UNCERTAIN DYNAMIC SYSTEMS

FRED C. SCHWEPPE

Department of Electrical Engineering
Massachusetts Institute of Technology
Cambridge, Massachusetts

PRENTICE-HALL, INC., *Englewood Cliffs, New Jersey*

Library of Congress Cataloging in Publication Data

SCHWEPPE, FRED C.
Uncertain dynamic systems.
(Prentice-Hall series in electrical engineering)

Bibliography: p.
1. System analysis. 2. Dynamics 3. Estimation theory. 4. Statistical hypothesis testing. I. Title.
QA402.S37 620′.72 72–11801
ISBN 0–13–935593–6

10 9 8 7 6 5 4 3 2 1

Printed in the United States of America

PRENTICE-HALL INTERNATIONAL, INC., *London*
PRENTICE-HALL OF AUSTRALIA, PTY. LTD., *Sydney*
PRENTICE-HALL OF CANADA, LTD., *Toronto*
PRENTICE-HALL OF INDIA PRIVATE LIMITED, *New Delhi*
PRENTICE-HALL OF JAPAN, INC., *Tokyo*

To Cindy, Edmund, Carl, and Fritz

"*Dignity at all costs*"

Contents

Preface

This book is designed to help the reader make certain (that is, not uncertain) statements about dynamic systems when uncertainty exists in a system's input and output and in the nature of the system itself. Interest centers on methods applicable to large-scale, multivariable systems. The problems of and interactions among modeling, analysis, and design are considered. Emphasis is placed on techniques which can be implemented on a digital computer.

Some new, previously unpublished results on unknown-but-bounded models are included, but most of the development is based on already existing and even classic concepts involving statistical-probabilistic models. The many different ideas are organized and presented so that their similarities and differences hopefully become clear. Emphasis is on applications rather than theory so discussions of how to and how not to use the theory are considered to be as important as the derivation of equations.

Concepts are first introduced for static (nondynamic) systems so that the basic ideas can be understood with a minimum of mathematical sophistication and complication. Extension to the complexities of dynamic systems then follows in a relatively straightforward fashion.

This book resulted from lecture notes for a control-system theory course in an electrical engineering curriculum. However, the material is widely applicable in many different areas of engineering as well as in economic and social systems. The book is suitable for independent study or can be used as a text for a one- or two-semester course.

The writing of the book was a part-time project spread over many years, and many people played a role in its evolution. Some of the earlier ideas were influenced by discussions with Harold Kushner, David Sakrison, and especially Leland Gardner. Various versions of the material have been presented in courses with the assistance of Edison Tse, Frank Galiana, J. Duncan Glover, Steve Hnylicza, and Robert Moore. Tom Kailath and Michael Athans provided critical and hence very useful comments which had a strong influence on the final draft. Barbara Smith, Helen Jackson, and Karen Keefe were extremely helpful in the many details associated with obtaining a final manuscript. Finally, but very definitely not least, Connie Spargo typed and retyped most of the manuscript. She demonstrated an uncanny ability to transform almost unreadable handwriting and rough sketches into clean, clear text, equations, and figures.

FRED C. SCHWEPPE

Cambridge, Massachusetts

1

Introduction

The three main topics of interest in this book are:

1. Modeling of uncertainty.
2. Analyzing the effects of uncertainty.
3. Designing systems to remove or compensate for uncertainty.

This introductory chapter is intended to remove some of the reader's uncertainty as to what he will encounter in later chapters.

1.1 Uncertainty

According to the dictionary, something is *uncertain* if it is indeterminate, indefinite, not reliable, untrustworthy, and not clearly defined. One of the main goals of this book is to make determinate, definite, reliable, trustworthy, clearly defined statements about uncertain dynamic systems. To make certain statements about an uncertain system, it is necessary to use precise, well-defined (i.e., certain) models for the uncertainty. Thus, in a very real sense, we are not interested in truly uncertain systems, but instead are interested in problems for which *certain* models are available. These certain models are chosen to represent what is considered to be uncertain about the system.

The following are examples of models for uncertainty. Possible models for a scalar x whose value is uncertain are

1. x is a random variable with specified probabilistic structure.
2. x is a random variable whose probabilistic structure contains uncertain parameters; for example, mean and variance.
3. x is a random variable whose probabilistic structure is unknown except for certain moments; for example, mean and variance.
4. x is completely unknown.
5. x is bounded; for example, $|x| \leq 1$.
6. x can take on only certain values; for example, $x = 1, 2$, or 3.

These six models can be combined into two basic types as follows:

1–3. x is a random variable whose probabilistic structure may be uncertain.

4–6. x belongs to some set.

Possible models for a K-dimensional vector $\mathbf{x}$ whose value is uncertain are

1. $\mathbf{x}$ is a random vector whose probabilistic structure may be uncertain.
2. $\mathbf{x}$ belongs to some set in a K-dimensional space; for example, $\mathbf{x}$ belongs to a closed set, $\mathbf{x}$ can take on only certain values, or $\mathbf{x}$ lies on some hypersurface.

Possible models for a scalar time function $x(t)$ over $0 \leq t \leq T$ whose value is uncertain are

1. $x(t)$ is a stochastic process whose probabilistic structure may be uncertain.
2. $x(t)$ is bounded for all t, $0 \leq t \leq T$; for example, $|x(t)| < 1$.
3. $x(t)$ must satisfy some integral constraints; for example,

$$\int_0^T x^2(t)\, dt \leq 1$$

4. The spectrum of $x(t)$ is constrained; for example,

$$|x(\omega)|^2 < \epsilon, \qquad \text{all } \omega > \omega_0,\ \omega < -\omega_0$$

$$x(\omega) = \int_0^T x(t)e^{-i\omega t}\, dt$$

5. $x(t)$ is parameterized; for example,

$$x(t) = a_0 + a_1 t + a_2 t^2$$

where the values of a_0, a_1, and a_2 are uncertain.

Models 2, 3, 4, and 5 can be viewed as methods of constraining $x(t)$ to lie in some set in the space of all possible time functions over $0 \leq t \leq T$. Similar types of models can be stated for uncertain vector time functions and functions of more than one independent variable.

The above examples illustrate the two main classes of uncertainty models that will be considered in this book:

1. Probabilistic models with possible uncertainty in the probability distributions.
2. Set-theoretic models (called unknown-but-bounded models).

Naturally not all possible combinations of such models will be studied. The actual subjects to be considered fall into four general categories:

1. *Bayesian:* Uncertainty is modeled by random variables and/or stochastic processes with either completely specified probability distributions or completely specified first and second moments (means and variances).
2. *Fisher:* Uncertainty is modeled by a combination of Bayesian models and completely unknown quantities.
3. *Unknown-but-bounded:* Uncertainty is modeled by set-theoretic models; the resulting quantities are called unknown-but-bounded processes.
4. *Weighted least squares:* "Reasonability" arguments are used in place of explicit models for the uncertainty.

1.2 Background Needed

The required prior background in probability theory is an understanding of the basic concepts of vector random variables; i.e., probability density, expectation, conditional density, and conditional expectation. However, few of the ideas (and none of the really basic concepts) require an understanding at a sophisticated mathematical level.

The book draws heavily on state space techniques for handling and analyzing deterministic dynamic systems. Thus the reader is assumed to have a prior understanding of such concepts and a thorough grasp of the basic linear algebra of vector and matrix theory and of methods for handling vector difference and differential equations.

Certain subjects to be covered are closely related to deterministic optimal control theory (calculus of variations, maximum principle, and dynamic programming). However, a prior background in these areas is not assumed of the reader.

This book evolved from lecture notes for a course in control-system theory and some of the discussions reflect this control-system theory orientation. However, the material is applicable to a much broader class of problems, and a reader without a control-system background should have little trouble in following the main ideas and developments.

1.3 Goals and Method of Approach

This book is not written to be an introduction, a survey, or a complete exposition. The book is written to take the reader from an understanding of certain basic ideas along a relatively narrow path which leads to a wide variety of advanced topics. The goal of the trip is to develop an understanding so the reader can successfully apply the ideas. The author personally feels that the route followed on this trip is limited primarily in the area of theory, not in the range of applications. An attempt is made to present the material so that the book can also be used as a reference by readers who already have some background and experience in the field.

The following pattern of presentation is continually repeated:

1. Concepts are first defined for the static case, i.e., a system defined by just a vector set of equations.
2. State space concepts are used to convert the static case into a discrete time dynamic system.
3. A discrete to continuous time limit is used to yield results for continuous time systems.

The theory for linear systems is presented before nonlinear systems are considered. This method of presentation has the disadvantage of being slow. However, four advantages to the chosen method of presentation are

1. Concepts are first introduced and discussed in simple situations (linear, static systems) where they can easily be understood.
2. No prior knowledge of stochastic processes is needed.
3. Many applications involve discrete time systems and analysis, i.e., digital computers.
4. The linear theory provides the basis for many of the practical techniques for handling nonlinear problems.

The use of state space concepts greatly simplifies the development of models and analysis techniques. State space techniques also provide results in a form directly suitable for implementation on analog or digital computers. This is very important, as the types of problems encountered in modern applications rarely yield simple, closed-form, analytic results. In fact, the underlying philosophy of this book is that a "solution" consists of an algorithm (equation) suitable for computer implementation. Discussions of computational problems are therefore included as an integral part of the overall presentation.

The examples presented as an aid to understanding, usually evolve a one- or two-dimensional time-invariant system. The exercises at the end of the chapters are designed to be done with the minimum amount of tedium, so they usually involve one- or two-dimensional time-invariant systems. However, these simple examples and problems should not be viewed as being typical of applications. The general theory is aimed toward developing techniques for solving complicated, high-dimension, and/or time-varying problems on a (usually digital) computer. If this were the "best of all possible worlds," the reader would have a computer console furnished with the book so the examples and problems could be more realistic.

This book assumes that the reader has a "reasonable degree" of mathematical sophistication. However, the material is presented in a discussion format rather than a theorem-proof format and results are often developed in an intuitive manner rather than as special cases of some more general, abstract mathematical concept. Two advantages to the chosen method of presentation are

1. In most (but not quite all) applications, intuitive understanding is more useful than general, abstract theorems.
2. A more sophisticated approach would either require the inclusion of background material or would restrict the book to a narrow class of readers.

The chosen method of presentation has the disadvantage that much excellent material available in the literature is written using more sophisticated mathematical formalization and terminology, and the book does not train the reader to handle such literature. The decision to write the book in its present form obviously reflects the author's personal interests.

1.4 Organization

The book is divided into five parts. Part I presents general discussions on system theory and types of mathematical models to be considered. Part II covers estimation theory for linear systems. Part III introduces certain aspects of decision theory (hypothesis testing), which is also viewed as a partial bridge between linear and nonlinear estimation theory. Part IV discusses estimation theory for nonlinear systems. Part V contains general discussions on system identification and stochastic control. The appendices contain background and supplementary material.

Except for Part V, the book is organized in a highly structured form.

The heart of structure is the basic sequence

Static system (vectors)

↓

Discrete-time dynamic system

↓

Continuous-time dynamic system

At each stage, consideration is given to the three basic uncertainty models (Bayesian, Fisher, unknown-but-bounded) and to the weighted-least-squares concept. This overall structure is evident in the table of contents.

The tight structural form does not force the reader to start at the beginning and go on a straight line to the end. On the contrary, the structure is intended to provide the reader with a choice of many possible routes through the book. The following general comments are intended to help one choose a route:

1. Chapter 2 contains a very general overview which is not essential to understanding the rest of the text.
2. The unknown-but-bounded model discussions can be viewed as a separate sequence and studied independent of a Bayesian, Fisher, weighted-least-squares sequence. Separation of the Bayesian, Fisher, and weighted least squares into individual sequences, is not recommended.
3. The linear and nonlinear model developments can be combined in different ways, for example, Chapters 4, 9, 11, and 12 can be studied as a sequence on the static model.
4. The basic static, discrete-time, continuous-time sequence should not be broken.
5. The survey discussions of Part V can be factored into a sequence. For example, Chapter 14 (system identification) and Chapter 15 (stochastic control) can follow Chapter 6, with only selected reading in other chapters.

Figure 1.4.1 summarizes the main prerequisite structure existing between the various chapters. This prerequisite structure is related to general overall understanding of concepts. Sometimes reference is made to specific points in earlier nonprerequisite chapters, but such references can either be skipped as being only "side points" or else the referenced material can be studied "out of context." As a final aid to choosing a route through this book, various sections and subsections are "daggered" (†) to indicate that they are not considered to be important to the main theme of the development.

The appendices play an ambiguous role which depends on the reader's

Chapter	Main Prerequisite Chapters	
1	None	
Part I		
2	None	
3	None	
Part II		
4	3	
5	None	
6	3 4 5	
7	3 6	
8	3 6	
Part III		
9	5	
10	3 6 7* 9	* implies chapter is
Part IV		recommended but is
11	5*	not essential
12	11	
13	3 6* 11	
Part V		
14	3 4 6	
15	3 4 6	

Figure 1.4.1. Prerequisite structure of chapters.

background and interests. The appendices contain definitions of basic mathematical concepts. Hopefully the reader is already familiar with many of these ideas, but it is expected that most readers will have to go to the appendices at various times.

1.5 References

The question of references presented a problem when writing this book as it is difficult to give credit where credit is due. Many of the basic principles can be traced back to the work on least squares by Gauss and Legendre. From this common base, the theory evolved through many different paths to its present state. Statisticians, economists, surveyors, natural scientists, and astronomers all developed various aspects, often independently of each other. The same basic results were often rederived many times for different applications. The engineers entered the picture with Wiener's work and proceeded with their own evolutionary cycle of deriving new results and rederiving old results which were thought to be new. The crediting problem is further complicated by the fact that the author has worked in this field for a relatively long time and many ideas have become so ingrained that he has forgotten where they came from originally.

In addition to giving credit, an ideal bibliography or set of references should prove to be a source of background reading and a guide to advanced studies. Unfortunately, this is also a difficult task. The basic concepts of uncertain systems apply and are used in many different fields, and the appropriate reference often depends on the reader's background.

The actual bibliography and references found in this book are the result of the following ground rules:

1. Credit is definitely given to all material that the author consciously used (whether it be book, paper, unpublished report, or personal discussion).
2. Background references are limited to only selected books, although in general many others may also be equally as good.
3. Selected references to papers and books where further or more advanced developments can be found are given.

A complete listing of all related papers and books is not attempted. However, a failure to reference some book or paper definitely does not constitute a nonrecommendation.

1.6 Notation

Half the job of reading any paper or book is usually the "figuring out of what the notation means." As a start in this direction, some of the conventions to be used are now stated.

A boldface symbol denotes a vector or matrix, i.e.,

a, A	scalars
$\mathbf{a}$, $\mathbf{A}$	vectors or matrices

Usually, boldface lowercase letters are vectors, while boldface capital letters are matrices. (This convention obviously is imprecise as a vector is also a matrix.) All vectors are column vectors unless explicitly stated otherwise. If $\mathbf{A}$ is a matrix, then

$\mathbf{A}^{-1}$	denotes inverse of $\mathbf{A}$ (assuming it exists)
$\lvert\mathbf{A}\rvert$	denotes determinant of $\mathbf{A}$
$\operatorname{tr}\mathbf{A}$	denotes trace of $\mathbf{A}$
$\mathbf{A}'$	denotes transpose of $\mathbf{A}$

The use of $\lvert\mathbf{A}\rvert$ for the determinant of $\mathbf{A}$ is somewhat ambiguous as $\lvert A\rvert$ is also used to denote the absolute value of the scalar A, but the meaning is always made clear in the text. A unit matrix is denoted by $\mathbf{I}$. The zero vector or matrix is noted by $\mathbf{0}$.

The symbol E always denotes expectation of a random variable or stochastic process. No notational convention is used to denote the difference between a random variable and a sample of a random variable.

If the random vector $\mathbf{x}$ has a Gaussian (or normal) probability density with mean $\mathbf{m}$ and covariance matrix $\mathbf{\Gamma}$,

$$E\{\mathbf{x}\} = \mathbf{m}$$
$$E\{[\mathbf{x} - \mathbf{m}][\mathbf{x} - \mathbf{m}]'\} = \mathbf{\Gamma}$$

this is often expressed as

$$\mathbf{x} \text{ is } N(\mathbf{m}, \mathbf{\Gamma})$$

No particular convention is used in determining whether an English or Greek symbol is to be used.

The following conventions are almost always followed (see also Fig. 2.1.1 of Chapter 2):

- $\mathbf{u}$ known system input or input control
- $\mathbf{w}$ uncertain disturbance "driving" the system
- $\mathbf{v}$ uncertain disturbance corrupting the observations
- $\mathbf{z}$ observations
- $\mathbf{x}$ system state vector

Time is usually the only independent variable considered. For continuous-time problems, t denotes time. In discrete-time problems, the integer n denotes time, the "event," or the sample number. When necessary, Δ denotes time between events, and so $n\Delta$ denotes discrete time.

An estimate of variable is denoted by a superscript caret; i.e., $\hat{\mathbf{x}}$ is an estimate of $\mathbf{x}$. The error in an estimate is denoted by $\boldsymbol{\delta}$, and so

$$\boldsymbol{\delta}_x = \mathbf{x} - \hat{\mathbf{x}}$$

A superscript tilde is often used to denote a dummy or related variable. Thus $\tilde{\mathbf{x}}$ may denote a variable similar in character to but different from $\mathbf{x}$.

Sets are denoted using notations such as

$$\mathbf{\Omega}_x = \{\mathbf{x} : g(\mathbf{x}) \leq 1\}$$

where this notation means $\mathbf{\Omega}_x$ is the set of all vectors $\mathbf{x}$ which satisfy the relation $g(\mathbf{x}) \leq 1$.

The equations and figures are numbered according to the chapter and section. Subsection numbering does not affect the equation and figure numbering. Thus (4.2.3) denotes the third numbered equation in the second section of Chapter 4, while Fig. 2.1.1 denotes the first figure in the first Section of Chapter 2. Not all equations are numbered. Equations of special importance are "boxed."

PART I

Framework

The reader who is planning to go through the whole book has a long journey in front of him. Part I is intended to provide a general framework which will make the actual journey more meaningful.

Chapter 2 is a combination road-map and tour-book description of the general nature of the country to be visited (system theory). Our actual journey will not cover all the roads and places of interest on the map, but a map helps one appreciate that which is actually seen. It may also stimulate the traveler to come back again on another trip.

In Chapter 3, the basic mathematical state space models to be investigated are discussed. The use of state space models can be viewed as a vehicle which makes the trip go faster. Chapter 3 discusses the generality of the models and shows that the state space model vehicle is very versatile and that little generality is sacrificed for speed provided that one knows how to use the models.

Before starting, it is worthwhile for the reader to glance over the background material contained in the appendices. Some of this is (hopefully) already known to the reader, but knowledge of which topics are discussed in the appendices will prove very useful for later reference. Reading these appendices is analogous to getting the car greased and full of gas, putting the family dog in the kennels, and making sure there are enough diapers for the baby. Neglect of such details can turn any trip into an agonizing experience.

2

System Theory

The chapters to follow present theory and techniques which can be considered part of the general field of *system theory*. Some of the interrelationships are briefly discussed in this chapter in an attempt to provide an overall perspective.

2.1 Terminology

It is necessary to define some basic terms. Figure 2.1.1 summarizes the needed definitions and illustrates their relationships. The variables **u, m, w, v, x, y,** and **z** of Fig. 2.1.1 may be of many types. They may be scalars or vectors. They may be functions of independent variables such as time and/or space. The range of the variables may be discrete or continuous, infinite or bounded. "Details" such as these do not effect the discussions to follow. The various parts of Fig. 2.1.1 are now discussed.

System: The system is represented by a mathematical model which may take many forms, such as algebraic equations, finite state machines, difference equations, ordinary differential equations, partial differential equations, and functional equations. The system model may be uncertain, as the mathematical model may not be known completely.

Symbolic sensor: The sensor box in Fig. 2.1.1 is a symbolic operation

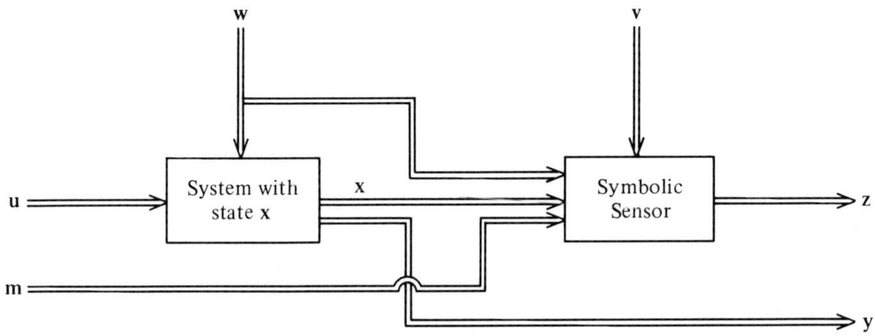

u Input Control
m Measurement Control
w System Disturbance
v Sensor Disturbance
x System State
y System Output
z Observation

Figure 2.1.1 Definitions of basic variables.

included to emphasize the fact that the internal workings and environment of a system are rarely observed directly. The sensor output is some, usually quite restricted, function of the system's state and disturbances. This symbolic sensor has no dynamic structure of its own. Dynamics of the physical sensors are included as part of the system model.

Disturbances, w, v: The disturbance **w** models the uncertainty (noise) in the inputs to the system. The disturbance **v** conceptually models the uncertainty (noise) introduced by the sensor (although in many cases, the input disturbance **w** ends up modeling some or all of the sensor uncertainty). Both types of disturbances are modeled as uncertain processes, i.e., processes whose actual values are not known exactly. The dynamic structures of the disturbances are included as part of the system model.

State x: The state of the system summarizes the effects of all the past inputs to the system. The future evolution of the system is completely specified (assuming that the system model is perfect) by the present value of the state and the future inputs. The state **x** can include "natural" states, sensor states, and disturbance states.

Observation z: The observation **z** is the output of the sensors. It is modeled as a known function of the system state **x** (and possibly of the input disturbance **w**) corrupted by the sensor disturbance **v** in some known way. Usually **z** is an explicit function of only a few components of **x**.

System output, y: The system output **y** is the quantity which interacts with the outside world so that system performance depends on how **y** behaves.

y is modeled as a known function of the system state, **x**, which may not be directly observable, i.e., part of **z**.

Control input, u: The control input **u** is the quantity which can be manipulated (within constraints) to control the system's state **x** and hence the output **y**.

Measurement control, m: The measurement control **m** is the quantity which can be manipulated (within constraints) to control how the observations **z** on the system are made. It should be noted that the input control **u** also affects the observations **z** through its influence on the states **x**. Hence **u** can also exercise some measurement control.

2.2 Basic Problems of System Theory

To relate the contents of this book to system theory, it is necessary to define system theory and the basic problem areas it encompasses. Unfortunately, system theory is an ill-defined generic term that means many different things to different people. The following should be viewed as an outline of the author's personal point of view, which has been colored to emphasize the subjects covered in this book.

System theory is a tool which engineers use to help them design the "best" system to do the job that must be done. A dominant characteristic of system theory is the interest in the analysis and design (synthesis) of systems from an input-output point of view. System theory uses mathematical manipulation of a mathematical model to help design the actual system. The computer (digital, analog, hybrid) is often the laboratory of the system theorist in which ideas are checked out.

One of many ways to classify the basic problems of system theory is in terms of *modeling*, *analysis*, *estimation*, and *control*. Figure 2.2.1 defines the four areas, various special cases, and their general interrelationships. The following gives more details on this classification.

Modeling

The problem of developing a mathematical model which adequately represents the physical situation is the most critical step. If the model is no good, the subsequent mathematical (and computer) manipulations are useless. On the other hand, the mathematical model should not be too complex, as this leads to unnecessary and sometimes insurmountable problems. The only general rules are

1. The modeling must be done by someone who thoroughly understands the nature and behavior of the actual system.
2. The model must be *usable*, i.e., capable of being analyzed.

	Mathematical Model	Controls	Observables	States	Problems to be Solved
Modeling	?	—	—	—	What is mathematical model?
Analysis	Known	Known	—	—	How does output behave?
Estimation					
State	Known	Known	To be Used	?	What is best estimate of state?
Identification	Incomplete	Known	To be Used	?	What is best estimate of system?
Adaptive	Incomplete	Known	To be Used	?	What is best estimate of state and system?
Control					
Open loop	Known	?	—	—	Which precomputed control gives best output?
Closed loop	Known	?	To be Used	?	Which control expressed as function of observables gives best output?
Adaptive	Incomplete	?	To be Used	?	Which control expressed as function of observables gives best output?

Figure 2.2.1. Basic problems of system theory.

The mathematical model specifies all the relationships discussed in Section 2.1, i.e., equations relating (1) the response of the states $\mathbf{x}$ to the control inputs $\mathbf{u}$ and disturbances $\mathbf{w}$; (2) the system's output $\mathbf{y}$ to the states $\mathbf{x}$, and (3) the observables $\mathbf{z}$ to the states $\mathbf{x}$, the sensor disturbances $\mathbf{v}$, and the measurement control $\mathbf{m}$. The mathematical model also specifies the uncertainties. For example, if an uncertain scalar is modeled as a random variable, the mathematical model may specify the probability density or some other related quantities such as the mean and variance.

Analysis

The basic question of analysis is, How does the system output behave? The two general types of analysis are qualitative and quantitative. A quantitative analysis determines precisely how the output behaves, for example, a plot of $\mathbf{y}$ versus time. A qualitative analysis determines general properties of the behavior of the output; for example, is the system stable?

In the case of disturbances modeled by stochastic processes, the analysis centers on the statistical or probabilistic properties of the system states rather than on the states themselves. Deterministic quantities that specify the statistical properties are defined (e.g., the probability distribution or various moments), and then the behavior of these deterministic quantities is studied. The quantities which define the statistics of the states can be modeled (or thought of) as the states of some "new" system. Thus the analysis of systems with stochastic disturbances can be viewed as a two-step process:

1. Develop a new system whose states determine the statistical properties of the states of the original system.
2. Analyze the new system as a deterministic system of interest in its own right.

In the case of disturbances known only to lie within sets, the analysis centers on the behavior of the resulting sets which contain the states. The two-step procedure used for the stochastic case can again be employed; i.e.,

1. Develop a new system whose states determine the sets which contain the states of the original system.
2. Analyze the new system as a deterministic system of interest in its own right.

Estimation

Estimation is the problem of using the observables $\mathbf{z}$ to estimate (calculate, deduce, infer, guess, etc.) properties of the actual system. Three types of estimation problems are defined in Fig. 2.2.1: state estimation, identification, and adaptive estimation.

In state estimation, the observable $\mathbf{z}$ is used to estimate the system's state $\mathbf{x}$ under the assumption that the system model is perfect. When time is the independent variable, three types of estimation problems are

1. Filtering: Estimate $\mathbf{x}(T)$ from $\mathbf{z}(t)$, $0 \leq t \leq T$.
2. Prediction: Estimate $\mathbf{x}(T + \tau)$ from $\mathbf{z}(t)$, $0 \leq t \leq T$, $\tau > 0$.
3. Smoothing: Estimate $\mathbf{x}(\tau)$, $0 \leq \tau \leq T$, from $\mathbf{z}(t)$, $0 \leq t \leq T$.

It is often impossible to develop a priori a completely specified mathematical model for the physical system. In the identification problem, the observable $\mathbf{z}$ is used to complete the model. The system to be identified is modeled in terms of some a priori structure and some uncertain variables $\boldsymbol{\alpha}$ whose values are to be "identified." These $\boldsymbol{\alpha}$ can be considered to be *new states* of the original system. Thus the identification problem can simply be considered to be a state estimation problem for a completely specified system where the new states $\boldsymbol{\alpha}$ are of interest instead of the natural system states $\mathbf{x}$. Adaptive estimation is the state estimation problem when the system itself requires identification, i.e., state estimation and system identification combined. It follows that adaptive estimation can be considered to be state estimation of a combined state containing both $\mathbf{x}$ and $\boldsymbol{\alpha}$.

Decision-making problems (hypothesis testing) involve situations where it is desired to make a choice among various alternative decisions (hypotheses). Such problems can be viewed as generalized state estimation problems where the definition of state has simply been expanded.

Hence all estimation and decision-making problems can be considered to be state estimation problems of varying degrees of complexity.

Control

The control problem is the choice (determination) of values for the control variables $\mathbf{u}$ and $\mathbf{m}$. In the classic view of control theory, these $\mathbf{u}$ and $\mathbf{m}$ are input *signals*. However, from a more general point of view, the $\mathbf{u}$ and $\mathbf{m}$ can also be system parameters which the engineer can vary to arrive at a final design. Three types of control problems are defined in Fig. 2.2.1: open-loop, closed-loop, and adaptive.

Open-loop control is precomputed in that the control is expressed in terms of the system's initial state and the independent variable(s), which is usually time. In closed-loop control, a law or function is desired which expresses the control variables as explicit functions of the observations $\mathbf{z}$. When uncertainty is present, closed-loop control usually results in better performance than open-loop control. A closed-loop-control law is often divided into two parts:

1. Estimation of the state or some function of the state from the observations.
2. Calculation of the control from the estimate.

Adaptive control is closed-loop control when the system is not completely specified, i.e., when the system must be identified. Thus, loosely speaking, the estimation portion of closed-loop control is replaced by adaptive estimation. Since adaptive estimation is just a more complicated version of state estimation, adaptive control is just a more complicated closed-loop-control problem.

The popular term *stochastic control* actually applies to many problems. Stochastic control may be either open- or closed-loop. In the closed-loop case, the observable **z** may or may not be the complete state **x**, with or without sensor errors.

2.3 Discussion

Four problems (modeling, analysis, estimation, and control) of system theory have been defined. Modeling is definitely the most important and critical problem. If the mathematical model is not valid, any subsequent analysis, estimation, or control study is meaningless. The development of the model in a convenient form can greatly reduce the complexity of the actual studies. Sometimes a "sloppy" model can be used by introducing pseudo-uncertain disturbances to account for modeling errors. Often the most important part of modeling is the specification of what problem is to be solved. The choice of model often depends on what question is being asked. The second most important area is analysis. Often the ability to analyze a situation is all that is needed—especially if the analysis is combined with good "seat of the pants" design. The relative importance of estimation and control depends on the degree of uncertainty. For situations involving only negligible uncertainty, control is obviously more important than estimation. However, as the degree of uncertainty increases, estimation becomes more and more important, until in many cases it is much more important than control.

In this book, modeling is discussed many times but usually only in general terms or by delineating the types of possible models and their implications. There are no general rules for modeling other than that the man developing the mathematical model must understand both the physical problem and the mathematics to be used.

The analysis of uncertainty plays the dominant role in this book. However, estimation theory provides many of the needed analysis tools, and most of the book appears to be devoted to estimation theory. (Estimation theory is taken to include identification, adaptive, and decision-making problems).

Control theory plays only a minor role in this book. Assume that it is desired to control some system. The biggest (and sometimes insurmountable) problem is usually to use the available data (information, measurements, etc.) to find out what the system is actually doing (i.e., to estimate its state). If the system's state can be estimated to some reasonable accuracy, the desired control is often obvious (or can be obtained by the use of deter-

1. State space structures.
2. White processes.

For linear systems, the resulting model is summarized in Fig. 3.1.1. More details are now given.

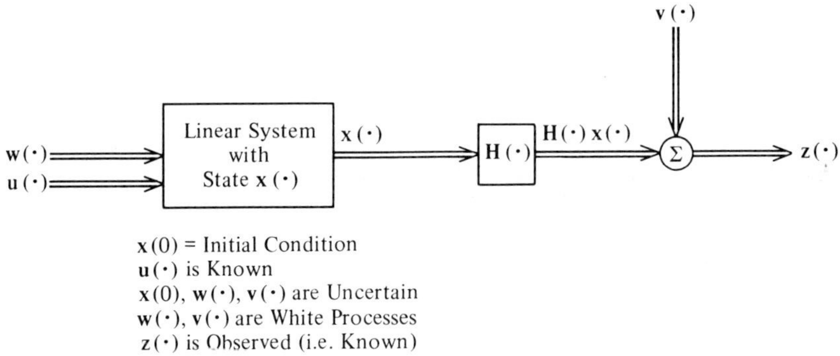

Figure 3.1.1 Basic structure of linear model.

3.1.1 State Space Structures

Consider the linear system of Fig. 3.1.1. The *input-output* behavior is modeled by a state space structure as follows:

Discrete time:
$$\mathbf{x}(n\Delta + \Delta) = \mathbf{\Phi}(n\Delta)\mathbf{x}(n\Delta) + \Delta\mathbf{G}(n\Delta)\mathbf{w}(n\Delta) + \Delta\mathbf{B}(n\Delta)\mathbf{u}(n\Delta)$$
$$\mathbf{z}(n\Delta) = \mathbf{H}(n\Delta)\mathbf{x}(n\Delta) + \mathbf{v}(n\Delta)$$

Continuous time:
$$\frac{d}{dt}\mathbf{x}(t) = \mathbf{F}(t)\mathbf{x}(t) + \mathbf{G}(t)\mathbf{w}(t) + \mathbf{B}(t)\mathbf{u}(t)$$
$$\mathbf{z}(t) = \mathbf{H}(t)\mathbf{x}(t) + \mathbf{v}(t)$$

In both cases

$\mathbf{x}(\cdot)$	state of system
$\mathbf{u}(\cdot)$	known (deterministic) input
$\mathbf{w}(\cdot)$	uncertain white input disturbance
$\mathbf{v}(\cdot)$	uncertain white observation disturbance
$\mathbf{z}(\cdot)$	observation
$\mathbf{x}(0)$	uncertain initial condition

Here and throughout this chapter the rather unconventional notation $\mathbf{x}(\cdot)$, etc., is used when the discussions apply equally to discrete- and continuous-time models. Thus $\mathbf{x}(\cdot)$ may be either $\mathbf{x}(n\Delta)$ or $\mathbf{x}(t)$. The Δ multiplying $\mathbf{G}(n\Delta)$ and $\mathbf{B}(n\Delta)$ is included to facilitate later manipulations involving a discrete to continuous time limit operation. (See Appendix B.3 and Section 4.2.) For the present, the Δ can be neglected if desired.

The corresponding nonlinear models are very similar in form, namely

Discrete time:
$$\mathbf{x}(n\Delta + \Delta) = \boldsymbol{\phi}[\mathbf{x}(n\Delta), \mathbf{w}(n\Delta), \mathbf{u}(n\Delta), n\Delta]$$
$$\mathbf{z}(n\Delta) = \mathbf{h}[\mathbf{x}(n\Delta), \mathbf{v}(n\Delta), n\Delta]$$

Continuous time:
$$\frac{d}{dt}\mathbf{x}(t) = \mathbf{f}[\mathbf{x}(t), \mathbf{w}(t), \mathbf{u}(t), t]$$
$$\mathbf{z}(t) = \mathbf{h}[\mathbf{x}(t), \mathbf{v}(t), t]$$

The matrices $\mathbf{\Phi}$, $\mathbf{F}$, $\mathbf{G}$, and $\mathbf{H}$ and the nonlinear functions $\mathbf{f}$, $\boldsymbol{\phi}$, and $\mathbf{h}$ are assumed to be completely known. Uncertainty enters via $\mathbf{w}(\cdot)$, $\mathbf{v}(\cdot)$, and $\mathbf{x}(0)$.

EXAMPLE: Define

$x(n)$	number of bottles of beer in store at time n
$u_1(n)$	number of bottles bought between n and $n + 1$ from wholesaler
$u_2(n)$	average number of bottles sold between n and $n + 1$ to consumers
$u_2(n) + w(n)$	actual number of bottles sold between n and $n + 1$ to consumers
$w(n)$	uncertainty in sales
Φ	"fraction" of bottles in store at time n that are not drunk between n and $n + 1$ by the proprietor

Then

$$x(n + 1) = \Phi x(n) + w(n) + u(n)$$
$$u(n) = u_1(n) - u_2(n)$$

EXAMPLE: Consider the following electrical network:

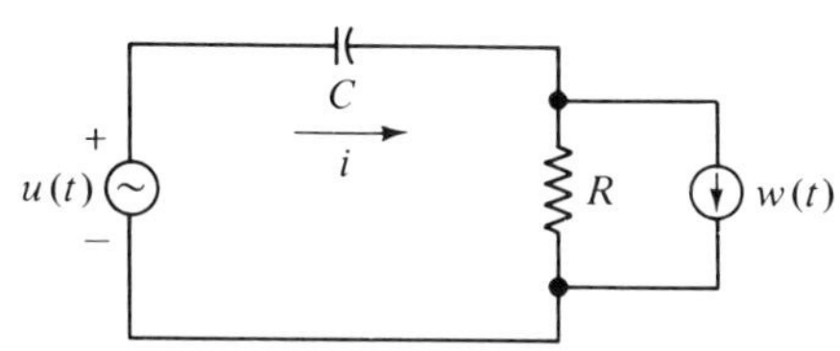

$u(t)$	known time-varying voltage source
$w(t)$	uncertain current generator representing a "noisy" resistor
$i(t)$	current flowing through capacitor

Assume no charge on capacitor at $t = 0$. Then

$$u(t) = \frac{1}{C}\int_0^t i(\tau)\, d\tau + R[i(t) - w(t)]$$

Define

$$x(t) = \text{voltage across capacitor}$$

$$x(t) = \frac{1}{C}\int_0^t i(\tau)\,d\tau$$

Then

$$\frac{d}{dt}x(t) = Fx(t) + Bu(t) + Gw(t)$$

$$F = -\frac{1}{RC} \qquad B = \frac{1}{RC} \qquad G = \frac{1}{C}$$

In the following developments of this book, the known input $u(\cdot)$ is often assumed to be equal to zero. This is usually done only to simplify the presentation, as (for linear systems) known inputs can often be handled independently of the uncertainties.

This book is written under the following basic assumption: *The reader is capable of developing state space structure models for deterministic systems.* Therefore detailed discussions on deterministic state space models are not given here. However, it should be emphasized that the development of such state space structures is not necessarily a simple task.

3.1.2 White Processes

In Fig. 3.1.1, the $\mathbf{w}(\cdot)$ and $\mathbf{v}(\cdot)$ are *white processes.* The general description of a white process is, "A white process has no time structure." In other words, knowledge of the white process's value at one instant of time provides no knowledge of what its value will be (or was) at any other time point. Three types of white processes are of interest:

1. Stochastic white processes.
2. Unknown-but-bounded white processes.
3. Completely unknown processes.

Consider the stochastic case first. A stochastic white process $\mathbf{w}(\cdot)$ is defined by

Discrete time, $\mathbf{w}(n\Delta)$:

$$E\{\mathbf{w}(n\Delta)\} = \mathbf{0}$$

$$E\{\mathbf{w}(n_1\Delta)\mathbf{w}'(n_2\Delta)\} = \begin{cases} \Delta^{-1}\mathbf{Q}(n_1\Delta), & n_1 = n_2 \\ 0, & n_1 \neq n_2 \end{cases}$$

Continuous time, $\mathbf{w}(t)$:

$$E\{\mathbf{w}(t)\} = \mathbf{0}$$

$$E\{\mathbf{w}(t_1)\mathbf{w}'(t_2)\} = \delta(t_1 - t_2)\mathbf{Q}(t)$$

where $\delta(t - \tau)$ is the Dirac delta function. Note that the probability densities of $\mathbf{w}(\cdot)$ are not specified; in particular it is not assumed in this book that white stochastic processes are Gaussian (or normal) processes. In various later developments, Gaussian models will be used, but in such cases $\mathbf{w}(\cdot)$ will be called a white Gaussian stochastic process.

The condition $E\{\mathbf{w}(n_1)\mathbf{w}'(n_2)\} = \mathbf{0}$, $n_1 \neq n_2$, implies that $\mathbf{w}(n_1)$ and $\mathbf{w}(n_2)$ are uncorrelated (also orthogonal because of the zero mean assumption). This does not imply that $\mathbf{w}(n_1)$ and $\mathbf{w}(n_2)$ are independent unless, for example, a Gaussian assumption is also used. (See Appendix *E*.) Therefore the stochastic white processes defined above may actually have some time structure, but this is a fine point which will be ignored. The processes could be called *wide sense* stochastic white processes if desired. (See Appendix *E*.)

The Δ dependence of the discrete-time covariance matrix is crucial when a discrete to continuous time limit is performed, and various ramifications of this dependence will be discussed in later chapters. For the present, it is simply noted that as $\Delta \longrightarrow 0$, $n \longrightarrow \infty$, $n\Delta \longrightarrow t$, it is not "unreasonable" to consider that the discrete-time white process approaches the continuous-time white process in some sense. For this chapter, the Δ dependence of the discrete-time white stochastic model can be ignored if desired.

Unfortunately the continuous-time stochastic white process is about the "worst-behaved" process imaginable. For fixed t, it is a zero mean random vector with infinite covariance matrix. It is not continuous anywhere. It does not exist in any physical sense. The continuous-time white stochastic concept yields formulas that are easy to study and implement, but its use also causes conceptual difficulties and fears as to the validity of any results. As will be discussed in later chapters, such fears are not important provided that one knows what one is doing.

Now consider the unknown-but-bounded case. An unknown-but-bounded white process $\mathbf{w}(\cdot)$ is defined by

Discrete time, $\mathbf{w}(n\Delta)$:	$\mathbf{w}(n\Delta)$ is completely unknown except for the constraint $\mathbf{w}(n\Delta) \in \mathbf{\Omega}_w(n\Delta)$, all n
Continuous time, $\mathbf{w}(t)$:	$\mathbf{w}(t)$ is completely unknown except for the constraint $\mathbf{w}(t) \in \mathbf{\Omega}_w(t)$, all t

where $\mathbf{\Omega}_w(n\Delta)$ and $\mathbf{\Omega}_w(t)$ are sets whose size and shape can vary with time. The sets usually considered in this book are ellipsoids, in which case (see Appendix H)

$$\mathbf{\Omega}_w(\cdot) = \{\mathbf{w} : \mathbf{w}'\mathbf{Q}^{-1}(\cdot)\mathbf{w} \leq 1\}$$

White unknown-but-bounded processes are like white stochastic processes in that they have no time structure. Their values at time t_1 or $n_1\Delta$ are not affected by their values at t_2 or $n_2\Delta$. The continuous-time versions need not be continuous functions but are well behaved in the sense that they have finite amplitudes, power, etc. Thus in some senses, they are a physically more satisfactory model than their continuous-time stochastic white process counterparts.

For both the stochastic and unknown-but-bounded ellipsoidal $\mathbf{w}(\cdot)$, the positive definite matrix $\mathbf{Q}(\cdot)$ measures the "size" (average or maximum possible value) of the uncertainties at the time of interest. The "larger" is $\mathbf{Q}$, the larger the uncertainty.

Finally, consider the completely unknown process, which is defined as follows: $\mathbf{w}(\cdot)$ *is a completely unknown process if nothing is known about its behavior.* At first glance this definition seems rather silly, but completely unknown processes are definitely a useful concept and it is necessary to define them. In one sense, the model for a completely unknown process is the fact that its behavior cannot be modeled. In later developments, results for completely unknown $\mathbf{w}(\cdot)$ are sometimes obtained as the limit of a white stochastic or unknown-but-bounded $\mathbf{w}(\cdot)$ as $\mathbf{Q}(\cdot)$ goes to infinity.

The above discussions considered the input disturbance $\mathbf{w}(\cdot)$, but the white observation disturbance $\mathbf{v}(\cdot)$ is defined in a similar fashion. The positive definite matrix $\mathbf{R}(\cdot)$ is used to measure the size of $\mathbf{v}(\cdot)$.

3.1.3 Relation Between x(0), w(·), and v(·)

To complete the specifications of Fig. 3.1.1, it is necessary to model $\mathbf{x}(0)$ and then consider the interrelations between $\mathbf{x}(0)$, $\mathbf{w}(\cdot)$, and $\mathbf{v}(\cdot)$.

The initial condition $\mathbf{x}(0)$ is modeled in three ways:

$\mathbf{x}(0)$, random vector: $E\{\mathbf{x}(0)\} = 0$, $E\{\mathbf{x}(0)\mathbf{x}'(0)\} = \boldsymbol{\psi}$

$\mathbf{x}(0)$, unknown-but-bounded: $\mathbf{x}(0) \in \boldsymbol{\Omega}_x(0)$ for the ellipsoidal model $\boldsymbol{\Omega}_x(0) = \{\mathbf{x} : \mathbf{x}'\boldsymbol{\psi}^{-1}\mathbf{x} \leq 1\}$

$\mathbf{x}(0)$, completely unknown: nothing is known about $\mathbf{x}(0)$

The model for the interrelationship between $\mathbf{x}(0)$, $\mathbf{w}(\cdot)$, and $\mathbf{v}(\cdot)$ can be summarized by

> $\mathbf{x}(0)$, $\mathbf{w}(\cdot)$, and $\mathbf{v}(\cdot)$ are not related in any way

To be more precise for stochastic models,

$$E\{\mathbf{x}(0)\mathbf{w}'(\cdot)\} = \mathbf{0}$$
$$E\{\mathbf{x}(0)\mathbf{v}'(\cdot)\} = \mathbf{0}$$
$$E\{\mathbf{w}(\cdot)\mathbf{v}'(\cdot)\} = \mathbf{0}$$

For unknown-but-bounded models, this "unrelated property" of $\mathbf{x}(0)$, $\mathbf{w}(\cdot)$, and $\mathbf{v}(\cdot)$ is inherent in the definition.

3.1.4 Bayesian, Fisher, and Unknown-but-Bounded Models

The three basic models to be considered in this book are the Bayesian, Fisher, and unknown-but-bounded models. These models are specific cases of the state space structure-white process models just discussed and are defined in Fig. 3.1.2. Note that there are three types of Fisher models. The names Bayesian and Fisher are used for historical reasons.

	White Input Disturbance	White Observation Disturbance	Initial Condition
Bayesian	Stochastic	Stochastic	Random Variable
Fisher 1)	Stochastic	Stochastic	Completely Unknown
2)	Completely Unknown	Stochastic	Random Variable
3)	Completely Unknown	Stochastic	Completely Unknown
Unknown-but-Bounded	Unknown-but-Bounded	Unknown-but-Bounded	Unknown-but-Bounded

Figure 3.1.2 Bayesian, Fisher, and unknown-but-bounded models.

For the sake of completeness and future reference, the complete definitions of the Bayesian and set-theoretic models for linear systems are now summarized. The Fisher models follow directly from Figure 3.1.2 and the Bayesian definitions.

Bayesian, discrete time:

$$\begin{aligned}
\mathbf{x}(n\Delta + \Delta) &= \mathbf{\Phi}(n\Delta)\mathbf{x}(n\Delta) + \Delta\mathbf{G}(n\Delta)\mathbf{w}(n\Delta) \\
\mathbf{z}(n\Delta) &= \mathbf{H}(n\Delta)\mathbf{x}(n\Delta) + \mathbf{v}(n\Delta) \\
&E\{\mathbf{x}(0)\} = 0 \qquad E\{\mathbf{v}(n\Delta)\} = \mathbf{0} \qquad E\{\mathbf{w}(n\Delta)\} = 0 \\
&E\{\mathbf{x}(0)\mathbf{v}'(n\Delta)\} = \mathbf{0} \qquad E\{\mathbf{x}(0)\mathbf{w}'(n\Delta)\} = 0 \\
&E\{\mathbf{v}(n_1\Delta)\mathbf{w}'(n_2\Delta)\} = \mathbf{0}, \qquad \text{all } n_1, n_2 \\
&E\{\mathbf{x}(0)\mathbf{x}'(0)\} = \boldsymbol{\psi} \\
&E\{\mathbf{w}(n_1\Delta)\mathbf{w}'(n_2\Delta)\} = \begin{cases} \Delta^{-1}\mathbf{Q}(n_1\Delta), & n_1 = n_2 \\ \mathbf{0}, & n_1 \neq n_2 \end{cases} \\
&E\{\mathbf{v}(n_1\Delta)\mathbf{v}'(n_2\Delta)\} = \begin{cases} \Delta^{-1}\mathbf{R}(n_1\Delta), & n_1 = n_2 \\ \mathbf{0}, & n_1 \neq n_2 \end{cases}
\end{aligned} \tag{3.1.1}$$

Bayesian, continuous time:

$$\begin{aligned}
&\frac{d}{dt}\mathbf{x}(t) = \mathbf{F}(t)\mathbf{x}(t) + \mathbf{G}(t)\mathbf{w}(t) \\
&\mathbf{z}(t) = \mathbf{H}(t)\mathbf{x}(t) + \mathbf{v}(t) \\
&E\{\mathbf{x}(0)\} = \mathbf{0} \qquad E\{\mathbf{v}(t)\} = \mathbf{0} \qquad E\{\mathbf{w}(t)\} = \mathbf{0} \\
&E\{\mathbf{x}(0)\mathbf{v}'(t)\} = \mathbf{0} \qquad E\{\mathbf{x}(0)\mathbf{w}'(t)\} = \mathbf{0} \\
&E\{\mathbf{v}(t_1)\mathbf{w}'(t_2)\} = \mathbf{0}, \qquad \text{all } t_1, t_2 \\
&E\{\mathbf{x}(0)\mathbf{x}'(0)\} = \boldsymbol{\psi} \\
&E\{\mathbf{w}(t_1)\mathbf{w}'(t_2)\} = \delta(t_1 - t_2)\mathbf{Q}(t_1) \\
&E\{\mathbf{v}(t_1)\mathbf{v}'(t_2)\} = \delta(t_1 - t_2)\mathbf{R}(t_1)
\end{aligned} \tag{3.1.2}$$

Unknown-but-bounded, discrete time:

$$\begin{aligned}
\mathbf{x}(n\Delta + \Delta) &= \boldsymbol{\Phi}(n\Delta)\mathbf{x}(n\Delta) + \Delta\mathbf{G}(n\Delta)\mathbf{w}(n\Delta) \\
\mathbf{z}(n\Delta) &= \mathbf{H}(n\Delta)\mathbf{x}(n\Delta) + \mathbf{v}(n\Delta) \\
\mathbf{x}(0) &\in \boldsymbol{\Omega}_x(0) \\
\mathbf{w}(n\Delta) &\in \boldsymbol{\Omega}_w(n\Delta) \\
\mathbf{v}(n\Delta) &\in \boldsymbol{\Omega}_v(n\Delta)
\end{aligned} \tag{3.1.3}$$

For ellipsoidal sets,

$$\begin{aligned}
\boldsymbol{\Omega}_x(0) &= \{\mathbf{x} : \mathbf{x}'\boldsymbol{\psi}^{-1}\mathbf{x} \le 1\} \\
\boldsymbol{\Omega}_w(n\Delta) &= \{\mathbf{w} : \mathbf{w}'\mathbf{Q}^{-1}(n\Delta)\mathbf{w} \le 1\} \\
\boldsymbol{\Omega}_v(n\Delta) &= \{\mathbf{v} : \mathbf{v}'\mathbf{R}^{-1}(n\Delta)\mathbf{v} \le 1\}
\end{aligned}$$

Unknown-but-bounded, continuous time:

$$\begin{aligned}
\frac{d}{dt}\mathbf{x}(t) &= \mathbf{F}(t)\mathbf{x}(t) + \mathbf{G}(t)\mathbf{w}(t) \\
\mathbf{z}(t) &= \mathbf{H}(t)\mathbf{x}(t) + \mathbf{v}(t) \\
\mathbf{x}(0) &\in \boldsymbol{\Omega}_x(0) \\
\mathbf{w}(t) &\in \boldsymbol{\Omega}_w(t) \\
\mathbf{v}(t) &\in \boldsymbol{\Omega}_v(t)
\end{aligned} \tag{3.1.4}$$

For ellipsoidal sets,

$$\begin{aligned}
\boldsymbol{\Omega}_x(0) &= \{\mathbf{x} : \mathbf{x}'\boldsymbol{\psi}^{-1}\mathbf{x} \le 1\} \\
\boldsymbol{\Omega}_w(t) &= \{\mathbf{w} : \mathbf{w}'\mathbf{Q}^{-1}(t)\mathbf{w} \le 1\} \\
\boldsymbol{\Omega}_v(t) &= \{\mathbf{v} : \mathbf{v}'\mathbf{R}^{-1}(t)\mathbf{v} \le 1\}
\end{aligned}$$

3.1.5 Discussion of Basic Model Form

The basic model form to be studied in this book combines a state space structure with uncertainties modeled by white processes. As will be seen in later chapters, the use of this model form simplifies the mathematical manipulation and provides a good basis for implementation. However, many readers with experience will be worried by the fact that many real-world models do not appear to be in this state space-white process form. For example, the real world is full of stochastic situations which involve things such as nonzero mean processes, correlated processes, bias errors, and dynamic systems which are not represented in state space form. The rest of the chapter is devoted to easing the reader's worries by showing that *the vast majority of real-world problems can be expressed in the state space-white process form.*

3.2 Generality of Basic Bayesian Model

The basic Bayesian model of (3.1.1) or (3.1.2) appears to be highly specialized. However, an extremely wide range of problems can be put into this form. Figure 3.2.1 tabulates 16 extensions or special cases of the basic Bayesian model (some of which can be viewed as Fisher models). These 16 cases are now discussed in more detail to illustrate the generality of the basic model.

3.2.1 Time-Invariant Systems

Consider (3.1.1) and (3.1.2). The corresponding time-invariant system models are obtained by making $\mathbf{\Phi}$, $\mathbf{F}$, $\mathbf{G}$, $\mathbf{H}$, $\mathbf{Q}$, and $\mathbf{R}$ constants.

A time-invariant Bayesian system model does not necessarily imply that the state $\mathbf{x}(\cdot)$ is a stationary stochastic process. The state may be a nonstationary process because of a *transient* caused by the initial conditions. If the model is a stable system, this transient may "die out" so that as time increases, the state becomes a stationary process.

EXAMPLE: The time-invariant Bayesian model

$$x(n+1) = 5x(n) + w(n)$$

results in an $x(n)$ which is never a stationary process.

3.2.2 No Dynamics

Consider (3.1.1) and (3.1.2) when

$$\mathbf{\Phi}(n\Delta) = \mathbf{I} \qquad \text{or} \qquad \mathbf{F}(t) = \mathbf{0}$$

$$\mathbf{G}(\cdot) = \mathbf{0}$$

Case	Special Case or Extension	Method of Handling
1	Time invariant system	Set $\mathbf{\Phi}(n\Delta) = \mathbf{\Phi}$, $\mathbf{G}(n\Delta) = \mathbf{G}$, etc.
2	No dynamics	Set $\mathbf{\Phi} = \mathbf{I}$ or $\mathbf{F} = \mathbf{0}$, $\mathbf{G} = \mathbf{0}$
3	Static model	Use $\mathbf{z} = \mathbf{Hx} + \mathbf{v}$
4	No observations	Set $\mathbf{H} = \mathbf{0}$
5	Perfect observations	Set $\mathbf{R} = \mathbf{0}$
6	Known input	Treat separately by addition and subtraction
7	No input uncertainty	Set $\mathbf{Q} = \mathbf{0}$
8	Known initial conditions	Set $\boldsymbol{\psi} = \mathbf{0}$; treat separately by addition and subtraction
9	Completely unknown initial conditions: Fisher	Set $\boldsymbol{\psi} = \infty\mathbf{I}$*
10	Nonzero mean initial conditions and disturbances	Treat separately by addition and subtraction
11	Completely unknown input: Fisher	Set $\mathbf{Q} = \infty\mathbf{I}$*
12	Nonwhite (time correlated) disturbance	Use state augmentation
13	Biases	Limiting case of nonwhite disturbance
14	Sensor dynamics	Use state augmentation
15	Observations of input disturbance	Use observations as known input plus disturbances
16	Gaussian assumption	Gives "stronger" interpretation to results

*True only in limited sense for certain applications.

Figure 3.2.1 Extensions and special cases of Bayesian linear model.

In such a case, the state is constant

$$\mathbf{x}(\cdot) = \mathbf{x}(0) = \mathbf{x}$$

so

$$\mathbf{z}(\cdot) = \mathbf{H}(\cdot)\mathbf{x} + \mathbf{v}(\cdot)$$

This is called a no-dynamics model.

EXAMPLE: Let x denote the length of a desk and $z(n)$, $n = 1, \ldots, 3$, represent three measurements with three different rulers, where $v(n)$ represents the errors in the rulers' calibrations. Then

$$z(n) = x + v(n), \qquad n = 1, 2, 3$$

3.2.3 *Static Case*

A further simplification of the no-dynamics model yields the static case defined by

$$\mathbf{z} = \mathbf{H}\mathbf{x} + \mathbf{v}$$

Time is not a factor. This static case is actually used as the starting point in the developments of later chapters.

EXAMPLE: Consider the no-dynamics "desk" example of Section 3.2.2. It can be viewed as a static case by writing

$$\mathbf{z} = \begin{bmatrix} z(1) \\ z(2) \\ z(3) \end{bmatrix} \qquad \mathbf{v} = \begin{bmatrix} v(1) \\ v(2) \\ v(3) \end{bmatrix}$$

$$\mathbf{H} = \begin{bmatrix} 1 \\ 1 \\ 1 \end{bmatrix}$$

$$\mathbf{z} = \mathbf{H}x + \mathbf{v}$$

3.2.4 *No Observations*

If there are no observations $\mathbf{z}(\cdot)$, the corresponding equations can simply be ignored. However, the case of no observations can also be viewed as a limiting case of the general model where $\mathbf{H}(\cdot) = \mathbf{0}$. In some sense, the case $\mathbf{R}(\cdot) = \infty\ \mathbf{I}$ also corresponds to no observations.

3.2.5 *Perfect Observations*

It can be argued that perfect observations are impossible, but in some applications it may be desired to model the observations as being perfect; i.e.,

$$\mathbf{z}(\cdot) = \mathbf{H}(\cdot)\mathbf{x}(\cdot)$$

The results for this case can be obtained as the limiting case when

$$\mathbf{R}(\cdot) = \mathbf{0}$$

In certain applications, some but not all of the observations may be viewed as being perfect. This can be modeled by allowing **R** to be a positive semi-definite matrix.

EXAMPLE: Consider

$$\mathbf{z}(n) = \begin{bmatrix} z_1(n) \\ z_2(n) \end{bmatrix}$$

$z_1(n)$ can be modeled as a perfect observation by using

$$\mathbf{R} = \begin{bmatrix} 0 & 0 \\ 0 & R_{22} \end{bmatrix}$$

It will be seen in later chapters that the case where $\mathbf{R}^{-1}$ does not exist often requires special handling, as many of the equations are derived assuming that $\mathbf{R}^{-1}$ exists and have terms involving $\mathbf{R}^{-1}$.

3.2.6 Known Inputs

In many applications, the system has known inputs as well as uncertain inputs. This is modeled by extending (3.1.1) and (3.1.2) to be

$$\mathbf{x}(n\Delta + \Delta) = \mathbf{\Phi}(n\Delta)\mathbf{x}(n\Delta) + \Delta\mathbf{G}(n\Delta)\mathbf{w}(n\Delta) + \Delta\mathbf{B}(n\Delta)\mathbf{u}(n\Delta)$$

$$\frac{d}{dt}\mathbf{x}(t) = \mathbf{F}(t)\mathbf{x}(t) + \mathbf{G}(t)\mathbf{w}(t) + \mathbf{B}(t)\mathbf{u}(t)$$

$$\left.\begin{matrix} \mathbf{u}(n\Delta) \\ \mathbf{u}(t) \end{matrix}\right\} \quad \text{known perfectly}$$

Such known inputs can conceptually be handled by treating them as a separate problem and "adding and subtracting" their effects.

In the discrete-time case, this can be done by defining

$$\mathbf{x}(n\Delta) = \mathbf{x}_1(n\Delta) + \mathbf{x}_2(n\Delta)$$

$$\mathbf{x}_1(n\Delta + \Delta) = \mathbf{\Phi}(n\Delta)\mathbf{x}_1(n\Delta) + \Delta\mathbf{G}(n\Delta)\mathbf{w}(n\Delta)$$

$$\mathbf{x}_1(0)\text{: uncertain}$$

$$\mathbf{x}_2(n\Delta + \Delta) = \mathbf{\Phi}(n\Delta)\mathbf{x}_2(n\Delta) + \Delta\mathbf{B}(n\Delta)\mathbf{u}(n\Delta)$$

$$\mathbf{x}_2(0) = \mathbf{0}$$

The behavior of the uncertain process $\mathbf{x}_1(n\Delta)$ and the certain process $\mathbf{x}_2(n\Delta)$ can be analyzed separately and then "added" together. Observations can be handled conceptually by defining

$$\mathbf{z}_1(n\Delta) = \mathbf{z}(n\Delta) - \mathbf{H}(n\Delta)\mathbf{x}_2(n\Delta)$$

Since $\mathbf{x}_2(n\Delta)$ is known, $\mathbf{z}_1(n\Delta)$ can be calculated by subtraction and viewed as a pseudo-observation modeled by

$$\mathbf{z}_1(n\Delta) = \mathbf{H}(n\Delta)\mathbf{x}_1(n\Delta) + \mathbf{v}(n\Delta)$$

The pseudo-observations $\mathbf{z}_1(n\Delta)$ can then be used to find out how $\mathbf{x}_1(n\Delta)$ is behaving. The continuous-time case follows in a similar fashion. The ideas are illustrated in Fig. 3.2.2.

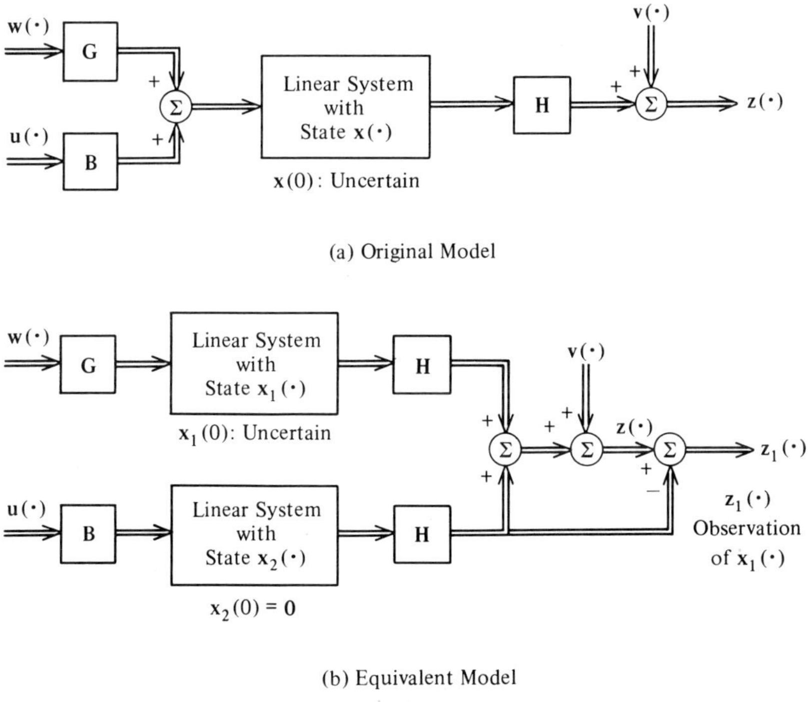

Figure 3.2.2 System with known inputs.

The above technique of handling known inputs by addition and subtraction depends heavily on the assumption that the model is linear.

The addition-subtraction technique is presented as a conceptual aid to show how the problem of known inputs can be handled. In practice, the known inputs can often be "blended" directly into the results so that addition and subtraction is not explicitly done.

EXAMPLE: The system

$$x(n\Delta + \Delta) = \Phi x(n\Delta) + w(n\Delta) + \sin(n\Delta)$$

$$z(n\Delta) = x(n\Delta) + v(n\Delta)$$

$$\mathbf{x}(0),\ w(n\Delta),\ v(n\Delta)\text{: zero mean, uncorrelated}$$

$$E\{x^2(0)\} = \psi \qquad E\{w(n_1\Delta)w(n_2\Delta)\} = \begin{cases} Q, & n_1 = n_2 \\ 0, & n_1 \neq n_2 \end{cases}$$

$$E\{v(n_1\Delta)v(n_2\Delta)\} = \begin{cases} R, & n_1 = n_2 \\ 0, & n_1 \neq n_2 \end{cases}$$

can be modeled as

$$\begin{aligned} x(n\Delta) &= x_1(n\Delta) + x_2(n\Delta) \\ x_1(n\Delta + \Delta) &= \Phi x_1(n\Delta) + w(n\Delta) \\ x_2(n\Delta + \Delta) &= \Phi x_2(n\Delta) + \sin(n\Delta) \\ E\{x_2^2(0)\} &= 0 \\ x_2(0) &= 0 \\ z_1(n\Delta) &= z(n\Delta) - x_2(n\Delta) = x_1(n\Delta) + v(n\Delta) \end{aligned}$$

3.2.7 No Input Uncertainty

The special case of (3.1.1) and (3.1.2) where there is no input uncertainty can be modeled by setting

$$\mathbf{Q}(\cdot) = \mathbf{0}$$

In such a case, $\mathbf{x}(\cdot)$ may still be an uncertain process because of uncertainty in $\mathbf{x}(0)$.

3.2.8 Known Initial Conditions

In some situations, the initial conditions $\mathbf{x}(0)$ can be assumed to be known perfectly. If the initial conditions are zero, $\mathbf{x}(0) = \mathbf{0}$, this can be handled by setting $\boldsymbol{\psi} = \mathbf{0}$. The case of nonzero but known initial conditions can conceptually be handled by setting $\boldsymbol{\psi} = \mathbf{0}$ and using addition and subtraction, as done with known inputs. For example, Fig. 3.2.2(b) can be used with $\mathbf{u}(\cdot) = \mathbf{0}$ but a nonzero initial condition on the system which generates $\mathbf{x}_2(\cdot)$.

3.2.9 Completely Unknown Initial Conditions: Fisher

In many applications, the initial conditions are completely unknown. A model where $\mathbf{x}(0)$ is completely unknown is a type of Fisher model. See Fig. 3.1.2. Conceptually such Fisher models have to be handled in a different fashion from Bayesian models where $\mathbf{x}(0)$ is viewed as a random vector with known covariance matrix $\boldsymbol{\psi}$. In later chapters the necessary logic is developed. However, it turns out that for many applications the results of a Fisher modeling and analysis are equivalent to a Bayesian model in which

$$\boldsymbol{\psi} = \infty\,\mathbf{I}$$

This is not surprising.

3.2.10 Nonzero Mean Initial Conditions and Disturbances

Consider a model where the initial condition, input, and sensor disturbance uncertainty are not zero mean. If the mean values are known, they can be treated as known initial conditions and inputs and can be handled by addition and subtraction, as in Sections 3.2.6 and 3.2.8.

EXAMPLE: Consider

$$\begin{aligned} x(n+1) &= 0.5x(n) + \tilde{w}(n) \\ z(n) &= x(n) + \tilde{v}(n) \\ E\{x(0)\} &= 1 \\ \tilde{w}(n) &= 2 + w(n) \\ \tilde{v}(n) &= 4 + v(n) \\ w(n), v(n)&: \quad \text{white processes} \end{aligned}$$

This model can be rewritten as

$$\begin{aligned} x(n) &= x_1(n) + x_2(n) \\ x_1(n+1) &= 0.5x_1(n) + w(n) \\ x_1(0)&: \text{zero mean random variable} \\ x_2(n+1) &= 0.5x_2(n) + 2 \\ x_2(0) &= 1 \\ z_1(n) &= z(n) - 4 - x_2(n) \end{aligned}$$

$z_1(n)$ can be viewed as a pseudo-observation modeled by

$$z_1(n) = x_1(n) + v(n)$$

When the mean values are not known, they are called biases. This case is discussed in Section 3.2.13.

3.2.11 Completely Unknown Input: Fisher

The input disturbance, $\mathbf{w}(\cdot)$ can be modeled as being completely unknown. This Fisher modeling of the input is analogous to that of Section 3.2.9 on initial conditions (see Fig. 3.1.2). As in Section 3.2.9, Fisher modeling of $\mathbf{w}(\cdot)$ actually requires a separate development, but for some applications it can be viewed as the limiting Bayesian case, where

$$\mathbf{Q}(\cdot) = \infty\, \mathbf{I}$$

Fisher models for the input are usually meaningful only when the dimension of the observation vector ($\mathbf{z}$) is not less than that of the input ($\mathbf{w}$). As will be seen in later chapters, special care is required in handling these models as many of the basic derivations result in equations having terms involving $\mathbf{Q}(\cdot)$.

A Fisher model of completely unknown inputs may initially seem to be of little interest, but it actually has many applications. A prime example occurs when a system's input is so erratic that sensors are used to observe the input. If these sensors have dynamics, the resulting model is Fisher with unknown inputs. This case is discussed further in Section 3.2.15.

3.2.12 Nonwhite (Time-Correlated) Disturbances

In many applications, the input and sensor disturbances are not white processes but rather *structured* processes; i.e., they are time-correlated.

Consider the effect of using a white process as an input to a driven dynamic system with state $\mathbf{x}(\cdot)$ and output $\mathbf{H}(\cdot)\mathbf{x}(\cdot)$. This output is also an uncertain process, but it has time structure in the sense that the value of $\mathbf{x}(t)$ or $\mathbf{x}(n\Delta)$ depends to some extent on the value of $\mathbf{x}(t-\tau)$ or $\mathbf{x}(n\Delta - m\Delta)$. This simple observation forms the basis for handling nonwhite disturbances.

Consider the model of Fig. 3.2.3 where the disturbances are not white

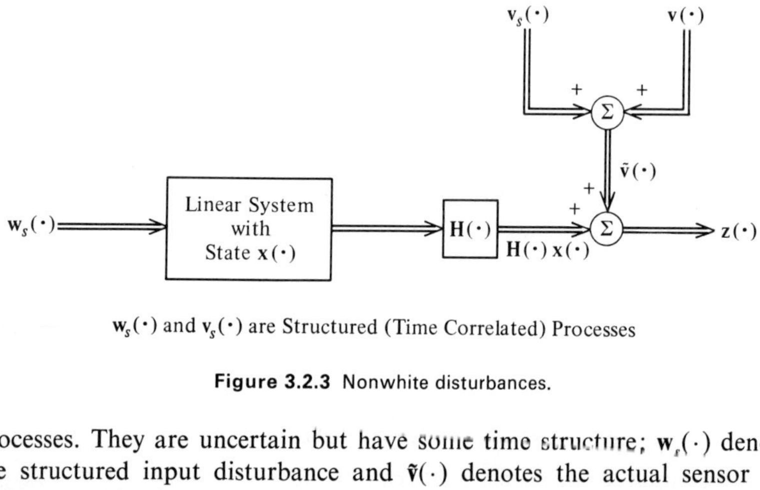

$\mathbf{w}_s(\cdot)$ and $\mathbf{v}_s(\cdot)$ are Structured (Time Correlated) Processes

Figure 3.2.3 Nonwhite disturbances.

processes. They are uncertain but have some time structure; $\mathbf{w}_s(\cdot)$ denotes the structured input disturbance and $\tilde{\mathbf{v}}(\cdot)$ denotes the actual sensor disturbance, which is the sum of a white process $\mathbf{v}(\cdot)$ and a structured process $\mathbf{v}_s(\cdot)$. Assume that the structured processes $\mathbf{w}_s(\cdot)$ and $\mathbf{v}_s(\cdot)$ can each be modeled as the output of a driven dynamic system with white inputs, $\mathbf{w}_w(\cdot)$ and $\mathbf{w}_v(\cdot)$, as illustrated in Fig. 3.2.4. Then a complete model can be obtained by combining Fig. 3.2.3 and Fig. 3.2.4 as in Fig. 3.2.5. Finally, since all the systems are linear, the various blocks of Fig. 3.2.5 can be combined to give Fig. 3.2.6. Figure 3.2.6 is back in the form of the basic model of Fig. 3.1.1, where the input and sensor disturbances are white. The combined state vector $\mathbf{x}_c(\cdot)$ is now larger, but the basic form is the same. The logic involved in the preceding steps is called augmenting the state vector or *state augmentation.*

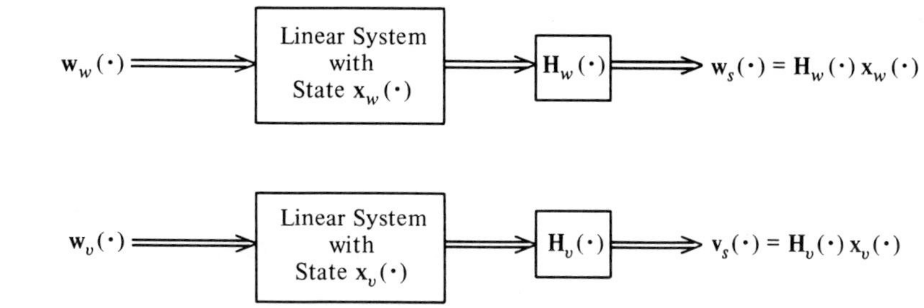

Figure 3.2.4 Models for two structured uncertain processes, $\mathbf{w}_s(\cdot)$ and $\mathbf{v}_s(\cdot)$.

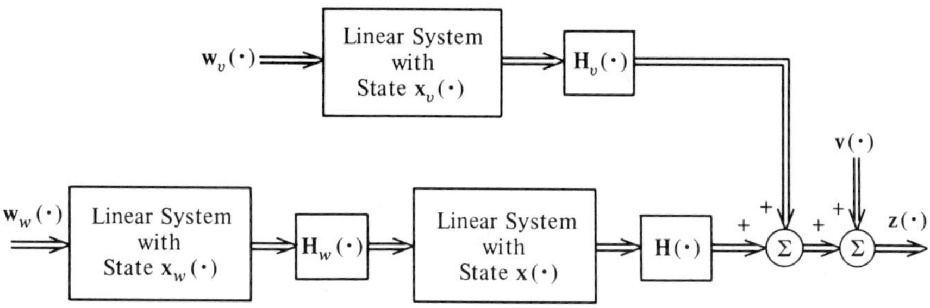

Figure 3.2.5 Complete model.

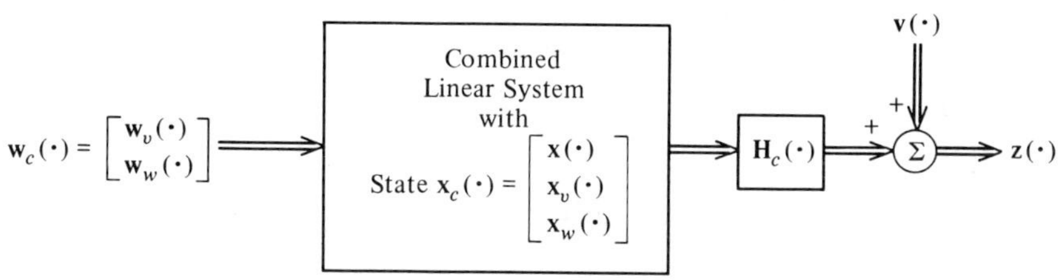

Figure 3.2.6 Combined form of complete model.

The crucial step is that of Fig. 3.2.4, where the structured processes are modeled as outputs of linear systems driven by white disturbances. In theory, not all structured processes can be modeled as in Fig. 3.2.4, but in practice such models can almost always provide satisfactory approximations.

Consider Fig. 3.2.3 and assume that the actual sensor disturbance $\tilde{\mathbf{v}}(\cdot)$ has no white component at all; i.e., $\mathbf{v}(\cdot) = \mathbf{0}$. Then the resultant model of Fig. 3.2.6 has $\mathbf{v}(\cdot) = \mathbf{0}$ and corresponds to the perfect observation case discussed in Section 3.2.5. This is a very important point which deserves emphasis; i.e., *the assumption of sensor disturbance with no white component results in a mathematical model with perfect observations.*

EXAMPLE: Consider

$$\frac{d}{dt} x_1(t) = F_1 x_1(t) + G_1 w_1(t)$$

$$\frac{d}{dt} v_s(t) = F_v v_s(t) + G_v w_v(t)$$

$$z(t) = x_1(t) + v_s(t)$$

where $w_1(t)$ and $w_v(t)$ are white stochastic processes. Define

$$\mathbf{x}(t) = \begin{bmatrix} x_1(t) \\ v_s(t) \end{bmatrix} \qquad \mathbf{w}(t) = \begin{bmatrix} w_1(t) \\ w_v(t) \end{bmatrix}$$

Then

$$\frac{d}{dt}\mathbf{x}(t) = \begin{bmatrix} F_1 & 0 \\ 0 & F_v \end{bmatrix}\mathbf{x}(t) + \begin{bmatrix} G_1 & 0 \\ 0 & G_v \end{bmatrix}\begin{bmatrix} w_1(t) \\ w_v(t) \end{bmatrix}$$

$$z(t) = [1 \quad 1]\mathbf{x}(t)$$

3.2.13 Biases

In this book, a bias is viewed as an uncertain but constant disturbancc. A bias can be considered a limiting case of a nonwhite disturbance as a constant is the *most time-correlated* process possible. To illustrate, consider a sensor disturbance $\tilde{\mathbf{v}}(\cdot)$ modeled as

$$\tilde{\mathbf{v}}(\cdot) = \mathbf{v}(\cdot) + \mathbf{b}$$

$$\mathbf{v}(\cdot): \ \text{white process}$$

$$\mathbf{b}: \ \text{bias}$$

where $\mathbf{b}$ is modeled as a random vector with

$$E\{\mathbf{b}\} = 0$$

$$E\{\mathbf{b}\mathbf{b}'\} = \boldsymbol{\psi}_b$$

Such a bias can be incorporated using the state augmentation ideas of Section 3.2.12 by introducing a new state $\mathbf{x}_b(\cdot)$, defined by

$$\mathbf{x}_b(n\Delta + \Delta) = \mathbf{x}_b(n\Delta) \qquad \mathbf{x}_b(n\Delta) = \mathbf{b}$$

or

$$\frac{d}{dt}\mathbf{x}_b(t) = \mathbf{0} \qquad \mathbf{x}_b(t) = \mathbf{b}$$

$$E\{\mathbf{x}_b(0)\} = 0$$

$$E\{\mathbf{x}_b(0)\mathbf{x}_b'(0)\} = \boldsymbol{\psi}_b$$

Biased input disturbances follow in a similar fashion.

Note that in this book a fundamental distinction is made between a biased disturbance and a nonzero mean disturbance, as discussed in Section 3.2.10. A disturbance with a known, nonzero, constant mean value is not called biased. A bias is a constant disturbance whose value is uncertain.

EXAMPLE: Let x denote the length of a table and let $z(n)$, $n = 1, \ldots,$ denote measurements made using the same ruler but by different people. If the ruler, is known to be 3 inches too long, then

$$z(n) = x + \tilde{v}(n)$$
$$E\{\tilde{v}(n)\} = 3$$

where $\tilde{v}(n)$ is nonzero mean disturbance. If the ruler were selected from a large pile of rulers whose calibration errors, b, were random with mean zero but variance 9, then

$$z(n) = x + v(n) + b$$
$$E\{b\} = 0$$
$$E\{b^2\} = 9$$

where b is a bias error.

3.2.14 Sensor Dynamics

Consider Fig. 3.2.7, which represents a linear system whose state $\mathbf{x}(\cdot)$ is being observed by sensors to yield $\mathbf{z}(\cdot)$. The model for the sensor implied by (3.1.1) and (3.1.2) is

$$\mathbf{z}(\cdot) = \mathbf{H}(\cdot)\mathbf{x}(\cdot) + \mathbf{v}(\cdot)$$

so that the sensor is an "instantaneous" device. In practice, sensors usually have dynamic response characteristics; for example, (1) a voltmeter may have mechanical moving parts, (2) a frequency meter usually measures only average frequency over some small time interval, (3) a thermocouple may have thermal inertia, and (4) a nationwide poll takes time to perform. Thus the model corresponding to Fig. 3.2.7 is often as shown in Fig. 3.2.8(a)

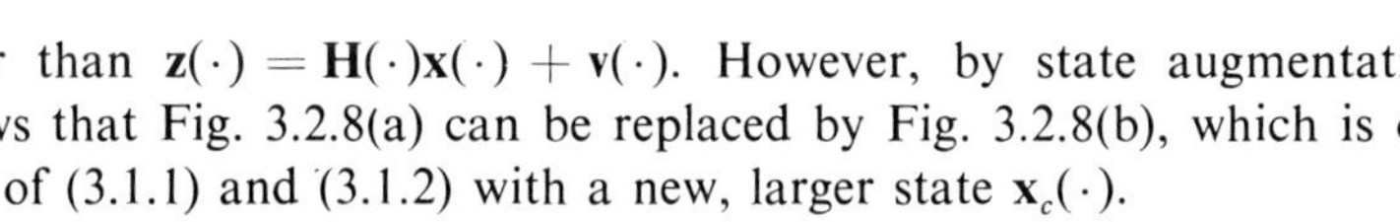

Figure 3.2.7

rather than $\mathbf{z}(\cdot) = \mathbf{H}(\cdot)\mathbf{x}(\cdot) + \mathbf{v}(\cdot)$. However, by state augmentation it follows that Fig. 3.2.8(a) can be replaced by Fig. 3.2.8(b), which is of the form of (3.1.1) and (3.1.2) with a new, larger state $\mathbf{x}_c(\cdot)$.

3.2.15 Observation of Input Disturbances

Consider a linear system with input disturbance $\tilde{\mathbf{w}}(\cdot)$. In many applications, observations may be made both on $\tilde{\mathbf{w}}(\cdot)$ and on the state $\mathbf{x}(\cdot)$, as illustrated in Fig. 3.2.9. In some cases only observations on $\tilde{\mathbf{w}}(\cdot)$ may be available. The method of handling a situation such as Fig. 3.2.9 depends on the nature of $\tilde{\mathbf{w}}(\cdot)$. Three cases are discussed.

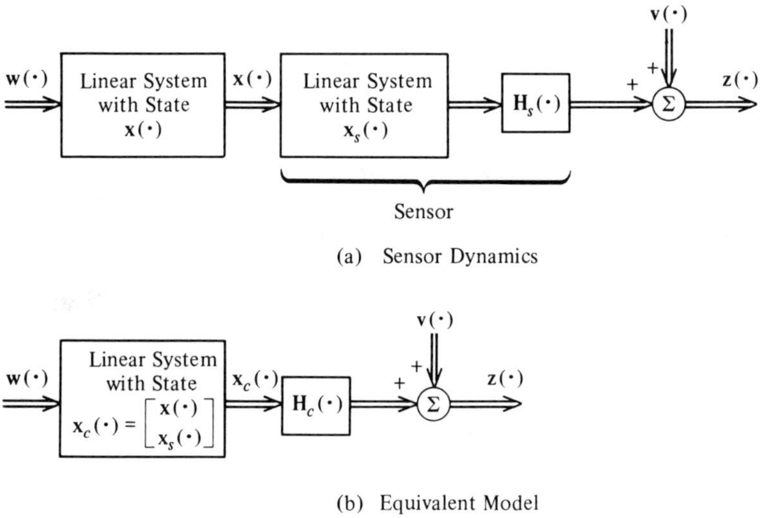

Figure 3.2.8 Sensor dynamics by state augmentation.

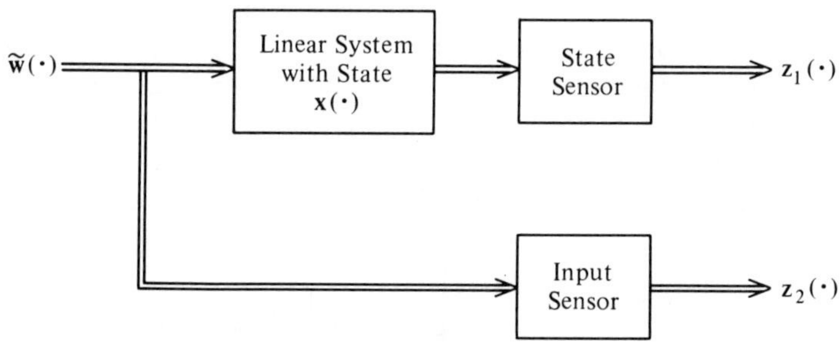

Figure 3.2.9 Observation of state and input disturbance.

Assume that $\tilde{\mathbf{w}}(\cdot)$ is a time-correlated (structured) nonwhite process. Then state augmentation as in Section 3.2.12 (and Section 3.2.14 if sensor dynamics are present) immediately yields a model of the basic form of (3.1.1) and (3.1.2).

Assume that the input and state sensors have no dynamics and that $\tilde{\mathbf{w}}(\cdot)$ is completely unknown, i.e., a Fisher model, as in Section 3.2.11. Then

$$\mathbf{z}_2(\cdot) = \mathbf{w}(\cdot) + \mathbf{v}_2(\cdot)$$

$$\mathbf{w}(\cdot)\text{: completely unknown}$$

as illustrated in Fig. 3.2.10(a). An equivalent model is shown in Fig. 3.2.10(b), where $\mathbf{z}_2(\cdot)$ is viewed as a known input and handled as in Section 3.2.6. Thus observation of the input can turn a Fisher model with completely

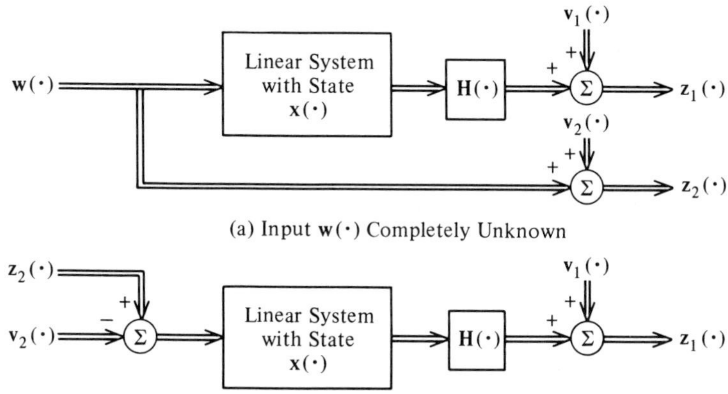

(a) Input w(·) Completely Unknown

(b) Equivalent Model, $z_2(\cdot)$ Viewed as Known Input

Figure 3.2.10 Observations of completely unknown input.

unknown inputs into a Bayesian model with known input plus a white input which is actually the negative of the input sensor disturbance.

Assume that $\mathbf{w}(\cdot)$ is completely unknown but that the input sensor has dynamics. State augmentation then yields a model like (3.1.1) and (3.1.2). However, in this case when the input sensor has dynamics, the resultant model is still Fisher, as in Section 3.2.11.

3.2.16 Gaussian Assumption

The stochastic models of Section 3.1.2 and all the preceding discussions have involved only the mean value and covariance matrices (first and second moments) of the random variables and processes. In later chapters, the additional assumption of Gaussian (or normal) probability densities is sometimes added to the model. The main effect of the Gaussian assumption is that stronger statements can be made on the meaning and nature of the results.

3.3 Generality of Basic Unknown-but-Bounded Models

The basic unknown-but-bounded models are summarized in (3.1.3) and (3.1.4). The Bayesian-Fisher models and unknown-but-bounded models are fundamentally different both in terms of physical assumptions and interpretations and in terms of type of mathematical concepts required. However, when ellipsoidal sets are used, the unknown-but-bounded and stochastic model equations are very similar in appearance. Another similarity is the fact that almost all the ideas of Section 3.2 on extensions and special cases of Bayesian models also apply to unknown-but-bounded models. (Section 3.2.16 on Gaussian models is an exception.) Therefore no attempt is made to duplicate the discussion of Section 3.2 for unknown-but-bounded models.

3.4 Generality of State Space Models

The basic state space structures of linear systems are ($\Delta = 1$)

$$\begin{aligned} \mathbf{x}(n+1) &= \mathbf{\Phi}(n)\mathbf{x}(n) + \mathbf{G}(n)\mathbf{w}(n) \\ z(n) &= \mathbf{H}(n)\mathbf{x}(n) + \mathbf{v}(n) \end{aligned} \tag{3.4.1}$$

or

$$\begin{aligned} \frac{d}{dt}\mathbf{x}(t) &= \mathbf{F}(t)\mathbf{x}(t) + \mathbf{G}(t)\mathbf{w}(t) \\ \mathbf{z}(t) &= \mathbf{H}(t)\mathbf{x}(t) + \mathbf{v}(t) \end{aligned} \tag{3.4.2}$$

The generality of these state space model forms are now discussed. The discussions center on structural form, and uncertainty plays little role.

3.4.1 Nonuniqueness of Model

A fundamental property of a state space model is that it is not unique in the sense that many different sets of values of $\mathbf{\Phi}$ (or $\mathbf{F}$), $\mathbf{G}$, and $\mathbf{H}$ can give the same input-output behavior, i.e., the same relation of $\mathbf{x}(0)$ and $\mathbf{w}(\cdot)$ to $\mathbf{z}(\cdot)$. The choice of a particular $\mathbf{\Phi}$ (or $\mathbf{F}$), $\mathbf{G}$, and $\mathbf{H}$ corresponds to the choice of a *coordinate system*. This choice can make a big difference in numerical studies. Two common forms are the *standard controllable form* and the *standard observable form*.

Further discussions on this important topic are not provided in this book.

3.4.2 Autoregressive and Moving-Average Models

Three model forms of interest are

1. Autoregressive models.
2. Moving-average models.
3. Mixed autoregressive and moving-average models.

These names historically arose in studies of stochastic models and statistical analysis of time series.

A discrete-time autoregressive model is ($\Delta = 1$)

$$\mathbf{y}(n+1) + \sum_{m=1}^{M_1} \mathbf{\Phi}_m(n)\mathbf{y}(n-m+1) = \mathbf{G}(n)\mathbf{w}(n)$$

$$\mathbf{z}(n) = \mathbf{y}(n)$$

Initial conditions are ignored for simplicity. Such an autoregressive model can be expressed in terms of a basic state space model such as (3.4.1). To illustrate, assume that $M_1 = 2$. Define

$$\mathbf{x}(n) = \begin{bmatrix} \mathbf{x}_1(n) \\ \mathbf{x}_2(n) \end{bmatrix} = \begin{bmatrix} \mathbf{y}(n-1) \\ \mathbf{y}(n) \end{bmatrix}$$

$$\mathbf{\Phi}(n) = \begin{bmatrix} \mathbf{0} & \mathbf{I} \\ -\mathbf{\Phi}_2(n) & -\mathbf{\Phi}_1(n) \end{bmatrix}$$

$$\tilde{\mathbf{G}}(n) = \begin{bmatrix} \mathbf{0} \\ \mathbf{G}(n) \end{bmatrix} \qquad \mathbf{H} = [\mathbf{0} \quad \mathbf{I}]$$

Then the state space model

$$\mathbf{x}(n+1) = \mathbf{\Phi}(n)\mathbf{x}(n) + \tilde{\mathbf{G}}(n)\mathbf{w}(n)$$

$$\mathbf{z}(n) = \mathbf{H}\mathbf{x}(n)$$

is completely equivalent to the original autoregressive model.

A continuous-time autoregressive model for $\mathbf{M}_1 = 2$ is

$$\mathbf{F}_0(t)\mathbf{y}(t) + \mathbf{F}_1(t)\frac{d}{dt}\mathbf{y}(t) + \frac{d^2}{dt^2}\mathbf{y}(t) = \mathbf{G}(t)\mathbf{w}(t)$$

$$\mathbf{z} = \mathbf{y}(t)$$

This can be expressed as a special case of (3.4.2) by choosing

$$\mathbf{x}(t) = \begin{bmatrix} \mathbf{x}_1(t) \\ \mathbf{x}_2(t) \end{bmatrix} = \begin{bmatrix} \mathbf{y}(t) \\ \frac{d}{dt}\mathbf{y}(t) \end{bmatrix}$$

$$\mathbf{F}(t) = \begin{bmatrix} \mathbf{0} & \mathbf{I} \\ -\mathbf{F}_0(t) & -\mathbf{F}_1(t) \end{bmatrix}$$

$$\tilde{\mathbf{G}}(t) = \begin{bmatrix} \mathbf{0} \\ \mathbf{G}(t) \end{bmatrix}$$

$$\mathbf{H}(t) = [\mathbf{I} \quad \mathbf{0}]$$

A discrete-time moving-average model ($\Delta = 1$) is

$$\mathbf{y}(n) = \sum_{m=1}^{M_2} \mathbf{B}_m(n)\mathbf{w}(n-m)$$

$$\mathbf{z}(n) = \mathbf{y}(n)$$

Such discrete-time moving-average models can also be expressed in terms of (3.4.1). To illustrate, assume that $M_2 = 3$. Then the following is an equivalent model:

$$\mathbf{x}(n) = \begin{bmatrix} \mathbf{x}_1(n) \\ \mathbf{x}_2(n) \\ \mathbf{x}_3(n) \end{bmatrix} = \begin{bmatrix} \mathbf{w}(n-3) \\ \mathbf{w}(n-2) \\ \mathbf{w}(n-1) \end{bmatrix}$$

$$\mathbf{\Phi} = \begin{bmatrix} \mathbf{0} & \mathbf{I} & \mathbf{0} \\ \mathbf{0} & \mathbf{0} & \mathbf{I} \\ \mathbf{0} & \mathbf{0} & \mathbf{0} \end{bmatrix}$$

$$\mathbf{G} = \begin{bmatrix} 0 \\ 0 \\ \mathbf{I} \end{bmatrix}$$

$$\mathbf{H}(n) = [\mathbf{B}_3(n) \quad \mathbf{B}_2(n) \quad \mathbf{B}_1(n)]$$

$$\mathbf{x}(n+1) = \mathbf{\Phi x}(n) + \mathbf{Gw}(n)$$

$$\mathbf{z}(n) = \mathbf{H}(n)\mathbf{x}(n)$$

Initial conditions are ignored for simplicity.

A continuous-time moving-average model is

$$\mathbf{y}(t) = \int_0^t \mathbf{B}(t, \tau)\mathbf{w}(t - \tau)\, d\tau$$

A related continuous time model is

$$\mathbf{y}(t) = \mathbf{B}_1(t)\mathbf{w}(t) + \mathbf{B}_2(t)\frac{d}{dt}\mathbf{w}(t) + \mathbf{B}_3(t)\frac{d^2}{dt^2}\mathbf{w}(t)$$

However, neither of these forms can be expressed in terms of (3.4.2).

In discrete time, a mixed autoregressive–moving-average model is

$$\mathbf{y}(n+1) + \sum_{m=1}^{M_1} \mathbf{\Phi}_m(n)\mathbf{y}(n - m + 1) = \sum_{m=1}^{M_2} \mathbf{B}_m(n)\mathbf{w}(n - m) + \mathbf{G}(n)\mathbf{w}(n)$$

$$\mathbf{z}(n) = \mathbf{y}(n)$$

This can be modeled in terms of (3.4.1) by simply combining the individual autoregressive and moving-average models. This straightforward modeling may result in a model with more state variables than are actually needed.

EXAMPLE: Consider

$$y(n+1) + \Phi_1 y(n) + \Phi_2 y(n-1) = w(n) + Bw(n-1)$$

One equivalent state space model is

$$\mathbf{x}(n) = \begin{bmatrix} x_1(n) \\ x_2(n) \end{bmatrix}$$

$$\begin{bmatrix} x_1(n+1) \\ x_2(n+1) \end{bmatrix} = \begin{bmatrix} 0 & -\Phi_2 \\ 1 & -\Phi_1 \end{bmatrix} \begin{bmatrix} x_1(n) \\ x_2(n) \end{bmatrix} + \begin{bmatrix} B \\ 1 \end{bmatrix} w(n)$$

$$y(n) = x_2(n)$$

A straightforward combination of individual state space models could yield

$$\begin{bmatrix} x_1(n+1) \\ x_2(n+1) \\ x_3(n+1) \end{bmatrix} = \begin{bmatrix} 0 & -\Phi_2 & 0 \\ 1 & -\Phi_1 & B \\ 0 & 0 & 0 \end{bmatrix} \begin{bmatrix} x_1(n) \\ x_2(n) \\ x_3(n) \end{bmatrix} + \begin{bmatrix} 0 \\ 1 \\ 1 \end{bmatrix} w(n)$$

$$y(n) = x_2(n)$$

A continuous-time mixed autoregressive–moving-average model is ($M_1 = 2$, $M_2 = 1$)

$$\frac{d^2\mathbf{y}(t)}{dt} + \mathbf{F}_1(t)\frac{d}{dt}\mathbf{y}(t) + \mathbf{F}_0(t)\mathbf{y}(t) = \mathbf{G}(t)\mathbf{w}(t) + \mathbf{B}(t)\frac{d}{dt}\mathbf{w}(t)$$

As long as $M_1 > M_2$, such continuous-time mixed models can be expressed in terms of the basic models of (3.4.2).

It is seen from the preceding that the autoregressive–moving-average models and the basic models of (3.4.1) and (3.4.2) have much in common. In the vast majority of problems of interest, either basic formulation can be used, but exceptions to this general rule exist.

3.5 Uncertain Model Parameters and System Identification

The basic linear model forms defined in Section 3.1,

1. Cover only systems defined by known $\mathbf{F}$, $\mathbf{\Phi}$, $\mathbf{G}$, and $\mathbf{H}$ matrices, i.e., allow for uncertainty only in the initial conditions, system input, and observation disturbances.
2. Require uncertainty models that are completely defined in terms of known $\boldsymbol{\psi}$, $\mathbf{Q}$, and $\mathbf{R}$ matrices.

The assumption that the $\mathbf{F}$, $\mathbf{\Phi}$, $\mathbf{G}$, $\mathbf{H}$, $\boldsymbol{\psi}$, $\mathbf{Q}$, and $\mathbf{R}$ matrices are known exactly is rarely (if ever) satisfied in practice. If this fact disturbs the reader, he should not despair too much, as there are three ways to handle the problem:

1. Ignore it and simply use an "engineering guess" of the model parameters (i.e., $\mathbf{F}$, $\mathbf{\Phi}$, etc., values).
2. Choose an approach which does not need the uncertain model parameters.
3. Use system identification.

The use of an engineering guess for the model parameters is the simplest and most common approach and it is often satisfactory (in the sense that the final result does the job). In later chapters, techniques for analyzing the effects of errors in the model parameters are discussed.

The weighted least squares estimation concept does not employ explicit models for the uncertainty in $\mathbf{w}$, $\mathbf{v}$, and $\mathbf{x}(0)$ and thus does not need values for the $\mathbf{\Phi}$, $\mathbf{R}$, and $\mathbf{Q}$ matrices. This approach requires one to choose other "similar" parameter values, but the logic underlying their choice is different.

System identification techniques are a more sophisticated approach wherein the observations are used to estimate the values of the model parameters as well as the state. It turns out that state augmentation concepts can often be used to convert a linear system with unknown (or uncertain) model parameters into a nonlinear system.

EXAMPLE: Consider the scalar system

$$x_1(n+1) = \Phi x_1(n) + w(n)$$
$$z(n) = x_1(n) + v(n)$$

The value of Φ is uncertain. The equivalent two-dimensional nonlinear system

$$x_1(n+1) = x_2(n)x_1(n) + w(n)$$
$$x_2(n+1) = x_2(n)$$
$$z(n) = x_1(n) + v(n)$$

does not have uncertain model parameters.

Thus the discussions of Part IV of this book on nonlinear systems can conceptually be applied to many linear systems with uncertain model parameters. However, Chapter 14 of Part V discusses system identification explicitly from a more fundamental point of view.

The study of models with completely known model parameters (as specified in Section 3.1) is important because they must be clearly understood before real-world solutions are attempted.

3.6 Discussion

The basic mathematical model forms of Section 3.1 are based on combining state space structures with white uncertain processes. The main point of this chapter is that an extremely wide range of problems can be put into this form. Sometimes this form is "natural"; in other cases the natural form has to be converted. A state space-white process model does not encompass all cases of possible interest, but it comes close, especially if approximations are allowed.

The role of this chapter relative to real-world problems and the rest of the book is summarized in Fig. 3.6.1. The figure is (hopefully) self-explanatory.

A reader with an interest in a class of physical problems whose natural models are not close to the basic state space form may find the need to go through Step II of Fig. 3.6.1 (i.e., use ideas of this chapter) rather frustrating. The book is written in its present manner primarily because it is not directed toward any particular class of physical problems. In the author's opinion, the basic state space-white process model forms of Section 3.1 are by far the most powerful framework on which to develop both conceptual understanding and practical, useful results. The book's length could double or triple if separate developments were attempted for the other natural mathematical models of possible interest.

Various types of models have been discussed: Bayesian, Fisher, and

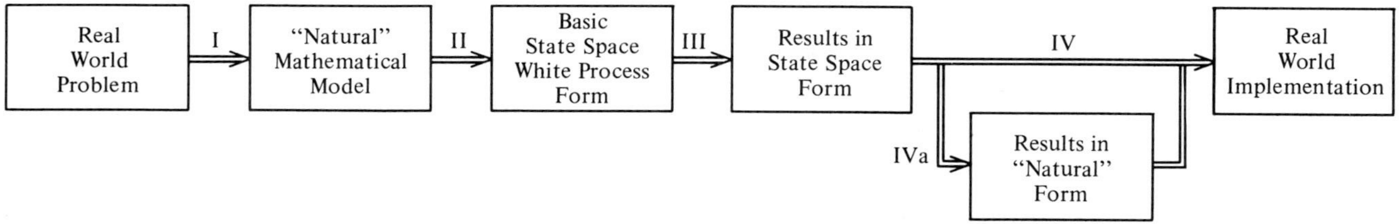

Step I: Up to Engineer
II: Discussed in this Chapter
III: } Discussed in Rest of Book
IV: }
IVa: Reverse Step II

Figure 3.6.1 Possible steps in engineering solution.

unknown-but-bounded. The question "Which model is best?" is meaningless and has no answer. The choice of model depends on the particular problem being studied. Various aspects of choosing a model will be discussed in Chapter 8.

Historical Notes and References: Chapter 3

Most of the ideas of this chapter are in the "folk-lore" category; that is, they are well known by workers in the field, but it is effectively impossible to say who originated them. Book List 3 on Deterministic Systems contains background material on state space concepts, while Book List 4 on Stochastic Systems contains many of the stochastic modeling ideas discussed here. Autoregressive-moving average models are more common in statistical time series references, as found in Mathematical Statistics, Book List 2. Frequency domain models tend to be emphasized in the Communications references of Book List 5. The terms Bayesian model and Fisher model are used in this book to try to categorize types of ideas, but the terminology is by no means universally used. Book List 4 on Stochastic Systems tends to emphasize Bayesian models, while Book List 2 on Mathematical Statistics emphasizes Fisher models. The unknown-but-bounded models are taken from Schweppe [9] and related works (see discussion on Section 6.6).

Exercises

3.1. Consider the no-dynamics scalar model

$$z(n) = x + v(n) + b, \qquad n = 1, \ldots, 3$$

$$E\{x\} = 0 \qquad E\{x^2\} = \psi_x$$

$$E\{v(n)\} = 0 \qquad E\{v^2(n)\} = R, \qquad n = 1, \ldots, 3$$

$$E\{v(n_1)v(n_2)\} = 0, \qquad n_1 \neq n_2$$

a. Describe in words physical (or economic or social, etc.) situations involving observations $z(n)$ on a constant state x with observation errors $v(n) + b$ for the following models for the constant b:

(1) b is a random variable:

$$E\{b\} = m \qquad E\{b^2\} = \psi_b$$

(2) The value of b is known.

(3) The value of b is completely unknown.

b. For each of the three cases of part a, express the model as a static case model:

$$\tilde{\mathbf{z}} = \mathbf{H}\tilde{\mathbf{x}} + \tilde{\mathbf{v}}$$

$$\tilde{\mathbf{z}} = \begin{bmatrix} z(1) \\ z(2) \\ z(3) \end{bmatrix} + \mathbf{c}$$

c: known vector

$$E\{\tilde{\mathbf{x}}\} = \mathbf{0} \qquad E\{\tilde{\mathbf{x}}\tilde{\mathbf{x}}'\} = \boldsymbol{\psi}$$

$$E\{\tilde{\mathbf{v}}\} = \mathbf{0} \qquad E\{\tilde{\mathbf{v}}\tilde{\mathbf{v}}'\} = \mathbf{R}$$

Specify $\mathbf{H}$, $\tilde{\mathbf{x}}$, $\tilde{\mathbf{v}}$, $\boldsymbol{\psi}$, $\mathbf{R}$, and $\mathbf{c}$.

3.2. Consider

$$z(t) = a_0 + a_1 t + a_2 t^2 + v(t)$$

$v(t)$: white stochastic process

$$\mathbf{a} = \begin{bmatrix} a_0 \\ a_1 \\ a_2 \end{bmatrix} \text{: completely unknown}$$

Express this model in the following two forms:

a. No-dynamics model:

$$z(t) = \mathbf{H}(t)\mathbf{x} + v(t)$$

b. Dynamic system with no input:

$$\frac{d}{dt}\mathbf{x}(t) = \mathbf{F}\mathbf{x}(t)$$

$$z(t) = \mathbf{H}\mathbf{x}(t) + v(t)$$

3.3. Let $y(t)$, $0 \le t \le T$, denote the uncertain position versus time of an intoxicated eagle flying in a gusty head wind. Assume that the eagle's path "looks" like a second-degree polynomial in time over short time spans; i.e.,

$$y(t) = a_0(t) + a_1(t)t + a_2(t)t^2, \qquad 0 \le t \le T$$

where for small time intervals, $\tau \ll T$, the $a_j(t)$, $t_0 < t < t_0 + \tau$, can be assumed to be constant.

a. Define one "reasonable" mathematical model for $y(t)$ based on a white stochastic process and/or random variables.

b. Define one "reasonable" model for $y(t)$ based on a white unknown-but-bounded process and/or uncertain constants which lie within known sets.

3.4. Consider a linear system with vector input $\mathbf{i}(t)$ and scalar output $y(t)$ modeled as

$$\frac{d}{dt}\mathbf{x}_1(t) = \mathbf{F}_1\mathbf{x}_1(t) + \mathbf{G}_1\mathbf{i}(t)$$

$$\mathbf{x}_1(0) = \mathbf{0}$$

$$y(t) = \mathbf{H}_1\mathbf{x}_1(t)$$

Assume that observations $z_1(t)$ of the output are modeled as

$$z_1(t) = y(t) + v(t)$$

$v(t)$: continuous-time white stochastic process

Assume that the input $\mathbf{i}(t)$ is the output of a linear system driven by a vector white stochastic process $\mathbf{w}_i(t)$ modeled as

$$\frac{d}{dt}\mathbf{i}(t) = \mathbf{F}_i\mathbf{i}(t) + \mathbf{w}_i(t)$$

$$E\{\mathbf{i}(0)\} = \mathbf{0} \qquad E\{\mathbf{i}(0)\mathbf{i}'(0)\} = \boldsymbol{\psi}_i$$

Assume that scalar observations $z_2(t)$ on $\mathbf{i}(t)$ are made by a sensor with dynamics but no uncertainty modeled as

$$\frac{d}{dt}z_2(t) = F_2 z_2(t) + \mathbf{G}_2\mathbf{i}(t)$$

$$z_2(0) = 0$$

Express the above in a state space-white process form like (3.1.2). Specify the resulting $\mathbf{F}$, $\mathbf{G}$, $\mathbf{H}$, $\boldsymbol{\psi}$, $\mathbf{Q}$, and $\mathbf{R}$ matrices.

3.5. Assume that $w(n)$ and $v(n)$ are scalar white unknown-but-bounded processes:

$$|w(n)| \leq Q^{1/2}$$

$$|v(n)| \leq R^{1/2}$$

Consider a sequence of observations $z(n)$, $n = 1, \ldots,$ modeled by

$$z(n) = y_1(n) + v(n)$$

$$y_1(n+1) + a_1 y_1(n) + a_2 y_1(n-1) = y_2(n)$$

$$y_2(n) = b_1 w(n) + b_2 w(n-1)$$

Express $z(n)$, $n = 1, \ldots,$ in the basic form

$$\mathbf{x}(n+1) = \mathbf{\Phi}\mathbf{x}(n) + \mathbf{G}w(n)$$

$$z(n) = \mathbf{H}\mathbf{x}(n) + v(n)$$

$$\mathbf{x}(0) \in \mathbf{\Omega}_x(0)$$

$$\mathbf{\Omega}_x(0)\text{: ellipsoid}$$

Try to keep the dimension of $\mathbf{x}$ as small as possible. Make whatever assumptions you feel are necessary to be able to specify $\mathbf{\Omega}_x(0)$.

3.6. Consider

$$\mathbf{x}(n+1) = \mathbf{\Phi}\mathbf{x}(n) + \mathbf{G}w(n)$$

$$z(n) = \mathbf{H}\mathbf{x}(n) + v(n)$$

$$\mathbf{x} = \begin{bmatrix} x_1 \\ x_2 \end{bmatrix}$$

$$\mathbf{x}(0) \in \mathbf{\Omega}_x(0) = \{\mathbf{x} : x_1 = 1\}$$

$$w(n) \in \mathbf{\Omega}_w = \{w : w < 2\}, \qquad \text{all } n$$

$$v(n) \in \mathbf{\Omega}_v = \{v : (v-3)^2 \leq 1\}, \qquad \text{all } n$$

Express this model in the form of an unknown-but-bounded model with ellipsoids as in (3.1.3).

3.7. The basic Bayesian model assumes that the input $\mathbf{w}(\cdot)$ and sensor $\mathbf{v}(\cdot)$ dis-

4

Analysis of Linear Systems

The behavior of the basic state space-white process linear system models defined in Section 3.1 is now analyzed and discussed. Bayesian and unknown-but-bounded models are considered. Fisher models are not discussed, as they enter primarily in estimation theory, which is considered in later chapters.

4.1 Bayesian

The basic Bayesian models are summarized in (3.1.1) and (3.1.2). The question of interest is now, *What are the properties of the resulting* $\mathbf{x}(n\Delta)$ *or* $\mathbf{x}(t)$ *and* $\mathbf{z}(n\Delta)$ *or* $\mathbf{z}(t)$? Since the $\mathbf{x}$ and $\mathbf{z}$ are random processes, the properties of interest are things such as probability densities or moments such as mean values and covariance matrices. The present discussions center on mean values, covariance matrices, and correlation functions, which, of course, do not completely specify the probability densities of the process except in the Gaussian case. General equations for the probability densities will be derived in Chapters 11 and 13.

4.1.1 Static Model

Consider the special Bayesian static model where

$$
\begin{aligned}
\mathbf{z} &= \mathbf{Hx} + \mathbf{v} \\
E\{\mathbf{x}\} = \mathbf{0} &\qquad E\{\mathbf{v}\} = \mathbf{0} \\
E\{\mathbf{xx}'\} &= \boldsymbol{\psi} \\
E\{\mathbf{vv}'\} &= \mathbf{R} \\
E\{\mathbf{xv}'\} &= \mathbf{0}
\end{aligned}
\tag{4.1.1}
$$

It is desired to determine the mean value and covariance matrix of $\mathbf{z}$.

Consider first the mean value. From (4.1.1)

$$E\{\mathbf{z}\} = \mathbf{H}E\{\mathbf{x}\} + E\{\mathbf{v}\}$$

so

$$\boxed{E\{\mathbf{z}\} = \mathbf{0}} \tag{4.1.2}$$

Thus $\mathbf{z}$ is a zero mean vector.

Let $\boldsymbol{\Gamma}_z$ denote the covariance matrix of $\mathbf{z}$:

$$\boldsymbol{\Gamma}_z = E\{[\mathbf{z} - E\{\mathbf{z}\}][\mathbf{z} - E\{\mathbf{z}\}]'\}$$

Since $\mathbf{z}$ is zero mean,

$$\boldsymbol{\Gamma}_z = E\{\mathbf{zz}'\}$$

From (4.1.1)

$$\boldsymbol{\Gamma}_z = E\{\mathbf{Hxx}'\mathbf{H}'\} + E\{\mathbf{vv}'\} + E\{H\mathbf{xv}'\} + E\{\mathbf{vx}'\mathbf{H}'\}$$

or since the *crossterms* are zero,

$$\boxed{\boldsymbol{\Gamma}_z = \mathbf{H}\boldsymbol{\psi}\mathbf{H}' + \mathbf{R}} \tag{4.1.3}$$

EXAMPLE: Consider

$$z = [1 \quad 1]\begin{bmatrix} x_1 \\ x_2 \end{bmatrix} + v$$

$$E\{\mathbf{x}\} = \mathbf{0} \qquad E\{v\} = 0$$

$$E\{\mathbf{xx}'\} = \begin{bmatrix} 2 & 1 \\ 1 & 4 \end{bmatrix} \qquad E\{v^2\} = 5$$

$$E\{\mathbf{x}v\} = \mathbf{0}$$

Then

$$E\{z^2\} = \boldsymbol{\Gamma}_z = [1 \quad 1]\begin{bmatrix} 2 & 1 \\ 1 & 4 \end{bmatrix}\begin{bmatrix} 1 \\ 1 \end{bmatrix} + 5 = 13$$

4.1.2 Discrete-Time Dynamic Systems

A Bayesian model for the behavior of the state $\mathbf{x}(n\Delta)$ is

$$\begin{aligned}
\mathbf{x}(n\Delta + \Delta) &= \mathbf{\Phi}(n\Delta)\mathbf{x}(n\Delta) + \Delta\mathbf{G}(n\Delta)\mathbf{w}(n\Delta) \\
E\{\mathbf{w}(n\Delta)\} &= \mathbf{0} \qquad E\{\mathbf{x}(0)\} = \mathbf{0} \\
E\{\mathbf{x}(0)\mathbf{w}'(n\Delta)\} &= \mathbf{0} \\
E\{\mathbf{x}(0)\mathbf{x}'(0)\} &= \boldsymbol{\psi} \\
E\{\mathbf{w}(n_1\Delta)\mathbf{w}'(n_2\Delta)\} &= \begin{cases} \Delta^{-1}\mathbf{Q}(n_1\Delta), & n_1 = n_2 \\ \mathbf{0}, & n_1 \neq n_2 \end{cases}
\end{aligned} \tag{4.1.4}$$

It is immediately clear that $\mathbf{x}(n\Delta)$ is a zero mean process, as

$$E\{\mathbf{x}(n\Delta + \Delta)\} = \mathbf{\Phi}(n\Delta)E\{\mathbf{x}(n\Delta)\}$$

and

$$E\{\mathbf{x}(0)\} = \mathbf{0}$$

so

$$\boxed{E\{\mathbf{x}(n\Delta)\} = \mathbf{0}, \qquad \text{all } n} \tag{4.1.5}$$

Let $\mathbf{\Gamma}(n\Delta)$ denote the covariance matrix of $\mathbf{x}(n\Delta)$:

$$\mathbf{\Gamma}(n\Delta) = E\{[\mathbf{x}(n\Delta) - E\{\mathbf{x}(n\Delta)\}][\mathbf{x}(n\Delta) - E\{\mathbf{x}(n\Delta)\}]'\}$$

Since $\mathbf{x}(n\Delta)$ is zero mean,

$$\mathbf{\Gamma}(n\Delta) = E\{\mathbf{x}(n\Delta)\mathbf{x}'(n\Delta)\}$$

Substituting (4.1.4) and manipulating gives

$$\begin{aligned}
\mathbf{\Gamma}(n\Delta + \Delta) &= E\{\mathbf{x}(n\Delta + \Delta)\mathbf{x}'(n\Delta + \Delta)\} \\
&= \mathbf{\Phi}(n\Delta)E\{\mathbf{x}(n\Delta)\mathbf{x}'(n\Delta)\}\mathbf{\Phi}'(n\Delta) \\
&\quad + \Delta^2\mathbf{G}(n\Delta)E\{\mathbf{w}(n\Delta)\mathbf{w}'(n\Delta)\}\mathbf{G}'(n\Delta) \\
&\quad + \Delta\mathbf{G}(n\Delta)E\{\mathbf{w}(n\Delta)\mathbf{x}'(n\Delta)\}\mathbf{\Phi}'(n\Delta) \\
&\quad + \Delta\mathbf{\Phi}(n\Delta)E\{\mathbf{x}(n\Delta)\mathbf{w}'(n\Delta)\}\mathbf{G}'(n\Delta)
\end{aligned}$$

Now $\mathbf{x}(n\Delta)$ depends only on $\mathbf{x}(0)$ and $\mathbf{w}(m\Delta)$, $m = 1 \ldots, n-1$, and since $\mathbf{w}(n\Delta)$ is a white process which is also uncorrelated with $\mathbf{x}(0)$,

$$E\{\mathbf{x}(n\Delta)\mathbf{w}'(n\Delta)\} = \mathbf{0}$$

Thus the crossterms are zero and the covariance matrix of $\mathbf{x}(n\Delta)$ satisfies

$$\boxed{\begin{aligned}
\mathbf{\Gamma}(n\Delta + \Delta) &= \mathbf{\Phi}(n\Delta)\mathbf{\Gamma}(n\Delta)\mathbf{\Phi}'(n\Delta) \\
&\quad + \Delta\mathbf{G}(n\Delta)\mathbf{Q}(n\Delta)\mathbf{G}'(n\Delta) \\
\mathbf{\Gamma}(0) &= \boldsymbol{\psi}
\end{aligned}} \tag{4.1.6}$$

This is a difference equation which can be solved starting with $\mathbf{\Gamma}(0)$ to give $\mathbf{\Gamma}(n\Delta)$ for all n.

In the special case of a time-invariant system model, (4.1.6) becomes

$$\begin{aligned}\mathbf{\Gamma}(n\Delta + \Delta) &= \mathbf{\Phi}\mathbf{\Gamma}(n\Delta)\mathbf{\Phi}' + \Delta\mathbf{G}\mathbf{Q}\mathbf{G}' \\ \mathbf{\Gamma}(0) &= \boldsymbol{\psi}\end{aligned} \tag{4.1.7}$$

If the dynamic system is stable (i.e., eigenvalues of $\mathbf{\Phi}$ are less than 1 in magnitude), $\mathbf{\Gamma}(n\Delta)$ of (4.1.7) will "go to" steady state. Define $\mathbf{\Gamma}_{ss}$ as a positive semidefinite matrix which satisfies

$$\boxed{\mathbf{\Gamma}_{ss} = \mathbf{\Phi}\mathbf{\Gamma}_{ss}\mathbf{\Phi}' + \Delta\mathbf{G}\mathbf{Q}\mathbf{G}'} \tag{4.1.8}$$

If the system is stable, $\mathbf{\Gamma}(n\Delta)$ goes to $\mathbf{\Gamma}_{ss}$ as n increases. When $\mathbf{\Gamma}(n\Delta) = \mathbf{\Gamma}_{ss}$, $\mathbf{x}(n\Delta)$ is a stationary process (in the wide sense). Note that if $\boldsymbol{\psi} = \mathbf{\Gamma}_{ss}$, then there is no transient and $\mathbf{\Gamma}(n\Delta) = \mathbf{\Gamma}_{ss}$ for all n.

EXAMPLE: Consider

$$\begin{bmatrix} x_1(n+1) \\ x_2(n+1) \end{bmatrix} = \begin{bmatrix} \Phi_{11} & 0 \\ 0 & \Phi_{22} \end{bmatrix} \begin{bmatrix} x_1(n) \\ x_2(n) \end{bmatrix} + \begin{bmatrix} 0 \\ 1 \end{bmatrix} w(n)$$

$|\Phi_{11}| < 1$ and $|\Phi_{22}| < 1$:

$$\mathbf{\Gamma}_{ss} = \begin{bmatrix} \Gamma_{11} & \Gamma_{12} \\ \Gamma_{12} & \Gamma_{22} \end{bmatrix}$$

Then from (4.1.8)

$$\Gamma_{11} = \Gamma_{21} = 0$$

$$\Gamma_{22} = \frac{Q}{1 - \Phi_{22}^2}$$

Let $\mathcal{P}(n_1\Delta, n_2\Delta)$ denote the correlation function of $\mathbf{x}(n\Delta)$. Since $\mathbf{x}(n\Delta)$ is zero mean,

$$\mathcal{P}(n_1\Delta, n_2\Delta) = E\{\mathbf{x}(n_1\Delta)\mathbf{x}'(n_2\Delta)\}$$

By definition

$$\mathcal{P}(n\Delta, n\Delta) = \mathbf{\Gamma}(n\Delta)$$

Consider first the case $n_1 = n + 1$, $n_2 = n$. From (4.1.4)

$$\begin{aligned}\mathcal{P}(n\Delta + \Delta, n\Delta) = \mathbf{\Phi}(n\Delta)&E\{\mathbf{x}(n\Delta)\mathbf{x}'(n\Delta)\} \\ &+ \Delta\mathbf{G}(n\Delta)E\{\mathbf{w}(n\Delta)\mathbf{x}'(n\Delta)\}\end{aligned}$$

or using the whiteness of $\mathbf{w}(n\Delta)$,

$$\begin{aligned}\mathcal{P}(n\Delta + \Delta, n\Delta) &= \mathbf{\Phi}(n\Delta)\mathcal{P}(n\Delta, n\Delta) \\ &= \mathbf{\Phi}(n\Delta)\mathbf{\Gamma}(n\Delta)\end{aligned}$$

In general, it follows that the correlation function of $\mathbf{x}(n\Delta)$ satisfies

$$\boxed{\begin{gathered}\mathcal{P}(n_1\Delta + \Delta, n_2\Delta) = \mathbf{\Phi}(n_1\Delta)\mathcal{P}(n_1\Delta, n_2\Delta) \\ n_1 \geq n_2 \\ \mathcal{P}(n_2\Delta, n_2\Delta) = \mathbf{\Gamma}(n_2\Delta)\end{gathered}} \tag{4.1.9}$$

Thus the correlation function $\mathcal{P}(n_3\Delta, n_2\Delta)$, $n_3 > n_2$, can be calculated by first solving (4.1.6) to get the covariance matrix $\mathbf{\Gamma}(n_2\Delta) = \mathcal{P}(n_2\Delta, n_2\Delta)$ and then using (4.1.9) repeated with $n_1 = n_2 + 1, \ldots, n_3 - 1$.

For a time-invariant system, (4.1.9) yields

$$\mathcal{P}(n_1\Delta, n_2\Delta) = \mathbf{\Phi}^{n_1 - n_2}\mathbf{\Gamma}(n_2\Delta) \qquad n_1 \geq n_2$$

Assume that the system is stable and that

$$\mathbf{\Gamma}(n_2\Delta) = \mathbf{\Gamma}_{ss}$$

so $\mathbf{x}(n\Delta)$ is a stationary (wide sense) process. Then the correlation function $\mathcal{P}(n_1\Delta, n_2\Delta)$ is a function only of $n_1 - n_2$.

EXAMPLE: Consider the scalar, time-invariant model

$$x(n\Delta + \Delta) = \Phi x(n\Delta) + \Delta G w(n\Delta)$$

From (4.1.6)

$$\Gamma(n\Delta) = \left(\frac{1 - \Phi^{2n}}{1 - \Phi^2}\right) G^2 Q\Delta + \Phi^{2n}\psi$$

If $|\Phi| \leq 1$, then

$$\Gamma_{ss} = \frac{\Delta G^2 Q}{1 - \Phi^2}$$

If $\Gamma(n\Delta) = \Gamma_{ss}$, then from (4.1.9)

$$\begin{gathered}\mathcal{P}(n_1\Delta, n_2\Delta) = \Gamma_{ss}\Phi^{n_1 - n_2} \\ n_1 > n_2\end{gathered}$$

Suppose that

$$\begin{aligned}\mathbf{z}(n\Delta) &= \mathbf{H}(n\Delta)\mathbf{x}(n\Delta) + \mathbf{v}(n\Delta) \\ E\{\mathbf{v}(n\Delta)\} &= 0 \\ E\{\mathbf{v}(n\Delta)\mathbf{v}'(n\Delta)\} &= \Delta^{-1}\mathbf{R}(n\Delta)\end{aligned}$$

Then using the logic leading to (4.1.3),

$$\begin{aligned}\mathbf{\Gamma}_z(n\Delta) &= E\{\mathbf{z}(n\Delta)\mathbf{z}'(n\Delta)\} \\ &= \mathbf{H}(n\Delta)\mathbf{\Gamma}(n\Delta)\mathbf{H}'(n\Delta) + \Delta^{-1}\mathbf{R}(n\Delta)\end{aligned}$$

4.1.3 Continuous-Time Dynamic System: Formal Limit

From (3.1.2) the Bayesian model for $\mathbf{x}(t)$ is

$$\begin{aligned}
\frac{d}{dt}\mathbf{x}(t) &= \mathbf{F}(t)\mathbf{x}(t) + \mathbf{G}(t)\mathbf{w}(t) \\
E\{\mathbf{w}(t)\} &= \mathbf{0} \qquad E\{\mathbf{x}(0)\} = \mathbf{0} \\
E\{\mathbf{x}(0)\mathbf{w}'(t)\} &= \mathbf{0} \\
E\{\mathbf{x}(0)\mathbf{x}'(0)\} &= \boldsymbol{\psi} \\
E\{\mathbf{w}(t_1)\mathbf{w}'(t_2)\} &= \delta(t_1 - t_2)\mathbf{Q}(t_1)
\end{aligned} \tag{4.1.10}$$

It follows directly that $\mathbf{x}(t)$ is a zero mean process

$$E\{\mathbf{x}(t)\} = \mathbf{0}$$

Let $\mathbf{\Gamma}(t)$ and $\boldsymbol{\mathcal{P}}(t_1, t_2)$ denote the covariance matrix and correlation function of $\mathbf{x}(t)$:

$$\mathbf{\Gamma}(t) = E\{\mathbf{x}(t)\mathbf{x}'(t)\}$$
$$\boldsymbol{\mathcal{P}}(t_1, t_2) = E\{\mathbf{x}(t_1)\mathbf{x}'(t_2)\}$$

The differential equations which $\mathbf{\Gamma}(t)$ and $\boldsymbol{\mathcal{P}}(t_1, t_2)$ obey are now derived by a formal discrete to continuous time limit. *Justification* is discussed in Section 4.2. (See Appendix B for discussion of deterministic case.)

Substitute

$$\mathbf{\Phi}(n\Delta) = \mathbf{I} + \Delta\mathbf{F}(n\Delta)$$

into (4.1.6) to give, after some manipulation,

$$\begin{aligned}
\frac{1}{\Delta}[\mathbf{\Gamma}(n\Delta + \Delta) - \mathbf{\Gamma}(n\Delta)] = \mathbf{F}(n\Delta)\mathbf{\Gamma}(n\Delta) &+ \mathbf{\Gamma}(n\Delta)\mathbf{F}'(n\Delta) \\
&+ \mathbf{G}(n\Delta)\mathbf{Q}(n\Delta)\mathbf{G}'(n\Delta) \\
&+ \Delta\mathbf{F}(n\Delta)\mathbf{\Gamma}(n\Delta)\mathbf{F}'(n\Delta)
\end{aligned}$$

In the limit as $n \longrightarrow \infty$, $\Delta \longrightarrow 0$, $n\Delta \longrightarrow t$, and assuming sufficient smoothness, this gives for the covariance matrix of $\mathbf{x}(t)$

$$\boxed{\begin{aligned}
\frac{d}{dt}\mathbf{\Gamma}(t) &= \mathbf{F}(t)\mathbf{\Gamma}(t) + \mathbf{\Gamma}(t)\mathbf{F}'(t) + \mathbf{G}(t)\mathbf{Q}(t)\mathbf{G}'(t) \\
\mathbf{\Gamma}(0) &= \boldsymbol{\psi}
\end{aligned}} \tag{4.1.11}$$

In a similar fashion, substitution into (4.1.9) yields

$$\frac{1}{\Delta}[\boldsymbol{\mathcal{P}}(n_1\Delta + \Delta, n_2\Delta) - \boldsymbol{\mathcal{P}}(n_1\Delta, n_2\Delta)] = \mathbf{F}(n_1\Delta)\boldsymbol{\mathcal{P}}(n_1\Delta, n_2\Delta)$$

or, in the limit, the correlation function of $\mathbf{x}(t)$ satisfies

$$\boxed{\frac{\partial}{\partial t_1}\boldsymbol{\mathcal{P}}(t_1, t_2) = \mathbf{F}(t_1)\boldsymbol{\mathcal{P}}(t_1, t_2), \qquad t_1 > t_2} \tag{4.1.12}$$

In the case of time-invariant systems, steady-state solutions and stationary processes follow, just as discussed for discrete time. For example, if the system is stable (eigenvalues of $\mathbf{F}$ have negative real parts), a steady-state covariance matrix $\mathbf{\Gamma}_{ss}$ corresponding to a stationary process is the positive semidefinite solution of

$$\boxed{\mathbf{0} = \mathbf{F}\mathbf{\Gamma}_{ss} + \mathbf{\Gamma}_{ss}\mathbf{F}' + \mathbf{G}\mathbf{Q}\mathbf{G}'} \tag{4.1.13}$$

EXAMPLE: Consider the scalar time-invariant system

$$\frac{d}{dt}x(t) = Fx(t) + Gw(t)$$

From 4.1.11

$$\Gamma(t) = \frac{G^2Q}{2F}(e^{2Ft} - 1) + \psi e^{2Ft}$$

so if $F < 0$, then

$$\Gamma_{ss} = -\frac{G^2Q}{2F}$$

If $\Gamma(t) = \Gamma_{ss}$, then from (4.1.12)

$$\mathcal{P}(t_1, t_2) = \Gamma_{ss} \exp\{F(t_1 - t_2)\}$$
$$t_1 \geq t_2$$

Consider

$$\mathbf{y}(t) = \mathbf{H}(t)\mathbf{x}(t)$$

Then

$$\begin{aligned}\mathbf{\Gamma}_y(t) &= E\{\mathbf{y}(t)\mathbf{y}'(t)\}\\ &= \mathbf{H}(t)\mathbf{\Gamma}(t)\mathbf{H}'(t)\end{aligned}$$

For

$$\mathbf{z}(t) = \mathbf{H}(t)\mathbf{x}(t) + \mathbf{v}(t)$$

$\mathbf{\Gamma}_z(t) = E\{\mathbf{z}(t)\mathbf{z}'(t)\}$ is infinite because of the continuous-time white process $\mathbf{v}(t)$.

4.1.4 Deterministic Inputs

Generalization of the zero mean stochastic models to include a deterministic (nonzero mean) input and/or nonzero mean initial conditions is straightforward (see Section 3.2). In continuous time, the new model is

$$\begin{aligned}&\frac{d}{dt}\mathbf{x}(t) = \mathbf{F}(t)\mathbf{x}(t) + \mathbf{G}(t)\mathbf{w}(t) + \mathbf{B}(t)\mathbf{u}(t)\\ &E\{\mathbf{x}(0)\} = \mathbf{m}\\ &\mathbf{w}(t)\text{: zero mean white stochastic process}\end{aligned} \tag{4.1.14}$$

The covariance matrix $\mathbf{\Gamma}(t)$ and correlation matrix $\mathcal{P}(t_1, t_2)$ of $\mathbf{x}(t)$ remain unchanged. However, $\mathbf{x}(t)$ is now a process with mean value

$$E\{\mathbf{x}(t)\} = \mathbf{x}_m(t)$$

where the mean value is determined by taking the expectation of (4.1.14):

$$\frac{d}{dt}\mathbf{x}_m(t) = \mathbf{F}(t)\mathbf{x}_m(t) + \mathbf{B}(t)\mathbf{u}(t)$$

$$\mathbf{x}_m(0) = \mathbf{m}$$

Note that

$$E\{\mathbf{x}(t)\mathbf{x}'(t)\} = \mathbf{\Gamma}(t) + \mathbf{x}_m(t)\mathbf{x}_m'(t)$$

Similar results follow in the discrete-time case.

4.1.5 Gaussian Models

The discussions thus far have not considered probability densities; only mean values, covariance matrices, and correlation functions have been dealt with. If the $\mathbf{x}(0)$, $\mathbf{w}(n\Delta)$ or $\mathbf{w}(t)$, and $\mathbf{v}(n\Delta)$ or $\mathbf{v}(t)$ are assumed to be Gaussian, the resulting equations can be interpreted in a much stronger sense.

Consider the static model

$$\mathbf{z} = \mathbf{H}\mathbf{x} + \mathbf{v}$$

Assume that $\mathbf{x}$ and $\mathbf{v}$ are Gaussian so that in shorthand notation

$$\mathbf{x} \text{ is } N(\mathbf{0}, \boldsymbol{\psi})$$

$$\mathbf{v} \text{ is } N(\mathbf{0}, \mathbf{R})$$

Then from Appendix *E*, $\mathbf{z}$ is also Gaussian, and from (4.1.3)

$$\mathbf{z} \text{ is } N(\mathbf{0}, \mathbf{H}\boldsymbol{\psi}\mathbf{H}' + \mathbf{R})$$

Thus when $\mathbf{x}$ and $\mathbf{v}$ are Gaussian, the equations of Section 4.1.1 can be used to determine the probability density of $\mathbf{z}$.

The same logic applies to both the discrete- and continuous-time dynamic systems. Thus if $\mathbf{x}(0)$ and $\mathbf{w}(n\Delta)$ or $\mathbf{w}(t)$ are assumed to be Gaussian $\mathbf{x}(t)$ is also Gaussian for all t, so for a zero mean model

$$\mathbf{x}(t) \text{ is } N(\mathbf{0}, \mathbf{\Gamma}(t))$$

In the zero mean Gaussian case, the condition that $\mathbf{\Gamma}(t)$ or $\mathbf{\Gamma}(n)$ is $\mathbf{\Gamma}_{ss}$ means the process is strict sense stationary (see Appendix *F*).

4.2 Continuous-Time White Stochastic Processes

Section 4.1 illustrates the basic approach used in this book, which is to start with vector concepts, extend them to discrete-time systems, and then use a limiting operation to obtain results for continuous-time systems. Appendix

B discusses the limit procedure for deterministic models. However, the concept of a continuous-time white stochastic process $\mathbf{w}(t)$ requires special attention.

Consider the Bayesian continuous time model

$$\frac{d}{dt}\mathbf{x}(t) = \mathbf{F}(t)\mathbf{x}(t) + \mathbf{G}(t)\mathbf{w}(t)$$
$$E\{\mathbf{w}(t_1)\mathbf{w}'(t_2)\} = \delta(t_1 - t_2)\mathbf{Q}(t_1) \tag{4.2.1}$$
$$\vdots$$

Discussion is definitely needed, as $\mathbf{w}(t)$ is extremely ill behaved in a mathematical sense and does not exist in any physical sense. For example, the variance of, say, a scalar white process $w(t)$ at time t is infinite:

$$E\{w^2(t)\} = \infty$$

The following questions are considered:

1. Why is Δ dependence of $\mathbf{w}(n\Delta)$ in (4.1.4) needed?
2. How can one justify the use of the "nonexisting" $\mathbf{w}(t)$ to model actual processes?
3. What special consequences (mathematical problems) arise from the use of $\mathbf{w}(t)$?

This section is intended only to convey an overall feeling, not general results. Hence the discussions are presented in terms of simple scalar, time-invariant examples instead of the usual vector, time-varying systems.

4.2.1 Need for Δ Behavior of w(nΔ)

The discrete-time model assumes that the variance of $\mathbf{w}(n\Delta)$ depends on Δ:

$$E\{\mathbf{w}(n_1\Delta)\mathbf{w}'(n_2\Delta)\} = \begin{cases} \Delta^{-1}\mathbf{Q}(n_1), & n_1 = n_2 \\ \mathbf{0}, & n_1 \neq n_2 \end{cases} \tag{4.2.2}$$

Such Δ dependence is needed to get meaningful continuous-time results in the sense that any other Δ dependence yields an $\mathbf{x}(t)$ process whose variance is either zero or infinite. Rather than prove this statement in general, it is merely shown to be true for a simple example.

Consider a "discrete time integrator"

$$x(n\Delta + \Delta) = x(n\Delta) + \Delta w(n\Delta), \qquad n = 0, \ldots, N-1$$
$$x(0) = 0$$

so that

$$x(N\Delta) = \Delta \sum_{n=0}^{N-1} w(n\Delta)$$

Assume that, instead of (4.2.2),

$$E\{w(n_1\Delta)w(n_2\Delta)\} = \begin{cases} \Delta^q Q, & n_1 = n_2 \\ 0, & n_1 \neq n_2 \end{cases}$$

Then

$$E\{x^2(N\Delta)\} = N\Delta^{q+2}Q$$

The limit $N \longrightarrow \infty$, $\Delta \longrightarrow 0$, $N\Delta = T$, gives

$$E\{x^2(T)\} = \begin{cases} \infty, & q > -1 \\ 0, & q < -1 \\ TQ, & q = -1 \end{cases}$$

Thus if the limit of the discrete-time integrator, i.e.,

$$\frac{d}{dt}x(t) = w(t)$$

$$x(0) = 0$$

is to be meaningful in the sense that $E\{x^2(t)\}$ is finite but nonzero, one must choose $q = -1$.

This argument only shows that the Δ behavior of (4.2.2) is necessary if the resulting continuous-time model is to make sense. It does not prove or justify the model itself in any way.

4.2.2 Justification for Use of w(t)

The $\mathbf{w}(t)$ of (4.2.1) is a physically impossible process, but there are many situations where the use of (4.2.1) can be justified.

One situation is as follows. Assume that it is desired to model some uncertain scalar time function $y(t)$ as a stochastic process. Assume that measurements (or other information) on $y(t)$ show that it can be viewed as a zero mean stochastic process with specified correlation function, $\mathcal{P}_{\text{measured}}(t, \tau)$ for some range of t and τ, where

$$\mathcal{P}(t, \tau) = E\{y(t)y(\tau)\}$$

Now consider

$$y_{\text{model}}(t) = \mathbf{H}(t)\mathbf{x}(t) \tag{4.2.3}$$

where $\mathbf{x}(t)$ is generated by (4.2.1). Such a model yields a $\mathcal{P}_{\text{model}}(t, \tau)$:

$$\mathcal{P}_{\text{model}}(t, \tau) = \mathbf{H}(t)\boldsymbol{\mathcal{P}}(t, \tau)\mathbf{H}'(\tau)$$

where $\boldsymbol{\mathcal{P}}(t, \tau)$ is given by (4.1.11) and (4.1.12). If $\mathbf{F}(t)$, $\mathbf{G}(t)$, $\mathbf{Q}(t)$, $\mathbf{H}(t)$, and $\boldsymbol{\psi}$ can be chosen so that

$$\mathcal{P}_{\text{model}}(t, \tau) \approx \mathcal{P}_{\text{measured}}(t, \tau)$$

over the *important range* of t and τ, it can be concluded that (4.2.1) combined with (4.2.3) provides a satisfactory model in some sense for $y(t)$. With this point of view, the ill behavior and nonexistence of the white stochastic process is of no consequence, as it is used only as a conceptual "fiction" to facilitate the formulation of a mathematical model which can easily be analyzed.

Another way to justify the use of (4.2.1) is as follows. Let

$$\frac{d}{dt}\mathbf{x}_c(t) = \mathbf{F}(t)\mathbf{x}_c(t) + \mathbf{G}(t)\mathbf{c}(t) \tag{4.2.4}$$

denote a mathematical model for a linear system with correlated (nonwhite) stochastic, zero mean, input $\mathbf{c}(t)$. Define

$$\mathbf{\mathcal{P}}_c(t, \tau) = E\{\mathbf{x}_c(t)\mathbf{x}_c'(\tau)\}$$

Now if (4.2.1) yields a $\mathbf{\mathcal{P}}(t, \tau)$ such that

$$\mathbf{\mathcal{P}}(t, \tau) \approx \mathbf{\mathcal{P}}_c(t, \tau)$$

over some desired range of t and τ, one can conclude that (4.2.1) is a reasonable approximate model for (4.2.4).

This concept is now illustrated by a simple example. Consider

$$\begin{aligned} \frac{d}{dt}x_c(t) &= Fx_c(t) + c(t) \\ F &< 0 \\ \frac{d}{dt}c(t) &= Ac(t) + w_c(t) \\ A &< 0 \\ E\{w_c(t_1)w_c(t_2)\} &= \delta(t_1 - t_2)Q_c \\ E\{x(0)c(0)\} &= 0 \end{aligned} \tag{4.2.5}$$

where $x_c(t)$ and $c(t)$ are stationary (see Exercise 4.12); i.e.,

$$E\{c^2(t)\} = -\frac{Q_c}{2A} = \Gamma_c, \qquad \text{all } t$$

$$E\{x_c^2(t)\} = \frac{\Gamma_c}{F(F + A)}, \qquad \text{all } t$$

Equation (4.2.5) involves a white stochastic process $w_c(t)$, but, for the sake of the present discussion, $w_c(t)$ is viewed as a mathematical fiction used to obtain a correlated $c(t)$ with

$$E\{c(t)c(\tau)\} = \Gamma_c e^{A(t-\tau)}, \qquad t > \tau$$

Define

$$\mathcal{P}_{x_c}(t, \tau) = E\{x_c(t)x_c(\tau)\}$$

Then it can be shown after some work that (see Exercise 4.12)

$$\mathcal{P}_{x_c}(t, \tau) = \frac{\Gamma_c}{(A + F)(A - F)}\left\{\frac{Ae^{F(t-\tau)}}{F} - e^{A(t-\tau)}\right\}, \qquad t > \tau \tag{4.2.6}$$

Now if

$$|A| \gg |F|$$

then

$$\mathcal{P}_{x_c}(t, \tau) \approx \frac{\Gamma_c}{AF} e^{F(t-\tau)}, \qquad t > \tau$$

Therefore if the condition $|A| \gg |F|$ is satisfied, the original model (4.2.5) with correlated input $c(t)$ can be replaced by the white process input model

$$\begin{aligned} \frac{d}{dt} x(t) &= Fx(t) + w(t) \\ E\{w(t_1)w(t_2)\} &= \delta(t_1 - t_2)Q \\ Q &= \frac{-2\Gamma_c}{A} = \frac{Q_c}{A^2} \end{aligned} \tag{4.2.7}$$

because for (4.2.7)

$$\mathcal{P}_x(t, \tau) = E\{x(t)x(\tau)\} = \frac{\Gamma_c}{AF} e^{F(t-\tau)}, \qquad t > \tau$$

The preceding example is now reinterpreted in terms of block diagrams. The original problem consists of a linear system with *system time constant* $|F|^{-1}$ driven by a physically realizable stochastic process $c(t)$ with *correlation time constant* $|A|^{-1}$. This is illustrated in Fig. 4.2.1(a). Figure 4.2.1(b) illustrates a model for $c(t)$ based on a fictional white process $w_c(t)$. Figure 4.2.1(c), the complete model, follows by combining Fig. 4.2.1(a) and (b). Figure 4.2.1(d) is the approximate model which results when $|A| \gg |F|$ holds so that the physical process $c(t)$ can be replaced by a white process $w(t)$ of (4.2.7).

To summarize, the following points on the use of the model (4.2.1) are important:

1. The model can be justified when it yields the same behavior as that which is experimentally obtained and/or when it yields approximately the same behavior as that calculated from a "real" input.

2. Justification of the model (4.2.2) depends on the nature of the system which is being driven.

3. When going from Fig. 4.2.1(c) to (d), the "time constant $|A|^{-1}$ system" is not simply dropped, as $w_c(t)$ and $w(t)$ are different white processes.

Although the discussion is based on the simple example of a scalar, time-invariant system, the basic ideas generalize to vector, time-varying systems.

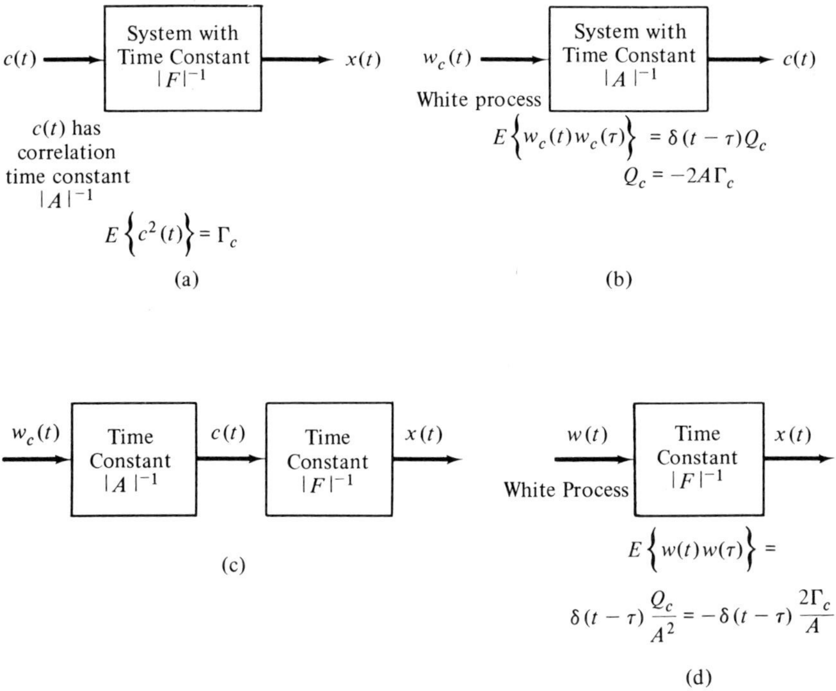

Figure 4.2.1 Block diagram for example.

4.2.3 $E\{x(t)w(t)\}$ = ?†

Consider (4.2.1). A question that sometimes arises is, What is $E\{x(t)w(t)\}$? This question can yield various answers. This is one of the unfortunate consequences of the use of a continuous-time white stochastic process model.

Assume that the continuous-time model is viewed as the limit of a discrete-time model. Consider the integrator of Section 4.2.1. Then two possible definitions and results are

$$E\{x(t)w(t)\} = \lim E\{x(n\Delta)w(n\Delta)\} = 0 \tag{4.2.8}$$

$$E\{x(t)w(t)\} = \lim E\left\{\left[\frac{x(n\Delta + \Delta) + x(n\Delta)}{2}\right]w(n\Delta)\right\} = \frac{Q}{2} \tag{4.2.9}$$

There are other possible definitions which yield still different answers, but discussion is restricted to (4.2.8) and (4.2.9).

Consider the example of Section 4.2.2. For the "physical" model of (4.2.5), it can be shown that (see Exercise 4.12)

$$E\{x_c(t)c(t)\} = \frac{Q_c}{2A(F + A)}$$

or when $|A| \gg |F|$

$$E\{x_c(t)c(t)\} = \frac{Q_c}{2A^2}$$

Thus in the equivalent "white" model (4.2.7) it is reasonable to assume that

$$E\{x(t)w(t)\} = \frac{Q}{2}$$

as this yields

$$E\{x(t)w(t)\} = \frac{Q_c}{2A^2}$$

Hence in this sense the definition of (4.2.9) is more appropriate than (4.2.8).

An alternative way to view the situation is to write

$$x(t) = \int_0^t e^{F(t-\tau)} w(\tau)\, d\tau$$

The question then turns on whether the upper limit of integration is t or $t - \epsilon$, where ϵ is an arbitrarily small positive number. It is often argued that "half of the delta function" is included so that

$$E\{x(t)w(t)\} = \frac{Q}{2}$$

For any given definition [i.e., (4.2.8) or (4.2.9)], the quantity $E\{x(t)w(t)\}$ is well specified in a mathematical sense. The problem is that different reasonable definitions lead to different results. Thus it is in a physical rather than mathematical sense that there is no general answer to the question, What is $E\{x(t)w(t)\}$? The choice of mathematical definition depends on the exact nature of the physical problem that is being modeled. Fortunately, in many applications, the value of $E\{w(t)x(t)\}$ is not important, so any convenient mathematical definition can be used and the problem is mostly of academic interest.

4.2.4 Input to Nonlinear Dynamic Systems†

When nonlinear systems are studied, some other confusing consequences of the white process model arise. Nonlinear systems will not be encountered until later in this book, but it is appropriate to consider an example here.

As a preliminary, a "strange" property of $w(t)$ is derived. Consider

$$s(N\Delta) = \sum_{n=1}^{N} \Delta^2 w^2(n\Delta)$$

If $w(n\Delta)$, $n = 1, \ldots, N$, is a well-defined process, it follows that in the limit $N \longrightarrow \infty$, $\Delta \longrightarrow 0$, $N\Delta \longrightarrow T$ that

$$s(N\Delta) \longrightarrow s(T) \longrightarrow 0$$

However, if $w(n\Delta)$, $n = 1, \ldots, N$, is a Gaussian zero mean, discrete-time white stochastic process with

$$E\{w(n_1\Delta)w(n_2\Delta)\} = \begin{cases} \dfrac{Q}{\Delta}, & n_1 = n_2 \\ 0, & n_1 \neq n_2 \end{cases}$$

then

$$E\{s(N\Delta)\} = \sum_{n=1}^{N} \Delta Q = N\Delta Q$$

Because of the Gaussian assumption (see Appendix E),

$$E\{\Delta w^2(n\Delta) - Q\}^2 = 2Q^2$$

After some manipulation it follows that for large N

$$E\{s^2(N\Delta)\} \approx (N\Delta)^2 Q^2 \qquad \text{(large } N\text{)}$$

Thus the variance of $s(N\Delta)$,

$$E\{[s(N\Delta) - E\{s(N\Delta)\}]^2\} = E\{s^2(N\Delta)\} - [E\{s(N\Delta)\}]^2$$

is for large N:

$$E\{[s(N\Delta) - E\{s(N\Delta)\}]^2\} \approx 0, \qquad \text{large } N$$

Hence for $w(n\Delta)$, a Gaussian white stochastic process,

$$\boxed{\begin{aligned} & s(N\Delta) = \sum_{n=1}^{N} \Delta^2 w^2(n\Delta) \\ & s(N\Delta) \longrightarrow s(T) \\ & s(T)\text{: random variable with mean } QT \text{ and zero variance} \end{aligned}} \tag{4.2.10}$$

Therefore when the white model is assumed, $s(T)$ can be viewed as a constant.

To illustrate the consequences of this property, consider the nonlinear continuous-time model

$$\frac{d}{dt}x_1(t) = w(t)$$

$$\frac{d}{dt}x_2(t) = x_1(t)w(t)$$

$$E\{w(t)w(\tau)\} = Q\delta(t - \tau)$$

$$w(t)\text{: zero mean Gaussian}$$

$$x_1(0) = x_2(0) = 0$$

One "reasonable" way to define $x_2(T)$ is as the limit of $\tilde{x}_2(N\Delta)$, $N \longrightarrow \infty$, $\Delta \longrightarrow 0$, $N\Delta \longrightarrow T$, where

$$\begin{aligned} \tilde{x}_2(N\Delta) &= \sum_{n=1}^{N} \Delta x_1(n\Delta)w(n\Delta) \\ x_1(n\Delta + \Delta) &= x_1(n\Delta) + \Delta w(n\Delta) \end{aligned} \tag{4.2.11}$$

Equation (4.2.11) can be manipulated into the form

$$\tilde{x}_2(N\Delta) = \frac{1}{2} \sum_{n=0}^{N} [x_1^2(n\Delta + \Delta) - x_1^2(n\Delta)] - \frac{1}{2} \sum_{n=1}^{N} \Delta^2 w^2(n\Delta)$$

so that

$$\tilde{x}_2(N\Delta) = \frac{x_1^2(N\Delta + \Delta)}{2} - \frac{1}{2} \sum_{n=1}^{N} \Delta^2 w^2(n\Delta)$$

Thus in the limit as $N \to \infty$, $\Delta \to 0$, $N\Delta \to T$, there are two possible answers:

$$\tilde{x}_2(T) = \frac{x_1^2(T)}{2}, \qquad \text{if } w(n\Delta) \text{ is well behaved} \tag{4.2.12}$$

$$\tilde{x}_2(T) = \frac{x_1^2(T)}{2} - \frac{1}{2}QT, \qquad \text{if } w(n\Delta) \text{ is a white stochastic process and (4.2.10) is used} \tag{4.2.13}$$

Is $x_2(T)$ given by (4.2.12) or (4.2.13)? If $w(n\Delta)$, $n = 1, \ldots,$ goes in the limit to some well-behaved (physically realizable) $w(t)$, then $\sum_{n=1}^{N} \Delta^2 w^2(n\Delta) \to 0$ and (4.2.12) is the result. Therefore if one wants answers which correspond to well-behaved processes, (4.2.12) is the best answer. Unfortunately, to obtain (4.2.12) from (4.2.11), it is necessary to ignore certain aspects of the mathematical model, and this does not satisfy our desire for mathematical consistency. However, (4.2.11) is only one of many ways to define $x_2(T)$. An equally reasonable definition is

$$\begin{aligned} \tilde{\tilde{x}}_2(N\Delta) &= \sum_{n=1}^{N} \Delta \left[\frac{x_1(n\Delta + \Delta) + x_1(n\Delta)}{2}\right] w(n\Delta) \\ x_1(n\Delta + \Delta) &= x_1(n\Delta) + \Delta w(n\Delta) \end{aligned} \tag{4.2.14}$$

so that after manipulation

$$\begin{aligned} \tilde{\tilde{x}}_2(N\Delta) &= \tilde{x}_2(N\Delta) + \frac{1}{2} \sum_{u=1}^{N} \Delta^2 w^2(n\Delta) \\ &= \frac{1}{2} \sum_{n=0}^{N} [x_1^2(n\Delta + \Delta) - x_1^2(n\Delta)] \end{aligned}$$

so in the limit

$$\tilde{\tilde{x}}_2(T) = \frac{x_1^2(T)}{2}$$

Thus by using the definition of (4.2.14) instead of (4.2.11) it is possible to derive (4.2.12) without violating the mathematics.

The problem of defining $x_2(T)$ is closely related to the problem of defining $E\{x(t)w(t)\}$. In particular, one can say that (4.2.8) and (4.2.11) are "similar" definitions and that (4.2.9) and (4.2.14) are "similar" definitions.

The various definitions yield results which differ only by a constant, i.e., in the mean value of $x_2(T)$. Thus if the mean value is not important in the problem of interest, either definition can be used and the whole problem is only of academic interest. If the mean value is important, it is necessary to examine the nature of the actual process approximated by $w(t)$ more closely and to remember that ideal integrators and ideal multipliers do not actually exist. In terms of frequency domain concepts, it may be necessary to worry about the bandwidth of the integrator and the multiplier relative to the bandwidth of $w(t)$.

4.2.5 Mathematical Formalism†

In this book no attempt is made either to define or to manipulate continuous-time stochastic processes in a precise mathematical fashion. However, as an aid to the reader who wants to go to the literature, some mathematical formalism and associated jargon is briefly discussed.

In a mathematical formulation, (4.2.1) would be written in the form [for scalar $w(t)$]

$$d\mathbf{x}(t) = \mathbf{F}(t)\mathbf{x}(t)\, dt + \mathbf{G}(t)\, db_w(t) \tag{4.2.15}$$

where

$$w(t) = \frac{db_w(t)}{dt}$$

$b(t)$ is called a Wiener process, or Brownian motion, or a process with independent increments. (See Appendix F.)

It has been shown that care is needed in defining exactly what is meant by a solution, i.e., in defining $E\{x(t)w(t)\}$, or a *stochastic integral*, or a *stochastic differential equation*. At the present time there are two main methods of approach, often called the *Itô calculus* and the *Stratonovich calculus*. Roughly speaking, (4.2.8) and (4.2.11) correspond to the Itô calculus, while (4.2.9) and (4.2.14) correspond to the Stratonovich calculus. In general, the Stratonovich calculus is more physically oriented. However, the Itô calculus actually has some mathematical advantages, and once its specialized properties are understood, the Itô calculus is sometimes easier to use. It is possible to change the system models in a compensating fashion so that one can get Stratonovich-calculus-type results using Itô calculus.

Any reader who is interested in doing mathematical research on continuous-time stochastic process problems must learn this mathematical formalism. Engineers who are interested in applying sophisticated results are also urged to "learn and understand the language" as it is used in most advanced papers and books. However, readers of all types must always remember that mathematical formalism does not replace engineering judgment with regard to when, where, or how the models should be used and what the results mean.

4.2.6 Discussion on White Stochastic Process

Continuous-time white stochastic process models are employed because they often greatly simplify the analysis. They are only approximations, but then all mathematical models are really only approximate representations of the real world. However, continuous-time white stochastic processes are a very special type of approximation. They have infinite variance and thus, in one sense, can never approximate a real process. In many situations, their use can be justified either as a mathematical artifice or because their effects are approximately the same as those of physical processes when they are both passed through some system or filter of interest. Nevertheless, the special nature of the approximation must be remembered; in particular, equally reasonable mathematical definitions can yield different mathematical results. Fortunately the differences are often only of academic interest, but they can lead to confusion. In the following chapters of this book, special discussions are given in those cases where these confusions arise.

The term *physically realizable* is often used when discussing stochastic process models. If $w(t)$ is a white stochastic process, it is clear that $w(t)$ is "not physically realizable" as $E\{w^2(t)\} = \infty$. It is often said that if

$$\frac{d}{dt}c(t) = Fc(t) + w(t)$$

$$E\{c(t)c(\tau)\} = \Gamma_c e^{A(t-\tau)}, \qquad t > \tau$$

then $c(t)$ is a physically realizable process. However,

$$E\left\{\frac{dc(t)}{dt}\right\}^2 = \infty$$

Since a physically realizable process cannot have a derivative with infinite variance, $c(t)$ is, in a very real sense, just as physically unrealizable as the white process $w(t)$. The conclusion is that *none of the continuous-time stochastic processes considered in this book are physically realizable.* This does not mean the equations to be derived are useless. It merely reemphasizes the point that all mathematical models are only approximate representations of the real world.

4.3 Unknown-but-Bounded

The output of a dynamic system driven by a white unknown-but-bounded process is now discussed. Instead of considering positive semidefinite covariance matrices as in the Bayesian stochastic model, interest centers on sets (which are usually ellipsoids defined by positive semidefinite matrices). Comparison of the set-theoretic concepts to be discussed with the stochastic theory reveals that

1. The detailed mathematical manipulations are very different.
2. The final equations look similar.
3. The final equations can behave quite differently.

It must always be remembered that the two models involve fundamentally different sets of physical assumptions.

It is recommended that Appendices G and H be read before proceeding, as the ideas therein are used extensively in the following.

4.3.1 Static Model

Consider the static model

$$\mathbf{z} = \mathbf{H}\mathbf{x} + \mathbf{v}$$
$$\mathbf{x} \in \mathbf{\Omega}_x$$
$$\mathbf{v} \in \mathbf{\Omega}_v$$

Let $\mathbf{\Omega}_z$ denote the set that contains all the possible resulting $\mathbf{z}$. Then

$$\mathbf{\Omega}_z = \{\mathbf{z}: \mathbf{z} = \mathbf{H}\mathbf{x} + \mathbf{v}, \mathbf{x} \in \mathbf{\Omega}_x, \mathbf{v} \in \mathbf{\Omega}_v\}$$

Define

$$\mathbf{\Omega}_{Hx} = \{\mathbf{z}: \mathbf{z} = \mathbf{H}\mathbf{x}, \mathbf{x} \in \mathbf{\Omega}_x\}$$

Then

$$\mathbf{\Omega}_z = \{\mathbf{z}: \mathbf{z} = \mathbf{z}_1 + \mathbf{v}, \mathbf{z}_1 \in \mathbf{\Omega}_{Hx}, \mathbf{v} \in \mathbf{\Omega}_v\}$$

or

$$\boxed{\mathbf{\Omega}_z = \text{vector sum of } \mathbf{\Omega}_{Hx} \text{ and } \mathbf{\Omega}_v} \tag{4.3.1}$$

Define

$$\begin{aligned} s_z(\boldsymbol{\eta}) &\quad \text{support function of } \mathbf{\Omega}_z \\ s_{Hx}(\boldsymbol{\eta}) &\quad \text{support function of } \mathbf{\Omega}_{Hx} \\ s_v(\boldsymbol{\eta}) &\quad \text{support function of } \mathbf{\Omega}_v \end{aligned}$$

Then (4.3.1) can be rewritten as the functional equation

$$\boxed{\begin{aligned} \mathbf{\Omega}_z &= \{\mathbf{z} : \mathbf{z}'\boldsymbol{\eta} \leq s_z(\boldsymbol{\eta}), \quad \text{all } \boldsymbol{\eta}, \boldsymbol{\eta}'\boldsymbol{\eta} = 1\} \\ s_z(\boldsymbol{\eta}) &= s_{Hx}(\boldsymbol{\eta}) + s_v(\boldsymbol{\eta}) \end{aligned}} \tag{4.3.2}$$

Now, assume that $\mathbf{\Omega}_x$ and $\mathbf{\Omega}_v$ are ellipsoids:

$$\mathbf{\Omega}_x = \{\mathbf{x}: \mathbf{x}'\boldsymbol{\psi}^{-1}\mathbf{x} \leq 1\}$$
$$\mathbf{\Omega}_v = \{\mathbf{v}: \mathbf{v}'\mathbf{R}^{-1}\mathbf{v} \leq 1\}$$

Then from (H.4) in Appendix H

$$\Omega_{Hx} = \{\mathbf{z} : \mathbf{z}'(\mathbf{H}\boldsymbol{\psi}\mathbf{H}')^{-1}\mathbf{z} \leq 1\}$$
$$s_v(\boldsymbol{\eta}) = (\boldsymbol{\eta}'\mathbf{R}\boldsymbol{\eta})^{1/2}$$
$$s_{Hx}(\boldsymbol{\eta}) = (\boldsymbol{\eta}'\mathbf{H}\boldsymbol{\psi}\mathbf{H}'\boldsymbol{\eta})^{1/2}$$

If this were the best of all possible worlds, the fact that Ω_{Hx} and Ω_v are ellipsoids would imply that Ω_z is an ellipsoid. Unfortunately the gods are not so kind, as

$$s_z(\boldsymbol{\eta}) = (\boldsymbol{\eta}'\mathbf{R}\boldsymbol{\eta})^{1/2} + (\boldsymbol{\eta}'\mathbf{H}\boldsymbol{\psi}\mathbf{H}'\boldsymbol{\eta})^{1/2} \qquad (4.3.3)$$

is not in general the support function of an ellipsoid. From a computational point of view, it is very desirable to work with ellipsoids, as they are defined by matrices which can be manipulated on a digital computer. Equations (4.3.2) and (4.3.3) are functional equations which are not well suited to numerical studies except in special cases. Therefore a bounding ellipsoid which contains Ω_z will be developed.

A special case of Holder's inequality is

$$(1-\gamma)^{-1}b_1^2 + \gamma^{-1}b_2^2 \geq (b_1 + b_2)^2, \qquad 0 \leq \gamma \leq 1$$

From (4.3.3), this gives

$$s_b(\boldsymbol{\eta}) \geq s_z(\boldsymbol{\eta}), \qquad \text{all } \boldsymbol{\eta}$$
$$s_b(\boldsymbol{\eta}) = [\boldsymbol{\eta}'[(1-\gamma)^{-1}\mathbf{R} + \gamma^{-1}\mathbf{H}\boldsymbol{\psi}\mathbf{H}']\boldsymbol{\eta}]^{1/2}$$

Thus the set $\Omega_{z,b}$ is a bounding ellipsoid for Ω_z, where

$$\boxed{\begin{aligned} \Omega_{z,b} &= \{\mathbf{z} : \mathbf{z}'\boldsymbol{\Gamma}_b^{-1}\mathbf{z} \leq 1\} \\ \boldsymbol{\Gamma}_b &= \tilde{\mathbf{R}} + \mathbf{H}\tilde{\boldsymbol{\psi}}\mathbf{H}' \\ \tilde{\mathbf{R}} &= (1-\gamma)^{-1}\mathbf{R} \\ \tilde{\boldsymbol{\psi}} &= \gamma^{-1}\boldsymbol{\psi} \\ \Omega_{z,b} &\supset \Omega_z \text{ for any } \gamma, \qquad 0 \leq \gamma \leq 1 \end{aligned}} \qquad (4.3.4)$$

Note the structural similarity between $\boldsymbol{\Gamma}_b$ of (4.3.4) and the covariance matrix (4.1.3).

It is very important to note the basic difference between (4.3.2), and the bounding ellipsoid (4.3.4). Equation (4.3.2) is simple in appearance, but it is a functional equation which is very difficult to study numerically using a digital computer; i.e., (4.3.2) must be satisfied for all values of the vector $\boldsymbol{\eta}$. The bounding ellipsoid (4.3.4), on the other hand, requires only matrix manipulation.

EXAMPLE: Consider

$$\mathbf{z} = \begin{bmatrix} 1 \\ 0 \end{bmatrix} x + \mathbf{v} \qquad \mathbf{z} = \begin{bmatrix} z_1 \\ z_2 \end{bmatrix} \qquad \mathbf{v} = \begin{bmatrix} v_1 \\ v_2 \end{bmatrix}$$

$$\mathbf{\Omega}_x = \{x : x^2 \leq 1\}$$

$$\mathbf{\Omega}_v = \{\mathbf{v} : v_1^2 + v_2^2 \leq 1\}$$

Using (4.3.4), a bounding ellipsoid $\mathbf{\Omega}_{z,b}$, is

$$\mathbf{\Omega}_{z,b} = \{\mathbf{z} : \mathbf{z}'\mathbf{\Gamma}_b^{-1}\mathbf{z} \leq 1\}$$

$$\mathbf{\Gamma}_b = \begin{bmatrix} \dfrac{1}{1-\gamma} + \dfrac{1}{\gamma} & 0 \\ 0 & \dfrac{1}{1-\gamma} \end{bmatrix}$$

$\mathbf{\Omega}_{z,b}$ is compared with $\mathbf{\Omega}_z$ in Fig. 4.3.1 for $\gamma = \frac{1}{2}$, so

$$\mathbf{\Gamma} = \begin{bmatrix} 4 & 0 \\ 0 & 2 \end{bmatrix}$$

In Fig. 4.3.1 the nature of $\mathbf{\Omega}_z$ was determined by inspection. However, $\mathbf{\Omega}_z$ could have been obtained from (4.3.2) as

$$s_v(\boldsymbol{\eta}) = (\eta_1^2 + \eta_2^2)^{1/2}$$

$$s_{Hx}(\boldsymbol{\eta}) = \eta_1$$

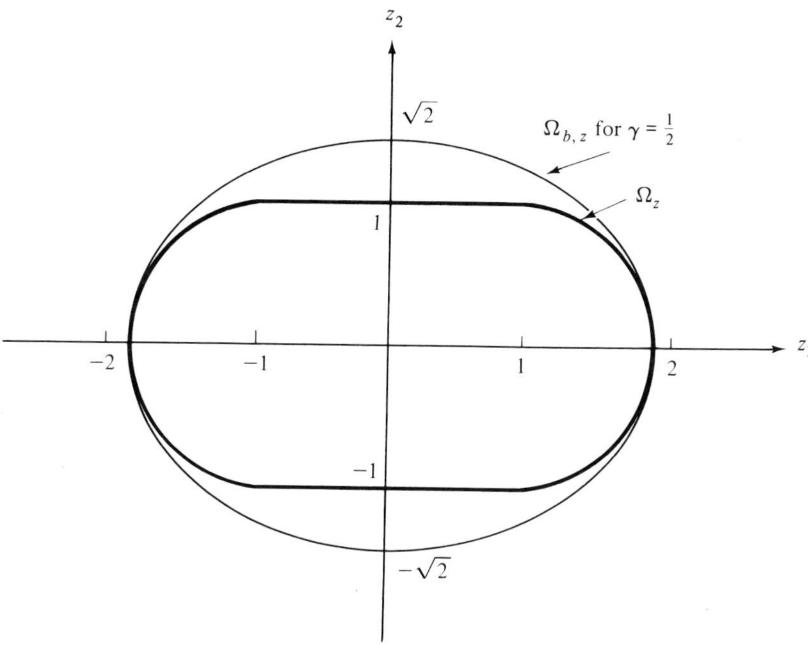

Figure 4.3.1 Example of bounding ellipsoid.

so

$$\Omega_z = \{\mathbf{z} : z_1\eta_1 + z_2\eta_2 \leq (\eta_1^2 + \eta_2^2)^{1/2} + \eta_1, \text{ all } \eta_1 \text{ and } \eta_2, \eta_1^2 + \eta_2^2 = 1\}$$

Determination of the size and shape of $\mathbf{\Omega}_z$ using this equation is awkward. If a digital computer were used, the computer would have to evaluate the equation for *all* possible values of η_1 and η_2 such that $\eta_1^2 + \eta_2^2 = 1$.

4.3.2 Discrete-Time Dynamic System

A discrete-time unknown-but-bounded process $\mathbf{x}(n\Delta)$, $n = 1, \ldots,$ can be modeled as

$$\mathbf{x}(n\Delta + \Delta) = \mathbf{\Phi}(n\Delta)\mathbf{x}(n\Delta) + \Delta\mathbf{G}(n\Delta)\mathbf{w}(n\Delta)$$
$$\mathbf{x}(0) \in \mathbf{\Omega}_x(0)$$
$$\mathbf{w}(n\Delta) \in \mathbf{\Omega}_w(n\Delta)$$

Let $\mathbf{\Omega}_x(n\Delta)$ denote the set containing all possible $\mathbf{x}(n\Delta)$. $\mathbf{\Omega}_x(n\Delta)$ is called the set of reachable states. It follows that

$$\mathbf{\Omega}_x(n\Delta + \Delta) = \{\mathbf{x} : \mathbf{x} - \mathbf{\Phi}(n\Delta)\mathbf{x}_1 + \Delta\mathbf{G}(n\Delta)\mathbf{w}, \ \mathbf{x}_1 \in \mathbf{\Omega}_x(n\Delta), \ \mathbf{w} \in \mathbf{\Omega}_w(n\Delta)\}$$

Define

$$\tilde{\mathbf{\Omega}}_x(n\Delta + \Delta \mid n\Delta) = \{\mathbf{x} : \mathbf{x} = \mathbf{\Phi}(n\Delta)\mathbf{x}_1, \ \mathbf{x}_1 \in \mathbf{\Omega}_x(n\Delta)\}$$
$$\mathbf{\Omega}_{Gw}(n\Delta) = \{\mathbf{x} : \mathbf{x} = \Delta\mathbf{G}(n\Delta)\mathbf{w}, \ \mathbf{w} \in \mathbf{\Omega}_w(n\Delta)\}$$

Then

$$\boxed{\mathbf{\Omega}_x(n\Delta + \Delta) = \text{vector sum of } \tilde{\mathbf{\Omega}}_x(n\Delta + \Delta \mid n\Delta) \text{ and } \mathbf{\Omega}_{Gw}(n\Delta)} \tag{4.3.5}$$

Define

$s_{x(n\Delta)}(\boldsymbol{\eta})$	support function of $\mathbf{\Omega}_x(n\Delta)$
$s_{w(n\Delta)}(\boldsymbol{\eta})$	support function of $\mathbf{\Omega}_w(n\Delta)$
$s_{Gw(n\Delta)}(\boldsymbol{\eta})$	support function of $\mathbf{\Omega}_{Gw}(n\Delta)$
$\tilde{s}(\boldsymbol{\eta})$	support function of $\tilde{\mathbf{\Omega}}_x(n\Delta + \Delta \mid n\Delta)$

Then

$$s_{Gw(n\Delta)}(\boldsymbol{\eta}) = s_{w(n\Delta)}[\Delta\mathbf{G}'(n\Delta)\boldsymbol{\eta}]$$
$$\tilde{s}(\boldsymbol{\eta}) = s_{x(n\Delta)}[\mathbf{\Phi}'(n\Delta)\boldsymbol{\eta}]$$

and (4.3.5) can be rewritten as

$$\boxed{\begin{aligned} s_{x(n\Delta+\Delta)}(\boldsymbol{\eta}) = {} & s_{w(n\Delta)}[\Delta\mathbf{G}'(n\Delta)\boldsymbol{\eta}] \\ & + s_{x(n\Delta)}[\mathbf{\Phi}'(n\Delta)\boldsymbol{\eta}] \end{aligned}} \tag{4.3.6}$$

Equation (4.3.6) is a functional equation relating the support functions of the sets of reachable states $\mathbf{\Omega}_x(n\Delta)$ and $\mathbf{\Omega}_x(n\Delta + \Delta)$.

Now assume that $\mathbf{\Omega}_x(0)$ and $\mathbf{\Omega}_w(n\Delta)$ are ellipsoids:

$$\mathbf{\Omega}_x(0) = \{\mathbf{x} : \mathbf{x}'\boldsymbol{\psi}^{-1}\mathbf{x} \leq 1\}$$

$$\mathbf{\Omega}_w(n\Delta) = \{\mathbf{w} : \mathbf{w}'\mathbf{Q}^{-1}(n\Delta)\mathbf{w} \leq 1\}$$

In general the resulting $\mathbf{\Omega}_x(n\Delta + \Delta)$ will not be an ellipsoid. However, equations for a bounding ellipsoid $\mathbf{\Omega}_{x,b}(n\Delta + \Delta)$ can be developed as follows:

1. Assume that a bounding ellipsoid $\mathbf{\Omega}_{x,b}(n\Delta)$, $\mathbf{\Omega}_{x,b}(n\Delta) \supset \mathbf{\Omega}_x(n\Delta)$, is given.
2. Find the equation defining $\mathbf{\Omega}_{x,b}(n\Delta + \Delta)$, $\mathbf{\Omega}_{x,b}(n\Delta + \Delta) \supset \mathbf{\Omega}_x(n\Delta + \Delta)$, assuming that $\mathbf{\Omega}_{x,b}(n\Delta) = \mathbf{\Omega}_x(n\Delta)$.

This logic leads to a difference equation whose solution always yields a bounding ellipsoid for $\mathbf{\Omega}_x(n\Delta)$ for all n. Define

$$\mathbf{\Omega}_{x,b}(n\Delta) = \{\mathbf{x} : \mathbf{x}'\mathbf{\Gamma}^{-1}(n\Delta)\mathbf{x} \leq 1\}$$

with support function

$$s_{x(n\Delta),b}(\boldsymbol{\eta}) = (\boldsymbol{\eta}'\mathbf{\Gamma}(n\Delta)\boldsymbol{\eta})^{1/2}$$

Then (4.3.6) becomes

$$s_{x(n\Delta+\Delta)}(\boldsymbol{\eta}) = (\Delta^2\boldsymbol{\eta}'\mathbf{G}(n\Delta)\mathbf{Q}(n\Delta)\mathbf{G}'(n\Delta)\boldsymbol{\eta})^{1/2} + (\eta'\mathbf{\Phi}(n\Delta)\mathbf{\Gamma}(n\Delta)\mathbf{\Phi}'(n\Delta)\boldsymbol{\eta})^{1/2}$$

Now, as in Section 4.3.1, a Holder-type inequality can be used to find the support function of a bounding ellipsoid which is larger than $s_{x(n\Delta+\Delta)}(\boldsymbol{\eta})$. The final result is

$$\begin{aligned}
\mathbf{\Omega}_{x,b}(n\Delta) &= \{\mathbf{x} : \mathbf{x}'\mathbf{\Gamma}^{-1}(n\Delta)\mathbf{x} \leq 1\} \\
\mathbf{\Omega}_{x,b}(n\Delta) &\supset \mathbf{\Omega}_x(n\Delta) \\
\mathbf{\Gamma}(n\Delta + \Delta) &= \left(\frac{1}{1 - \Delta\beta(n\Delta)}\right)\mathbf{\Phi}(n\Delta)\mathbf{\Gamma}(n\Delta)\mathbf{\Phi}'(n\Delta) \\
&\quad + \Delta\mathbf{G}(n\Delta)\tilde{\mathbf{Q}}(n\Delta)\mathbf{G}'(n\Delta) \\
\tilde{\mathbf{Q}}(n\Delta) &= \frac{\mathbf{Q}(n\Delta)}{\beta(n\Delta)} \\
\frac{1}{\Delta} &\geq \beta(n\Delta) \geq 0 \\
\mathbf{\Gamma}(0) &= \boldsymbol{\psi}
\end{aligned} \tag{4.3.7}$$

Note the structural similarity between the $\mathbf{\Gamma}$ equation of (4.3.7) and (4.1.6) for the Bayesian model.

For time-invariant stable systems, the set of reachable states $\mathbf{\Omega}_x(n\Delta)$ will reach a steady-state value $\mathbf{\Omega}_x(ss)$. The corresponding bounding ellipsoid $\mathbf{\Omega}_{x,b}(ss)$ is given by

$$\mathbf{\Omega}_{x,b}(ss) = \{\mathbf{x} : \mathbf{x}'\mathbf{\Gamma}_{ss}^{-1}\mathbf{x} \leq 1\}$$

$$\mathbf{\Gamma}_{ss} = \left(\frac{1}{1-\Delta\beta}\right)\mathbf{\Phi}\mathbf{\Gamma}_{ss}\mathbf{\Phi}' + \frac{\Delta}{\beta}\mathbf{G}\mathbf{Q}\mathbf{G}' \tag{4.3.8}$$

$$\frac{1}{\Delta} \geq \beta \geq 0$$

EXAMPLE: Consider

$$x(n+1) = \Phi x(n) + w(n), \qquad |\Phi| < 1$$

$$|w(n)| \leq Q^{1/2}$$

Then from (4.3.8)

$$\Gamma_{ss} = \frac{(1-\beta)Q}{\beta(1-\beta-\Phi^2)}$$

Thus a choice of $\beta = 1 - \Phi^2$ yields $\Gamma_{ss} = \infty$, while $1 > \beta > 1 - \Phi^2$ yields a negative Γ_{ss}. This is not inconsistent with the theory; it merely says that only the β such that

$$0 \leq \beta < 1 - \Phi^2$$

are of any interest. The best value of β is

$$\beta = 1 - \Phi$$

as this yields a minimum Γ_{ss}

$$\Gamma_{ss} = \frac{Q}{(1-\Phi)^2}$$

which in this case actually defines the true set of reachable states.

As in Section 4.3.1, note the difference in computational problems between the general functional equation (4.3.6) and the matrix equations of (4.3.7), which give the bounding ellipsoid.

4.3.3 Continous-Time Dynamic System

A continuous-time unknown-but-bounded process $\mathbf{x}(t)$ can be modeled as

$$\frac{d}{dt}\mathbf{x}(t) = \mathbf{F}(t)\mathbf{x}(t) + \mathbf{G}(t)\mathbf{w}(t)$$

$$\mathbf{x}(0) \in \mathbf{\Omega}_x(0) = \{\mathbf{x} : \mathbf{x}'\boldsymbol{\psi}^{-1}\mathbf{x} \leq 1\}$$

$$\mathbf{w}(t) \in \mathbf{\Omega}_w(t) = \{\mathbf{w} : \mathbf{w}'\mathbf{Q}^{-1}(t)\mathbf{w} \leq 1\}$$

Let $\mathbf{\Omega}_x(t)$ denote the set of reachable states.

Let $\mathbf{\Omega}_{x,b}(t)$ denote a bounding ellipsoid defined by $\mathbf{\Gamma}(t)$ for $\mathbf{\Omega}_x(t)$. A differential equation for $\mathbf{\Gamma}(t)$ can be obtained by applying the discrete to continuous time limit, $\Delta \longrightarrow 0$, $n \longrightarrow \infty$, $n\Delta \longrightarrow t$. Substituting

$$\mathbf{\Phi}(n\Delta) = \mathbf{I} + \Delta\mathbf{F}(n\Delta)$$

into the difference equation (4.3.7) gives

$$\begin{aligned}\mathbf{\Gamma}(n\Delta + \Delta) = (1 + \Delta\beta(n\Delta) + \Delta^2\beta^2(n\Delta) + \cdots)[&\mathbf{\Gamma}(n\Delta + \Delta\mathbf{F}(n\Delta)\mathbf{\Gamma}(n\Delta) \\ &+ \Delta\mathbf{\Gamma}(n\Delta)\mathbf{F}'(n\Delta) + \Delta^2\mathbf{F}(n\Delta)\mathbf{\Gamma}(n\Delta)\mathbf{F}'(u\Delta)] \\ &+ \frac{\Delta}{\beta(n\Delta)}\mathbf{G}(n\Delta)\mathbf{Q}(n\Delta)\mathbf{G}'(n\Delta)\end{aligned}$$

or

$$\begin{aligned}\frac{1}{\Delta}[\mathbf{\Gamma}(n\Delta + \Delta) - \mathbf{\Gamma}(n\Delta)] = \mathbf{F}(n\Delta)\mathbf{\Gamma}(n\Delta) &+ \mathbf{\Gamma}(n\Delta)\mathbf{F}'(n\Delta) \\ &+ \beta(n\Delta)\mathbf{\Gamma}(n\Delta) + \frac{1}{\beta(n\Delta)}\mathbf{G}(n\Delta)\mathbf{Q}(n\Delta)\mathbf{G}'(n\Delta) \\ &+ (\text{higher-order terms in } \Delta)\end{aligned}$$

$$\frac{1}{\Delta} \geq \beta(n\Delta) \geq 0$$

Thus in the limit $\Delta \longrightarrow 0$, $n \longrightarrow \infty$, $n\Delta \longrightarrow t$,

$$\boxed{\begin{aligned}&\mathbf{\Omega}_{x,b}(t) = \{\mathbf{x} : \mathbf{x}'\mathbf{\Gamma}^{-1}(t)\mathbf{x} \leq 1\} \\ &\mathbf{\Omega}_{x,b}(t) \supset \mathbf{\Omega}_x(t) \\ &\frac{d}{dt}\mathbf{\Gamma}(t) = \mathbf{F}(t)\mathbf{\Gamma}(t) + \mathbf{\Gamma}(t)\mathbf{F}'(t) + \beta(t)\mathbf{\Gamma}(t) + \mathbf{G}(t)\tilde{\mathbf{Q}}(t)\mathbf{G}'(t) \\ &\tilde{\mathbf{Q}}(t) = \frac{\mathbf{Q}(t)}{\beta(t)} \\ &\mathbf{\Gamma}(0) = \boldsymbol{\psi} \\ &\infty \geq \beta(t) \geq 0\end{aligned}} \tag{4.3.9}$$

Equation (4.3.9) can also be derived directly without using the discrete to continuous time limit. However, the proof uses Pontryagin's maximum principle, so it is not given here.

4.3.4 Choice of Free Parameter, β

For simplicity, the discussions are restricted to the continuous-time case, but similar discussions also apply in the discrete-time case.

The matrix differential equation (4.3.9) defines the matrix $\mathbf{\Gamma}(t)$, which in turn defines a bounding ellipsoid $\mathbf{\Omega}_{x,b}(t)$ which contains all possible $\mathbf{x}(t)$. However, the differential equation contains a free parameter (time function) $\beta(t)$, and a bounding ellipsoid is obtained for any nonnegative $\beta(t)$. Obviously it is desired to choose a $\beta(t)$ which is optimum in some sense. There is no simple answer, but various possibilities are now discussed.

One "reasonable" approach is to choose the $\beta(t)$ so as to minimize $(d/dt)\mathbf{\Gamma}(t)$ in some sense. The most mathematically convenient criterion is

$$\text{Choose } \beta(t) \text{ to minimize } \mathrm{tr}\left[\frac{d}{dt}\mathbf{\Gamma}(t)\right]$$

If $\tilde{\beta}(t)$ denotes the minimum value,

$$\tilde{\beta}(t) = \left[\frac{\mathrm{tr}[\mathbf{G}(t)\mathbf{Q}(t)\mathbf{G}'(t)]}{\mathrm{tr}[\mathbf{\Gamma}(t)]}\right]^{1/2} \tag{4.3.10}$$

Substitution of $\tilde{\beta}(t)$ into (4.3.9) yields a completely defined differential equation for $\mathbf{\Gamma}(t)$. Unfortunately the equation is now nonlinear in $\mathbf{\Gamma}(t)$.

If interest centered on $\mathbf{\Gamma}(T)$ for some fixed T, the "ultimate" approach would be to formulate the choice of $\beta(t)$, $0 \leq t \leq T$, as the calculus of variations problem of choosing the particular $\beta(t)$, $0 \leq t \leq T$, which minimizes $\mathbf{\Gamma}(T)$ in some sense. Because of the constraint on $\beta(t)$, Pontryagin's maximum principle of optimal control theory is an appropriate tool to use.

From a practical point of view, it would be nice to have a constant value, $\beta(t) = \beta_0$. For a stable time-invariant system it would be reasonable to choose β_0 such that the $\mathbf{\Gamma}_{ss}$ which satisfies

$$\mathbf{F}\mathbf{\Gamma}_{ss} + \mathbf{\Gamma}_{ss}\mathbf{F}' + \beta_0\mathbf{\Gamma}_{ss} + \frac{1}{\beta_0}\mathbf{G}\mathbf{Q}\mathbf{G}' = \mathbf{0}$$

is minimum in some sense (see example in Section 4.3.2).

More insight into the effect of $\beta(t)$ may be obtained by rewriting (4.3.9) in the form

$$\begin{aligned} \frac{d}{dt}\mathbf{\Gamma}(t) &= \tilde{\mathbf{F}}(t)\mathbf{\Gamma}(t) + \mathbf{\Gamma}(t)\tilde{\mathbf{F}}'(t) + \mathbf{G}(t)\tilde{\mathbf{Q}}(t)\mathbf{G}'(t) \\ \tilde{\mathbf{F}}(t) &= \mathbf{F}(t) + \left(\frac{\beta(t)}{2}\right)\mathbf{I} \\ \tilde{\mathbf{Q}}(t) &= \left(\frac{1}{\beta(t)}\right)\mathbf{Q}(t) \end{aligned} \tag{4.3.11}$$

This shows that a large $\beta(t)$ tends to make the system more unstable while a small $\beta(t)$ tends to amplify the effect of the input bound $\mathbf{Q}(t)$.

4.3.5 Deterministic Inputs

Generalization of the unknown-but-bounded models to include a deterministic input and/or an initial ellipsoid which is not centered at the origin is straightforward. In continuous time the new model is

$$\frac{d}{dt}\mathbf{x}(t) = \mathbf{F}(t)\mathbf{x}(t) + \mathbf{G}(t)\mathbf{w}(t) + \mathbf{B}(t)\mathbf{u}(t)$$

$$\mathbf{x}(0) \in \mathbf{\Omega}_x(0) = \{\mathbf{x} : [\mathbf{x} - \mathbf{m}]'\boldsymbol{\psi}^{-1}[\mathbf{x} - \mathbf{m}] \leq 1\}$$

The bounding ellipsoid $\mathbf{\Omega}_{x,b}(t)$ now has its center at $\mathbf{x}_m(t)$ instead of at the origin, where $\mathbf{x}_m(t)$ is given by

$$\frac{d}{dt}\mathbf{x}_m(t) = \mathbf{F}(t)\mathbf{x}_m(t) + \mathbf{B}(t)\mathbf{u}(t)$$

$$\mathbf{x}_m(0) = \mathbf{m}$$

The equation for $\mathbf{\Gamma}(t)$ which defines the size and shape of the bounding ellipsoid remains unchanged.

Similar results are obtained in the discrete-time case.

4.4 Frequency Domain Discussion†

Frequency domain techniques play a minor role in this book. However, they are important in certain types of applications. They also furnish insights which can be valuable. Thus a few brief observations are now made. No attempt is made to be rigorous or complete. For example, the question of existence of transforms is ignored entirely. The discussions assume that the reader already has a frequency domain background.

Let $\mathfrak{F}_y(j\omega)$ denote the Fourier transform of some time function $y(t)$, $0 \le t \le T$, as given by

$$\mathfrak{F}_y(j\omega) = \frac{1}{\sqrt{2\pi}} \int_0^T y(t)e^{-j\omega t}\, dt$$

Define

$$G_y(\omega) = \lim_{T\to\infty} \frac{1}{T} \left| \frac{1}{\sqrt{2\pi}} \int_0^T y(t)e^{-j\omega t}\, dt \right|^2$$

Thus

$$G_y(\omega) = \lim_{T\to\infty} \frac{1}{T} \left| \mathfrak{F}_y(j\omega) \right|^2$$

Let $P_y(\omega_1, \omega_2)$ denote the "amount of power" contained in $y(t)$ in the frequency range ω_2 to ω_1, $\omega_1 > \omega_2 \ge 0$. Then

$$P_y(\omega_1, \omega_2) = 2 \int_{\omega_2}^{\omega_1} G_y(\omega)\, d\omega$$

The "total power" in $y(t)$ can be defined as

$$P_y = \frac{1}{T} \int_0^T y^2(t)\, dt$$

or by Parseval's theorem,

$$P_y = \int_{-\infty}^{\infty} G_y(\omega)\, d\omega$$

4.4.1 White Processes†

Scalar stochastic white processes, $w(t)$, defined by

$$E\{w(t)\} = 0$$
$$E\{w(t)w(\tau)\} = \delta(t - \tau)Q$$

and scalar unknown-but-bounded white processes, $w(t)$, defined by

$$|w(t)| \leq Q^{1/2}$$

are discussed.

Consider first a stochastic white process. Interest centers on expected values. It can be shown that

$$E\{G_w(\omega)\} = Q$$

Thus

$$E\{P_w(\omega_1, \omega_2)\} = 2|\omega_1 - \omega_2|Q$$

and

$$E\{P_w\} = \infty$$

so the average total power of $w(t)$ is infinite. This is to be expected as $w(t)$ has infinite variance; i.e.,

$$E\{w^2(t)\} = \delta(t - t)Q = \infty$$

Now consider an unknown-but-bounded white process. Interest centers on bounds. There does not exist a finite upper bound on $G_w(\omega)$, as $w(t)$ can be a pure sine wave which yields an impluse. This, of course, does not imply that $w(t)$ can have infinite total power. The total power of an unknown-but-bounded white process is obviously bounded by

$$P_w \leq Q$$

The power contained in some narrow range of frequencies $\omega_1 \leq \omega \leq \omega_2$, $|\omega_1 - \omega_2|$ small, is probably bounded by

$$P_w(\omega_1, \omega_2) \leq \frac{Q}{2}, \qquad |\omega_1 - \omega_2| \text{ small}$$

as $Q/2$ is the power in $Q^{1/2}\sin(\omega_1 t)$.

4.4.2 Time-Invariant, Stable Dynamic Systems†

Consider a time-invariant, stable linear system with scalar input $w(t)$ and scalar output $y(t)$ defined by

$$\frac{d}{dt}\mathbf{x}(t) = \mathbf{F}\mathbf{x}(t) + \mathbf{G}w(t)$$

$$y(t) = \mathbf{H}\mathbf{x}(t)$$

Let $A(j\omega)$ denote the frequency response of the system so that

$$A(j\omega) = \mathbf{H}[\mathbf{I}j\omega - \mathbf{F}]^{-1}\mathbf{G}$$

Then

$$\mathfrak{F}_y(j\omega) = A(j\omega)\mathfrak{F}_w(j\omega)$$
$$G_y(\omega) = |A(j\omega)|^2 G_w(\omega)$$

Assume that $w(t)$ is a white stochastic process. Then

$$E\{G_y(\omega)\} = |A(j\omega)|^2 E\{G_w(\omega)\} = |A(j\omega)|^2 Q$$

$$E\{P_y(\omega_1, \omega_2)\} = 2Q \int_{\omega_2}^{\omega_1} |A(j\omega)|^2 \, d\omega$$

$$E\{P_y\} = Q \int_{-\infty}^{\infty} |A(j\omega)|^2 \, d\omega$$

so that $E\{P_y\}$ is finite.

Assume that $w(t)$ is a white unknown-but-bounded process. Then $G_y(\omega)$ is unbounded, as it can contain impulses. For sufficiently small $|\omega_1 - \omega_2|$, it is probably true that

$$P_y(\omega_1, \omega_2) \leq \frac{Q}{2} |A(j\omega_1)|^2, \qquad |\omega_1 - \omega_2| \text{ small}$$

It also follows that P_y is bounded, where the bound depends on the structure of the linear system.

4.4.3 Comparison†

An unknown-but-bounded process can be a sine wave. Thus care must be exercised when working with quantities such as $G_y(\omega)$ as they can contain impulses (i.e., infinities). White stochastic processes have, on the average, an infinite total power. The infinite power of a white stochastic process means that it can never exist in practice. These infinities, however, do not mean that the basic models are meaningless, as both white unknown-but-bounded and stochastic processes have only a finite amount of power in any finite (neither zero nor infinite) band of frequencies. The stochastic model enables one to calculate the average power, while the unknown-but-bounded model enables one to calculate an upper bound on the power in the band.

The discussions were given in terms of continuous-time models, but similar results can be obtained for the discrete-time case. In discrete time, the annoying complications of infinities do not arise.

4.5 Role of White Process Model

The basic dynamic models considered in this chapter are of the form

$$\mathbf{x}(n+1) = \mathbf{\Phi}(n+1)\mathbf{x}(n) + \mathbf{G}(n)\mathbf{w}(n)$$

$$\frac{d\mathbf{x}(t)}{dt} = \mathbf{F}(t)\mathbf{x}(t) + \mathbf{G}\mathbf{w}(t)$$

$\mathbf{w}(n)$ or $\mathbf{w}(t)$: white process (stochastic or unknown but bounded)

The whiteness of the input $\mathbf{w}(n)$ or $\mathbf{w}(t)$ is what made the analysis easy and made the equations [e.g., (4.1.6)] come out in such a convenient form. This emphasizes the importance of the ideas of Chapter 3 on how to convert "natural" models into the state space-white process model form.

Consider a case where some physical system has an input which is an uncertain but nonwhite process. Since white processes have no time structure, one might be tempted to consider a white process as some sort of *worst case*. Thus in the interest of expediency, one might be tempted to simply replace the actual nonwhite input by a white input of the same size (variance or bound) in the hopes of obtaining a worst case analysis. This logic is valid for unknown-but-bounded models but does *not* work for stochastic models. For stochastic models it all depends on how the time structure of the nonwhite process "interacts" with the dynamics of the physical system itself.

EXAMPLE: Consider the zero mean Bayesian model

$$x_1(n+1) = \Phi_1 x_1(n) + w_s(n)$$

$$w_s(n+1) = \Phi_s w_s(n) + w(n)$$

$$E\{w(n_1)w(n_2)\} = \begin{cases} Q, & n_1 = n_2 \\ 0, & n_1 \neq n_2 \end{cases}$$

$$|\Phi_1| < 1 \qquad |\Phi_s| < 1$$

where $w_s(n)$ is a stationary process, so from (4.1.8)

$$E\{w_s^2(n)\} = \Gamma_s = \frac{Q}{1 - \Phi_s^2}$$

Define

$$\mathbf{x}(n) = \begin{bmatrix} x_1(n) \\ w_s(n) \end{bmatrix} \qquad \mathbf{\Phi} = \begin{bmatrix} \Phi_1 & 1 \\ 0 & \Phi_s \end{bmatrix}$$

$$\mathbf{\Gamma}(n) = E\{\mathbf{x}(n)\mathbf{x}'(n)\}$$

Assume that $\mathbf{x}(n)$ is also a stationary process so

$$\mathbf{\Gamma}(n) = \mathbf{\Gamma} = \begin{bmatrix} \Gamma_{11} & \Gamma_{12} \\ \Gamma_{12} & \Gamma_{22} \end{bmatrix}$$

where from (4.1.8)

$$\mathbf{\Gamma} = \mathbf{\Phi\Gamma\Phi}' + \begin{bmatrix} 0 & 0 \\ 0 & Q \end{bmatrix}$$

Solving for $\mathbf{\Gamma}_{11} = E\{x_1^2(n)\}$ gives

$$\mathbf{\Gamma}_{11} = \frac{(1 + \mathbf{\Phi}_1\mathbf{\Phi}_s)\mathbf{\Gamma}_s}{(1 - \mathbf{\Phi}_1\mathbf{\Phi}_s)(1 - \mathbf{\Phi}_1^2)}$$

Now consider the effects of varying $\mathbf{\Phi}_1$ while keeping $\mathbf{\Gamma}_s$ constant. Let $\mathbf{\Gamma}_{11,\text{white}}$ denote the result when $w_s(n)$ is white. Since $w_s(n)$ is white when $\mathbf{\Phi}_s = 0$,

$$\mathbf{\Gamma}_{11,\text{white}} = \frac{\mathbf{\Gamma}_s}{1 - \mathbf{\Phi}_1^2}$$

Thus for $0 < |\mathbf{\Phi}_1| < 1$, $0 < |\mathbf{\Phi}_s| < 1$,

$$\mathbf{\Gamma}_{11} > \mathbf{\Gamma}_{11,\text{white}}, \qquad \mathbf{\Phi}_1, \mathbf{\Phi}_s \text{ have same sign}$$

$$\mathbf{\Gamma}_{11} < \mathbf{\Gamma}_{11,\text{white}}, \qquad \mathbf{\Phi}_1, \mathbf{\Phi}_s \text{ have different signs}$$

4.6 Discussion

Two fundamentally different uncertainty models were analyzed; the Bayesian, stochastic, random-variable model based on a white stochastic process and the unknown-but-bounded model based on a white unknown-but-bounded process. The Bayesian model deals with covariances, means, etc., while the unknown-but-bounded model deals with sets (bounds). However, both models can be analyzed by deterministic matrix equations of very similar forms; i.e., for continuous-time models,

$$\frac{d}{dt}\mathbf{\Gamma}(t) = \mathbf{F\Gamma}(t) + \mathbf{\Gamma}(t)\mathbf{F}' + \beta_1(t)\mathbf{\Gamma}(t) + \frac{\mathbf{GQG}'}{\beta_2(t)}$$

where for the Bayesian model

$$\beta_1(t) = 0 \qquad \beta_2(t) = 1$$

$\mathbf{\Gamma}(t)$: covariance matrix

and for the unknown-but-bounded model

$$\beta_1(t) = \beta_2(t) = \beta(t) \qquad \infty \geq \beta(t) \geq 0$$

$\mathbf{\Gamma}(t)$: the matrix which defines the ellipsoid

The free $\beta(t)$ of the unknown-but-bounded model can make its analysis somewhat more difficult, as one is tempted to try to find optimum values for $\beta(t)$. Freedom is both a blessing and a curse.

In the Bayesian model, one can talk about a "most probable" value of the state, but (for Gaussian processes) the difference between the true and most probable values can be arbitrarily large. In the unknown-but-bounded model, there is no such thing as a "most probable" or "most likely" value, but it is always known for certain that the true value lies somewhere within some set. The Bayesian model enables statements on the average amount of power contained in a process in a given frequency range. The unknown-but-bounded model provides upper bounds on the power contained in a given frequency range.

The derivations for the unknown-but-bounded models were much more complicated, as it was first necessary to develop general solutions (in terms of support functions) for arbitrary sets before specializing to ellipsoids defined by positive definite matrices. The Bayesian model derivations were "simple," as it was possible to work directly with covariance matrices without first developing general formulas for the probability densities (as will be done in Chapters 11 and 13).

Historical Notes and References: Chapter 4

Section 4.1

This material is widely available in many forms. See Book List 4 on Stochastic Systems.

Section 4.2

This discussion also applies to Section 7.1. The Book Lists 1, 4, and 5 on Probability, Stochastic Systems, and Communications contain various approaches to the continuous-time white stochastic process question. Some of many papers related to the discrete-continuous time limit approach are Kalman [2]; Mehra [1], Swerling [1]. For more details on stochastic integrals and differentials, see Itô [1]; Dynkin [1]; Wong and Zakai [1], [2]; Wong [1]; and the various works of Kushner, Wonham, Strantovich, and Kailath.

Exercises

4.1. Consider the uncertain model

$$x(n + 1) = \Phi x(n) + w(n)$$

Describe in words some physical (or social or economic, etc.) process to which this model might be applicable when

a. $x(0)$ random variable
   $w(n)$ white stochastic process

b. $|x(0)| < \psi^{1/2}$
   $|w(n)| \leq Q^{1/2}$

4.2. Consider the unstable system

$$\begin{bmatrix} x_1(n+1) \\ x_2(n+1) \end{bmatrix} = \begin{bmatrix} 2 & 0 \\ 0 & \Phi_{22} \end{bmatrix} \begin{bmatrix} x_1(n) \\ x_2(n) \end{bmatrix} + \begin{bmatrix} 0 \\ 1 \end{bmatrix} w(n)$$

$$|\Phi_{22}| < 1$$

Under what condition can $\mathbf{x}(n)$ have a steady-state covariance matrix, $\boldsymbol{\Gamma}_{ss}$?

4.3. Consider a (wide sense) stationary process $x(t)$ with correlation function

$$\mathcal{P}(t_1, t_2) = 2e^{-3(t_1 - t_2)}, \qquad t_1 > t_2$$

and mean value

$$E\{x(t)\} = 4$$

$x(t)$ can be modeled in the form

$$\frac{d}{dt}x(t) = Fx(t) + w(t) + b, \qquad b\text{: constant}$$

What are the values of F, Q, ψ, and b?

4.4. Consider the Bayesian model

$$x(n+1) = \Phi x(n) + w(n) + b$$

$$E\{x(0)\} = 0 \qquad E\{w(n)\} = 0 \qquad E\{b\} = 0$$

$$E\{x^2(0)\} = \psi$$

$$E\{w(n_1)w(n_2)\} = \begin{cases} 0, & n_1 \neq n_2 \\ Q, & n_1 = n_2 \end{cases}$$

$$E\{b^2\} = B$$

$$x(0),\ w(n),\ b\text{: uncorrelated}$$

Write the equations you would use to evaluate $E\{x^2(N)\}$. You do not have to solve them.

4.5. Consider

$$x(n+1) = 0.5x(n) + u(n) + w(n) + 1$$

$$z(n) = x(n) + v(n)$$

$$x(0) = 0$$

$$w(n),\ v(n)\text{: zero mean, uncorrelated}$$

$$E\{w(n_1)w(n_2)\} = \begin{cases} 1, & n_1 = n_2 \\ 0, & n_1 \neq n_2 \end{cases}$$

$$E\{v(n_1)v(n_2)\} = \begin{cases} 1, & n_1 = n_2 \\ 0, & n_1 \neq n_2 \end{cases}$$

The input $u(n)$ is a control signal which is to be obtained from the observations by a control law of the form

$$u(n) = Kz(n)$$

$$K\text{: constant}$$

It is desired to choose the control so as to minimize $E\{x^2(N)\}$ for large N. Which of the following two values would you choose?

$$K = -0.5$$

$$K = 0$$

4.6. Consider the two scalar time-invariant Bayesian models

$$x(n\Delta + \Delta) = \Phi x(n\Delta) + \Delta G w(n\Delta)$$

$$\frac{d}{dt}x(t) = Fx(t) + Gw(t)$$

Consider the corresponding $\Gamma(n\Delta)$ and $\Gamma(t)$ and show directly that $\Gamma(n\Delta) \longrightarrow \Gamma(t)$ as $n \longrightarrow \infty$, $\Delta \longrightarrow 0$, $n\Delta \longrightarrow t$.

4.7. Consider the scalar Bayesian model

$$\frac{d}{dt}x(t) = Fx(t) + w(t)$$

$$\Gamma(t) = E\{x^2(t)\}$$

One way to calculate a differential equation which $\Gamma(t)$ satisfies is to use

$$\begin{aligned}\frac{d}{dt}\Gamma(t) &= \frac{d}{dt}E\{x^2(t)\}\\ &= 2E\left\{x(t)\frac{dx}{dt}\right\}\end{aligned}$$

Carry out the details and compare the results with (4.1.11). *Hint:* Dirac delta functions are often defined so that

$$\int_0^{t_1} \delta(t_1 - t_2)\, dt_2 = \frac{1}{2}$$

4.8. Consider the unknown-but-bounded model

$$\begin{aligned}x(n+1) &= \Phi x(n) + b + w(n)\\ x(0) &= 0\\ |w(n)| &\leq 1\\ \Phi &= 0.5\end{aligned}$$

Define

$$\Omega_x = [x: 8 \leq x \leq 12]$$

What value or values of b will make $x(N) \in \Omega_x$ for large N?

4.9. Consider the unknown-but-bounded process

$$\begin{aligned}x(n+1) &= \Phi x(n) + w(n)\\ |w(n)| &\leq Q^{1/2}\\ |x(0)| &\leq \psi^{1/2}, \qquad 0 < \Phi < 1\end{aligned}$$

a. For large n, find $\Omega(n)$, the smallest set such that

$$x(n) \in \Omega(n)$$

b. Let $\Omega_b(n)$ denote a bounding ellipsoid such that

$$x(n) \in \Omega_b(n)$$

For large n, compare $\Omega_b(n)$ with $\Omega(n)$ (choose the best value for β).

4.10. Consider the uncertain process

$$\begin{bmatrix} x_1(n+1) \\ x_2(n+1) \end{bmatrix} = \begin{bmatrix} \frac{1}{2} & 1 \\ 0 & \frac{1}{2} \end{bmatrix} \begin{bmatrix} x_1(n) \\ x_2(n) \end{bmatrix} + \begin{bmatrix} 0 \\ 1 \end{bmatrix} w(n)$$

$$x_1(0) = x_2(0) = 0$$

a. Assume that

$$w(n) \text{ is } N(0, 1)$$

$$E\{w(n_1)w(n_2)\} = 0, \qquad n_1 \neq n_2$$

Let $p_n(\mathbf{x})$ denote the probability density of $\mathbf{x}(n)$. Sketch the curve $p_n(\mathbf{x}) = c$ for some c, $n = 0, 1, 2$ and $n = \infty$.

b. Assume that

$$|w(n)| \leq 1$$

Let $\mathbf{\Omega}(n)$ denote the smallest set such that $\mathbf{x}(n) \in \mathbf{\Omega}(n)$. Sketch $\mathbf{\Omega}(n)$ for $n = 0, 1, 2$ and $n = \infty$.

4.11. Derive (4.1.11) in a direct fashion using formal manipulation of the Dirac delta function for a time-invariant system. *Hint:* If $\boldsymbol{\theta}(t)$ is the fundamental matrix of $\mathbf{F}$, then

$$\mathbf{x}(t) = \boldsymbol{\theta}(t)\mathbf{x}(0) + \boldsymbol{\theta}(t)\int_0^t \boldsymbol{\theta}^{-1}(s)\mathbf{G}\mathbf{w}(s)\,ds$$

$$\begin{aligned}\mathbf{\Gamma}(t) &= \boldsymbol{\theta}(t)\boldsymbol{\psi}\boldsymbol{\theta}'(t) \\ &\quad + \boldsymbol{\theta}(t)E\left\{\int_0^t \boldsymbol{\theta}^{-1}(s_1)\mathbf{G}\mathbf{w}(s_1)\,ds_1 \int_0^t \mathbf{w}'(s_2)\mathbf{G}'\boldsymbol{\theta}^{-1\prime}(s_2)\,ds_2\right\}\boldsymbol{\theta}'(t)\end{aligned}$$

Use of the Dirac delta function behavior of a stochastic white process gives

$$\mathbf{\Gamma}(t) = \boldsymbol{\theta}(t)\boldsymbol{\psi}\boldsymbol{\theta}'(t) + \boldsymbol{\theta}(t)\left[\int_0^t \boldsymbol{\theta}^{-1}(s)\mathbf{G}\mathbf{Q}\mathbf{G}'\boldsymbol{\theta}^{-1\prime}(s)\,ds\right]\boldsymbol{\theta}'(t)$$

Differentiation with respect to t eventually yields (4.1.11).

4.12. Derive (4.2.6). *Hint:*

$$\boldsymbol{\mathcal{P}}_c(t, \tau) = E\left\{\begin{bmatrix} x_c(t) \\ c(t) \end{bmatrix}[x_c(\tau)c(\tau)]\right\}$$

From (4.1.13)

$$\mathbf{\Gamma}_c(t) = \boldsymbol{\mathcal{P}}_c(t, t) = \mathbf{\Gamma}_c \begin{bmatrix} \dfrac{1}{F(F+A)} & -\dfrac{1}{F+A} \\ -\dfrac{1}{F+A} & 1 \end{bmatrix}, \qquad \text{all } t$$

Let $\boldsymbol{\theta}(t)$ be a fundamental matrix defined by

$$\frac{d}{dt}\boldsymbol{\theta}(t) = \begin{bmatrix} F & I \\ 0 & A \end{bmatrix}\boldsymbol{\theta}(t), \qquad \boldsymbol{\theta}(0) = \mathbf{I}$$

so that

$$\boldsymbol{\theta}(t) = \begin{bmatrix} e^{Ft} & \frac{1}{F - A}(e^{Ft} - e^{At}) \\ 0 & e^{At} \end{bmatrix}$$

Then

$$\mathcal{P}_c(t, \tau) = \boldsymbol{\theta}(t - \tau)\mathcal{P}_c(\tau, \tau)$$

4.13. Consider (4.3.4). Prove that the bounding ellipsoid $\boldsymbol{\Omega}_{2,b}$ can be made to "touch" $\boldsymbol{\Omega}_z$ at any desired "point" by appropriate choice of γ. *Hint:* Consider the support functions $s_z(\boldsymbol{\eta})$ and $s_b(\boldsymbol{\eta})$ and show that for any $\boldsymbol{\eta} = \boldsymbol{\eta}_1$ there exists a γ such that $s_z(\boldsymbol{\eta}_1) = s_b(\boldsymbol{\eta}_1)$. *Comment:* The proof can be generalized to discrete and continuous-time dynamic systems.

4.14. Consider the ellipsoids

$$\boldsymbol{\Omega}_k = \{\mathbf{x} : (\mathbf{x} - \mathbf{m}_k)'\boldsymbol{\Gamma}^{-1}(\mathbf{x} - \mathbf{m}_k) \leq 1\}, \qquad k = 1, \ldots, K$$

Define $\boldsymbol{\Omega}_{vs}$ as the vector sum of all K ellipsoids

$$\boldsymbol{\Omega}_{vs} = \left\{\mathbf{x} : \mathbf{x} = \sum_{k=1}^{K} \mathbf{x}_k,\ \mathbf{x}_k \in \boldsymbol{\Omega}_k\right\}$$

Prove that $\boldsymbol{\Omega}_b$ is a bounding ellipsoid for $\boldsymbol{\Omega}_{vs}$, where

$$\boldsymbol{\Omega}_b = \{\mathbf{x} : (\mathbf{x} - \mathbf{m}_b)'\boldsymbol{\Gamma}_b^{-1}(\mathbf{x} - \mathbf{m}_b) \leq 1\}$$

$$\boldsymbol{\Gamma}_b = \sum_{k=1}^{K} \alpha_k^{-1}\boldsymbol{\Gamma}_k$$

$$\mathbf{m}_b = \sum_{k=1}^{K} \mathbf{m}_K$$

$$\text{for any } \alpha_k > 0, \qquad \sum_{k=1}^{K} \alpha_k \leq 1$$

Hint: Use support functions and the special form of Holder's inequality

$$\sum_{k=1}^{K} \alpha_k^{-1} b_k^2 \geq \left(\sum_{k=1}^{K} \alpha_k\right)^{-1}\left(\sum_{k=1}^{K} b_k\right)^2, \qquad \alpha_k > 0$$

where

$$b_k = \sqrt{\boldsymbol{\eta}'\boldsymbol{\Gamma}_k\boldsymbol{\eta}}$$

5

Estimation: Static Linear Systems

This is the first of many chapters on estimation theory, practice, and applications. Therefore it is appropriate to try to convey some general concepts before getting down to specific details.

From the point of view of this book, there are three important quantities of interest in estimation:

1. The variables to be estimated.
2. The observations.
3. The mathematical model describing (a) how the observations are related to the variables of interest which are to be estimated and (b) the uncertainties present.

In this chapter these quantities become

1. Variables to be estimated: state vector $\mathbf{x}$
2. Observations: $\mathbf{z}$.
3. Mathematical model:
   a. $\mathbf{z} = \mathbf{H}\mathbf{x} + \mathbf{v}$
      $\mathbf{H}$: a known matrix
      $\mathbf{v}$: an uncertain observation disturbance
   b. Model for uncertainty in $\mathbf{x}$ and $\mathbf{v}$.

In later chapters much more complicated models will be considered, but it is important first to get the basic ideas clear.

To understand what estimation is all about, define

$\mathbf{x}_{\text{true}}$ true value of the state $\mathbf{x}$
$\mathbf{z}_{\text{actual}}$ actual value of observation
$\mathbf{v}_{\text{actual}}$ actual value of observation uncertainty

Then

$$\mathbf{z}_{\text{actual}} = \mathbf{H}\mathbf{x}_{\text{true}} + \mathbf{v}_{\text{actual}}$$

In this equation

$\left.\begin{matrix}\mathbf{z}_{\text{actual}} \\ \mathbf{H}\end{matrix}\right\}$ numerical values are known

$\left.\begin{matrix}\mathbf{x}_{\text{true}} \\ \mathbf{v}_{\text{actual}}\end{matrix}\right\}$ numerical values are not known.

Although numerical values for $\mathbf{x}_{\text{true}}$ and $\mathbf{v}_{\text{actual}}$ are not known, the mathematical model conveys some information on their values; i.e., there is a model for their *uncertainty*. Now define

$$\hat{\mathbf{x}}\text{: estimate of value of } \mathbf{x}_{\text{true}}$$

The estimate $\hat{\mathbf{x}}$ can depend only on what is known; i.e., $\hat{\mathbf{x}}$ can depend only on $\mathbf{z}_{\text{actual}}$ and the mathematical model (**H** and the uncertainty models for $\mathbf{x}_{\text{true}}$ and $\mathbf{v}_{\text{actual}}$). Usually it is desired to view the estimate $\hat{\mathbf{x}}$ as some specified function of the observation $\mathbf{z}_{\text{actual}}$. This function is called the estimator. The nature of this estimator (function) can be determined from the mathematical model (**H** and the uncertainty models for $\mathbf{x}_{\text{true}}$ and $\mathbf{v}_{\text{actual}}$) and hence can be specified before the observation is actually made. In this chapter, the estimator is often a linear matrix *operator* **W**, in which case

$\hat{\mathbf{x}} = \mathbf{W}\mathbf{z}_{\text{actual}}$
W: an estimator (determined by **H** and uncertainty models for $\mathbf{x}_{\text{true}}$ and $\mathbf{v}_{\text{actual}}$)

In general $\hat{\mathbf{x}}$ will not equal $\mathbf{x}_{\text{true}}$. Thus one problem is to try to specify the best estimator (the best **W**) which minimizes, in some sense, the error $\mathbf{x}_{\text{true}} - \hat{\mathbf{x}}$. Now assume that an estimator (such as **W**) has been chosen in some fashion, that $\mathbf{z}_{\text{actual}}$ has been obtained, and that an estimate $\hat{\mathbf{x}}$ has been calculated. Another problem is to determine how close $\hat{\mathbf{x}}$ is to $\mathbf{x}_{\text{true}}$. Since the numerical value of the error $\mathbf{x}_{\text{true}} - \hat{\mathbf{x}}$ is not known, the problem is to develop an uncertainty model for $\mathbf{x}_{\text{true}} - \hat{\mathbf{x}}$. The uncertainty in $\mathbf{x}_{\text{true}} - \hat{\mathbf{x}}$ depends on the mathematical model (**H** and the uncertainty in $\mathbf{x}_{\text{true}}$ and $\mathbf{v}_{\text{actual}}$), the chosen estimator (**W**), and, in some cases, $\mathbf{z}_{\text{actual}}$. It is usually desirable and often possible to develop an uncertainty model for $\mathbf{x}_{\text{true}} - \hat{\mathbf{x}}$ which depends only on the mathematical model and not $\mathbf{z}_{\text{actual}}$. This is desired because then an *error analysis* can be done before the actual observation is made to determine how close $\hat{\mathbf{x}}$ will be to $\mathbf{x}_{\text{true}}$. Hence in general terms, the *basic estimation problem*: is

1. Find the estimator (function)

$$\mathbf{z}_{\text{actual}} \longrightarrow \boxed{\text{estimator}} \longrightarrow \text{estimate } \hat{\mathbf{x}}$$

   such that the estimate $\hat{\mathbf{x}}$ is as close to $\mathbf{x}_{\text{true}}$ as possible; i.e., the uncertainty in $\mathbf{x}_{\text{true}} - \hat{\mathbf{x}}$ is minimized in some sense. The estimator (function) can be specified from the mathematical model ($\mathbf{H}$, uncertain models for $\mathbf{x}_{\text{true}}$ and $\mathbf{v}_{\text{actual}}$).

2. Determine a model for the uncertainty in $\mathbf{x}_{\text{true}} - \hat{\mathbf{x}}$, where this model depends on the chosen estimator, the mathematical model, and possibly $\mathbf{z}_{\text{actual}}$.

There are really two models for the uncertainty in $\mathbf{x}_{\text{true}}$:

1. A priori model: the model for $\mathbf{x}_{\text{true}}$ which models the uncertainty before the observation is made.
2. A posteriori model: the model for $\mathbf{x}_{\text{true}} - \hat{\mathbf{x}}$ which models the uncertainty in $\mathbf{x}_{\text{true}}$ after the observation has been made and processed to yield the estimate $\hat{\mathbf{x}}$.

The choice of estimator (such as $\mathbf{W}$) depends on the a priori model. The a posteriori model depends on which estimator (which $\mathbf{W}$) is chosen. A more general view of estimation theory (than that being used here) views the problem as follows: *Given the a priori model for the uncertainty in* $\mathbf{x}_{\text{true}}$, $\mathbf{v}_{\text{actual}}$, *and the observation* $\mathbf{z}_{\text{actual}}$, *determine the particular a posteriori model for the uncertainty in* $\mathbf{x}_{\text{true}}$ *which involves the "least" uncertainty.* With this approach, a separate estimate $\hat{\mathbf{x}}$ and a separate uncertainty model for $\mathbf{x}_{\text{true}} - \hat{\mathbf{x}}$ do not exist.

EXAMPLE: Consider $\mathbf{z} = \mathbf{Hx}$:

$$\begin{bmatrix} z_1 \\ z_2 \end{bmatrix} = \begin{bmatrix} 1 & 1 \\ 0 & 1 \end{bmatrix} \begin{bmatrix} x_1 \\ x_2 \end{bmatrix}$$

i.e., there is no observation uncertainty $\mathbf{v}$ (or $\mathbf{v}_{\text{actual}}$ is known to be zero). Then

$$\mathbf{W} = \mathbf{H}^{-1} = \begin{bmatrix} 1 & -1 \\ 0 & 1 \end{bmatrix}$$

is obviously the best estimator as it gives $\hat{\mathbf{x}} = \mathbf{W}\mathbf{z}_{\text{actual}} = \mathbf{x}_{\text{true}}$. In this case the estimator $\mathbf{W}$ depends only on $\mathbf{H}$ and is independent of the (a priori) uncertainty model for $\mathbf{x}$. Since $\hat{\mathbf{x}} = \mathbf{x}_{\text{true}}$, there is no (a posteriori) uncertainty model for the value of $\mathbf{x}$ if $\mathbf{W} = \mathbf{H}^{-1}$.

EXAMPLE: Consider

$$z_1 = x + v_1$$
$$z_2 = x + v_2$$

Assume that

1. The value of x_{true} is completely unknown; i.e., there is no "a priori knowledge" of the value of x_{true}.
2. $v_{1,\text{true}}$ and $v_{2,\text{true}}$ can be considered to be samples of random variables v_1 and v_2, where

$$E\{v_1\} = E\{v_2\} = E\{v_1 v_2\} = 0$$
$$E\{v_1^2\} = 1$$
$$E\{v_2^2\} = 4$$

If the variances of v_1 and v_2 were not considered, a "reasonable" estimator would be

$$\mathbf{W}'_{\text{reasonable}} = \begin{bmatrix} \frac{1}{2} \\ \frac{1}{2} \end{bmatrix}$$

so

$$\hat{x} = \frac{z_1 + z_2}{2}$$

In this case

$$x_{\text{true}} - \hat{x} = -\frac{v_1 + v_2}{2}$$

so the (a posteriori) uncertainty in $x_{\text{true}} - \hat{x}$ can be modeled as a random variable with

$$E\{x_{\text{true}} - \hat{x}\} = 0$$
$$E\{[x_{\text{true}} - \hat{x}]^2\} = \tfrac{1}{4}[1 + 4] = \tfrac{5}{4}$$

The above estimate is reasonable, but in Section 5.3 it will be shown that the "best" estimator is

$$\mathbf{W}'_{\text{best}} = \begin{bmatrix} \frac{4}{5} \\ \frac{1}{5} \end{bmatrix}$$

so

$$\hat{x} = \tfrac{4}{5} z_1 + \tfrac{1}{5} z_2$$

In this case

$$E\{x_{\text{true}} - \hat{x}\} = 0$$
$$E\{[x_{\text{true}} - {}^2\hat{x}]\} = \tfrac{4}{5}$$

EXAMPLE: Consider

$$z = x + v$$

Assume that x and v are both unknown-but-bounded in that it is known only that

$$|x_{\text{true}}| \leq 1$$

$$|v_{\text{actual}}| \leq 2$$

After z_{actual} has been observed, it then follows from elementary geometry that x_{true} lies within some set (range)

$$x_{\min} \leq x_{\text{true}} \leq x_{\max}$$

and that a reasonable estimator is defined by

$$\hat{x} = \frac{x_{\max} + x_{\min}}{2}$$

where $x_{\max}$, $x_{\min}$, and $\hat{x}$ are nonlinear functions of z_{actual} as follows:

	$\lvert z_{\text{actual}} \rvert < 1$	$z_{\text{actual}} > 1$	$z_{\text{actual}} < -1$
$x_{\max}$	$+1$	$+1$	$z + 2$
$x_{\min}$	-1	$z - 2$	-1
$\hat{x}$	0	$\frac{z-1}{2}$	$\frac{z+1}{2}$

In this case the estimator cannot be expressed as $\hat{x} = Wz$. The (a posteriori) uncertainty model for x_{true} is

$$-\left(\frac{x_{\max} - x_{\min}}{2}\right) \leq x_{\text{true}} - \hat{x} \leq \frac{x_{\max} - x_{\min}}{2}$$

In most of the further discussion in this and later chapters, the $\mathbf{z}_{\text{actual}}$, $\mathbf{x}_{\text{true}}$, etc., notation is not used, and only $\mathbf{x}$, $\mathbf{z}$, etc., are written.

The above constitutes a general introduction to estimation theory concepts. The specific nature of this chapter is now discussed. Consider

$$\begin{aligned} \mathbf{z} &= \mathbf{H}\mathbf{x} + \mathbf{v} \\ \mathbf{x} &= K_1 \text{ vector} \\ \mathbf{z} &= K_2 \text{ vector} \\ \mathbf{v} &= K_2 \text{ vector} \\ \mathbf{H} &= K_2 \text{ by } K_1 \text{ matrix} \end{aligned} \tag{5.1}$$

where $\mathbf{H}$ is known, the value of the state $\mathbf{x}$ is uncertain, $\mathbf{v}$ is the observation uncertainty, and $\mathbf{z}$ is observed. It is desired to use the observation $\mathbf{z}$ to obtain an estimate of $\mathbf{x}$. This is called the state estimation problem for a static linear system. There are many different ways to model the uncertainty in $\mathbf{x}$ and $\mathbf{v}$. Thus many different estimators and estimates are possible. The following situations are discussed in this chapter:

1. Bayesian model: $\mathbf{x}$ and $\mathbf{v}$ are random vectors.

2. Fisher model: $\mathbf{x}$ is completely unknown; $\mathbf{v}$ is a random vector.
3. Weighted least squares: no models for $\mathbf{x}$ and $\mathbf{v}$.
4. Unknown-but-bounded model: $\mathbf{x}$ and $\mathbf{v}$ are constrained to lie in specified sets.

Sections 5.1–5.3 consider estimators for Bayesian and Fisher models. Section 5.4 considers weighted-least-squares estimators. Weighting matrices are discussed in Section 5.5. Unknown-but-bounded models are considered in Section 5.6. A dichotomy or two-step estimation procedure is discussed in Section 5.7. Section 5.8 considers the case of perfect observations. Non-optimum estimators and modeling errors are discussed in Sections 5.9 and 5.10. Some computational aspects are considered in Section 5.11. A final discussion is given in Section 5.12.

5.1 Bayesian

5.1.1 General

Consider the Bayesian model

$$\begin{aligned} \mathbf{z} &= \mathbf{H}\mathbf{x} + \mathbf{v} \\ E\{\mathbf{x}\} &= \mathbf{m}_x \\ E\{(\mathbf{x} - \mathbf{m}_x)(\mathbf{x} - \mathbf{m}_x)'\} &= \boldsymbol{\psi} \\ E\{\mathbf{v}\} &= \mathbf{0} \\ E\{\mathbf{v}\mathbf{v}'\} &= \mathbf{R} \\ E\{\mathbf{x}\mathbf{v}'\} &= \mathbf{0} \end{aligned}$$

Then

The estimator

$$\hat{\mathbf{x}} = \mathbf{W}\mathbf{z} + \mathbf{W}_0$$

yields the minimum error covariance matrix for all $\mathbf{W}$ and $\mathbf{W}_0$ when

$$\begin{aligned} \mathbf{W} &= \boldsymbol{\Sigma}\mathbf{H}'\mathbf{R}^{-1} \\ \mathbf{W}_0 &= \boldsymbol{\Sigma}\boldsymbol{\psi}^{-1}\mathbf{m}_x \\ \boldsymbol{\Sigma} &= (\mathbf{H}'\mathbf{R}^{-1}\mathbf{H} + \boldsymbol{\psi}^{-1})^{-1} \end{aligned} \tag{5.1.1}$$

The minimum error covariance matrix is

$$\begin{aligned} \boldsymbol{\Sigma} &= E\{(\mathbf{x} - \hat{\mathbf{x}})(\mathbf{x} - \hat{\mathbf{x}})'\} \\ E\{\hat{\mathbf{x}}\} &= \mathbf{m}_x \end{aligned}$$

By matrix manipulation using (A.3) of Appendix A, (5.1.1) can be rewritten as

$$\boxed{\begin{aligned} \hat{\mathbf{x}} &= \mathbf{m}_x + \boldsymbol{\psi}\mathbf{H}'[\mathbf{H}\boldsymbol{\psi}\mathbf{H}' + \mathbf{R}]^{-1}[\mathbf{z} - \mathbf{H}\mathbf{m}_x] \\ \boldsymbol{\Sigma} &= \mathrm{E}\{(\mathbf{x} - \hat{\mathbf{x}})(\mathbf{x} - \hat{\mathbf{x}})'\} \\ \boldsymbol{\Sigma} &= \boldsymbol{\psi} - \boldsymbol{\psi}\mathbf{H}'(\mathbf{H}\boldsymbol{\psi}\mathbf{H}' + \mathbf{R})^{-1}\mathbf{H}\boldsymbol{\psi} \end{aligned}} \tag{5.1.2}$$

The result (5.1.1) has important theoretical and practical implications. The condition $E\{\hat{\mathbf{x}}\} = \mathbf{m}_x$ means that $E\{\hat{\mathbf{x}}\} = E\{\mathbf{x}\}$, so $\hat{\mathbf{x}}$ is unbiased. Define

$$\tilde{\boldsymbol{\Sigma}} = E\{[\mathbf{x} - \tilde{\mathbf{x}}][\mathbf{x} - \tilde{\mathbf{x}}]'\}$$
$$\tilde{\mathbf{x}} = \tilde{\mathbf{W}}\mathbf{z} + \tilde{\mathbf{W}}_0$$
$$\tilde{\mathbf{W}}, \tilde{\mathbf{W}}_0: \text{not as in (5.1.1)}$$

Then (5.1.1) means that

$$\boldsymbol{\Sigma} \leq \tilde{\boldsymbol{\Sigma}}$$

in that $\tilde{\boldsymbol{\Sigma}} - \boldsymbol{\Sigma}$ is positive semidefinite (see Appendix A.2). This means that $\boldsymbol{\Sigma}$ is a global minimum in almost any reasonable sense of the world. To illustrate why this global minimization is important, suppose that

$$\mathbf{x} = \begin{bmatrix} \mathbf{x}_1 \\ \mathbf{x}_2 \end{bmatrix}$$

where $\mathbf{x}_2$ is of no interest and it is desired only to estimate $\mathbf{x}_1$. The global minimization means that the best estimate of $\mathbf{x}_1$ is still $\hat{\mathbf{x}}_1$, where

$$\hat{\mathbf{x}} = \begin{bmatrix} \hat{\mathbf{x}}_1 \\ \hat{\mathbf{x}}_2 \end{bmatrix}$$

and $\hat{\mathbf{x}}$ is given by (5.1.1). Thus one cannot make further improvements on $\hat{\mathbf{x}}_1$ at the expense of $\hat{\mathbf{x}}_2$. Similarly, the best estimate of

$$\mathbf{y} = \mathbf{C}\mathbf{x}$$

is given by

$$\hat{\mathbf{y}} = \mathbf{C}\hat{\mathbf{x}}$$
$$\boldsymbol{\Sigma}_y = E\{(\mathbf{y} - \hat{\mathbf{y}})(\mathbf{y} - \hat{\mathbf{y}})'\} = \mathbf{C}\boldsymbol{\Sigma}\mathbf{C}'$$

The proof of (5.1.1) is now discussed. Assume for simplicity that $\mathbf{m}_x = \mathbf{0}$. Consider the class of random vectors $\tilde{\mathbf{x}}$ defined by

$$\tilde{\mathbf{x}} = \mathbf{A}\mathbf{z}$$

Define

$$\tilde{\boldsymbol{\Sigma}} = E\{(\mathbf{x} - \tilde{\mathbf{x}})(\mathbf{x} - \tilde{\mathbf{x}})'\}$$

which can be rewritten

$$\tilde{\boldsymbol{\Sigma}} = E\{[\mathbf{x} - \mathbf{W}\mathbf{z} - (\mathbf{A} - \mathbf{W})z][\mathbf{x} - \mathbf{W}\mathbf{z} - (\mathbf{A} - \mathbf{W})z]'\}$$

where $\hat{\mathbf{x}} = \mathbf{Wz}$ as in (5.1.1). Now after some manipulation, it can be shown that

$$E\{[\mathbf{x} - \mathbf{Wz}]\mathbf{z}'\} = \mathbf{0} \tag{5.1.3}$$

Then after more manipulation

$$\begin{aligned} \tilde{\mathbf{\Sigma}} &= \mathbf{\Sigma} + \mathbf{B} \\ \mathbf{\Sigma}&: \text{given by (5.1.1)} \\ \mathbf{B} &= [\mathbf{A} - \mathbf{W}]E\{\mathbf{zz}'\}[\mathbf{A} - \mathbf{W}]' \end{aligned} \tag{5.1.4}$$

Now $\mathbf{B}$ is a positive semidefinite matrix. Thus for any $\mathbf{A}$,

$$\tilde{\mathbf{\Sigma}} \geq \mathbf{\Sigma} \tag{5.1.5}$$

The minimum $\tilde{\mathbf{\Sigma}}$ occurs when $\mathbf{B} = \mathbf{0}$, and this happens when

$$\mathbf{A} = \mathbf{W}$$

The general case of nonzero mean follows in a similar fashion.

One important property of $\hat{\mathbf{x}}$ is (5.1.3), which says that the random vector $\hat{\mathbf{x}} - \mathbf{x}$ is *orthogonal* to $\mathbf{z}$ or any linear function of $\mathbf{z}$, so

$$\boxed{\begin{aligned} E\{[\mathbf{x} - \hat{\mathbf{x}}]\mathbf{z}'\} = \mathbf{0} \\ E\{[\mathbf{x} - \hat{\mathbf{x}}]\hat{\mathbf{x}}'\} = \mathbf{0} \end{aligned}} \tag{5.1.6}$$

It also follows that

$$\begin{aligned} E\{\mathbf{x}\hat{\mathbf{x}}'\} &= E\{\hat{\mathbf{x}}\hat{\mathbf{x}}'\} \\ \mathbf{\Sigma} &= E\{\mathbf{xx}'\} - E\{\hat{\mathbf{x}}\hat{\mathbf{x}}'\} \\ E\{\mathbf{xx}'\} &\geq E\{\hat{\mathbf{x}}\hat{\mathbf{x}}'\} \end{aligned} \tag{5.1.7}$$

EXAMPLE: Consider

$$\begin{pmatrix} z_1 \\ z_2 \end{pmatrix} = \begin{pmatrix} 1 \\ 1 \end{pmatrix} x + \begin{pmatrix} v_1 \\ v_2 \end{pmatrix}$$

$$\mathbf{R} = \begin{pmatrix} 1 & 0 \\ 0 & 4 \end{pmatrix} \qquad \psi = 8 \qquad m_x = 0,$$

Then from (5.1.1)

$$\begin{aligned} \Sigma &= (1 + \tfrac{1}{4} + \tfrac{1}{8})^{-1} = \tfrac{8}{11} \\ \hat{x} &= \tfrac{8}{11}z_1 + \tfrac{2}{11}z_2 \end{aligned}$$

The error uncertainty model (the error covariance matrix $\mathbf{\Sigma}$) is independent of $\mathbf{z}$ and can be evaluated before the observations are made. This is extremely important in practice as it enables one to do an *error analysis* to decide how accurate the estimate will be when (if) it is actually calculated. Such error analysis can be very useful in design.

EXAMPLE: Consider

$$\begin{pmatrix} z_1 \\ z_2 \end{pmatrix} = \begin{pmatrix} 1 \\ 1 \end{pmatrix} x + \begin{pmatrix} v_1 \\ v_2 \end{pmatrix}$$

$$\mathbf{R} = \begin{bmatrix} 1 & 0 \\ 0 & R_{22} \end{bmatrix} \qquad \boldsymbol{\psi} = 8 \qquad m_x = 0$$

Assume that the "meter" which yields z_2 has not yet been designed so that the value of R_{22} is still free. Assume that "system specifications" require that the final estimate have an error with variance less than or equal to $\frac{1}{2}$. Since accurate meters cost money, it is reasonable to try to find the maximum value of R_{22} that is acceptable. For the model,

$$\Sigma = [1 + R_{22}^{-1} + \tfrac{1}{8}]^{-1}$$

so the maximum allowable R_{22} can be obtained by setting $\Sigma = \frac{1}{2}$ to give

$$R_{22} = \tfrac{8}{7}$$

This design can be done before any actual observations are made. Obviously a meter with $R_{22} = \frac{8}{7}$ will enable the specifications to be met only if the "best" estimator is used.

5.1.2 Gaussian

The discussions and results of Section 5.1.1 are valid for non-Gaussian as well as Gaussian models. However if the model is specialized to the Gaussian case so $\mathbf{x}$ and $\mathbf{v}$ are jointly normal, then $\hat{\mathbf{x}}$ given by (5.1.1) or (5.1.2) is also the *conditional expectation* estimate,

$$\hat{\mathbf{x}} = E\{\mathbf{x} \mid \mathbf{z}\} \tag{5.1.8}$$

This can be proven by rewriting (E.8) of Appendix E as

$$E\{\mathbf{x} \mid \mathbf{z}\} = \mathbf{m}_x + \mathbf{\Gamma}_{xz}\mathbf{\Gamma}_{zz}^{-1}[\mathbf{z} - \mathbf{m}_z]$$
$$\mathbf{m}_z = E\{\mathbf{z}\} = \mathbf{H}\mathbf{m}_x$$
$$\mathbf{\Gamma}_{zz} = E\{[\mathbf{z} - \mathbf{m}_z][\mathbf{z} - \mathbf{m}_z]'\}$$
$$\mathbf{\Gamma}_{xz} = E\{[\mathbf{x} - \mathbf{m}_x][\mathbf{z} - \mathbf{m}_z]'\}$$

and then noting that

$$\mathbf{\Gamma}_{xz} = \boldsymbol{\psi}\mathbf{H}'$$
$$\mathbf{\Gamma}_{zz} = \mathbf{H}\boldsymbol{\psi}\mathbf{H}' + \mathbf{R}$$

so that (5.1.2) yields (5.1.8).

This conditional expectation interpretation in the Gaussian case is not needed until Part III, Chapter 9, and is discussed in more detail in Part IV, Chapter 11. The Gaussian case is discussed here mostly for later reference and completeness.

The fact that the estimator (5.1.1) or (5.1.2) is *optimum* without a

Gaussian assumption is very important in practice. The Gaussian assumption primarily effects only the interpretation that can be given to the results. This situation is analogous to the discussions in Chapter 4 on $\mathbf{\Gamma}(t)$ or $\mathbf{\Gamma}(n)$, the covariance matrix of a stochastic process, $\mathbf{x}(t)$ or $\mathbf{x}(n)$. In the Gaussian case, knowledge of $\mathbf{\Sigma}$ enables one to specify the probability distribution of $\mathbf{x} - \hat{\mathbf{x}}$, i.e.

$$\mathbf{x} - \hat{\mathbf{x}} \text{ is } N(\mathbf{0}, \mathbf{\Sigma})$$

5.2 Fisher

5.2.1 General

Consider

$$\mathbf{z} = \mathbf{H}\mathbf{x} + \mathbf{v}$$

$$\mathbf{x}\text{: completely unknown}$$

$$E\{\mathbf{v}\} = 0$$

$$E\{\mathbf{v}\mathbf{v}'\} = \mathbf{R}$$

Then

> The estimate
>
> $$\hat{\mathbf{x}} = \mathbf{W}\mathbf{z}$$
>
> yields the minimum error covariance matrix for all $\mathbf{W}$ under the constraint
>
> $$E\{\hat{\mathbf{x}}\} = \mathbf{x}, \qquad \text{all } \mathbf{x} \tag{5.2.1}$$
>
> when
>
> $$\mathbf{W} = \mathbf{\Sigma}\mathbf{H}'\mathbf{R}^{-1}$$
>
> $$\mathbf{\Sigma} = E\{[\mathbf{x} - \hat{\mathbf{x}}][\mathbf{x} - \hat{\mathbf{x}}]\} = [\mathbf{H}'\mathbf{R}^{-1}\mathbf{H}]^{-1}$$
>
> if $\mathbf{\Sigma}$ exists.

For $\mathbf{\Sigma}$ to exist, it is necessary that the dimension of $\mathbf{z}$ not be smaller than the dimension of $\mathbf{x}$; i.e., $K_2 \geq K_1$.

A proof of (5.2.1) is as follows. Consider an estimator

$$\tilde{\mathbf{x}} = \mathbf{A}\mathbf{z} + \mathbf{A}_0$$

where $\mathbf{A}$ and $\mathbf{A}_0$ are free. The unbiased constraint

$$E\{\tilde{\mathbf{x}}\} = \mathbf{x}$$

requires that

$$\mathbf{A}_0 + \mathbf{A}\mathbf{H}\mathbf{x} = \mathbf{x}, \qquad \text{all } \mathbf{x}$$

Since $\mathbf{A}_0$ and $\mathbf{A}$ cannot depend on $\mathbf{x}$, it is necessary that

$$\begin{aligned} \mathbf{A}_0 &= \mathbf{0} \\ \mathbf{A}\mathbf{H}\mathbf{x} &= \mathbf{x}, \qquad \text{all } \mathbf{x} \end{aligned} \tag{5.2.2}$$

Define

$$\tilde{\mathbf{\Sigma}} = E\{[\mathbf{x} - \mathbf{A}\mathbf{z}][\mathbf{x} - \mathbf{A}\mathbf{z}]'\}$$

for any $\mathbf{A}$. Using (5.2.2),

$$\tilde{\mathbf{\Sigma}} = E\{\mathbf{A}\mathbf{v}\mathbf{v}'\mathbf{A}'\} = \mathbf{A}\mathbf{R}\mathbf{A}'$$

Let

$$\begin{aligned} &\mathbf{A} = \mathbf{\Sigma}\mathbf{H}'\mathbf{R}^{-1} + \mathbf{C} \\ &\mathbf{C}\text{: unspecified} \\ &\mathbf{\Sigma} = [\mathbf{H}'\mathbf{R}^{-1}\mathbf{H}]^{-1} \end{aligned} \tag{5.2.3}$$

Then

$$\tilde{\mathbf{\Sigma}} = \mathbf{\Sigma} + \mathbf{\Sigma}\mathbf{H}'\mathbf{C}' + \mathbf{C}\mathbf{H}\mathbf{A} + \mathbf{C}\mathbf{R}\mathbf{C}'$$

Substitution of (5.2.3) into (5.2.2) gives

$$[\mathbf{\Sigma}\mathbf{H}'\mathbf{R}^{-1} + \mathbf{C}]\mathbf{H}\mathbf{x} = \mathbf{x}, \qquad \text{all } \mathbf{x}$$

so that the unbiased constraint requires $\mathbf{C}\mathbf{H} = \mathbf{0}$. Thus

$$\tilde{\mathbf{\Sigma}} = \mathbf{\Sigma} + \mathbf{C}\mathbf{R}\mathbf{C}'$$

Since $\mathbf{C}\mathbf{R}\mathbf{C}'$ is positive semidefinite, $\tilde{\mathbf{\Sigma}}$ is minimized by choosing

$$\mathbf{C} = \mathbf{0}, \qquad \mathbf{A} = \mathbf{\Sigma}\mathbf{H}'\mathbf{R}^{-1} = \mathbf{W}$$

which proves (5.2.1).

EXAMPLE: Consider

$$\begin{bmatrix} z_1 \\ z_2 \end{bmatrix} = \begin{bmatrix} 1 \\ 2 \end{bmatrix} x + \begin{bmatrix} v_1 \\ v_2 \end{bmatrix}$$

$$\mathbf{R} = \begin{bmatrix} 1 & 0 \\ 0 & 4 \end{bmatrix}$$

Then from (5.2.1)

$$\begin{aligned} &\Sigma = (1 + \tfrac{1}{4})^{-1} = \tfrac{4}{5} \\ &\hat{x} = \tfrac{4}{5}z_1 + \tfrac{1}{5}z_2 \end{aligned}$$

For the Fisher model, as in the Bayesian model, the fact that the error covariance matrix $\mathbf{\Sigma}$ is independent of $\mathbf{z}$ and can be precomputed is of major practical importance.

It is necessary to assume that $K_2 \geq K_1$, i.e., that the dimension of $\mathbf{z}$ is not smaller than the dimension of $\mathbf{x}$. Mathematically, this condition is necessary (not sufficient) if $\mathbf{\Sigma} = [\mathbf{H}'\mathbf{R}^{-1}\mathbf{H}]^{-1}$ is to exist. The condition is logical from a physical point of view, as $\mathbf{x}$ is assumed to be completely unknown. Thus the observation $\mathbf{z}$ can be viewed as defining K_2 equations in K_1 unknowns. If $K_1 > K_2$, a unique solution cannot be expected; i.e., there are many values of $\mathbf{x}$ for which $\mathbf{Hx} = \mathbf{z}$. For $K_1 = K_2$ (and assuming that $\mathbf{H}$ is invertible) $\hat{\mathbf{x}} = \mathbf{H}^{-1}\mathbf{z}$, so that $\hat{\mathbf{x}}$ is the unique vector for which $\mathbf{H}\hat{\mathbf{x}} = \mathbf{z}$. In this case, the modeling of the uncertainty in $\mathbf{v}$ has no effect on the estimate but of course still affects the error covariance matrix $\mathbf{\Sigma}$.

In the case where $K_1 > K_2$ and $\mathbf{\Sigma}$ does not exist, some authors like to invoke the so-called pseudoinverse. Loosely speaking, this approach says to pick the particular value $\mathbf{x}$ which satisfies $\mathbf{Hx} = \mathbf{z}$ and which is of minimum length; i.e., $\mathbf{x}'\mathbf{x}$ is minimum. Although the pseudoinverse has nice mathematical properties, the author feels that it cannot be glibly used without careful justification in each case. Any decision to choose a *minimum length* solution should be made consciously and not solely because it is convenient. Therefore pseudoinverses are not discussed further in this book.

As in the Bayesian case, $\hat{\mathbf{x}}$ is global optimum in the sense that $\tilde{\mathbf{\Sigma}} - \mathbf{\Sigma}$ is positive semidefinite. Thus, to estimate $\mathbf{y}$,

$$\mathbf{y} = \mathbf{Cx}$$

one can use

$$\hat{\mathbf{y}} = \mathbf{C}\hat{\mathbf{x}}$$

$$E\{\hat{\mathbf{y}}\} = \mathbf{y}$$

$$E\{[\hat{\mathbf{y}} - \mathbf{y}][\hat{\mathbf{y}} - \mathbf{y}]'\} = \mathbf{\Sigma}_y = \mathbf{C\Sigma C}'$$

5.2.2 Gaussian

The discussions and results of Section 5.2.1 are valid for non-Gaussian as well as Gaussian models. However if the model is specialized to the Gaussian case where $\mathbf{v}$ is normal while $\mathbf{x}$ is still unknown, the estimate of (5.2.1) is also the *maximum likelihood* estimate. (See Exercise 5.9.) The maximum likelihood interpretation is not needed until Chapter 9 and is discussed in more detail in Chapter 11. It is mentioned here primarily for the sake of completeness.

5.3 Relationship Between Fisher and Bayesian

The estimators resulting from Bayesian and Fisher models are related in various ways, but it is important to emphasize that *Bayesian and Fisher models of uncertainty are fundamentally different concepts which, in general, yield different results.* This basic point is easy to forget, as, in the rest of this section, the close relationships between Bayesian and Fisher model estimators are discussed.

The estimators for the Bayesian model of Section 5.1 and the Fisher model of Section 5.2 are as follows. From (5.1.1)

$$\begin{aligned}\hat{\mathbf{x}}_{\text{Bayesian}} &= [\boldsymbol{\psi}^{-1} + \mathbf{H}'\mathbf{R}^{-1}\mathbf{H}]^{-1}[\mathbf{H}'\mathbf{R}^{-1}\mathbf{z} + \boldsymbol{\psi}^{-1}\mathbf{m}_x] \\ \boldsymbol{\Sigma}_{\text{Bayesian}} &= [\boldsymbol{\psi}^{-1} + \mathbf{H}'\mathbf{R}^{-1}\mathbf{H}]^{-1}\end{aligned} \tag{5.3.1}$$

while from (5.2.1)

$$\begin{aligned}\hat{\mathbf{x}}_{\text{Fisher}} &= [\mathbf{H}'\mathbf{R}^{-1}\mathbf{H}]^{-1}\mathbf{H}'\mathbf{R}^{-1}\mathbf{z} \\ \boldsymbol{\Sigma}_{\text{Fisher}} &= [\mathbf{H}'\mathbf{R}^{-1}\mathbf{H}]^{-1}\end{aligned} \tag{5.3.2}$$

In both the Bayesian and Fisher models, the $\hat{\mathbf{x}}$ are unbiased but in different ways. $\hat{\mathbf{x}}_{\text{Bayesian}}$ is unbiased in the sense that

$$E\{\hat{\mathbf{x}}_{\text{Bayesian}}\} = \mathbf{m}_x = E\{\mathbf{x}\}$$

where E on the left-hand side denotes expectation over $\mathbf{x}$ and $\mathbf{v}$, while E on the right-hand side denotes expectation over just $\mathbf{x}$. This unbiased property arose as a natural consequence of the definition of the "best" estimate. On the other hand, $\hat{\mathbf{x}}_{\text{Fisher}}$ is unbiased in the sense that

$$E\{\hat{\mathbf{x}}_{\text{Fisher}}\} = \mathbf{x}$$

for any value of $\mathbf{x}$ where the expectation is over just $\mathbf{v}$. This unbiased property was directly imposed as an important part of the definition of the best estimate.

Now consider how to go back and forth between Bayesian and Fisher results. In the Bayesian model, the random vector $\mathbf{x}$ is known (before any observations) to have a covariance matrix $\boldsymbol{\psi}$. As $\boldsymbol{\psi} \longrightarrow \infty\,\mathbf{I}$ (in some sense or other), the a priori distribution of $\mathbf{x}$ spreads out, and the a priori knowledge of the *value* of $\mathbf{x}$ becomes *more and more uncertain*. In the limit,

$$\boxed{\hat{\mathbf{x}}_{\text{Bayesian}} = \hat{\mathbf{x}}_{\text{Fisher}} \qquad \text{when } \boldsymbol{\psi}^{-1} = \mathbf{0}} \tag{5.3.3}$$

Thus the Fisher model estimator equations can be derived as a limiting case of the Bayesian estimator. This fact is used in later chapters.

The reverse route from Fisher to Bayesian is done by introducing a new measurement. Assume that in a Fisher formulation two observations are made; the original one

$$\mathbf{z} = \mathbf{H}\mathbf{x} + \mathbf{v}$$

and a new one given by

$$\begin{aligned}\mathbf{z}_1 &= \mathbf{x} + \mathbf{v}_1 \\ E(\mathbf{v}_1\mathbf{v}_1') &= \boldsymbol{\psi} \\ E(\mathbf{v}\mathbf{v}_1') &= \mathbf{0} \\ E(\mathbf{v}_1) &= \mathbf{0}\end{aligned}$$

These two observations can be combined into one by defining an augmented vector as follows:

$$\tilde{\mathbf{z}} = \tilde{\mathbf{H}}\mathbf{x} + \tilde{\mathbf{v}}$$

$$\tilde{\mathbf{z}} = \begin{bmatrix} \mathbf{z} \\ -- \\ \mathbf{z}_1 \end{bmatrix} \qquad \tilde{\mathbf{v}} = \begin{bmatrix} \mathbf{v} \\ -- \\ \mathbf{v}_1 \end{bmatrix}$$

$$\tilde{\mathbf{H}} = \begin{bmatrix} \mathbf{H} \\ -- \\ \mathbf{I} \end{bmatrix}$$

$$\tilde{\mathbf{R}} = E\{\tilde{\mathbf{v}}\tilde{\mathbf{v}}'\} = \left[\begin{array}{c|c} \mathbf{R} & \mathbf{0} \\ \hline \mathbf{0} & \boldsymbol{\psi} \end{array}\right]$$

From the Fisher logic of Section 5.2,

$$\hat{\mathbf{x}}_{\text{Fisher}} = [\tilde{\mathbf{H}}'\tilde{\mathbf{R}}^{-1}\tilde{\mathbf{H}}]^{-1}\tilde{\mathbf{H}}'\tilde{\mathbf{R}}^{-1}\tilde{\mathbf{z}}$$

or

$$\hat{\mathbf{x}}_{\text{Fisher}} = [\boldsymbol{\psi}^{-1} + \mathbf{H}'\mathbf{R}^{-1}\mathbf{H}]^{-1}\{\mathbf{H}'\mathbf{R}^{-1}\mathbf{z} + \boldsymbol{\psi}^{-1}\mathbf{z}_1\} \tag{5.3.4}$$

Comparison of (5.3.4) with (5.3.1) shows that the Fisher model estimator with the extra observation $\mathbf{z}_1$ equals the Bayesian model estimator if $\mathbf{z}_1 = \mathbf{m}_x$, the mean value of $\mathbf{x}$ (when $\mathbf{x}$ is a random vector). However, this equality of estimator equations does not mean that the Fisher and Bayesian models have somehow become equal. In the Bayesian model, $\boldsymbol{\psi}$ is associated with the random vector $\mathbf{x}$. In the Fisher model, $\boldsymbol{\psi}$ is associated with measurement noise $\mathbf{v}_1$ of the extra observation $\mathbf{z}_1$ and $\mathbf{x}$ is still considered to be completely unknown. The point is that a priori measurements made on an unknown vector $\mathbf{x}$ can have the same effect on the estimate as the assumption that $\mathbf{x}$ is a random vector.

An important property of the Bayesian model (5.3.1) is, from (5.1.6),

$$E\{\mathbf{x} - \hat{\mathbf{x}}_{\text{Bayesian}}]\mathbf{z}'\} = \mathbf{0}$$

However, for the Fisher model (5.3.2), a little manipulation shows that

$$E\{\mathbf{x} - \hat{\mathbf{x}}_{\text{Fisher}}]\mathbf{z}'\} = -[\mathbf{H}'\mathbf{R}^{-1}\mathbf{H}]^{-1}\mathbf{H}'$$

This illustrates once again that despite their interrelationships, Fisher and Bayesian models are different. One mathematical reason for this particular difference is that the proof of (5.1.6) uses (5.1.1), which contains $[\mathbf{H}\boldsymbol{\psi}\mathbf{H}' + \mathbf{R}]^{-1}$, and when $K_2 > K_1$ and $\boldsymbol{\psi} = \infty\,\mathbf{I}$, the inverse does not exist.

5.4 Weighted Least Squares

In a Bayesian version of (5.1) both $\mathbf{x}$ and $\mathbf{v}$ are modeled as random vectors. In a Fisher model, $\mathbf{v}$ is modeled as a random vector but $\mathbf{x}$ is considered to be completely unknown. In weighted-least-squares theory, it is important to emphasize that *no model for the uncertainty in* **x** *or* **v** *is used.* Weighted-least-squares estimation theory replaces modeling and optimality arguments by the *intuitive* judgment that given **z**, a "reasonable" estimate of **x** would be obtained by choosing the value of **x** that minimizes

$$J(\mathbf{x}) = (\mathbf{z} - \mathbf{H}\mathbf{x})'\mathbf{R}^{-1}(\mathbf{z} - \mathbf{H}\mathbf{x}) \tag{5.4.1}$$

where $\mathbf{R}^{-1}$ is now a positive definite weighting matrix chosen on the basis of *engineering judgment.*

Let $\hat{\mathbf{x}}$ denote the **x** that minimizes (5.4.1). Then

$$\boxed{\begin{aligned}\hat{\mathbf{x}} &= (\mathbf{H}'\mathbf{R}^{-1}\mathbf{H})^{-1}\mathbf{H}'\mathbf{R}^{-1}\mathbf{z} \\ J(\hat{\mathbf{x}}) &= \mathbf{z}'[\mathbf{R}^{-1} - \mathbf{R}^{-1}\mathbf{H}(\mathbf{H}'\mathbf{R}^{-1}\mathbf{H})^{-1}\mathbf{H}'\mathbf{R}^{-1}]\mathbf{z}\end{aligned}} \tag{5.4.2}$$

To prove (5.4.2), manipulate (5.4.1) into the form

$$\begin{aligned}J(\mathbf{x}) &= \mathbf{z}'[\mathbf{R}^{-1} - \mathbf{R}^{-1}\mathbf{H}\mathbf{\Sigma}\mathbf{H}'\mathbf{R}^{-1}]\mathbf{z} \\ &\quad + [\mathbf{x} - \mathbf{\Sigma}\mathbf{H}'\mathbf{R}^{-1}\mathbf{z}]'\mathbf{\Sigma}^{-1}[\mathbf{x} - \mathbf{\Sigma}\mathbf{H}'\mathbf{R}^{-1}\mathbf{z}] \\ \mathbf{\Sigma} &= [\mathbf{H}'\mathbf{R}^{-1}\mathbf{H}]^{-1}\end{aligned}$$

Assume that $\mathbf{\Sigma}$ exists. Since the second term is nonnegative and is the only term involving **x**, it is obvious that $\hat{\mathbf{x}}$ is as in (5.4.2). A direct minimization of $J(\mathbf{x})$ yields, of course, the same result; i.e.,

$$\frac{\partial J(\mathbf{x})}{\partial \mathbf{x}} = 2\mathbf{H}'\mathbf{R}^{-1}[\mathbf{z} - \mathbf{H}\mathbf{x}] = \mathbf{0}$$

has a solution $\hat{\mathbf{x}}$ given by (5.4.2), providing that $\mathbf{\Sigma}$ exists.

The estimator of (5.4.2) is identical in form to the Fisher estimator of (5.2.1). The difference between the two approaches is

Fisher: **R** is a covariance matrix of zero mean observation error **v** in the model $\mathbf{z} = \mathbf{H}\mathbf{x} + \mathbf{v}$.

Weighted least squares: **R** is a positive definite weighting matrix in (5.4.1) chosen by judgment.

In many cases, this difference can be viewed as just a play on words. However, the importance of the difference can be illustrated by the following example.

EXAMPLE: Assume that you are a consultant and that a prospective client comes to you with a problem that you see can be formulated as a $\mathbf{z} = \mathbf{H}\mathbf{x} + \mathbf{v}$ estimation problem. Assume that the client is intelligent but completely ignorant of all probability and/or statistical concepts. You will probably make a sale (of your services) much easier if you use a weighted-least-squares argument, as the client can readily understand what you will do for him and why. You may lose him if you insist on using "big words" such as covariance matrix, random vector, and unbiased.

A weighted-least-squares estimator which has the same form as the Bayesian estimator of (5.1.1) can be obtained by redefining (5.4.1) as

$$J(\mathbf{x}) = [\mathbf{z} - \mathbf{Hx}]'\mathbf{R}^{-1}[\mathbf{z} - \mathbf{Hx}] + [\mathbf{x} - \mathbf{m}_x]'\boldsymbol{\psi}^{-1}[\mathbf{x} - \mathbf{m}_x]$$

By definition, weighted-least-squares estimation theory requires no models for the uncertainty in **x** and **v**. This has the disadvantage that there is no way to evaluate the performance of the estimate, i.e., to analyze its behavior. In this sense, the weighted-least-squares logic does not provide a complete answer to the basic estimation problem specified in the introduction to this chapter. This is the penalty one pays for not "needing" models for the uncertainty in **x** and **v**.

In the literature (including some of the author's own papers), the term weighted least squares is often used even when a Fisher or Bayesian model is being assumed. The approach being used in this book was chosen to emphasize that uncertainty models are not an essential part of the weighted-least-squares concept.

5.5 Weighting Patterns

The estimators of Sections 5.1–5.4 are of the form

$$\hat{\mathbf{x}} = \mathbf{Wz} + \mathbf{W}_0 \tag{5.5.1}$$

where the matrix **W** and the vector $\mathbf{W}_0$ specify the estimator. The matrix **W** is called the weighting pattern matrix. If x is scalar and if $W_0 = 0$,

$$\hat{x} = \sum_{k=1}^{K_2} \mathbf{W}_k z_k$$

and the W_k, $k = 1, \ldots, K_2$, determine the relative weights (importance) that are given to each observation z_k in obtaining $\hat{x}$. These weights are determined by the "size" of the uncertainties ($\mathbf{R}$, $\boldsymbol{\psi}$) and the structure of **H**.

Consideration of the weighting pattern matrix **W** can aid in understanding how the estimator works. The author finds that the following informal discussion provides a motivation for why the basic estimator forms of the preceding sections are reasonable. Consider the Fisher model

$$\mathbf{z} = \mathbf{H}x + \mathbf{v}$$

$$\mathbf{z} = \begin{bmatrix} z_1 \\ z_2 \end{bmatrix} \qquad \mathbf{v} = \begin{bmatrix} v_1 \\ v_2 \end{bmatrix} \qquad x\text{: scalar}$$

$$E\{\mathbf{v}\mathbf{v}'\} = \begin{bmatrix} R_{11} & 0 \\ 0 & R_{22} \end{bmatrix}$$

$$\mathbf{H} = \begin{bmatrix} H_1 \\ H_2 \end{bmatrix}$$

so

$$\mathbf{W} = [\mathbf{H}'\mathbf{R}^{-1}\mathbf{H}]^{-1}\mathbf{H}'\mathbf{R}^{-1} = [W_1 W_2]$$

In the special case

$$R_{11} = R_{22}$$

$\mathbf{W}$ is proportional to $\mathbf{H}'$, which, after a little thought, is a very reasonable answer. Similarly, in the special case

$$H_1 = H_2$$

$\mathbf{W}$ is proportional to $[\mathbf{R}_{11}^{-1}\ \mathbf{R}_{22}^{-1}]$, which again makes a lot of sense. In general, the estimator matrix $\mathbf{W}$ is determined by the *interaction* of $\mathbf{H}$ and $\mathbf{R}$.

EXAMPLE: Consider the Fisher model

$$\mathbf{z} = \mathbf{H}x + \mathbf{v}$$

$$\mathbf{z} = \begin{bmatrix} z_1 \\ z_2 \end{bmatrix} \qquad \mathbf{v} = \begin{bmatrix} v_1 \\ v_2 \end{bmatrix}$$

$$\mathbf{H} = \begin{bmatrix} 1 \\ 8 \end{bmatrix} \qquad \mathbf{R} = \begin{bmatrix} 1 & 0 \\ 0 & 4 \end{bmatrix}$$

Then

$$\mathbf{W} = [\tfrac{1}{17}\ \ \tfrac{2}{17}]$$

Thus although z_2 is a noisier observation than z_1 ($R_{22} > R_{11}$), z_2 receives more weight ($W_2 > W_1$) because of the effect of $\mathbf{H}$ in determining $\mathbf{W}$.

In the above discussion of the Fisher model, x was a scalar and $[\mathbf{H}'\mathbf{R}^{-1}\mathbf{H}]^{-1}$ did not effect the ratio of W_1 to W_2. In the case of vector $\mathbf{x}$, the $[\mathbf{H}'\mathbf{R}^{-1}\mathbf{H}]^{-1}$ term can play a more important role.

In the weighted-least-squares approach, $\mathbf{R}^{-1}$ was viewed as a weighting matrix. This weighting should not be confused with the weighting pattern matrix $\mathbf{W}$ of (5.5.1).

The above discussions can also be modified for application to a Bayesian model.

5.6 Unknown-but-Bounded

Consider

$$\mathbf{z} = \mathbf{H}\mathbf{x} + \mathbf{v}$$

$$\mathbf{x} \in \mathbf{\Omega}_x$$

$$\mathbf{v} \in \mathbf{\Omega}_v$$

where $\mathbf{\Omega}_x$ and $\mathbf{\Omega}_v$ are sets in K_1- and K_2-dimensional spaces, respectively. The particular sets to be discussed are ellipsoids.

5.6.1 Arbitrary Sets

Since $\mathbf{v}$ must lie in a set, it follows that a given observation $\mathbf{z} = \mathbf{z}_{\text{actual}}$ combines with the set Ω_v in K_2 space to define a new set in K_1 space which must contain $\mathbf{x}$. Thus even if $\mathbf{x}$ itself is not bounded a priori, the observation $\mathbf{z} = \mathbf{z}_{\text{actual}}$ specifies a set $\Omega_{x|z_{\text{actual}}}$ which must contain $\mathbf{x}$; i.e.,

$$\mathbf{x} \in \Omega_{x|z_{\text{actual}}}$$

$$\Omega_{x|z} = \{\mathbf{x}: \mathbf{z} - \mathbf{H}\mathbf{x} \in \Omega_v\}$$

Consider the two sets, Ω_x and $\Omega_{x|z}$. Each set contains $\mathbf{x}$. Thus $\mathbf{x}$ must lie in their intersection. Let Ω_{est} denote this intersection. Then

$$\boxed{\Omega_{\text{est}} = \Omega_x \cap \Omega_{x|z}} \tag{5.6.1}$$

The estimation problem is now conceptually complete. The modeling of the uncertainties implies that, given the observation $\mathbf{z}$, the value of $\mathbf{x}$ must lie within Ω_{est}. Furthermore, Ω_{est} is the smallest set, which must contain $\mathbf{x}$ and which can be calculated from the available information. Thus the set Ω_{est} is the "best" estimate set.

Note that the estimate of the vector $\mathbf{x}$ is defined as a set—not as a vector as in the Sections 5.1–5.4. An unknown-but-bounded model does not provide any specific way of determining which vector within Ω_{est} is the best estimate of $\mathbf{x}$. Naturally, a reasonable choice for a vector estimate is to define an $\hat{\mathbf{x}}$ as the *center* of Ω_{est}, where the center can be defined in any convenient way. Such center-defined $\hat{\mathbf{x}}$ will often be used in the following, but it must always be remembered that the basic logic leads to an estimate which is a set, not a vector.

EXAMPLE: Consider

$$\begin{bmatrix} z_1 \\ z_2 \end{bmatrix} = \begin{bmatrix} x_1 \\ x_2 \end{bmatrix} + \begin{bmatrix} v_1 \\ v_2 \end{bmatrix}$$

$$\Omega_x = \{\mathbf{x}: |x_1| \leq 2, |x_2| \leq 2\}$$

$$\Omega_v = \{\mathbf{v}: v_1^2 + v_2^2 \leq 1\}$$

The sets Ω_x, $\Omega_{x|z}$, and Ω_{est} are shown in Fig. 5.6.1 for

$$\mathbf{z} = \begin{bmatrix} 2 \\ 2 \end{bmatrix}$$

If a vector estimate $\hat{\mathbf{x}}$ is desired, it can be chosen to be the center of Ω_{est} of Fig. 5.6.1(c).

5.6.2 Ellipsoids

Now assume that Ω_x and Ω_v are ellipsoids:

$$\Omega_x = \{\mathbf{x}: [\mathbf{x} - \mathbf{m}_x]'\boldsymbol{\psi}^{-1}[\mathbf{x} - \mathbf{m}_x] \leq 1\}$$

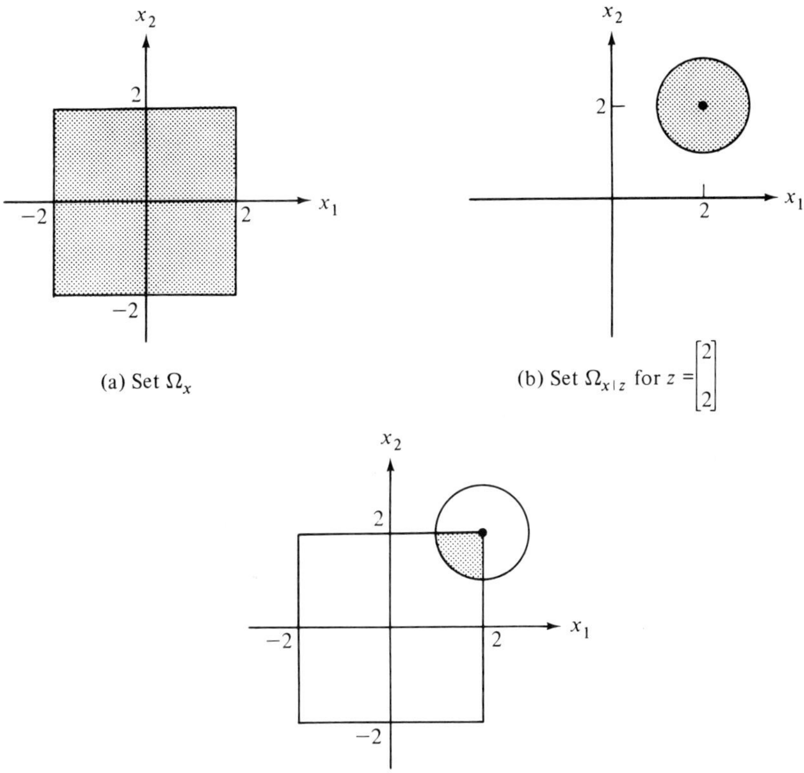

(a) Set Ω_x

(b) Set $\Omega_{x|z}$ for $z = \begin{bmatrix} 2 \\ 2 \end{bmatrix}$

(c) $\Omega_{\text{est}} = \Omega_x \cap \Omega_{x|z}$

Figure 5.6.1 Sets of example.

$$\boldsymbol{\Omega}_v = \{\mathbf{v} : \mathbf{v}'\mathbf{R}^{-1}\mathbf{v} \leq 1\}$$

Then

$$\boldsymbol{\Omega}_{x|z} = \{\mathbf{x} : [\mathbf{z} - \mathbf{H}\mathbf{x}]'\mathbf{R}^{-1}[\mathbf{z} - \mathbf{H}\mathbf{x}] \leq 1\}$$

$$\boldsymbol{\Omega}_{\text{est}} = \{\mathbf{x} : [\mathbf{x} - \mathbf{m}_x]'\boldsymbol{\psi}^{-1}[\mathbf{x} - \mathbf{m}_x] \leq 1 \text{ and } [\mathbf{z} - \mathbf{H}\mathbf{x}]'\mathbf{R}^{-1}[\mathbf{z} - \mathbf{H}\mathbf{x}] \leq 1\}$$

As already noted in Section 4.3, we do not live in the best of all possible worlds, so $\boldsymbol{\Omega}_{\text{est}}$ is not an ellipsoid and is not well suited to computer study. However, again as in Section 4.3, all is not lost, as a bounding ellipsoid can be found which contains the intersection of ellipsoids.

Consider the set

$$\boldsymbol{\Omega}_{\hat{x}} = \{\mathbf{x} : [1 - \rho][\mathbf{x} - \mathbf{m}_x]'\boldsymbol{\psi}^{-1}[\mathbf{x} - \mathbf{m}_x] + \rho[\mathbf{z} - \mathbf{H}\mathbf{x}']\mathbf{R}^{-1}[\mathbf{z} - \mathbf{H}\mathbf{x}] \leq 1\} \tag{5.6.2}$$

It is obvious that

$$\boldsymbol{\Omega}_{\hat{x}} \supset \boldsymbol{\Omega}_{\text{est}}, \qquad \text{any } \rho, \qquad 0 \leq \rho \leq 1$$

$\boldsymbol{\Omega}_{\hat{x}}$ is an ellipsoid. It turns out after a lot of manipulation that

$$\begin{aligned}
\boldsymbol{\Omega}_{\hat{x}} &\supset \boldsymbol{\Omega}_{\text{est}} \\
\boldsymbol{\Omega}_{\hat{x}} &= \{\mathbf{x}: [\mathbf{x} - \hat{\mathbf{x}}]'\boldsymbol{\Sigma}^{-1}[\mathbf{x} - \hat{\mathbf{x}}] \leq 1\} \\
\hat{\mathbf{x}} &= \mathbf{W}\mathbf{z} + \mathbf{W}_0 \\
\boldsymbol{\Sigma} &= (1 - \delta^2)\boldsymbol{\Sigma}_b \\
\mathbf{W} &= \boldsymbol{\Sigma}_b\mathbf{H}'\tilde{\mathbf{R}}^{-1} \\
\mathbf{W}_0 &= \boldsymbol{\Sigma}_b\tilde{\boldsymbol{\psi}}^{-1}\mathbf{m}_x \\
\boldsymbol{\Sigma}_b &= [\tilde{\boldsymbol{\psi}}^{-1} + \mathbf{H}'\tilde{\mathbf{R}}^{-1}\mathbf{H}]^{-1} \\
\tilde{\mathbf{R}} &= \rho^{-1}\mathbf{R} \\
\tilde{\boldsymbol{\psi}} &= (1 - \rho)^{-1}\boldsymbol{\psi} \\
\delta^2 &= [\mathbf{z} - \mathbf{H}\mathbf{m}_x]'[\mathbf{H}\tilde{\boldsymbol{\psi}}\mathbf{H}' + \tilde{\mathbf{R}}]^{-1}[\mathbf{z} - \mathbf{H}\mathbf{m}_x] \\
0 &\leq \rho \leq 1
\end{aligned} \tag{5.6.3}$$

The center of the bounding ellipsoid estimate set $\boldsymbol{\Omega}_{\hat{x}}$ is $\hat{\mathbf{x}}$, and the size and shape of the set are determined by $\boldsymbol{\Sigma}$. Although an unknown-but-bounded model yields an estimate which is a set rather than a single vector, it is quite reasonable to consider $\hat{\mathbf{x}}$ to be the best estimate. For example, $\hat{\mathbf{x}}$ is a *min-max* estimate; it minimizes the maximum error between $\hat{\mathbf{x}}$ and the true but unknown value of $\mathbf{x}$ (if the difference between $\boldsymbol{\Omega}_{\hat{x}}$ and $\boldsymbol{\Omega}_{\text{est}}$ is ignored).

If $\hat{\mathbf{x}}$ is viewed as *the* estimate, then the size and shape of $\boldsymbol{\Omega}_{\hat{x}}$ determines the error so that the matrix $\boldsymbol{\Sigma}$ provides the desired error analysis (i.e., model of uncertainty in $\mathbf{x} - \hat{\mathbf{x}}$). The $\boldsymbol{\Sigma}$ of (5.6.3) depends on δ^2 and $\mathbf{z}$ and hence cannot be evaluated before the observation is taken. However, it is obvious by setting $\delta^2 = 0$ in (5.6.3) that

$$\boldsymbol{\Sigma}_b \geq \boldsymbol{\Sigma} \tag{5.6.4}$$

Thus $\boldsymbol{\Sigma}_b$ can be used to determine a bound on the maximum possible error that can occur.

The results (5.6.3) can be considered to be analogous to the Bayesian estimation case in the sense that a priori bounds are available on both $\mathbf{x}$ and $\mathbf{v}$. The analogous Fisher case occurs when $\mathbf{x}$ is completely unknown and only $\mathbf{v}$ is constrained to lie within an ellipsoid. After setting $\boldsymbol{\psi}^{-1} = \mathbf{0}$ and choosing $\rho = 1$, (5.6.3) becomes

$$\begin{aligned}
\boldsymbol{\Omega}_{\hat{x}} &= \{\mathbf{x}: [\mathbf{x} - \hat{\mathbf{x}}]'\boldsymbol{\Sigma}^{-1}[\mathbf{x} - \hat{\mathbf{x}}] \leq 1\} \\
\hat{\mathbf{x}} &= [\mathbf{H}'\mathbf{R}^{-1}\mathbf{H}]^{-1}\mathbf{H}'\mathbf{R}^{-1}\mathbf{z} \\
\boldsymbol{\Sigma} &= (1 - \delta^2)[\mathbf{H}'\mathbf{R}^{-1}\mathbf{H}]^{-1} \\
\delta^2 &= \mathbf{z}'[\mathbf{R}^{-1} - \mathbf{R}^{-1}\mathbf{H}(\mathbf{H}'\mathbf{R}^{-1}\mathbf{H})^{-1}\mathbf{H}'\mathbf{R}^{-1}]\mathbf{z}
\end{aligned} \tag{5.6.5}$$

$[\mathbf{H}'\mathbf{R}^{-1}\mathbf{H}]^{-1}$ is assumed to exist so that it is necessary that $K_2 \geq K_1$; i.e., the dimension of $\mathbf{z}$ is not smaller than the dimension of $\mathbf{x}$. In this case, where $\mathbf{x}$ is completely unknown, there are no free parameters such as ρ, and the ellipsoid $\mathbf{\Omega}_{\hat{x}}$ is $\mathbf{\Omega}_{\text{est}}$.

At first it appears that $1 - \delta^2$ may go negative, so $\mathbf{\Sigma}$ of (5.6.3) may become negative definite. However, it can be shown that this cannot happen provided that the model itself is valid. This is easily proved in the special case $\boldsymbol{\psi}^{-1} = \mathbf{0}$ when (5.6.5) applies. Substitution of $\mathbf{z} = \mathbf{Hx} + \mathbf{v}$ into (5.6.5) gives

$$\mathbf{\Sigma} = \{1 + \mathbf{v}'\mathbf{R}^{-1}\mathbf{H}(\mathbf{H}'\mathbf{R}^{-1}\mathbf{H})^{-1}\mathbf{H}'\mathbf{R}^{-1}\mathbf{v} - \mathbf{v}'\mathbf{R}^{-1}\mathbf{v}\}[\mathbf{H}'\mathbf{R}^{-1}\mathbf{H}]^{-1}$$

Since $\mathbf{v}'\mathbf{R}^{-1}\mathbf{v} \leq 1$, $\mathbf{\Sigma}$ is always positive semidefinite. $\mathbf{\Sigma}$ may be $\mathbf{0}$ if $\mathbf{v}'\mathbf{R}^{-1}\mathbf{v} = 1$ because $\mathbf{R}^{-1}\mathbf{H}(\mathbf{H}'\mathbf{R}^{-1}\mathbf{H})^{-1}\mathbf{H}'\mathbf{R}^{-1}$ is only positive semidefinite (if $K_2 > K_1$).

EXAMPLE: Consider

$$\begin{bmatrix} z_1 \\ z_2 \end{bmatrix} = \begin{bmatrix} 1 \\ 1 \end{bmatrix} x + \begin{bmatrix} v_1 \\ v_2 \end{bmatrix}$$

$$\mathbf{R} = \begin{bmatrix} 1 & 0 \\ 0 & 4 \end{bmatrix} \qquad \psi^{-1} = 0 \qquad m_x = 0$$

Then from (5.6.5)

$$\Sigma = (1 - \delta^2)\tfrac{4}{5}$$
$$\delta^2 = \tfrac{1}{5}(z_1 - z_2)^2$$

so

$$\Sigma = [5 - (z_1 - z_2)^2]\tfrac{4}{25}$$
$$\hat{x} = \tfrac{4}{5}z_1 + \tfrac{1}{5}z_2$$

There is a whole family of bounding ellipsoid estimate sets $\mathbf{\Omega}_{\hat{x}}$ depending on the choice of ρ, $1 \geq \rho \geq 0$. The lack of uniqueness arises because the ellipsoidal set $\mathbf{\Omega}_{\hat{x}}$ only bounds $\mathbf{\Omega}_{\text{est}}$ and there are many such bounding ellipsoids. An optimum choice of value for ρ is possible if a scalar measure of the size of the matrix $\mathbf{\Sigma}$ is chosen such as: tr $\mathbf{\Sigma}$, maximum eigenvalue of $\mathbf{\Sigma}$, $|\mathbf{\Sigma}|$, or $\mathbf{d}'\mathbf{\Sigma}\mathbf{d}$, where $\mathbf{d}$ is some chosen *direction* of special importance. Unfortunately, the resulting equations are nontrivial to solve. A somewhat easier (but still nontrivial) approach to choosing ρ is to minimize some function of $\mathbf{\Sigma}_b$ of (5.6.4), as $\mathbf{\Sigma}_b$ does not depend on the observation $\mathbf{z}$. This also has the advantage that ρ can be precalculated before the observation is made.

The bounding ellipsoid estimate for the unknown-but-bounded model is a linear operation as specified in (5.6.3). The matrix $\mathbf{\Sigma}$ which measures the uncertainty in $\mathbf{x} - \hat{\mathbf{x}}$ (the size and shape of $\mathbf{\Omega}_{\hat{x}}$) is a function of the observation $\mathbf{z}$ and hence cannot be precomputed. However, by using (5.6.4), a "larger" matrix $\mathbf{\Sigma}_b$ (which specified a larger ellipsoid) which is independent of $\mathbf{z}$ can be obtained and used for an error analysis.

The bounding ellipsoid estimate set $\mathbf{\Omega}_{\hat{x}}$ is closely related to the Bayesian, Fisher, and weighted-least-squares results. Consider first the Fisher model estimates of (5.2.1) and the "corresponding" one of (5.6.5). The equations for $\hat{\mathbf{x}}$ are identical (equal estimators). For the Fisher model, the covariance matrix of the error in the estimate is given by $(\mathbf{H}'\mathbf{R}^{-1}\mathbf{H})^{-1}$. The corresponding concept for the unknown-but-bounded model is the size of the ellipsoid defined by the positive definite matrix $\mathbf{\Sigma}$ of (5.6.5). These two *error measures* are related but definitely not equal, as $\mathbf{\Sigma}$ of (5.6.5) depends on the value of the observation $\mathbf{z}$. In fact, it was shown using (5.6.5) that for certain observations it is possible to have $\mathbf{\Sigma} = \mathbf{0}$ so that the estimate $\hat{\mathbf{x}}$ is equal to the unknown $\mathbf{x}$; i.e., there is no error at all. No such phenomenon is encountered in the Fisher model estimator. Similar comparisons can be drawn between the results of the Bayesian theory of Section 5.2 and the unknown-but-bounded results of (5.6.3). The presence of the free parameter ρ complicates the use of (5.6.3), as a separate study is needed to specify a value. The Bayesian and Fisher models yielded globally optimum estimates in the sense that to estimate $\mathbf{y} = \mathbf{C}\mathbf{x}$ one uses $\hat{\mathbf{y}} = \mathbf{C}\hat{\mathbf{x}}$. In the case of bounding ellipsoids this is not quite true, as it is conceivable to choose ρ so as to minimize the size of the estimate set in some direction.

Unknown-but-bounded models actually yield an estimate which is a set. No particular vector in the set is a more likely estimate than any other vector. In general, any needed choice of a particular vector within the estimate set should be treated as a separate problem. For example, min-max estimation theory follows as a special case. The use of ellipsoidal sets tends to obscure the set nature of the problem, as the ellipsoid center forms a natural vector estimate and for ease of discussion, the center, $\hat{\mathbf{x}}$, is usually called (or at least viewed as) *the* estimate.

The basic unknown-but-bounded model estimation concept is not restricted to ellipsoidal sets. For example, it is possible and may be desirable to work directly with nonellipsoidal sets such as polyhedra. In certain applications, the exact $\mathbf{\Omega}_{\text{est}}$ may be calculable on a computer, while in more general cases, some new $\mathbf{\Omega}_{\hat{x}} \supset \mathbf{\Omega}_{\text{est}}$ will have to be defined.

5.7 A Dichotomy

Suppose that two separate observations of the same state are made so that

$$\mathbf{z}_1 = \mathbf{H}_1\mathbf{x} + \mathbf{v}_1$$
$$\mathbf{z}_2 = \mathbf{H}_2\mathbf{x} + \mathbf{v}_2$$

Obviously these observations could be combined into a single model

$$\mathbf{z} = \mathbf{H}\mathbf{x} + \mathbf{v}$$
$$\mathbf{H} = \begin{bmatrix}\mathbf{H}_1\\ \mathbf{H}_2\end{bmatrix} \qquad \mathbf{z} = \begin{bmatrix}\mathbf{z}_1\\ \mathbf{z}_2\end{bmatrix} \qquad \mathbf{v} = \begin{bmatrix}\mathbf{v}_1\\ \mathbf{v}_2\end{bmatrix} \tag{5.7.1}$$

and processed as a single vector observation. However, in some applications, a two-step process or dichotomy is desired; i.e.,

1. Make separate estimates of $\mathbf{x}$, one from $\mathbf{z}_1$ and one from $\mathbf{z}_2$.
2. Combine the two estimates to get a final estimate.

It turns out that this dichotomy can result in little or no loss in accuracy.

Consider first a Fisher model where $\mathbf{x}$ is unknown and $\mathbf{v}_1$ and $\mathbf{v}_2$ are zero mean, uncorrelated random vectors so that

$$
\begin{aligned}
E\{\mathbf{v}_1\mathbf{v}_1'\} &= \mathbf{R}_1 \\
E\{\mathbf{v}_2\mathbf{v}_2'\} &= \mathbf{R}_2 \\
E\{\mathbf{v}_1\mathbf{v}_2'\} &= \mathbf{0}
\end{aligned}
$$

Then by combining $\mathbf{z}_1$ and $\mathbf{z}_2$ into a single vector, it follows from (5.2.1) that

$$
\begin{aligned}
\hat{\mathbf{x}} &= \mathbf{\Sigma}[\mathbf{H}_1'\mathbf{R}_1^{-1}\mathbf{z}_1 + \mathbf{H}_2'\mathbf{R}_2^{-1}\mathbf{z}_2] \\
\mathbf{\Sigma} &= [\mathbf{H}_1'\mathbf{R}_1^{-1}\mathbf{H}_1 + \mathbf{H}_2'\mathbf{R}_2^{-1}\mathbf{H}_2]^{-1}
\end{aligned}
\tag{5.7.2}
$$

Now define

$$
\left.
\begin{aligned}
\hat{\mathbf{x}}(j) &= \mathbf{\Sigma}(j)\mathbf{H}_j'\mathbf{R}_j^{-1}\mathbf{z}_j \\
\mathbf{\Sigma}(j) &= [\mathbf{H}_j'\mathbf{R}_j^{-1}\mathbf{H}_j]^{-1}, \qquad j = 1, 2
\end{aligned}
\right\}: \text{Estimate using } \mathbf{z}_j,\ j = 1, 2, \text{ assuming Fisher model}
$$

The estimate of (5.7.2) can be rewritten as

$$
\hat{\mathbf{x}} = [\mathbf{\Sigma}(1)^{-1} + \mathbf{\Sigma}(2)^{-1}]^{-1}\{\mathbf{\Sigma}(1)^{-1}\hat{\mathbf{x}}(1) + \mathbf{\Sigma}(2)^{-1}\hat{\mathbf{x}}(2)\} \tag{5.7.3}
$$

Thus $\hat{\mathbf{x}}(1)$ and $\hat{\mathbf{x}}(2)$ can be calculated first and then combined to get $\hat{\mathbf{x}}$. To carry this argument a bit further, define

$$
\boldsymbol{\delta}_x(j) = \mathbf{x} - \hat{\mathbf{x}}(j), \qquad j = 1, 2
$$

so that

$$
\mathbf{\Sigma}(j) = E\{\boldsymbol{\delta}_x(j)\boldsymbol{\delta}_x'(j)\}
$$

Then (5.7.3) can be interpreted as a Fisher model estimate obtained by viewing $\hat{\mathbf{x}}(1)$ and $\hat{\mathbf{x}}(2)$ as "observations" on a completely unknown $\mathbf{x}$ with *observation uncertainty* $\boldsymbol{\delta}_x(j)$. Figure 5.7.1 is meant to summarize these concepts.

Consider a Bayesian version of the preceding Fisher model where $\mathbf{x}$ is now a random vector uncorrelated with $\mathbf{v}_1$ and $\mathbf{v}_2$ with

$$
\begin{aligned}
E\{\mathbf{x}\} &= \mathbf{m}_x \\
E\{[\mathbf{x} - \mathbf{m}_x][\mathbf{x} - \mathbf{m}_x]'\} &= \boldsymbol{\psi}
\end{aligned}
$$

In this case the Bayesian model equation (5.1.1) yields

$$
\begin{aligned}
\hat{\mathbf{x}} &= \mathbf{\Sigma}[\mathbf{H}_1'\mathbf{R}_1^{-1}\mathbf{z}_1 + \mathbf{H}_2'\mathbf{R}_2^{-1}\mathbf{z}_2 + \boldsymbol{\psi}^{-1}\mathbf{m}_x] \\
\mathbf{\Sigma} &= [\mathbf{H}_1'\mathbf{R}_1^{-1}\mathbf{H}_1 + \mathbf{H}_2'\mathbf{R}_2^{-1}\mathbf{H}_2 + \boldsymbol{\psi}^{-1}]^{-1}
\end{aligned}
\tag{5.7.4}
$$

The dichotomy illustrated by Fig. 5.7.1 still applies if the a priori distribution is associated with, say, $\mathbf{z}_1$, while $\mathbf{z}_2$ is first processed assuming that $\mathbf{x}$ is com-

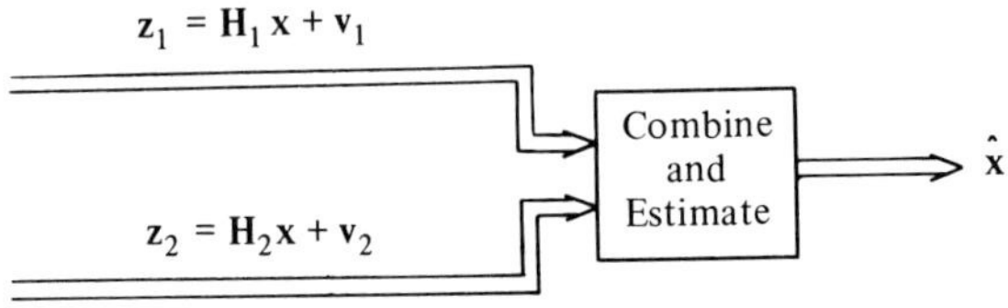

(a) **Straightforward Estimation**

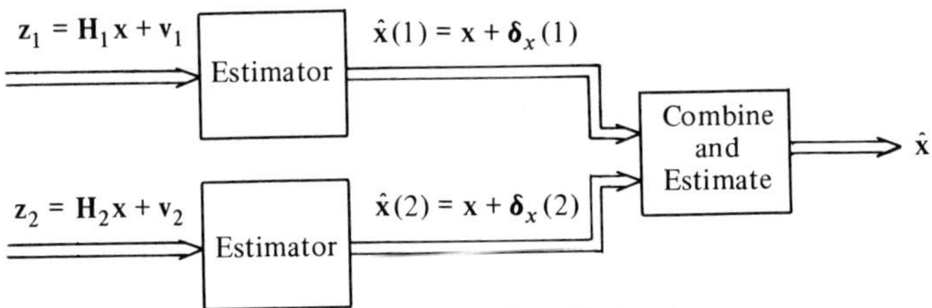

(b) **Two Step Estimation**

Figure 5.7.1 A dichotomy.

pletely unknown (i.e., as a Fisher model). Define

$$\left.\begin{aligned}\hat{\mathbf{x}}(1) &= \boldsymbol{\Sigma}(1)[\mathbf{H}_1'\mathbf{R}_1^{-1}\mathbf{z}_1 + \boldsymbol{\psi}^{-1}\mathbf{m}_x] \\ \boldsymbol{\Sigma}(1) &= [\mathbf{H}_1'\mathbf{R}_1^{-1}\mathbf{H}_1 + \boldsymbol{\psi}]^{-1}\end{aligned}\right\}: \text{Estimate using } \mathbf{z}_1 \text{ assuming Bayesian model}$$

$$\left.\begin{aligned}\hat{\mathbf{x}}(2) &= \boldsymbol{\Sigma}(2)[\mathbf{H}_2'\mathbf{R}_2^{-1}\mathbf{z}_2] \\ \boldsymbol{\Sigma}(2) &= [\mathbf{H}_2'\mathbf{R}_2^{-1}\mathbf{H}_2]^{-1}\end{aligned}\right\}: \text{Estimate using } \mathbf{z}_2 \text{ assuming Fisher model}$$

Then (5.7.4) can be rewritten as

$$\left.\begin{aligned}\hat{\mathbf{x}} &= [\boldsymbol{\Sigma}^{-1}(1) + \boldsymbol{\Sigma}^{-1}(2)]^{-1} \\ &\quad \times\{\boldsymbol{\Sigma}^{-1}(1)\hat{\mathbf{x}}(1) + \boldsymbol{\Sigma}^{-1}(2)\hat{\mathbf{x}}(2)\}\end{aligned}\right\}: \text{Combination of } \hat{\mathbf{x}}(1) \text{ and } \hat{\mathbf{x}}(2) \text{ viewing } \hat{\mathbf{x}}(1) \text{ and } \hat{\mathbf{x}}(2) \text{ as observations on completely unknown } \mathbf{x} \text{ (Fisher model)}$$

Thus the "trick" is to use the Bayesian a priori distribution only once and then use Fisher model concepts the rest of the way. This is closely related to the idea in Section 5.3 of viewing a Bayesian model as a Fisher model with an extra measurement corresponding to the a priori distribution. Figure 5.7.2 summarizes the relationship between the dichotomy for the Fisher and Bayesian case.

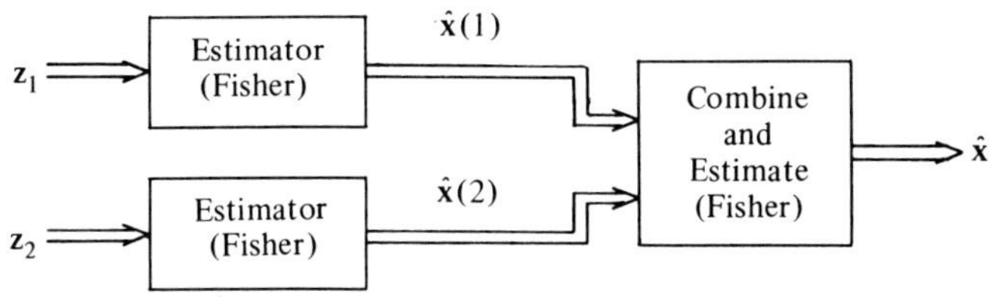

(a) Fisher

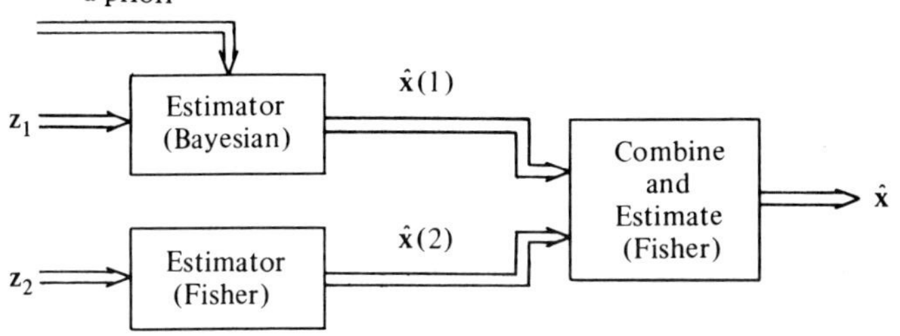

(b) Bayesian

Figure 5.7.2 Dichotomy for Bayesian and Fisher models.

Finally, consider an unknown-but-bounded model where

$$\mathbf{x} \in \mathbf{\Omega}_x$$
$$\mathbf{v}_1 \in \mathbf{\Omega}_{v_1}$$
$$\mathbf{v}_2 \in \mathbf{\Omega}_{v_2}$$

Let $\mathbf{\Omega}_{\hat{x}}$ denote the exact estimate set calculated using both $\mathbf{z}_1$ and $\mathbf{z}_2$ and let $\mathbf{\Omega}_{\hat{x}}(j), j = 1, 2$ denote the individual exact estimate sets calculated from just $\mathbf{z}_j, j = 1, 2$. It follows directly from the set theory arguments that

$$\mathbf{\Omega}_{\hat{x}} = \mathbf{\Omega}_{\hat{x}}(1) \cap \mathbf{\Omega}_{\hat{x}}(2)$$

so a dichotomy like that of Fig. 5.7.1 yields exact answers as long as exact sets are considered. However, when working with bounding ellipsoids, the use of a two-step estimate procedure may result in a larger estimate set. Note that in the special case where $\mathbf{v}_1$ and $\mathbf{v}_2$ lie in ellipsoids,

$$\mathbf{\Omega}_{v_1} = \{\mathbf{v}_1 : \mathbf{v}_1'\mathbf{R}_1^{-1}\mathbf{v}_2 \leq 1\}$$
$$\mathbf{\Omega}_{v_2} = \{\mathbf{v}_2 : \mathbf{v}_2'\mathbf{R}_2^{-1}\mathbf{v}_2 \leq 1\}$$

a combined model such as (5.7.1) involving $\mathbf{v}$

$$\mathbf{v} = \begin{bmatrix} \mathbf{v}_1 \\ \mathbf{v}_2 \end{bmatrix}$$

yields a $\mathbf{\Omega}_v$ which is not an ellipsoid.

5.8 Perfect Observations†

In some applications, the sensors are so accurate that one is tempted to assume that they are perfect. Thus instead of

$$\mathbf{z} = \mathbf{H}\mathbf{x} + \mathbf{v}$$

the model might become

$$\mathbf{z} = \mathbf{H}\mathbf{x}$$

Consider a Bayesian model. The desired estimator results in the limiting case when $\mathbf{R} \longrightarrow \mathbf{0}$. From (5.1.2),

$$\begin{aligned} \hat{\mathbf{x}} &= \mathbf{m}_x + \boldsymbol{\psi}\mathbf{H}'[\mathbf{H}\boldsymbol{\psi}\mathbf{H}']^{-1}[\mathbf{z} - \mathbf{H}\mathbf{m}_x] \\ \mathbf{\Sigma} &= \boldsymbol{\psi} - \boldsymbol{\psi}\mathbf{H}'(\mathbf{H}\boldsymbol{\psi}\mathbf{H}')^{-1}\mathbf{H}\boldsymbol{\psi} \end{aligned} \tag{5.8.1}$$

assuming that $(\mathbf{H}\boldsymbol{\psi}\mathbf{H}')^{-1}$ exists. Note that

$$\mathbf{H}\hat{\mathbf{x}} = \mathbf{z}$$
$$\mathbf{H}\mathbf{\Sigma}\mathbf{H}' = \mathbf{0}$$

as expected. If $\mathbf{H}^{-1}$ exists (or if $\mathbf{H}$ is of full rank when $K_2 > K_1$), $\hat{\mathbf{x}}$ becomes the true value of $\mathbf{x}$. $\mathbf{\Sigma}^{-1}$ does not exist.

Consider a Fisher model where $\mathbf{x}$ is completely unknown. A meaningful problem results when

$$\begin{aligned} \mathbf{z}_1 &= \mathbf{H}_1\mathbf{x} + \mathbf{v}_1 \\ \mathbf{z}_2 &= \mathbf{H}_2\mathbf{x} \\ \mathbf{z} &= \mathbf{H}\mathbf{x} + \mathbf{v} = \begin{bmatrix} \mathbf{H}_1 \\ \mathbf{H}_2 \end{bmatrix}\mathbf{x} + \begin{pmatrix} \mathbf{v}_1 \\ 0 \end{pmatrix} \\ E\{\mathbf{v}_1\mathbf{v}_1\} &= \mathbf{R}_1 \\ E\{\mathbf{v}\mathbf{v}'\} &= \begin{pmatrix} \mathbf{R}_1 & \mathbf{0} \\ \mathbf{0} & \mathbf{0} \end{pmatrix} = \mathbf{R} \end{aligned}$$

i.e., when some of the sensors are assumed to be perfect. $\mathbf{R}^{-1}$ does not exist. A general method of approach is to make a clever change of variables, but the details are messy. There exists an alternative "proof" which is much simpler and yields the same answer if $[\mathbf{H}_1'\mathbf{R}_1'^{-1}\mathbf{H}_1]^{-1}$ exists. The alternative proof is based on the trick of viewing $\mathbf{z}_2$ as

$$\mathbf{z}_2 = \mathbf{H}_2\mathbf{x} + \mathbf{v}_2$$

$$\mathbf{z}_2\text{: zero mean random vector}$$

$$E\{\mathbf{v}_2\mathbf{v}_2'] = \mathbf{R}_2$$

The Fisher theory of Section 5.2 is then applied to

$$\tilde{\mathbf{z}} = \begin{bmatrix} \mathbf{z}_1 \\ \mathbf{z}_2 \end{bmatrix} = \begin{bmatrix} \mathbf{H}_1 \\ \mathbf{H}_2 \end{bmatrix} \mathbf{x} + \begin{bmatrix} \mathbf{v}_1 \\ \mathbf{v}_1 \end{bmatrix}$$

and the limit is then taken as $\mathbf{R}_2 \to \mathbf{0}$. The result is (assuming that $[\mathbf{H}_1'\mathbf{R}_1^{-1}\mathbf{H}_1]^{-1}$ exists)

$$\begin{aligned} \hat{\mathbf{x}} &= \mathbf{\Sigma}\mathbf{H}_1'\mathbf{R}_1^{-1}\mathbf{z}_1 + \boldsymbol{\theta}\mathbf{H}_2'[\mathbf{H}_2\boldsymbol{\theta}\mathbf{H}_2']^{-1}\mathbf{z}_2 \\ \mathbf{\Sigma} &= \boldsymbol{\theta} - \boldsymbol{\theta}\mathbf{H}_2'[\mathbf{H}_2\boldsymbol{\theta}\mathbf{H}_2']^{-1}\mathbf{H}_2\boldsymbol{\theta} = E\{(\mathbf{x} - \hat{\mathbf{x}})(\mathbf{x} - \hat{\mathbf{x}})'\} \\ \boldsymbol{\theta} &= [\mathbf{H}_1'\mathbf{R}_1^{-1}\mathbf{H}_1]^{-1} \end{aligned} \tag{5.8.2}$$

Note that

$$\mathbf{H}_2\hat{\mathbf{x}} = \mathbf{z}_2$$

as expected. $\boldsymbol{\theta}$ is the error covariance matrix that would exist if the perfect observation $\mathbf{z}_2$ did not exist. Hence the second term in $\mathbf{\Sigma}$ can be viewed as the amount the perfect observation reduces the error. $\mathbf{\Sigma}^{-1}$ does not exist.

Finally, consider an unknown-but-bounded model, in particular (5.6.3). This equation first has to be manipulated into a general form such as (5.1.2). After such manipulations, setting $\mathbf{R} = \mathbf{0}$ and then choosing $\rho = 0$ yields

$$\begin{aligned} \hat{\mathbf{x}} &= \mathbf{m}_x + \boldsymbol{\psi}\mathbf{H}'[\mathbf{H}\boldsymbol{\psi}\mathbf{H}']^{-1}[\mathbf{z} - \mathbf{Hm}] \\ \mathbf{\Sigma} &= (1 - \delta^2)\{\boldsymbol{\psi} - \boldsymbol{\psi}\mathbf{H}'[\mathbf{H}\boldsymbol{\psi}\mathbf{H}']^{-1}\mathbf{H}\boldsymbol{\psi}\} \\ \delta^2 &= [\mathbf{z} - \mathbf{Hm}_x]'[\mathbf{H}\boldsymbol{\psi}\mathbf{H}']^{-1}[\mathbf{z} - \mathbf{Hm}_x] \end{aligned} \tag{5.8.3}$$

assuming that $(\mathbf{H}\boldsymbol{\psi}\mathbf{H}')^{-1}$ exists. In this case there is no free parameter, and the true estimate set is the ellipsoid, $\Omega_{\text{est}} = \Omega_{\hat{x}}$. $\mathbf{\Sigma}^{-1}$ does not exist.

A Fisher model with perfect observation could be viewed as a combination of a Fisher and an unknown-but-bounded model such as

$$\mathbf{z}_1 = \mathbf{H}_1\mathbf{x} + \mathbf{v}_1$$

$$E\{\mathbf{v}_1\mathbf{v}_1'\} = R$$

$$\mathbf{x} \in \Omega_x = \{\mathbf{x}\colon \mathbf{H}_2\mathbf{x} = \mathbf{z}_2\}$$

5.9 Nonoptimum Estimators

The preceding sections have concentrated on optimum estimators for different models and definitions of optimum. In many applications, nonoptimum estimators may be used for economic reasons or because the model is not known perfectly (see Section 5.10). The analysis of such nonoptimum estimators is now discussed.

Consider a Bayesian model $\mathbf{z} = \mathbf{Hx} + \mathbf{v}$, where

$$E\{\mathbf{x}\} = \mathbf{m}_x \qquad E\{\mathbf{v}\} = \mathbf{0}$$
$$E\{(\mathbf{x} - \mathbf{m}_x)(\mathbf{x} - \mathbf{m}_x)'\} = \boldsymbol{\psi}$$
$$E\{\mathbf{vv}'\} = \mathbf{R}$$
$$E\{\mathbf{xv}'\} = \mathbf{0}$$

Assume that an estimator of the form

$$\hat{\mathbf{x}} = \mathbf{Wz} + \mathbf{W}_0 \tag{5.9.1}$$

is to be used, where the estimator matrix $\mathbf{W}$ and the vector $\mathbf{W}_0$ are not optimum. Define

$\mathbf{b}$: bias error in estimate

$$\mathbf{b} = \mathbf{m}_x - E\{\hat{\mathbf{x}}\}$$

$\boldsymbol{\Sigma}$: estimate error covariance matrix

$$\boldsymbol{\Sigma} = E\{[\mathbf{x} - \hat{\mathbf{x}} - \mathbf{b}][\mathbf{x} - \hat{\mathbf{x}} - \mathbf{b}]'\}$$

Then it follows that

$$\boxed{\begin{aligned} \mathbf{b} &= [\mathbf{I} - \mathbf{WH}]\mathbf{m}_x - \mathbf{W}_0 \\ \boldsymbol{\Sigma} &= [\mathbf{I} - \mathbf{WH}]\boldsymbol{\psi}[\mathbf{I} - \mathbf{WH}]' + \mathbf{WRW}' \end{aligned}} \tag{5.9.2}$$

The "size" of $\mathbf{b}$ and the difference between the $\boldsymbol{\Sigma}$ of (5.9.2) and the $\boldsymbol{\Sigma}$ of, say, (5.1.1) (the minimum possible estimate error covariance matrix) provide a measure of how much extra uncertainty is introduced by using a non-optimum estimator.

Consider a Fisher model $\mathbf{z} = \mathbf{Hx} + \mathbf{v}$, where $\mathbf{x}$ is completely unknown and

$$E\{\mathbf{v}\} = \mathbf{0}$$
$$E\{\mathbf{vv}'\} = \mathbf{R}$$

Consider an estimator such as (5.9.1). Define the estimate bias error $\mathbf{b}$

$$\mathbf{b} = \mathbf{x} - E\{\hat{\mathbf{x}}\}$$

Then

$$\boxed{\begin{aligned} \mathbf{b} &= [\mathbf{I} - \mathbf{WH}]\mathbf{x} - \mathbf{W}_0 \\ \boldsymbol{\Sigma} &= \mathbf{WRW}' \end{aligned}} \tag{5.9.3}$$

EXAMPLE: Consider the Fisher model

$$\begin{pmatrix} z_1 \\ z_2 \end{pmatrix} = \begin{pmatrix} 1 \\ 1 \end{pmatrix} x + \begin{pmatrix} v_1 \\ v_2 \end{pmatrix}$$

$$\mathbf{R} = \begin{pmatrix} 1 & 0 \\ 0 & 4 \end{pmatrix}$$

$$x\text{: unknown}$$

Assume that the following reasonable estimator is chosen:

$$\hat{x} = \mathbf{Wz}$$

$$\mathbf{W} = [\tfrac{1}{2} \quad \tfrac{1}{2}]$$

Then from (5.9.3)

$$b = \left\{1 - [\tfrac{1}{2} \quad \tfrac{1}{2}]\begin{bmatrix} 1 \\ 1 \end{bmatrix}\right\}x = 0$$

$$\Sigma = [\tfrac{1}{2} \quad \tfrac{1}{2}]\begin{bmatrix} 1 & 0 \\ 0 & 4 \end{bmatrix}\begin{bmatrix} \tfrac{1}{2} \\ \tfrac{1}{2} \end{bmatrix} = \tfrac{5}{4}$$

For this case, it was shown in Section 5.2 that the optimum estimator $\hat{x}_{\text{opt}}$ and its error variances Σ_{opt} are

$$\hat{x}_{\text{opt}} = \tfrac{4}{5}z_1 + \tfrac{1}{5}z_2$$

$$\Sigma_{\text{opt}} = \tfrac{4}{5}$$

Comparison of (5.9.2) and (5.9.3) shows that for the same estimator a Bayesian model may yield a larger $\mathbf{\Sigma}$ than the Fisher model. However, great care is needed in comparing (5.9.2) and (5.9.3), as Bayesian and Fisher models are basically different and the definitions of the two **b**s are not equivalent. In a Bayesian model the bias of (5.9.2) is finite, while in the Fisher model the bias of (5.9.3) depends on **x**. Since **x** is completely unknown in a Fisher model, the bias error of (5.9.3) can be arbitrarily large (unless $\mathbf{WH} = \mathbf{I}$).

Consider an unknown-but-bounded model. In this case the estimate is actually a set rather than a vector as in (5.9.1). A nonoptimum estimator should have the property that it always contains the true state **x**. There are various ways that a nonoptimum estimator can be defined, but details are not given here. Note that the bounding ellipsoid is actually a nonoptimum estimator in the sense that it is usually larger than the minimum possible estimate set.

5.10 Modeling Errors

One of the main reasons nonoptimum estimators are used is because the model is not known perfectly.

Consider a Fisher model where the true model is

$$
\begin{aligned}
\mathbf{z} &= \mathbf{H}\mathbf{x} + \mathbf{v} \\
E\{\mathbf{v}\} &= \mathbf{0} \\
E\{\mathbf{v}\mathbf{v}'\} &= \mathbf{R} \\
\mathbf{x}&: \text{unknown}
\end{aligned}
\tag{5.10.1}
$$

Assume that the values of **H** and/or **R** are not known precisely. The observation **z** may be processed using an "optimum" estimator which is designed using the assumed but faulty model

$$
\begin{aligned}
\mathbf{z} &= \tilde{\mathbf{H}}\mathbf{x} + \mathbf{v} \\
E\{\mathbf{v}\} &= 0 \\
E\{\mathbf{v}\mathbf{v}'\} &= \tilde{\mathbf{R}} \\
\mathbf{x}&: \text{unknown}
\end{aligned}
\tag{5.10.2}
$$

so the estimator is

$$
\begin{aligned}
\hat{\mathbf{x}} &= \mathbf{W}\mathbf{z} \\
\mathbf{W} &= [\tilde{\mathbf{H}}\tilde{\mathbf{R}}^{-1}\tilde{\mathbf{H}}']^{-1}\tilde{\mathbf{H}}\tilde{\mathbf{R}}^{-1}
\end{aligned}
\tag{5.10.3}
$$

The ideas of Section 5.9 can be used to see what effects the modeling errors $\mathbf{H} - \tilde{\mathbf{H}}$ and $\mathbf{R} - \tilde{\mathbf{R}}$ have on performance. This approach obviously also applies to Bayesian models.

Consider a Fisher or Bayesian model. An interesting question is, Are modeling errors in **H** or in **R** (and $\boldsymbol{\psi}$) more important? In other words, which is more important, structural models (**H**) or uncertainty models ($\mathbf{R}, \boldsymbol{\psi}$)? Obviously an exact answer depends on the specific problem. However, a good general rule of thumb is that *structural modeling errors are more important than statistical modeling errors*. This rule is based primarily on experience. However, to illustrate the rule, consider a Fisher model such as (5.10.1) where the structure **H** is known (so $\tilde{\mathbf{H}} = \mathbf{H}$) while the covariance matrix **R** is not known perfectly (so $\tilde{\mathbf{R}} \neq \mathbf{R}$). Then combining (5.10.3) with (5.9.3) gives $\mathbf{b} = \mathbf{0}$ so that the unbiased property of the Fisher estimator remains valid even when $\tilde{\mathbf{R}}$ is completely different from **R**.

Now consider an unknown-but-bounded model. Let the true model be

$$
\begin{aligned}
\mathbf{z} &= \mathbf{H}\mathbf{x} + \mathbf{v} \\
\mathbf{v} \in \boldsymbol{\Omega}_v &= \{\mathbf{v}: \mathbf{v}'\mathbf{R}^{-1}\mathbf{v} \leq 1\} \\
\mathbf{x}&: \text{completely unknown}
\end{aligned}
$$

and let the assumed faulty model be

$$
\begin{aligned}
\mathbf{z} &= \tilde{\mathbf{H}}\mathbf{x} + \mathbf{v} \\
\mathbf{v} \in \boldsymbol{\Omega}_v &= \{\mathbf{v}: \mathbf{v}'\tilde{\mathbf{R}}^{-1}\mathbf{v} \leq 1\} \\
\mathbf{x}&: \text{completely unknown}
\end{aligned}
$$

Assume that $\mathbf{H} = \tilde{\mathbf{H}}$. Then if $\mathbf{R} < \tilde{\mathbf{R}}$, it follows that the resulting estimate set will simply be too large. However, if $\mathbf{R} > \tilde{\mathbf{R}}$, the resulting estimate set may not actually contain the true value of $\mathbf{x}$. Similarly, if $\tilde{\mathbf{H}} \neq \mathbf{H}$, the estimate set may not contain the true $\mathbf{x}$. Thus an unknown-but-bounded model is in some sense more sensitive to modeling errors than a Fisher or Bayesian model. This is not surprising. Set models yield precise statements; i.e., $\mathbf{x}$ is contained in some set. Fisher and Bayesian models yield only best estimates with no absolute guarantee that the true value is close to the estimate. Note that for an unknown-but-bounded model with $\tilde{\mathbf{H}} \neq \mathbf{H}$ or $\tilde{\mathbf{R}} < \mathbf{R}$ the resulting estimate set may disappear; i.e., there may be no values of $\mathbf{x}$ that are consistent with both the observation $\mathbf{z}$ and the assumed models $\tilde{\mathbf{H}}$ and $\tilde{\mathbf{R}}$. If this happens, it is obvious that a modeling error exists; i.e., the modeling error is "detected." This redeeming feature is discussed in more detail in Part III.

5.11 Computational Aspects

The estimators of this chapter are defined in terms of matrix and vector operations and hence are "easy" to implement on a digital computer using standard subroutines. However, there are various computational aspects which should be mentioned.

A "typical" estimator of this chapter has the form

$$\hat{\mathbf{x}} = [\mathbf{H}'\mathbf{R}^{-1}\mathbf{H}]^{-1}\mathbf{H}'\mathbf{R}^{-1}\mathbf{z}$$

or

$$\hat{\mathbf{x}} = \boldsymbol{\Sigma}\boldsymbol{\chi}$$
$$\boldsymbol{\Sigma} = \boldsymbol{\mathcal{I}}^{-1}$$
$$\boldsymbol{\mathcal{I}} = \mathbf{H}'\mathbf{R}^{-1}\mathbf{H}$$
$$\boldsymbol{\chi} = \mathbf{H}'\mathbf{R}^{-1}\mathbf{z}$$

Hence a straightforward computer implementation would be

1. Calculate $\boldsymbol{\chi}$ and $\boldsymbol{\mathcal{I}}$.
2. Invert $\boldsymbol{\mathcal{I}}$ to get $\boldsymbol{\Sigma}$.
3. Calculate $\hat{\mathbf{x}} = \boldsymbol{\Sigma}\boldsymbol{\chi}$.

However, a computationally more efficient and accurate procedure is

1. Calculate $\boldsymbol{\chi}$ and $\boldsymbol{\mathcal{I}}$.
2. Solve the linear equation $\boldsymbol{\mathcal{I}}\hat{\mathbf{x}} = \boldsymbol{\chi}$, for $\hat{\mathbf{x}}$ using special techniques for positive definite matrices such as "square root factorization."

It often seems in practice that $\boldsymbol{\mathcal{I}}$ and hence $\boldsymbol{\Sigma}$ is *ill-conditioned*, which means numerical accuracy problems arise due to lack of enough significant figures in $\mathcal{J}$ and/or roundoff errors. After several "bad experiences," the author has made it a general rule to *always* assume that $\mathcal{J}$ is ill-conditioned until it is proved otherwise. A good practical way to check the numerical validity of a $\mathbf{\Sigma}$ obtained by numerically inverting $\mathcal{J}$ is

1. Define

$$\tilde{\mathcal{J}} = \mathcal{J} + \mathbf{D}$$

$\mathbf{D}$: small random perturbating positive definite matrix

$$\mathcal{J} \gg \mathbf{D}$$

2. Compute numerically

$$\mathbf{\Sigma} = \mathcal{J}^{-1}$$
$$\tilde{\mathbf{\Sigma}} = [\tilde{\mathcal{J}}]^{-1}$$

3. If $\mathbf{\Sigma} \approx \tilde{\mathbf{\Sigma}}$, then one can trust $\mathbf{\Sigma}$.

When $\mathcal{J}$ is ill-conditioned and it is difficult to obtain $\mathbf{\Sigma}$, the advantages of computing $\hat{\mathbf{x}}$ by solving $\mathcal{J}\hat{\mathbf{x}} = \boldsymbol{\chi}$ for $\hat{\mathbf{x}}$ become even greater.

5.12 Discussion

An awful lot of words and matrix manipulations were used in this chapter to derive a lot of equations which look very similar. For example, the equation

$$\hat{\mathbf{x}} = \{\mathbf{H}'\mathbf{R}^{-1}\mathbf{H}\}^{-1}\mathbf{H}'\mathbf{R}^{-1}\mathbf{z} \tag{5.12.1}$$

and its variations appeared many times. However, the reader must not view this chapter as an academic exercise in which the author demonstrates his cleverness by rederiving the same result many different ways. Each derivation starts off from a different set of physical assumptions (models for uncertainty) and proceeds along different paths (definition of "best"). Thus, although the final equations are all roughly the same in appearance, each proof constitutes a separate result of value in its own right. Even more important, the physical interpretation to be given to the results depends heavily on the model. For example,

Bayesian, $\hat{\mathbf{x}}$: estimate of value of random vector
$\mathbf{\Sigma}$: covariance matrix
Fisher, $\hat{\mathbf{x}}$: estimate of unknown constant vector
$\mathbf{\Sigma}$: covariance matrix
Weighted least squares, $\hat{\mathbf{x}}$: "reasonable" estimate
Unknown-but-bounded, $\hat{\mathbf{x}}$: center of ellipsoid
$\mathbf{\Sigma}$: defines size and shape of ellipsoid

In spite of the many roads leading to equations such as (5.12.1), it cannot be concluded that such equations apply to all static linear system estimation problems. There exist other methods of handling and modeling uncertainty which can yield different results. If the reader studies the various methods and derivations of this chapter, he will see various points where "reasonable" assumptions were made not only because they were reasonable but also because they lead to equations such as (5.12.1). Examples are the unbiased constraint, the minimization of variance, the use of least squares (instead of, say, least fourths), and the use of ellipsoids. However, the assumptions are reasonable, and the resulting equations are easy to implement on a computer. In many practical problems, such criteria are sufficient.

Figure 5.12.1 is an attempt to summarize some of the ideas presented in this chapter. Much of the terminology of Fig. 5.12.1 has not been discussed, but it is included here to help the reader wade through the related literature. Section 11.6 contains a more general but related overview.

	Models for Uncertainty		
	Noise $\mathbf{v}$	State $\mathbf{x}$	Reasons for Choosing Estimate
Bayesian	Random vector	Random vector	Gaussian: 1) Conditional expectation 2) Regression General: 1) Wide sense conditional expectation 2) Projection 3) Least-squares linear regression 4) Minimum variance linear estimate (unbiased)
Fisher	Random vector	Unknown	Gaussian: Maximum likelihood General: Minimum variance unbiased linear estimate
Weighted least squares	No model		Minimize weighted quadratic distance between $\mathbf{z}$ and $\mathbf{Hx}$
Unknown-but-bounded	Element of set	Element of set	General: Estimate is smallest set which contains $\mathbf{x}$ Ellipsoid: If desired, center of ellipsoid can be viewed as a point estimate.

Figure 5.12.1. Comparison of different approaches and jargon: $\mathbf{z} = \mathbf{Hx} + \mathbf{v}$.

Historical Notes and References: Chapter 5

A complete exposition on the history of estimation theory is beyond the scope of this book. However, the names of Gauss and Legendre have to be mentioned. Gauss is usually given credit for weighted least squares, although Legendre's work also occurred at the same time.

Sections 5.1 to 5.4

The basic ideas are widely available in many forms and many different methods of proof are used. See Book Lists 2, 4, and 5 on Mathematical Statistics, Stochastic Systems, and Communications. Fisher concepts are based on the paper by R. A. Fisher [1] and its subsequent developments. Much has been written on the philosophical differences between Bayesian and Fisher models and weighted-least-squares estimators. Such discussions were not given here as it is the author's opinion that it is up to the reader to decide for his own particular problem which approach to use.

For material on pseudoinverses, see Penrose [1], [2]. See Magness and McQuise [1] for comparison of minimum variance estimators.

Section 5.6

See discussion of Section 6.6.

Section 5.11

See also discussion of Section 6.14. Forsythe and Moler [1] discuss square root factorization which is also called Cholesky's method.

Exercises

Problems 5.1–5.5 are based on the following electrical network of one battery and two resistors of known values $r_1 = 1$ and $r_2 = 2$ and two voltmeters: The meter readings z_1 and z_2 are modeled as

$$z_j = \text{true value} + \text{uncertainty}, \qquad j = 1, 2$$

The battery voltage V is uncertain. For readers who do not know (or have forgotten) how networks behave, perfect meter readings (no observation uncertainty) would be

$$z_1 = \frac{Vr_1}{r_1 + r_2}$$

$$z_2 = \frac{Vr_2}{r_1 + r_2}$$

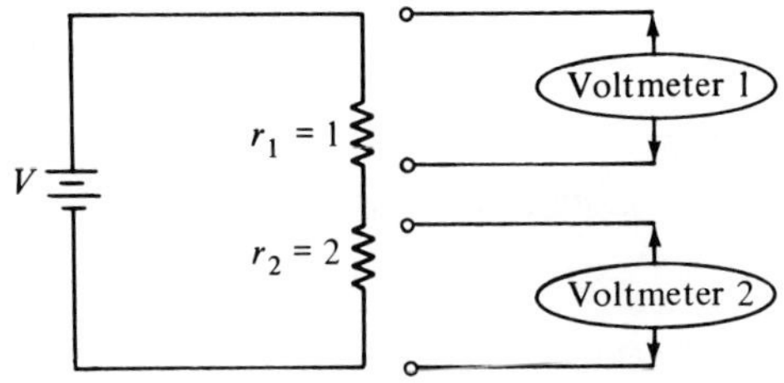

Figure P5.1

5.1. Assume that the meter uncertainties can be modeled as zero mean, uncorrelated random variables with variances $R_{11} = 1$ and $R_{22} = 9$. Assume that no a priori knowledge on V is available.

a. Express a "best" estimate $\hat{V}$ of V as a function of z_1 and z_2.
b. What is the variance of the error in this best estimate?
c. An unbiased estimate of V could be obtained by

$$\hat{V} = z_1 + z_2$$

What is the variance of the error in this estimate?

5.2. Assume that the meter uncertainties can be modeled as zero mean, independent random variables with variance $R_{11} = 1$ and $R_{22} = 9$. For the two following cases, find equations for the best estimate of V and the resulting error variance.

a. A third voltmeter is placed across the battery to yield

$$z_3 = V + \text{uncertainty}$$

where the uncertainty is a zero mean, random variable with variance $R_{33} = 4$.

b. The battery was drawn at random from a large collection of batteries whose average voltage is 8 volts with a standard derivation of $\sqrt{R_{33}} = 2$.

5.3. Assume that no models for the uncertainties in the meter readings and battery voltage are available. Assume that you therefore exercise your engineering judgment and decide to estimate V by the principle of "least $2m$ths," which is as follows: Given z_k, $k = 1, \ldots, K$, $\hat{x}$ is value of x which minimizes

$$J(x) = \sum_{k=1}^{k} [z_k - H_k x]^{2m}$$

a. What is the equation for the estimated battery voltage?
b. What happens as $m \longrightarrow \infty$?

5.4. Assume that V is completely unknown. Assume that the meter reading uncertainties are bounded so that they lie in the rectangular box

$$|\text{uncertainty in 1st meter}| \leq 1$$

$$|\text{uncertainty in 2nd meter}| \leq 3$$

a. Find Ω, the smallest set in which V must lie after the meter readings have been made.
b. Specify one set of meter readings which determine the value of V exactly.

5.5. Assume that the meter uncertainties are such that

$$|\text{uncertainty in } j\text{th meter}| < 2, \qquad j = 1, 2$$

Assume that engineering judgment says that this "box" can be approximated by a circle of radius $\sqrt{2}$. Assume that the battery was drawn from a collection of batteries whose voltages ranged from 6 to 10 volts.

a. Using ellipsoid bounds, find the "smallest" set which must contain V after the meter readings are made.
b. What is the "best" choice of the free parameter ρ if a value must be chosen before an observation?

5.6. Assume that it is desired to estimate the value of some zero mean random variable x using the best linear combination of samples of three other zero mean random variables: $z(n)$, $n = 1, 2, 3$. Prove that the following two "reasonable-appearing" statements are actually false:
a. If $E\{xz(1)\} \neq 0$, then $z(1)$ is always part of the best estimate of x.
b. If $E\{xz(1)\} = 0$, then $z(1)$ is never of use in estimating x.
Hint: Look for simple counterexamples.

5.7. Consider the zero mean Bayesian model

$$\mathbf{z} = \mathbf{Hx} + \mathbf{v}$$

$$E\{\mathbf{x}\} = 0 \qquad E\{\mathbf{v}\} = \mathbf{0} \qquad E\{\mathbf{xv}'\} = \mathbf{0}$$

$$E\{\mathbf{xx}'\} = \boldsymbol{\psi} \qquad E\{\mathbf{vv}'\} = \mathbf{R}$$

Assume that $\mathbf{v}$ denotes observation errors made by some meters. Assume that you are the meter's manufacturer and that you want to use the observation $\mathbf{z}$ to obtain an estimate of $\mathbf{v}$. What equations would you use to define the estimate $\hat{\mathbf{v}}$?

5.8. Consider the zero mean Bayesian model

$$\mathbf{z} = \mathbf{Hx} + \mathbf{v}$$

Define for arbitrary $\mathbf{W}$

$$\hat{\mathbf{x}} = \mathbf{Wz}$$

$$J(\mathbf{W}) = E\{[\mathbf{x} - \hat{\mathbf{x}}]'\mathbf{C}[\mathbf{x} - \hat{\mathbf{x}}]\}$$

$$\mathbf{C}\text{: arbitrary positive definite matrix}$$

View $J(\mathbf{W})$ is a scalar function of the matrix $\mathbf{W}$. Determine the particular $\mathbf{W}$ which minimizes $J(\mathbf{W})$; i.e., which satisfies

$$\frac{\partial J(\mathbf{W})}{\partial \mathbf{W}} = \mathbf{0}$$

(See Appendix C.) Show that the minimizing $\mathbf{W}$ is independent of $\mathbf{C}$. *Comment:* This is an alternative way to prove (5.1.1) in the zero mean case.

5.9. Consider a Fisher model where $\mathbf{v}$ is Gaussian:

$$\mathbf{z} = \mathbf{Hx} + \mathbf{v}$$

$$\mathbf{x}\text{: unknown}$$

$$\mathbf{v} \text{ is } N(\mathbf{0}, \mathbf{R})$$

Let $p(\mathbf{z} : \mathbf{x})$ denote the probability density of $\mathbf{z}$, where the dependence of the density on the value of the vector $\mathbf{x}$ is explicitly included in the notation. Assume that an observation of $\mathbf{z}$ is made and let $\mathbf{z}_{\text{actual}}$ denote the value of this observation. The function $p(\mathbf{z}_{\text{actual}} : \mathbf{x})$, viewed as a function of $\mathbf{x}$, is called the likelihood function. The maximum likelihood estimate of $\mathbf{x}$ is defined to be the value of $\mathbf{x}$ that maximizes $p(\mathbf{z} : \mathbf{x})$ for $\mathbf{z} = \mathbf{z}_{\text{actual}}$. Show that if $\hat{\mathbf{x}}$ maximizes $p(\mathbf{z} : \mathbf{x})$, then $\hat{\mathbf{x}}$ is given by (5.2.1). See also Chapter 11.

5.10. Derive (5.1.2) starting with (5.1.1).

5.11. Derive (5.1.3) and (5.1.4).

5.12. Derive (5.6.3) from (5.6.2) by starting with (5.6.3) and working backward.

5.13. Consider the K ellipsoids

$$\mathbf{\Omega}_k = \{\mathbf{x} : [\mathbf{x} - \mathbf{c}_k]'\mathbf{\Gamma}_k^{-1}[\mathbf{x} - \mathbf{c}_k] \leq 1\}, \qquad k = 1, \ldots, K$$

Let $\mathbf{\Omega}_{\text{int}}$ denote the intersection of these K ellipsoids so that

$$\begin{aligned}\mathbf{\Omega}_{\text{int}} &= \mathbf{\Omega}_1 \cap \mathbf{\Omega}_2 \cdots \cap \mathbf{\Omega}_k \\ &= \{\mathbf{x} : [\mathbf{x} - \mathbf{c}_x]'\mathbf{\Gamma}_k^{-1}[\mathbf{x} - \mathbf{c}_k] \leq 1,\ k = 1, \ldots, K\}\end{aligned}$$

Show that

$$\begin{aligned}\mathbf{\Omega}_b &\supset \mathbf{\Omega}_{\text{int}} \\ \mathbf{\Omega}_b &= \{\mathbf{x} : [\mathbf{x} - \mathbf{c}_b]'\mathbf{\Gamma}_b^{-1}[\mathbf{x} - \mathbf{c}_b] \leq 1\} \\ \mathbf{c}_b &= \tilde{\mathbf{\Gamma}}_b\left(\sum_{k=1}^{K} \gamma_k \mathbf{\Gamma}_k^{-1}\mathbf{c}_k\right) \\ \mathbf{\Gamma}_b^{-1} &= \left(\frac{1}{1 - \delta^2}\right)\tilde{\mathbf{\Gamma}}_b^{-1} \\ \tilde{\mathbf{\Gamma}}_b^{-1} &= \sum_{k=1}^{K} \gamma_k \mathbf{\Gamma}_k^{-1} \\ \delta^2 &= \sum_{k=1}^{K} \gamma_k \mathbf{c}_k'\mathbf{\Gamma}_k^{-1}\mathbf{c}_k - \mathbf{c}_b'\tilde{\mathbf{\Gamma}}_b^{-1}\mathbf{c}_b \\ \sum_{k=1}^{K} \gamma_k &\leq 1 \\ \gamma_k &> 0\end{aligned}$$

Show that in the special case $K = 2$

$$\delta^2 = [\mathbf{c}_1 - \mathbf{c}_2]'[\gamma_1^{-1}\mathbf{\Gamma}_1 + \gamma_2^{-1}\mathbf{\Gamma}_2]^{-1}[\mathbf{c}_1 - \mathbf{c}_2]$$

Hint: Expand

$$\begin{aligned}\sum_{k=1}^{K} \gamma_k[\mathbf{x} - \mathbf{c}_k]'\mathbf{\Gamma}_k^{-1}[\mathbf{x} - \mathbf{c}_k] &= \mathbf{x}'\left[\sum_{k=1}^{K} \gamma_k \mathbf{\Gamma}_k^{-1}\right]\mathbf{x} \\ &\quad - 2\mathbf{x}'\left[\sum_{k=1}^{K} \gamma_k \mathbf{\Gamma}_k^{-1}\right]\mathbf{c}_k + \sum_{k=1}^{K} \gamma_k \mathbf{c}_k'\mathbf{\Gamma}_k^{-1}\mathbf{c}_k\end{aligned}$$

and "complete the square":

$$\sum_{k=1}^{K} \gamma_k[\mathbf{x} - \mathbf{c}_k]'\mathbf{\Gamma}_k^{-1}[\mathbf{x} - \mathbf{c}_k] = [\mathbf{x} - \mathbf{c}_b]'\mathbf{\Gamma}_b^{-1}[\mathbf{x} - \mathbf{c}_b] + \delta^2$$

6

Estimation: Discrete-Time Linear Dynamic Systems

A discrete-time linear system model is (for time increment between samples $\Delta = 1$)

$$\mathbf{x}(n+1) = \mathbf{\Phi}(n)\mathbf{x}(n) + \mathbf{G}(n)\mathbf{w}(n)$$
$$\mathbf{z}(n) = \mathbf{H}(n)\mathbf{x}(n) + \mathbf{v}(n), \qquad n = 1, \ldots$$

$\mathbf{x}(n)$	state, a K_1 vector	
$\mathbf{z}(n)$	observation, a K_2 vector	(6.1)
$\mathbf{v}(n)$	white observation uncertainty, a K_2 vector	
$\mathbf{w}(n)$	white system driving uncertainty, a K_3 vector	
$\mathbf{x}(0)$	initial condition which may be uncertain	
n	time	

The matrices, $\mathbf{H}(n)$, $\mathbf{\Phi}(n)$, and $\mathbf{G}(n)$, are assumed to be known functions of time n. The problem to be considered is how to use the observations up to time n_2, $\mathbf{z}(1) \cdots \mathbf{z}(n_2)$, to estimate the state $\mathbf{x}(n_1)$ at some time n_1.

This estimation problem is obviously related to the static case discussed in Chapter 5. The model is now much more complicated but the state space-white process structures of (6.1), enables the present estimation problem to be solved by merely combining the ideas of Chapters 4 and 5. This emphasizes the importance of Chapter 3, which discussed how a wide variety of "natural" mathematical models could be put in the form of (6.1). Hence this chapter should not be read unless Chapters 3, 4, and 5 are already understood.

A wide range of physical (or social or economic, etc.) problems can be formulated as an estimation problem which can be solved using a model such as (6.1). A few examples (highly oversimplified) are summarized in Fig. 6.1. Two comments on Fig. 6.1 are in order. First, in all cases an estimate of the state $\mathbf{x}(n)$ itself is not of prime interest, but it turns out that an estimate of the state enables one to estimate the desired quantity. Second, if the author really knew how to handle the stock market problem, he wouldn't have written this book.

The following notation is used:

$\hat{\mathbf{x}}(n_1 \mid n_2)$ "best" estimate of state $\mathbf{x}(n_1)$ using observations $\mathbf{z}(1) \cdots \mathbf{z}(n_2)$

and for Bayesian and Fisher models

$\mathbf{\Sigma}(n_1 \mid n_2)$ covariance matrix of the estimate errors $\mathbf{x}(n_1) - \hat{\mathbf{x}}(n_1 \mid n_2)$

The three types of state estimation problems to be considered are

$n_1 = n_2$ filtering
$n_1 > n_2$ prediction
$1 \leq n_1 < n_2$ smoothing

In Section 6.1, a general matrix manipulation approach is discussed which reduces all models to the static case model of Chapter 5. In Sections 6.2 and 6.3, the Bayesian case, where $\mathbf{x}(0)$, $\mathbf{v}(n)$, and $\mathbf{w}(n)$ are modeled stochastically, is analyzed. The resulting equations are often called the Kalman or Kalman-Bucy filter. Fisher models and weighted least squares are discussed in Sections 6.4 and 6.5. In Section 6.6, estimators based on unknown-but-bounded models are discussed. In Section 6.7, the concept of finite memory estimation is presented. Smoothing is discussed in Section 6.8. The limiting case of perfect observations is discussed in Section 6.9. Weighting patterns are discussed in Section 6.10. The analysis and design of nonoptimum estimators are discussed in Sections 6.11 and 6.12. Modeling errors are considered in Section 6.13. Some problems associated with actual implementation on a digital computer are discussed in Section 6.14. Section 6.15 considers the common problem of sampled data systems where discrete-time observations are made on a continuous-time system. Section 6.16 considers an important area of application, curve fitting. The concept of asymptotic efficiency is briefly discussed in Section 6.17. The role of the white process assumption is discussed in Section 6.18. A final discussion is given in Section 6.19.

6.1 Matrix Manipulation

The dynamic system model (6.1) can be reduced to the static case model by defining suitable matrices. The desired estimation equations can then be found by using the ideas of Chapter 5 and matrix manipulation.

Problem	Observations $\mathbf{z}(n)$ Are	$\mathbf{v}(n)$ Represents	$\mathbf{w}(n)$ Represents	$\mathbf{\Phi}(n)$ Models	Desired Quantity
Radar Tracking an Airplane	Range, Elevation and Azimuth	Radar Errors	Wind Gusts and Pilot's Maneuvers	Airplane Dynamics	Prediction of Future Position of Airplane
Load Forecasting for Electric Power System	Total Demand in Megawatts	Meter Errors (Usually ≈ 0)	Randomness of People's Desires for Power	Dependence of Load Demand on Time of Day, Month, etc.	Prediction of Future Demand in Megawatts
Detection of Underground Nuclear Explosions	Output of Array of Seismometers	Noise Caused by Ocean Waves, Rail-road Trains, etc.	"Signal" from Explosion	Propagation of Signals Through Earth	Shape of Signal from Explosion
Making Money on Stock Market	Stock Prices	$v(n) = 0$	Uncertainty in Economy and People's Buying Whims	Dynamics of Economy	Is Stock Going Up or Down?

Figure 6.1. Oversimplified examples of state estimation problems.

Define

$$\mathbf{z}_N = \begin{bmatrix} \mathbf{z}(1) \\ \cdot \\ \cdot \\ \cdot \\ \mathbf{z}(N) \end{bmatrix} \qquad \mathbf{v}_N = \begin{bmatrix} \mathbf{v}(1) \\ \cdot \\ \cdot \\ \cdot \\ \mathbf{v}(N) \end{bmatrix} \qquad \mathbf{w}_N = \begin{bmatrix} \mathbf{w}(0) \\ \cdot \\ \cdot \\ \cdot \\ \mathbf{w}(N-1) \end{bmatrix}$$

Then it follows from (6.1) that there exist matrices $\mathbf{A}_N$ and $\mathbf{B}_N$ such that

$$\begin{aligned} \mathbf{z}_N &= \mathbf{A}_N \mathbf{x}(N) + \tilde{\mathbf{v}}_N \\ \tilde{\mathbf{v}}_N &= \mathbf{B}_N \mathbf{w}_N + \mathbf{v}_N \end{aligned} \tag{6.1.1}$$

Now (6.1.1) is in the form of a static system model, so $\mathbf{x}(N)$ can be estimated from $\mathbf{z}(1) \cdots \mathbf{z}(N)$ to yield $\hat{\mathbf{x}}(N|N)$ by combining (6.1.1) with the results of Chapter 5 (for any method of uncertainty modeling).

EXAMPLE: Consider

$$\begin{aligned} x(n+1) &= \Phi x(n) + w(n) \\ z(n) &= x(n) + v(n), \qquad n = 1, 2, 3 \end{aligned}$$

Then

$$\begin{bmatrix} z(1) \\ z(2) \\ z(3) \end{bmatrix} = \begin{bmatrix} \Phi \\ \Phi^2 \\ \Phi^3 \end{bmatrix} x(0) + \begin{bmatrix} 1 & 0 & 0 \\ \Phi & 1 & 0 \\ \Phi^2 & \Phi & 1 \end{bmatrix} \begin{bmatrix} w(0) \\ w(1) \\ w(2) \end{bmatrix} + \begin{bmatrix} v(1) \\ v(2) \\ v(3) \end{bmatrix}$$

or in the form of (6.1.1),

$$\begin{bmatrix} z(1) \\ z(2) \\ z(3) \end{bmatrix} = \begin{bmatrix} \Phi^{-2} \\ \Phi^{-1} \\ 1 \end{bmatrix} x(3) + \begin{bmatrix} -\Phi^{-1} & -\Phi^{-2} \\ 0 & -\Phi^{-1} \\ 0 & 0 \end{bmatrix} \begin{bmatrix} w(1) \\ w(2) \end{bmatrix} + \begin{bmatrix} v(1) \\ v(2) \\ v(3) \end{bmatrix}$$

The above matrix manipulation approach is not pursued in this book because the required manipulations are tedious and yield little insight (at least to the author). However, it is a very general technique, which can prove useful in special cases. Furthermore, the fact that (6.1) can be viewed as a static system model is used in certain later conceptual arguments.

6.2 Bayesian: Prediction and Filtering, Basic Theory

In the Bayesian model, stochastic-probabilistic models are used for all the uncertainties. Thus for (6.1), $\mathbf{x}(n+1) = \mathbf{\Phi}(n)\mathbf{x}(n) + \mathbf{G}(n)\mathbf{w}(n)$, $\mathbf{z}(n) = \mathbf{H}(n)\mathbf{x}(n) + \mathbf{v}(n)$, and the $\mathbf{x}(0)$, $\mathbf{v}(n)$, and $\mathbf{w}(n)$ are modeled as zero mean uncorrelated random vectors with

$$E\{\mathbf{v}(n_1)\mathbf{v}'(n_2)\} = \begin{cases} \mathbf{R}(n_1), & n_1 = n_2 \\ \mathbf{0}, & n_1 \neq n_2 \end{cases}$$

$$E\{\mathbf{w}(n_1)\mathbf{w}'(n_2)\} = \begin{cases} \mathbf{Q}(n_1), & n_1 = n_2 \\ \mathbf{0}, & n_1 \neq n_2 \end{cases}$$

$$\mathbf{R}(n_1) > \mathbf{0}$$

$$E\{\mathbf{x}(0)\mathbf{x}'(0)\} = \boldsymbol{\psi}$$

The case of perfect observations, $\mathbf{R} = \mathbf{0}$, is discussed in Section 6.9.

To simplify some of the notation, define

$$\mathbf{z}_N = \begin{bmatrix} \mathbf{z}(1) \\ \cdot \\ \cdot \\ \cdot \\ \mathbf{z}(N) \end{bmatrix} \tag{6.2.1}$$

so $\mathbf{z}_N$ is the vector containing all observations up to time N. Also define

$$\boldsymbol{\delta}_x(N\,|\,m) = \mathbf{x}(N) - \hat{\mathbf{x}}(N\,|\,m) \tag{6.2.2}$$

so $\boldsymbol{\delta}_x(N\,|\,m)$ is the error in the estimate $\hat{\mathbf{x}}(N\,|\,m)$. Usually $\hat{\mathbf{x}}(N\,|\,m)$ is an unbiased estimate ($E\{\hat{\mathbf{x}}(N\,|\,m)\} = E\{\mathbf{x}(N)\}$) so the error covariance matrix is

$$\boldsymbol{\Sigma}(N\,|\,m) = E\{\boldsymbol{\delta}_x(N\,|\,m)\boldsymbol{\delta}'_x(N\,|\,m)\}$$

6.2.1 Prediction

The basic *one-step* prediction problem is to calculate $\hat{\mathbf{x}}(N+1\,|\,N)$, the estimate of $\mathbf{x}(N+1)$ using $\mathbf{z}(1) \cdots \mathbf{z}(N)$ (or, more succinctly, $\mathbf{z}_N$).

The solution logic proceeds as follows. From the model (6.1)

$$\mathbf{x}(N+1) = \boldsymbol{\Phi}(N)\mathbf{x}(N) + \mathbf{G}(N)\mathbf{w}(N)$$

Assume that $\hat{\mathbf{x}}(N\,|\,N)$, the best estimate of $\mathbf{x}(N)$ for $\mathbf{z}_N$, is available, say, by using the ideas of Section 6.1 so the formulas of Section 5.1 are used. The global optimality of the estimators of Section 5.1 proves that the best estimate of $\boldsymbol{\Phi}(N)\mathbf{x}(N)$ is given by $\boldsymbol{\Phi}(N)\hat{\mathbf{x}}(N\,|\,N)$. In the model, $\mathbf{w}(N)$ is a white process so no "information" on it is contained in $\mathbf{z}_N$ and thus the best prediction of $\mathbf{w}(N)$ that can be made from $\mathbf{z}_N$ is its mean value; i.e., $\mathbf{0}$. Thus

$$\boxed{\hat{\mathbf{x}}(N+1\,|\,N) = \boldsymbol{\Phi}(N)\hat{\mathbf{x}}(N\,|\,N)} \tag{6.2.3}$$

Now

$$\boldsymbol{\delta}_x(N+1\,|\,N) = \boldsymbol{\Phi}(N)\boldsymbol{\delta}_x(N\,|\,N) + \mathbf{G}(N)\mathbf{w}(N)$$

so that the error covariance matrix of the one-step prediction is given by

$$\boxed{\boldsymbol{\Sigma}(N+1\,|\,N) = \boldsymbol{\Phi}(N)\boldsymbol{\Sigma}(N\,|\,N)\boldsymbol{\Phi}'(N) + \mathbf{G}(N)\mathbf{Q}(N)\mathbf{G}'(N)} \tag{6.2.4}$$

Note the relationship between (6.2.4) and the Chapter 4 equations for $\boldsymbol{\Gamma}(n)$, the covariance matrix for $\mathbf{x}(n)$.

Extension of the single-step prediction concept to multiple steps yields, $m > 0$,

$$\begin{aligned}
\hat{\mathbf{x}}(N+m+1\,|\,N) &= \boldsymbol{\Phi}(N+m)\cdots\boldsymbol{\Phi}(N)\hat{\mathbf{x}}(N\,|\,N)\\
\hat{\mathbf{x}}(N+m+1\,|\,N) &= \boldsymbol{\Phi}(N+m)\hat{\mathbf{x}}(N+m\,|\,N)\\
\boldsymbol{\Sigma}(N+m+1\,|\,N) &= \boldsymbol{\Phi}(N+m)\boldsymbol{\Sigma}(N+m\,|\,N)\boldsymbol{\Phi}'(N+m)\\
&\quad + \mathbf{G}(N+m)\mathbf{Q}(N+m)\mathbf{G}'(N+m)
\end{aligned}$$

6.2.2 Filtering

The desired form of the filtering solution is a difference equation (recursive relationship) expressing $\hat{\mathbf{x}}(N+1\,|\,N+1)$ in terms of $\hat{\mathbf{x}}(N\,|\,N)$ and $\mathbf{z}(N+1)$.

The logic which yields the desired equations can be summarized in the following steps:

1. Assume that $\hat{\mathbf{x}}(N+1\,|\,N+1)$ is to be calculated from just $\hat{\mathbf{x}}(N\,|\,N)$ and $\mathbf{z}(N+1)$.
2. Use the one-step prediction logic to change the problem to calculating $\hat{\mathbf{x}}(N+1\,|\,N+1)$ from $\hat{\mathbf{x}}(N+1\,|\,N)$ and $\mathbf{z}(N+1)$.
3. Solve a Fisher estimation problem where $\hat{\mathbf{x}}(N+1\,|\,N)$ and $\mathbf{z}(N+1)$ are considered to be two independent observations on an unknown vector $\mathbf{x}(N+1)$.

At first glance, this logic seems rather ridiculous as it involves using Fisher estimation on a Bayesian problem; i.e., $\mathbf{x}(N+1)$ is a random vector, not an unknown vector. The key step lies in interpreting $\hat{\mathbf{x}}(N+1\,|\,N)$ as a vector which contains all the "information" in $\mathbf{z}_N$ and the probabilistic structure about $\mathbf{x}(N+1)$. Thus, to combine $\hat{\mathbf{x}}(N+1\,|\,N)$ and $\mathbf{z}(N+1)$, $\mathbf{x}(N+1)$ is considered to be completely unknown, as all probabilistic structure is contained in $\hat{\mathbf{x}}(N+1\,|\,N)$. This heuristic argument provides a "feel" for the equations which will be exploited several times in later discussions.

This argument can be made rigorous by combining the dichotomy of Section 5.7 with the matrix manipulations of Section 6.1. By matrix manipulation

$$\mathbf{z}_N = \boldsymbol{\Lambda}_{N+1}\mathbf{x}(N+1) + \tilde{\mathbf{v}}(N)$$

$$\mathbf{z}(N+1) = \mathbf{Hx}(N+1) + \mathbf{v}(N+1)$$

where $\tilde{\mathbf{v}}(N)$ is some complicated random vector, which is, however, uncorrelated with $\mathbf{v}(N+1)$. Given this formulation, the dichotomy of Section 5.7 justifies the basic logic. The logic is summarized in Fig. 6.2.1, which is intended to be compared with Fig. 5.7.2.

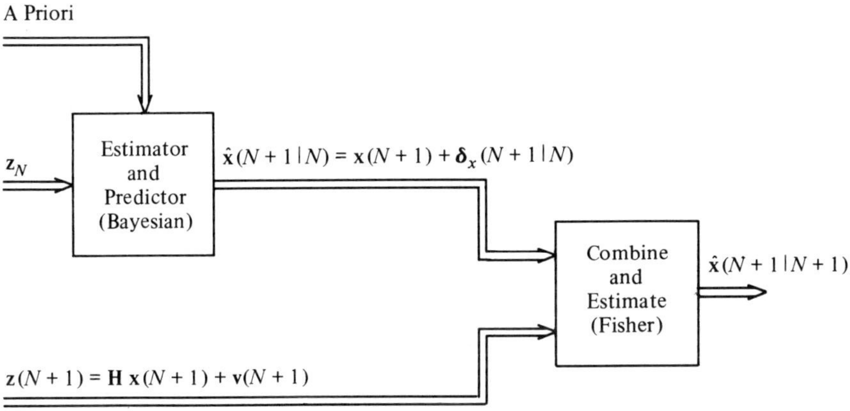

Figure 6.2.1 Basis of argument.

The detailed manipulations are now given. Define

$$\tilde{\mathbf{z}} = \tilde{\mathbf{H}}\mathbf{x}(N+1) + \tilde{\mathbf{v}}$$

$$\tilde{\mathbf{z}} = \begin{bmatrix} \mathbf{z}(N+1) \\ \hat{\mathbf{x}}(N+1 \mid N) \end{bmatrix}$$

$$\tilde{\mathbf{H}} = \begin{bmatrix} \mathbf{H}(N+1) \\ \mathbf{I} \end{bmatrix}$$

$$\tilde{\mathbf{v}} = \begin{bmatrix} \mathbf{v}(N+1) \\ -\boldsymbol{\delta}_x(N+1 \mid N) \end{bmatrix}$$

$$\boldsymbol{\delta}_x(N+1 \mid N) = \mathbf{x}(N+1) - \hat{\mathbf{x}}(N+1 \mid N)$$

This is viewed as a Fisher estimation formulation involving the unknown vector $\mathbf{x}(N+1)$. Define

$$\tilde{\mathbf{R}} = E\{\tilde{\mathbf{v}}\tilde{\mathbf{v}}'\} = \begin{bmatrix} \mathbf{R}(N+1) & 0 \\ 0 & \boldsymbol{\Sigma}(N+1 \mid N) \end{bmatrix}$$

Then from Section 5.2,

$$\hat{\mathbf{x}}(N+1 \mid N+1) = [\tilde{\mathbf{H}}'\tilde{\mathbf{R}}^{-1}\tilde{\mathbf{H}}]^{-1}\tilde{\mathbf{H}}'\tilde{\mathbf{R}}^{-1}\tilde{\mathbf{z}}$$

or after substituting using (6.2.3) and (6.2.4) and manipulating,

$$\begin{aligned}
\hat{\mathbf{x}}(N+1|N+1) &= \mathbf{\Sigma}(N+1|N+1) \\
&\quad \times \{\mathbf{H}'(N+1)\mathbf{R}^{-1}(N+1)\mathbf{z}(N+1) \\
&\quad + \mathbf{\Sigma}^{-1}(N+1|N)\mathbf{\Phi}(N)\hat{\mathbf{x}}(N|N)\} \\
\mathbf{\Sigma}(N+1|N) &= \mathbf{\Phi}(N)\mathbf{\Sigma}(N|N)\mathbf{\Phi}'(N) + \mathbf{G}(N)\mathbf{Q}(N)\mathbf{G}'(N) \\
\mathbf{\Sigma}(N+1|N+1) &= [\mathbf{H}'(N+1)\mathbf{R}^{-1}(N+1)\mathbf{H}(N+1) \\
&\quad + \mathbf{\Sigma}^{-1}(N+1|N)]^{-1} \\
\mathbf{\Sigma}(0|0) &= \boldsymbol{\psi} \\
\hat{\mathbf{x}}(0|0) &= \mathbf{0}
\end{aligned} \tag{6.2.5}$$

Thus (6.2.5) is an equation for the filter estimate $\hat{\mathbf{x}}(N|N)$ of $\mathbf{x}(N)$ given $\mathbf{z}(1) \cdots \mathbf{z}(N)$ for the Bayesian model version of (6.1).

In (6.2.5), the estimator is viewed as the vector difference equation (recursion) which specifies the estimate $\hat{\mathbf{x}}(N|N)$, $N = 1, \ldots$. The matrix difference equation for $\mathbf{\Sigma}(N|N)$ has to be solved to obtain the estimator. The matrix difference equation also provides the model for the uncertainty in the error $\mathbf{x}(N) - \hat{\mathbf{x}}(N|N)$, as

$$\begin{aligned}
E\{[\mathbf{x}(N) - \hat{\mathbf{x}}(N|N)][\mathbf{x}(N) - \hat{\mathbf{x}}(N|N)']\} &= \mathbf{\Sigma}(N|N) \\
E\{\hat{\mathbf{x}}(N|N)\} &= E\{\mathbf{x}(N)\}
\end{aligned} \tag{6.2.6}$$

$\mathbf{\Sigma}(N|N)$ is the error covariance matrix. In this book the matrix difference equation for $\mathbf{\Sigma}(N|N)$ is called a matrix Riccati equation (discrete time). This terminology is not necessarily standard.

Use of the matrix identity (A.3) in Appendix A and some manipulation shows that (6.2.5) can be rewritten in the form

$$\begin{aligned}
\hat{\mathbf{x}}(N+1|N+1) &= \mathbf{\Phi}(N)\hat{\mathbf{x}}(N|N) + \mathbf{K}(N+1) \\
&\quad \times \{\mathbf{z}(N+1) - \mathbf{H}(N+1)\mathbf{\Phi}(N)\hat{\mathbf{x}}(N|N)\} \\
\mathbf{K}(N+1) &= \mathbf{\Sigma}(N+1|N)\mathbf{H}'(N+1)[\mathbf{R}(N+1) \\
&\quad + \mathbf{H}(N+1)\mathbf{\Sigma}(N+1|N)\mathbf{H}'(N+1)]^{-1} \\
\hat{\mathbf{x}}(0|0) &= \mathbf{0} \\
\mathbf{\Sigma}(N+1|N), \mathbf{\Sigma}(N+1|N+1) &\qquad \text{as in (6.2.5)}
\end{aligned} \tag{6.2.7}$$

6.3 Bayesian: Interpretation, Alternative Forms, Properties

The basic relationships of Section 6.2 can be expressed in terms of many different-looking difference equations and viewed in many different lights.

6.3.1 Predict, Observe, Predict, etc.

Implementation of (6.2.5) can be interpreted as a two-step cycle:

1. Predict $\mathbf{x}(N+1)$ from $\hat{\mathbf{x}}(N \mid N)$.
2. Estimate $\mathbf{x}(N+1)$ from $\hat{\mathbf{x}}(N+1 \mid N)$ and $\mathbf{z}(N+1)$.

Step 1 is closely related to the analysis in Chapter 4. Step 2 is based on the ideas of Chapter 5. Thus the two-step cycle could be viewed as

1. Predict using equations of Chapter 4.
2. Estimate using equations of Chapter 5.

6.3.2 Linear Feedback or Differential Corrections

Equation (6.2.5) or (6.2.7) can be rewritten by matrix manipulation as [see Appendix A, (A.3)]

$$\boxed{\begin{aligned}
\hat{\mathbf{x}}(N+1 \mid N+1) &= \mathbf{\Phi}(N)\hat{\mathbf{x}}(N \mid N) + \mathbf{K}(N+1) \\
&\quad \times \{\mathbf{z}(N+1) - \mathbf{H}(N+1)\mathbf{\Phi}(N)\hat{\mathbf{x}}(N \mid N)\} \\
\mathbf{K}(N+1) &= \mathbf{\Sigma}(N+1 \mid N+1)\mathbf{H}'(N+1)\mathbf{R}^{-1}(N+1) \\
\mathbf{\Sigma}(N+1 \mid N+1) &= \mathbf{\Sigma}(N+1 \mid N) - \mathbf{\Sigma}(N+1 \mid N)\mathbf{H}'(N+1) \\
&\quad \times [\mathbf{R}(N+1) + \mathbf{H}(N+1)\mathbf{\Sigma}(N+1 \mid N) \\
&\quad \times (\mathbf{H}'(N+1)]^{-1}\mathbf{H}(N+1)\mathbf{\Sigma}(N+1 \mid N) \\
\mathbf{\Sigma}(N+1 \mid N) &= \mathbf{\Phi}(N)\mathbf{\Sigma}(N \mid N)\mathbf{\Phi}'(N) + \mathbf{G}(N)\mathbf{Q}(N)\mathbf{G}'(N) \\
\mathbf{\Sigma}(0 \mid 0) &= \boldsymbol{\psi} \\
\hat{\mathbf{x}}(0 \mid 0) &= \mathbf{0}
\end{aligned}} \tag{6.3.1}$$

In the equations for $\hat{\mathbf{x}}$, the second term on the right can be viewed as a *differential correction* based on $\mathbf{z}(N+1)$ to the old estimate (given by the first term). The equation for $\mathbf{\Sigma}(N+1 \mid N+1)$ is another form of the matrix Riccati equation. The second term on the right of this Riccati equation shows how much the error covariance matrix resulting from the old estimate [the first term $\mathbf{\Sigma}(N+1 \mid N)$] is reduced by the new observation. The estimator of (6.3.1) can be viewed as a linear feedback system with time-varying gains $\mathbf{K}(N+1)$. This is illustrated as a block diagram in Fig. 6.3.1, where $\mathbf{H}(N+1)\mathbf{\Phi}(N)\hat{\mathbf{x}}(N \mid N)$ is viewed as $\hat{\mathbf{z}}(N+1 \mid N)$, the estimate of $\mathbf{z}(N+1)$ made from $\hat{\mathbf{x}}(N \mid N)$. Thus the input to the time-varying gain $\mathbf{K}(N+1)$ is $\boldsymbol{\delta}_z(N+1 \mid N)$, defined as in (6.2.2). The basic dynamic model can be expressed as in Fig. 6.3.2. Thus the discrete-time filter of Fig. 6.3.1 can also be viewed as in Fig. 6.3.3. Figure 6.3.3 views the optimum filter as feedback system around a model of the dynamic system where the observations are used to try to keep the model's output $[\hat{\mathbf{x}}(N \mid N)]$ as close to $\mathbf{x}(N)$ as possible.

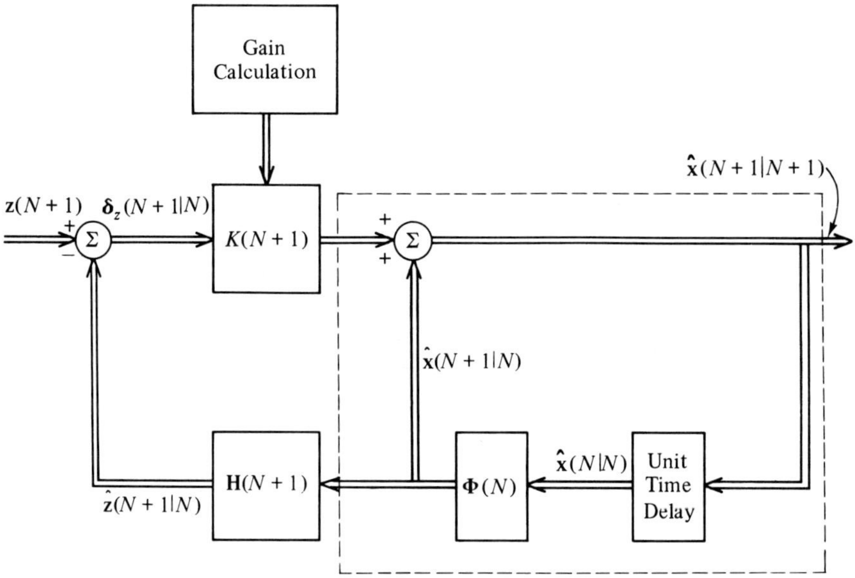

Figure 6.3.1 Bayesian discrete-time filter.

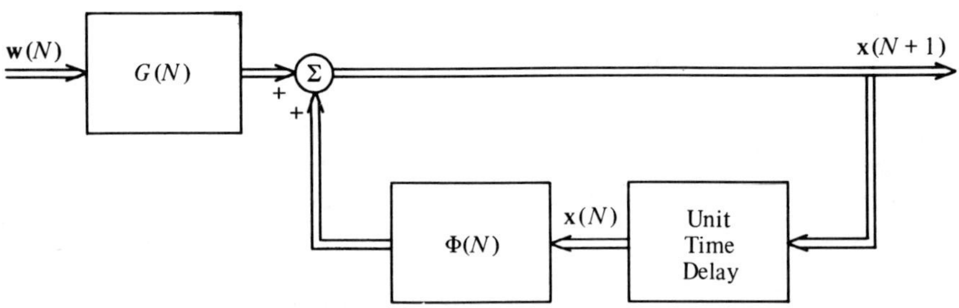

Figure 6.3.2 Discrete-time dynamic system.

This is often a very useful interpretation in practical application and is discussed again in later chapters (Chapter 8 and Chapter 13 on nonlinear models).

6.3.3 Whiteness of Residual Process

An important property of the residuals $\boldsymbol{\delta}_z(N+1|N) = \mathbf{z}(N+1) - \hat{\mathbf{z}}(N+1|N)$ in Fig. 6.3.1 and 6.3.3 is that $\boldsymbol{\delta}_z(N+1|N)$ *is a stochastic*

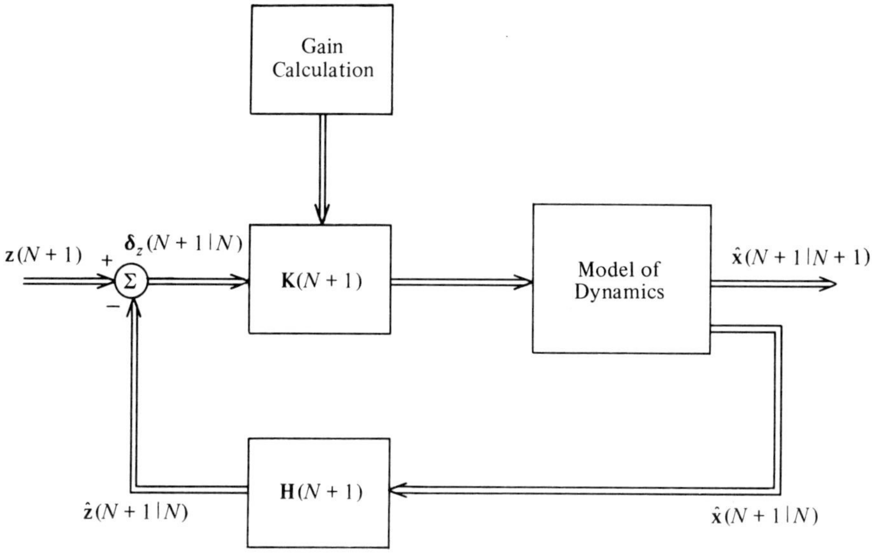

Figure 6.3.3 Bayesian discrete-time filter.

white process; i.e.,

$$E\{\boldsymbol{\delta}_z(N+1\,|\,N)\boldsymbol{\delta}_z'(m+1\,|\,m)\} = \mathbf{0}, \qquad N \neq m$$

This property can be deduced as follows. For the $\mathbf{z} = \mathbf{Hx} + \mathbf{v}$ Bayesian model, (5.1.6) stated that

$$E\{(\mathbf{x} - \hat{\mathbf{x}})\mathbf{z}'\} = \mathbf{0}$$

In the present case, this gives

$$E\{(\mathbf{x}(N+1) - \hat{\mathbf{x}}(N+1\,|\,N))\mathbf{z}_N'\} = \mathbf{0}$$

which, because $E\{\mathbf{v}(N+1)\mathbf{z}_N'\} = \mathbf{0}$, gives

$$E\{(\mathbf{z}(N+1) - \hat{\mathbf{z}}(N+1\,|\,N))\mathbf{z}_N'\} = \mathbf{0}$$

Now $\mathbf{z}(m+1) - \mathbf{z}(m+1\,|\,m)$ can be written as a linear function of $\mathbf{z}_{m+1}$, so for $m < N$,

$$\mathbf{z}(m+1) - \hat{\mathbf{z}}(m+1\,|\,m) = \mathbf{C}\mathbf{z}_N$$

where $\mathbf{C}$ is some complicated matrix. The whiteness property of $\boldsymbol{\delta}_z(N+1\,|\,N)$ follows immediately.

After carrying out the detailed manipulations,

$$\boxed{\begin{aligned} &\boldsymbol{\delta}_z(N+1\,|\,N) = \mathbf{z}(N+1) - \mathbf{H}(N+1)\boldsymbol{\Phi}(N)\hat{\mathbf{x}}(N\,|\,N) \\ &E\{\boldsymbol{\delta}_z(N+1\,|\,N)\boldsymbol{\delta}_z'(m+1\,|\,m)\} \\ &\quad = \begin{cases} \mathbf{0}, & N \neq m \\ \mathbf{R}(N+1) + \mathbf{H}(N+1)\boldsymbol{\Sigma}(N+1\,|\,N)\mathbf{H}'(N+1), & N = m \end{cases} \end{aligned}} \tag{6.3.2}$$

The "whiteness" of $\boldsymbol{\delta}_z(n+1 \mid n)$ is sometimes viewed as a basic property of an optimum estimator. It means the *correction* $\boldsymbol{\delta}_z(N+1 \mid N)$ made using $\mathbf{z}(N+1)$ cannot be predicted (except to be zero) from $\boldsymbol{\delta}_z(N \mid N-1)$, which depends on $z(1) \cdots z(N)$.

6.3.4 Information Flow†

Another rewriting of (6.2.5) yields an interpretation in terms of *information flow*.

Define

$$\mathscr{I}(N \mid N) = \boldsymbol{\Sigma}^{-1}(N \mid N) \qquad \text{information matrix}$$

$$\boldsymbol{\chi}(N \mid N) = \mathscr{I}(N \mid N)\hat{\mathbf{x}}(N \mid N) \qquad \text{information vector}$$

The matrix $\mathscr{I}(N \mid N)$ is also called the *Fisher information matrix* or the *normal matrix*. It can be shown from information theory concepts that $\mathscr{I}(N \mid N)$ is the amount of information contained in $\hat{\mathbf{x}}(N \mid N)$, i.e., in $\mathbf{z}(1) \cdots \mathbf{z}(N)$ about $\mathbf{x}(N)$. The association of the term *information vector* with $\boldsymbol{\chi}(N \mid N)$ is not standard terminology. $\boldsymbol{\chi}(N \mid N)$ is considered to be the actual information in $\hat{\mathbf{x}}(N \mid N)$ [i.e., in $\mathbf{z}(1) \cdots \mathbf{z}(N)$] about $\mathbf{x}(N)$. After a fair amount of manipulation, (6.2.5) can be rewritten as

$$\boxed{\begin{aligned}
\boldsymbol{\chi}(N+1 \mid N+1) &= \mathbf{H}'(N+1)\mathbf{R}^{-1}(N+1)\mathbf{z}(N+1) \\
&\quad + [\mathbf{I} - \boldsymbol{\Xi}(N+1 \mid N)]\boldsymbol{\Phi}'(N)^{-1}\boldsymbol{\chi}(N \mid N) \\
\mathscr{I}(N+1 \mid N+1) &= \mathbf{H}'(N+1)\mathbf{R}^{-1}(N+1)\mathbf{H}(N+1) \\
&\quad + [\mathbf{I} - \boldsymbol{\Xi}(N+1 \mid N)]\boldsymbol{\Phi}'(N)^{-1} \\
&\quad \times \mathscr{I}(N \mid N)\boldsymbol{\Phi}^{-1}(N) \\
\hat{\mathbf{x}}(N \mid N) &= \mathscr{I}^{-1}(N \mid N)\boldsymbol{\chi}(N \mid N) \\
\boldsymbol{\Xi}(N+1 \mid N) &= [\boldsymbol{\Phi}(N)\mathscr{I}^{-1}(N \mid N)\boldsymbol{\Phi}'(N) \\
&\quad + \mathbf{G}(N)\mathbf{Q}(N)\mathbf{G}'(N)]^{-1}[\mathbf{G}(N)\mathbf{Q}(N)\mathbf{G}'(N)] \\
\boldsymbol{\chi}(0 \mid 0) &= \mathbf{0} \\
\mathscr{I}(0 \mid 0) &= \boldsymbol{\psi}^{-1}
\end{aligned}} \tag{6.3.3}$$

The following interpretations can be made from the equation for $\boldsymbol{\chi}(N+1 \mid N+1)$:

1. $\mathbf{H}'(N+1)\mathbf{R}^{-1}(N+1)\mathbf{z}(N+1)$ is the new information contained in $\mathbf{z}(N+1)$ about $\mathbf{x}(N+1)$.
2. $[\mathbf{I} - \boldsymbol{\Xi}(N+1 \mid N)]\boldsymbol{\Phi}'(N)^{-1}\boldsymbol{\chi}(N \mid N)$ is the information contained in $\mathbf{z}(1) \cdots \mathbf{z}(N)$ about $\mathbf{x}(N+1)$.
3. Information is additive.

The following interpretations can be made from the equation (a matrix Riccati equation) for $\mathcal{I}(N+1 \mid N+1)$:

1. $\mathbf{H}'(N+1)\mathbf{R}^{-1}(N+1)\mathbf{H}(N+1)$ is the amount of information contained in $\mathbf{z}(N+1)$ about $\mathbf{x}(N+1)$.
2. $[\mathbf{I} - \mathbf{\Xi}(N+1 \mid N)]\mathbf{\Phi}'(N)^{-1}\mathcal{I}(N \mid N)\mathbf{\Phi}^{-1}(N)$ is the amount of information contained in $\mathbf{z}(1) \cdots \mathbf{z}(N)$ about $\mathbf{x}(N+1)$.
3. Information matrices are additive.

Note that information and information matrices propagate in *inverse time*, i.e., in terms of $\mathbf{\Phi}^{-1}(N)$, which is assumed to exist. $\mathbf{\Xi}(N+1 \mid N)$ is considered to be a measure of how much information is lost by the presence of the uncertain system input, $\mathbf{w}(N)$.

In general, (6.3.3) presents no particular computational advantage over (6.2.5). However, when $\mathbf{Q}(N) = \mathbf{0}$ there may be a great advantage, as then $\mathbf{\Xi}(N+1 \mid N) = \mathbf{0}$, and $\boldsymbol{\chi}(N+1 \mid N+1)$ and $\mathcal{I}(N+1 \mid N+1)$ are given by linear difference equations which require no matrix inversions (except for $\mathbf{R}$ and $\mathbf{\Phi}$). This can be a big help if an estimate $\hat{\mathbf{x}}(N \mid N)$ is wanted only for some final value of N and not for $n = 1, \ldots, N-1$, as only one matrix inversion at time N is needed (see Section 6.14).

6.3.5 Time-Invariant Models and Steady-State Behavior

The preceding equations apply to time-varying models. Simplifications result when the system models are time-invariant; i.e., when the $\mathbf{\Phi}$, $\mathbf{G}$, $\mathbf{H}$, $\mathbf{Q}$, and $\mathbf{R}$ matrices are constants.

Assume that the model values $\mathbf{\Phi}$, $\mathbf{G}$, $\mathbf{H}$, $\mathbf{Q}$, and $\mathbf{R}$ are time-invariant. In general the difference equations for the estimate $\hat{\mathbf{x}}(N \mid N)$ are time-varying because of the time-varying error covariance matrix $\mathbf{\Sigma}(N \mid N)$ [a time-varying gain $\mathbf{K}(N+1)$ in the feedback loop]. However, in many applications $\mathbf{\Sigma}(N+1 \mid N+1)$ converges to a constant matrix for large N; i.e.,

$$\mathbf{\Sigma}(N \mid N) \longrightarrow \mathbf{\Sigma}_\infty \tag{6.3.4}$$

When this occurs, the estimator itself becomes a time-invariant system, the steady-state filter (estimator). Obviously there are many possible forms in which this steady-state estimator can be expressed. To illustrate, from (6.2.5)

$$\boxed{\begin{aligned} \hat{\mathbf{x}}(N+1 \mid N+1) &= \mathbf{\Sigma}_\infty\{\mathbf{H}'\mathbf{R}^{-1}\mathbf{z}(N+1) \\ &\qquad + \mathbf{\Sigma}_\infty^{-1}(1 \mid 0)\mathbf{\Phi}\hat{\mathbf{x}}(N \mid N)\} \\ \mathbf{\Sigma}_\infty &= [\mathbf{H}'\mathbf{R}^{-1}\mathbf{H} + \mathbf{\Sigma}_\infty^{-1}(1 \mid 0)]^{-1} \\ \mathbf{\Sigma}_\infty(1 \mid 0) &= \mathbf{\Phi}\mathbf{\Sigma}_\infty\mathbf{\Phi}' + \mathbf{G}\mathbf{Q}\mathbf{G}' \end{aligned}} \tag{6.3.5}$$

while from (6.3.1)

$$\boxed{\begin{aligned} \hat{\mathbf{x}}(N+1|N+1) &= \mathbf{\Phi}\hat{\mathbf{x}}(N|N) \\ &\quad + \mathbf{K}_\infty\{\mathbf{z}(N+1) - \mathbf{H}\mathbf{\Phi}\hat{\mathbf{x}}(N|N)\} \\ \mathbf{K}_\infty &= \mathbf{\Sigma}_\infty\mathbf{H}'\mathbf{R}^{-1} \\ \mathbf{\Sigma}_\infty &= \mathbf{\Sigma}_\infty(1|0) - \mathbf{\Sigma}_\infty(1|0) \\ &\quad \times \mathbf{H}'[\mathbf{R} + \mathbf{H}\mathbf{\Sigma}_\infty(1|0)\mathbf{H}']^{-1}\mathbf{H}\mathbf{\Sigma}_\infty(1|0) \\ \mathbf{\Sigma}_\infty(1|0) &= \mathbf{\Phi}\mathbf{\Sigma}_\infty\mathbf{\Phi}' + \mathbf{G}\mathbf{Q}\mathbf{G}' \end{aligned}} \tag{6.3.6}$$

In both (6.3.5) and (6.3.6) $\mathbf{\Sigma}_\infty(1|0)$ represents the steady-state, error covariance matrix associated with predicting one time step into the future. Other forms for a steady-state estimator can be obtained from (6.2.7) and (6.3.3). Equation (6.3.5) or (6.3.6) defines a linear, time-invariant system with input $\mathbf{z}(n)$, $n = 1, \ldots, N+1$, and output $\hat{\mathbf{x}}(N+1|N+1)$. Note that if the initial condition matrix $\boldsymbol{\psi}$ is such that

$$\boldsymbol{\psi} = \mathbf{\Sigma}_\infty$$

then (6.3.5) and (6.3.6) hold for all N for time invariant models.

The steady-state filter of (6.3.5) or (6.3.6) is easy to implement once $\mathbf{\Sigma}_\infty$ has been determined. $\mathbf{\Sigma}_\infty$ can be obtained by solving the matrix difference equation for $\mathbf{\Sigma}(N|N)$ until (if) it converges to a steady-state value. Alternatively one can try to find a positive definite solution to an algebraic equation such as [from (6.3.5)]

$$\mathbf{\Sigma}_\infty = \{\mathbf{H}'\mathbf{R}^{-1}\mathbf{H} + [\mathbf{\Phi}\mathbf{\Sigma}_\infty\mathbf{\Phi}' + \mathbf{G}\mathbf{Q}\mathbf{G}']^{-1}\}^{-1} \tag{6.3.7}$$

This leads to the obvious question, What are the conditions under which $\mathbf{\Sigma}(N|N)$ converges to a steady-state value and for which (6.3.7) has a unique positive definite solution? Such conditions exist and are closely related to the concepts of observability and controllability (see Appendix B). In particular it can be shown that

> If the model
>
> $$\begin{aligned} \mathbf{x}(n+1) &= \mathbf{\Phi}\mathbf{x}(n) + \mathbf{G}\mathbf{w}(n) \\ \mathbf{z}(n) &= \mathbf{H}\mathbf{x}(n) \end{aligned} \tag{6.3.8}$$
>
> is controllable and observable and if $\mathbf{Q} > \mathbf{0}$ and $\mathbf{R} > \mathbf{0}$, then $\mathbf{\Sigma}(N|N)$ *converges to a unique positive definite* $\mathbf{\Sigma}_\infty$, *given by* (6.3.7).

If (6.3.7) has a unique, positive definite solution, then $\mathbf{\Sigma}_\infty$ must be independent of the initial error covariance matrix $\boldsymbol{\psi}$.

The following heuristic argument is presented instead of a formal mathematical proof of (6.3.8). If $\mathbf{\Sigma}_\infty$ is to be unique, it must be independent of the initial condition $\boldsymbol{\psi}$. This occurs if the input uncertainty $\mathbf{w}(n)$ "excites" all the states, as then, for large N, the effect of $\mathbf{w}(n)$ has to dominate the initial uncertainty, $\mathbf{x}(0)$. The conditions of a controllable system and $\mathbf{Q} > \mathbf{0}$ guarantee that $\mathbf{w}(n)$ excites all the states. If $\mathbf{R} = \mathbf{0}$ (see Section 6.9) $\mathbf{\Sigma}^{-1}(N|N)$ does not exist, while if the system is not observable, $\mathbf{\Sigma}(N|N)$ may "blow up" (some components go to infinity). Hence the conditions of $\mathbf{R} > \mathbf{0}$ and an observable system prevent $\mathbf{\Sigma}(N|N)$ from going to zero or infinity for large N.

Note that (6.3.8) says nothing about necessary conditions. Note also that the stability of the model $\mathbf{\Phi}$ does not affect the argument.

EXAMPLE: Consider a scalar time-invariant model

$$x(n+1) = \Phi x(n) + w(n)$$

$$z(n) = x(n) + v(n)$$

$$E\{v^2(n)\} = R \qquad E\{w^2(n)\} = Q \qquad E\{x^2(0)\} = \psi$$

From (6.3.5) or (6.3.6)

$$\Sigma(n+1|n+1) = \frac{R\Phi^2\Sigma(n) + RQ}{\Phi^2\Sigma(n) + Q + R}$$

$$\Sigma(0|0) = \psi$$

If steady-state error covariances Σ_∞ exist, they are given by nonnegative solutions to

$$\Sigma = \frac{-[Q + R(1-\Phi^2)] \pm \sqrt{[Q + R(1-\Phi^2)]^2 + 4\Phi^2 RQ}}{2\Phi^2}$$

Consider three cases.

Case 1: If $Q = 1$ and $R = 1$, Φ arbitrary, then

$$\Sigma_\infty = \frac{\Phi^2 - 2 + \sqrt{\Phi^4 + 4}}{2\Phi^2}$$

Case 2: If $Q = 0$, $R = 1$, and $|\Phi| < 1$, then

$$\Sigma_\infty = 0$$

Case 3: If $Q = 0$, $R = 1$, and $|\Phi| > 1$, then

$$\Sigma_\infty = \frac{\Phi^2 - 1}{\Phi^2}$$

or

$$\Sigma_\infty = 0$$

The general behavior of $\Sigma(n|n)$ versus n is sketched in Fig. 6.3.4 for each of the three cases for different initial conditions ψ. The behavior is sketched as a continuous-time function for ease of presentation.

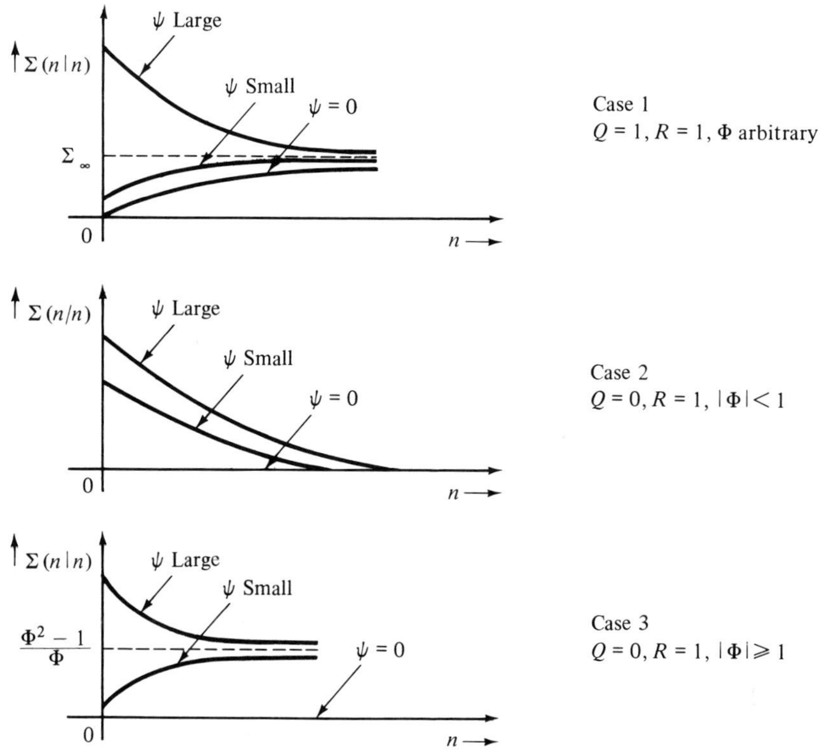

Figure 6.3.4 Behavior of $\Sigma(n|n)$ of example.

6.3.6 Stability

An important question to consider in many applications is, *Is the estimator a stable dynamic system?* The answer is usually yes but may be no. The estimator (6.3.1) can be rewritten as

$$
\begin{aligned}
\hat{\mathbf{x}}(N+1|N+1) &= \tilde{\mathbf{\Phi}}(N)\hat{\mathbf{x}}(N|N) + \mathbf{K}(N+1)\mathbf{z}(N+1) \\
\tilde{\mathbf{\Phi}}(N) &= [\mathbf{I} - \mathbf{K}(N+1)\mathbf{H}(N+1)]\mathbf{\Phi}(N)
\end{aligned}
\tag{6.3.9}
$$

The stability of the dynamic system (6.3.9) depends on the properties of $\tilde{\mathbf{\Phi}}(N)$.

Consider the case of a time-invariant system model so that $\mathbf{\Phi}$, $\mathbf{H}$, $\mathbf{G}$, $\mathbf{Q}$, and $\mathbf{R}$ are all time-invariant. Assume that $\mathbf{\Sigma}(N|N)$ goes to a steady-state value $\mathbf{\Sigma}_\infty$, so (6.3.9) becomes, for large N,

$$
\begin{aligned}
\hat{\mathbf{x}}(N+1|N+1) &= \tilde{\mathbf{\Phi}}\hat{\mathbf{x}}(N|N) + \mathbf{K}_\infty\mathbf{z}(N+1) \\
\tilde{\mathbf{\Phi}} &= [\mathbf{I} - \mathbf{K}_\infty\mathbf{H}]\tilde{\mathbf{\Phi}} \\
\mathbf{K}_\infty &= \mathbf{\Sigma}_\infty\mathbf{H}'\mathbf{R}^{-1}
\end{aligned}
\tag{6.3.10}
$$

Equation (6.3.10) is a stable dynamic system if the eigenvalues of $\tilde{\mathbf{\Phi}}$ are all less than 1 in magnitude. It turns out for this time-invariant case that

The filter (6.3.10) is stable if $\boldsymbol{\psi}^{-1}$ exists and if $\alpha\mathbf{I} \leq \boldsymbol{\Sigma}_\infty \leq \beta\mathbf{I}$, $0 < \alpha \quad \beta < \infty$	(6.3.11)

In other words, the steady-state filter is stable if the steady-state error covariance matrix is a finite matrix of full rank. From (6.3.8) it thus follows that

The filter (6.3.10) is stable if the model is controllable and observable and $\mathbf{Q} > \mathbf{0}$, $\mathbf{R} > \mathbf{0}$.	(6.3.12)

An indirect argument will be used to justify (6.3.11). Formal proofs are also possible. Consider

$$\boldsymbol{\delta}_x(N|N) = \mathbf{x}(N) - \hat{\mathbf{x}}(N|N)$$

It follows that

$$\boldsymbol{\delta}_x(N+1|N+1) = \tilde{\boldsymbol{\Phi}}\boldsymbol{\delta}_x(N|N) + \tilde{\mathbf{w}}(N)$$
$$\tilde{\mathbf{w}}(N) = \mathbf{G}\mathbf{w}(N) - \mathbf{K}_\infty\mathbf{v}(N+1)$$

Now by definition, in steady state

$$E\{\boldsymbol{\delta}_x(N|N)\boldsymbol{\delta}'_x(N|N)\} = \boldsymbol{\Sigma}_\infty$$

It follows that if $\boldsymbol{\Sigma}_\infty \leq \beta\mathbf{I}$, $\beta < \infty$, then $\tilde{\boldsymbol{\Phi}}$ must correspond to a stable system if $\boldsymbol{\psi}^{-1}$ exists ($\boldsymbol{\Sigma}(0|0) = \boldsymbol{\psi}$). To see why the condition $\alpha\mathbf{I} \leq \boldsymbol{\Sigma}_\infty$, $0 < \alpha$, was imposed, assume that $\boldsymbol{\Sigma}_\infty = \mathbf{0}$. In such a case, $\mathbf{K}_\infty = \mathbf{0}$ and $\tilde{\boldsymbol{\Phi}} = \boldsymbol{\Phi}$ so that if the basic model is unstable, so is the filter. Equation (6.3.11) says only that the condition "$\boldsymbol{\Sigma}_\infty$ is finite and of full rank" is sufficient to guarantee stability. Nothing has been said about necessary conditions.

It is important to note that (6.3.11) and (6.3.12) hold even if the model itself is unstable.

More elaborate discussions about stability, which include the time-varying case, can be found in the literature, although the derivations become very tedious. The key conditions are still observability, controllability, and $\mathbf{Q} > \mathbf{0}$, $\mathbf{R} > \mathbf{0}$.

EXAMPLE: Consider a scalar time-invariant version of (6.1):

$$x(n+1) = \Phi x(n) + w(n)$$
$$z(n) = x(n) + v(n)$$
$$E\{v^2(n)\} = R \qquad E\{w^2(n) = Q \qquad E\{x^2(0)\} = \psi$$

Consider (6.3.10). Then from the example of Section 6.3.5

Case 1: If $Q = 1$ and $R = 1$, $\mathbf{\Phi}$ arbitrary, then

$$\tilde{\mathbf{\Phi}} = \frac{\mathbf{\Phi}^2 + 2 - \sqrt{\mathbf{\Phi}^4 + 4}}{\mathbf{\Phi}}$$

so that $|\tilde{\mathbf{\Phi}}| \leq 1$ and the filter is stable, as predicted by (6.3.11) and (6.3.12).

Case 2: If $Q = 0$, $R = 1$, and $|\mathbf{\Phi}| \leq 1$, then

$$\tilde{\mathbf{\Phi}} = \mathbf{\Phi}$$

and the filter is stable. The conditions (6.3.11) and (6.3.12) are not satisfied, but then they are only sufficient, not necessary.

Case 3: If $Q = 0$, $R = 1$, and $|\mathbf{\Phi}| \geq 1$, then there are two $\mathbf{\Sigma}_\infty$ and hence two K_∞ and $\tilde{\mathbf{\Phi}}$, where

$$\tilde{\mathbf{\Phi}} = \frac{1}{\mathbf{\Phi}}, \qquad \text{if } \mathbf{\Sigma}_\infty = \frac{\mathbf{\Phi}^2 - 1}{\mathbf{\Phi}^2}$$

$$\tilde{\mathbf{\Phi}} = \mathbf{\Phi}, \qquad \text{if } \mathbf{\Sigma}_\infty = 0$$

When $\tilde{\mathbf{\Phi}} = 1/\mathbf{\Phi}$, the system is stable, as predicted by (6.3.11) [but not by (6.3.12)]. When $\tilde{\mathbf{\Phi}} = \mathbf{\Phi}$, the system is unstable and neither (6.3.11) or (6.3.12) are satisfied.

6.3.7 Nonzero Mean Inputs and Initial Conditions

Consider

$$\mathbf{x}(n + 1) = \mathbf{\Phi}(n)\mathbf{x}(n) + \mathbf{G}(n)\mathbf{w}(n) + \mathbf{B}(n)\mathbf{u}(n)$$
$$\mathbf{z}(n) = \mathbf{H}(n)\mathbf{x}(n) + \mathbf{v}(n)$$
$$\mathbf{w}(n), \mathbf{v}(n)\text{: as before}$$
$$E\{\mathbf{x}(0)\} = \mathbf{m}$$
$$E\{[\mathbf{x}(0) - \mathbf{m}][\mathbf{x}(0) - \mathbf{m}]'\} = \boldsymbol{\psi}$$
$$\mathbf{u}(n)\text{: known time function}$$

This can be viewed as a Bayesian model (6.1) with nonzero mean input uncertainty and nonzero mean initial conditions. Methods for handling such nonzero mean models were discussed in Chapter 3, Section 3.2. From (6.2.5) the resulting estimator is

$$\hat{\mathbf{x}}(N + 1 \mid N + 1) = \mathbf{\Sigma}(N + 1 \mid N + 1)\{\mathbf{H}'(N + 1)\mathbf{R}^{-1}(N + 1)\mathbf{z}(N + 1) + \mathbf{\Sigma}^{-1}(N + 1 \mid N)[\mathbf{\Phi}(N)\hat{\mathbf{x}}(N \mid N) + \mathbf{B}(N)\mathbf{u}(N)]\}$$
$$\hat{\mathbf{x}}(0 \mid 0) = \mathbf{m}$$

where $\mathbf{\Sigma}$ is as in (6.2.5). In a feedback form, the estimator is

$$\hat{\mathbf{x}}(N + 1 \mid N + 1) = \mathbf{\Phi}(N)\hat{\mathbf{x}}(N \mid N) + \mathbf{B}(N)\mathbf{u}(N) + \mathbf{K}(N + 1) \times \{\mathbf{z}(N + 1) - \mathbf{H}(N + 1)[\mathbf{\Phi}(N)\hat{\mathbf{x}}(N \mid N) + \mathbf{B}(N)\mathbf{u}(N)]\}$$
$$\hat{\mathbf{x}}(0 \mid 0) = \mathbf{m}$$

where the gain $\mathbf{K}$ is as in (6.3.1).

6.3.8 *Effect of Initial Conditions*

The initial conditions for the difference equation defining the estimator for the Bayesian model are

$$\hat{\mathbf{x}}(0 \mid 0) = E\{\mathbf{x}(0)\}$$
$$\mathbf{\Sigma}(0 \mid 0) = \boldsymbol{\psi}$$

One of the important aspects of the discussions of Sections 6.3.4 and 6.3.5 is that if the filter becomes a time-invariant system that is stable and/or if the filter is stable, the effects of the initial conditions decay with time.

6.3.9 *Discussion of Bayesian*

The error covariance matrix $\mathbf{\Sigma}(n \mid n)$ is obtained by solving a matrix Riccati equation (discrete-time version) which does not depend on the observations. The estimate $\hat{\mathbf{x}}(n \mid n)$ is obtained from a linear [in the observation $\mathbf{z}(n)$] vector difference equation. This recursive form wherein $\hat{\mathbf{x}}(N+1 \mid N+1)$ is expressed in terms of $\hat{\mathbf{x}}(N \mid N)$, $\mathbf{\Sigma}(N+1 \mid N+1)$, and $\mathbf{z}(N+1)$ [and $\mathbf{\Sigma}(N+1 \mid N+1)$ in terms of $\mathbf{\Sigma}(N \mid N)$] is extremely useful as the past $\mathbf{z}(n)$, $n = 1, \ldots,$ do not have to be stored. In fact, if the basic argument of Section 6.2.2 did not actually yield the optimum, the equations would still often be used because of their practical value, i.e., ease of implementation.

The global optimality of $\hat{\mathbf{x}}(N \mid N)$ has not been explicitly demonstrated, but it holds by direct analogy with Chapter 5. To be specific $\hat{\mathbf{x}}(N \mid N)$ yields the minimum error covariance matrix of any linear estimator in the sense that

$$\mathbf{\Sigma}(N \mid N) \leq \tilde{\mathbf{\Sigma}}(N \mid N)$$

where $\tilde{\mathbf{\Sigma}}(N \mid N)$ is the error covariance matrix for any other linear estimator. All the discussions of Section 5.1 apply here, so they are not repeated. For example, to estimate $\mathbf{y} = \mathbf{C}\mathbf{x}(N)$ from $\mathbf{z}(1) \cdots \mathbf{z}(N)$, one uses $\hat{\mathbf{y}} = \mathbf{C}\hat{\mathbf{x}}(N \mid N)$.

Consider (6.2.5), (6.2.7), (6.3.1), and (6.3.3). They all look different, but they are equivalent provided that the indicated matrix inversions exist. No one form is best, and they all have their place both in theory and practice. In Section 6.14, the choice of algorithm for actual computation is discussed.

The interpretation of $\mathbf{\Sigma}(N \mid N)$ as the estimate error covariance matrix is, as in Chapter 5, important, as it provides an error analysis which determines how well the estimator performs. Since $\mathbf{\Sigma}(n \mid n)$ can be computed before the observations are made, this error analysis can be used to study what observation accuracies are needed, etc.

The fact that $\mathbf{\Sigma}(N \mid N)$ can be precomputed is also important in implementation. For example, in the feedback from of (6.3.1) and Fig. 6.3.1, the gains $\mathbf{K}(N+1)$ can be precomputed and stored.

It is important not to let all the manipulations obscure the fact that no new estimation theory beyond that of Chapter 5 was employed. The important aspects of the discussion are the nature and interpretation of the resulting difference equations.

The basic difference equations for the Bayesian model estimator are often called the *Kalman filter* or the *Kalman-Bucy filter*.

At no point in the discussion has a Gaussian assumption been used. If the model is specialized to the Gaussian case, the estimates, become *conditional expectation* estimates. This follows directly from the discussion of Section 5.1.2 and the fact that only the estimation theory of Chapter 5 was used. When $\mathbf{x}(n)$, $\mathbf{w}(n)$ and $\mathbf{v}(n)$ are all jointly normal, then $\hat{\mathbf{x}}(n \mid n)$ is also Gaussian, and

$$\mathbf{x}(n) - \hat{\mathbf{x}}(n \mid n) \text{ is } N[\mathbf{0}, \boldsymbol{\Sigma}(n \mid n)]$$

See Exercise 6.14.

6.4 Fisher: Filtering and Prediction

Two Fisher-type models for (6.1) are

1. $\mathbf{x}(0)$ is completely unknown, and $\mathbf{w}(n)$ is a white stochastic process.
2. $\mathbf{x}(0)$ is completely unknown, and $\mathbf{w}(n)$ is completely unknown,

where in both cases $\mathbf{v}(n)$ is modeled as a white stochastic process. To expedite the derivations, direct proofs such as in Section 6.2 are not presented. Instead, the arguments of Section 5.3 are used; i.e., Fisher estimators are obtained from the Bayesian estimators by letting the appropriate covariance matrices go to infinity. This approach is not very elegant, but it is rigorous in the sense that the matrix manipulation concepts of Section 6.1 can be used to turn the models of (6.1) into the form discussed in Sections 5.1–5.3, where it was proved that the limit procedure does yield the "right" equations defining the estimators.

6.4.1 Unknown Initial Conditions

Consider (6.1), where $\mathbf{x}(0)$ is completely unknown but $\mathbf{w}(n)$ and $\mathbf{v}(n)$ are white stochastic processes. The Bayesian model results are converted into this Fisher model by letting $\boldsymbol{\psi} \longrightarrow \infty \mathbf{I}$. From (6.2.5), it is seen that this implies that $\boldsymbol{\Sigma}^{-1}(1 \mid 0) = \mathbf{0}$, so that in the limit

$$\boldsymbol{\Sigma}^{-1}(1 \mid 1) = [\mathbf{H}'(1)\mathbf{R}^{-1}(1)\mathbf{H}(1)]$$

If $\mathbf{H}'\mathbf{R}^{-1}\mathbf{H}$ actually has an inverse, $\boldsymbol{\Sigma}(1 \mid 1)$ exists and the difference equations (6.2.5), (6.3.1), etc., can be applied directly. However, if the dimension of $\mathbf{z}$ is less than the dimension of $\mathbf{x}$(i.e., if $K_1 > K_2$), $\mathbf{H}'\mathbf{R}^{-1}\mathbf{H}$ is singular and $\boldsymbol{\Sigma}(1 \mid 1)$ does not exist. This difficulty is solved by waiting until m observations have been made and forming a single vector from $\mathbf{z}(1) \cdots \mathbf{z}(m)$, which is then treated as one observation. If m is large enough, this yields a $\mathbf{\Sigma}(m \mid m)$, which exists (usually the condition $mK_2 > K_1$ suffices). For $n > m$, (6.2.5), etc., can then be applied directly. Thus the only problem that arises when $\mathbf{x}(0)$ is completely unknown is the "start up" of the estimation equations. The details when $K_1 > K_2$ are messy but straightforward. The matrix manipulation ideas of Section 6.1 are helpful.

6.4.2 Unknown Input†

Consider (6.1), where $\mathbf{x}(0)$ and $\mathbf{w}(n)$, $n = 1, \ldots,$ are completely unknown, while $\mathbf{v}(n)$ is a white stochastic process. Such a problem can make physical sense if the dimension of the observation $\mathbf{z}$ exceeds the dimension of the completely unknown input $\mathbf{w}(n)$.

If a Fisher model with unknown input makes sense, a solution can be found from the Bayesian results by letting $\mathbf{Q}(n) \longrightarrow \infty\, \mathbf{I}$. To illustrate how to proceed, consider the equation for $\mathbf{\Sigma}(N+1 \mid N)$ of (6.2.5). Using the matrix identity (A.3) in Appendix A, this can be rewritten as

$$\begin{aligned}\mathbf{\Sigma}^{-1}(N+1 \mid N) = {} & [\mathbf{\Phi}(N)\mathbf{\Sigma}(N \mid N)\mathbf{\Phi}'(N)]^{-1} - [\mathbf{\Phi}(N)\mathbf{\Sigma}(N \mid N)\mathbf{\Phi}'(N)]^{-1}\mathbf{G}(N) \\ & \times \{\mathbf{Q}^{-1}(N) + \mathbf{G}'(N)[\mathbf{\Phi}(N)\mathbf{\Sigma}(N \mid N)\mathbf{\Phi}'(N)]^{-1}\mathbf{G}(N)\}^{-1} \\ & \times \mathbf{G}'(N)[\mathbf{\Phi}(N)\mathbf{\Sigma}(N \mid N)\mathbf{\Phi}'(N)]^{-1}\end{aligned}$$

Thus in the limit $\mathbf{Q}^{-1}(N) = \mathbf{0}$, one possible estimator for the Fisher model with completely unknown inputs is

Use (6.2.5) with

$$\begin{aligned}\mathbf{\Sigma}^{-1}(N+1 \mid N) = {} & [\mathbf{\Phi}(N)\mathbf{\Sigma}(N \mid N)\mathbf{\Phi}'(N)]^{-1} \\ & - [\mathbf{\Phi}(N)\mathbf{\Sigma}(N \mid N)\mathbf{\Phi}'(N)]^{-1}\mathbf{G}(N) \\ & \times \{\mathbf{G}'(N)[\mathbf{\Phi}(N)\mathbf{\Sigma}(N \mid N)\mathbf{\Phi}'(N)]^{-1}\mathbf{G}(N)\}^{-1} \\ & \times \mathbf{G}'(N)[\mathbf{\Phi}(N)\mathbf{\Sigma}(N \mid N)\mathbf{\Phi}'(N)]^{-1}\end{aligned} \qquad (6.4.1)$$

Note that $\mathbf{\Sigma}(N+1 \mid N)$ does not exist since

$$\mathbf{G}'(N)\mathbf{\Sigma}^{-1}(N+1 \mid N) = \mathbf{0}$$
$$\mathbf{\Sigma}^{-1}(N+1 \mid N)\mathbf{G}(N) = \mathbf{0}$$

It is, of course, assumed here that the resulting $\mathbf{\Sigma}(N+1 \mid N+1)$ exists. If it does not exist, more manipulation is required.

The estimate $\hat{\mathbf{x}}(N \mid N)$ given by (6.4.1) is an unbiased, minimum variance estimate (in the sense of Section 5.2). Thus

$$E\{\hat{\mathbf{x}}(N \mid N\} = \mathbf{x}(N)$$
$$E\{[\mathbf{x}(N) - \hat{\mathbf{x}}(N \mid N)][\mathbf{x}(N) - \hat{\mathbf{x}}(N \mid N)]'\} = \mathbf{\Sigma}(N \mid N)$$

where the expectation is over $\mathbf{v}(n)$, $n = 1, \ldots, N$.

In certain applications it is sometimes desired to obtain an estimate of $\mathbf{w}(N)$ itself. Let $\hat{\mathbf{w}}(N|N+1)$ denote the estimate of $\mathbf{w}(N)$ made from $\mathbf{z}(1) \cdots \mathbf{z}(N+1)$. It is obvious from (6.1) that $\mathbf{z}(1) \cdots \mathbf{z}(N)$ provides no information on $\mathbf{w}(N)$, as $\mathbf{w}(N)$ is completely unknown and is "first observed" by $\mathbf{z}(N+1)$. From (6.1), one can write

$$\mathbf{z}(N+1) = \mathbf{H}(N+1)[\mathbf{\Phi}(N)\mathbf{x}(N) + \mathbf{G}(N)\mathbf{w}(N)] + \mathbf{v}(N+1)$$

Employing arguments such as those used in Section 6.2.2, write $\hat{\mathbf{x}}(N|N)$ as $\hat{\mathbf{x}}(N|N) = \mathbf{x}(N) - \boldsymbol{\delta}_x(N|N)$. The estimate $\hat{\mathbf{w}}(N|N+1)$ can be calculated by combining $\mathbf{z}(N+1)$ and $\hat{\mathbf{x}}(N|N)$ into a single vector observation of the "unknown" vector formed by combining $\mathbf{x}(N)$ and $\mathbf{w}(N)$.

The condition that the dimension of $\mathbf{z}$ exceeds the dimension of $\mathbf{w}$ can often occur. A prime example is the case where both the input and output of a system are observed, as discussed in Section 3.2.15.

6.4.3 Discussion of Fisher

Given the filtered estimate $\hat{\mathbf{x}}(N|N)$, a "good" one-step prediction is given by

$$\hat{\mathbf{x}}(N+1|N) = \mathbf{\Phi}(N)\hat{\mathbf{x}}(N|N)$$

The corresponding error covariance matrix $\mathbf{\Sigma}(N+1|N)$, of course, does not exist when the input $\mathbf{w}(n)$ is unknown, as $\mathbf{\Sigma}^{-1}(N+1|N)$ of (6.4.1) does not have an inverse. This is as expected from the nature of the model.

For a time-invariant system model, a steady-state filter may result, just as in the Bayesian case. The discussions of Section 6.3.5 still apply, so they are not repeated.

The derivation of the estimator for the Fisher model as a limit of Bayesian model estimator is based on the concepts of Section 5.3. Section 5.3 also states that the *properties* of $\hat{\mathbf{x}}(N|N)$ for a Fisher model are different from those of $\hat{\mathbf{x}}(N|N)$ for a Bayesian model even in the limit.

The use of the limit argument does lead to an interesting question. Can one simply use the Bayesian model equations with $\boldsymbol{\psi}$ and/or $\mathbf{Q}(n)$ chosen to be very large? In practice the answer may be yes. In fact it is the author's opinion (based on intuition, not experience) that the possibility of obtaining a Fisher model estimator by using Bayesian equations with large $\boldsymbol{\psi}$ and/or $\mathbf{Q}(n)$ should always be considered as an alternative to the explicit equations which were derived. With this approach, it will be necessary to determine "large enough" values for $\boldsymbol{\psi}$ and/or $\mathbf{Q}(n)$ by numerical studies wherein the values of $\boldsymbol{\psi}$ and $\mathbf{Q}(n)$ are increased until the filter's input-output behavior becomes independent of $\boldsymbol{\psi}$ and/or $\mathbf{Q}(n)$.

A Gaussian assumption was not used. If the world is specialized to the Gaussian case, the estimates become *maximum likelihood* estimates (see Section 5.2.2).

6.5 Weighted Least Squares: Filtering

The weighted-least-squares estimation logic of Section 5.4 is now applied to the dynamic case.

6.5.1 General Case

For the static model of Section 5.4, the estimator was defined by,

Choose the value of $\mathbf{x}$ which minimizes

$$J(\mathbf{x}) = [\mathbf{z} - \mathbf{Hx}]'\mathbf{R}^{-1}[\mathbf{z} - \mathbf{Hx}]$$

where $\mathbf{R}$ is a positive definite matrix chosen by engineering judgement.

For the dynamic model of (6.1), the corresponding estimator is defined by

Choose the $\mathbf{x}(N)$ and $\mathbf{w}(n)$, $n = 0, \ldots, N - 1$, which minimize

$$\begin{aligned} J[\mathbf{x}(N), \mathbf{w}(0) \cdots \mathbf{w}(N-1)] &= \sum_{n=1}^{N} [\mathbf{z}(n) - \mathbf{H}(n)\mathbf{x}(n)]'\mathbf{R}^{-1}(n)[\mathbf{z}(n) - \mathbf{H}(n)\mathbf{x}(n)] \\ &+ \sum_{n=0}^{N-1} \mathbf{w}'(n)\mathbf{Q}^{-1}(n)\mathbf{w}(n) + \mathbf{x}'(0)\boldsymbol{\psi}^{-1}\mathbf{x}(0) \end{aligned} \tag{6.5.1}$$

subject to the constraint that

$$\mathbf{x}(n+1) = \mathbf{\Phi}(n)\mathbf{x}(n) + \mathbf{G}(n)\mathbf{w}(n)$$

where $\mathbf{R}(n)$, $\mathbf{Q}(n)$, and $\boldsymbol{\psi}$ are positive definite matrices chosen by engineering judgement.

Let $\hat{\mathbf{x}}(N|N)$ denote the resulting value of $\mathbf{x}(N)$. If one actually performs the minimization using Lagrange multipliers, the resulting equations for $\hat{\mathbf{x}}(N|N)$ are the same as those of Sections 6.2 and 6.3.

6.5.2 Exponential Smoothing

A special case of (6.5.1) is called exponential smoothing. For exponential smoothing, $\mathbf{R}(n)$, $\boldsymbol{\psi}$, and $\mathbf{Q}(n)$ in (6.5.1) are chosen to be

$$\begin{aligned} &\mathbf{R}(n) = e^{-\alpha(N-n)}\,\mathbf{I}, \qquad n = 1, \ldots, N \\ &\mathbf{Q}^{-1}(n) = \mathbf{0} \qquad \boldsymbol{\psi}^{-1} = \mathbf{0} \\ &\alpha > 0 \end{aligned} \tag{6.5.2}$$

so (6.5.1) becomes

Choose $\mathbf{x}(N)$ which minimizes

$$J[\mathbf{x}(N)] = \sum_{n=1}^{N} e^{-\alpha(N-n)}[\mathbf{z}(n) - \mathbf{H}(n)\mathbf{x}(n)]'[\mathbf{z}(n) - \mathbf{H}(n)\mathbf{x}(n)]$$

subject to the constraint

$$\mathbf{x}(n+1) = \mathbf{\Phi}(n)\mathbf{x}(n) \tag{6.5.3}$$

where α is chosen by engineering judgment, $\alpha > 0$.

Let $\hat{\mathbf{x}}(N|N)$ denote the solution to (6.5.3) with $\hat{\mathbf{x}}(n|N)$, $n = 1, \ldots, N$, the resulting backward time solution to $\mathbf{x}(n+1) = \mathbf{\Phi}(n)\mathbf{x}(n)$. The motivation behind (6.5.3) is that $\hat{\mathbf{x}}(n|N)$ is the solution to $\mathbf{x}(n+1) = \mathbf{\Phi}(n)\mathbf{x}(n)$ which "best fits" the observations $\mathbf{z}(n)$, $n = 1, \ldots, N$, where the most recent (i.e., n close to N) observations are given the most weight and the old observations are "weighted out" (or exponentially smoothed out).

The necessary estimator equations follow directly from the results of Sections 6.2 and 6.3. However, since $\mathbf{R}(n)$ now depends on N, some manipulation is required and some of the steps will be outlined.

The easiest place to start is with the information flow formulation of (6.3.3). For simplicity, assume that $\mathbf{H}(n) = \mathbf{H}$ and $\mathbf{\Phi}(n) = \mathbf{\Phi}$. Assume that N is fixed. Then for (6.5.2), (6.3.3) becomes

$$\begin{aligned}
\tilde{\boldsymbol{\chi}}(n|n) &= \mathbf{\Phi}'^{-1}\tilde{\boldsymbol{\chi}}(n-1|n-1) + \mathbf{H}'\mathbf{z}(n)e^{-\alpha(N-n)} \\
\tilde{\mathfrak{J}}(n|n) &= \mathbf{\Phi}'^{-1}\tilde{\mathfrak{J}}(n-1|n-1)\mathbf{\Phi}^{-1} + \mathbf{H}'\mathbf{H}e^{-\alpha(N-n)} \\
\tilde{\boldsymbol{\chi}}(0|0) &= \mathbf{0} \\
\tilde{\mathfrak{J}}(0|0) &= \mathbf{0} \\
n &= 1, \ldots, N
\end{aligned}$$

where superscript tildes have been added to emphasize that N is considered fixed. The desired $\hat{\mathbf{x}}(N|N)$ for fixed N is given by

$$\hat{\mathbf{x}}(N|N) = \tilde{\mathfrak{J}}^{-1}(N|N)\tilde{\boldsymbol{\chi}}(N|N)$$

The following closed-form expressions follow:

$$\tilde{\mathfrak{J}}(N|N) = \sum_{n=1}^{N} [\mathbf{\Phi}']^{n-N}\mathbf{H}'\mathbf{H}[\mathbf{\Phi}]^{n-N}e^{-\alpha(N-n)}$$

$$\tilde{\boldsymbol{\chi}}(N|N) = \sum_{n=1}^{N} [\mathbf{\Phi}']^{n-N}\mathbf{H}'\mathbf{z}(n)e^{-\alpha(N-n)}$$

Now let $\tilde{\mathfrak{J}}(N+1|N+1)$, $\tilde{\boldsymbol{\chi}}(N+1|N+1)$, and $\hat{\mathbf{x}}(N+1|N+1) = \tilde{\mathfrak{J}}^{-1}(N+1|N+1)\tilde{\boldsymbol{\chi}}(N+1|N+1)$ denote the corresponding results for $N+1$. It then follows that

$$\begin{aligned}
\tilde{\mathfrak{J}}(N+1|N+1) &= e^{-\alpha}\mathbf{\Phi}'^{-1}\tilde{\mathfrak{J}}(N|N)\mathbf{\Phi}^{-1} + \mathbf{H}'\mathbf{H} \\
\tilde{\boldsymbol{\chi}}(N+1|N+1) &= e^{-\alpha}\mathbf{\Phi}'^{-1}\tilde{\boldsymbol{\chi}}(N|N) + \mathbf{H}'\mathbf{z}(N+1)
\end{aligned} \tag{6.5.4}$$

Define

$$\tilde{\mathbf{\Sigma}}(N|N) = \tilde{\mathfrak{J}}(N|N)$$

Then after some manipulation

$$\boxed{\begin{aligned}\hat{\mathbf{x}}(N+1|N+1) &= \tilde{\mathbf{\Sigma}}(N+1|N+1)[e^{-\alpha}[\mathbf{\Phi}\tilde{\mathbf{\Sigma}}(N|N)\mathbf{\Phi}']^{-1} \\ &\qquad \times \mathbf{\Phi}\hat{\mathbf{x}}(N|N) + \mathbf{H}'\mathbf{z}(N+1)] \\ \tilde{\mathbf{\Sigma}}(N+1|N+1) &= [e^{-\alpha}[\mathbf{\Phi}\tilde{\mathbf{\Sigma}}(N|N)\mathbf{\Phi}']^{-1} + \mathbf{H}'\mathbf{H}]^{-1} \\ \tilde{\mathbf{\Sigma}}^{-1}(0|0) &= \mathbf{0} \qquad \hat{\mathbf{x}}(0|0) = \mathbf{0}\end{aligned}} \tag{6.5.5}$$

which is in a form analogous to (6.2.5). After more manipulation a feedback form analogous to (6.3.1) can be obtained:

$$\hat{\mathbf{x}}(N+1|N+1) = \mathbf{\Phi}\hat{\mathbf{x}}(N|N) + \mathbf{K}(N+1)[\mathbf{z}(N+1) - \mathbf{H\Phi}\hat{\mathbf{x}}(N|N)]$$
$$K(n+1) = \tilde{\mathbf{\Sigma}}(N+1|N+1)\mathbf{H}'$$

It should be emphasized that $\tilde{\mathbf{\Sigma}}(N+1|N+1)$ in (6.5.5) cannot be interpreted as an error covariance matrix, as exponential smoothing is a weighted-least-squares formulation.

EXAMPLE: Consider a scalar case where (6.5.3) becomes

$$J[x(N)] = \sum_{n=1}^{N} e^{-\alpha(N-n)}[z(n) - x(n)]^2$$
$$x(n+1) = \Phi x(n)$$

Then from (6.5.5)

$$\hat{x}(N+1|N+1) = \Phi\hat{x}(N|N) + K(N+1)[z(N+1) - \Phi\hat{x}(N|N)]$$
$$K(N+1) = \tilde{\Sigma}(N+1|N+1)$$
$$\tilde{\Sigma}(N+1|N+1) = \frac{\Phi^2\tilde{\Sigma}(N|N)}{\Phi^2\tilde{\Sigma}(N|N) + e^{-\alpha}}$$

Assume that steady state is reached so that

$$\tilde{\Sigma}(N+1|N+1) = \tilde{\Sigma}(N|N) = \tilde{\Sigma}_\infty$$

Then

$$\tilde{\Sigma}_\infty = \begin{cases} 0 \\ \dfrac{\Phi^2 - e^{-\alpha}}{\Phi^2} \end{cases}$$

If $e^{-\alpha} < \Phi^2$, then the nonzero $\tilde{\Sigma}_\infty$ will be obtained. It is interesting to compare this example with the example of Section 6.3.5.

The motivation for using exponential smoothing is discussed in Chapter 8; see especially Section 8.2.3. The reader can actually read most of Chapter 8 after finishing this chapter, so a brief jump ahead might be appropriate if motivation is of any interest.

Note that $J[\hat{\mathbf{x}}(N)]$ of (6.5.3) could be generalized to a form more like (6.5.1), i.e., with $\mathbf{Q} \neq \mathbf{0}$. This was not done because it complicates the mathematics and, in the author's opinion, has limited application.

6.6 Unknown-but-Bounded: Filtering

Unknown-but-bounded models are now considered for (6.1)

$$\mathbf{x}(n+1) = \mathbf{\Phi}(n)\mathbf{x}(n) + \mathbf{G}(n)\mathbf{w}(n), \qquad \mathbf{z}(n+1) = \mathbf{H}(n)\mathbf{x}(n) + \mathbf{v}(n)$$

where $\mathbf{w}(n)$ and $\mathbf{v}(n)$ are white unknown-but-bounded processes and $\mathbf{x}(0)$ is an unknown-but-bounded vector such that

$$\begin{aligned} \mathbf{x}(0) &\in \mathbf{\Omega}_x(0) = \{\mathbf{x}: \mathbf{x}'\boldsymbol{\psi}^{-1}\mathbf{x} \leq 1\} \\ \mathbf{v}(n) &\in \mathbf{\Omega}_v(n) = \{\mathbf{v}: \mathbf{v}'\mathbf{R}^{-1}(n)\mathbf{v} \leq 1\}, \qquad \text{all } n \\ \mathbf{w}(n) &\in \mathbf{\Omega}_w(n) = \{\mathbf{w}: \mathbf{w}'\mathbf{Q}^{-1}(n)\mathbf{w} \leq 1\}, \qquad \text{all } n \end{aligned}$$

6.6.1 General Case

Define

$\mathbf{\Omega}_{\text{est}}(n_1 \mid n_2)$	minimum set containing $\mathbf{x}(n_1)$ that can be calculated from $\mathbf{z}(1) \cdots \mathbf{z}(n_2)$

It is desired to find a recursive relation expressing $\mathbf{\Omega}_{\text{est}}(N+1 \mid N+1)$ in terms of $\mathbf{\Omega}_{\text{est}}(N \mid N)$ and $\mathbf{z}(N+1)$.

The Bayesian model yielded results which could be viewed as the two-step cycle: predict, estimate, predict, etc. This same cycle applies to the unknown-but-bounded model.

Consider first the prediction step. Define

$$\begin{aligned} \tilde{\mathbf{\Omega}}_{\text{est}}(N+1 \mid N) &= \{\mathbf{x}: \mathbf{x} = \mathbf{\Phi}(N)\mathbf{x}_1, \mathbf{x}_1 \in \mathbf{\Omega}_{\text{est}}(N \mid N)\} \\ \mathbf{\Omega}_{\mathbf{G}\mathbf{w}}(N) &= \{\mathbf{x}: \mathbf{x} = \mathbf{G}(N)\mathbf{w}, \mathbf{w} \in \mathbf{\Omega}_w(N)\} \end{aligned}$$

Then following the ideas of Chapter 4, the one-step prediction $\mathbf{\Omega}_{\text{est}}(N+1 \mid N)$ is the vector sum of $\tilde{\mathbf{\Omega}}_{\text{est}}(N+1 \mid N)$ and $\mathbf{\Omega}_{\mathbf{G}\mathbf{w}}(N)$ so

$$\boxed{\begin{aligned} \mathbf{\Omega}_{\text{est}}(N+1 \mid N) = \{\mathbf{x}: \mathbf{x} = \mathbf{x}_1 + \mathbf{x}_2, \mathbf{x}_1 \in \tilde{\mathbf{\Omega}}_{\text{est}}(N+1 \mid N), \\ \mathbf{x}_2 \in \mathbf{\Omega}_{\mathbf{G}\mathbf{w}}(N)\} \end{aligned}} \tag{6.6.1}$$

Now consider the estimation step. Define $\mathbf{\Omega}_{\mathbf{x}|\mathbf{z}}(N)$ as the set of $\mathbf{x}$ which could have caused $\mathbf{z}(N)$. Then

$$\mathbf{\Omega}_{\mathbf{x}|\mathbf{z}}(N) = \{\mathbf{x}: \mathbf{z}(N) - \mathbf{H}(N)\mathbf{x} \in \mathbf{\Omega}_v(N)\}$$

From the estimation logic of Chapter 5 $\mathbf{\Omega}_{\text{est}}(N+1 \mid N+1)$ is the intersection of $\Omega_{est}(N+1|N)$ and $\Omega_{x|z}(N+1)$, so

$$\boxed{\Omega_{est}(N+1|N+1) = \{\mathbf{x}: \mathbf{x} \in \Omega_{est}(N+1|N) \text{ and } \mathbf{x} \in \Omega_{x|z}(N+1)\}} \tag{6.6.2}$$

Equations (6.6.1) and (6.6.2) form the desired recursive relationship which can theoretically be solved for all N once $\Omega_x(0)$ is given. $\Omega_x(0)$ can be the whole space if desired.

EXAMPLE: Consider

$$\mathbf{x}(n+1) = \begin{bmatrix} 1 & 1 \\ 0 & 1 \end{bmatrix}\mathbf{x}(n) + \begin{bmatrix} 0 \\ 1 \end{bmatrix} w(n)$$

$$z(n) = [1 \quad 0]\mathbf{x}(n) + v(n)$$

$$\mathbf{x}(0) \in \Omega_x(0)$$

$$\Omega_x(0) = \{\mathbf{x}: \mathbf{x}'\mathbf{x} \le 4\}$$

$$|v(n)| \le \tfrac{1}{2}$$

$$|w(n)| \le 1$$

Assume that

$$z(1) = 1.5$$

The sets $\Omega_{est}(0|0)$, $\tilde{\Omega}_{est}(1|0)$, $\Omega_{Gw}(0)$, and $\Omega_{est}(1|0)$ are shown in Fig. 6.6.1. Figure 6.6.2 shows $\Omega_{est}(1|0)$, $\Omega_{x|z}(1)$, and the resulting $\Omega_{est}(1|1)$.

The shortcomings of set relationships such as (6.6.1) and (6.6.2) have been noted before; they are difficult to implement on a computer, as $\Omega_{est}(N|N)$ is not an ellipsoid (or other nice shape) even though the basic sets are ellipsoids. Hence bounding ellipsoids are used.

Note that the general arguments of (6.6.1) and (6.6.2) actually apply for arbitrary sets. The ellipsoid nature of the uncertainty models was not used.

6.6.2 Bounding Ellipsoid

Let $\Omega_{\hat{x}}(N|N)$ denote a bounding ellipsoid for $\Omega_{est}(N|N)$

$$\Omega_{\hat{x}}(N|N) \supset \Omega_{est}(N|N) \tag{6.6.3}$$

of the form

$$\Omega_{\hat{x}}(N|N) = \{\mathbf{x}: [\mathbf{x} - \hat{\mathbf{x}}(N|N)]'\mathbf{\Sigma}^{-1}(N|N)[\mathbf{x} - \hat{\mathbf{x}}(N|N)] \le 1\}$$

Difference equations for the desired $\hat{\mathbf{x}}(N|N)$ and $\mathbf{\Sigma}(N|N)$ can be obtained as follows: Assume that (6.6.3) holds. Then

1. Calculate an ellipsoid $\Omega_{\hat{x}}(N+1|N)$ such that $\Omega_{\hat{x}}(N+1|N) \supset$ vector sum of $\tilde{\Omega}_{\hat{x}}(N+1|N)$ and $\Omega_{Gw}(N)$, where $\tilde{\Omega}_{\hat{x}}(N+1|N)$ is calculated from $\Omega_{\hat{x}}(N|N)$.
2. Calculate an ellipsoid $\Omega_{\hat{x}}(N+1|N+1)$ such that $\Omega_{\hat{x}}(N+1|N+1)$ $\subset$ intersection of $\Omega_{\hat{x}}(N+1|N)$ and $\Omega_{x|z}(N+1)$.

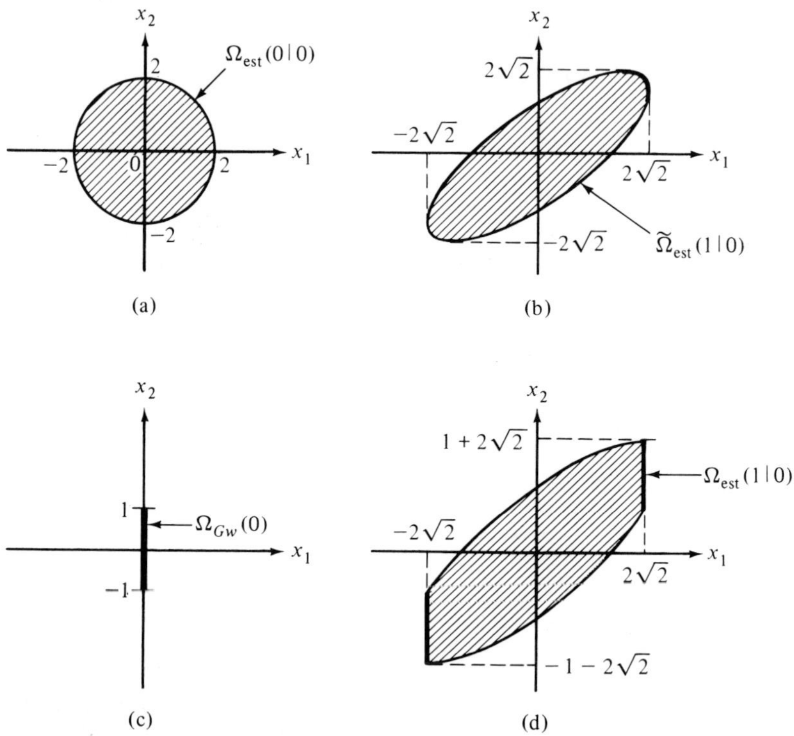

Figure 6.6.1 Evolution of $\Omega_{\text{est}}(0|0)$ into $\Omega_{\text{est}}(1|0)$ for example.

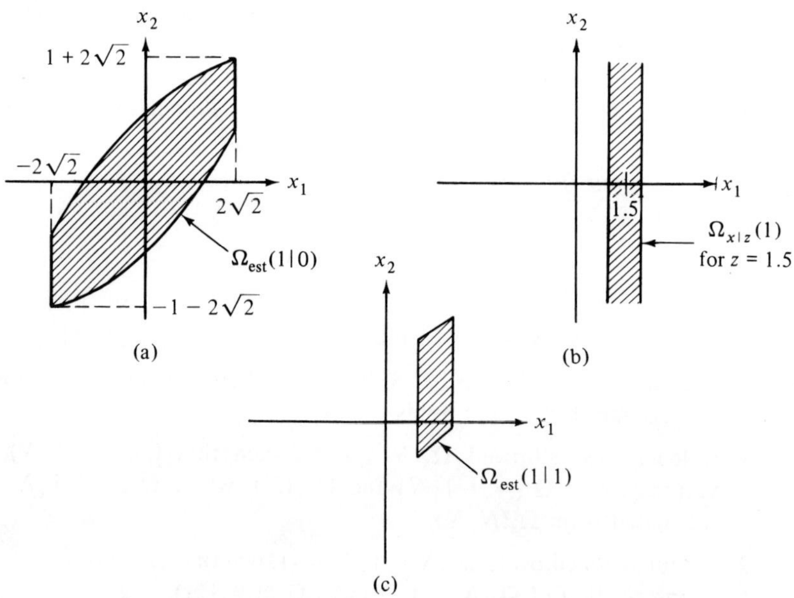

Figure 6.6.2 Relation of $\Omega_{\text{est}}(1|1)$ to $\Omega_{\text{est}}(1|0)$ for example.

The first step was done in Chapter 4, Section 4.3.2, while the second step was effectively done in Chapter 5, Section 5.6.2. Hence it is only necessary to put the appropriate equations together with the right notation.

The results are as follows: For the model (6.1) with unknown-but-bounded uncertainty modeled by ellipsoids, $\mathbf{x}(N)$ is contained in $\mathbf{\Omega}_{\hat{x}}(N|N)$, where

$$
\begin{aligned}
\mathbf{\Omega}_{\hat{x}}(N+1|N+1) &= \{\mathbf{x}: [\mathbf{x} - \hat{\mathbf{x}}(N+1|N+1)]' \\
&\quad \times \mathbf{\Sigma}^{-1}(N+1|N+1) \\
&\quad \times [\mathbf{x} - \hat{\mathbf{x}}(N+1|N+1)] \leq 1\} \\
\hat{\mathbf{x}}(N+1|N+1) &= \mathbf{\Sigma}_b(N+1|N+1)\{\tilde{\mathbf{\Sigma}}^{-1}(N+1|N) \\
&\quad \times \mathbf{\Phi}(N)\hat{\mathbf{x}}(N|N) + \mathbf{H}'(N+1) \\
&\quad \times \tilde{\mathbf{R}}^{-1}(N+1)\mathbf{z}(N+1)\} \\
\mathbf{\Sigma}(N+1|N+1) &= [1 - \delta^2(N+1)]\mathbf{\Sigma}_b(N+1|N+1) \\
\mathbf{\Sigma}_b(N+1|N+1) &= \{\tilde{\mathbf{\Sigma}}^{-1}(N+1|N) \\
&\quad + \mathbf{H}'(N+1)\tilde{\mathbf{R}}^{-1}(N+1)\mathbf{H}(N+1)\}^{-1} \\
\mathbf{\Sigma}(N+1|N) &= \left[\frac{1}{1-\beta(N)}\right]\mathbf{\Phi}(N)\mathbf{\Sigma}(N|N)\mathbf{\Phi}'(N) \\
&\quad + \mathbf{G}(N)\tilde{\mathbf{Q}}(N)\mathbf{G}'(N) \\
\tilde{\mathbf{\Sigma}}(N+1|N) &= \left(\frac{1}{1-\rho(N+1)}\right)\mathbf{\Sigma}(N+1|N) \\
\tilde{\mathbf{Q}}(N) &= \frac{1}{\beta(N)}\mathbf{Q}(N) \\
\tilde{\mathbf{R}}(N) &= \frac{1}{\rho(N)}\mathbf{R}(N) \\
\delta^2(N+1) &= [\mathbf{z}(N+1) - \mathbf{H}(N+1)\mathbf{\Phi}(N)\hat{\mathbf{x}}(N|N)]' \\
&\quad \times \{\mathbf{H}(N+1)\tilde{\mathbf{\Sigma}}(N+1|N)\mathbf{H}'(N+1) \\
&\quad + \tilde{\mathbf{R}}(N+1)\}^{-1}[\mathbf{z}(N+1) \\
&\quad - \mathbf{H}(N+1)\mathbf{\Phi}(N)\hat{\mathbf{x}}(N|N)]
\end{aligned}
\tag{6.6.4}
$$

for any $\rho(N+1)$, $\beta(N)$

$$
\begin{aligned}
1 \geq \rho(N+1) &\geq 0 \\
1 \geq \beta(N) &\geq 0 \\
\hat{\mathbf{x}}(0|0) &= \mathbf{0} \\
\mathbf{\Sigma}(0|0) &= \boldsymbol{\psi}
\end{aligned}
$$

Equation (6.6.4) can, of course, be rewritten in various ways to give interpretations analogous to the linear feedback or information flow concepts discussed in Section 6.3 for the Bayesian model. In the linear feedback form (6.6.4) becomes

$$
\begin{aligned}
\hat{\mathbf{x}}(N+1 \mid N+1) &= \boldsymbol{\Phi}(N)\hat{\mathbf{x}}(N \mid N) + \mathbf{K}(N+1) \\
&\quad \times \{\mathbf{z}(N+1) - \mathbf{H}(N+1)\boldsymbol{\Phi}(N)\hat{\mathbf{x}}(N \mid N)\} \\
\boldsymbol{\Sigma}_b(N+1 \mid N+1) &= \tilde{\boldsymbol{\Sigma}}(N+1 \mid N) - \tilde{\boldsymbol{\Sigma}}(N+1 \mid N)\mathbf{H}'(N+1) \\
&\quad \times \{\tilde{\mathbf{R}}(N+1) + \mathbf{H}(N+1)\tilde{\boldsymbol{\Sigma}}(N+1 \mid N) \\
&\quad \times \mathbf{H}'(N+1)\}^{-1}\mathbf{H}(N+1)\tilde{\boldsymbol{\Sigma}}(N+1 \mid N) \\
\mathbf{K}(N+1) &= \boldsymbol{\Sigma}_b(N+1 \mid N+1)\mathbf{H}'(N+1)\tilde{\mathbf{R}}^{-1}(N+1) \\
&\left.\begin{aligned}
\boldsymbol{\Sigma}_b(N+1 \mid N+1) \\
\tilde{\boldsymbol{\Sigma}}(N+1 \mid N) \\
\delta^2(N+1) \\
\rho(N+1) \\
\beta(N)
\end{aligned}\right\} \text{as in (6.6.4)} \\
\hat{\mathbf{x}}(0) &= \mathbf{0} \\
\boldsymbol{\Sigma}(0 \mid 0) &= \boldsymbol{\psi}
\end{aligned}
\tag{6.6.5}
$$

The feedback implementation of (6.6.5) is illustrated in block diagram form in Fig. 6.6.3.

Equations (6.6.4) and (6.6.5) are "like" the Bayesian model equations (6.2.5) and (6.3.1) except for the $\beta(N)$, $\rho(N+1)$, and $\delta^2(N+1)$ terms. Compare Fig. 6.6.3 with Fig. 6.3.3. The discussions in Sections 4.3.4 and 5.6.2 on the choice of the free parameters $\beta(N)$ and $\rho(N+1)$ apply equally

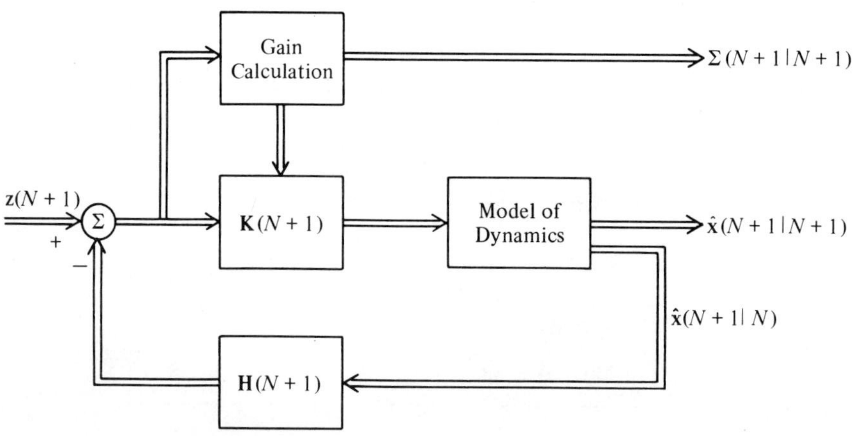

Figure 6.6.3 Unknown-but-bounded discrete-time filter.

here. The $\beta(N)$ and $\rho(N+1)$ can be prespecified or special logics can be developed so that they are computed as a function of $\hat{\mathbf{x}}(N|N)$ and $\mathbf{\Sigma}(N|N)$. The $\delta^2(N+1)$ term depends on the observations, so $\mathbf{\Sigma}(N+1|N+1)$ is a function of the observations. For the Bayesian model, the corresponding $\mathbf{\Sigma}(N+1|N+1)$ is independent of the observations. Thus the unknown-but-bounded estimator, viewed as a system with input $\mathbf{z}(N+1)$ and output $\hat{\mathbf{x}}(N+1|N+1)$, is a time-varying nonlinear system. For the Bayesian model, the corresponding estimator is a time-varying linear system.

6.6.3 Bounding Ellipsoid: Special Cases†

The fact that in (6.6.4) or (6.6.5) the estimate $\hat{\mathbf{x}}(N|N)$ is a nonlinear function of the observation $\mathbf{z}(1) \cdots \mathbf{z}(N+1)$ is not in itself undesirable. However, the fact that $\mathbf{\Sigma}(N|N)$ depends on the observation means that it cannot be precomputed, so all of (6.6.4) [or (6.6.5)] must be computed *on-line*. Furthermore, since $\mathbf{\Sigma}(N|N)$ determines the "size" of the ellipsoid, it specifies "how well" the estimator is working, and the observation dependence means that error analysis cannot be done before the observations are made.

Both these problems can be solved by considering a "bounding" bounding ellipsoid $\Omega_{\hat{x},b}(N|N)$ with $\hat{\mathbf{x}}_b(N|N)$ and $\mathbf{\Sigma}_b(N|N)$ as defined by

$$\boxed{\begin{aligned} \Omega_{\hat{x},b}(N+1|N+1) = \{\mathbf{x}: [\mathbf{x} - &\hat{\mathbf{x}}_b(N+1|N+1)]' \\ &\times \mathbf{\Sigma}_b^{-1}(N+1|N+1) \\ &\times [\mathbf{x} - \hat{\mathbf{x}}_b(N+1|N+1)] \leq 1\} \\ \left.\begin{matrix}\hat{\mathbf{x}}_b(N+1|N+1) \\ \mathbf{\Sigma}_b(N+1|N+1)\end{matrix}\right\} &\begin{matrix}\text{given by (6.6.4) or (6.6.5)} \\ \text{with } \delta^2(N+1) = 0\end{matrix} \end{aligned}} \tag{6.6.6}$$

It is felt to be obvious that

$$\Omega_{\hat{x},b}(N|N) \supset \Omega_{\hat{x}}(N|N)$$

so that (6.6.6) defines an ellipsoid which always contains the state $\mathbf{x}(N)$. Equation (6.6.6) is a linear estimator in that the ellipsoid center $\hat{\mathbf{x}}(N|N)$ is a linear function of the observations $\mathbf{z}(1) \cdots \mathbf{z}(N)$. More important, the matrix $\mathbf{\Sigma}_b(N|N)$ which specifies the size of the ellipsoid can be precomputed by solving a matrix Riccati equation which is independent of the observations. The idea of setting $\delta^2(n+1) = 0$ is closely related to the ideas of Section 5.6.

The $\mathbf{\Sigma}_b(N|N)$ of (6.6.6) can also be used as a precomputable error analysis for the nonlinear estimate (6.6.4) or (6.6.5) as $\mathbf{\Sigma}_b(N|N) \geq \mathbf{\Sigma}(N|N)$ provided that the same values for $\rho(N+1)$ and $\beta(N)$ are used; i.e., the $\rho(N+1)$ and $\beta(N)$ have to be prespecified.

Another special case of (6.6.4) or (6.6.5) of interest occurs when the $\beta(N)$ and $\rho(N+1)$ are chosen in a special way. Let $\beta(N)$ and $\rho(N+1)$ denote some prespecified set of values used to determine a bounding ellipsoid as in (6.6.6). Let $\tilde{\beta}(N)$ and $\tilde{\rho}(N+1)$ denote the values to be used in (6.6.4), where

$$\tilde{\beta}(N) = \frac{\beta(N)}{1 - [1 - \beta(N)]q(N)}$$

$$\tilde{\rho}(N+1) = \frac{\rho(N+1)}{1 - [1 - \beta(N)][1 - \rho(N+1)]q(N)}$$

$$\begin{aligned} q(N+1) = {} & [1 - \beta(N)][1 - \rho(N+1)]q(N) \\ & + [\mathbf{z}(N+1) - \mathbf{H}(N+1)\mathbf{\Phi}(N)\hat{\mathbf{x}}(N \mid N)]' \\ & \times \{\mathbf{H}(N+1)\tilde{\mathbf{\Sigma}}(N+1 \mid N)\mathbf{H}'(N+1) + \tilde{\mathbf{R}}(N+1)\}^{-1} \\ & \times [\mathbf{z}(N+1) - \mathbf{H}(N+1)\mathbf{\Phi}(N)\hat{\mathbf{x}}(N \mid N)] \end{aligned}$$

It turns out that for these $\tilde{\beta}(N)$ and $\tilde{\rho}(N+1)$

$$\begin{aligned} \mathbf{\Sigma}(N+1 \mid N+1) &= [1 - q(N+1)]\mathbf{\Sigma}_b(N+1 \mid N+1) \\ \hat{\mathbf{x}}(N+1 \mid N+1) &= \hat{\mathbf{x}}_b(N+1 \mid N+1) \end{aligned}$$

Therefore the ellipsoid

$$\begin{aligned} \tilde{\mathbf{\Omega}}_{\hat{x},b}(N+1 \mid N+1) = {} & \{\mathbf{x}: [\mathbf{x} - \hat{\mathbf{x}}_b(N+1 \mid N+1)]' \\ & \times \tilde{\mathbf{\Sigma}}_b^{-1}(N+1 \mid N+1) \\ & \times [\mathbf{x} - \hat{\mathbf{x}}_b(N+1 \mid N+1)] \leq 1\} \end{aligned}$$

$$\left.\begin{aligned} & \hat{\mathbf{x}}_b(N+1 \mid N+1) \\ & \mathbf{\Sigma}_b(N+1 \mid N+1) \end{aligned}\right\} \begin{aligned} & \text{given by (6.6.4) or (6.6.5)} \\ & \text{with } \delta^2(N+1) = 0 \end{aligned} \tag{6.6.7}$$

$$\mathbf{\Sigma}_b(N+1 \mid N+1) = [1 - q(N+1)]\mathbf{\Sigma}_b(N+1 \mid N+1)$$

is such that

$$\mathbf{\Omega}_{\hat{x},b}(N \mid N) \supset \tilde{\mathbf{\Omega}}_{\hat{x},b}(N \mid N) \supset \mathbf{\Omega}_{\hat{x}}(N \mid N)$$

The ellipsoid $\mathbf{\Omega}_{\hat{x},b}(N \mid N)$ of (6.6.7) has the same center as $\mathbf{\Omega}_{\hat{x},b}(N \mid N)$ of (6.6.6) but is smaller (or the same). Hence (6.6.7) constitutes an improvement over (6.6.6) which does not cost much in the way of computation, as $q(N)$ is relatively easy to obtain.

6.6.4 Time-Invariant Models and Steady-State Behavior

Consider a time-invariant model so that the $\mathbf{\Phi}$, $\mathbf{G}$, $\mathbf{H}$, $\mathbf{Q}$, and $\mathbf{R}$ matrices are constant. Even if constant values ρ and β are used for the free parameters $\rho(N+1)$ and $\beta(N)$, a steady-state concept corresponding to the $\mathbf{\Sigma}_\infty$ of the Bayesian-Fisher model does not occur for the bounding ellipsoids of Section 6.6.2, as $\mathbf{\Sigma}(N \mid N)$ depends on the observations and thus does not go to any steady-state value. Therefore it is still necessary to implement equations such as (6.6.4) and (6.6.5), which can be viewed as a time-invariant nonlinear system with input $\mathbf{z}(n)$, $n = 1, \ldots, N$, and output $\hat{\mathbf{x}}(N|N)$.

A linear time-invariant estimator can be obtained by removing the $\boldsymbol{\Sigma}$ dependence on the observations by setting $\delta^2(N+1) = 0$ to get the bounding ellipsoid of Section 6.6.3. In this case, (6.6.6) yields

$$
\boxed{
\begin{aligned}
\hat{\mathbf{x}}(N+1|N+1) &= \boldsymbol{\Phi}\hat{\mathbf{x}}(N|N) + \mathbf{K}_\infty\{\mathbf{z}(N+1) - \mathbf{H}\boldsymbol{\Phi}\hat{\mathbf{x}}(N|N)\} \\
\mathbf{K}_\infty &= \boldsymbol{\Sigma}_\infty\mathbf{H}'\tilde{\mathbf{R}}^{-1} \\
\boldsymbol{\Sigma}_\infty &= \{(1-\rho)\boldsymbol{\Sigma}_\infty^{-1}(1|0) + \mathbf{H}'\tilde{\mathbf{R}}^{-1}\mathbf{H}\}^{-1} \\
\boldsymbol{\Sigma}_\infty(1|0) &= \left(\frac{1}{1-\beta}\right)\boldsymbol{\Phi}\boldsymbol{\Sigma}_\infty\boldsymbol{\Phi}' + \mathbf{G}\tilde{\mathbf{Q}}\mathbf{G}' \\
\tilde{\mathbf{R}} &= \frac{1}{\rho}\mathbf{R} \\
\tilde{\mathbf{Q}} &= \frac{1}{\beta}\mathbf{Q} \\
0 &\leq \rho \leq 1 \\
0 &\leq \beta \leq 1
\end{aligned}
}
\tag{6.6.8}
$$

Discussions on existence of a unique positive definite $\boldsymbol{\Sigma}_\infty$ and the stability of (6.6.8) follow as in the Bayesian model case of Section 6.3 except that the presence of the free parameters, ρ and β, must be considered.

6.6.5 Discussion of Unknown-but-Bounded

The bounding ellipsoid concept yields equations which can be implemented as easily as those from Bayesian and Fisher models. However, there is the extra problem of choosing values for the free parameters $\beta(N)$ and $\rho(N+1)$. This is not a simple problem and no general "rules of thumb" or "proved" formulas are presently available.

An unknown-but-bounded model is potentially a very useful tool but an important point should be made. At the time this book was written, only very limited numerical experience was available (to the author) on the behavior of the bounding ellipsoid algorithm. Hence it should not be viewed as a proved technique.

6.7 Finite Memory Filtering

The estimators discussed in the preceding sections are of the growing memory type; i.e., $\hat{\mathbf{x}}(N|N)$ is an estimate of $\mathbf{x}(N)$ calculated from observations, $\mathbf{z}(1) \cdots \mathbf{z}(N)$. As N increases, the estimate always "remembers" the observations back to the time observations began ($N = 1$). In some applications this is not satisfactory. For example, assume that the model is not valid over time intervals longer than m units of time (see Section 8.2 for further discussion). In such cases, one approach is to use a finite memory estimator, i.e., to find $\hat{\mathbf{x}}(N|N, N-m)$, the estimate of $\mathbf{x}(N)$ given observations $\mathbf{z}(n)$, $n = N - m, \ldots, N$, where m is fixed while N is increasing. Such finite memory estimators are sometimes called *sliding arc* estimators.

The basic ideas of the derivation for a Bayesian model are as follows:

1. Assume that $\hat{\mathbf{x}}(N|N, N-m)$ is known. Calculate $\hat{\mathbf{x}}(N+1|N+1, N-m)$ from $\hat{\mathbf{x}}(N|N, N-m)$ and $\mathbf{z}(N+1)$ as in Section 6.2.2.
2. Assume that $\hat{\mathbf{x}}(N+1|N+1, N+1-m)$ is known. Calculate $\hat{\mathbf{x}}(N+1|N+1, N-m)$ from $\hat{\mathbf{x}}(N+1|N+1, N+1-m)$ and $\mathbf{z}(N-m)$ using a variation of the ideas in Section 6.2.2.
3. Equate the equations from steps 1 and 2 to obtain a relationship between $\hat{\mathbf{x}}(N|N, N-m)$ and $\hat{\mathbf{x}}(N+1|N+1, N+1-m)$, $\mathbf{z}(N+1)$, $\mathbf{z}(N-m)$.

This procedure has a nice interpretation in terms of the information flow ideas of Section 6.3.4. The information vector $\boldsymbol{\chi}(N+1|N+1, N+1-m)$ is calculated by adding to $\boldsymbol{\chi}(N|N, N-m)$ the information contained in $\mathbf{z}(N+1)$ and subtracting the information contained in $\mathbf{z}(N-m)$. The information matrices are calculated in the same way, i.e., by adding the new and subtracting the old.

The estimator resulting from the above steps can be written in various ways. One general form is

$$\hat{\mathbf{x}}(N+1|N+1, N+1-m) = \mathbf{C}_1(N)\hat{\mathbf{x}}(N|N, N-m) + \mathbf{C}_2(N)\mathbf{z}(N+1) + \mathbf{C}_3(N)\mathbf{z}(N-m) \tag{6.7.1}$$

This form however can exhibit instability problems which are not encountered in the corresponding growing memory estimator.

The preceding discussions have assumed a Bayesian modeling of the uncertainty, but similar results are possible for Fisher and unknown-but-bounded models and weighted-least-squares filters.

For a time-invariant system model, a finite memory estimator can reach a steady state just as growing memory estimators can.

EXAMPLE: Consider the Fisher model

$$z(n) = x + v(n)$$

$$x\text{: unknown}$$

$$E\{v(n_1)v(n_2)\} = \begin{cases} R, & n_1 = n_2 \\ 0, & n_1 \neq n_2 \end{cases}$$

The *growing memory* estimate $\hat{x}(N|N)$ is given by

$$\hat{x}(N+1|N+1) = \frac{1}{N+1}\sum_{n=1}^{N+1} z(n)$$

or

$$\hat{x}(N+1|N+1) = \frac{1}{N+1}z(N+1) + \frac{N}{N+1}\hat{x}(N|N)$$

Similarly, the finite memory estimate $\hat{x}(N|N, N-m)$ is obviously given by (for $N > m$)

$$\hat{x}(N+1|N+1, N+1-m) = \frac{1}{m+1}\sum_{n=N+1-m}^{N+1} z(n)$$

which can be written in the form of (6.7.1) as

$$\hat{x}(N+1|N+1, N+1-m) = \hat{x}(N|N, N-m) + \frac{1}{m+1}z(N+1) - \frac{1}{m+1}z(N-m)$$

$$C_1(N) = 1$$

$$C_2(N) = \frac{1}{m+1}$$

$$C_3(N) = -\frac{1}{m+1}$$

which is *not* a stable system.

6.8 Smoothing

Thus far, interest has centered on prediction, $\hat{\mathbf{x}}(m|N)$, $m > N$, and filtering, $\hat{\mathbf{x}}(N|N)$, relationships. The smoothing problem of calculating $\hat{\mathbf{x}}(m|N)$, $N > m > 1$, is now considered. Three classes of smoothing problems can be defined, as in Fig. 6.8.1. Still more classes can be defined if, for example, a finite memory smoother is considered.

	m	n
Fixed Interval (signal reconstruction)	varies from 1 to n	fixed
Fixed Point (growing lag)	both increasing such that $n - m$ is fixed	
Fixed Lag (moving point)	fixed	increasing

It is desired to calculate $\hat{\mathbf{x}}(n-m|n)$.

Figure 6.8.1. Classes of smoothing problems.

Explicit equations for smoothing estimators are often very complicated and hence appear to be difficult to understand. However, the smoothing concept becomes an almost trivial extension of filtering when the estimating problem is divided into two parts involving the *past* and *future* observations, respectively. Define

$$\mathbf{z}_m = \begin{bmatrix} \mathbf{z}(1) \\ \cdot \\ \cdot \\ \cdot \\ \mathbf{z}(m) \end{bmatrix} \qquad \mathbf{z}_{N-m} = \begin{bmatrix} \mathbf{z}(m+1) \\ \cdot \\ \cdot \\ \cdot \\ \mathbf{z}(N) \end{bmatrix}$$

$$\mathbf{v}_m = \begin{bmatrix} \mathbf{v}(1) \\ \cdot \\ \cdot \\ \cdot \\ \mathbf{v}(m) \end{bmatrix} \qquad \mathbf{v}_{N-m} = \begin{bmatrix} \mathbf{v}(m+1) \\ \cdot \\ \cdot \\ \cdot \\ \mathbf{v}(N) \end{bmatrix}$$

$$\mathbf{w}_m = \begin{bmatrix} \mathbf{w}(0) \\ \cdot \\ \cdot \\ \cdot \\ \mathbf{w}(m-1) \end{bmatrix} \qquad \mathbf{w}_{N-m} = \begin{bmatrix} \mathbf{w}(m) \\ \cdot \\ \cdot \\ \cdot \\ \mathbf{w}(N-1) \end{bmatrix}$$

$\hat{\mathbf{x}}(m \mid \mathbf{z}_m)$ — estimate of $\mathbf{x}(m)$ made from $\mathbf{z}_m$ (i.e., the past and present observations)

$\hat{\mathbf{x}}(m \mid \mathbf{z}_{N-m})$ — estimate of $\mathbf{x}(m)$ made from $\mathbf{z}_{N-m}$ (i.e., the future observations)

Using the matrix manipulation concepts of Section 6.1, it is possible to find matrices $\mathbf{C}_1 \cdots \mathbf{C}_6$ such that

$$\begin{aligned} \mathbf{z}_m &= \mathbf{C}_1\mathbf{x}(m) + \tilde{\mathbf{v}}_m \\ \tilde{\mathbf{v}}_m &= \mathbf{C}_2\mathbf{w}_m + \mathbf{C}_3\mathbf{v}_m \\ \mathbf{z}_{N-m} &= \mathbf{C}_4\mathbf{x}(m) + \tilde{\mathbf{v}}_{N-m} \\ \tilde{\mathbf{v}}_{N-m} &= \mathbf{C}_5\mathbf{w}_{N-m} + \mathbf{C}_6\mathbf{v}_{N-m} \end{aligned}$$

Now this formulation is like the dichotomy of Section 5.7, so that $\mathbf{x}(m)$ can be estimated in three steps as follows:

1. Calculate $\hat{\mathbf{x}}(m \mid \mathbf{z}_m)$ as a filtering problem running forward in time from 1 to m.
2. Calculate $\hat{\mathbf{x}}(m \mid \mathbf{z}_{N-m})$ as a filtering problem that is run backward in time from N to $m + 1$ (plus a one-step prediction backward to m).
3. Combine $\hat{\mathbf{x}}(m \mid \mathbf{z}_m)$ and $\hat{\mathbf{x}}(m \mid \mathbf{z}_{N-m})$ into the desired $\hat{\mathbf{x}}(m \mid N)$ by viewing them as separate observations of $\mathbf{x}(m)$.

This three-step solution applies equally to Bayesian, Fisher, and unknown-but-bounded models. For a Bayesian model, the a priori $\boldsymbol{\psi}$ associated with $\mathbf{x}(0)$ is used on the forward filter, while the backward filter (from N backward to $m + 1$) views $\mathbf{x}(N)$ as a completely unknown vector. For an unknown-but-bounded model, an initial set $\boldsymbol{\Omega}_x(0)$ constraining $\mathbf{x}(0)$ is handled in the same way. For a Bayesian model, the third step of combination is done using the concepts of Section 6.2.2, i.e., viewing $\mathbf{x}(m)$ itself as completely unknown.

The weighted-least-squares logic of Section 6.5 was presented in terms of the filtering problem. However, a little thought shows that minimization of $J[\mathbf{x}(N), \mathbf{w}(0) \cdots \mathbf{w}(N + 1)]$ actually yields an $\hat{\mathbf{x}}(m|N)$, all m, $1 \leq m \leq N$, i.e., a smoothing solution.

The three-step method for obtaining a smoothed solution can be implemented directly or it may be more convenient to manipulate the equations into a recursive form. However, some of these recursive forms can exhibit instability so care is needed. There are many possible forms which can be derived.

6.9 Perfect Observations†

The models considered in Sections 6.1–6.8 have assumed (6.1) with $\mathbf{z}(n) = \mathbf{H}(n)\mathbf{x}(n) + \mathbf{v}(n)$, where $\mathbf{v}(n)$ is a white uncertain process. The case of perfect observation where $\mathbf{v}(n) = 0$ is now considered.

It is important at this point for the reader to reread Section 3.2, as it discusses how $\mathbf{v}(n) = \mathbf{0}$ models result from problems with observation uncertainties that are nonwhite. Thus.

$$\mathbf{z}(n) = \mathbf{H}(n)\mathbf{x}(n) + \mathbf{v}_s(n)$$

$$\mathbf{v}_s(n)\text{: nonwhite, structured process}$$

is modeled in state space form by

$$\mathbf{z}(n) = \mathbf{H}_c\mathbf{x}_c(n)$$

Hence the "perfect observation" case does not necessarily imply that the original problem is being modeled as having perfect observations.

For the sake of discussion, the case of perfect observations is viewed as the limiting case of the earlier results when there is no observation uncertainty; i.e., $\mathbf{R}(n) \longrightarrow \mathbf{0}$. This point of view can be made rigorous if desired.

Consider a Bayesian model version of (6.1). To set $\mathbf{R}(n) = \mathbf{0}$, it is merely necessary to choose some set of equations which do not involve $\mathbf{R}^{-1}(n)$.

From (6.2.7) and (6.3.1) it follows that for $\mathbf{R}(n) = \mathbf{0}$

$$
\boxed{
\begin{aligned}
\hat{\mathbf{x}}(N+1|N+1) &= \mathbf{\Phi}(N)\hat{\mathbf{x}}(N|N) + \mathbf{\Sigma}(N+1|N)\mathbf{H}'(N+1) \\
&\quad \times [\mathbf{H}(N+1)\mathbf{\Sigma}(N+1|N)\mathbf{H}'(N+1)]^{-1} \\
&\quad \times [\mathbf{z}(N+1) - \mathbf{H}(N+1)\mathbf{\Phi}(N)\hat{\mathbf{x}}(N|N)] \\
\mathbf{\Sigma}(N+1|N+1) &= \mathbf{\Sigma}(N+1|N) - \mathbf{\Sigma}(N+1|N)\mathbf{H}'(N+1) \\
&\quad \times [\mathbf{H}(N+1)\mathbf{\Sigma}(N+1|N)\mathbf{H}'(N+1)]^{-1} \\
&\quad \times \mathbf{H}(N+1)\mathbf{\Sigma}(N+1|N) \\
\mathbf{\Sigma}(0|0) &= \boldsymbol{\psi} \\
\hat{\mathbf{x}}(0|0) &= \mathbf{0} \\
\mathbf{\Sigma}(N+1|N) &= \mathbf{\Phi}(N)\mathbf{\Sigma}(N|N)\mathbf{\Phi}'(N) + \mathbf{G}(N)\mathbf{Q}(N)\mathbf{G}'(N)
\end{aligned}
}
\tag{6.9.1}
$$

provided that the indicated inverse exists. From (6.9.1), it follows that

$$
\begin{aligned}
&\mathbf{H}(N+1)\hat{\mathbf{x}}(N+1|N+1) = \mathbf{z}(N+1) = \mathbf{H}(N+1)\mathbf{x}(N+1) \\
&\mathbf{H}(N+1)\mathbf{\Sigma}(N+1|N+1) = \mathbf{0} \\
&\mathbf{\Sigma}(N+1|N+1)\mathbf{H}'(N+1) = \mathbf{0}
\end{aligned}
\tag{6.9.2}
$$

As expected $\mathbf{\Sigma}^{-1}(N+1|N+1)$ does not exist.

EXAMPLE: Consider the Bayesian case of observing a constant x_1 with structured observation noise $v_s(n)$:

$$
\begin{gathered}
z(n) = x_1 + v_s(n) \\
v_s(n+1) = \Phi v_s(n) + w_s(n) \\
E\{x_1^2\} = \psi_1 \qquad E\{v_s^2(0)\} = \psi_s \\
E\{w_s(n_1)w_s(n_2)\} = \begin{cases} Q, & n_1 = n_2 \\ 0, & n_1 \neq n_2 \end{cases} \\
x_1,\ v_s(0),\ w_s(n)\text{: zero mean, uncorrelated}
\end{gathered}
$$

Define

$$
\mathbf{x}(n) = \begin{bmatrix} x_1(n) \\ x_2(n) \end{bmatrix} = \begin{bmatrix} x_1 \\ v_s(n) \end{bmatrix}
$$

$$
\mathbf{H} = [1 \quad 1] \qquad \mathbf{\Phi} = \begin{bmatrix} 1 & 0 \\ 0 & \Phi \end{bmatrix}
$$

$$
\mathbf{G} = \begin{bmatrix} 0 \\ 1 \end{bmatrix}
$$

Consider first the error covariance matrix:

$$
\mathbf{\Sigma}(N|N) = \begin{bmatrix} \Sigma_{11}(N|N) & \Sigma_{12}(N|N) \\ \Sigma_{12}(N|N) & \Sigma_{22}(N|N) \end{bmatrix}
$$

From (6.9.2)

$$
\Sigma_{22}(N|N) = \Sigma_{11}(N|N) = -\Sigma_{12}(N|N)
$$

so

$$\boldsymbol{\Sigma}(N|N) = \boldsymbol{\Sigma}_{11}(N|N)\begin{bmatrix} 1 & -1 \\ -1 & 1 \end{bmatrix}$$

Then from (6.9.1)

$$\boldsymbol{\Sigma}_{11}(N+1|N+1) = \boldsymbol{\Sigma}_{11}(N|N) - \frac{(1-\boldsymbol{\Phi})^2\boldsymbol{\Sigma}_{11}^2(N|N)}{(1-\boldsymbol{\Phi})^2\boldsymbol{\Sigma}_{11}(N|N)+Q}$$

or for large N

$$\boldsymbol{\Sigma}_{11}(N|N) = \frac{Q}{N(1-\boldsymbol{\Phi})^2}$$

Now consider the estimate of the constant x_1. From (6.9.1)

$$\begin{aligned}\hat{x}_1(N+1|N+1) &= \hat{x}_1(N|N) + K(N+1)[z(N+1) \\ &\quad - \hat{x}_1(N|N) - \boldsymbol{\Phi}\hat{x}_2(N|N)] \\ K(N+1) &= \frac{(1-\boldsymbol{\Phi})\boldsymbol{\Sigma}_{11}(N|N)}{(1-\boldsymbol{\Phi})^2\boldsymbol{\Sigma}_{11}(N|N)+Q}\end{aligned}$$

From (6.9.2)

$$\hat{x}_1(N|N) + \hat{x}_2(N|N) = z(N)$$

so

$$\begin{aligned}\hat{x}_1(N+1|N+1) &= \hat{x}_1(N|N) + K(N+1)[z(N+1) \\ &\quad - (1-\boldsymbol{\Phi})\hat{x}_1(N|N) - \boldsymbol{\Phi}z(N)]\end{aligned}$$

For large N,

$$K(N+1) = \frac{1}{(1-\boldsymbol{\Phi})(1+N)}$$

$$\hat{x}_1(N+1|N+1) = \frac{N}{N+1}\hat{x}_1(N|N) + \frac{z(N+1)-\boldsymbol{\Phi}z(N)}{(1-\boldsymbol{\Phi})(1+N)}$$

Equation (6.9.1) is a K_1-dimensional vector and matrix difference equation (K_1 is a dimension of $\mathbf{x}$). However, not all these dimensions are "needed," as, for example, $\mathbf{H}(N+1)\boldsymbol{\Sigma}(N+1|N+1) = \mathbf{0}$. It is possible to rewrite the equations in a reduced dimensional form which will usually have computational advantages. The above example illustrated this point.

Some further understanding of the Bayesian form (6.9.1) can be obtained by considering the special case

$$\mathbf{x}(n) = \begin{bmatrix} \mathbf{x}_1(n) \\ \mathbf{x}_2(n) \end{bmatrix} \qquad \mathbf{w}(n) = \begin{bmatrix} \mathbf{w}_1(n) \\ \mathbf{w}_2(n) \end{bmatrix}$$

$$\boldsymbol{\Phi}(n) = \begin{bmatrix} \boldsymbol{\Phi}_{11}(n) & \boldsymbol{\Phi}_{12}(n) \\ \boldsymbol{\Phi}_{21}(n) & \boldsymbol{\Phi}_{22}(n) \end{bmatrix} \qquad \mathbf{H}(n) = [\mathbf{0} \quad \mathbf{I}] \qquad \mathbf{G}(n) = \mathbf{I}$$

$$\mathbf{Q}(n) = \begin{bmatrix} \mathbf{Q}_{11}(n) & \mathbf{0} \\ 0 & \mathbf{Q}_{22}(n) \end{bmatrix}$$

Instead of substituting into (6.9.1), one can proceed as follows to deduce the answer. Since $\mathbf{z}(n) = \mathbf{x}_2(n)$, the model (6.1) can be viewed as

$$\mathbf{x}_1(n+1) = \mathbf{\Phi}_{11}(n)\mathbf{x}_1(n) + \mathbf{w}_1(n) + \mathbf{\Phi}_{12}(n)\mathbf{z}(n) \tag{6.9.3}$$

where $\mathbf{\Phi}_{12}(n)\mathbf{z}(n)$ is to be viewed as a deterministic input. Now define

$$\tilde{\mathbf{z}}(N+1) = \mathbf{z}(N+1) - \mathbf{\Phi}_{22}(N)\mathbf{z}(N)$$

Since $\mathbf{z}(n) = \mathbf{x}_2(n)$

$$\tilde{\mathbf{z}}(N+1) = \mathbf{\Phi}_{21}(N)\mathbf{x}_1(N) + \mathbf{w}_2(N)$$

Now view $\tilde{\mathbf{z}}(N+1)$ as an observation on $\mathbf{x}_1(N)$ with "an $\mathbf{H} = \mathbf{\Phi}_{21}(N)$" and observation uncertainty $\mathbf{w}_2(N)$, where $\mathbf{x}_1(N)$ is modeled by a system (6.9.3) with a deterministic input. Hence the equations of Section 6.2 and 6.3 apply; in particular see Section 6.3.7.

In certain applications, it may be desired to model some of the sensors as being perfect while others have uncertainty. This can be modeled by using an $\mathbf{R}(n)$ where $\mathbf{R}^{-1}(n)$ does not exist. The desired estimator equations are obviously exactly those from which (6.9.1) was obtained, as they involved only $\mathbf{R}(n)$. However, caution is needed, as the fact that one sensor is several orders of magnitude more accurate than some other sensor does not necessarily imply that the accurate sensor can be safely assumed to be perfect. It all depends on the nature of the problem.

The author recommends against the general use of a perfect observation model unless great care is exercised as "strange things" can happen. In the special case where the perfect observation model is the result of modeling nonwhite observation errors, less care is needed.

Similar results are obtained from weighted least squares, Fisher models, and unknown-but-bounded models with bounding ellipsoids. For bounding ellipsoids, when $\mathbf{R}(n) = \mathbf{0}$, one also chooses $\rho(n) = 0$.

6.10 Weighting Patterns

The estimators of the preceding sections have been expressed in terms of difference equations because such difference equations are usually most convenient to implement on a digital computer. However, when the estimator is a linear operator, the difference equation can be solved to give

$$\hat{\mathbf{x}}(N|N) = \sum_{n=1}^{N} \mathbf{W}(n, N)\mathbf{z}(n) + \mathbf{W}_0(N) \tag{6.10.1}$$

The $\mathbf{W}(n, N)$, $n = 1, \ldots, N$, matrices are called the *weighting pattern*. Section 5.5 contains related discussions.

A weighting pattern implementation is usually awkward to implement if $\hat{\mathbf{x}}(N|N)$, $N = 1, \ldots,$ is desired, but it can be very useful if only $\hat{\mathbf{x}}(N|N)$ for some fixed N is desired. One main advantage of the weighting pattern concept is in the insight it provides. The relative "sizes" of $\mathbf{W}(n, N)$, $n = 1, \ldots, N$, indicate how much "importance" (weight) is attached to the past observations. A weighting pattern implementation such as (6.10.1) can also be needed when the corresponding recursive equations are unstable as can occur with finite memory and some smoothing estimators.

6.11 Analysis of Nonoptimum Estimators

The preceding sections have discussed "optimum" estimators and methods of analyzing their performance [$\mathbf{\Sigma}(N|N)$ is a measure of performance]. In many applications, annoying real-world considerations such as cost and lack of knowledge of the model (see also Section 6.13) force the use of a nonoptimum estimator. Section 5.9 contains related discussions.

Consider the model

$$\begin{aligned}
\mathbf{x}_1(n+1) &= \mathbf{\Phi}_1(n)\mathbf{x}_1(n) + \mathbf{G}_1(n)\mathbf{w}_1(n) \\
\mathbf{z}(n) &= \mathbf{y}(n) + \mathbf{v}_1(n) \\
\mathbf{y}(n) &= \mathbf{H}_1(n)\mathbf{x}_1(n)
\end{aligned} \tag{6.11.1}$$

Assume that an estimate $\hat{\mathbf{y}}(n|n)$ of $\mathbf{y}(n)$ is to be obtained as the output of a nonoptimum estimator represented by the linear dynamic system

$$\begin{aligned}
\mathbf{x}_2(n+1) &= \mathbf{\Phi}_2(n)\mathbf{x}_2(n) + \mathbf{G}_2(n)\mathbf{z}(n+1) \\
\hat{\mathbf{y}}(n|n) &= \mathbf{H}_2(n)\mathbf{x}_2(n)
\end{aligned} \tag{6.11.2}$$

It is desired to investigate the behavior of the estimate error

$$\boldsymbol{\delta}_y(n|n) = \mathbf{y}(n) - \hat{\mathbf{y}}(n|n)$$

Since (6.11.1) and (6.11.2) are linear, they can be combined to give

$$\begin{aligned}
\mathbf{x}(n+1) &= \mathbf{\Phi}(n)\mathbf{x}(n) + \mathbf{G}(n)\mathbf{w}(n) \\
\boldsymbol{\delta}_y(n|n) &= \mathbf{H}(n)\mathbf{x}(n) \\
\mathbf{x}(n) &= \begin{bmatrix} \mathbf{x}_1(n) \\ \mathbf{x}_2(n) \end{bmatrix} \qquad \mathbf{w}(n) = \begin{bmatrix} \mathbf{w}_1(n) \\ \mathbf{v}_1(n+1) \end{bmatrix} \\
\mathbf{\Phi}(n) &= \begin{bmatrix} \mathbf{\Phi}_1(n) & \mathbf{0} \\ \mathbf{G}_2(n)\mathbf{H}_1(n+1)\mathbf{\Phi}_1(n) & \mathbf{\Phi}_2(n) \end{bmatrix} \\
\mathbf{G}(n) &= \begin{bmatrix} \mathbf{G}_1(n) & \mathbf{0} \\ \mathbf{G}_2(n)\mathbf{H}_1(n+1)\mathbf{G}_1(n) & \mathbf{G}_2(n) \end{bmatrix} \\
\mathbf{H}(n) &= [\mathbf{H}_1(n) \quad -\mathbf{H}_2(n)] \\
\mathbf{Q}(n) &= \begin{bmatrix} \mathbf{Q}(n) & \mathbf{0} \\ \mathbf{0} & \mathbf{R}_1(n+1) \end{bmatrix}
\end{aligned} \tag{6.11.3}$$

Equation (6.11.3) is a linear system of equations and the behavior of $\boldsymbol{\delta}_y(n \mid n)$ can be investigated using the results of Chapter 4. The basic ideas are summarized in Fig. 6.11.1. This approach obviously applies directly to Bayesian, Fisher, and unknown-but-bounded models. The general construction leading to (6.11.3) also applies to nonlinear estimators (as might be encountered for unknown-but-bounded models), but in such cases the theory of Chapter 4 cannot be applied directly.

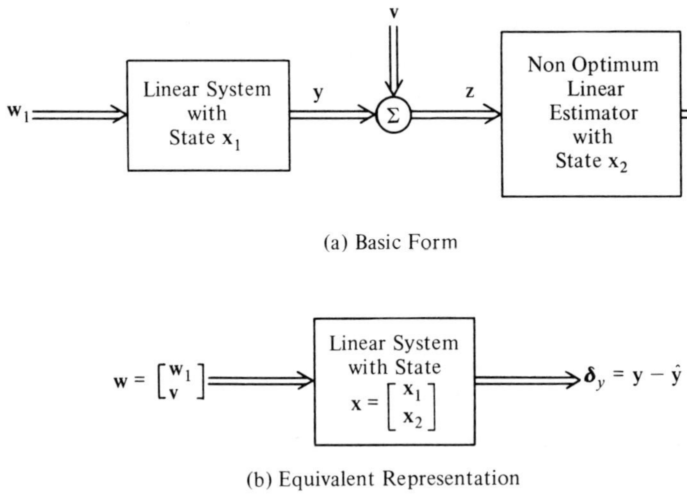

Figure 6.11.1 Analysis of nonoptimum estimators.

The desired performance analysis for a Bayesian model is obtained by solving a linear matrix equation for the corresponding covariance matrix $\boldsymbol{\Gamma}(n)$. If both $\mathbf{x}_1$ and $\mathbf{x}_2$ are K-dimensional, $\boldsymbol{\Gamma}(n)$ is a $2K$ by $2K$ matrix, so a $2K$ by $2K$ linear system of equations must be solved. The corresponding performance analysis for the optimum estimator is given by the nonlinear matrix Ricceti equation [see (6.3.1)] for the K by K matrix $\boldsymbol{\Sigma}$. It can be shown (see Exercise 6.15) that the K by K nonlinear Riccati equation for $\boldsymbol{\Sigma}$ can be (partly) replaced by a $2K$ by $2K$ linear equation, so, in this sense, the performance analysis of optimum and nonoptimum Bayesian (Fisher) estimators are of the same order of difficulty. However, when using a digital computer, the K by K nonlinear equations for $\boldsymbol{\Sigma}$ may be a preferable form.

6.12 Design of Constrained Estimators

The use of a nonoptimum estimator may be forced by "cost" considerations such as limited computer capacity (speed and/or memory). General discussions on the design of practical (nonoptimum) estimators are deferred to Chapter 8, but one particular approach to the design of "optimum" nonoptimum estimators is now considered.

In the estimation theory of the earlier sections, there are few or no a priori restrictions on the class of estimators, so the optimum estimator's complexity is dictated by the model. In practice, it may be desired to constrain the class of allowable estimators and then to choose the optimum estimator lying in this constrained class.

Consider a Bayesian model where the allowable estimators are constrained to lie within some class of linear systems. The choice of the best estimator from the class is, conceptually, a simple extension of the ideas of Section 6.11 on the analysis of nonoptimum Bayesian filters. The approach is as follows:

1. Let the nonoptimum filter of Fig. 6.11.1 belong to the class of allowable estimators.
2. Do a performance analysis as in Section 6.11.
3. Vary the nonoptimum estimator *over the class* of allowable estimators to see which yields the best performance.

As an example, assume that the estimator is restricted to be of the form of (6.11.2) for specified $\mathbf{G}_2$ and $\mathbf{H}_2$ and dimension of $\mathbf{x}_2$. $\mathbf{\Phi}_2$ is to be time-invariant but is otherwise free. Then a particular $\mathbf{\Phi}_2$ is to be chosen to minimize $\mathbf{\Gamma}_{\delta_v}(n)$ in some sense. In general, *global optimization* is not possible, and it is necessary to choose some particular scalar criterion such as $J = \text{tr}\{\mathbf{\Gamma}_{\delta_v}(N)\}$ for specified N or $J = \sum_{n=1}^{N} \text{tr}\{\mathbf{\Gamma}_{\delta_v}(n)\}$ and then solve for the $\mathbf{\Phi}_2$ which gives a minimum.

The class of possible estimators can, of course, be characterized in various ways depending on the problem. For example, $\mathbf{\Phi}_2$ may be constrained to have many "zero terms" so that only a few variables are left to be optimized. If the estimator is allowed to have time-varying coefficients, the calculus of variations can be used to solve for the best values. In fact, the optimum estimator can be derived by applying the above design philosophy to a class of time-varying estimators which happens to include the true optimum.

The constrained estimator design concept also applies to Fisher models. In Fisher problems, it is usually desirable to exercise care so that only unbiased estimators are considered. The design concept also applies to unknown-but-bounded models, especially if the equations are constrained to be linear (although a linear constraint is not needed conceptually).

6.13 Modeling Errors

Modeling errors such as uncertainty in the values of $\mathbf{\Phi}$, $\mathbf{G}$, $\mathbf{H}$, $\mathbf{Q}$, $\mathbf{R}$, and $\boldsymbol{\psi}$ are always present to some degree. In some applications, the effects of modeling errors dominate the behavior. Modeling errors will be discussed several times again in later chapters. For the present, it is merely noted that the discussions of Section 5.10 on modeling errors can be extended directly to dynamic system models such as (6.1). The effect of modeling errors on an optimum filter can be analyzed using the ideas of Section 6.11. It also follows that, as in the special sense discussed in Section 5.10, estimator performance is more sensitive to structural model errors in $\mathbf{\Phi}$, $\mathbf{G}$, and $\mathbf{H}$ than to uncertainty model errors in $\mathbf{Q}$, $\mathbf{R}$, and $\boldsymbol{\psi}$.

The phrase, *divergence of the Kalman filter*, has been used in the literature to denote poor performance (errors, much larger than indicated by $\mathbf{\Sigma}$) which results from modeling errors (sometimes called "approximations"). This use of the term divergence should not be confused with the distance measure concept to be discussed in Part III.

6.14 Choice of Algorithms

The difference equations for the various discrete-time estimators can be written in many different ways. For example, a Bayesian filter can be expressed as (6.2.5), (6.2.7), (6.3.1), or (6.3.3). And many other equations which look different but which express the same basic relationship can be written. Implementation on a digital computer requires the choice of some particular equation (algorithm), and hopefully this choice minimizes computation time, computer storage, and computation error. Unfortunately it is very difficult (if not impossible) to give any general rules on how to choose the best equation. However, some of the factors which effect the final decision can be discussed.

The choice of algorithm (or equation form) depends to a great extent on the computer to be used. In some computers, addition and multiplication require about the same amount of time, while in other computers multiplication is much slower. In some computers, more time may be spent in bookkeeping and internal data transferring than in actual arithmetic operations. The nature of computational error buildup depends on computer word length and on whether fixed-point or floating-point computations are done. The skill of the programmer who converts the equations to a computer program and the "language" he writes in are also very important.

In many problems, the $\mathbf{\Phi}$, $\mathbf{G}$, $\mathbf{H}$, $\mathbf{R}$, and $\mathbf{Q}$ matrices are sparse, i.e., have few nonzero entries. Such sparsity can be exploited in various ways, and the cleverness with which it is done can affect the performance of any particular algorithm. In some cases, the sparsity can be removed automatically by choice of model and method of analysis. For example, an autoregression model (see Section 3.4.2) results in a sparse $\mathbf{\Phi}$, but it is possible to analyze it directly without converting to the form of (6.1). The relative advantages often depend on the computer and the programmer.

The global optimality of $\hat{\mathbf{x}}(N|N)$ enables one to estimate $\mathbf{y}(N) = \mathbf{Cx}(N)$ by using $\hat{\mathbf{y}}(N|N) = \mathbf{C}\hat{\mathbf{x}}(N|N)$. If $\mathbf{y}(N)$ is a scalar, it appears at first that computation of $\hat{\mathbf{x}}(N|N)$ in order to get $\hat{\mathbf{y}}(N|N)$ may be wasteful of computer time, etc. It is often possible to rewrite the equations so that $\hat{\mathbf{x}}(N|N)$ does not appear explicitly, but such rewriting may or may not actually be helpful. It depends on the form of $\mathbf{\Phi}$, $\mathbf{G}$, etc., on the computer, and on the programmer.

To illustrate some of the explicit problems encountered in the choice of an algorithm, consider the two different forms, (6.2.5) and (6.3.1), for the matrix Riccati equation for $\mathbf{\Sigma}(N+1|N+1)$. Equation (6.2.5) requires inversion of K_1 by K_1 matrices (K_1 is the dimension of $\mathbf{x}$), while (6.3.1) requires inversion of a K_2 by K_2 matrix (K_2 is the dimension of $\mathbf{z}$). Since K_1 is often greater than K_2, (6.3.1) may seem to be superior in terms of computer time. However, (6.3.1) requires more matrix multiplications and may actually be slower. Furthermore, (6.3.1) requires subtraction, which can cause loss of accuracy.

The positive definite matrix $\mathbf{\Sigma}$ is symmetric. Thus, if $\mathbf{\Sigma}$ is K by K, it is really necessary to calculate only $(K/2)(K+1)$ terms. If $\mathbf{\Sigma}$ is calculated recursively by ignoring its symmetry, it is usually good practice to force it to be symmetrical by averaging, the off-diagonal terms at each step of the recursion.

Computational errors of many types can cause trouble. Matrix inversion can yield meaningless numbers if the dimension is large and "standard prepackaged" matrix inversion subroutines are used. The nonlinear difference equations for $\mathbf{\Sigma}$ can yield meaningless numbers for large N if care in programming is not exercised. The cause of such meaningless numbers are roundoff errors which result from the computer's finite word lengths. There are many ways to combat such roundoff error buildup, and a complete discussion is beyond the scope of this book. However the *square root filtering* technique must be mentioned as it can be very effective. See also the discussion in Section 5.11.

The $\mathbf{\Sigma}(n|n)$ and the gain matrices $K(n)$, $n = 1, \ldots, N$, can be precomputed before the observations are made (except in some unknown-but-bounded formulations). It may be more advantageous to precompute, store, and "look up" rather than compute in real time (on-line). Also, if computational error buildup is causing trouble, it may be possible to use an accurate computer to precalculate the gains, etc., while the estimator itself is implemented on a smaller, less accurate machine.

The above arguments on precomputing gains also apply to weighting patterns. If computer error buildup is causing trouble in the estimator difference equations, a weighting pattern implementation can be used instead as a weighting pattern implementation is less sensitive to computer error buildup.

One final point should be made on the question of computation errors. An estimate $\hat{\mathbf{x}}(N|N)$ is always in error because of the uncertainties in $\mathbf{x}(0)$,

w, and **v**. Additional errors introduced by the computations are not important unless they approach (or exceed) the magnitude of the inherent error in the estimate.

A decision on choice of algorithm should actually include many other aspects such as modeling errors and possibilities of using a nonoptimum estimator. Often such aspects are of much more importance. The discussion in Chapter 8 considers the problem from this more general point of view.

6.15 Sampled Data Systems

Many discrete-time estimator problems arise in *sampled data* systems when discrete-time observations are made on a continuous-time system. Two versions of the situation are illustrated in Fig. 6.15.1. The difference lies in the way the observation uncertainty is assumed to enter.

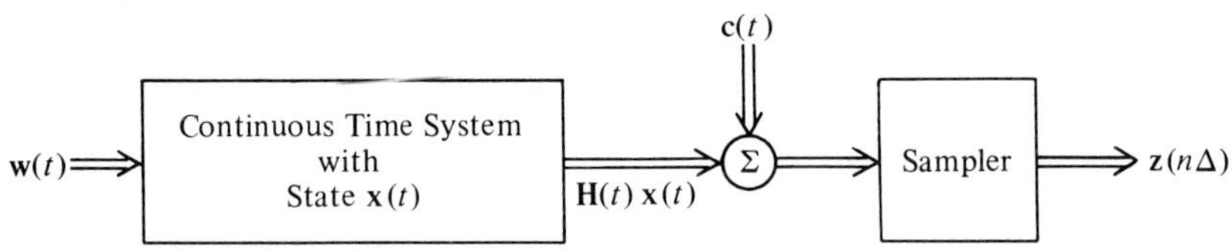

$\mathbf{w}(t)$: Continuous Time White Process
$\mathbf{c}(t)$: Continuous Time Uncertain Process

(a)

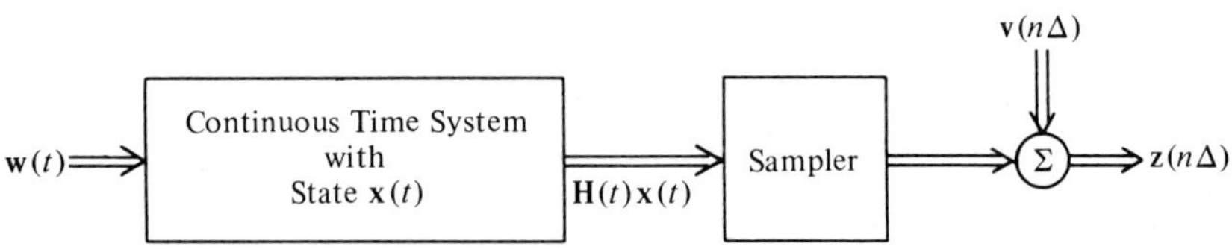

$\mathbf{v}(n\Delta)$: White Discrete Time Uncertain Process

(b)

Figure 6.15.1 Sampled data observation models.

A crucial factor in the analysis of sampled data estimation problems is the behavior of the sampler. Let $\mathbf{y}(t)$ denote the continuous-time input to the sampler and $\mathbf{z}(n\Delta)$ denote the discrete-time output of the sampler. Two sampler types will be considered:

$$\text{Instantaneous sampler:} \quad \mathbf{z}(n\Delta) = \mathbf{y}(t)\Big|_{t=n\Delta}$$

$$\text{Finite memory, averaging sampler:} \quad \mathbf{z}(n\Delta) = \frac{1}{\epsilon}\int_{n\Delta}^{n\Delta+\epsilon} \mathbf{y}(t)\,dt$$

There are many other possible sampler models but these two are sufficient to illustrate the basic ideas. It is assumed that $\epsilon < \Delta$.

A basic method of approach for all cases is as follows:

1. Develop a discrete-time model for the actual observations $\mathbf{z}(n\Delta)$ of the form

$$\mathbf{x}(n\Delta + \Delta) = \mathbf{\Phi}(n\Delta)\mathbf{x}(n\Delta) + \Delta\mathbf{G}(n\Delta)\mathbf{w}(n\Delta)$$
$$\mathbf{z}(n\Delta) = \mathbf{H}(n\Delta)\mathbf{x}(n\Delta) + \mathbf{v}(n\Delta)$$

where

$$\mathbf{x}(n\Delta) = \mathbf{x}(t)\Big|_{t=n\Delta}$$

$\mathbf{x}(t)$: state of continuous-time system at time t

2. Use discrete-time estimation theory to get $\hat{\mathbf{x}}(N\Delta \mid N\Delta)$, which yields the desired $\hat{\mathbf{x}}(T \mid T)$ for $T = N\Delta$.
3. If $\hat{\mathbf{x}}(T_1 \mid T)$ for $T_1 \neq N\Delta$ is desired, calculate it by prediction (or extrapolation or interpolation).

The discussions of this section concentrate on the first step.

To simplify the discussions, only Bayesian models will be considered. Thus

$$\frac{d}{dt}\mathbf{x}(t) = \mathbf{F}(t)\mathbf{x}(t) + \mathbf{G}(t)\mathbf{w}(t)$$
$$E\{\mathbf{w}(t)\mathbf{w}(\tau)\} = \delta(t - \tau)\mathbf{Q}(t)$$
$$E\{\mathbf{x}(0)\mathbf{x}'(0)\} = \boldsymbol{\psi}$$

$\mathbf{w}(t)$ and $\mathbf{x}(0)$ are zero mean, uncorrelated. Results for other uncertainty models follow in a similar fashion.

6.15.1 Instantaneous Sampler

Consider case (b) of Fig. 6.15.1. A model for an instantaneous sampler output is

$$\begin{aligned}
\mathbf{z}(n\Delta) &= \mathbf{H}(n\Delta)\mathbf{x}(n\Delta) + \mathbf{v}(n\Delta) \\
\mathbf{x}(n\Delta + \Delta) &= \mathbf{\Phi}(n\Delta)\mathbf{x}(n\Delta) + \Delta\mathbf{G}(n\Delta)\mathbf{w}(n\Delta) \\
E\{\mathbf{w}(n_1\Delta)\mathbf{w}'(n_2\Delta)\} &= \begin{cases} \mathbf{Q}(n_1\Delta), & n_1 = n_2 \\ \mathbf{0}, & n_1 \neq n_2 \end{cases} \\
E\{\mathbf{x}(0)\mathbf{x}'(0)\} &= \boldsymbol{\psi} \\
\mathbf{\Phi}(n\Delta) &= \boldsymbol{\theta}(t, \tau)\Big|_{\substack{t=n\Delta+\Delta \\ \tau=n\Delta}} \qquad (6.15.2) \\
\Delta\mathbf{G}(n\Delta) &= \mathbf{I} \\
\mathbf{w}(n\Delta) &= \left[\int_{n\Delta}^{t} \boldsymbol{\theta}(t, \tau)\mathbf{G}(\tau)\mathbf{w}(\tau)\, d\tau\right]\Bigg|_{t=n\Delta+\Delta} \\
\mathbf{Q}(n\Delta) &= \left[\int_{n\Delta}^{t} \boldsymbol{\theta}(t, \tau)\mathbf{G}(\tau)\mathbf{Q}(\tau)\mathbf{G}'(\tau)\boldsymbol{\theta}'(t, \tau)\, d\tau\right]\Bigg|_{t=n\Delta+\Delta}
\end{aligned}$$

$\boldsymbol{\theta}(t, \tau)$: fundamental matrix for $\mathbf{F}(t)$

For this model

$$\mathbf{x}(t)\Big|_{t=n\Delta} = \mathbf{x}(n\Delta)$$

so the results of Chapter 6 can be used to get $\hat{\mathbf{x}}(N\Delta/N\Delta) = \hat{\mathbf{x}}(T\,|\,T)\Big|_{T=N\Delta}$.
If $\hat{\mathbf{x}}(t\,|\,N\Delta)$, $N\Delta + \Delta > t \geq N\Delta$, is desired, it can be obtained by

$$\hat{\mathbf{x}}(t\,|\,N\Delta) = \boldsymbol{\theta}(t, \tau)\Big|_{\tau=N\Delta} \hat{\mathbf{x}}(N\Delta\,|\,N\Delta)$$

Now consider case (a) of Fig. 6.15.1. Equation (6.15.2) still applies for an instantaneous sampler except that

$$\begin{aligned} \mathbf{z}(n\Delta) &= \mathbf{H}(n\Delta)\mathbf{x}(n\Delta) + \tilde{\mathbf{v}}(n\Delta) \\ \tilde{\mathbf{v}}(n\Delta) &= \mathbf{c}(t)\Big|_{t=n\Delta} \end{aligned} \tag{6.15.3}$$

If $\mathbf{c}(t)$ is a correlated process (nonwhite), (6.15.3) is meaningful. $\tilde{\mathbf{v}}(n\Delta)$ is not white, so the ideas of Section 3.2 have to be used. However, if the *correlation time constant* of $\mathbf{c}(t)$ is small compared to the time between samples Δ, $\tilde{\mathbf{v}}(n\Delta)$ can be assumed to be white. If $\mathbf{c}(t)$ in Fig. 6.15.1(a) is assumed to be a white process, (6.15.3) is not meaningful, as it yields $E\{\tilde{\mathbf{v}}(n\Delta)\tilde{\mathbf{v}}'(n\Delta)\} = \infty\,\mathbf{I}$. In such a case, the assumption of an instantaneous sampler cannot be used. (See Section 6.15.2.)

6.15.2 Finite Memory, Averaging Sampler

Consider case (a) of Fig. 6.15.1. Equation (6.15.2) still applies for an averaging sampler except that

$$\begin{aligned} \mathbf{z}(n\Delta) &= \tilde{\mathbf{H}}(n\Delta)\mathbf{x}(n\Delta) + \tilde{\mathbf{v}}(n\Delta) \\ \tilde{\mathbf{H}}(n\Delta) &= \frac{1}{\epsilon}\int_{n\Delta}^{n\Delta+\epsilon} \mathbf{H}(t)\boldsymbol{\theta}(t, \tau)\Big|_{\tau=n\Delta} dt \\ \tilde{\mathbf{v}}(n\Delta) &= \frac{1}{\epsilon}\int_{n\Delta}^{n\Delta+\epsilon}\left[\mathbf{c}(t) + \int_{n\Delta}^{t} \boldsymbol{\theta}(t, \tau)\mathbf{G}(\tau)\mathbf{w}(\tau)\,d\tau\right] dt \end{aligned} \tag{6.15.4}$$

In this case $\tilde{\mathbf{v}}(n\Delta)$ and $\mathbf{w}(n\Delta)$ in (6.15.2) are correlated. Hence if an optimum estimate of $\mathbf{x}(N\Delta)$ from $\mathbf{z}(n\Delta)$, $n = 1, \ldots,$ is desired, it is necessary to use a formulation, as discussed in Exercise 3.8, which allows for such correlation. However, for many practical situations the correlation between $\tilde{\mathbf{v}}(n\Delta)$ and $\mathbf{w}(n\Delta)$ can be ignored.

Note that in (6.15.4) if the observation uncertainty $\mathbf{c}(t)$ of Fig. 6.15.1(a) is a white process then $\tilde{\mathbf{v}}(n\Delta)$ is a white zero mean process with a finite covariance matrix. However, when $\mathbf{c}(t) = \mathbf{v}(t)$, $\tilde{\mathbf{v}}(n\Delta)$ of (6.15.4) is still correlated with $\mathbf{w}(n\Delta)$ of (6.15.2).

6.15.3 Discussion of Sampled Data

The ideas and concepts of how to handle discrete-time observations of continuous-time systems (sampled data) are very important, as a very wide range of practical problems are of this form. Obviously not all possibilities were discussed, but hopefully the reader can handle other situations on his own.

The discussions point out an important moral about sampled data systems: *Never glibly ignore the dynamic behavior of the sampling device.* Instantaneous samplers do not exist in the real world.

6.16 Curve Fitting†

Often one is given some time function $z(n)$, $n = 1, \ldots, N$, and is asked to find a *curve* of the form

$$\sum_{k=1}^{K} a_k \eta_k(n), \qquad n = 1, \ldots, N$$

which best fits $z(n)$ in some sense where the $\eta_k(n)$ are specified. The problem is to find the best values for the a_k. Common curve-fitting problems involve

$$\left.\begin{aligned} \eta_k(n) &= \alpha_k^n \\ \eta_k(n) &= \sin(\omega_k n) \\ \eta_k(n) &= n^k \end{aligned}\right\} k = 1, \ldots, K \qquad \begin{aligned} (6.16.1) \\ (6.16.2) \\ (6.16.3) \end{aligned}$$

6.16.1 Least Squares

A least-squares curve-fitting problem formulation is as follows: Given $z(n)$, $n = 1, \ldots, N$, find the a_k, $k = 1, \ldots, K$, which minimize

$$J = \sum_{n=1}^{N} \left[z(n) - \sum_{k=1}^{K} a_k \eta_k(n) \right]^2 \tag{6.16.4}$$

Equation (6.16.4) can be reexpressed in the state estimation notation by writing

$$\sum_{k=1}^{K} a_k \eta_k(n) = \mathbf{H}(n)\mathbf{x}$$

$$\mathbf{x} = \begin{bmatrix} a_1 \\ \cdot \\ \cdot \\ \cdot \\ a_K \end{bmatrix} \tag{6.16.5}$$

$$\mathbf{H}'(n) = \begin{bmatrix} \eta_1(n) \\ \cdot \\ \cdot \\ \cdot \\ \eta_K(n) \end{bmatrix}$$

Because of the close relation between weighted-least-squares estimators and Fisher model estimators, the curve-fitting problem of (6.16.4) can sometimes be viewed as being equivalent to a Fisher estimation problem.

Sometimes it is possible and more convenient to view a curve-fitting problem as an estimation problem for a model such as

$$\begin{aligned} z(n) &= \mathbf{H}\mathbf{x}(n) + v(n) \\ \mathbf{x}(n+1) &= \mathbf{\Phi}\mathbf{x}(n) \\ \mathbf{x}(0) &= \begin{bmatrix} a_1 \\ \cdot \\ \cdot \\ \cdot \\ a_K \end{bmatrix} \end{aligned} \tag{6.16.6}$$

This is possible if $\mathbf{\Phi}$ and $\mathbf{H}$ can be chosen so that

$$\begin{bmatrix} \eta_1(n) \\ \cdot \\ \cdot \\ \cdot \\ \eta_K(n) \end{bmatrix} = \mathbf{H}\mathbf{\Phi}^n$$

This can be done for the $\eta_k(n)$ of (6.16.1)–(6.16.3).

6.16.2 Min-Max

Define

$$\delta(n) = z(n) - \sum_{k=1}^{K} a_k \eta_k(n)$$

It may be desired to choose the a_k, $k = 1, \ldots, K$, to minimize the maximum $|\delta(n)|$, $n = 1, \ldots, N$. This min-max curve-fitting formulation can be attacked using unknown-but-bounded model estimation theory. Define

$$\mathbf{z}_N = \begin{bmatrix} z(1) \\ \cdot \\ \cdot \\ \cdot \\ z(N) \end{bmatrix}$$

$$\mathbf{x} = \begin{bmatrix} a_1 \\ \cdot \\ \cdot \\ \cdot \\ a_K \end{bmatrix}$$

$$\mathbf{H}'(n) = \begin{bmatrix} \eta_1(n) \\ \cdot \\ \cdot \\ \cdot \\ \eta_K(n) \end{bmatrix}$$

$$\mathbf{H}_N = \begin{bmatrix} \mathbf{H}(1) \\ \cdot \\ \cdot \\ \cdot \\ \mathbf{H}(N) \end{bmatrix}$$

$$\delta_N = \begin{bmatrix} \delta(1) \\ \cdot \\ \cdot \\ \cdot \\ \delta(N) \end{bmatrix}$$

so that

$$\delta_N = \mathbf{z}_N - \mathbf{H}_N\mathbf{x}$$

Consider the unknown-but-bounded estimator model

$$\mathbf{z}_N = \mathbf{H}_N\mathbf{x} + \mathbf{v}_N$$

$$\mathbf{v}_N = \begin{bmatrix} v(1) \\ \cdot \\ \cdot \\ \cdot \\ v(N) \end{bmatrix}$$

$$|v(n)| \leq R^{1/2}, \qquad n = 1, \ldots, N$$

$$\mathbf{x}\text{:} \quad \text{completely unknown}$$

Let $\mathbf{\Omega}_{\text{est}}(N)$ denote the resulting K-dimensional set which must contain $\mathbf{x}$. If R is fixed, this estimate set will not exist for some $\mathbf{z}_N$. However, for a given $\mathbf{z}_N$, there exists an R_0 such that for $R \geq R_0$, $\mathbf{\Omega}_{\text{est}}(N)$ exists. For $R = R_0$, $\mathbf{\Omega}_{\text{est}}(N)$ contains a single vector, and this vector is the vector that minimizes the $\max|\delta(n)|$, $n = 1, \ldots, N$. Hence the min-max curve-fitting problem can be solved by estimation theory by varying R until the estimate set contains only a single vector.

There exists a variety of variations on the above ideas.

6.16.3 Polynomial Curve Fitting

A polynomial curve-fitting problem involves use of (6.16.3); i.e.,

$$\eta_k(n) = n^k$$

The minimizing a_k of (6.16.4) can be found by solving the Fisher model estimation problem

$$
\begin{aligned}
z(n) &= y(n) + v(n) \\
y(n) &= \mathbf{H}(n)\mathbf{x} \\
\mathbf{x} &= \begin{bmatrix} a_0 \\ \cdot \\ \cdot \\ \cdot \\ a_K \end{bmatrix} \\
E\{v(n_1)v(n_2)\} &= \begin{cases} R, & n_1 = n_2 \\ 0, & n_1 \neq n_2 \end{cases} \\
E\{v(n)\} &= 0 \\
\mathbf{II}'(n) &= \begin{bmatrix} 1 \\ n \\ n^2 \\ \cdot \\ \cdot \\ \cdot \\ n^K \end{bmatrix}
\end{aligned}
\tag{6.16.7}
$$

Let $\hat{\mathbf{x}}(N\,|\,N)$ denote the Fisher estimate of $\mathbf{x}$ using $z(1) \cdots z(N)$. Then

$$
\begin{aligned}
\hat{\mathbf{x}}(N\,|\,N) &= \left[\sum_{n=1}^{N} \mathbf{H}'(n)\mathbf{H}(n)\right]^{-1} \sum_{n=1}^{N} \mathbf{H}'(n)z(n) \\
\mathbf{\Sigma}_x(N\,|\,N) &= R\left[\sum_{k=1}^{N} \mathbf{H}'(n)\mathbf{H}(n)\right]^{-1} \\
\mathbf{\Sigma}_x(N\,|\,N) &= E\{[\mathbf{x} - \hat{\mathbf{x}}(N\,|\,N)][\mathbf{x} - \hat{\mathbf{x}}(N\,|\,N)]'\} \\
\hat{y}(n\,|\,N) &= \mathbf{H}(n)\hat{\mathbf{x}}(N\,|\,N) \\
\mathbf{\Sigma}_y(n\,|N) &= E\{[y(n) - \hat{y}(n\,|\,N)]^2\}
\end{aligned}
\tag{6.16.8}
$$

The matrix inversion associated with $\mathbf{\Sigma}_x(N\,|\,N)$ can be accomplished by introducing orthonormal polynomials $\gamma_k(n, N)$, $k = 1, \ldots, K$, $n = 1, \ldots, N$. These polynomials have the following properties:

$$
\gamma_k(n, N) = \sum_{j=0}^{k} c_j n^j
$$

$$
\sum_{n=1}^{N} \gamma_k(n, N)\gamma_j(n, N) = \begin{cases} 0, & j \neq k \\ 1, & j = k \end{cases}
$$

Such orthonormal polynomials can be generated by the following formulas:

$$
\begin{aligned}
\gamma_k(n, N) &= \frac{1}{k!}\left(\frac{2k+1}{N(N^2-1)\cdots(N^2-k^2)}\right)^{1/2} \xi_k(n, N) \\
\xi_{k+1}(n, N) &= (2k+1)(2n-k-1)\xi_k(n, N) \\
&\quad - (N^2-k^2)k^2\xi_{k-1}(n, N)
\end{aligned}
\tag{6.16.9}
$$

For $k = 0, 1, 2,$

$$\gamma_0(n, N) = \left(\frac{1}{N}\right)^{1/2}$$

$$\gamma_1(n, N) = \left(\frac{12}{N(N^2 - 1)}\right)^{1/2}\left[n - \frac{N+1}{2}\right]$$

$$\gamma_2(n, N) = \left(\frac{180}{N(N^2 - 1(N^2 - 4)}\right)^{1/2}\left[n^2 - (N + 1)n + \frac{(N + 1)(N + 2)}{6}\right]$$

The behavior of $\gamma_0(n, N)$ through $\gamma_5(n, N)$ is indicated in Fig. 6.16.1 for the case where N is very large.

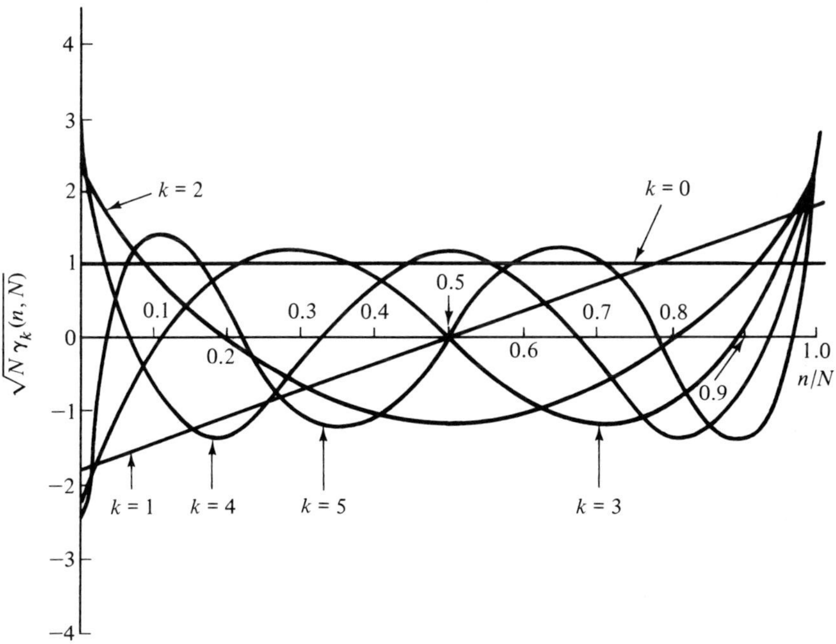

Figure 6.16.1 Orthonormal polynomials, $\gamma_k\ (n, | N)$ for large N.

Now return to the Fisher model estimation problem. The polynomial $y(n)$ can be expressed as

$$y(n) = \sum_{k=0}^{K} b_k \gamma_k(n, N)$$

so

$$z(n) = \tilde{\mathbf{H}}(n)\tilde{\mathbf{x}} + v(n)$$

$$\tilde{\mathbf{x}} = \begin{bmatrix} b_0 \\ \cdot \\ \cdot \\ \cdot \\ b_K \end{bmatrix}$$

$$\tilde{\mathbf{H}}'(n) = \begin{bmatrix} \gamma_0(n, N) \\ \cdot \\ \cdot \\ \cdot \\ \gamma_K(n, N) \end{bmatrix}$$

Because of the orthonormal properties,

$$\sum_{n=1}^{N} \tilde{\mathbf{H}}'(n)\tilde{\mathbf{H}}(n) = \mathbf{I}$$

so that

$$\hat{\tilde{\mathbf{x}}}(N \mid N) = \sum_{n=1}^{N} \tilde{\mathbf{H}}'(n)z(n)$$

$$\mathbf{\Sigma}_{\tilde{x}}(N \mid N) = R\,\mathbf{I}$$

These results in terms of $\tilde{\mathbf{x}}$ (i.e., the b_k) can now be converted back to $\mathbf{x}$ (i.e., the a_k) and a wide variety of interesting equations follows.

Plots of $\Sigma_y(n \mid N)$ for $K = 0, \ldots, 5$ are presented in Fig. 6.16.2 for the case of large N. Let $W_{K,K}(n, N)$ denote the weighting function for estimating the highest degree coefficient of a polynomial so that

$$\hat{a}_K(N) = W_{K,K}(n, N)z(n)$$

The shape of these weighting functions can be seen from Fig. 6.16.1, as $W_{K,K}(n, N)$ is proportional to $\gamma_K(n, N)$. For many polynomial curve-fitting

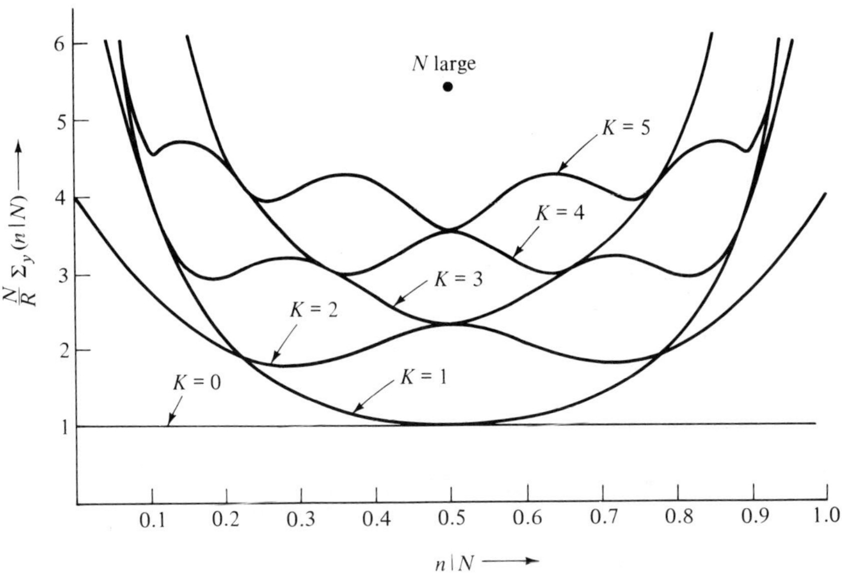

Figure 6.16.2 Error variance of polynomial curve fit.

applications it is more convenient to consider polynomials *referenced to the midpoint*. Thus instead of (6.16.7), consider

$$z(n) = y(n) + v(n)$$

First degree: $y(n) = x_1 + x_2\left(n - \frac{N+1}{2}\right)$

Second degree: $y(n) = x_1 + x_2\left(n - \frac{N+1}{2}\right) + \frac{x_3}{2}\left(n - \frac{N+1}{2}\right)^2$

Let $\mathbf{\Sigma}_x(N \mid N)$ denote the resulting error covariance matrix. Then

First degree:

$$\mathbf{\Sigma}_x(N \mid N) = R\begin{bmatrix} \frac{1}{N} & 0 \\ 0 & \frac{12}{N(N^2-1)} \end{bmatrix}$$

Second degree:

$$\mathbf{\Sigma}_x(N \mid N) = R\begin{bmatrix} \frac{3(3N^2-1)}{4N(N^2-4)} & 0 & \frac{-30}{N(N^2-4)} \\ 0 & \frac{12}{N(N^2-1)} & 0 \\ \frac{-30}{N(N^2-4)} & 0 & \frac{720}{N(N^2-1)(N^2-4)} \end{bmatrix}$$

Many more similar results can be relatively easily derived using the orthonormal polynomials.

The preceding equations result from a Fisher model. This Fisher formulation yields the coefficients which minimize the J of (6.16.4). However, care is needed in interpreting the Fisher model result in terms of the original curve-fitting problem. The curve-fitting problem is for arbitrary $z(n)$, $n = 1, \ldots, N$, while the Fisher model assumes that $z(n)$ is the sum of a polynomial and a white stochastic process of known variance R.

6.16.4 Adaptive Curve Fitting

Consider the polynomial curve-fitting problem. An adaptive curve-fitting problem is how to use the $z(1) \cdots z(N)$ to specify the degree of polynomial, K, to be used.

At first thought, one might attempt to find the optimum value K by minimizing J of (6.16.4) over both the a_k and the value of K. However, such a minimization yields $K = N - 1$, with the corresponding minimum J being zero. Thus something else must be done (unless $K = N - 1$ is actually desired).

To find an optimum value for K ($K < N$), some modeling of $z(n)$, $n = 1, \ldots, N$, is needed. One reasonable approach is as follows. Assume that $z(n)$, $n = 1, \ldots, N$, is the sum of some *smooth* time function $g(n)$, $n = 1, \ldots, N$, and a discrete-time white process $v(n)$, $n = 1, \ldots, N$:

$$z(n) = g(n) + v(n), \qquad n = 1, \ldots, N$$
$$E\{v(n)\} = 0$$
$$E\{v(n_1)v(n_2)\} = \begin{cases} R, & n_1 = n_2 \\ 0, & n_1 \neq n_2 \end{cases}$$

Further assume that $g(n)$ can be "closely approximated" by a K_0th degree polynomial, i.e.,

$$g(n) = \sum_{k=0}^{K_0} a_k n^k, \qquad n = 1, \ldots, N$$

but that the values of K_0 and the a_k are not known. Then, the value of K_0 can be "estimated" by hypothesizing that K_0 actually takes on values, $K_0 = 0$, $K_0 = 1, \ldots$, and then using hypothesis testing techniques to choose one of the hypotheses. The needed hypothesis-testing theory is given in Chapters 9 and 10. When choosing the hypothesized value, one can start with $K_0 = 0$ and increase the hypothesized value of K_0 until a hypothesis is accepted. If an upper bound $K_{\max}$ on the true value of K_0 is available, one can start with the hypothesis that $K_0 = K_{\max}$ and "work down" until a hypothesis is rejected.

The above concept can obviously be applied to $\eta_k(n)$ which are not polynomials. It can also be applied to problems such as deciding whether it is "better" to fit a given $z(1) \cdots z(N)$ by polynomials or Fourier series.

6.16.5 Stochastic Polynomials

The polynomial curve-fitting problem of Section 6.16.4 can be viewed as the Fisher estimation of the state of an undriven dynamic system ($Q = 0$). The estimation problem resulting when this same dynamic system is driven by a white stochastic process is now discussed. The resulting signal to be estimated is called a stochastic polynomial (for the lack of a better term).

A second-degree polynomial curve-fitting problem can be formulated as the Fisher model

$$\begin{aligned}
\mathbf{x}(n+1) &= \mathbf{\Phi}\mathbf{x}(n) \\
z(n) &= \mathbf{H}\mathbf{x}(n) + v(n) \\
\mathbf{\Phi} &= \begin{bmatrix} 1 & 1 & \frac{1}{2} \\ 0 & 1 & 1 \\ 0 & 1 & 1 \end{bmatrix} \\
\mathbf{H} &= [1 \quad 0 \quad 0] \\
E\{v(n_1)v(n_2)\} &= \begin{cases} R, & n_1 = n_2 \\ 0, & n_1 \neq n_2 \end{cases} \\
\mathbf{x}(0)&: \text{ completely unknown}
\end{aligned} \tag{6.16.10}$$

where

$$\mathbf{H}\mathbf{x}(n) = x_1(0) + x_2(0)n + \frac{x_2(0)}{2}n^2$$

A stochastic polynomial curve-fitting problem results when (6.16.10) is modified to the Fisher model (unknown initial conditions)

$$\mathbf{x}(n+1) = \mathbf{\Phi}\mathbf{x}(n) + \mathbf{G}w(n)$$

$$\mathbf{G} = \begin{bmatrix} 0 \\ 0 \\ 1 \end{bmatrix} \tag{6.16.11}$$

$$E\{w(n_1)w(n_2)\} = \begin{cases} Q, & n_1 = n_2 \\ 0, & n_1 \neq n_2 \end{cases}$$

Some explicit numerical results for this problem obtained using a digital computer are now summarized.

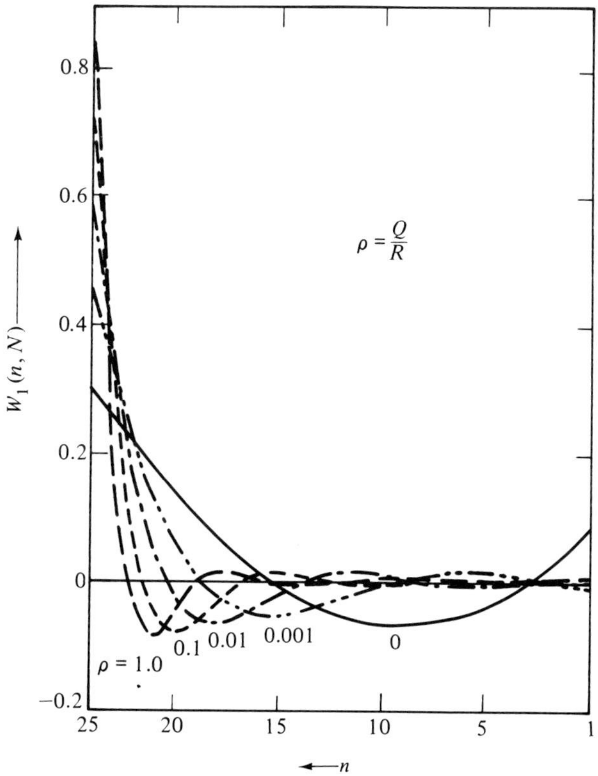

Figure 6.16.3 Weighting function for estimating position $x_1(N)$, $N = 25$.

The studies were motivated by the problem of using position measurements ($z(n)$) made on a moving vehicle to estimate the vehicle's position velocity and acceleration, where

$$\mathbf{x}(N) = \begin{bmatrix} x_1(N) \\ x_2(N) \\ x_3(N) \end{bmatrix} = \begin{bmatrix} \text{position} \\ \text{velocity} \\ \text{acceleration} \end{bmatrix}$$

With this interpretation, the undriven dynamic model of (6.16.10) is a constant acceleration model, while the driven dynamic model of (6.16.11) can be considered a random acceleration model.

Let $\hat{\mathbf{x}}(N|N)$ denote the estimate using $z(1) \cdots z(N)$. Then in terms of weighting patterns

$$\hat{x}_k(N \mid N) = \sum_{n=1}^{N} W_k(n, N) z(n), \qquad k = 1, 2, 3$$

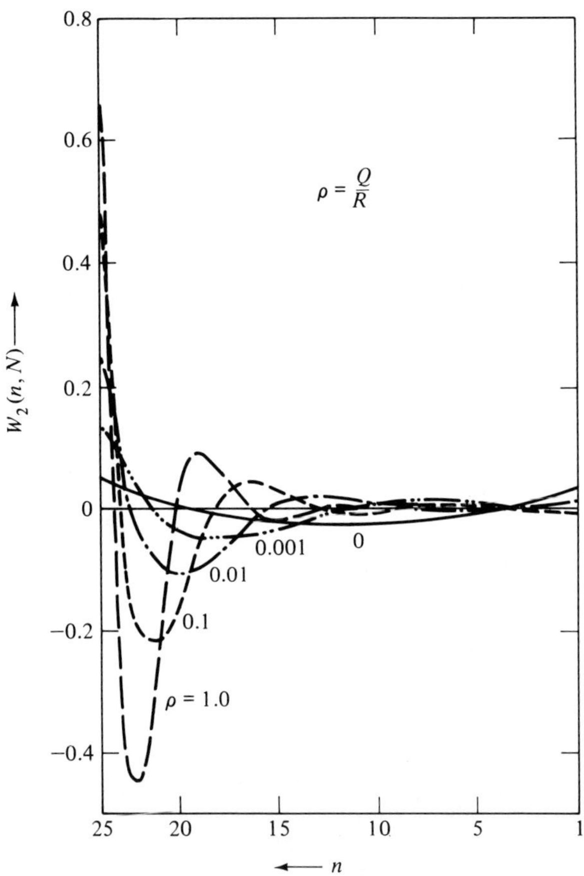

Figure 6.16.4 Weighting function for estimating velocity $x_2(N)$, $N = 25$.

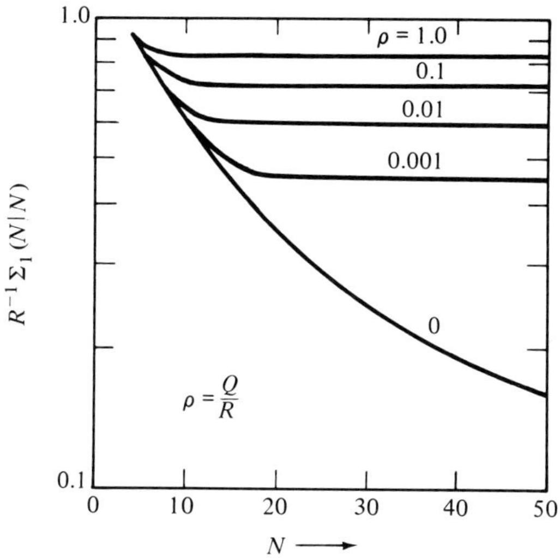

Figure 6.16.5 Error variance in position versus filter length.

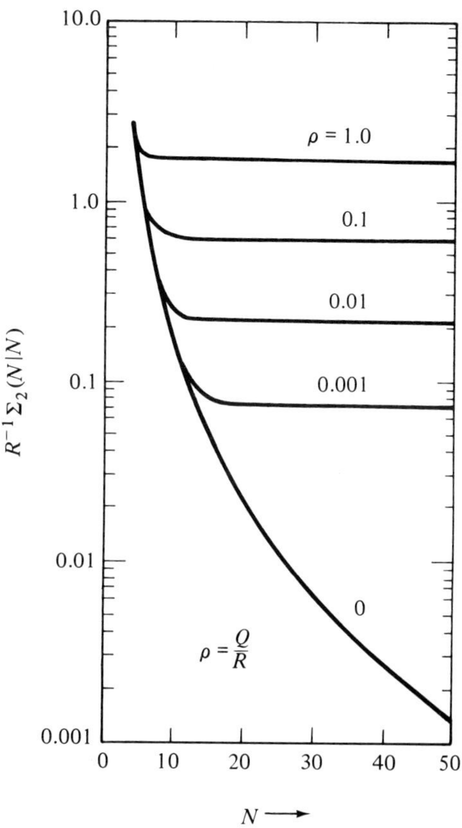

Figure 6.16.6 Error variance in velocity versus filter length.

The position weighting pattern $W_1(n, N)$ and velocity weighting pattern $W_2(n, N)$ are given in Fig. 6.16.3 and 6.16.4 for $N = 25$ and various values of ρ where $\rho = Q/R$. Let $\Sigma_k(N|N)$ denote the error variance of $\hat{\mathbf{x}}_k(N|N)$, $k = 1, 2, 3$, so $\Sigma_1(N|N)$ is the error variance of the position estimate, etc. $\Sigma_1(N|N)$, $\Sigma_2(N|N)$, and $\Sigma_3(N|N)$ versus N is plotted in Fig. 6.16.5–6.16.7 for various values of ρ. In all these figures, $\rho = 0$ corresponds to results of Section 6.16.3 for a deterministic second-degree polynomial ($K = 2$). These results illustrate the large differences that can result when a relatively small amount of driving white process is introduced.

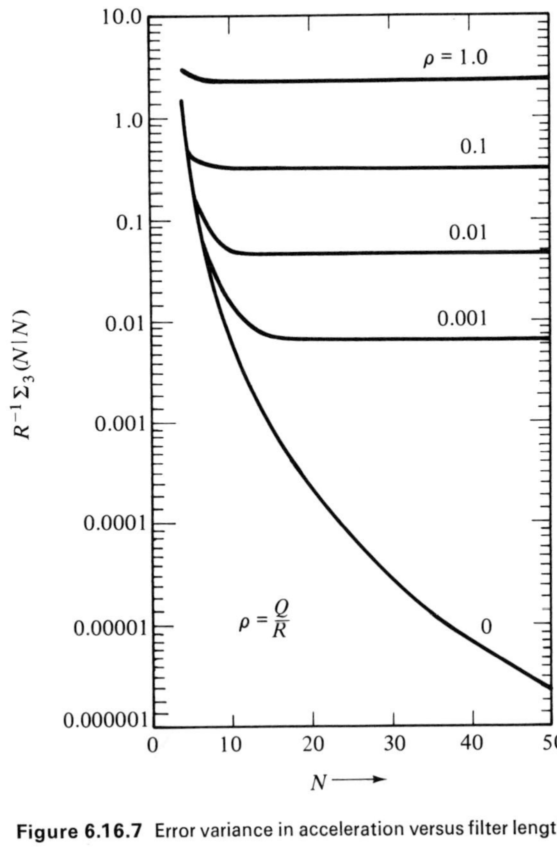

Figure 6.16.7 Error variance in acceleration versus filter length.

6.17 Asymptotic Efficiency: Fisher Estimation†

For certain types of Fisher models, a nonoptimum filter can be just as effective as an optimum filter for long observation time intervals. Such a nonoptimum filter is said to be asymptotically efficient.

Consider a Fisher estimation problem of the form

$$z_{\text{corr}}(n) = y(n) + v_s(n), \qquad n = 1, \ldots, N$$

$$y(n) = \mathbf{H}(n)\mathbf{x} = \sum_{k=1}^{K} a_k \eta_k(n)$$

$$\mathbf{H}'(n) = \begin{bmatrix} \eta_1(n) \\ \cdot \\ \cdot \\ \cdot \\ \eta_K(n) \end{bmatrix} \tag{6.17.1}$$

$$\mathbf{x} = \begin{bmatrix} a_1 \\ \cdot \\ \cdot \\ \cdot \\ a_K \end{bmatrix} : \quad \mathbf{x} \text{ is unknown}$$

$v_s(n)$: zero mean stationary stochastic process with correlation function

$$E\{v_s(n)v_s(m)\} = \mathcal{P}(n-m)$$

The optimum estimator for this model can be expressed in terms of a weighting pattern as

$$\hat{\mathbf{x}}_{\text{corr}}(N \mid N) = \sum_{n=1}^{N} \mathbf{W}_{\text{corr}}(n, N) z_{\text{corr}}(n)$$

Let $\mathbf{\Sigma}_{\text{corr}}(N|N)$ denote the error covariance matrix of $\mathbf{x}_{\text{corr}}(N|N)$. Now consider a different Fisher problem with the same $y(n)$ but with white instead of correlated noise:

$$z_{\text{white}}(n) = y(n) + v(n) \tag{6.17.2}$$
$$v(n): \quad \text{white stochastic process}$$

For this case the optimum estimator can also be expressed in terms of a weighting pattern as

$$\hat{\mathbf{x}}_{\text{white}}(N \mid N) = \sum_{n=1}^{N} \mathbf{W}_{\text{white}}(n, N) z_{\text{white}}(n)$$

Let $\mathbf{\Sigma}_{\text{white}}(N|N)$ denote the error covariance matrix of $\hat{\mathbf{x}}_{\text{white}}(N|N)$.

Assume that the actual observations involve correlated noise [i.e., (6.17.1)] but that the estimation is to be done using a filter that is designed to be optimum for white noise. If $\hat{\mathbf{x}}_{\text{w}|\text{c}}(N|N)$ denotes the resulting estimate,

$$\hat{\mathbf{x}}_{\text{w}|\text{c}}(N \mid N) = \sum_{n=1}^{N} \mathbf{W}_{\text{white}}(n, N) z_{\text{corr}}(n)$$

Clearly $\hat{\mathbf{x}}_{\text{w}|\text{c}}(N|N)$ is the output of a nonoptimum estimator. Let $\mathbf{\Sigma}_{\text{w}|\text{c}}(N|N)$ denote the error covariance matrix of $\hat{\mathbf{x}}_{\text{w}|\text{c}}(N|N)$.

In general, it follows that

$$\mathbf{\Sigma}_{\text{white}}(N \mid N) \neq \mathbf{\Sigma}_{\text{corr}}(N \mid N) \tag{6.17.3}$$
$$\mathbf{\Sigma}_{\text{w}|\text{c}}(N \mid N) > \mathbf{\Sigma}_{\text{corr}}(N \mid N)$$

However, for certain classes of $\eta_k(n)$, it turns out that for large N

$$\mathbf{\Sigma}_{\text{w}|\text{c}}(N \mid N) \longrightarrow \mathbf{\Sigma}_{\text{corr}}(N \mid N) \tag{6.17.4}$$

which means that the *white-noise-designed* estimator is asymptotically (large

N) as efficient as the optimum estimator designed using the correlated noise properties. Classes of the $\eta_k(n)$ for which (6.17.3) often happens are polynomials $\eta_k(n) = n^k$ and sine functions $\eta_k(n) = \sin(\omega_k n)$. Such asymptotic efficiency is also encountered in continuous-time problems.

EXAMPLE: Consider

$$\begin{aligned} z_{\text{corr}}(n) &= x_1 + v_s(n), \qquad n = 1, \ldots, N \\ E\{v_s(n_1)v_s(n_2)\} &= \Gamma\Phi^{|n_1 - n_2|} \\ z_{\text{white}}(n) &= x_1 + v(n), \qquad n = 1, \ldots, N \\ E\{v(n_1)v(n_2)\} &= \begin{cases} \Gamma, & n_1 = n_2 \\ 0, & n_1 \neq n_2 \end{cases} \\ x_1&: \text{unknown} \qquad |\Phi| < 1 \end{aligned}$$

The correlated noise estimation problem can be handled as in the example of Section 6.9 or by using the matrix manipulation approach of Section 6.1 (see Exercise 6.9). The performance of the nonoptimum estimator can be calculated directly from

$$W_{\text{white}}(n, N) = \frac{1}{N}$$

The results are

$$\begin{aligned} \Sigma_{\text{white}}(N|N) &= \frac{\Gamma}{N} \\ \Sigma_{\text{corr}}(N|N) &= \frac{\Gamma(1 + \Phi)}{N(1 - \Phi) + 2\Phi} \\ \Sigma_{w|c}(N|N) &= \frac{\Gamma(1 + \Phi)}{N(1 - \Phi)} + \frac{\Gamma 2\Phi(1 - \Phi^N)}{N^2(1 - \Phi)^2} \end{aligned}$$

It can be seen that

$$\Sigma_{w|c}(N|N) \geq \Sigma_{\text{corr}}(N|N)$$

but as N increases,

$$\Sigma_{\text{corr}}(N|N) \longrightarrow \Sigma_{w|c}(N|N) \longrightarrow \frac{\Gamma(1 + \Phi)}{N(1 - \Phi)}$$

6.18 Role of White Process Model

The basic models considered in this chapter involve white uncertain processes, $\mathbf{w}(n)$ and $\mathbf{v}(n)$. The whiteness of these processes is what made the analysis "easy" and what yielded such "simple" equations. This emphasizes the importance of the ideas of Chapter 3 on how to convert "natural" models into a state space-white process form. This is the same statement made in Section 4.5, but it bears repeating.

In Section 4.5, the point was also made that for stochastic models, an input nonwhite process should not be arbitrarily replaced by a white process of the same variance in the hopes of obtaining a worst case analysis. This point applies equally to the case of observation noise. A white stochastic observation noise model can yield higher or lower errors than a nonwhite model with the same variance.

EXAMPLE: Consider

$$z(n) = x_1 + v_s(n)$$

$$x_1 : \text{unknown}$$

$$v_s(n+1) = \Phi v_s(n) + w_s(n)$$

$$E\{w_s(n)\} = 0$$

$$E\{w_s(n_1)w_s(n_2)\} = \begin{cases} Q, & n_1 = n_2 \\ 0, & n_1 \neq n_2 \end{cases}$$

$$|\Phi| < 1$$

Assume that $v_s(n)$ is stationary so that

$$\Gamma = E\{v_s^2(n)\} = \frac{Q}{1 - \Phi^2}$$

Define

$$\Sigma_{11}(N|N) = \text{error covariance matrix of } \hat{\mathbf{x}}_1(N/N)$$

Then using the example of Section 6.9 or 6.17 it follows that for large N

$$\Sigma_{11}(N|N) = \frac{Q}{N(1-\Phi)^2} = \frac{\Gamma(1+\Phi)}{N(1-\Phi)}$$

Now consider varying Φ with Γ held constant. Define $\Sigma_{11,\text{white}}(N|N)$ to be the error variance in estimating x when $v_s(n)$ is white. Since $v_s(n)$ is white when $\Phi = 0$,

$$\Sigma_{11,\text{white}} = \frac{\Gamma}{N}$$

Thus for $|\Phi| < 1$,

$$\Sigma_{11} > \Sigma_{11,\text{white}} \qquad \Phi > 0$$

$$\Sigma_{11} < \Sigma_{11,\text{white}} \qquad \Phi < 0$$

Exercise 7.9 is also related to this example.

6.19 Discussion

This chapter illustrates how a wide variety of complex formulas can be derived from a "simple" model such as (6.1). By no means all possible cases of interest were considered, and many more pages of equations and discussion could easily be added. However, it is hoped that sufficient material is provided so the reader can do his own extensions as needed.

A point made at the end of Chapter 5 should be reemphasized. Many of the equations are very similar in appearance but if the derivations started with different models for uncertainty, the results have different physical interpretations.

The complexity of the various equations is actually misleading. All the equations have simple interpretations such as the two-step cycle: predict, estimate. The complexity is actually just bookkeeping. This bookkeeping is the type that can easily be implemented on a digital computer.

The Bayesian model, Fisher model, and weighted-least-squares estimators are old, well-established techniques that have a long history of successful applications (plus a less well-published history of failures). The unknown-but-bounded model estimator is a new technique which has not yet been applied to other than academic problems.

Historical Notes and References: Chapter 6

Mendel and Giesling [1] is a bibliography of over 900 references, many of which are related to (mostly Bayesian) aspects of this chapter. Some of these references are discussed below, but many are not.

Section 6.2, 6.3

This discussion also applies to Sections 7.1 and 7.2.

The difference equations (or differential equations) which specify the optimum linear filter estimate are often called the *Kalman filter* or the *Kalman-Bucy filter* because of the fundamental papers Kalman [1], [2], and Kalman and Bucy [1]. These beautiful papers presented and analyzed the techniques and stimulated countless other theoretical investigations and practical applications. However, it turns out that many of the same basic ideas can also be found in other works which did not receive the attention they deserved. The author was first introduced to the concepts by the work of Swerling [2], [3]. The basic results can also be viewed as extensions of the Weiner filter for continuous-time processes which originated with Weiner [1], and which Shinbrot [1] extended to the time varying case. Singer and Frost [1] compare the performance of Kalman and Weiner filters.

The basic argument of Section 6.2.2 and the information flow concepts are drawn from the mathematical statistical literature as characterized in particular by Kullbeck [1].

The whiteness of the residuals is related to the innovations approach of Kailath [2], [4], Kailath and Frost [1] which is, in turn, related to the whitening filter concept used in the Weiner theory.

The matrix Riccati equation, conditions for unique steady state solutions, and stability are all closely coupled and many related papers are available; for example, see Kalman [2], Deyst and Price [1], and Levin [1].

The references of Book List 4 on Stochastic Systems develop these Bayesian results in various ways.

Section 6.4

The completely unknown input case is taken from Glover [1]. Friendland [1] considers estimation of bias errors which can be viewed as unknown initial conditions.

Section 6.5

The method of approach used for the exponential smoothing derivation is the direct result of discussions with R. Curry. However, the concept is old.

Section 6.6

This discussion also applies to Sections 4.3, 5.6, and 7.5.

The basic discrete time results are taken from Schweppe [9]. However, certain of the ideas are taken from or motivated by the work of Witsenhausen [1], [2], [3], and Kahan [1]. If the work of these two men had not been available, it is very doubtful the unknown-but-bounded theory would exist in its present form. A related paper which provided background for the development is Schweppe and Knudsen [1]. Certain aspects of the discrete time theory resulted from discussions with L. Gardner. Discussions with F. Schlaepfer helped crystallize many of the concepts. Schlaepfer [1] contains much more material on continuous-time results and also discusses the corresponding distributed parameter (partial differential equation model) results. See also Schlaepfer and Schweppe [1]. The special algorithm of (6.6.7) is taken from Bersakaus and Rhodes [1], who derived it in a different way. Hnylicza [1] develops a bounding set algorithm based on hyperplanes and polyhedron rather than ellipsoids.

Section 6.7

Blum [1], [2], considers the special case of "polynomial-like" signals. The discussions are taken from Schweppe [3]. Jazwinski [2] develops the concept in detail.

Section 6.8

Explicit smoothing equations are available in Book List 4 on Stochastic Systems; see especially Meditch [3]. Related papers are Bryson and Frasier [1]; Meditch [1], [2]; Rauch, Tung and Striebel [1]; Rauch [1]. Frasier and Potter [1] employ the basic logic used here.

Section 6.12

Athans and Tse [1] contains a derivation of the optimum estimator which follows this philosophy for the continuous-time cases.

Section 6.13

The ideas have been discussed in more detail in various papers. See, for example, Heffs [1]; Nash and Tuteur [1]; Neal [1]; Nishimuru [1]; Sorenson [1]. The divergence problem is discussed many places, of which Price [1] and Fitzgerald [1] are examples.

Section 6.14

Numerical methods such as matrix inversion techniques are discussed in many books such as Bodewig [1]; Gantmacher [1]; Householder [1]; Lanczos [1]; Forsythe and Moler [1]; Franklin [1]. Kaminski, Bryson and Schmidt [1] provide a survey of square root filtering.

Section 6.16

Extensive work on polynomials has been done by Blum [1], [2]. The orthonormal polynomial equations are taken from Chitterden, et al [1], although they should also be available elsewhere. The curves are taken from Schweppe [1], [3]; Jones and Schweppe [1].

Section 6.17

Grenader and Rosenblatt [1] discuss this subject in much more detail.

Exercises

6.13 on "cloud tracking" is taken from Schweppe and Knudsen [1]. The observer concept of 6.17 was introduced by Luenberger [1], [2]. Spatial recursions as in 6.18 are discussed in Schweppe [11].

Exercises

6.1. Consider the zero mean Bayesian model

$$\mathbf{z}(n) = \mathbf{H}(n)\mathbf{x} + \mathbf{v}(n)$$

$$E\{\mathbf{x}\mathbf{x}'\} = \boldsymbol{\psi}$$

$$E\{\mathbf{v}(n_1)\mathbf{v}'(n_2)\} = \begin{cases} \mathbf{R}(n_1), & n_1 = n_2 \\ \mathbf{0}, & n_1 \neq n_2 \end{cases}$$

$$E\{\mathbf{x}\mathbf{v}(n)\} = \mathbf{0}$$

Derive difference equations for $\hat{\mathbf{x}}(N|N)$ and $\boldsymbol{\Sigma}(N|N)$ using the matrix manipulation concept of Section 6.1.

6.2. Consider the zero mean Bayesian model

$$z(n) = x + v(n)$$

$$E\{v(n_1)v(n_2)\} = \begin{cases} R, & n_1 = n_2 \\ 0, & n_1 \neq n_2 \end{cases}$$

$$E\{x^2\} = \psi$$

$$E\{xv(n)\} = 0$$

Write explicit equations for the estimate $\hat{x}(N|N)$ of x given $z(1) \cdots z(N)$ by starting with four different (but equivalent) equations: (6.2.5), (6.2.7), (6.3.1), and (6.3.3). Show that these four equations all yield the same result, which for large N is

$$\hat{x}(N\,|\,N) \approx \frac{1}{N}\sum_{n=1}^{N} z(n)$$

$$\Sigma(N\,|\,N) \approx \frac{R}{N}$$

6.3. Show by matrix manipulation that the matrix Riccati equations for $\Sigma(N\,|\,N)$ in (6.2.5) and (6.3.1) are equivalent.

6.4. Consider the zero mean Bayesian model

$$x(n+1) = \Phi x(n) + w(n)$$

$$z(n) = x(n) + v(n)$$

$$E\{w(n_1)w(n_2)\} = \begin{cases} Q, & n_1 = n_2 \\ 0, & n_1 \neq n_2 \end{cases}$$

$$E\{v(n_1)v(n_2)\} = \begin{cases} R, & n_1 = n_2 \\ 0, & n_1 \neq n_2 \end{cases}$$

$$E(x^2(0)\} = \psi$$

Assume that it is desired to estimate the value of $x(N+10) - x(N+5)$ using $z(1) \cdots z(N)$ for increasing N. Write the general form of the difference equations which specify the best estimate and the variance of the error in the estimate. Do not worry about a detailed solution.

6.5. Consider the dynamic system

$$x_1(n+1) = \Phi_1 x_1(n) + w(n)$$

$$E\{x_1^2(0)\} = \psi$$

with completely unknown input $w(n)$. Assume that the state $x_1(n)$ is observed:

$$z_1(n) = x_1(n) + v_1(n)$$

$$E\{v_1(n_1)v_1(n_2)\} = \begin{cases} R_1, & n_1 = n_2 \\ 0, & n_1 \neq n_2 \end{cases}$$

Assume that the unknown input $w(n)$ is also observed by a sensor with output $z_2(n)$, where the sensor has "memory" modeled by

$$x_2(n+1) = \Phi_2 x_2(n) + w(n)$$

$$z_2(n) = x_2(n) + v_2(n)$$

$$x_2(0) = 0$$

$$E\{v_2(n_1)v_2(n_2)\} = \begin{cases} R_2, & n_1 = n_2 \\ 0, & n_1 \neq n_2 \end{cases}$$

Assume that $x_1(0)$, $v_1(n)$, and $v_2(n)$ are uncorrelated zero mean. Assume that it is desired to use the observations to estimate what the input is doing. Write difference equations which specify $\hat{w}(N\,|\,N+1)$, the best estimate of $w(N)$, using $z_1(1) \cdots z_1(N+1)$ and $z_2(1) \cdots z_2(N+1)$.

6.6. Consider the unknown-but-bounded model

$$x(n+1) = \tfrac{1}{2}x(n) + w(n)$$
$$z(n) = x(n) + v(n)$$
$$|w(n)| \leq 1 \qquad |v(n)| \leq 1$$
$$|x(0)| \leq [-2 + \sqrt{8}]^{1/2}$$

Assume that a bounding ellipsoid algorithm is used to calculate the ellipsoid

$$\Omega_{\hat{x}}(N) = \{x: (x - \hat{x}(N|N)^2 \leq \Sigma(N|N)\}$$

which contains $x(N)$. Assume that it was decided to use

$$\rho = \tfrac{1}{2} \qquad \beta = \tfrac{1}{2}$$

Evaluate $\Sigma(N|N)$ for $N = 0, 1, 2, 3$ assuming that the following measurements are obtained:

$$z(1) = 0$$
$$z(2) = 1$$
$$z(3) = 10$$

What can you say about the measurement $z(3) = 10$?

6.7. Consider the zero mean Bayesian model with no input uncertainty:

$$x(n+1) = \Phi x(n)$$
$$z(n) = x(n) + v(n)$$
$$E\{x^2(0)\} = \psi$$
$$E\{v(n_1)v(n_2)\} = \begin{cases} R, & n_1 = n_2 \\ 0, & n_1 \neq n_2 \end{cases}$$

When $n > m$, define $\hat{x}(n|n, n-m)$ to be the estimate of $x(n)$ given $z(n) \cdots z(n-m)$ where m is a fixed integer. Derive a set of difference equations for $\hat{x}(n|n, n-m)$ for $n = m, \ldots$.

6.8. Consider the Fisher model

$$\mathbf{x}(n\Delta + \Delta) = \begin{bmatrix} 1 & \Delta \\ 0 & 1 \end{bmatrix} \mathbf{x}(n\Delta)$$
$$\mathbf{z}(n\Delta) = \mathbf{x}(n\Delta) + \mathbf{v}(n\Delta)$$
$$E\{\mathbf{v}(n_1\Delta)\mathbf{v}'(n_2\Delta)\} = \begin{cases} \mathbf{R}, & n_1 = n_2 \\ \mathbf{0}, & n_1 \neq n_2 \end{cases}$$
$$\mathbf{R} = \begin{pmatrix} R_1 & 0 \\ 0 & R_2 \end{pmatrix}$$

$\mathbf{x}(0)$: completely unknown

$$R_1 = 1 \qquad R_2 = 100$$
$$\Delta = 0.001$$

Since $R_2 \gg R_1$, is it reasonable to model the observation z_1 as coming from a perfect sensor ($R_1 = 0$)? *Hint:* Calculate $\Sigma(N\Delta \mid N\Delta)$ for $N = 2$ for $R_1 = 1$, $R_2 = 100$ and $R_1 = 0$, $R_2 = 100$.

6.9. Consider the Bayesian model with $x(0)$, w, v zero mean, uncorrelated:

$$x(n+1) = \Phi x(n) + w(n), \qquad 0 < \Phi < 1$$

$$z(n) = x(n) + v(n)$$

$$E\{x^2(0)\} = \psi$$

$$E\{w(n_1)w(n_2)\} = \begin{cases} Q, & n_1 = n_2 \\ 0, & n_1 \neq n_2 \end{cases}$$

$$E\{v(n_1)v(n_2)\} = \begin{cases} R, & n_1 = n_2 \\ 0, & n_1 \neq n_2 \end{cases}$$

Assume that computer restrictions constrain the filter to be of the form

$$\hat{x}(N \mid N) = Kz(N)$$

What value of K yields the minimum estimate error covariance in steady state?

6.10. Consider the zero mean Bayesian model

$$\frac{d}{dt}x(t) = (-1)x(t) + w(t)$$

$$y(t) = x(t) + v(t)$$

$$E\{w(t)w(\tau)\} = \delta(t - \tau)2$$

$$E\{v(t)v(\tau)\} = \delta(t - \tau)4$$

$$E\{x^2(0)\} = 0$$

Assume that $y(t)$ goes through a sampler to yield $z(n)$, $n = 1, \ldots,$ where the sampler is modeled by

$$\frac{d}{dt}z(t) = (-50)z(t) + y(t)$$

$$z(n) = z(t)|_{t=n}$$

What equations would you use to obtain an estimate of $y(t)|_{t=N}$ from $z(1) \cdots z(N)$ for large N? *Comment:* Don't be afraid to make an engineering approximation.

6.11. Consider the following situation. There is a linear dynamic system

$$x(n+1) = \Phi x(n) + u(n)$$

$$x(0) = 0$$

whose input $u(n)$ is a correlated, zero mean, stochastic process defined by

$$u(n+1) = Au(n) + w(n)$$

$$u(0) = 0$$

$$E\{w(n_1)w(n_2)\} = \begin{cases} 0, & n_1 \neq n_2 \\ Q, & n_1 = n_2 \end{cases}$$

Assume that only the input $u(n)$ is observed so that

$$z(n) = u(n) + v(n)$$
$$E(v(n)\} = 0$$
$$E\{v(n)w'(n)\} = 0$$
$$E\{v(n_1)v'(n_2)\} = \begin{cases} 0, & n_1 \neq n_2 \\ R, & n_1 = n_2 \end{cases}$$

It is desired to estimate $x(n)$ using $z(1) \cdots z(n)$. An engineer interested in saving money proposes the following simple approach:

a. Build an optimum filter to yield $\hat{u}(n|n)$, $n = 1$, as computed from $z(m)$, $m = 1, \ldots, n$.

b. Calculate $\hat{x}(N|N)$ by solving

$$\hat{x}(n+1|n+1) = \mathbf{\Phi}\hat{x}(n|n) + \hat{u}(n|n)$$
$$\hat{x}(0|0) = 0$$

recursively until $n = N$. The engineer has sacrificed some accuracy for the sake of economics (simplicity). Why?

6.12. Consider

$$x(n+1) = \mathbf{\Phi}x(n) + w(n) + bu(n)$$
$$z(n) = x(n) + v(n)$$
$$x(0) = 0$$

$w(n)$, $v(n)$: zero mean, uncorrelated

$$E\{w(n_1)w(n_2)\} = \begin{cases} Q, & n_1 = n_2 \\ 0, & n_1 \neq n_2 \end{cases}$$

$$E\{v(n_1)v(n_2)\} = \begin{cases} R, & n_1 = n_2 \\ 0, & n_1 \neq n_2 \end{cases}$$

$\mathbf{\Phi}$, Q, R: known
$u(n)$, $n = 1, \ldots$: known time function
b: completely unknown constant

Let $\hat{b}(N)$ denote an *optimum* estimate of b using $z(1) \cdots z(N)$. Write a recursive estimate which relates $\hat{b}(N+1)$ to $\hat{b}(N)$ and $z(N+1)$. Do not worry about startup problems. *Hint:* Use state augmentation and make b a state.

6.13. Consider a "cloud" of many individual chaff particles in orbit around the earth above the atmosphere. Let $\mathbf{x}^i(t)$ denote the state vector of the ith particle, where $\mathbf{x}$ is a six vector:

$$\mathbf{x} = \begin{bmatrix} \mathbf{x}_p \\ \mathbf{x}_v \end{bmatrix} \qquad \mathbf{x}_p = \begin{bmatrix} x_1 \\ x_2 \\ x_3 \end{bmatrix}: \text{ position coordinates}$$

$$\frac{d}{dt}\mathbf{x}_p = \mathbf{x}_v: \text{ velocity coordinates}$$

Assume that the particles are close enough to each other so that a valid approximation for the dynamics is

$$\mathbf{x}^i(t_2) = \mathbf{\Phi}(t_2, t_1)\mathbf{x}^i(t_1), \qquad \text{all } i$$

where $\mathbf{\Phi}$ is a known matrix; i.e., all chaff trajectories are governed by the same linear equations of motion. Assume that $\mathbf{\Phi}^{-1}(t_2, t_1)$ exists. Assume that all chaff trajectories are different; i.e.,

$$\mathbf{x}^i(t) \neq \mathbf{x}^j(t), \qquad i \neq j$$

a. Assume at time 0 that all trajectories are contained in some six-dimensional ellipsoid, $\mathbf{\Omega}_x(0)$:

$$\mathbf{x}^i(0) \in \mathbf{\Omega}_x(0), \qquad \text{all } i$$

$$\mathbf{\Omega}_x(0) = \{\mathbf{x}: [\mathbf{x} - \mathbf{x}_c(0)]'\mathbf{\Gamma}^{-1}(0)[\mathbf{x} - \mathbf{x}_c(0)] \leq 1\}$$

Write the equations which define the set $\mathbf{\Omega}_x(t)$ which is the smallest set such that

$$\mathbf{x}^i(t) \in \mathbf{\Omega}_x(t), \qquad \text{all } i$$
$$\text{all } t$$

b. Assume that the chaff particles are dense so that when the cloud is "looked" at or mapped a well-defined boundary is seen. (In practice the looking is done by radar but for the present discussion, "eyeballing" can be assumed.) Assume that the cloud is mapped at discrete times, t_n, $n = 1, \ldots, N$. Each map consists of the three-dimensional surface in position space which is the boundary of the cloud. To be more explicit, define

$$\mathbf{\Omega}_z(t_n) = \{\mathbf{x}_p: \mathbf{x}_p \in \mathbf{\Omega}_{\mathbf{x}}(t_n)\}$$

Then a cloud map is the boundary of $\mathbf{\Omega}_z(t_n)$. Assume that the $\mathbf{\Omega}_z(t_n)$ are ellipsoids:

$$\mathbf{\Omega}_z(t_n) = \{\mathbf{x}_p: [\mathbf{x}_p - \mathbf{z}(t_n)]'\mathbf{R}^{-1}(t_n)[\mathbf{x}_p - \mathbf{z}(t_n)] \leq 1\}$$

Assume that there is no a priori information available on the shape of the cloud; i.e., $\mathbf{\Omega}_x(0)$ cannot be assumed to be an ellipsoid. Show how you would use the maps $\mathbf{\Omega}_z(t_n)$, $n = 1, \ldots, N$, to calculate a six-dimensional bounding ellipsoid estimate set $\mathbf{\Omega}_{\hat{x}}(t_N)$ such that

$$\mathbf{\Omega}_{\hat{x}}(t_N) \supset \mathbf{\Omega}_x(t_N)$$

Give explicit equations.

6.14. Consider (6.1). Assume that all random vectors and processes are Gaussian. Assume that $\hat{\mathbf{x}}(N|N)$ and $\hat{\mathbf{x}}(N+1|N)$ now denote conditional expectations; i.e.,

$$\hat{\mathbf{x}}(n|m) = E\{\mathbf{x}(n)|\mathbf{Z}_m\}$$

In the following, use the equations and concepts of Appendix E.

a. Prove that (6.2.3) holds.
b. Prove that (6.2.7) holds.

Hint: The derivation consists of two main steps:

1. Show that $\hat{\mathbf{x}}(N+1 \mid N+1) = \hat{\mathbf{x}}(N+1 \mid N)$ + linear operation on $\boldsymbol{\delta}_z(N+1 \mid N)$.
2. Manipulate to get explicit equations.

The first step begins by writing

$$E\{\mathbf{x}(N+1) \mid \mathbf{z}_{N+1}\} = E\{\mathbf{x}(N+1) \mid \mathbf{z}_N, \boldsymbol{\delta}_z(N+1 \mid N)\}$$

Now

$$E\{\boldsymbol{\delta}_z(N+1 \mid N)\mathbf{z}'_N\} = 0$$

which by the Gaussian assumption means $\boldsymbol{\delta}_z(N+1 \mid N)$ and $\mathbf{z}_N$ are independent. Thus

$$E\{\mathbf{x}(N+1) \mid \mathbf{z}_{N+1}\} = E\{\mathbf{x}(N+1) \mid \mathbf{z}_N\} + E\{\mathbf{x}(N+1) \mid \boldsymbol{\delta}_z(N+1 \mid N)$$

This completes the first step. The second step is much more tedious. Define

$$\mathbf{A}(N+1) = \begin{bmatrix} \mathbf{A}_{11} & \mathbf{A}_{12} \\ \mathbf{A}'_{12} & \mathbf{A}_{22} \end{bmatrix} = E\left\{\begin{bmatrix} \mathbf{x}(N+1) \\ \boldsymbol{\delta}_z(N+1 \mid N) \end{bmatrix}\begin{bmatrix} \mathbf{x}(N+1) \\ \boldsymbol{\delta}_z(N+1 \mid N) \end{bmatrix}'\right\}$$

Now

$$E\{\mathbf{x}(N+1) \mid \boldsymbol{\delta}_z(N+1 \mid N)\} = \mathbf{A}_{12}\mathbf{A}_{22}^{-1}\boldsymbol{\delta}_z(N+1 \mid N)$$

so $\mathbf{A}_{12}$ and $\mathbf{A}_{22}$ must be evaluated. Now

$$\boldsymbol{\delta}_z(N+1 \mid N) = \mathbf{H}(N+1)\boldsymbol{\delta}_x(N+1 \mid N) + \mathbf{v}(N+1)$$

which yields

$$\mathbf{A}_{22} = \mathbf{R}(N+1) + \mathbf{H}(N+1)\boldsymbol{\Sigma}(N+1 \mid N)\mathbf{H}'(N+1)$$

To evaluate $\mathbf{A}_{12}$, $\mathbf{x}(N+1)$ is written as

$$\mathbf{x}(N+1) = \boldsymbol{\Phi}(N)E\{\mathbf{x}(N) \mid \mathbf{z}_N\} + \boldsymbol{\delta}_x(N+1 \mid N)$$

which yields

$$\mathbf{A}_{12} = \boldsymbol{\Sigma}(N+1 \mid N)\mathbf{H}'(N+1)$$

Putting all these equations together yields (6.2.7).

6.15. Consider the nonlinear matrix Riccati difference equation

$$\boldsymbol{\Sigma}(n+1) = [[\boldsymbol{\Phi}(n)\boldsymbol{\Sigma}(n)\boldsymbol{\Phi}'(n) + \mathbf{Q}(n)]^{-1} + \mathbf{R}^{-1}(n)]^{-1}$$
$$\boldsymbol{\Sigma}(0) = \boldsymbol{\psi}$$

a. Show that if

$$\boldsymbol{\Gamma}_1(n+1) = \boldsymbol{\Phi}(n)\boldsymbol{\Gamma}_1(n)\boldsymbol{\Phi}'(n) + \mathbf{Q}(n)$$
$$\boldsymbol{\Gamma}_1(0) = \boldsymbol{\psi}$$
$$\boldsymbol{\Gamma}_2^{-1}(n+1) = \boldsymbol{\Phi}'^{-1}(n)\boldsymbol{\Gamma}_2^{-1}(n)\boldsymbol{\Phi}^{-1}(n) + \mathbf{R}^{-1}(n)$$
$$\boldsymbol{\Gamma}_2(0) = \boldsymbol{\psi}$$

then

$$\boldsymbol{\Gamma}_2(n) \leq \boldsymbol{\Sigma}(n) \leq \boldsymbol{\Gamma}_1(n), \qquad \text{all } n$$

Hint: The upper bound $\mathbf{\Gamma}_1(n)$ is $\mathbf{\Sigma}(n)$ for $\mathbf{R}^{-1}(n) = \mathbf{0}$, which implies an "infinite" uncertainty in the observations. The lower bound $\mathbf{\Gamma}_2(n)$ is $\mathbf{\Sigma}(n)$ for $\mathbf{Q}(n) = \mathbf{0}$, which implies that there is no uncertainty associated with the input to the dynamic system. Use these two comments to argue "physically" why the result is true.

b. Let $\mathbf{x}(n)$ and $\mathbf{p}(n)$ denote two K-dimensional vectors defined by

$$\begin{bmatrix} \mathbf{x}(n+1) \\ \mathbf{p}(n+1) \end{bmatrix} = \begin{bmatrix} \mathbf{\Phi}(n) & \mathbf{Q}(n)\mathbf{\Phi}'^{-1}(n) \\ \mathbf{R}^{-1}(n)\mathbf{\Phi}(n) & [\mathbf{I} + \mathbf{R}^{-1}(n)\mathbf{Q}(n)]\mathbf{\Phi}'^{-1}(n) \end{bmatrix} \begin{bmatrix} \mathbf{x}(n) \\ \mathbf{p}(n) \end{bmatrix}$$

$$\mathbf{x}(0) = \boldsymbol{\psi}\mathbf{p}(0)$$

Then show that

$$\mathbf{x}(n) = \mathbf{\Sigma}(n)\mathbf{p}(n)$$

Define $\boldsymbol{\theta}(N)$ to be such that

$$\begin{bmatrix} \mathbf{x}(N) \\ \mathbf{p}(N) \end{bmatrix} = \begin{bmatrix} \boldsymbol{\theta}_{11}(N) & \boldsymbol{\theta}_{12}(N) \\ \boldsymbol{\theta}_{21}(N) & \boldsymbol{\theta}_{22}(N) \end{bmatrix} \begin{bmatrix} \mathbf{x}(0) \\ \mathbf{p}(0) \end{bmatrix}$$

Then show that

$$\mathbf{\Sigma}(N) = [\boldsymbol{\theta}_{11}(N)\boldsymbol{\psi} + \boldsymbol{\theta}_{12}(N)][\boldsymbol{\theta}_{21}(N)\boldsymbol{\psi} + \boldsymbol{\theta}_{22}(N)]^{-1}$$

6.16. For discrete time, the definition of observability (see Appendix B) can be stated as follows: The system

$$\mathbf{x}(n+1) = \mathbf{\Phi}\mathbf{x}(n)$$
$$\mathbf{y}(n) = \mathbf{H}\mathbf{x}(n)$$

is observable if the initial state $\mathbf{x}(0)$ can be calculated from $\mathbf{y}(n)$, $n = 1, \ldots,$ N, for some finite N. Consider the Fisher model

$$\mathbf{x}(n+1) = \mathbf{\Phi}\mathbf{x}(n)$$
$$\mathbf{z}(n) = \mathbf{H}\mathbf{x}(n) + \mathbf{v}(n)$$
$$\mathbf{x}(0)\text{: completely unknown}$$
$$E\{\mathbf{v}(n)\} = \mathbf{0}$$
$$E\{\mathbf{v}(n_1)\mathbf{v}'(n_2)\} = \begin{cases} \alpha\,\mathbf{I}, & n_1 = n_2 \\ \mathbf{0}, & n_1 \neq n_2 \end{cases}$$
$$\boldsymbol{\theta}(n+1) = \mathbf{\Phi}\boldsymbol{\theta}(n)$$
$$\boldsymbol{\theta}(0) = \mathbf{I}$$
$$\boldsymbol{\theta}(n)\text{: transition matrix}$$

Show that

$$\hat{\mathbf{x}}(0\,|\,N) = \mathcal{J}^{-1}(0\,|\,N) \sum_{n=1}^{N} \boldsymbol{\theta}'(n)\mathbf{H}'\mathbf{z}(n)$$

$$\mathcal{J}(0\,|\,N) = \sum_{n=1}^{N} \boldsymbol{\theta}'(n)\mathbf{H}'\mathbf{H}\boldsymbol{\theta}(n)$$

$$\mathbf{\Sigma}(0\,|\,N) = \alpha[\mathcal{J}(0\,|\,N)]^{-1}$$

Discuss why the system is observable if $\mathcal{J}(0\,|\,N)$ is of full rank for some finite N.

6.17. Consider the problem of calculating (reconstructing) the state from perfect observations when the system input is known perfectly. Such a state calculator (reconstructor) is called an *observer*. Since the input $\mathbf{u}(n)$ is known perfectly, it can be assumed to be zero. Consider the Bayesian model of Section 6.1 with

$$\mathbf{Q}(n) = \mathbf{0}$$
$$\mathbf{R}(n) = \mathbf{0}$$
$$\boldsymbol{\psi} = \alpha\,\mathbf{I}$$

where α is arbitrarily large. Show that

$$\begin{aligned}
\hat{\mathbf{x}}(N+1\,|\,N+1) &= \mathbf{\Phi}(N)\hat{\mathbf{x}}(N\,|\,N) + \mathbf{K}(N+1)[\mathbf{z}(N+1) \\
&\quad - \mathbf{H}(N+1]\mathbf{\Phi}(N)\hat{\mathbf{x}}(N\,|\,N)] \\
\hat{\mathbf{x}}(0\,|\,0) &= \mathbf{0} \\
\mathbf{K}(N+1) &= \mathbf{\Sigma}(N+1\,|\,N)\mathbf{H}'(N+1) \\
&\quad \times [\mathbf{H}(N+1)\mathbf{\Sigma}(N+1\,|\,N)\mathbf{H}'(N+1)]^{-1} \\
\mathbf{\Sigma}(N+1\,|\,N+1) &= \mathbf{\Sigma}(N+1\,|\,N) - \mathbf{\Sigma}(N+1\,|\,N)\mathbf{H}'(N+1) \\
&\quad \times [\mathbf{H}(N+1)\mathbf{\Sigma}(N+1\,|\,N)\mathbf{H}'(N+1)]^{-1} \\
&\quad \times \mathbf{H}(N+1)\mathbf{\Sigma}(N+1\,|\,N) \\
\mathbf{\Sigma}(N+1\,|\,N) &= \mathbf{\Phi}(N)\mathbf{\Sigma}(N\,|\,N)\mathbf{\Phi}'(N) \\
\mathbf{\Sigma}(0\,|\,0) &= \alpha\,\mathbf{I}
\end{aligned}$$

Note that the gain $\mathbf{K}(N+1)$ is independent of the value of α. Analysis of these equations shows that (if the system is observable) for some finite N_0, $\mathbf{\Sigma}(N\,|\,N) = \mathbf{0}$, $N \geq N_0$, and thus the state is determined exactly: $\hat{\mathbf{x}}(N\,|\,N) = \mathbf{x}(N)$ for $N \geq N_0$. For $N \geq N_0$, any gain $\mathbf{K}(N+1)$ can be used. This is called a *deadbeat observer*.

6.18. Consider a Fisher model of the form

$$\mathbf{z} = \mathbf{H}\mathbf{x} + \mathbf{v}$$

$$\mathbf{z} = \begin{bmatrix} \mathbf{z}_1 \\ \cdot \\ \cdot \\ \cdot \\ \mathbf{z}_N \end{bmatrix} \qquad \mathbf{v} = \begin{bmatrix} \mathbf{v}_1 \\ \cdot \\ \cdot \\ \cdot \\ \mathbf{v}_N \end{bmatrix}$$

$$\mathbf{x} = \begin{bmatrix} \mathbf{x}_1 \\ \mathbf{x}_{12} \\ \mathbf{x}_2 \\ \mathbf{x}_{23} \\ \mathbf{x}_3 \\ \cdot \\ \cdot \\ \cdot \\ \mathbf{x}_{N-1,N} \\ \mathbf{x}_{N,N} \end{bmatrix}$$

where $\mathbf{H}$ is such that

$$\mathbf{z}_1 = \mathbf{H}_1 \begin{bmatrix} \mathbf{x}_1 \\ \mathbf{x}_{12} \end{bmatrix} + \mathbf{v}_1$$

$$\mathbf{z}_2 = \mathbf{H}_2 \begin{bmatrix} \mathbf{x}_{12} \\ \mathbf{x}_2 \\ \mathbf{x}_{23} \end{bmatrix} + \mathbf{v}_2$$

$$\vdots$$

$$\mathbf{z}_N = \mathbf{H}_N \begin{bmatrix} \mathbf{x}_{N-1,N} \\ \mathbf{x}_N \end{bmatrix} + \mathbf{v}_N$$

$\mathbf{x}$ is unknown, and $\mathbf{v}$ is zero mean and such that

$$E\{\mathbf{v}_k\mathbf{v}_j'\} = \begin{cases} \mathbf{R}_k, & k = j \\ \mathbf{0}, & k \neq j \end{cases}$$

$$E\{\mathbf{v}\mathbf{v}'\} = \mathbf{R}$$

Models like this can arise in surveying and electrical networks where the $\mathbf{z}_k$ correspond to observations made in different *spatial regions*. Each observation $\mathbf{z}_k$ is a function of coordinates (state, $\mathbf{x}_k$) belonging just to the *local region* and of coordinates (states, $\mathbf{x}_{k-1,k}$ and $\mathbf{x}_{k,k+1}$) which are common to adjacent regions. Define

$$\left.\begin{matrix} \hat{\mathbf{x}}_1(\mathbf{z}_1) \\ \hat{\mathbf{x}}_{12}(\mathbf{z}_1) \end{matrix}\right\} \text{ optimum estimates of } \mathbf{x}_1 \text{ and } \mathbf{x}_{12} \text{ made just from } \mathbf{z}_1$$

$$\left.\begin{matrix} \hat{\mathbf{x}}_{12}(\mathbf{z}_1, \mathbf{z}_2) \\ \hat{\mathbf{x}}_2(\mathbf{z}_1, \mathbf{z}_2) \\ \hat{\mathbf{x}}_{23}(\mathbf{z}_1, \mathbf{z}_2) \end{matrix}\right\} \text{ optimum estimates of } \mathbf{x}_{12}, \mathbf{x}_2, \text{ and } \mathbf{x}_{23} \text{ made from } \mathbf{z}_1 \text{ and } \mathbf{z}_2$$

$$\left.\begin{matrix} \hat{\mathbf{x}}_{12}[\hat{\mathbf{x}}_{12}(\mathbf{z}_1), \mathbf{z}_2] \\ \hat{\mathbf{x}}_2[\hat{\mathbf{x}}_{12}(\mathbf{z}_1), \mathbf{z}_2] \\ \hat{\mathbf{x}}_{23}[\hat{\mathbf{x}}_{12}(\mathbf{z}_1), \mathbf{z}_2] \end{matrix}\right\} \text{ optimum estimates of } \mathbf{x}_{12}, \mathbf{x}_2, \text{ and } \mathbf{x}_{23} \text{ made from } \hat{\mathbf{x}}_{12}(\mathbf{z}_1) \text{ and } \mathbf{z}_2$$

Show that

$$\hat{\mathbf{x}}_{12}[\mathbf{z}_1, \mathbf{z}_2] = \hat{\mathbf{x}}_{12}[\hat{\mathbf{x}}_{12}(\mathbf{z}_1), \mathbf{z}_2]$$

$$\hat{\mathbf{x}}_3\{\mathbf{z}_1, \mathbf{z}_2, \mathbf{z}_3\} = \hat{\mathbf{x}}_3\{\hat{\mathbf{x}}_{23}[\hat{\mathbf{x}}_{12}(\mathbf{z}_1), \mathbf{z}_2], \mathbf{z}_3\}$$

$$\vdots$$

Thus a single spatial recursion through the data yields the desired estimate of $\mathbf{x}_N$, i.e., $\hat{\mathbf{x}}_N(\mathbf{z}_1 \cdots \mathbf{z}_N)$. Of course it is then necessary to go back through the data to get, for example, $\hat{\mathbf{x}}_1(\mathbf{z}_1 \cdots \mathbf{z}_N)$ instead of just $\hat{\mathbf{x}}_1(\mathbf{z}_1)$. This problem is analogous to the smoothing problem.

7

Estimation: Continuous-Time Linear Dynamic Systems

The continuous-time system model to be discussed is

$$\frac{d}{dt}\mathbf{x}(t) = \mathbf{F}(t)\mathbf{x}(t) + \mathbf{G}(t)\mathbf{w}(t)$$

$$\mathbf{z}(t) = \mathbf{H}(t)\mathbf{x}(t) + \mathbf{v}(t)$$

$\mathbf{z}(t)$	observation, a K_2 vector	
$\mathbf{x}(t)$	state, a K_1 vector	(7.1)
$\mathbf{v}(t)$	white observation uncertainty, a K_2 vector	
$\mathbf{w}(t)$	white system driving uncertainty, a K_3 vector	
$\mathbf{x}(0)$	initial condition, which may be uncertain	
t	time	

The matrices $\mathbf{F}(t)$, $\mathbf{G}(t)$, and $\mathbf{H}(t)$ are all assumed to be known. It is desired to determine an estimator to process the observations so as to yield an estimate of the state.

By analogy with the discrete-time case of Chapter 6, define

$\hat{\mathbf{x}}(t_1 \mid t_2)$	estimate of $\mathbf{x}(t_1)$ using observations
$\mathbf{z}(t)$	$0 \leq t \leq t_2$

and for Bayesian and Fisher models

$\boldsymbol{\Sigma}(t_1 \mid t_2)$: covariance matrix of the estimate errors $[\mathbf{x}(t_1) - \hat{\mathbf{x}}(t_1 \mid t_2)]$

where

$$\begin{aligned} t_1 &= t_2 && \text{filtering} \\ t_1 &> t_2 && \text{prediction} \\ t_1 &< t_2 && \text{smoothing} \end{aligned}$$

The model (7.1) is the continuous-time version of the discrete-time model (6.1) of Chapter 6. Some rather interesting arguments sometimes result when one asks the question, Are the discrete- or the continuous-time results more useful? The answer is, of course, that the question is stupid. Comparing discrete- and continuous-time models is like comparing red and white wines; both can be very good, but neither is superior in all cases. Of course, some people tend to favor one wine type, and in this sense the author favors the discrete-time model of Chapter 6 over the continuous-time results to be developed in this chapter. The reason lies in the fact that almost all estimation problems the author has worked on have required discrete- rather than continuous-time estimation theory. As a result of this bias, the reader will find that this chapter tends to be rather cryptic and often refers back to Chapter 6 for more explanation on concepts. The main role of this chapter is to report the results of using the discrete to continuous limit on the solutions of Chapter 6.

The discrete to continuous time limit for Bayesian models is considered in Section 7.1. Bayesian, Fisher, weighted-least-squares and unknown-but-bounded filters are discussed in Sections 7.2–7.5. Weighting patterns and implementation are discussed in Sections 7.6 and 7.7. A final discussion is given in Section 7.8.

7.1 Discrete to Continuous Time Limit: Bayesian

The "recipe" to be used to convert the Bayesian discrete-time results of Chapter 6 into continuous-time results is as follows: Assume that results are available for the zero mean model

$$\begin{aligned} \mathbf{x}(n\Delta + \Delta) &= \boldsymbol{\Phi}(n\Delta)\mathbf{x}(n\Delta) + \Delta\mathbf{G}(n\Delta)\mathbf{w}(n\Delta) \\ \mathbf{z}(n\Delta) &= \mathbf{H}(n\Delta)\mathbf{x}(n\Delta) + \mathbf{v}(n\Delta) \\ E\{\mathbf{x}(0)\mathbf{x}'(0)\} &= \boldsymbol{\psi} \\ E\{\mathbf{w}(n_1\Delta)\mathbf{w}'(n_2\Delta)\} &= \begin{cases} \Delta^{-1}\mathbf{Q}(n_1), & n_1 = n_2 \\ \mathbf{0}, & n_1 \neq n_2 \end{cases} \\ E\{\mathbf{v}(n_1\Delta)\mathbf{v}'(n_2\Delta)\} &= \begin{cases} \Delta^{-1}\mathbf{R}(n_1\Delta), & n_1 = n_2 \\ \mathbf{0}, & n_1 \neq n_2 \end{cases} \\ &\mathbf{x}(0),\ \mathbf{w}(n\Delta),\ \mathbf{v}(n\Delta)\text{: uncorrelated.} \end{aligned} \tag{7.1.1}$$

Then the corresponding continuous-time results for (7.1) follow from the two step recipe;

1. Set $\mathbf{\Phi}(n\Delta) = \mathbf{I} + \Delta\mathbf{F}(n\Delta)$
2. Let $n \longrightarrow \infty \qquad \Delta \longrightarrow 0 \qquad n\Delta \longrightarrow t$

The Δ dependence of $\mathbf{w}(n\Delta)$ has been discussed in Section 4.2. A similar discussion is now given to motivate and justify the Δ dependence of $\mathbf{v}(n\Delta)$.

7.1.1 Need for Δ Behavior of v(nΔ)

Consider the Fisher model

$$z(n\Delta) = x + v(n\Delta)$$

where x is an unknown constant and $v(n\Delta)$ is a discrete white process. Then the estimate $\hat{x}(N\Delta \mid N\Delta)$ of x given $z(\Delta) \cdots z(N\Delta)$ is

$$\hat{x}(N\Delta \mid N\Delta) = \frac{1}{N}\sum_{n=1}^{N} z(n\Delta)$$

so

$$x - \hat{x}(N\Delta \mid N\Delta) = \frac{1}{N}\sum_{n=1}^{N} v(n\Delta)$$

Now assume that

$$E\{v(n_1\Delta)v(n_2\Delta)\} = \begin{cases} \Delta^q R, & n_1 = n_2 \\ 0, & n_1 \neq n_2 \end{cases}$$

Then the estimate error variance is

$$E\{[x - \hat{x}(N\Delta \mid N\Delta)]^2\} = \frac{\Delta^q R}{N}$$

The limit $N \longrightarrow \infty$, $\Delta \longrightarrow 0$, $N\Delta \longrightarrow T$ gives

$$E\{[x - \hat{x}(T \mid T)]^2\} = \begin{cases} \infty, & q > -1 \\ 0, & q < -1 \\ \dfrac{R}{T}, & q = -1 \end{cases}$$

Hence if the model

$$z(t) = x + v(t) \tag{7.1.2}$$

is to be meaningful in the sense that the error variance is nonzero but finite, one must choose $q = -1$. This argument shows only that the $\mathbf{v}(n\Delta)$ Δ dependence of (7.1.1) is necessary. It does not justify the use of the model.

7.1.2 Justification for Use of v(t)

The model (7.1.2) yields

$$E\{z^2(t)\} = \infty\, R$$

Hence justification is definitely needed.

In practice the actual situations of interest are

$$\mathbf{z}(t) = \mathbf{H}(t)\mathbf{x}(t) + \mathbf{c}(t)$$

where $\mathbf{c}(t)$ is some physically realizable process. However, a white process $\mathbf{v}(t)$ can be used in place of $\mathbf{c}(t)$ to design and evaluate estimators provided that

1. The estimator designed assuming $\mathbf{v}(t)$ approximates the estimator designed using $\mathbf{c}(t)$.
2. The performance (error covariance matrix) predicted on the basis of $\mathbf{v}(t)$ approximates the performance that will be obtained from $\mathbf{c}(t)$.

This often happens because the observations $\mathbf{z}(t)$ are passed through a linear system (the estimator) to get the estimate. Hence the arguments of Section 4.2 can be used to justify the use of $\mathbf{v}(t)$.

A simple example is used to illustrate what happens. Consider the following two versions of a continuous-time problem where it is desired to estimate a scalar constant, x.

White observation process: $z(t) = x + v(t), \qquad 0 \leq t \leq T$

$$\begin{aligned} &v(t)\text{: white stochastic process} \\ &E\{v(t)v(\tau)\} = \delta(t-\tau)R \end{aligned} \tag{7.1.3}$$

Correlated observation process: $z(t) = x + c(t), \qquad 0 \leq t \leq T$

$$\begin{aligned} &\frac{d}{dt}c(t) = Ac(t) + w(t) \\ &A < 0 \\ &w(t)\text{: white stochastic process} \\ &E\{w(t)w(\tau)\} = \delta(t-\tau)Q \\ &E\{c^2(t)\} = -\frac{Q}{2A} = \Gamma_c, \qquad \text{all } t \\ &E\{c(0)\} = 0 \end{aligned} \tag{7.1.4}$$

In both versions, x is modeled as a random variable (Bayesian model) where

$$E\{x\} = 0 \qquad E\{x^2\} = \psi$$

It can be shown (see Exercise 7.9) that for the white model (7.1.3)

$$\begin{aligned} \hat{x}(T \mid T) &= \left[\left(\frac{R}{T}\right)^{-1} + \psi^{-1}\right]^{-1} \int_0^T \frac{z(t)\,dt}{R} \\ \Sigma(T \mid T) &= \left[\left(\frac{R}{T}\right)^{-1} + \psi^{-1}\right]^{-1} \end{aligned} \tag{7.1.5}$$

while for the correlated model (7.1.4)

$$\begin{aligned} \hat{x}(T \mid T) &= \frac{\Sigma(T \mid T)}{2\Gamma_c}\left\{z(T) + z(0) - A\int_0^T z(t)\,dt\right\} \\ \Sigma(T \mid T) &= \left[\left(\frac{2\Gamma_c}{-TA + 2}\right)^{-1} + \psi^{-1}\right]^{-1} \end{aligned} \tag{7.1.6}$$

If it is assumed that $|TA| \gg 1$, (7.1.6) reduces to

$$\hat{x}(T \mid T) \approx -\frac{\Sigma(T \mid T)A}{2\Gamma_c} \int_0^T z(t)\,dt$$

$$\Sigma(T \mid T) \approx \left[\left(\frac{2\Gamma_c}{-TA}\right)^{-1} + \psi^{-1}\right]^{-1}$$

which corresponds to the white model results (7.1.5) if R in (7.1.3) is chosen to be

$$R = \frac{2\Gamma_c}{-A}$$

Hence when the condition $|TA| \gg 1$ is satisfied, the use of $v(t)$ instead of $c(t)$ is justified in the sense that both yield the same estimator and performance provided that R in (7.1.3) is chosen correctly. The condition $|TA| \gg 1$ means the observation time interval must be long relative to the *noise time constant.*

The above example was done for the case of estimating a constant x. If it is generalized to the case

$$\frac{d}{dt}x(t) = Fx(t) + w(t)$$

$$z(t) = Hx(t) + c(t)$$

the two conditions under which $c(t)$ could be replaced by $v(t)$ are

$$|F| \ll |A|$$

$$|AT| \gg 1$$

The condition $|F| \ll |A|$ was encountered in Section 4.2. The arguments can be generalized to vector time-varying systems.

To summarize, the following points on justification of use of a white $\mathbf{v}(t)$ model are important:

1. Justification may be possible when $\mathbf{z}(t)$ is to be used as an input to an estimator.
2. Justification depends on the nature of $\mathbf{x}(t)$.
3. Justification depends on the length of the observation interval.
4. The corresponding value of $\mathbf{R}$ depends on the nature of the actual observation error process.

7.2 Bayesian : Filtering and Prediction

For a Bayesian model version of (7.1), $\mathbf{x}(0)$, $\mathbf{w}(t)$, and $\mathbf{v}(t)$ are zero mean, uncorrelated, and

$$\begin{aligned} E\{\mathbf{x}(0)\mathbf{x}'(0)\} &= \boldsymbol{\psi} \\ E\{\mathbf{w}(t)\mathbf{w}'(\tau)\} &= \delta(t-\tau)\mathbf{Q}(t) \\ E\{\mathbf{v}(t)\mathbf{v}'(\tau)\} &= \delta(t-\tau)\mathbf{R}(t) \end{aligned} \tag{7.2.1}$$

The corresponding discrete-time model is (7.1.1) and the corresponding filtering equations are in Sections 6.2 and 6.3. Introducing Δ into (6.3.1) gives

$$
\begin{aligned}
\hat{\mathbf{x}}(N\Delta + \Delta \mid N\Delta + \Delta) &= \mathbf{\Phi}(N\Delta)\hat{\mathbf{x}}(N\Delta \mid N\Delta) + \Delta\mathbf{K}(N\Delta + \Delta)\{\mathbf{z}(N\Delta + \Delta) \\
&\quad - \mathbf{H}(N\Delta + \Delta)\mathbf{\Phi}(N\Delta)\hat{\mathbf{x}}(N\Delta \mid N\Delta)\} \\
\mathbf{K}(N\Delta + \Delta) &= \mathbf{\Sigma}(N\Delta + \Delta \mid N\Delta + \Delta)\mathbf{H}'(N\Delta + \Delta)\mathbf{R}^{-1}(N\Delta + \Delta) \\
\mathbf{\Sigma}(N\Delta + \Delta \mid N\Delta + \Delta) &= \mathbf{\Sigma}(N\Delta + \Delta \mid N\Delta) - \mathbf{\Sigma}(N\Delta + \Delta \mid N\Delta)\mathbf{H}'(N\Delta + \Delta) \\
&\quad \times \{\Delta^{-1}\mathbf{R}(N\Delta + \Delta) + \mathbf{H}(N\Delta + \Delta)\mathbf{\Sigma}(N\Delta + \Delta \mid N\Delta) \\
&\quad \times \mathbf{H}'(N\Delta + \Delta)\}^{-1}\mathbf{H}(N\Delta + \Delta)\mathbf{\Sigma}(N\Delta + \Delta \mid N\Delta) \\
\mathbf{\Sigma}(N\Delta + \Delta \mid N\Delta) &= \mathbf{\Phi}(N\Delta)\mathbf{\Sigma}(N\Delta \mid N\Delta)\mathbf{\Phi}'(N\Delta) + \Delta\mathbf{G}(N\Delta)\mathbf{Q}(N\Delta)\mathbf{G}'(N\Delta) \\
\mathbf{\Sigma}(0 \mid 0) &= \boldsymbol{\psi} \\
\hat{\mathbf{x}}(0 \mid 0) &= \mathbf{0}
\end{aligned}
$$

Substituting

$$\mathbf{\Phi}(N\Delta) = \mathbf{I} + \Delta\mathbf{F}(N\Delta)$$

gives

$$
\begin{aligned}
\mathbf{\Sigma}(N\Delta + \Delta \mid N\Delta) &= \mathbf{\Sigma}(N\Delta \mid N\Delta) + \Delta\{\mathbf{F}(N\Delta)\mathbf{\Sigma}(N\Delta \mid N\Delta) + \mathbf{\Sigma}(N\Delta \mid N\Delta)\mathbf{F}'(N\Delta) \\
&\quad + \mathbf{G}(N\Delta)\mathbf{Q}(N\Delta)\mathbf{G}'(N\Delta)\} + \Delta^2\mathbf{F}(N\Delta)\mathbf{\Sigma}(N\Delta \mid N\Delta)\mathbf{F}'(N\Delta)
\end{aligned} \tag{7.2.2}
$$

Taking the limit $N \longrightarrow \infty$, $\Delta \longrightarrow 0$, $N\Delta \longrightarrow t$,

$$
\boxed{
\begin{aligned}
\frac{d}{dt}\hat{\mathbf{x}}(t \mid t) &= \mathbf{F}(t)\hat{\mathbf{x}}(t \mid t) + \mathbf{K}(t)\{\mathbf{z}(t) - \mathbf{H}(t)\hat{\mathbf{x}}(t \mid t)\} \\
\mathbf{K}(t) &= \mathbf{\Sigma}(t \mid t)\mathbf{H}'(t)\mathbf{R}^{-1}(t) \\
\frac{d}{dt}\mathbf{\Sigma}(t \mid t) &= \mathbf{F}(t)\mathbf{\Sigma}(t \mid t) + \mathbf{\Sigma}(t \mid t)\mathbf{F}'(t) \\
&\quad - \mathbf{\Sigma}(t \mid t)\mathbf{H}'(t)\mathbf{R}^{-1}(t)\mathbf{H}(t)\mathbf{\Sigma}(t \mid t) + \mathbf{G}(t)\mathbf{Q}(t)\mathbf{G}'(t) \\
\hat{\mathbf{x}}(0 \mid 0) &= \mathbf{0} \\
\mathbf{\Sigma}(0 \mid 0) &= \boldsymbol{\psi}
\end{aligned}
} \tag{7.2.3}
$$

Equation (7.2.3) defines the time-varying linear filter which processes $\mathbf{z}(t)$ to obtain the filter estimate $\hat{\mathbf{x}}(t \mid t)$. The nonlinear matrix differential equation for $\mathbf{\Sigma}(t \mid t)$ is the matrix Riccati equation. The basic dynamic model and the continuous-time filter of (7.2.3) can be represented as in Figs. 7.2.1 and 7.2.2. Hence the basic structure of the continuous-time filter is of the form shown in Fig. 7.2.3. Note the similarity of these block diagrams to the discrete-time versions of Fig. 6.3.1–6.3.3.

As in the discrete-time case the residual process $\boldsymbol{\delta}_z(t \mid t)$ is a white

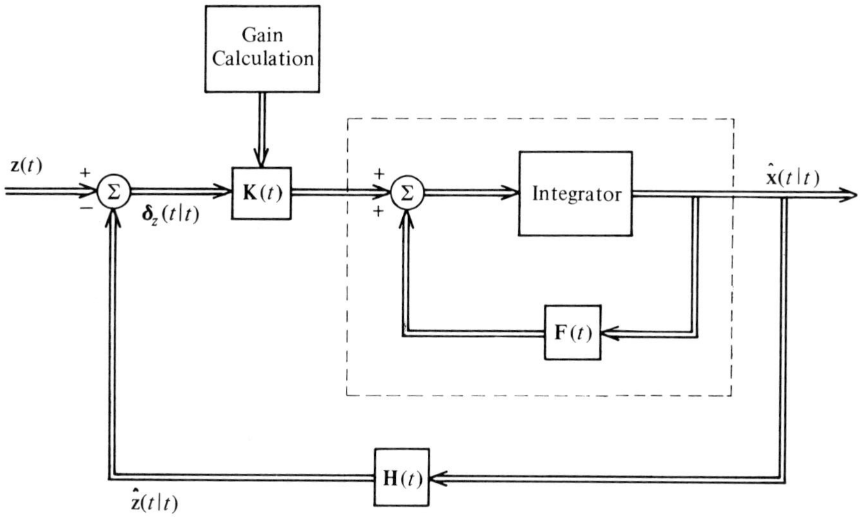

Figure 7.2.1 Bayesian continuous-time filter.

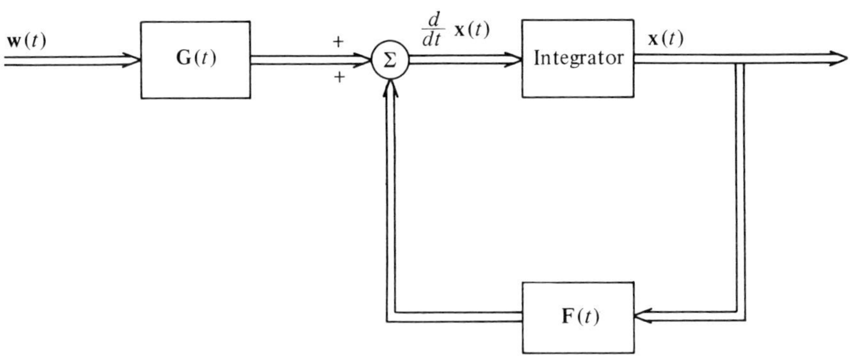

Figure 7.2.2 Dynamic model.

stochastic process:

$$\boxed{\begin{aligned} \boldsymbol{\delta}_z(t \mid t) &= \mathbf{z}(t) - \mathbf{H}(t)\hat{\mathbf{x}}(t \mid t) \\ E\{\boldsymbol{\delta}_z(t \mid t)\boldsymbol{\delta}_z'(\tau \mid \tau)\} &= \boldsymbol{\delta}(t - \tau)\mathbf{R}(t) \end{aligned}} \tag{7.2.4}$$

The fact that $\boldsymbol{\delta}_z(t \mid t)$ and $\mathbf{v}(t)$ have the same statistics (i.e., $\mathbf{R}$) does not imply that they are the same process.

The global optimality of $\hat{\mathbf{x}}(N \mid N)$ in discrete time implies similar global optimality of $\hat{\mathbf{x}}(T \mid T)$.

The basic equation (7.2.3) can be written in various forms. Consider

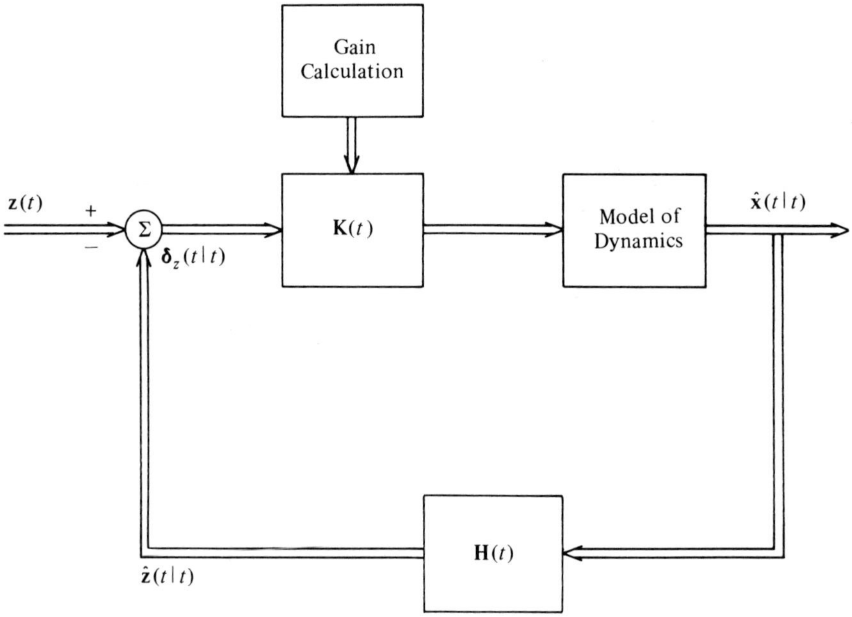

Figure 7.2.3 Bayesian continuous-time filter.

an information flow formulation. By analogy with Section 6.3.4, define

$$\mathcal{I}(t \mid t) = \boldsymbol{\Sigma}^{-1}(t \mid t) \qquad \text{information matrix}$$

$$\boldsymbol{\chi}(t \mid t) = \mathcal{I}(t \mid t)\hat{\mathbf{x}}(t \mid t) \qquad \text{information vector}$$

Then (7.2.3) can be rewritten as [assume that $\mathcal{I}(t \mid t)$ exists]

$$\begin{aligned}
\frac{d}{dt}\boldsymbol{\chi}(t \mid t) &= -[\mathbf{F}'(t) + \mathcal{I}(t \mid t)\mathbf{G}(t)\mathbf{Q}(t)\mathbf{G}'(t)]\boldsymbol{\chi}(t \mid t) \\
&\quad + \mathbf{H}'(t)\mathbf{R}^{-1}(t)\mathbf{z}(t) \\
\frac{d}{dt}\mathcal{I}(t \mid t) &= -\mathbf{F}'(t)\mathcal{I}(t \mid t) - \mathcal{I}(t \mid t)\mathbf{F}(t) \\
&\quad + \mathbf{H}(t)\mathbf{R}^{-1}(t)\mathbf{H}(t) - \mathcal{I}(t \mid t)\mathbf{G}(t)\mathbf{Q}(t)\mathbf{G}'(t)\mathcal{I}(t \mid t) \\
\hat{\mathbf{x}}(t \mid t) &= \mathcal{I}^{-1}(t \mid t)\boldsymbol{\chi}(t \mid t) \\
\boldsymbol{\chi}(0 \mid 0) &= \mathbf{0} \\
\mathcal{I}(0 \mid 0) &= \boldsymbol{\psi}^{-1}
\end{aligned} \tag{7.2.5}$$

The discussions in Chapter 6 on the interpretation of these equations can be carried over directly. The special case $\mathbf{Q}(t) = \mathbf{0}$ (no input uncertainty) does not overly simplify (7.2.3), but it does help in the information formulation as the equation for $\boldsymbol{\chi}(t \mid t)$ becomes independent of $\mathcal{I}(t \mid t)$ and $\mathcal{I}(t \mid t)$ is given by a linear matrix differential equation.

The performance of the Bayesian filter is, of course, determined by $\boldsymbol{\Sigma}(t \mid t)$, the error covariance matrix.

Equation (7.2.3) yields the filtering solution, i.e., $\hat{\mathbf{x}}(t_1 | t_2)$ for $t_1 = t_2$. The prediction solution follows directly from the ideas of Section 6.2.1. Let $\boldsymbol{\theta}(t_1, t_2)$ denote the fundamental matrix associated with $\mathbf{F}(t)$. Then for $\tau > 0$,

$$
\boxed{
\begin{aligned}
&\hat{\mathbf{x}}(t + \tau \mid t) = \boldsymbol{\theta}(t + \tau, t)\hat{\mathbf{x}}(t, t) \\
&\text{or} \\
&\frac{d}{d\tau}\hat{\mathbf{x}}(t + \tau \mid t) = \mathbf{F}(t + \tau)\hat{\mathbf{x}}(t + \tau \mid t) \\
&\frac{d}{d\tau}\boldsymbol{\Sigma}(t + \tau \mid t) = \mathbf{F}(t + \tau)\boldsymbol{\Sigma}(t + \tau \mid t) + \boldsymbol{\Sigma}(t + \tau \mid t)\mathbf{F}'(t + \tau) \\
&\qquad\qquad + \mathbf{G}(t + \tau)\mathbf{Q}(t + \tau)\mathbf{G}'(t + \tau) \\
&\qquad\qquad \tau > 0
\end{aligned}
}
\tag{7.2.6}
$$

Consider the special case of a time-invariant model (constant **F**, **G**, **H**, **Q**, and **R**). In many cases $\boldsymbol{\Sigma}(t \mid t)$ of (7.2.3) goes to some steady-state value $\boldsymbol{\Sigma}_\infty$ for large t, where $\boldsymbol{\Sigma}_\infty$ is the positive definite solution of

$$
\mathbf{0} = \mathbf{F}\boldsymbol{\Sigma} + \boldsymbol{\Sigma}\mathbf{F}' - \boldsymbol{\Sigma}\mathbf{H}'\mathbf{R}^{-1}\mathbf{H}\boldsymbol{\Sigma} + \mathbf{G}\mathbf{Q}\mathbf{G}'
$$

Then for large t, or if $\boldsymbol{\psi} = \boldsymbol{\Sigma}_\infty$, the estimator becomes a linear time-invariant system with input $\mathbf{z}(t)$,

$$
\frac{d}{dt}\hat{\mathbf{x}}(t \mid t) = [\mathbf{F} - \boldsymbol{\Sigma}_\infty\mathbf{H}'\mathbf{R}^{-1}\mathbf{H}]\hat{\mathbf{x}}(t \mid t) + \boldsymbol{\Sigma}_\infty\mathbf{H}'\mathbf{R}^{-1}\mathbf{z}(t) \tag{7.2.7}
$$

The discussions of Section 6.3 on when $\boldsymbol{\Sigma}_\infty$ exists and is unique also apply here.

The discussion in Section 6.3 on stability of the filter also applies to (7.2.3).

EXAMPLE: Consider a scalar time-invariant model

$$
\frac{d}{dt}x(t) = Fx(t) + w(t)
$$

$$
z(t) = x(t) + v(t)
$$

If steady-state error variances Σ_∞ exist, they are given by nonnegative solutions to

$$
\Sigma = FR \pm \sqrt{F^2R^2 + RQ}
$$

Consider three cases.

Case 1: If $Q = 1$ and $R = 1$, F arbitrary, then

$$
\Sigma_\infty = F + \sqrt{F^2 + 1}
$$

and the steady-state filter is

$$\frac{d}{dt}\hat{x}(t\,|\,t) = -\sqrt{F^2+1}\,\hat{x}(t\,|\,t) + \Sigma_\infty z(t).$$

Case 2: If $Q = 0$, $R = 1$, and $F < 0$, then

$$\Sigma_\infty = 0$$

and the steady-state estimate is $\hat{x}(t\mid t) = 0$.

Case 3: If $Q = 0$, $R = 1$, and $F > 0$, then

$$\Sigma_\infty = 2F$$

or

$$\Sigma_\infty = 0$$

where the value to be used depends on ψ. $\Sigma_\infty = 0$ results from $\psi = 0$. The steady-state filter is stable for $\Sigma_\infty = 2F$.

The equations of this section are often called the Kalman filter or the Kalman-Bucy filter.

7.3 Weighted Least Squares

7.3.1 General Case

A discrete-time weighted-least-squares criterion was given in Section 6.5.1. Application of the discrete to continuous time limit to (6.5.1) yields the following continuous time criterion:

Choose the $\mathbf{x}(T)$ and $\mathbf{w}(t)$, $0 \le t \le T$, which minimize

$$J = \int_0^T \{[\mathbf{z}(t) - \mathbf{y}(t)]'\mathbf{R}^{-1}(t)[\mathbf{z}(t) - \mathbf{y}(t)] + \mathbf{w}'(t)\mathbf{Q}^{-1}(t)\mathbf{w}(t)\}\,dt + \mathbf{x}'(0)\boldsymbol{\psi}^{-1}\mathbf{x}(0) \tag{7.3.1}$$

subject to the constraint

$$\frac{d}{dt}\mathbf{x}(t) = \mathbf{F}(t)\mathbf{x}(t) + \mathbf{G}(t)\mathbf{w}(t)$$

$$\mathbf{y}(t) = \mathbf{H}(t)\mathbf{x}(t)$$

Application of the calculus of variations to this problem yields, after much cleverness, the Bayesian solution of Section 7.2. It turns out that in the course of the solution, one also obtains formulas for the smoothed estimate $\hat{\mathbf{x}}(t\,|\,T)$, $0 \le t \le T$. (See Exercise 7.8).

7.3.2 Exponential Smoothing

A special case of weighted least squares called exponential smoothing results when $\mathbf{R}(t)$, $\boldsymbol{\psi}$, and $\mathbf{Q}(t)$ in (7.3.1) are chosen to be

$$\begin{aligned} \mathbf{R}(t) &= e^{-\alpha[T-t]}\,I, \qquad 0 \le t \le T \\ \mathbf{Q}^{-1}(t) &= \mathbf{0} \qquad \boldsymbol{\psi}^{-1} = \mathbf{0} \\ \alpha &> 0 \end{aligned} \tag{7.3.2}$$

so (7.3.1) becomes

Choose $\mathbf{x}(T)$ which minimizes

$$J[\mathbf{x}(T)] = \int_0^T e^{-\alpha(T-t)}[\mathbf{z}(t) - \mathbf{H}(t)\mathbf{x}(t)]'[\mathbf{z}(t) - \mathbf{H}(t)\mathbf{x}(t)]\,dt \tag{7.3.3}$$

subject to the constraint

$$\frac{d}{dt}\mathbf{x}(t) = \mathbf{F}(t)\mathbf{x}(t)$$

Let $\hat{\mathbf{x}}(T|T)$ denote the solution to (7.3.3). The desired equations for $\hat{\mathbf{x}}(T|T)$ follow directly as a special case of Section 7.2. However, the complications arising from $\mathbf{R}(t)$ depending on T as in (7.3.2) make it worthwhile to outline the steps. The general approach followed in Section 6.5.2 for discrete time is repeated.

For simplicity assume that $\mathbf{F}(t) = \mathbf{F}$ and $\mathbf{H}(t) = \mathbf{H}$. Substitution of (7.3.2) into the information flow formulation of (7.2.5) yields equations for $\tilde{x}(T|T)$ and $\tilde{\mathfrak{J}}(T|T)$ which are viewed as solutions for fixed T. Differentiating with respect to T then yields

$$\frac{d}{dT}\tilde{x}(T|T) = -[\mathbf{F}' + \alpha\mathbf{I}]\tilde{x}(T|T) + \mathbf{H}'\mathbf{z}(T)$$

$$\frac{d}{dT}\tilde{\mathfrak{J}}(T|T) = -\alpha\tilde{\mathfrak{J}}(T|T) - \mathbf{F}'\tilde{\mathfrak{J}}(T|T) - \tilde{\mathfrak{J}}(T|T)\mathbf{F} + \mathbf{H}'\mathbf{H}$$

$$\tilde{x}(0|0) = \mathbf{0}$$

$$\tilde{\mathfrak{J}}(0|0) = \mathbf{0}$$

where the desired $\hat{\mathbf{x}}(T|T)$ is given by

$$\hat{\mathbf{x}}(T|T) = \mathbf{\Sigma}(T|T)\tilde{x}(T|T)$$

$$\tilde{\mathbf{\Sigma}}(T|T) = \tilde{\mathfrak{J}}^{-1}(T|T)$$

More differentiation with respect to T then yields, after some manipulation,

$$\boxed{\begin{aligned}
\frac{d\hat{\mathbf{x}}(T|T)}{dT} &= \mathbf{F}\hat{\mathbf{x}}(T|T) + \mathbf{K}(T)[\mathbf{z}(T) - \mathbf{H}\hat{\mathbf{x}}(T|T)] \\
\mathbf{K}(T) &= \tilde{\mathbf{\Sigma}}(T|T)\mathbf{H}' \\
\frac{d}{dT}\tilde{\mathbf{\Sigma}}(T|T) &= \mathbf{F}\tilde{\mathbf{\Sigma}}(T|T) + \tilde{\mathbf{\Sigma}}(T|T)\mathbf{F}' + \alpha\tilde{\mathbf{\Sigma}}(T|T) \\
&\quad -\tilde{\mathbf{\Sigma}}(T|T)\mathbf{H}'\mathbf{H}\tilde{\mathbf{\Sigma}}(T|T) \\
\hat{\mathbf{x}}(0|0) &= \mathbf{0} \\
\tilde{\mathbf{\Sigma}}^{-1}(0|0) &= \mathbf{0}
\end{aligned}} \tag{7.3.4}$$

This result could, of course, also have been derived by applying the discrete to continuous time limit directly to (6.5.5).

EXAMPLE: Consider the special case where x is a scalar and $H = 1$. Then from (7.3.4)

$$\frac{d\hat{x}(T|T)}{dT} = F\hat{x}(T|T) + K(T)[z(T) - \hat{x}(T|T)]$$

$$K(T) = \tilde{\Sigma}(T|T)$$

$$\frac{d\tilde{\Sigma}(T|T)}{dT} = (2F + \alpha)\tilde{\Sigma}(T|T) - \tilde{\Sigma}^2(T|T)$$

Assume steady state is reached so that

$$\frac{d\tilde{\Sigma}(T|T)}{dT} = 0$$

Then

$$\tilde{\Sigma}_\infty = \begin{cases} 0 \\ 2F + \alpha \end{cases}$$

The zero $\tilde{\Sigma}_\infty$ results if F is negative and $\alpha \leq |2F|$.

7.4 Fisher: Filtering†

Extension of the discrete-time Fisher model of Section 6.4 to continuous time yields a model where $\mathbf{v}(t)$ is a continuous white stochastic process while either or both $\mathbf{x}(0)$ and $\mathbf{w}(t)$ are completely unknown.

Consider first the case where $\mathbf{w}(t)$ and $\mathbf{v}(t)$ are modeled as continuous white stochastic processes and only $\mathbf{x}(0)$ is considered to be unknown. As in Section 6.4.1 this is handled by setting $\boldsymbol{\psi}^{-1} = \mathbf{0}$ in the Bayesian theory. Thus the initial condition for the matrix Riccati equation of (7.2.3) is $\boldsymbol{\Sigma}(0|0) = \infty\,\mathbf{I}$. This is not overly desirable, but one can simply choose $\boldsymbol{\Sigma}(0|0)$ to be "very big" or use the information flow equations where the initial condition for the matrix Riccati equation (7.2.5) is $\mathcal{I}(0|0) = \mathbf{0}$.

Now consider the case where $\mathbf{w}(t)$ is completely unknown, while $\mathbf{v}(t)$ is a continuous-time stochastic white process and $\mathbf{x}(0)$ is modeled either as a random vector or completely unknown. Unfortunately attempts to apply the discrete to continuous time limit to the results of Section 6.4.2 and/or to set $Q(t) = \infty\,\mathbf{I}$ in (7.2.3) have failed to yield "nice" general answers. The author feels that fairly general results are possible, but he also feels that the problem is primarily only of academic interest so it is left open. [One approach is to consider the weighted-least-squares criterion of (7.3.1) with $\mathbf{Q}^{-1}(t) = \mathbf{0}$. In the jargon of optimum control theory, the answer lies on singular arcs.]

An engineering approximation to the estimator for an unknown $\mathbf{w}(t)$ model can, of course, be obtained by simply using the Bayesian model equations with $\mathbf{Q}(t)$ very large.

7.5 Unknown-but-Bounded: Filtering

For (7.1) with unknown-but-bounded ellipsoid models, $\mathbf{v}(t)$ and $\mathbf{w}(t)$ are continuous-time, white unknown-but-bounded processes, while $\mathbf{x}(0)$ is an unknown-but-bounded vector.

The corresponding discrete-time model is (3.1.3) and the corresponding discrete-time bounding ellipsoid filtering equations are (6.6.5). Including Δ, (6.6.5) can be written as

$$\begin{aligned}
\hat{\mathbf{x}}(N\Delta + \Delta \mid N\Delta + \Delta) &= \mathbf{\Phi}(N\Delta)\hat{\mathbf{x}}(N\Delta \mid N\Delta) + \Delta\rho(N+1)[1 - \delta^2(N\Delta + \Delta)]^{-1} \\
&\quad \times \mathbf{\Sigma}(N\Delta + \Delta \mid N\Delta + \Delta)\mathbf{H}'(N\Delta + \Delta)\mathbf{R}^{-1}(N\Delta + \Delta) \\
&\quad \times \{\mathbf{z}(N\Delta + \Delta) - \mathbf{H}(N\Delta + \Delta)\mathbf{\Phi}(N\Delta)\hat{\mathbf{x}}(N\Delta \mid N\Delta)\} \\
\mathbf{\Sigma}(N\Delta + \Delta \mid N\Delta + \Delta) &= [1 - \delta^2(N\Delta + \Delta)]\{[1 - \Delta\rho(N\Delta + \Delta)]^{-1} \\
&\quad \times \mathbf{\Sigma}(N\Delta + \Delta \mid N\Delta) - [1 - \Delta\rho(N\Delta + \Delta)]^{-2} \\
&\quad \times \mathbf{\Sigma}(N\Delta + \Delta \mid N\Delta)\mathbf{H}'(N\Delta + \Delta)[[\Delta\rho(N\Delta + \Delta)]^{-1} \\
&\quad \times \mathbf{R}(N+1) + [1 - \Delta\rho(N\Delta + \Delta)]^{-1}\mathbf{H}(N\Delta + \Delta) \\
&\quad \times \mathbf{\Sigma}(N\Delta + \Delta \mid N\Delta)\mathbf{H}'(N\Delta + \Delta)]^{-1}\mathbf{H}(N\Delta + \Delta) \\
&\quad \times \mathbf{\Sigma}(N\Delta + \Delta \mid N\Delta)\}
\end{aligned} \tag{7.5.1}$$

$$\frac{1}{\Delta} \geq \rho(N\Delta + \Delta) \geq 0$$

$$\begin{aligned}
\mathbf{\Sigma}(N\Delta + \Delta \mid N\Delta) &= [1 - \Delta\beta(N\Delta)]^{-1}\mathbf{\Phi}(N\Delta)\mathbf{\Sigma}(N\Delta \mid N\Delta)\mathbf{\Phi}'(N\Delta) \\
&\quad + \Delta^2[\Delta\beta(N\Delta)]^{-1}\mathbf{G}(N\Delta)\mathbf{Q}(N\Delta)\mathbf{G}'(N\Delta)
\end{aligned}$$

$$\frac{1}{\Delta} \geq \beta(n\Delta) \geq 0$$

$$\mathbf{\Sigma}(0 \mid 0) = \boldsymbol{\psi}$$

$$\hat{\mathbf{x}}(0 \mid 0) = \mathbf{0}$$

$$\begin{aligned}
\delta^2(N\Delta + \Delta) &= [\mathbf{z}(N\Delta + \Delta) - \mathbf{H}(N\Delta + \Delta)\mathbf{\Phi}(N\Delta)\hat{\mathbf{x}}(N\Delta \mid N\Delta)]' \\
&\quad \times \{(1 - \Delta\rho(N\Delta + \Delta))^{-1}\mathbf{H}(N\Delta + \Delta)\mathbf{\Sigma}(N\Delta + \Delta \mid N\Delta) \\
&\quad \times \mathbf{H}'(N\Delta + \Delta) + (\Delta\rho(N\Delta + \Delta))^{-1}\mathbf{R}(N\Delta + \Delta)\}^{-1} \\
&\quad \times [\mathbf{z}(N\Delta + \Delta) - \mathbf{H}(N\Delta + \Delta)\mathbf{\Phi}(N\Delta)\hat{\mathbf{x}}(N\Delta \mid N\Delta)]
\end{aligned}$$

The $\beta(N+1)$ and $\rho(N+1)$ of Chapter 6 have been replaced by $\Delta\beta(N\Delta + \Delta)$ and $\Delta\rho(N\Delta + \Delta)$ so that the constraints have been changed accordingly. Substitution of $\mathbf{\Phi}(n\Delta) = \mathbf{I} + \Delta\mathbf{F}(n\Delta)$ and taking the limit $N \longrightarrow \infty$, $\Delta \longrightarrow 0$, $N\Delta \longrightarrow t$ yields the desired result. For the dynamic system (7.1) with unknown-but-bounded uncertainties modeled by

$$\begin{aligned}
\mathbf{x}(t) &\in \{\mathbf{x} : \mathbf{x}'\boldsymbol{\psi}^{-1}\mathbf{x} \leq 1\} \\
\mathbf{w}(t) &\in \{\mathbf{w} : \mathbf{w}'\mathbf{Q}^{-1}(t)\mathbf{w} \leq 1\} \\
\mathbf{v}(t) &\in \{\mathbf{v} : \mathbf{v}'\mathbf{R}^{-1}(t)\mathbf{v} \leq 1\}
\end{aligned}$$

the true $\mathbf{x}(t)$ must be contained within the time-varying bounding ellipsoid $\boldsymbol{\Omega}_{\hat{x}}(t\,|\,t)$ defined by

$$
\begin{aligned}
\boldsymbol{\Omega}_{\hat{x}}(t\,|\,t) &= \{\mathbf{x}\colon [\mathbf{x} - \hat{\mathbf{x}}(t\,|\,t)]'\boldsymbol{\Sigma}^{-1}(t\,|\,t)[\mathbf{x} - \hat{\mathbf{x}}(t\,|\,t)] \le 1\} \\
\frac{d}{dt}\hat{\mathbf{x}}(t\,|\,t) &= \mathbf{F}(t)\hat{\mathbf{x}}(t\,|\,t) + \mathbf{K}(t)[\mathbf{z}(t) - \mathbf{H}(t)\hat{\mathbf{x}}(t\,|\,t)] \\
\mathbf{K}(t) &= \boldsymbol{\Sigma}(t\,|\,t)\mathbf{H}'(t)\tilde{\mathbf{R}}^{-1}(t) \\
\frac{d}{dt}\boldsymbol{\Sigma}(t\,|\,t) &= \mathbf{F}(t)\boldsymbol{\Sigma}(t\,|\,t) + \boldsymbol{\Sigma}(t\,|\,t)\mathbf{F}'(t) \\
&\quad -\boldsymbol{\Sigma}(t\,|\,t)\mathbf{H}'(t)\tilde{\mathbf{R}}^{-1}(t)\mathbf{H}(t)\boldsymbol{\Sigma}(t\,|\,t) + \mathbf{G}(t)\tilde{\mathbf{Q}}(t)\mathbf{G}'(t) \\
&\quad + \{\beta(t) + \rho(t) - \delta^2(t)\}\boldsymbol{\Sigma}(t\,|\,t) \\
\delta^2(t) &= [\mathbf{z}(t) - \mathbf{H}(t)\hat{\mathbf{x}}(t\,|\,t)]'\tilde{\mathbf{R}}^{-1}(t)[\mathbf{z}(t) - \mathbf{H}(t)\hat{\mathbf{x}}(t\,|\,t)] \\
\tilde{\mathbf{R}}(t) &= \left(\frac{1}{\rho(t)}\right)\mathbf{R}(t) \\
\tilde{\mathbf{Q}}(t) &= \left(\frac{1}{\beta(t)}\right)\mathbf{Q}(t) \\
\boldsymbol{\Sigma}(0\,|\,0) &= \boldsymbol{\psi} \\
\hat{\mathbf{x}}(0\,|\,0) &= \mathbf{0} \\
\infty &> \rho(t) \ge 0 \\
\infty &> \beta(t) \ge 0
\end{aligned}
\tag{7.5.2}
$$

Any nonnegative $\rho(t)$ and $\beta(t)$ can be used. The estimator can be viewed in terms of a block diagram as in Fig. 7.5.1.

The discussions in the earlier chapters on the choice of values for β and ρ also apply in general terms to the continuous-time case.

As in the discrete-time case, these unknown-but-bounded results bear many resemblances to the corresponding Bayesian results for stochastic models. This similarity can be emphasized by defining

$$\tilde{\mathbf{F}}(t) = \mathbf{F}(t) + \frac{[\beta(t) + \rho(t) - \delta^2(t)]}{2}\mathbf{I}$$

to yield from (7.5.2)

$$
\begin{aligned}
\frac{d}{dt}\boldsymbol{\Sigma}(t\,|\,t) &= \tilde{\mathbf{F}}(t)\boldsymbol{\Sigma}(t\,|\,t) + \boldsymbol{\Sigma}(t\,|\,t)\tilde{\mathbf{F}}'(t) \\
&\quad -\boldsymbol{\Sigma}(t\,|\,t)\mathbf{H}'(t)\tilde{\mathbf{R}}^{-1}(t)\mathbf{H}(t)\boldsymbol{\Sigma}(t\,|\,t) + \mathbf{G}(t)\tilde{\mathbf{Q}}(t)\mathbf{G}'(t)
\end{aligned}
$$

The estimator of (7.5.2) is a nonlinear system, as $\hat{\mathbf{x}}(t\,|\,t)$ is a nonlinear function of $\mathbf{z}$. To remove this coupling and nonlinearity, set $\delta^2(t) = 0$ to get

$$
\begin{aligned}
&\boldsymbol{\Omega}_{\hat{x},b}(t\,|\,t) = \{\mathbf{x}\colon [\mathbf{x} - \hat{\mathbf{x}}_b(t\,|\,t)]'\boldsymbol{\Sigma}_b^{-1}(t\,|\,t)[\mathbf{x} - \hat{\mathbf{x}}_b(t\,|\,t)] \le 1\} \\
&\left.\begin{aligned}\hat{\mathbf{x}}_b(t\,|\,t)\\ \boldsymbol{\Sigma}_b(t\,|\,t)\end{aligned}\right\}\text{given by (7.5.2) with } \delta^2(t) = 0
\end{aligned}
\tag{7.5.3}
$$

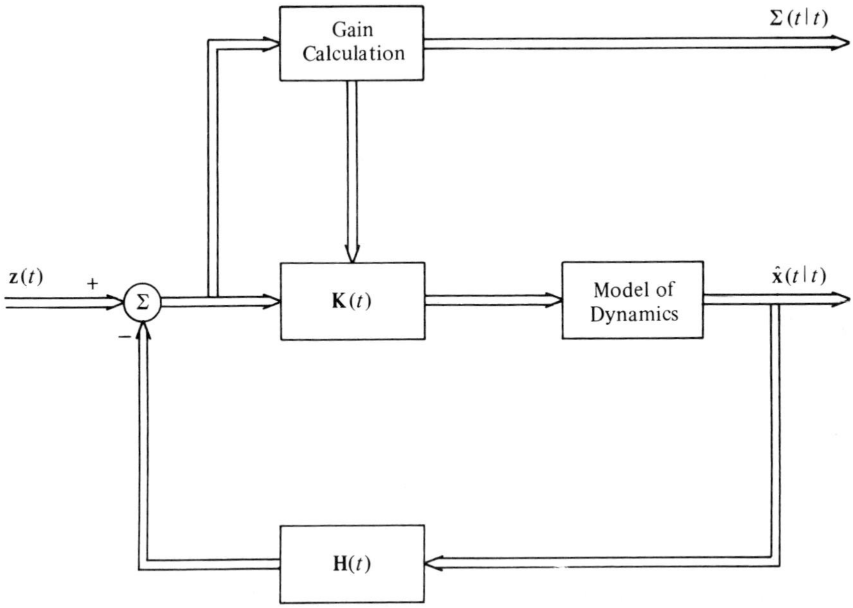

Figure 7.5.1 Unknown-but-bounded continuous-time filter.

This defines a bounding bounding ellipsoid in that

$$\mathbf{\Omega}_{\hat{x},b}(t \mid t) \supset \mathbf{\Omega}_{\hat{x}}(t \mid t)$$

The size of $\mathbf{\Sigma}(t \mid t)$ is a measure of the performance of the unknown-but bounded model filter. $\mathbf{\Sigma}(t \mid t)$ of (7.5.2) is a function of the observations. If it is desired to calculate a bound on the performance that is independent of the actual observations, $\mathbf{\Sigma}_b(t \mid t)$ of (7.5.3) can be used.

As in (6.6.7) an improvement on (7.5.3) can be obtained by considering

$$\boxed{\begin{aligned} \tilde{\mathbf{\Omega}}_{\hat{x},b}(t \mid t) &= \{\mathbf{x}: [\mathbf{x} - \hat{\mathbf{x}}_b(t \mid t)]'\tilde{\mathbf{\Sigma}}_b^{-1}(t \mid t)[\mathbf{x} - \hat{\mathbf{x}}_b(t \mid t) \leq 1\} \\ &\left.\begin{matrix} \hat{\mathbf{x}}_b(t \mid t) \\ \tilde{\mathbf{\Sigma}}_b(t \mid t) \end{matrix}\right\} \text{given by (7.5.2) with } \delta^2(t) = 0 \\ \tilde{\mathbf{\Sigma}}_b(t \mid t) &= [1 - q(t)]\mathbf{\Sigma}_b(t \mid t) \\ \frac{dq}{dt}(t) &= -[\beta(t) + \rho(t)]q(t) \\ &\quad + [\mathbf{z}(t) - \mathbf{H}(t)\hat{\mathbf{x}}(t \mid t)]'\tilde{\mathbf{R}}^{-1}(t)[\mathbf{z}(t) - \mathbf{H}(t)\hat{\mathbf{x}}(t \mid t)] \end{aligned}} \qquad (7.5.4)$$

as

$$\mathbf{\Omega}_{\hat{x},b}(t \mid t) \supset \tilde{\mathbf{\Omega}}_{\hat{x},b}(t \mid t) \supset \mathbf{\Omega}_{\hat{x}}(t \mid t)$$

For a time-invariant model, a steady-state $\mathbf{\Sigma}_\infty$ does not exist for (7.5.2) but does exist for the bounding bounding ellipsoid of (7.5.3).

7.6 Weighting Patterns

The estimators have been expressed in terms of differential equations. By analogy with the discrete-time case, these same estimators can often be expressed in terms of their weighting pattern.

In terms of its weighting pattern $\mathbf{W}(t, T)$ an estimator is expressed as

$$\hat{\mathbf{x}}(T|T) = \int_0^T \mathbf{W}(t, T)\mathbf{z}(t)\,dt + \mathbf{W}_0(T) \tag{7.6.1}$$

Calculation of such weighting patterns is conceptually straightforward. The differential equation (7.2.3) for a continuous-time Bayesian filter can be rewritten as

$$\frac{d}{dt}\hat{\mathbf{x}}(t|t) = \tilde{\mathbf{F}}(t)\hat{\mathbf{x}}(t|t) + \tilde{\mathbf{G}}(t)\mathbf{z}(t)$$

$$\tilde{\mathbf{F}}(t) = \mathbf{F}(t) - \mathbf{\Sigma}(t|t)\mathbf{H}'(t)\mathbf{R}^{-1}(t)\mathbf{H}(t)$$

$$\mathbf{G}(t) = \mathbf{\Sigma}(t|t)\mathbf{H}'(t)\mathbf{R}^{-1}(t)$$

Let $\boldsymbol{\theta}(t_1, t_2)$ denote the fundamental matrix associated with $\tilde{\mathbf{F}}(t)$. Then it follows that the weighting pattern of (7.6.1) is given by

$$\mathbf{W}(t, T) = \boldsymbol{\theta}(T, t)\tilde{\mathbf{G}}(t) \tag{7.6.2}$$

Similar results are obtained for Fisher models, for finite memory filters, for smoothing problems, and for bounding bounding ellipsoids.

A weighting pattern is often a useful quantity to "look at," as it provides a "feel" for how the filter behaves. Thus one may want to calculate a weighting pattern such as $\mathbf{W}(t, T)$ even though the actual implementation is done using differential equations.

EXAMPLE: Consider the Bayesian model

$$\frac{d}{dt}x(t) = Fx(t) + w(t)$$

$$z(t) = x(t) + v(t)$$

$$R = 1 \qquad Q = 1$$

with the corresponding steady-state filter given by

$$\frac{d}{dt}\hat{x}(t|t) = -\sqrt{F^2 + 1}\hat{x}(t|t) + (F + \sqrt{F^2 + 1})z(t)$$

(see example of Section 7.2). Thus

$$\hat{x}(T|T) = \int_0^T W(t, T)z(t)\,dt$$

$$W(t, T) = (F + \sqrt{F^2 + 1})e^{-(\sqrt{F^2+1})(T-t)}$$

and for $F = -1$ (a stable model)

$$W(t, T) = (-1 + \sqrt{2})e^{-\sqrt{2}(T-t)}$$

while for $F = 1$ (an unstable model)

$$W(t, T) = (1 + \sqrt{2})e^{-\sqrt{2}(T-t)}$$

which illustrates again that an unstable model can yield a stable filter.

7.7 Implementation

The continuous-time estimators and analysis equations of this chapter are expressed in terms of differential equations or weighting patterns (integral equations). In general these are time-varying equations, and exact implementation is often very difficult if not completely impractical. One practical approach is to use a digital computer to "solve" the differential equations by numerical integration. If numerical integration is viewed as a method of replacing a differential equation by a difference equation, one logical approach is to simply use the difference equations of Chapter 6. However, it may be more convenient in certain cases to apply a standard numerical integration technique directly to the continuous-time equations. It all depends on the particular application.

In the case of a linear time-invariant estimator, direct implementation in terms of electronic circuitry becomes practical in many cases.

7.8 Discussion

Many of the preceding discussions on the continuous-time case have been cryptic, as the basic ideas were already developed for the discrete-time case in Chapter 6. Note, however, that the final continuous-time equations are somewhat simpler in appearance and, for many people at least, more readily visualized. Thus it is sometimes convenient to use the continuous-time case as a vehicle for investigating behavior even if a discrete-time case is of interest.

In Chapter 6 discussions were provided on smoothing, perfect observations, finite memory filters, analysis and design of nonoptimum filters, and the effects of modeling errors. These discrete-time concepts all carry over to the continuous-time case.

Historical Notes and References: Chapter 7

Section 7.1

See discussion on Section 4.2.

Section 7.2

The discussions on Sections 6.2 and 6.3 also apply here. A discrete to continuous time limit procedure is not the only possible method of derivation.

For example, the same results can be obtained by the projection theorem or by the Weiner Hopf integral equation as in Kalman and Bucy [1]. Kyskeransk [1] considers linear models with pure time delays.

Section 7.4

Schweppe [10] discusses a different approach to the unknown input case which uses frequency domain techniques.

Section 7.5

See discussion of Section 6.6.

Exercises

Perfect observations, as in 7.12, are discussed in more detail in Bryson and Johansen [1] for stochastic, and Schleapfer and Schweppe [1] for unknown-but-bounded.

Exercises

7.1. Consider the zero mean Bayesian model

$$\frac{d}{dt}x(t) = Fx(t) + w(t)$$

$$z(t) = Hx(t) + v(t)$$

$$E\{x^2(0)\} = \psi$$

where $w(t)$ and $v(t)$ are white stochastic processes with corresponding Q and R.

a. Find the complete time-varying solution to the filtering problem; i.e., find a differential equation for $\hat{x}(t \mid t)$ and a closed-form expression for $\Sigma(t \mid t)$.
b. Find the steady-state solution by direct methods and then verify that it is the limit of the time-varying solution.
c. Discuss the behavior of the time-varying solution as $\psi \longrightarrow \infty$ so that the model becomes Fisher with respect to the initial condition.

Hint: The differential equation

$$\frac{d}{dt}\Sigma(t) = 2F\Sigma(t) + Q - \Sigma^2(t)\frac{H^2}{R}$$

$$\Sigma(0) = \psi, \qquad F < 0$$

$$Q > 0,\ R > 0$$

$$\psi > 0$$

has the solution

$$\Sigma(t) = \frac{Q\sinh(\tilde{F}t) + \psi\{\tilde{F}\cosh(\tilde{F}t) + F\sinh(\tilde{F}t)\}}{\tilde{F}\cosh(\tilde{F}t) - F\sinh(\tilde{F}t) + (\psi H^2/R)\sinh(\tilde{F}t)}$$

$$\tilde{F} = \sqrt{F^2 + \frac{QH^2}{R}}$$

Also,

$$R(F + \tilde{F}) = \frac{QH^2}{\tilde{F} - F}$$

7.2. Consider the Bayesian model, $0 \leq t \leq T$,

$$\begin{aligned} z(t) &= x(t) + v(t) \\ E\{v(t)v(\tau)\} &= \delta(t - \tau)2 \\ E\{v(t)\} &= 0 \\ E\{x(t)\} &= u(t) \\ E\{[x(t) - u(t)][x(\tau) - u(\tau)]\} &= e^{-3|t-\tau|}, \\ v(t),\ x(t)&\text{: uncorrelated} \end{aligned}$$

Let $\hat{x}(T \mid T)$ denote a best estimate of $x(T)$ made from $z(t)$, $0 \leq t \leq T$. For large T, specify a differential equation for $\hat{x}(T \mid T)$. Determine the variance of this estimate for large T.

7.3. Carry out the details of the discrete to continuous time limit which yields (7.2.3).

7.4. Consider the zero mean Bayesian model

$$\begin{aligned} \frac{d}{dt}x(t) &= Fx(t) + w(t) \\ z(t) &= x(t) + v(t) \\ E\{w(t)w(\tau)\} &= \delta(t - \tau)Q \\ E\{v(t)v(\tau)\} &= \delta(t - \tau)R \\ E\{x^2(0)\} &= \psi \end{aligned}$$

In steady state, the filter which generates $\hat{x}(t, t)$ is of the form

$$\frac{d}{dt}\hat{x}(t \mid t) = A\hat{x}(t \mid t) + Bz(t)$$

Sketch the behavior of A and B versus F for

$$\begin{aligned} R &= 1 & Q &= 1 \\ R &= 100 & Q &= 1 \\ R &= 1 & Q &= 100 \\ R &= 100 & Q &= 100 \end{aligned}$$

7.5. Consider a Fisher model with no input uncertainty:

$$\begin{aligned} \frac{dx}{dt}(t) &= Fx(t) \\ z(t) &= x(t) + v(t) \\ x(0)&\text{: completely unknown} \\ E\{v(t)v(\tau)\} &= \delta(t - \tau)R \end{aligned}$$

Define

$$y(t) = x(t) - x(0)$$

$$\hat{y}(T \mid T)\text{: estimate of } y(T) \text{ from } z(t),\ 0 \leq t \leq T$$

Express the estimator which yields $\hat{y}(T \mid T)$ in terms of a weighting pattern; i.e., if

$$\hat{y}(T \mid T) = \int_0^T W(t, T) z(t)\, dt$$

what is the best $W(t, T)$?

7.6. Consider the Bayesian model

$$\frac{d}{dt}\begin{pmatrix} x_1(t) \\ x_2(t) \end{pmatrix} = \begin{pmatrix} F_{11} & F_{12} \\ F_{21} & F_{22} \end{pmatrix}\begin{pmatrix} x_1(t) \\ x_2(t) \end{pmatrix} + \begin{pmatrix} w_1(t) \\ w_2(t) \end{pmatrix}$$

$$E[\mathbf{w}(t)\mathbf{w}'(t)] = \delta(t - \tau)\begin{pmatrix} Q_{11} & 0 \\ 0 & Q_{22} \end{pmatrix}$$

$$E[\mathbf{x}(0)\mathbf{x}'(0)] = \boldsymbol{\psi}$$

$$z_1(t) = x_1(t) + v_1(t)$$

$$E[v_1(t)v_1(\tau)] = \delta(t - \tau)R_1$$

$$z_2(t) = x_2(t)$$

$$\mathbf{x}(0), \mathbf{w}(t), v_1(\tau)\text{: zero mean uncorrelated}$$

One approach to handling this problem is to define

$$u(t) = F_{12} z_2(t)$$

and view $u(t)$ as a deterministic input for the following Bayesian model:

$$\frac{d}{dt} x_1(t) = F_{11} x_1(t) + u(t) + w_1(t)$$

$$z_1(t) = x_1(t) + v_1(t)$$

Let $\hat{x}_1(t \mid t)$ denote an estimate of $x_1(t)$ obtained by applying (7.2.3) to the above model. Explain in words why $\hat{x}_1(T \mid T)$ is not the best estimate of $x_1(T)$ given $z_1(t)$ and $z_2(t)$, $0 \leq t \leq T$.

7.7. Consider a Bayesian model with $x(0)$, w, v zero mean, uncorrelated:

$$\frac{d}{dt} x(t) = Fx(t) + w(t)$$

$$z(t) = x(t) + v(t)$$

$$E\{w(t)w(\tau)\} = \delta(t - \tau)Q$$

$$E\{v(t)v(\tau)\} = \delta(t - \tau)R$$

$$E\{x^2(0)\} = \psi$$

$$Q = 1 \qquad R = 1 \qquad \psi = 1$$

Assume that the true value of F is $\tilde{F}$ but that this value is not known and an optimum filter is designed for $F = 0.5$. Let $\Sigma_\infty(\tilde{F})$ denote the steady-state estimate error variance resulting from the filter designed for $F = -0.5$ when the input is the $z(t)$ resulting from $F = \tilde{F}$. Sketch $\Sigma_\infty(\tilde{F})$ versus $\tilde{F}$ for $-10 \leq \tilde{F} < 0$. What happens for $\tilde{F} \geq 0$?

7.8. Consider the weighted-least-squares formulation of Section 7.3. Prove that it yields an answer of the form of the Bayesian solution (7.2.3). *Comment:* Do not attempt this exercise unless you have a good background in the calculus of variations or, equivalently, Pontryagin's maximum principle. *Hint:* If $\mathbf{p}(t)$ denotes the costate or adjoint or Lagrange variables, the transversality conditions are $\mathbf{p}(T) = \mathbf{0}$, $\mathbf{p}(0) = -\boldsymbol{\psi}^{-1}\hat{\mathbf{x}}(0 \mid T)$. As a first step, derive the following smoothing equations:

$$\frac{d}{dt}\hat{\mathbf{x}}(t \mid T) = \mathbf{F}(t)\hat{\mathbf{x}}(t \mid T) + \mathbf{G}(t)\mathbf{Q}(t)\mathbf{G}'(t)[x(t \mid t) + \mathcal{J}(t \mid t)\hat{\mathbf{x}}(t \mid T)]$$

$$\hat{\mathbf{x}}(T \mid T) = \mathcal{J}^{-1}(T \mid T)x(T \mid T)$$

$$\left.\begin{matrix} x(t \mid t) \\ \mathcal{J}(t \mid t) \end{matrix}\right\}\text{as in (7.2.5)}$$

$$0 \leq t \leq T$$

Then introduce a change of variables

$$\hat{\mathbf{x}}(t \mid t) = \boldsymbol{\Sigma}(t \mid t)x(t \mid t)$$

to get (7.2.3).

7.9. Derive (7.1.5) and (7.1.6) by starting with a discrete-time model and taking the limit. *Hint:* The discrete-time equivalent to (7.1.3) is

$$z(n\Delta) = x + v(n\Delta), \qquad n = 1, \ldots, N$$

$$E\{v(n_1\Delta)v(n_2\Delta)\} = \begin{cases} \dfrac{R}{\Delta}, & n_1 = n_2 \\ 0, & n_1 \neq n_2 \end{cases}$$

so

$$\hat{x}(N\Delta \mid N\Delta) = \Sigma(N\Delta \mid N\Delta) \sum_{n=1}^{N} \frac{\Delta}{R} z(n\Delta)$$

$$\Sigma(N\Delta \mid N\Delta) = \left[\left[\frac{R}{N\Delta}\right]^{-1} + \psi^{-1}\right]^{-1}$$

The discrete-time equivalent to (7.1.4) is

$$z(n\Delta) = x + c(n\Delta), \qquad n = 1, \ldots, N$$

$$c(n\Delta + \Delta) = \Phi c(n\Delta) + \Delta w(n\Delta)$$

$$\Phi = 1 + \Delta A$$

$$E\{w(n\Delta)\} = 0$$

$$E\{w(n_1\Delta)w(n_2\Delta)\} = \begin{cases} \Delta^{-1}Q, & n_1 = n_2 \\ 0, & n_1 \neq n_2 \end{cases}$$

$$E\{c(0)\} = 0$$

$$E\{c^2(n\Delta)\} = \frac{Q\Delta}{1 - \Phi^2} = \Gamma_c, \qquad \text{all } n$$

Define

$$\mathbf{z}_N = \begin{bmatrix} z(\Delta) \\ \cdot \\ \cdot \\ \cdot \\ z(N\Delta) \end{bmatrix} \qquad \mathbf{H}_N = \begin{bmatrix} 1 \\ \cdot \\ \cdot \\ \cdot \\ 1 \end{bmatrix}$$

$$\mathbf{c}_N = \begin{bmatrix} c(\Delta) \\ \cdot \\ \cdot \\ \cdot \\ c(N\Delta) \end{bmatrix}$$

$$\mathbf{R}_N = E\{c_N c_N'\}$$

so that

$$\hat{x}(N\Delta \mid N\Delta) = \Sigma(N\Delta \mid N\Delta)\mathbf{H}_N'\mathbf{R}_N^{-1}\mathbf{z}_N$$

$$\Sigma(N\Delta \mid N\Delta) = E\{\hat{x}(N\Delta \mid N\Delta) - x\}^2 = [(\mathbf{H}_N'\mathbf{R}_N^{-1}\mathbf{H}_N) + \psi^{-1}]^{-1}$$

Now

$$\mathbf{R}_N = \frac{Q\Delta}{1 - \Phi^2}\begin{bmatrix} 1 & \Phi & \Phi^2 & \cdots & \Phi^{N-1} \\ \Phi & 1 & \Phi & \cdots & \Phi^{N-2} \\ \Phi^2 & \Phi & 1 & \cdots & \Phi^{N-3} \\ \cdots & \cdots & \cdots & \cdots & \cdots \\ \Phi^{N-1} & \Phi^{N-2} & \Phi^{N-3} & \cdots & 1 \end{bmatrix}$$

so that

$$\mathbf{R}_N^{-1} = (Q\Delta)^{-1}\begin{bmatrix} 1 & -\Phi & 0 & 0 & \cdots & 0 & 0 \\ -\Phi & 1+\Phi^2 & -\Phi & 0 & \cdots & 0 & 0 \\ 0 & -\Phi & 1+\Phi^2 & -\Phi & \cdots & 0 & 0 \\ \cdots & \cdots & \cdots & \cdots & \cdots & \cdots & \cdots \\ 0 & 0 & 0 & 0 & \cdots & -\Phi & 1 \end{bmatrix}$$

Thus

$$\Sigma(N\Delta/N\Delta) = \left[\left[\frac{(2 + \Delta A)\mathbf{\Gamma}_c}{(-N + 2)\Delta A + 2}\right]^{-1} + \psi^{-1}\right]^{-1}$$

This exercise is closely related to the examples of Sections 6.9, 6.17, and 6.18.

7.10. Consider the curve-fitting problem where given $z(t)$, $0 \le t \le T$, it is desired to find the a_k and b_k which minimize

$$J = \int_0^T [z(t) - y(t)]^2\, dt$$

$$y(t) = \sum_{k=1}^{K} a_k \cos(\omega_k t) + b_k \sin(\omega_k t)$$

$$\omega_k,\ k = 1, \ldots, K\text{: specified}$$

A Fourier series expansion of a time function $z(t)$, $0 \le t \le T$, is given by

$\tilde{y}(t)$, where

$$\tilde{y}(t) = \sum_{k=1}^{K} \alpha_{k,T} \cos(\omega_k t) + \beta_{k,T} \sin(\omega_k t)$$

$$\alpha_{k,T} = \frac{2}{\pi} \int_0^T z(t) \cos(\omega_k t)\, dt$$

$$\beta_{k,T} = \frac{2}{\pi} \int_0^T z(t) \sin(\omega_k t)\, dt$$

a. Show that the Fourier coefficients $\alpha_{k,T}$ and $\beta_{k,T}$ become equal to the a_k and b_k which minimize J for large T and that for all J that $\alpha_{k,T}$ and $\beta_{k,T}$ are equal to the minimizing a_k and b_k if the ω_k are chosen so that $T\omega_k = 2\pi k$, $k = 1, \ldots, K$.

b. Consider the dynamic system

$$\frac{d}{dt}\mathbf{x}_k(t) = \mathbf{F}_k \mathbf{x}_k(t) + \mathbf{G}z(t)$$

$$\mathbf{F}_k = \begin{bmatrix} 0 & \omega_k \\ -\omega_k & 0 \end{bmatrix} \qquad \mathbf{G} = \begin{bmatrix} 0 \\ 1 \end{bmatrix}$$

$$\mathbf{x}_k(t) = \begin{bmatrix} x_{1k}(t) \\ x_{2k}(t) \end{bmatrix} \qquad \mathbf{x}_k(0) = \mathbf{0}$$

Show that for the Fourier coefficients of part a,

$$\begin{bmatrix} -\beta_{k,T} \\ \alpha_{k,T} \end{bmatrix} = \begin{bmatrix} \cos(\omega_k T) & -\sin(\omega_k T) \\ \sin(\omega_k T) & \cos(\omega_k T) \end{bmatrix} \mathbf{x}(T)$$

Also show that for all T

$$a_{k,T}^2 + b_{k,T}^2 = x_1^2(T) + x_2^2(T)$$

7.11. Consider the nonlinear matrix Ricatti differential equation

$$\frac{d}{dt}\mathbf{\Sigma}(t) = \mathbf{F}(t)\mathbf{\Sigma}(t) + \mathbf{\Sigma}(t)\mathbf{F}'(t) + \mathbf{Q}(t) - \mathbf{\Sigma}(t)\mathbf{R}^{-1}(t)\mathbf{\Sigma}(t)$$

$$\mathbf{\Sigma}(0) = \boldsymbol{\psi}$$

a. Show that if

$$\frac{d}{dt}\mathbf{\Gamma}_1(t) = \mathbf{F}(t)\mathbf{\Gamma}_1(t) + \mathbf{\Gamma}_1(t)\mathbf{F}(t) + \mathbf{Q}(t)$$

$$\mathbf{\Gamma}_1(0) = \boldsymbol{\psi}$$

$$\frac{d}{dt}\mathbf{\Gamma}_2^{-1}(t) = -\mathbf{F}'(t)\mathbf{\Gamma}_2^{-1}(t) - \mathbf{\Gamma}_2^{-1}(t)\mathbf{F}(t) + \mathbf{R}^{-1}(t)$$

$$\mathbf{\Gamma}_2(0) = \boldsymbol{\psi}$$

then

$$\mathbf{\Gamma}_2(t) \leq \mathbf{\Sigma}(t) \leq \mathbf{\Gamma}_1(t)$$

Hint: See Exercise 6.15.

b. Consider $\mathbf{x}(t)$ and $\mathbf{p}(t)$ defined by

$$\frac{d}{dt}\begin{bmatrix}\mathbf{x}(t)\\ \mathbf{p}(t)\end{bmatrix} = \begin{bmatrix}\mathbf{F}(t) & \mathbf{Q}(t)\\ \mathbf{R}^{-1}(t) & -\mathbf{F}'(t)\end{bmatrix}\begin{bmatrix}\mathbf{x}(t)\\ \mathbf{p}(t)\end{bmatrix}$$

$$\mathbf{x}(0) = \boldsymbol{\psi}\mathbf{p}(0)$$

Show that

$$\mathbf{x}(t) = \boldsymbol{\Sigma}(t)\mathbf{p}(t)$$

Let $\boldsymbol{\theta}(t_1, t_2)$ denote a fundamental matrix such that

$$\mathbf{x}(t) = \boldsymbol{\theta}_{11}(t, 0)\mathbf{x}(0) + \boldsymbol{\theta}_{12}(t, 0)\mathbf{p}(0)$$

$$\mathbf{p}(t) = \boldsymbol{\theta}_{21}(t, 0)\mathbf{x}(0) + \boldsymbol{\theta}_{22}(t, 0)\mathbf{p}(0)$$

Then show that

$$\boldsymbol{\Sigma}(t) = [\boldsymbol{\theta}_{11}(t, 0)\boldsymbol{\psi} + \boldsymbol{\theta}_{12}(t, 0)][\boldsymbol{\theta}_{21}(t, 0)\boldsymbol{\psi} + \boldsymbol{\theta}_{22}(t, 0)]^{-1}$$

7.12. Consider a zero mean Bayesian model with perfect observations:

$$\frac{d}{dt}\mathbf{x}(t) = F\mathbf{x}(t) + \mathbf{G}\mathbf{w}(t)$$

$$\mathbf{z}(t) = \mathbf{H}\mathbf{x}(t)$$

$$E\{\mathbf{w}(t)\mathbf{w}'(\tau)\} = \delta(t - \tau)\mathbf{Q}$$

$$E\{\mathbf{x}(0)\mathbf{x}'(0)\} = \boldsymbol{\psi}$$

a. Show that if $[\mathbf{HGQG'H'}]^{-1}$ exists, then

$$\frac{d}{dt}\hat{\mathbf{x}}(t\,|\,t) = \mathbf{F}\hat{\mathbf{x}}(t\,|\,t) + [\boldsymbol{\Sigma}(t\,|\,t)\mathbf{F}' + \mathbf{GQG'}]\mathbf{H}'[\mathbf{HGQG'H'}]^{-1}\left\{\frac{d}{dt}\mathbf{z}(t) - \mathbf{HF}\hat{\mathbf{x}}(t\,|\,t)\right\}$$

$$\frac{d}{dt}\boldsymbol{\Sigma}(t\,|\,t) = \mathbf{F}\boldsymbol{\Sigma}(t\,|\,t) + \boldsymbol{\Sigma}(t\,|\,t)\mathbf{F}' - [\boldsymbol{\Sigma}(t\,|\,t)\mathbf{F}' + \mathbf{GQG'}]\mathbf{H}'[\mathbf{HGQG'H'}]^{-1}\mathbf{H}[\mathbf{F}\boldsymbol{\Sigma}(t\,|\,t) + \mathbf{GQG'}] + \mathbf{GQG'}$$

$$\hat{\mathbf{x}}(0\,|\,0) = \mathbf{0}$$

$$\boldsymbol{\Sigma}(0\,|\,0) = \tilde{\boldsymbol{\psi}}$$

where $\tilde{\boldsymbol{\psi}}$ is such that

$$\mathbf{H}\tilde{\boldsymbol{\psi}} = \mathbf{0}$$

$$\tilde{\boldsymbol{\psi}}\mathbf{H}' = \mathbf{0}$$

The initial conditions are written under the assumption that $\mathbf{z}(0) = \mathbf{0}$. Also show that from the assumed initial condition

$$\mathbf{H}\boldsymbol{\Sigma}(t\,|\,t) = \mathbf{0}$$

$$\mathbf{H}\hat{\mathbf{x}}(t\,|\,t) = \mathbf{z}(t)$$

Hint: Apply a discrete to continuous time limit to (6.9.1). Combining (7.2.2) with (6.9.1) and (6.9.2) gives

$$\boldsymbol{\Sigma}(N\Delta + \Delta\,|\,N\Delta)\mathbf{H}' = \Delta\{\boldsymbol{\Sigma}(N\Delta\,|\,N\Delta)\mathbf{F}'(N\Delta) + \mathbf{G}(N\Delta)\mathbf{Q}(N\Delta)\mathbf{G}'(N\Delta)\}\mathbf{H}' + \Delta^2\mathbf{F}(N\Delta)\boldsymbol{\Sigma}(N\Delta\,|\,N\Delta)\mathbf{F}'(N\Delta)\mathbf{H}'$$

$$\mathbf{H\Sigma}(N\Delta + \Delta \mid N\Delta) = \Delta\mathbf{H}\{\mathbf{F}(N\Delta)\mathbf{\Sigma}(N\Delta \mid N\Delta) + \mathbf{G}(N\Delta)\mathbf{Q}(N\Delta)\mathbf{G}'(N\Delta)\} + \Delta^2\mathbf{HF}(N\Delta)\mathbf{\Sigma}(N\Delta \mid N\Delta)\mathbf{F}'(N\Delta)$$

$$\mathbf{H\Sigma}(N\Delta + \Delta \mid N\Delta)\mathbf{H}' = \Delta\mathbf{HG}(N\Delta)\mathbf{Q}(N\Delta)\mathbf{G}'(N\Delta)\mathbf{H}' + \Delta^2\mathbf{HF}(N\Delta)\mathbf{\Sigma}(N\Delta \mid N\Delta)\mathbf{F}'(N\Delta)\mathbf{H}'$$

$$\mathbf{z}(N\Delta + \Delta) - \mathbf{H}[I + \Delta\mathbf{F}(N\Delta)]\hat{\mathbf{x}}(N\Delta \mid N\Delta) = \mathbf{z}(N\Delta + \Delta) - \mathbf{z}(N\Delta) - \Delta\mathbf{HF}(N\Delta)\hat{\mathbf{x}}(N\Delta \mid N\Delta)$$

b. Consider the special case

$$\mathbf{x} = \begin{bmatrix} x_1 \\ x_2 \end{bmatrix} \qquad \mathbf{w} = \begin{bmatrix} w_1 \\ w_2 \end{bmatrix}$$

$$\frac{d}{dt}\mathbf{x}(t) = \mathbf{Fx}(t) + \mathbf{w}(t)$$

$$z(t) = x_2(t)$$

$$\mathbf{F} = \begin{bmatrix} F_{11} & F_{12} \\ F_{21} & F_{22} \end{bmatrix}$$

$$\mathbf{E}\{\mathbf{w}(t)\mathbf{w}'(\tau)\} = \delta(t - \tau)\mathbf{Q}$$

$$\mathbf{Q} = \begin{bmatrix} Q_{11} & Q_{12} \\ Q_{12} & Q_{22} \end{bmatrix}$$

Show that

$$\frac{d}{dt}\hat{x}_1(t \mid t) = F_{11}\hat{x}_1(t \mid t) + F_{12}\hat{x}_2(t \mid t) + [(F_{21}\Sigma_{11}(t \mid t) + Q_{12})Q_{22}^{-1}]\left[-F_{21}\hat{x}_1(t \mid t) - F_{22}\hat{x}_2(t \mid t) + \frac{d}{dt}z(t)\right]$$

$$\hat{x}_2(t \mid t) = z(t)$$

$$\frac{d}{dt}\Sigma_{11}(t \mid t) = 2F_{11}\Sigma_{11}(t \mid t) + Q_{11} - (F_{21}\Sigma_{11}(t \mid t) + Q_{12})^2 Q_{22}^{-1}$$

$$\Sigma_{22}(t \mid t) = \Sigma_{12}(t \mid t) = 0$$

c. The results of part a involve $d\mathbf{z}(t)/dt$. However, consider an equation of the form

$$\frac{d}{dt}\hat{\mathbf{x}}(t \mid t) = \mathbf{A}_1(t)\hat{\mathbf{x}}(t \mid t) + \mathbf{A}_2(t)\mathbf{z}(t) + \mathbf{A}_3(t)\frac{d}{dt}\mathbf{z}(t)$$

Show that $\hat{\mathbf{x}}(t \mid t)$ can be obtained without actually implementing $[d\mathbf{z}(t)/dt]$ by defining

$$\mathbf{y}(t) = \hat{\mathbf{x}}(t \mid t) - \mathbf{A}_3(t)\mathbf{z}(t)$$

and writing

$$\frac{d}{dt}\mathbf{y}(t) = \mathbf{A}_1(t)\mathbf{y}(t) + \left[\mathbf{A}_2(t) - \frac{d\mathbf{A}_3(t)}{dt} + \mathbf{A}_3(t)\right]\mathbf{z}(t).$$

8

Application of Linear Model Estimation Theory

This chapter is a "soft" chapter in that it is mostly general discussion with few or no "hard" equations or derivations. However, the author feels that the concepts are as important in practice as the actual equations. A whole book could be written by combining the concepts of this chapter with the "hard" equations of the earlier chapters. The writing of this new book is left as an exercise for the reader.

In Section 8.1, the problems of obtaining models are discussed. In Section 8.2, the design of practical estimators is considered. In Section 8.3, the use of estimation theory in general system and signal design problems is discussed.

8.1 Development of Models

Many possible mathematical models are listed in Chapter 3 and analyzed in Chapter 4. In Chapters 5–7, estimation logics are based on these models. A big question is, How is the model chosen? There is no easy answer, but the question is now discussed.

It is obvious that if the model is bad, the resulting analysis and system design is bad. However, a designer can defeat himself if he insists on having a "perfect" model, as it may take too long to develop and may involve too much work to analyze. Although not very philosophically satisfying, there do exist situations where a poor result is better than no result (provided that the poor answer is not too poor). For example, in practice an answer may be needed in 10 days, and if it is not available by then, it is too late. Hence *overmodeling* can be as bad a sin as *undermodeling*. The designer must rely on his own physical insights and engineering judgment. It also pays to have luck on one's side.

A mathematical model can be considered to consist of two aspects:

1. Structure.
2. Numerical values (of parameters).

In the best of all possible worlds both the structure and the numerical values are either given (by someone else) or can easily be obtained from well-known and proved *basic principles*. Usually, however, life is more difficult. A general approach to model development is the following three-step procedure:

1. Hypothesize the structure of the model.
2. Estimate (guess, determine) numerical values.
3. Test the validity of the model.

When past observations are available on the system to be modeled, the second two steps can be done directly using the ideas of Part III of this book on hypothesis testing and Chapter 14 on system identification. The next two subsections deal with aspects of the first and second steps.

8.1.1 Uncertainty

Three approaches to the modeling of uncertainty are

1. Empirical.
2. Theoretical.
3. Pseudo.

Although the logic underlying these approaches differs, the actual mathematical models may be similar or identical in form.

An empirical approach to modeling uncertainty is based on a history of past behavior. To illustrate, assume that a model for an uncertain time function $y(t)$ is desired for $t > 0$. Assume that a record of the past behavior of $y(t)$, say, from $-T \leq t \leq 0$ (where T is positive), is available. Four possible models for the future $t > 0$ behavior of $y(t)$ are

1. Assume that $y(t)$ is an uncertain constant x which is modeled as a random variable with

$$\text{Mean} = \frac{1}{T}\int_{-T}^{0} y(t)\,dt$$

$$\text{Variance} = \frac{1}{T}\int_{-T}^{0} [y(t) - \text{mean}]^2\,dt$$

2. Assume that $y(t)$ is a white unknown-but-bounded process $w(t)$ defined by

$$y_{\min} \leq w(t) \leq y_{\max}$$

where

$$y_{\min} = \min y(t), \qquad -T \leq t \leq 0$$

$$y_{\max} = \max y(t), \qquad -T \leq t \leq 0$$

3. Assume that $y(t)$ is a process $x(t)$ generated by

$$\frac{d}{dt} x(t) = Fx(t) + w(t)$$

$$E\{w(t)w(\tau)\} = \delta(t - \tau)Q$$

where F, Q, and initial conditions are chosen to yield a correlation function (4.1.12) which approximates

$$\mathcal{P}(t, \tau) = \frac{1}{T + t} \int_{-T}^{t} y(s)y(s - \tau)\, ds$$

in some reasonable fashion.

4. Assume that $y(t)$ is an unknown-but-bounded process $x(t)$ generated by

$$\frac{d}{dt} x(t) = Fx(t) + w(t)$$

$$|w(t)|^2 \leq Q$$

where F, Q, and initial conditions are chosen to yield the minimum Q such that $\mathbf{\Omega}_x(t)$ contains $y(t)$, $-T \leq t \leq 0$.

There are, of course, many other possibilities. For example, the scalar $y(t)$ might be modeled as in model 3 or 4 by $y(t) = \mathbf{H}\mathbf{x}(t)$, where $\mathbf{x}(t)$ is a vector. This empirical logic is obviously based on some sort of stationarity assumption; i.e., past and future behaviors are assumed to be similar. Chapter 14 on system identification can be viewed as being a much more extensive discussion on empirical modeling.

Five examples of the theoretical approach to modeling uncertainty are

1. Thermal noise power levels can be calculated for use in white stochastic process models.
2. Manufacturer listed sensor or component accuracies can be converted into the variances of a stochastic model or into the bounds of an unknown-but-bounded model.
3. Knowledge of calibration or operating procedures used on a system can be converted into variances or bounds.
4. Analog to digital quantization effects can be modeled as an unknown-but-bounded uncertainty.

5. A maneuvering vehicle can be handled as an unknown-but-bounded model if a bound on maximum acceleration is known.

These basic uncertainty models can often combine with deterministic dynamic models to yield more complex models. For example, thermal noise "acting through" an electronic circuit can be modeled as a correlated process where the **F**, **G**, and **H** matrices are determined by the structure and values of the circuit.

The pseudo approach to uncertainty modeling refers to the use of an uncertainty model instead of using a deterministic model which is actually (or at least theoretically) available. The uncertainty model may be desired because it results in a simpler analysis, because it is too difficult (expensive) to gather all the data necessary for an exact model, or because the exact model is too complex to be included in the computer. Examples of pseudo uncertainty models are

1. A recording of Beethoven's Fifth Symphony yields a time function which is actually a deterministic time function. However, for the purposes of designing a high-fidelity set, it may be more convenient to model the symphony as an uncertain time function modeled as a stochastic process.
2. A deterministic dynamic system may be exactly described by a nonlinear differential equation or by a high-order set of linear differential equations. However, for purposes of analysis, it may be convenient to use a lower-order linear model with an uncertain input to account for the *neglected terms.*
3. A man 100 miles away may be making a series of rational decisions. However, instead of installing a telephone line to find out what he is doing, it may be cheaper to model him as an uncertain process.

The scope of possible pseudo uncertainty models is limited only by one's imagination.

It is obvious that the empirical, theoretical, and pseudo approaches can be combined in various ways. It is also obvious that the difference between them is often in the eye of the beholder. The actual mathematical models may be the same. For example, arguments for modeling Beethoven's Fifth as a stochastic process can be based on empirical and theoretical considerations as well as pseudo arguments. (The author prefers not to talk about Beethoven's work as random noise, although the term does seem to apply to some of the more modern compositions.) It is the model, not the name, that is important. The three approaches to modeling were defined only for the sake of discussion.

None of the preceding examples involves a case where probability distributions are given. The author has rarely encountered a problem where a complete probabilistic model was actually available. This fact underlies the reasons why the preceding chapters have devoted much effort to non-Bayesian formulations and to Bayesian theory which uses only first and second moments (means and variances). On the other hand, a Bayesian model is easy to analyze and study, so one is often tempted to assume a probabilistic structure to make the job easier. The use of an assumed probabilistic structure might be called a pseudo-Bayesian model.

8.1.2 Dynamics

The basic models used in this book are of the form of uncertain functions driving dynamic systems and corrupting the observations. The development of the needed dynamic models can be divided into two steps:

1. Determine the basic equations for the dynamic input and output relationships.
2. Convert the equations into the state space form used in this book.

The first step is usually the most difficult and critical.

Unfortunately the only general rule for the first step is that the engineer must really understand the physical mechanisms underlying the problem. As noted in Section 6.13 on modeling errors, the model for the dynamics is usually "more important" than the uncertainty models but it is difficult to generalize. For example, in the stochastic polynomial example of Section 6.16, the introduction of a little uncertainty (the stochastic white process input) could completely change the nature of the results.

8.2 Practical Filter Design

It is easy to become overly obsessed with mathematical optimality. Properties such as minimum error are usually less important than properties such as cost and reliability. *Practical* filters and the role theory plays in their design are now discussed. Most of the discussions also apply to prediction and smoothing.

Ideally, the term *practical* should be defined but no really viable definition is available. One design may be practical if a large digital computer is available but impractical if only four resistors, three capacitors, and one abacus are available. One approach is to define the "most practical" filter to be the "simplest" one that performs the necessary job satisfactorily. Of course, this definition does not lead to precise design techniques; for example, try to take the partial derivative of performance with respect to simplicity. Another approach is to define the most practical filter as the best-performing filter that can be built with the available equipment (analog components or digital computer software). This definition can lead to explicit design methods.

Before discussing design, it is necessary to decide how to determine that a filter performs satisfactorily. In practice, explicit criteria are rarely available, so a "seat of the pants" decision is usually made. Three ways that data on performance can be obtained are

1. Field test: Implement the filter and test performance against real data.
2. Simulation: Implement the filter on a general-purpose digital or analog computer and see how it performs for simulated inputs.
3. Analysis: Evaluate performance against some mathematical model for the real data.

8.2.1 Philosophies of Design: Good Model Available

Assume that a good mathematical model for the real world is available. Four philosophies of practical design can then be defined as follows:

1. Method I: Choose a simple filter which performs satisfactorily.
2. Method II: Choose a simple mathematical model which results in an optimum filter which is simple and which performs satisfactorily.
3. Method III: Choose a simple filter structure and then use mathematics to obtain optimum parameter values.
4. Method IV: Choose the most complex filter structure which can be implemented and then use mathematics to obtain optimum parameter values.

Method I is excellent if the designer has a lot of past experience in the particular application of interest. If such past experience is not available, method I can be used by designing the optimum for the good model and then seeing how the optimum can be simplified.

In method II there are really two levels of models. One is the good model used for performance evaluation. The second is the model to be used for the actual design. The designer looks at the good model and then removes the "less important" physical phenomena to get a design model. For example, he may ignore correlations, biases, and dynamics with small time constants. The designer then determines the optimum filter for the design model. If this filter is not "simple enough" to be practical, he returns to his design model and attempts to simplify it further.

In method III, the designer chooses a simple filter structure on the basis of his own insights and past experiences (as in method I) but uses mathematics to complete the design procedure (as in method II) by determining optimum values for the estimator parameters. This is the technique of constrained estimator design discussed in Section 6.12.

Method IV is a variation of method III in which the designer chooses the most complex filter structure which can be implemented with available hardware (analog components or digital computer capability). The free parameters are then optimized as in method III.

The first three design methods may yield a simple filter whose performance is not satisfactory. In such a case it is necessary to go back and try again. In method IV, the performance had better be satisfactory, as it is the best that can be obtained with available equipment. In the first three design philosophies, the performance of the optimum estimator (for the good model) is a valuable quantity to know, as it provides a *bench mark* against which the performance of practical estimators can be measured. If a particular estimator design performs *almost as well as* the optimum, there is little reason to attempt to improve it. If the optimum performance were not known, much effort might be wasted in a futile search to find a nonexistent way to make a major improvement. This logic leads to the following very important principle of practical filter design: *If filter performance is specified, the first step for any design is to evaluate the performance of the optimum filter (for the good model)*. This is a critical step, as it may turn out that the desired performance is not attainable by any filter. If so, the filter designer should complain to the man who designed the sensors (unless, of course, the same man has both jobs).

It is impossible to say that any one philosophy of estimator design is the most powerful. The "best" choice depends on both the nature of the problem and the way the designer likes to think.

There is an obvious weak point in the above discussions. A good model of the physical process is assumed to be available to use to evaluate estimator performance. In theory, this good model can be obtained by turning method II around and adding more and more physical phenomena to the model until any neglected phenomena have no noticeable effect. In practice the designer usually relies on his judgment and hopes he chooses a good model.

8.2.2 Feedback Tracking Systems

Discrete- and continuous-time, Bayesian, and unknown-but-bounded filters are represented by block diagrams in Fig. 6.3.3, 6.6.3, 7.2.3, and 7.5.1. All these block diagrams are summarized in Fig. 8.2.1.

The basic feedback form of Fig. 8.2.1 comes as no surprise. A good control engineer who never heard of estimation theory would probably choose a system like Fig. 8.2.1 if he had to use the observations $\mathbf{z}$ to track the time variations of the state $\mathbf{x}$. Such a control engineer would try to pick his gains

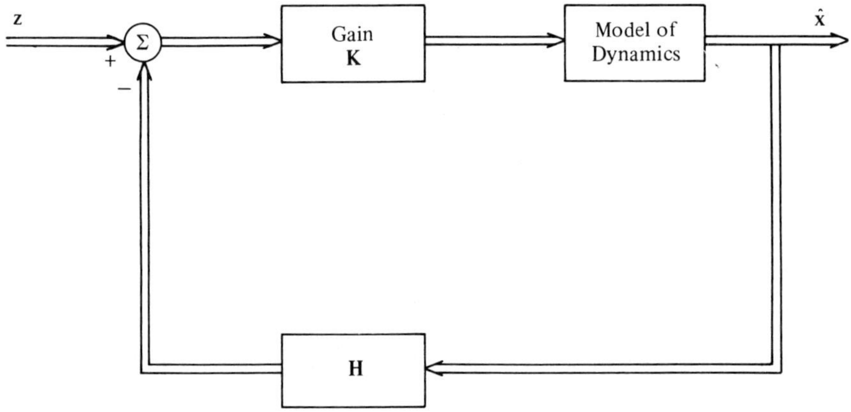

Figure 8.2.1 General form of filter.

K to give "satisfactory" performance, i.e., a stable system that is not too sluggish to follow variations in **x** and yet is not too sensitive to high frequency noise in **z**.

From this point of view, a reasonable method of estimator design is to choose the form of Fig. 8.2.1 and then use the equations for the optimum gains to guide the choice of simpler gains. For example, since **K** is a matrix, one might try to set some of the *small coefficients* equal to zero. Alternately, one might try to approximate an optimum time-varying gain **K** with a constant or piecewise constant gain.

This feedback tracking interpretation points out a crucial shortcoming of the definition of *optimum* used in the preceding chapters. Optimum was defined in terms of quantities such as minimum error variance or minimum set size. However, from the point of view of building a system to track **x**, the system's response to *unit step functions* as measured in terms of *rise time*, *overshoot*, and *damping* may be more important criteria. Hence when evaluating a filter's performance, a simulation may be required to see how it behaves when viewed as a feedback tracking system.

The following approach to estimator design is often used:

1. Determine and implement (or simulate) the optimum estimator (from Chapters 5–7) with **Q**, **R**, and $\boldsymbol{\psi}$ viewed as free parameters.
2. Vary **Q**, **R**, and $\boldsymbol{\psi}$ until the feedback tracker (estimator) with the best rise time, overshoot, etc., is obtained.

At first glance, this design technique may appear rather silly, but it can be justified in varying ways depending on the nature of the problem. If a good model is available (with specified **Q**, **R**, and $\boldsymbol{\psi}$), the above design procedure can be viewed as a method of obtaining a compromise between performance as measured by, say, error variance and performance as measured by, say, rise time and overshoot. In the other extreme, when model $\mathbf{Q}$, $\mathbf{R}$, and $\boldsymbol{\psi}$ values are not known, the above design procedure enables design in terms of performance criteria which can be measured, i.e., overshoot, etc. When model $\mathbf{Q}$, $\mathbf{R}$, and $\boldsymbol{\psi}$ values are not known, it might seem to be more reasonable to design the entire filter structure directly for the best overshoot, etc., response. However, such design is very difficult when the model structure itself is not simple, i.e., when complications such as time-varying dynamics or vector observations are present. The above design philosophy first uses the estimation theory to get an optimum filter structure and then resorts to seat of the pants engineering to find the best values for $\mathbf{Q}$, $\mathbf{R}$, and $\boldsymbol{\psi}$.

When designing by the preceding technique, $\mathbf{R}$ is often held fixed while $\boldsymbol{\psi}$ and $\mathbf{Q}$ are varied, as "good" model values for $\mathbf{R}$ may be available. In many applications the effect of initial uncertainty as measured by $\boldsymbol{\psi}$ is not critical, so only $\mathbf{Q}$ is varied.

Performance criteria such as rise time and overshoot were discussed, but another important concept is stability. As indicated in Chapters 6 and 7, the optimum filter is "almost always" stable as long as $\mathbf{Q}$ is not zero. In general, as $\mathbf{Q}$ increases, the gain $\mathbf{K}$ of the optimum filter increases, and this has the effect of *stabilizing* the feedback system. Thus if the actual model of interest has $\mathbf{Q} = \mathbf{0}$ (no uncertain input), it may be desirable to change the model to incorporate a finite $\mathbf{Q}$ for the sole purpose of obtaining a stable feedback system. Similarly, if the actual model $\mathbf{Q}$ is small, it may be desirable to use an artificially large value of $\mathbf{Q}$ to obtain "more" stability. This can be viewed as the use of a pseudo uncertain input which is introduced to improve filter stability.

Sometimes the filter's design can be modified after actual implementation in the field. In such a case a reasonable approach is to vary the gains $\mathbf{K}$ until the "best" performance results. Rather than vary $\mathbf{K}$ directly, it is often better to use the optimum theory to express $\mathbf{K}$ in terms of, say, $\mathbf{Q}$ and thus vary $\mathbf{K}$ by varying $\mathbf{Q}$. This is especially helpful when $\mathbf{K}$ is a large dimensional matrix while $\mathbf{Q}$ is a small dimensional matrix (hopefully a scalar).

8.2.3 Effective Memory Span

The feedback tracking interpretation of a filter provides one type of insight and philosophy of design. A related but different point of view is obtained by considering the effective memory time span of the filter. The effective memory span, τ, of a filter determines how much of the past data is being used to obtain the present estimate; i.e., $\hat{\mathbf{x}}(T|T)$ depends effectively only on $\mathbf{z}(t)$, $T - \tau \leq t \leq T$. In terms of a linear filter with a weighting function, $\mathbf{W}(t, T)$, τ is defined to be such that $\mathbf{W}(t, T)$ is effectively zero for $0 \leq t \leq T - \tau$.

Memory span is not an overly important concept in an idealized world where mathematical models are perfect and optimum filters are implemented. Old data are automatically used or rejected by the filter as required. However, in many applications, mathematical models are valid for only a short period of time into the past. For example, an airplane's motion might be modeled as a straight-line constant velocity path for short time intervals, but the model obviously cannot be extended back in time to before the airplane left the ground. If the airplane's motion can be modeled only for time spans of 15 seconds or less, any filter used to process radar observations of the airplane should have a memory span of 15 seconds or less.

A filter's memory span can be reduced in several ways. An obvious method is to use a true finite memory filter. An alternative approach is the introduction of an extra fictitious white process with "size" **Q** as an input into the dynamics of the mathematical model used to design the filter. The larger the value of **Q**, the less the present depends on the past. Thus if a filter is designed assuming a model with a large (pseudo) uncertain system input, the filter automatically forgets much of the past, i.e., reduces its memory span. This concept is related to the concepts of Section 8.2.2, where stabilization is obtained by adding a pseudo uncertainty of size **Q**.

To illustrate the concepts, consider a problem when it is known that $y(n)$, $n = 1, \ldots,$ can be modeled as a second-degree polynomial in time for time intervals of $\tau = 15$ or less so that

$$y(n) = x_0 + nx_1 + \frac{n^2 x_2}{2}, \qquad n = N - \tau, \ldots, N$$

or

$$\begin{aligned} \mathbf{x}(n+1) &= \mathbf{\Phi}\mathbf{x}(n) \\ \mathbf{\Phi} &= \begin{bmatrix} 1 & 1 & \frac{1}{2} \\ 0 & 1 & 1 \\ 0 & 0 & 1 \end{bmatrix} \\ y(n) &= [1 \quad 0 \quad 0]\mathbf{x}(n) \end{aligned} \tag{8.2.1}$$

Assume that

$$z(n) = y(n) + v(n)$$

$$E\{v(n_1)v(n_2)\} = \begin{cases} R, & n_1 = n_2 \\ 0, & n_1 \neq n_2 \end{cases}$$

In a true finite memory approach, (8.2.1) would be combined with a finite memory filter (Fisher model) of length 15. In the pseudowhite process approach, (8.2.1) might be modeled to

$$\begin{aligned} \mathbf{x}(n+1) &= \mathbf{\Phi}x(n) + \begin{pmatrix} 0 \\ 0 \\ 1 \end{pmatrix} w(n) \\ E\{w(n_1)w(n_2)\} &= \begin{cases} Q, & n_1 = n_2 \\ 0, & n_1 \neq n_2 \end{cases} \end{aligned} \tag{8.2.2}$$

and then the filter would be designed using (8.2.2) as the model for some value of Q. Equation (8.2.2) corresponds to the stochastic polynomial of Section 6.16.5. Figure 8.2.2 compares some weighting patterns obtained by these two methods when one is attempting to estimate the *velocity* of $y(n)$ at $n = N$. Both weighting patterns in Fig. 8.2.2 have an effective memory span of $\tau = 15$, but they have very different characteristics. The true finite memory filter weights data that are 15 units old almost as much as the most recent data. The pseudowhite process filter applies, effectively, no weight to data 15 units old.

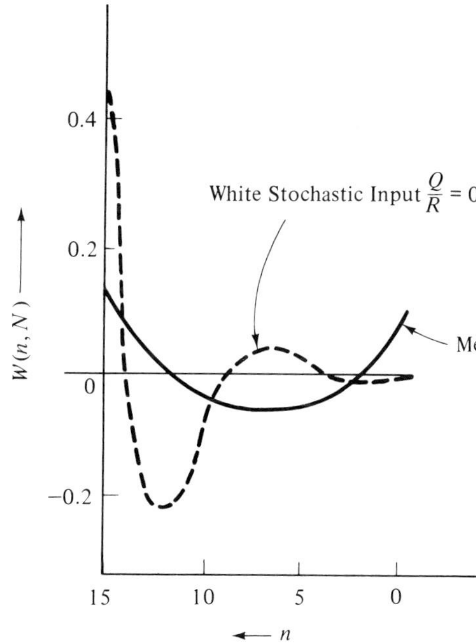

Figure 8.2.2 Two weighting functions with effective memory of $\tau = 15$: polynomial motion.

It is clear that effective memory span itself is not necessarily the only important property of a filter. However, in practice, it is often very desirable to know the memory span and sometimes to actually include it as part of the filter specifications.

In rough general terms, the feedback tracking and limited memory span points of view are related by "an increase in gain reduces effective memory length and increases stability." An increase in $\mathbf{Q}$, the amount of white process driving the system model, increases feedback gain, reduces effective memory span, and increases stability.

Two ways have been discussed to control effective memory span: the use of $\mathbf{Q}$ in a Bayesian model and the use of a true finite memory filter.

A third and conceptually different way is to use the special weighted-least-squares approach called exponential smoothing, as discussed in Sections 6.5.2 and 7.3.2. The parameter α in these sections plays an analogous role to $\mathbf{Q}$; i.e., increasing α decreases the effective memory span. No general rules are available as to whether it is better to adjust $\mathbf{Q}$ in a Bayesian model or to adjust α in exponential smoothing. The resulting estimators are very similar but not identical, and it is a matter for engineering judgment. In many cases, it probably makes little difference which approach is used. The exponential smoothing weighted-least-squares approach has the advantage that the relationship between the parameter α and the filter's effective memory span is "more intuitively obvious."

8.2.4 Philosophies of Design: Good Model Not Available

When a good model is not available (as is often the case), there are various ways to proceed with the design. As already discussed, the feedback tracking point of view can be used to design in terms of overshoot, stability, etc., or the finite memory point of view can be used when the available model is valid only for a limited portion of the past. Alternatively, a feedback tracker can simply be implemented in the field and the gain varied until the best performance is obtained.

Another approach is to use the observations to try to determine a good model by using system identification. An estimator which simultaneously estimates and identifies is sometimes called an adaptive estimator. This identification—adaption approach can definitely be of great value. However, explicit discussion will be deferred to Chapter 14, as some more theory has to be developed first.

8.2.5 Time-Invariant Models and Filters

The general emphasis of the preceding chapters and the preceding discussions in this chapter has been on time-varying system models, so naturally the time-invariant system is included as a special case. A further specialization is a time-invariant, stationary system model where the observation intervals are long and the filter itself is time-invariant. In this very special case, one can sometimes employ frequency domain techniques. In some classes of practical filter design problems, such frequency domain techniques and methods of thinking can be superior to the use of the general techniques of this book and the ideas of the preceding sections.

The reader who is not familiar with frequency domain methods of design (and some who are) is about to become frustrated, as the above statement is not going to be backed up with a discussion on when, where, or how. Such a discussion is beyond the scope of this book. The author merely wants to state that he does not believe that the techniques considered in this book are the best for all problems (just most).

8.3 Sensor, Signal, Measurement, etc., Design

Estimation theory is a powerful tool for much more than just designing estimators. The performance analysis part of estimation theory is a basic vehicle for making many overall system design decisions and formulating many signal and measurement design problems. In the following discussion a good model is assumed to be available.

Consider the design of some complete system of the form of Fig. 8.3.1.

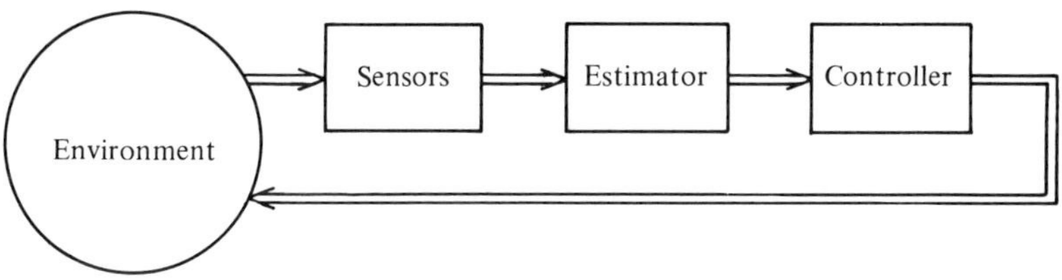

Figure 8.3.1

The *environment* is to be controlled. In this picture, the complete system has three components:

1. The sensors, which gather information about the state of the environment.
2. The estimator, which converts the sensor measurements into an estimate of the state (or some function thereof) of the environment.
3. The controller, which uses the estimated state (or some function thereof) as the basis for control decisions and actions.

The controller's decisions and effectiveness depend on the uncertainty in the output of the estimator. The uncertainty of the estimator's output depends on the quality and quantity of the measurements obtained from the sensors. Because of these ties, estimation theory is important in the design of both the sensors and the controller. (The general discussions of Chapter 2 are closely related to the present discussions.)

Often some hypothesized set of sensors and their accuracies are specified and the critical question is whether or not these sensors are sufficient to "do the job"; i.e., will the resulting state estimate have a small enough uncertainty? Such questions emphasize the importance of the theory, as the minimum possible error covariance matrix Σ can be obtained without ever designing the estimator itself.

Sometimes the performance analysis of estimation theory can be used to formulate a sensor system or signal design problem as a deterministic optimization problem. The basic concept is very simple. The error covariance matrix Σ for the optimum estimate is evaluated as a function of the "free" aspects of the signal or sensor system. The specific values of these free aspects are then chosen to minimize some desired scalar function of Σ. Examples of such free aspects are

1. A radar wave form must be determined subject to constraints on energy, peak power, bandwidth, etc.

2. A limited number of meters must be located on an electric power EHV network so as to give the best state estimate.

Unfortunately, in most design problems, the choice of a scalar measure of the size of Σ is not obvious. This is not a weakness of the theory. It is merely a fact of life that an engineer must often use his own judgment to choose a final design in the face of conflicting criteria.

The sensor, signal, measurement, etc., design problem is related to the concept of designing constrained estimators, as in Section 6.12. In both cases, an error covariance matrix Σ is evaluated as a function of free parameters, and then the values of the free parameters are optimized to minimize Σ in some sense. In the case of constrained estimator design, the free parameters are part of the estimator itself. In the case of sensor, signal, measurement design, etc., the estimator is fixed, and the free parameters specify the sensor measurement system or the *input* signal. Obviously, the two ideas can be combined if desired.

8.4 Summary Discussion

The choice of model is often the most critical aspect of a design and development engineering job, but it is impossible to give explicit rules or techniques. This book discusses a wide set of possible models and shows how to analyze and use them. The actual choice of model is the reader's problem.

Optimum theory is more than an academic exercise used to write papers for learned journals. It is a very powerful tool but must be used by an engineer with judment.

Estimation theory is used for much more than the design of estimators. It forms the basis for solving many system design problems.

Historical Notes and References: Chapter 8

Section 8.3

The signal design concepts are taken from Schweppe [2]; Athans and Schweppe [2]. Some explicit results for particular cases can be found in Schweppe [5]; Schweppe and Gray [1]. More discussions related to the meter placement problem area can be found in Schweppe and Wildes [1]; Schweppe and Rom [1]; and Schweppe [11]. Other discussions on signal design can be found in Meier and Peschon [1] and Kushner [5].

PART III

Hypothesis Testing

The next stage of our journey takes us along a relatively narrow strip of terrain which leads across the broad fields of hypothesis testing, detection theory, and decision making. Many interesting places are not on the itinerary, but those that are visited yield very pretty views. In a way we are traveling on an isthmus which connects the lands of linear models and nonlinear models as we draw on the results of Part II and as our answers have applications in Part IV.

In Chapters 9 and 10 we repeat the same type of straight-line path followed in Chapters 5 and 8; start with vectors, proceed to discrete-time dynamic systems, go on to continuous-time dynamic systems, and then attempt to indicate how the theory can be used in the design of practical hypothesis-testing systems, sensor systems, and signals. As in Chapters 5–8, the models are still linear. However, the Bayesian theory of Chapters 5–8 was developed without a Gaussian assumption, while in Chapters 9 and 10, a Gaussian assumption is used repeatedly.

9

Hypothesis Testing: Static Systems

Estimation theory and hypothesis testing are related in broad general terms as follows:

1. Estimation theory: Given observations made on some specified process, what is the state of the process?
2. Hypothesis testing: Given observations made on one of M specified processes, which of the M processes is observed?

In estimation theory (see the introduction to Chapter 5) there is only one model for the observations, and the problem is to use this model to determine some way of processing the actual observations to yield an estimate of the state of the model. In a hypothesis-testing problem with M hypotheses, there are M different possible models for the observations, and the problem is to use these M models to determine some way of processing the actual observations to decide which of the M models corresponds to the observed values.

In this chapter, Gaussian assumption is often used. The notational convention is

$$\mathbf{z} \text{ is } N[\mathbf{m}_z, \mathbf{\Gamma}_z]$$

meaning that $\mathbf{z}$ is a normal random vector with mean $\mathbf{m}_z$ and covariance matrix $\mathbf{\Gamma}_z$.

In this chapter, the observations are assumed to be a single vector $\mathbf{z}$. Assume that there are M hypotheses (possible models for $\mathbf{z}$) denoted by $\mathcal{H}_j$, $j = 1, \ldots, M$, where

$$\mathcal{H}_j: \quad \mathbf{z} = \mathbf{z}_j$$
$$\mathbf{z}_j: \quad \text{uncertain vector defined by some model}$$

The problem is to use the actual observations to decide which hypothesis $\mathcal{H}_j$ is true.

EXAMPLE: Let z denote a perfect measurement on the length of a desk. It is desired to determine whether the desk is made of wood or metal. Assume that it is known that all wood desks are 5 feet long, while all metal desks are 6 feet long. The two models or hypotheses $\mathcal{H}_1$ and $\mathcal{H}_2$ are

$$\mathcal{H}_1: \quad \text{desk is wood}$$
$$\mathcal{H}_2: \quad \text{desk is metal}$$

These hypotheses can be made more precise by writing

$$\mathcal{H}_1: \quad z = 5$$
$$\mathcal{H}_2: \quad z = 6$$

Thus given the actual observation z_{actual}, a decision can be made as follows:

If $z_{\text{actual}} = 5$, the desk is wood
If $z_{\text{actual}} = 6$, the desk is metal

Now assume that there is uncertainty in the measurements modeled by v, a zero mean Gaussian random variable with variance R. Then the two hypotheses are

$$\mathcal{H}_1: \quad z = 5 + v$$
$$v \text{ is } N(0, R)$$
$$\mathcal{H}_2: \quad z = 6 + v$$
$$v \text{ is } N(0, R)$$

A reasonable decision rule for using an actual observation $z = z_{\text{actual}}$ is now

If $z_{\text{actual}} < 5.5$, decide that the desk is wood
If $z_{\text{actual}} > 5.5$, decide that the desk is metal

In this case, the decision may, of course, be wrong; i.e., the wrong hypothesis may be chosen.

For the purpose of this book, there are two basic quantities in hypothesis testing, the decision function and the decision rule. Loosely speaking, they are

1. Decision function: a function (equation) which converts the actual observations into one or several numbers.
2. Decision rule: a rule or law which uses the (output of the) decision function to decide between the hypotheses.

The process of decision making (hypothesis testing) is illustrated in Fig. 9.1.

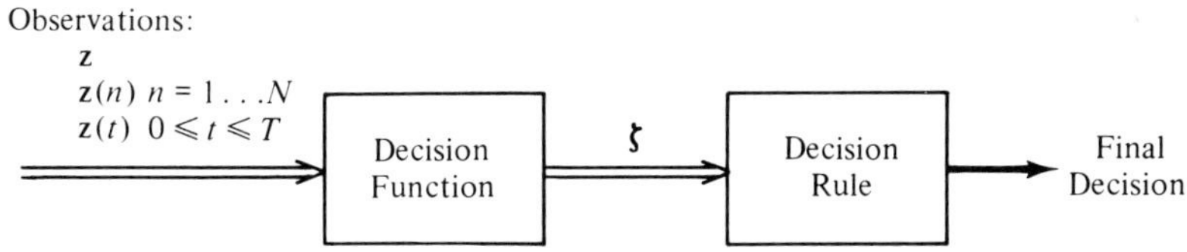

ζ, the Output of Decision Function, is a Scalar or Vector

Figure 9.1 Basic steps of hypothesis testing.

EXAMPLE: Consider

$$\mathcal{H}_1: \quad z(n) = 5 + v(n), \qquad n = 1, \ldots, N$$

$$\mathcal{H}_2: \quad z(n) = 6 + v(n), \qquad n = 1, \ldots, N$$

where for both hypotheses, the observation uncertainty $v(n)$ is modeled as

$$E\{v(n_1)v(n_2)\} = \begin{cases} R, & n_1 = n_2 \\ 0, & n_1 \neq n_2 \end{cases}$$

$$E\{v(n)\} = 0$$

Let $\xi(N)$ denote a decision function defined by

$$\xi(N) = \frac{1}{N} \sum_{n=1}^{N} z(n)$$

Then a reasonable deicision rule is

$$\text{Choose } \mathcal{H}_1 \text{ if } \xi(N) < 5.5$$
$$\text{Choose } \mathcal{H}_2 \text{ if } \xi(N) > 5.5$$

With this approach there is only one decision function. A different but equivalent approach is to define two decision functions $\xi_1(N)$ and $\xi_2(N)$,

$$\xi_1(N) = \frac{1}{N} \sum_{n=1}^{N} [z(n) - 5]$$

$$\xi_2(N) = \frac{1}{N} \sum_{n=1}^{N} [z(n) - 6]$$

and use the decision rule

Choose $\mathcal{H}_1$ if $\xi_1(N)$ is closer to zero than $\xi_2(N)$
Choose $\mathcal{H}_2$ otherwise

The term hypothesis testing arises because the choice as to which process is observed is based on hypothesized models. Thus hypothesis testing could also be called model testing. Hypothesis testing is sometimes called decision theory. The detection theory of communication theory is a special case. There are many interrelationships between estimation theory and hypothesis testing. For example, the following points of view can be taken:

1. Estimation theory: A special case of hypothesis testing in which the decision is made between a process with state $\mathbf{x}$ and a process with state $\mathbf{x} + \delta\mathbf{x}$, where $\delta\mathbf{x}$ is a small perturbation.

2. Hypothesis testing: A special case of estimation theory in which the observations $\mathbf{z}$ are the sum of the M possible processes, $\mathbf{z}_j$, $j = 1, \ldots, M$,

$$\mathbf{z} = \sum_{j=1}^{M} a_j \mathbf{z}_j$$

where $M - 1$ of the a_j are zero and the problem is to estimate which of the a_j is 1.

Sides will not be taken on the question of which theory is more important or comes first.

In this chapter, the static case is considered where a single vector observation $\mathbf{z}$ is made. Thus this chapter is analogous to Chapter 5 on linear estimation theory and the same basic pattern is repeated. A reader who has had no prior exposure to hypothesis testing may not appreciate its wide range of applications. For this reason Section 9.1 contains brief descriptions of various applications. Many of these applications involve dynamic models which will not be discussed until Chapter 10, but they serve to help the reader understand the broad scope of hypothesis testing. Section 9.2 discusses the general method of approach to be used. Bayesian, Fisher, weighted-least-squares, and unknown-but-bounded theories are considered in Sections 9.3, 9.4, 9.5, and 9.6, respectively. A brief general discussion is given in Section 9.7.

9.1 Areas of Application

Hypothesis testing is a very versatile tool which can be employed in a wide variety of problems. To illustrate the scope of applications, a few examples are briefly described.

9.1.1 Signal Detection

A common problem in radar, sonar, and seismic systems is to decide whether a signal is present. This is called a detection problem and has two hypotheses:

$\mathcal{H}_1$: observe signal plus noise
$\mathcal{H}_2$: observe noise only

The noise may be caused by receiver noise and background noise of many sorts (background radiation, jammers, schools of fish, multipath, trains). A stochastic model is usually used for the noise. The signal may have many forms. "Active" radar and sonar systems bouncing signals off of hard, point targets yield deterministic signals, often with unknown parameters such as time of arrival and amplitude. Stochastic signal models are encountered in passive sonar tracking and in passive radars observing radio stars. Stochastic signal models are also encountered in active radar-sonar systems where the targets are large "soft" targets such as a chaff cloud, or a school of fish. In seismic systems and in some radar-sonar problems, it is sometimes desirable to consider the signal to be completely unknown or unknown-but-bounded in some fashion. In all three types of systems, arrays of sensors are often used so that vector observations are made.

9.1.2 Communication Systems

In a communication system, the transmitter often transmits one of M time functions $u_j(t)$, $0 \leq t \leq T$, $j = 1, \ldots, M$, over a channel, and the receiver uses the observed signal to decide which signal was transmitted. The hypotheses to be tested by the receiver are

$$\mathcal{H}_j: \quad \text{observe } s_j(t) \text{ plus noise}, \qquad j = 1, \ldots, M, 0 \leq t$$

where $s_j(t)$ is the response of the channel to $u_j(t)$, $0 \leq t \leq T$. For a simple additive noise channel, $s_j(t) = u_j(t)$. Some channels with memory can be represented by a linear dynamic system with $u_j(t)$ as an input and $s_j(t)$ as output so that the $s_j(t)$ are still deterministic. In channels with multipath (random channels), the $s_j(t)$ are often modeled as stochastic processes whose statistics are determined by the $u_j(t)$.

9.1.3 Maneuvering Vehicles

Hypothesis testing is helpful when tracking a vehicle (say, by radar or sonar) which may execute abrupt maneuvers at unknown times. The vehicle motion without maneuvering is modeled as some linear dynamic system usually with an uncertain input. An additional driving input $\delta(t)$ to the dynamic system is hypothesized where the maneuver at time t_1 is modeled by

$$\delta(t) = 0, \qquad t < t_1$$

$$\delta(t) = \text{large value}, \qquad t \geq t_1$$

where $\delta(t)$ is viewed as the *acceleration* which causes the maneuver. This is an event detection problem where decisions on the following hypotheses are continually being made for all time t:

$\mathcal{H}_1$: maneuver has not taken place up to present time, t
$\mathcal{H}_2$: maneuver has taken place sometime in past

After the maneuver has been detected, it may be desired to estimate (identify) the time it began.

9.1.4 Component Failure

Large, multiple-component systems which operate over long periods of time can expect to experience some type of component failure. Hypothesis testing provides a method of using observations made on the system to detect these component failures. One example is an electrical power generation and transmission system where it is desired to determine when one of the transmission lines has been lost, say, from lightning or some inadvertent (or advertent) human action. Let $\mathbf{Y}_0$ denote the complex admittance matrix of the network when all lines are in operation and let $\mathbf{Y}_j$, $j = 1, \ldots, M$, denote the admittance matrix when the jth line has been removed. Then one might test the hypotheses

$\mathcal{H}_1$: observations made on network with admittance matrix $\mathbf{Y}_0$
$\mathcal{H}_2$: observations made on network with an admittance matrix that is not $\mathbf{Y}_0$

If the computational facilities are available, it would be more desirable to test the $M + 1$ hypothesis:

$\mathcal{H}_j$: observations made on network with admittance matrix $\mathbf{Y}_j$, $j = 0, \ldots, M$

This $M + 1$ hypothesis test has the advantage of also deciding (estimating) which line was lost.

9.1.5 Bad Data Point and Sensor Failure Detection

Another example of an event detection problem is *bad data point* detection. This common problem is often encountered when a sequence of observations is fed into a digital computer for processing. Completely erroneous data may be mixed in with the good observations. These bad data points can be caused by many things, such as temporary sensor failure, noise bursts on communication lines, or temporary failure of A to D conversion equipment. Bad data can often be detected using hypothesis-testing concepts. If the sensors are redundant, hypothesis testing can be used to cross-check them to decide which data (or sensor) is bad.

9.1.6 Deghosting

In certain radar systems, observations are made at different times on many different targets. To predict the future trajectories of any one target, it is first necessary to decide which of the observations were actually made on the same target. This is the deghosting problem and can be formulated as a hypothesis-testing problem in various ways.

9.1.7 Pattern Recognition

Pattern recognition can be viewed as a special case of hypothesis testing. In pattern recognition, an observation $\mathbf{z}$ is to be used to decide what pattern caused it. Each possible pattern can be viewed as one hypothesis. The main problem in pattern recognition is the development of models for the $\mathbf{z}$ corresponding to each pattern (hypothesis).

9.1.8 Parameter Estimation

Hypothesis testing can be used as a method to do state estimation for nonlinear systems. Assume that a vector observation

$$\mathbf{z} = \mathbf{h}(\boldsymbol{\alpha}) + \mathbf{v}$$

is made, where $\mathbf{h}$ is a known nonlinear function of an unknown parameter vector $\boldsymbol{\alpha}$. It is desired to use $\mathbf{z}$ to estimate $\boldsymbol{\alpha}$. Assume that α is a scalar. Then one way to make the estimate is to assume that α can take on only one of M possible values, α_j, $j = 1, \ldots, M$. Hypothesis tests are then made for

$$\mathcal{H}_j: \quad \mathbf{z} = \mathbf{z}_j$$

$$\mathbf{z}_j = \mathbf{h}(\alpha_j) + \mathbf{v}, \qquad j = 1, \ldots, M$$

This technique is often used for practical reasons even if α is not actually restricted to M values. The concept extends to the case of vector $\boldsymbol{\alpha}$ but rapidly loses its practical appeal as the dimension of $\boldsymbol{\alpha}$ increases. One common example of a scalar parameter is a radar doppler filter bank which is used to estimate the frequency shift (i.e., velocity) of a moving target. In an oversimplified version of the problem, the hypotheses are

$$\mathcal{H}_j: \quad z(t) = \sin(\omega_j t) + v(t), \qquad j = 1, \ldots, M$$

The doppler filter bank consists of M *matched filters*, one for each value of ω_j, and these can be viewed as generating M decision functions.

9.1.9 Model Verification; System Identification

In many design and analysis problems, one chooses a mathematical model by combining limited information with a lot of judgment and hopefully some luck (see Chapter 8). If actual observations are available, an attempt to verify the model can be made by testing the hypotheses:

$\mathcal{H}_1$: observations came from model
$\mathcal{H}_2$: observations did not come from model

System identification, which is discussed extensively in Chapter 14, involves the use of the observations to help determine a model (i.e., to identify the system). As will be seen in Chapter 14, the mathematics to be developed for hypothesis testing is very useful in system identification.

9.2 Philosophy of Presentation

When one makes a decision in the presence of uncertainty, there is the possibility of making a wrong decision. Therefore one logical way to develop decision functions and decision rules is to choose them to minimize some suitably defined cost or risk function. This approach is pursued extensively in the literature, but not in this book.

In this book, the choice of decision functions and rules is based mostly on *reasonability arguments*, with no real attempt to define general cost or risk functions. As expected, the final results are the same in that the decision functions and rules to be discussed are by far the most common ones used in practice. There are two main reasons the mathematical approach of minimum cost or risk is not emphasized. First, a good treatment requires the introduction and extensive discussion of a lot of concepts if all cases of interest are to be covered. Second, and more important, the author often finds the minimization of some cost or risk function to be "unsatisfying," as it obscures the basic trade-offs which are present. Also, in many cases there is no one decision function and rule which minimizes a *global* cost or risk function (as was possible with the estimation theory of Part II).

The above is not meant to imply that risk or cost function theory is useless. It is interesting and does add insight once it is understood. The author merely feels that its full development is simply not worth the time and effort, relative to the goals of this book.

9.3 Bayesian

Hypothesis testing is discussed for a Bayesian model. A Bayesian model is one which is completely defined in a probabilistic sense. The M hypotheses (models) have given probability densities and there is a specified probability that a given hypothesis is true.

Consider the jth model for a vector $\mathbf{z}$ given by the jth hypothesis:

$$\mathcal{H}_j: \quad \mathbf{z} \text{ is random vector with probability density } p_j(\mathbf{z}) \qquad j = 1, \ldots, M. \tag{9.3.1}$$

Define

$$\rho_j: \quad \text{a priori probability that hypothesis } \mathcal{H}_j \text{ is true,} \qquad j = 1, \ldots, M$$

$$\sum_{j=1}^{M} \rho_j = 1$$

9.3.1 Log Likelihood Function Decision Making

The most common decision functions for a Bayesian model are based on likelihood functions. Assume that $\mathbf{z} = \mathbf{z}_{\text{actual}}$ is observed. In the present case, where $p_j(\mathbf{z})$ is the probability density of $\mathbf{z}$ for the jth hypothesis, the likelihood function is simply $p_j(\mathbf{z})$ evaluated for $\mathbf{z} = \mathbf{z}_{\text{actual}}$. [Why isn't $p_j(\mathbf{z}_{\text{actual}})$ called the probability that $\mathbf{z} = \mathbf{z}_{\text{actual}}$?] If $\mathbf{z}_{\text{actual}}$ is observed and there are M hypothesized models (probability densities) with a priori probabilities of occurring, ρ_j, it is *reasonable* to choose a decision rule which picks the hypothesis with the maximum value of $\rho_j p_j(\mathbf{z})|_{\mathbf{z}=\mathbf{z}_{\text{actual}}}$. Thus in the case $M = 2$ a reasonable decision rule is

$$\begin{aligned} &\text{Choose } \mathcal{H}_1 \text{ if } \rho_1 p_1(\mathbf{z})|_{\mathbf{z}=\mathbf{z}_{\text{actual}}} > \rho_2 p_2(\mathbf{z})|_{\mathbf{z}=\mathbf{z}_{\text{actual}}} \\ &\text{Choose } \mathcal{H}_2 \text{ if } \rho_1 p_1(\mathbf{z})|_{\mathbf{z}=\mathbf{z}_{\text{actual}}} < \rho_2 p_2(\mathbf{z})|_{\mathbf{z}=\mathbf{z}_{\text{actaul}}} \end{aligned} \tag{9.3.2}$$

In the general case where $M > 2$, the decision rule is

$$\boxed{\text{Choose } \mathcal{H}_k \text{ if } j = k \text{ yields maximum } \rho_j p_j(\mathbf{z})|_{\mathbf{z}=\mathbf{z}_{\text{actual}}}, \quad j = 1, \ldots, M} \tag{9.3.3}$$

In the author's opinion, the best justification for the use of (9.3.3) is its "reasonability." However, some "optimal" properties of (9.3.3) are now discussed. When $M = 2$, there are two ways an error can be made. Define

$\alpha_{1,2}$: probability $\mathcal{H}_1$ is chosen when $\mathcal{H}_2$ is true
$\alpha_{2,1}$: probability $\mathcal{H}_2$ is chosen when $\mathcal{H}_1$ is true

The $\alpha_{1,2}$ and $\alpha_{2,1}$ are given various names in various contexts. Sometimes they are the *type I* and *type II* error probabilities. Sometimes they are the *failure to detect* and *false alarm* probabilities. Sometimes $\alpha_{1,2}$ is the *significance level* and $1 - \alpha_{2,1}$ is the *power of the test*. The total *probability of error*, P_e, is defined as

$$P_e = \rho_2 \alpha_{1,2} + \rho_1 \alpha_{2,1}$$

Then it will be shown that

$$\boxed{\begin{array}{l}\text{The use of the decision rule (9.3.2) minimizes} \\ \text{the total probability of error } P_e \ (M = 2)\end{array}} \tag{9.3.4}$$

This total probability of error can be viewed as one possible cost or risk function. To prove (9.3.4), consider an arbitrary decision function $d(\mathbf{z})$ and decision rule

$$\begin{aligned} &\text{Choose } \mathcal{H}_1 \text{ if } \mathbf{z} \in \Omega_1 \\ &\text{Choose } \mathcal{H}_2 \text{ if } \mathbf{z} \in \Omega_2 \\ &\Omega_1 = \{\mathbf{z}: d(\mathbf{z}) > 0\} \\ &\Omega_2 = \{\mathbf{z}: d(\mathbf{z}) < 0\} \end{aligned} \tag{9.3.5}$$

Then the probabilities of the two types of error are

$$\alpha_{1,2} = \int_{\mathbf{z} \in \Omega_1} p_2(\mathbf{z})\, d\mathbf{z}$$

$$\alpha_{2,1} = \int_{\mathbf{z} \in \Omega_2} p_1(\mathbf{z})\, d\mathbf{z}$$

It also follows that

$$\alpha_{1,2} = 1 - \int_{\mathbf{z}\in\Omega_2} p_2(\mathbf{z})\, d\mathbf{z}$$

Thus the total probability of error P_e can be written

$$P_e = \rho_2 + \int_{\mathbf{z}\in\Omega_2} [-\rho_2 p_2(\mathbf{z}) + \rho_1 p_1(\mathbf{z})]\, d\mathbf{z}$$

The minimum P_e occurs if the set Ω_2 contains all the $\mathbf{z}$ for which the integrand is negative, i.e., if

$$\Omega_2 = \{\mathbf{z}\colon -\rho_2 p_2(\mathbf{z}) + \rho_1 p_1(\mathbf{z}) < 0\}$$

Equation (9.3.4) follows directly. Actually there exist possible decision rules (such as *randomized* rules) which cannot be expressed in the form of (9.3.5), but the above proof is sufficient for our purposes.

EXAMPLE: Consider the two hypotheses

$$\mathcal{H}_1\colon \quad z = z_1, \quad z_1 \text{ has probability density } p_1(z)$$
$$\mathcal{H}_2\colon \quad z = z_2, \quad z_2 \text{ has probability density } p_2(z)$$

where $p_1(z)$ and $p_2(z)$ are as in Fig. 9.3.1(a) and (b). Assume that the a priori probabilities are equal, $\rho_1 = \rho_2 = \frac{1}{2}$. Then from Fig. 9.1.1(c), it is obvious that (9.3.2) yields

$$\text{Choose } \mathcal{H}_1 \text{ if } 0.5 < z_{\text{actual}} \leq 1.5$$
$$\text{Choose } \mathcal{H}_2 \text{ if } -1 \leq z_{\text{actual}} \leq 0.5$$

For this rule, $\alpha_{1,2} = \frac{1}{4}$, $\alpha_{2,1} = 0$, and $P_e = \frac{1}{8}$.

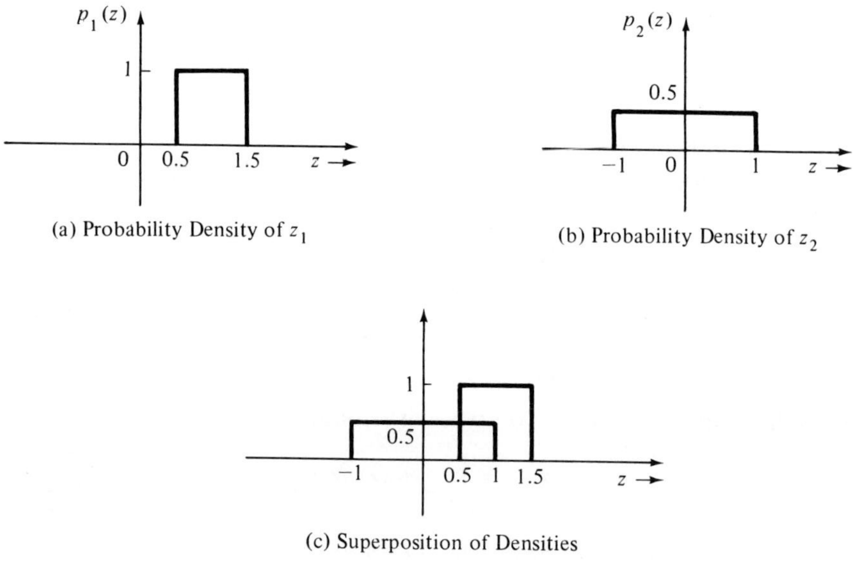

Figure 9.3.1 Probability density for example.

Instead of working with the likelihood function $p_j(\mathbf{z})$, it is usually more convenient to work with logarithms. Define

$$\xi_j(\mathbf{z}): \text{ log likelihood function}$$

$$\xi_j(\mathbf{z}) = \ln\{p_j(\mathbf{z})\}, \qquad j = 1, \ldots, m$$

The decision rule (9.3.3) now becomes

$$\boxed{\begin{array}{c}\text{Choose } \mathcal{H}_k \text{ if } j = k \text{ yields maximum } \xi_j(\mathbf{z}_{\text{actual}}) + \ln\{\rho_j\} \\ j = 1, \ldots, M\end{array}} \tag{9.3.6}$$

It is often convenient to view the decision functions $\xi_j(\mathbf{z})$, $j = 1, \ldots, M$, as a vector $\boldsymbol{\xi}(\mathbf{z})$:

$$\boldsymbol{\xi}(\mathbf{z}) = \begin{bmatrix} \xi_1(\mathbf{z}) \\ \cdot \\ \cdot \\ \cdot \\ \xi_M(\mathbf{z}) \end{bmatrix}$$

The decision-making process is illustrated in Fig. 9.3.2.

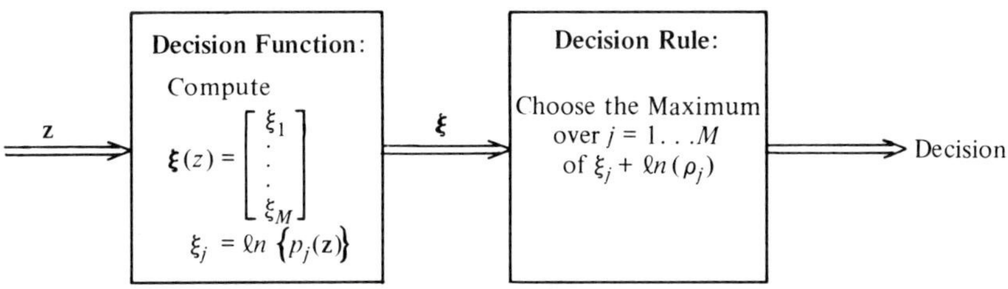

Figure 9.3.2 Decision making for M hypothesis problem.

Equation (9.3.6) defines a decision rule directly in terms of the a priori probabilities, ρ_j, and has been shown when $M = 2$ to minimize the total probability of error, P_e. However, in practice a different rule might be desired. Assume that $M = 2$. Then the probability of one type of error $\alpha_{1,2}$ can be made as small as desired at the expense of increasing the probability of the other type of error $\alpha_{2,1}$. In many applications, one may feel that one type of error is far more critical than the other and hence be willing to trade a small $\alpha_{1,2}$ for a large $\alpha_{2,1}$.

EXAMPLE: If you are a man taking a woman to a dance and the hypotheses are

$\mathcal{H}_1$: She does not expect you to bring flowers
$\mathcal{H}_2$: She expects you to bring flowers

a rule which yields a small $\alpha_{1,2}$ is definitely desired, where $\alpha_{1,2}$ is the probability that you will decide not to bring flowers when she expects them.

Therefore a more useful version of (9.3.6) is

$$\boxed{\begin{array}{l}\text{Choose } \mathcal{H}_k \text{ if } j = k \text{ yields maximum } \xi_j(\mathbf{z}) + l_j \\ \text{where the } l_j \text{ are chosen to give the desired} \\ \text{error performance}\end{array}} \tag{9.3.7}$$

Often hypothesis-testing theory is developed using the likelihood ratio instead of the likelihood function. Consider the case of two hypotheses ($M = 2$) and define

$$\lambda(\mathbf{z})\text{: log likelihood ratio}$$

$$\lambda(\mathbf{z}) = \ln\left\{\frac{p_1(\mathbf{z})}{p_2(\mathbf{z})}\right\}$$

$$\lambda(\mathbf{z}) = \xi_1(\mathbf{z}) - \xi_2(\mathbf{z})$$

Since

$$\ln\left\{\frac{\rho_1 p_1(\mathbf{z})}{\rho_2 p_2(\mathbf{z})}\right\} = \lambda(\mathbf{z}) + \ln\left\{\frac{\rho_1}{\rho_2}\right\}$$

the decision rule (9.3.6) can be expressed as ($M = 2$)

$$\text{Choose } \mathcal{H}_1 \text{ if } \lambda(\mathbf{z}_{\text{actual}}) > d_0$$

$$\text{Choose } \mathcal{H}_2 \text{ if } \lambda(\mathbf{z}_{\text{actual}}) < d_0$$

$$d_0 = -\ln\left(\frac{\rho_1}{\rho_2}\right)$$

The use of the likelihood ratio gives the same final answer as that obtained from the likelihood function. The likelihood function is emphasized in this book, as the author feels it is more natural when one considers all cases of interest.

9.3.2 Log Likelihood Functions: Gaussian

Consider the case

$$\mathcal{H}_j\text{: } \mathbf{z} = \mathbf{z}_j, \qquad \mathbf{z}_j \text{ is } N(\mathbf{m}_j, \mathbf{\Gamma}_j), \qquad j = 1, \dots, M$$

$$\mathbf{z}\text{: } K \text{ vector}$$

$$\mathbf{\Gamma}_j, \mathbf{\Gamma}_j^{-1}\text{: exist}$$

The probability density (or likelihood function) assuming that $\mathcal{H}_j$ is true is

$$p_j(\mathbf{z}) = [(2\pi)^K |\mathbf{\Gamma}_j|]^{-1/2} \exp(-\tfrac{1}{2})\{[\mathbf{z} - \mathbf{m}_j]'\mathbf{\Gamma}_j^{-1}[\mathbf{z} - \mathbf{m}_j]\}, \qquad j = 1, 2$$

Thus the log likelihood function ξ_j for the jth hypothesis is

$$\boxed{2\xi_j(\mathbf{z}) = -K\ln(2\pi) - \ln|\mathbf{\Gamma}_j| - [\mathbf{z} - \mathbf{m}_j]'\mathbf{\Gamma}_j^{-1}[\mathbf{z} - \mathbf{m}_j]} \tag{9.3.8}$$

In the case where

$$\mathcal{H}_j\colon \mathbf{z} = \mathbf{H}_j\mathbf{x}_j + \mathbf{v}_j, \qquad j = 1, \ldots, M$$

$$\left.\begin{array}{l}\mathbf{x}_j \text{ is } N(\mathbf{0}, \boldsymbol{\psi}_j)\\ \mathbf{v}_j \text{ is } N(\mathbf{0}, \mathbf{R}_j)\end{array}\right\}\mathbf{x}, \mathbf{v} \text{ independent}$$

then

$$\mathcal{H}_j\colon \mathbf{z} \text{ is } N(\mathbf{0}, \boldsymbol{\Gamma}_j), \qquad j = 1, \ldots, M$$

$$\boldsymbol{\Gamma}_j = \mathbf{H}_j\boldsymbol{\Psi}_j\mathbf{H}_j' + \mathbf{R}_j$$

and

$$2\xi_j(\mathbf{z}) = c_j - \mathbf{z}'[\mathbf{H}_j\boldsymbol{\psi}_j\mathbf{H}_j' + \mathbf{R}_j]^{-1}\mathbf{z}, \qquad j = 1, \ldots, M$$

$$c_j = -K\ln(2\pi) - \ln|\mathbf{H}_j\boldsymbol{\psi}_j\mathbf{H}_j' + \mathbf{R}_j|$$

Let $\hat{\mathbf{x}}_j$ be the best estimate of $\mathbf{x}$ given $\mathbf{z}$, assuming that $\mathcal{H}_j$ is true. From Chapter 5 [Eq. (5.1.2)], one form for $\hat{\mathbf{x}}_j$ is

$$\hat{\mathbf{x}}_j = \boldsymbol{\psi}_j\mathbf{H}_j'[\mathbf{H}_j\boldsymbol{\psi}_j\mathbf{H}_j' + \mathbf{R}_j]^{-1}\mathbf{z}$$

which illustrates the close tie between hypothesis testing and estimation theory.

9.3.3 Equal Covariance Matrices: Gaussian

Consider the Gaussian case where the hypotheses differ only in mean values; i.e., they have equal covariance matrices, so

$$\mathcal{H}_j\colon \mathbf{z} = \mathbf{z}_j, \qquad \mathbf{z}_j \text{ is } N(\mathbf{m}_j, \boldsymbol{\Gamma})$$

For this model (9.3.8) can be written

$$2\xi_j(\mathbf{z}) = \gamma + 2\tilde{\xi}_j(\mathbf{z})$$

$$\gamma = -K\ln(2\pi) - \ln|\boldsymbol{\Gamma}| - \mathbf{z}'\boldsymbol{\Gamma}^{-1}\mathbf{z} - \mathbf{m}_j'\boldsymbol{\Gamma}^{-1}\mathbf{m}_j$$

$$\tilde{\xi}_j|\mathbf{z}| = \mathbf{z}'\boldsymbol{\Gamma}^{-1}\mathbf{m}_j$$

where $\boldsymbol{\Gamma}^{-1}$ is assumed to exist. Now γ has no terms involving *both* $\mathbf{z}$ and $\mathbf{m}_j$ so γ does not contain any interaction between the observations $\mathbf{z}$ and the $\mathbf{m}_j$ which determine the differences between the hypothesized models. Thus γ can be neglected in decision making and

> The modified likelihood function $\tilde{\xi}_j(\mathbf{z})$
>
> $$\tilde{\xi}_j(\mathbf{z}) = \mathbf{z}'\boldsymbol{\Gamma}^{-1}\mathbf{m}_j \qquad (9.3.9)$$
>
> can be used as a decision function in place of the log likelihood function $\xi_j(\mathbf{z})$

The choice of decision rule, of course, depends on whether $\xi_j(\mathbf{z})$ or $\tilde{\xi}_j(\mathbf{z})$ is used as a decision function. The important point is that if appropriate decision rules are used, both decision functions $\xi_j(\mathbf{z})$ and $\tilde{\xi}_j(\mathbf{z})$ give the same performance.

EXAMPLE: Consider

$$\mathcal{H}_j: \mathbf{z} = \mathbf{z}_j, \qquad \mathbf{z}_j \text{ is } N(\mathbf{m}_j, \mathbf{I}), \qquad j = 1, 2$$

$$p_1 = p_2 = 0.5$$

Combining (9.3.8) with (9.3.6) yields the decision rule:

$$\text{Choose } \mathcal{H}_1 \text{ if } [\mathbf{z} - \mathbf{m}_1]'[\mathbf{z} - \mathbf{m}_1] < [\mathbf{z} - \mathbf{m}_2]'[\mathbf{z} - \mathbf{m}_2]$$
$$\text{Choose } \mathcal{H}_2 \text{ if } [\mathbf{z} - \mathbf{m}_1]'[\mathbf{z} - \mathbf{m}_1] > [\mathbf{z} - \mathbf{m}_2]'[\mathbf{z} - \mathbf{m}_2]$$

However, the same decision will be made by the rule

$$\text{Choose } \mathcal{H}_1 \text{ if } \tilde{\xi}_1(\mathbf{z}) > \tilde{\xi}_2(\mathbf{z}) + l$$
$$\text{Choose } \mathcal{H}_2 \text{ if } \tilde{\xi}_1(\mathbf{z}) < \tilde{\xi}_2(\mathbf{z}) + l$$

$$l = \frac{\mathbf{m}_1'\mathbf{m}_1 - \mathbf{m}_2'\mathbf{m}_2}{2}$$

$$\tilde{\xi}_j(\mathbf{z}) = \mathbf{z}'\mathbf{m}_j, \qquad j = 1, 2$$

9.3.4 Probability Density of Log Likelihood Function: Gaussian

The log likelihood function $\xi_j(\mathbf{z})$ is a random variable when it is viewed as a function of the random variable $\mathbf{z}$. The probability density of $\xi_j(\mathbf{z})$ for Gaussian models is now discussed.

Assume that

$$\mathcal{H}_j: \mathbf{z} = \mathbf{z}_j, \qquad \mathbf{z}_j \text{ is } N(\mathbf{m}_j, \mathbf{\Gamma}_j), \qquad j = 1, \ldots, M$$

Define

$\xi_j(\mathbf{z}_k)$: random variable obtained when (1) log likelihood function is evaluated for hypothesis $\mathcal{H}_j$ and (2) $\mathbf{z}$ is a random vector corresponding to hypothesis $\mathcal{H}_k$.

If $\mathbf{z}$ is a K-dimensional vector, (9.3.8) gives

$$2\xi_j(\mathbf{z}_j) = -K\ln(2\pi) - \ln|\mathbf{\Gamma}_j| - \chi_K^2$$
$$\chi_K^2 = [\mathbf{z}_j - \mathbf{m}_j]'\mathbf{\Gamma}_j^{-1}[\mathbf{z}_j - \mathbf{m}_j]$$

Thus if $\mathbf{\Gamma}_j^{-1}$ exists, the random variable χ_K^2 is a *chi-squared* random variable with K *degrees of freedom*, where

$$E\{\chi_K^2\} = K$$
$$E\{[\chi_K^2 - K]^2\} = 2K$$

As K becomes large, the probability density of χ_K^2 becomes Gaussian; i.e., *χ_K^2 is approximately $N(K, 2K)$ for large K.* See Appendix E.5 for more details. Since the probability density of χ_K^2 is thc known, the probability density of $2\xi_j(z_j)$ can be calculated. For example,

$$\boxed{\begin{array}{l} \text{For large } K, \\ \qquad 2\xi_j(\mathbf{z}_j) \text{ is } N(c, 2K) \\ \qquad\qquad c = -K[1 + \ln(2\pi)] - \ln|\mathbf{\Gamma}_j| \\ K \text{ is dimension of } \mathbf{z} \end{array}} \tag{9.3.10}$$

The probability density of $\xi_j(\mathbf{z}_j)$, the jth log likelihood function when the jth hypothesis is true, has been obtained. Unfortunately, the probability density of $\xi_j(\mathbf{z}_k), j \neq k$, is in general more complicated. However, nice results arise in the special case of Section 9.3.3 when the models differ only in the mean values and the covariance matrices are the same. Only the modified log likelihood function $\tilde{\xi}_j(\mathbf{z})$ has to be considered, and $\tilde{\xi}_j(z)$ is Gaussian for any hypothesis. Since $\mathbf{z}_j$ is $N(\mathbf{m}_j, \mathbf{\Gamma}), j = 1, \ldots, M$,

$$\boxed{\begin{array}{l} \tilde{\xi}_j(\mathbf{z}_k) = \mathbf{z}_k'\mathbf{\Gamma}^{-1}\mathbf{m}_j \\ \tilde{\xi}_j(\mathbf{z}_k) \text{ is } N(\mu_{jk}, \mu_{jj}) \\ \mu_{jk} = \mathbf{m}_k'\mathbf{\Gamma}^{-1}\mathbf{m}_j \end{array}} \tag{9.3.11}$$

9.3.5 Distance Measures†

The probabilities of the different types of error ($\alpha_{1,2}$ and $\alpha_{2,1}$) are a good measure of performance. Unfortunately they are often difficult to evaluate. Therefore a simpler approach to the evaluation of performance based on the *distance* between random vectors is discussed. These distance measures generalize nicely to the discrete- and continuous-time dynamic systems to be discussed in Chapter 10.

The distance between two random vectors is a measure of how far apart or different the corresponding probability densities are. If the two probability densities $p_1(\mathbf{z})$ and $p_2(\mathbf{z})$ are "very different", it should be easy to decide between them, i.e., to have a small probability of error. Thus distances provide, in some sense, a measure of decision performance. Two of many possible distance measures, the divergence and the Bhattacharyya distance, are discussed.

The divergence J between two random vectors $\mathbf{z}_1$ and $\mathbf{z}_2$ with probability densities $p_1(\mathbf{z})$ and $p_2(\mathbf{z})$ is defined as

$$J = \int [p_1(\mathbf{z}) - p_2(\mathbf{z})] \ln\left(\frac{p_1(\mathbf{z})}{p_2(\mathbf{z})}\right) d\mathbf{z} \tag{9.3.12}$$

The divergence has a simple interpretation: $J =$ *average value of* $\lambda(\mathbf{z})$ *when*

$\mathcal{H}_1$ is true minus average value of $\lambda(\mathbf{z})$ when $\mathcal{H}_2$ is true:

$$\lambda(\mathbf{z}) = \ln\left(\frac{p_1(\mathbf{z})}{p_2(\mathbf{z})}\right), \qquad \text{log likelihood ratio}$$

The Bhattacharyya distance between two random vectors $\mathbf{z}_1$ and $\mathbf{z}_2$ is defined as

$$B = -\ln \int [p_1(\mathbf{z})p_2(\mathbf{z})]^{1/2}\, d\mathbf{z} \tag{9.3.13}$$

Bhattacharyya also called this measure a divergence, but for sake of discussion, the term divergence is reserved for J of (9.3.12). One motivation for (9.3.13) is based on a geometrical argument, which is easiest to see if for scalar random variables z_1 and z_2 which can take on only K possible values $z_{k,j}$, $k = 1, \ldots, K$, $j = 1, 2$, with corresponding probabilities $p_{k,j}$ where

$$\sum_{k=1}^{K} p_{k,j} = 1$$

(The index k is running over possible values of z_j.) The probability distribution of z_j can then be viewed as a *vector* in a K-dimensional space with components $p_{k,j}^{1/2}$, $k = 1, \ldots, K$. Since the *length* of this vector is 1, the $(p_{k,j})^{1/2}$, $k = 1, \ldots, K$, can be viewed as *direction cosines*. Let γ denote the *angle* between these two vectors so that

$$\cos\gamma = \sum_{k=1}^{K} (p_{k,1}p_{k,2})^{1/2}$$

If z can actually take on continuous values, the corresponding result in terms of probability densities is

$$\cos\gamma = \int [p_1(z)p_2(z)]^{1/2}\, dz$$

Introducing the logarithm, ignoring the cosine, and going to the case of vector $\mathbf{z}$ leads to (9.3.13). Another way to motivate (9.3.13) is to note that

$$\int [p_1^{1/2}(\mathbf{z}) - p_2^{1/2}(\mathbf{z})]^2\, d\mathbf{z} = 2 - 2\int [p_1(\mathbf{z})p_2(\mathbf{z})]^{1/2}\, d\mathbf{z}$$

so that B of (9.3.13) can be directly related to the *quadratic distance* between $p_1^{1/2}(\mathbf{z})$ and $p_2^{1/2}(\mathbf{z})$.

It should be stated that the term distance is applied to J and B in a descriptive rather than mathematical sense. For example, J does not satisfy the triangle inequality.

The two distance measures can be compared by defining λ_j to be the random variable obtained from $\mathbf{z}_j$ so that

$$\lambda_j = \lambda(\mathbf{z}) \qquad \text{when } \mathbf{z} \text{ is a random vector } \mathbf{z}_j, \qquad j = 1, 2$$

Then

$$J = E\{\lambda_1\} - E\{\lambda_2\} \tag{9.3.14}$$

and after some manipulation

$$2B = -\ln E\{e^{-\lambda_1/2}\} - \ln E\{e^{\lambda_2/2}\}$$

Thus the measures differ in the order of taking the expectation and logarithm. The inequality

$$\ln E\{x\} \geq E\{\ln x\}$$

yields

$$J \geq 4B \tag{9.3.15}$$

The distance measures do not necessarily correspond directly to the probabilities of error, but there are many interrelationships. The Bhattacharyya distance can be used to calculate both an upper and lower bound for the probability of error. For example, consider the total probability of error P_e when the decision rule (9.3.2) is used so that

$$P_e = \rho_2 \int_{\mathbf{z} \in \Omega_1} p_2(z)\, dz + \rho_2 \int_{\mathbf{z} \in \Omega_2} p_1(\mathbf{z})\, d\mathbf{z}$$

$$\Omega_1 = \{\mathbf{z}\colon \rho_1 p_1(\mathbf{z}) > \rho_2 p_2(\mathbf{z})\}$$

$$\Omega_2 = \{\mathbf{z}\colon \rho_1 p_1(\mathbf{z}) < \rho_2 p_2(\mathbf{z})\}$$

Then after a little (or a lot of) thought,

$$P_e \leq \int [\rho_1 \rho_2 p_1(\mathbf{z}) p_2(\mathbf{z})]^{1/2}\, d\mathbf{z}$$

so

$$\boxed{P_e \leq (\rho_1 \rho_2)^{1/2} e^{-B}} \tag{9.3.16}$$

The Bhattacharyya distance is closely related to quantities of importance in coding and information theory.

The two distance measures are not always equivalent in the sense that they can lead to different conclusions. For example, assume that there are three hypotheses:

$$\mathcal{H}_j\colon \mathbf{z} \text{ has probability density } p_j(\mathbf{z}), \qquad j = 1, 2, 3$$

Use of the divergence might lead to the conclusion that

$$\mathcal{H}_1 \text{ is closer to } \mathcal{H}_2 \text{ than to } \mathcal{H}_3$$

while the Bhattacharyya distance might lead to the conclusion that

$$\mathcal{H}_1 \text{ is closer to } \mathcal{H}_3 \text{ than to } \mathcal{H}_2$$

In such cases the choice of distance measure is not obvious. In general, the Bhattacharyya distance seems to be the "better" measure.

Note that the distance between two random vectors or hypotheses is defined independently of any a priori probability on which hypothesis is true.

9.3.6 Distance Measures: Gaussian†

Consider the Gaussian case where

$$\mathcal{H}_j\colon \mathbf{z} = \mathbf{z}_j, \qquad \mathbf{z}_j \text{ is } N(\mathbf{m}_j, \boldsymbol{\Gamma}_j), \qquad j = 1, 2$$

After integration and manipulation,

$$\boxed{\begin{aligned} 2J &= \operatorname{tr}\{[\boldsymbol{\Gamma}_1 - \boldsymbol{\Gamma}_2][\boldsymbol{\Gamma}_2^{-1} - \boldsymbol{\Gamma}_1^{-1}]\} + [\mathbf{m}_1 - \mathbf{m}_2]'[\boldsymbol{\Gamma}_1^{-1} \\ &\quad + \boldsymbol{\Gamma}_2^{-1}][\mathbf{m}_1 - \mathbf{m}_2] \\ 4B &= 2\ln|\tfrac{1}{2}(\boldsymbol{\Gamma}_1 + \boldsymbol{\Gamma}_2)| - \ln|\boldsymbol{\Gamma}_1| - \ln|\boldsymbol{\Gamma}_2| \\ &\quad + (\mathbf{m}_1 - \mathbf{m}_2)'[\boldsymbol{\Gamma}_1 + \boldsymbol{\Gamma}_2]^{-1}(\mathbf{m}_1 - \mathbf{m}_2) \end{aligned}} \tag{9.3.17}$$

or, equivalently,

$$4B = \ln|\tfrac{1}{4}(\boldsymbol{\Gamma}_1 + \boldsymbol{\Gamma}_2)(\boldsymbol{\Gamma}_1^{-1} + \boldsymbol{\Gamma}_2^{-1})| + (\mathbf{m}_1 - \mathbf{m}_2)'[\boldsymbol{\Gamma}_1 + \boldsymbol{\Gamma}_2]^{-1}(\mathbf{m}_1 - \mathbf{m}_2)$$

The inequality (9.3.15) can be strengthened to $J \geq 8B$.

EXAMPLE: Consider

$$z_1 \text{ is } N(0, 1)$$

$$z_2 \text{ is } N(m_2, 3)$$

The distance between these random variables is

$$J = \frac{2}{3} + \frac{2}{3}m_2^2$$

$$8B = 2\ln\left(\frac{4}{3}\right) + \frac{m_2^2}{2}$$

The case of equal covariance matrices when $\boldsymbol{\Gamma}_1 = \boldsymbol{\Gamma}_2 = \boldsymbol{\Gamma}$ yields

$$J = 8B = [\mathbf{m}_1 - \mathbf{m}_2]'\boldsymbol{\Gamma}^{-1}|\mathbf{m}_1 - \mathbf{m}_2| \tag{9.3.18}$$

Note that $J = 8B$ contains the term $\mathbf{m}_1'\boldsymbol{\Gamma}^{-1}\mathbf{m}_2$, which specifies the probability density of the modified log likelihood function $\tilde{\xi}_j(\mathbf{z}_k)$ of (9.3.11). Thus where the hypotheses have equal covariance matrices, the various probabilities of error can be calculated from knowledge of the distance measures (after, of course, a decision rule has been chosen). When there are unequal covariance matrices, this is not, in general, true.

9.3.7 Role of Gaussian Assumption

Section 5.1 discussed estimation theory for a static model with Bayesian uncertainty. Hence it corresponds to the preceding discussions. However the theory of Section 5.1 yielded the same estimator for both Gaussian and non-Gaussian models, where the estimator depended only on the first and second moments, the mean $\mathbf{m}$, and covariance matrix $\boldsymbol{\Gamma}$. Therefore it is reasonable to ask, Can the preceding formulas based on Gaussian models yield good results when applied to non-Gaussian models with specified means and covariance matrices? Two cases will be considered: first, equal covariance matrices, and then unequal covariance matrices. The answer is yes in both cases but for different reasons.

Consider the case of equal covariance matrices:

$$\begin{aligned} \mathcal{H}_j\colon \mathbf{z} &= \mathbf{z}_j, \qquad j = 1, 2 \\ \mathbf{z}_j &= \mathbf{m}_j + \mathbf{y} \\ E(\mathbf{y}) &= 0 \qquad E\{\mathbf{y}\mathbf{y}'\} = \mathbf{\Gamma} \end{aligned} \tag{9.3.19}$$

If $\mathbf{y}$ were Gaussian, the decision functions would be the modified log likelihood functions $\tilde{\xi}_j(\mathbf{z})$ of (9.3.9),

$$\tilde{\xi}_j(\mathbf{z}) = \mathbf{z}'\mathbf{\Gamma}^{-1}\mathbf{m}_j$$

and since $M = 2$, a simple decision function

$$\tilde{\lambda}(\mathbf{z}) = \tilde{\xi}_1(\mathbf{z}) - \tilde{\xi}_2(\mathbf{z}) = \mathbf{z}'\mathbf{\Gamma}^{-1}[\mathbf{m}_1 - \mathbf{m}_2] \tag{9.3.20}$$

could be used. It will now be shown how to argue that such a decision function is also reasonable for (9.3.19) when $\mathbf{y}$ is not Gaussian. Assume that a decision function $d(\mathbf{z})$ of the following form is to be used:

$$d(\mathbf{z}) = \mathbf{z}'\mathbf{c}$$

The problem is to choose $\mathbf{c}$. Define

$$d(\mathbf{z}_j) = \mathbf{z}_j'\mathbf{c}, \qquad j = 1, 2$$

so that $d(\mathbf{z}_j)$ is $d(\mathbf{z})$ when $\mathcal{H}_j$ is true. Then a reasonable approach is

> Choose $\mathbf{c}$ so that $d(\mathbf{z}_1)$ and $d(\mathbf{z}_2)$ are *as far apart as possible* in some sense (9.3.21)

To use this approach, consider

$$\begin{aligned} E\{d(\mathbf{z}_j)\} &= \mathbf{m}_j'\mathbf{c}, \qquad j = 1, 2 \\ E\{(d(\mathbf{z}) - \mathbf{m}_j'\mathbf{c})^2\} &= \mathbf{c}'\mathbf{\Gamma}\mathbf{c} \end{aligned}$$

Since the variances of $d(\mathbf{z}_1)$ and $d(\mathbf{z}_2)$ are the same, it is reasonable choose $\mathbf{c}$ to maximize the difference between the mean values of $d(\mathbf{z}_1)$ and $d(\mathbf{z}_2)$. After a little thought one might also decide that it is reasonable to add a constraint that $\mathbf{c}'\mathbf{\Gamma}\mathbf{c}$ be constant. Hence (9.3.21) could be interpreted as *choose* $\mathbf{c}$ *to maximize* $(\mathbf{m}_1 - \mathbf{m}_2)'\mathbf{c}$ *subject to the constraint that* $\mathbf{c}'\mathbf{\Gamma}\mathbf{c}$ *is constant.* The result is $\mathbf{c} = \alpha\mathbf{\Gamma}^{-1}(\mathbf{m}_1 - \mathbf{m}_2)$, where the scalar α is unimportant, so that $d(\mathbf{z}) = \mathbf{z}'\mathbf{\Gamma}^{-1}(\mathbf{m}_1 - \mathbf{m}_2)$. This $d(\mathbf{z})$ is the same as the $\tilde{\lambda}(\mathbf{z})$ of (9.3.20). Thus the Gaussian model result has been derived by a "reasonable argument" for the non-Gaussian case.

Now consider the non-Gaussian case where the covariance matrices are not equal (and for simplicity the random vectors are zero mean) so that

$$\begin{aligned} \mathcal{H}_j\colon \mathbf{z} &= \mathbf{z}_j, \qquad j = 1, 2 \\ E\{\mathbf{z}_j\mathbf{z}_j'\} &= \mathbf{\Gamma}_j \\ E\{\mathbf{z}_j\} &= \mathbf{0} \end{aligned} \tag{9.3.22}$$

If $\mathbf{z}_j$ were Gaussian the decision function using (9.3.8) is

$$\lambda(\mathbf{z}) = 2\xi_1(\mathbf{z}) - 2\xi_2(\mathbf{z}) = \mathbf{z}'(\mathbf{\Gamma}_2^{-1} - \mathbf{\Gamma}_1^{-1})\mathbf{z} - \ln\left\{\frac{|\mathbf{\Gamma}_1|}{|\mathbf{\Gamma}_2|}\right\} \tag{9.3.23}$$

where the $\ln\{|\mathbf{\Gamma}_1|/|\mathbf{\Gamma}_2|\}$ term is unimportant. Therefore if one wants to duplicate the argument used in the equal covariance matrix case, one might choose a decision function

$$d(\mathbf{z}) = \mathbf{z}'\mathbf{D}\mathbf{z}$$

and then try to find the matrix $\mathbf{D}$ which makes $d(\mathbf{z}_1)$ and $d(\mathbf{z}_2)$ as far apart as possible. The difference in mean values is easy to compute:

$$E\{d(\mathbf{z}_1)\} - E\{d(\mathbf{z}_2)\} = \text{tr}[\mathbf{D}(\mathbf{\Gamma}_1 - \mathbf{\Gamma}_2)]$$

Unfortunately the rest of the equal covariance matrix argument cannot be duplicated, as the variance of $d(\mathbf{z}_j)$ depends on the *higher-order* moments of $\mathbf{z}$, which by assumption are not known. At this point one could modify the arguments but unfortunately there are many reasonable approaches and they yield different answers, only one of which is $\mathbf{D} = \mathbf{\Gamma}_1^{-1} - \mathbf{\Gamma}_2^{-1}$ which corresponds to (9.3.23). This does not imply that the Gaussian formulas should not be applied in a case such as (9.3.22). If only means and covariance matrices are known, the author definitely recommends that the Gaussian formulas be used to give decision functions. (In many cases the probability distributions of Section 9.3.4 can also be used as approximations, especially if the dimension of $\mathbf{z}$ is large.) The main point is that for non-Gaussian hypotheses with unequal covariance matrices, there are many reasonable decision functions which depend only on the mean and covariance matrices. Thus the Gaussian model formulas such as (9.3.23) do not play the unique role they played in the cases of estimation theory and hypotheses with equal covariance matrices.

9.3.8 Analysis and Design of Nonoptimum Decision Functions: Gaussian

For Gaussian hypotheses, the decision functions are of two basic types. In the case of equal covariance matrices, the modified log likelihood function is a linear operation on the observations [$\tilde{\xi}_j(\mathbf{z})$ is of form $\mathbf{z}'\mathbf{c}$], while for the general case of unequal covariance matrices, the log likelihood function combines linear and quadratic operations [$\xi_j(\mathbf{z})$ is of form $(\mathbf{z} - \mathbf{c})'\mathbf{D}(\mathbf{z} - \mathbf{c})$]. In practice the actual decision functions may be of these forms but be nonoptimum in the sense that the wrong values for $\mathbf{c}$ and/or $\mathbf{D}$ are used. This could occur because of modeling errors or because of computation restrictions on the form of decision functions which can be used.

The performance of such nonoptimum decision functions can be studied by analyzing their probability densities under the various hypotheses. This is quite simple for linear decision functions but presents problems for the nonlinear quadratic type of decision functions of the general, nonequal covariance matrix case. For the general case, only the mean value of the quadratic decision function is easy to obtain.

The design of the best decision functions from some specified class can be done by using variational theory to find the specific values which yield the best performance. One example of such a procedure is to find the best linear decision functions $\mathbf{c}_j'\mathbf{z}$ for use when the $\mathcal{H}_j$ involve unequal covariance matrices.

The above discussions are analogous in spirit to those of Sections 5.9, 6.11, and 6.12 on the design and analysis of nonoptimum estimators. The discussions are also closely related to the ideas of Section 9.3.7.

9.3.9 Discussion of Bayesian

The important points of the preceding discussions are now summarized.

The log likelihood function is a very "reasonable" decision function. A combination of a log likelihood decision function with certain decision rules (based on a priori probabilities) can be shown to be optimum in the sense of minimizing the total probability of error. However, it is often better to choose decision rules on the basis of engineering judgment on which types of error are more important.

For Gaussian models, the log likelihood function is easy to evaluate.

Knowledge of the probability density of the decision functions $\xi_j(\mathbf{z})$ when the various hypotheses are true is important, as it enables one to evaluate the various types of probability of error. For Gaussian models,

1. In general, the probability density of $\xi_j(\mathbf{z})$ when $\mathcal{H}_k$ is true is easy to evaluate only when $j = k$.

2. When the hypotheses differ only in mean value (equal covariance matrices), a modified function $\tilde{\xi}_j(\mathbf{z})$ can be used to make decisions, and the probability density of $\tilde{\xi}_j(\mathbf{z})$ when $\mathcal{H}_k$ is true is easy to evaluate for all j and k.

Distance measures provide a method of evaluating in some sense how well a decision can be made. They are useful when the various types of probability of error are difficult to evaluate, but they are not perfect. For the Gaussian case with equal covariance matrices, knowledge of the distances enables one to evaluate the various types of probability of error.

In the non-Gaussian case when only the first and second moments (means and covariance matrices) are specified, one might as well use the formulas for the Gaussian case.

9.4 Fisher

Fisher hypothesis-testing problems are Bayesian problems in which some of the probabilistic models for uncertainty are removed by assuming that the quantities are completely unknown. Such variations on the Bayesian model of Section 9.3 are now discussed. The use of the term Fisher for the following problems is not standard jargon, but it does follow by analogy with the estimation theory of Part II.

9.4.1 No a priori Probabilities

One Fisher-type hypothesis-testing problem occurs when it is completely unknown a priori which hypothesis is true, i.e., when there are no a priori probabilities ρ_j, $j = 1, \ldots, M$. This change has no effect on either the log likelihood functions or the divergence and Bhattacharyya distance concepts. The various types of error probabilities such as $\alpha_{1,2}$ are also still defined, but the total probability of error P_e is no longer defined. The main effect of dropping a priori probabilities on the hypothesis is in the choice of decision rule.

One way to choose a decision rule is to simply evaluate the probabilities of error such as $\alpha_{1,2}$ and $\alpha_{2,1}$ as functions of the decision rule and then use engineering judgment.

Another approach to this Fisher problem is philosophically different from the Bayesian case. The log likelihood function $\xi_j(\mathbf{z})$ is a scalar function of the random vector $\mathbf{z}$, and the probability density of ξ_j can be evaluated if the probability density of $\mathbf{z}$ is known. Define

$p_{j|j}(\xi)$: probability density of $\xi_j(\mathbf{z})$ assuming that $\mathbf{z} = \mathbf{z}_j$; i.e., $p_{j|j}(\xi)$ is the probability density of $\xi_j(\mathbf{z}_j)$

The probability density $p_{j|j}(\xi)$ was evaluated in Section 9.3.4 for the Gaussian case where $\mathbf{z} = \mathbf{z}_j$, $\mathbf{z}_j$ is $N(\mathbf{m}_j, \mathbf{\Gamma}_j)$. Assume that $p_{j|j}(\xi)$ is *smooth* and *unimodal* so that it is reasonable to talk about the "tails of the distribution." Define

$$\mathbf{\Omega}_{\text{tail}, j} = \{\xi : \xi < d_j\}$$

where d_j is chosen so that

$$\int_{-\infty}^{d_j} p_{j|j}(\xi)\, d\xi \ll 1$$

Thus the probability of ξ_j falling within the "tail" is very small if $\mathcal{H}_j$ is true. This is illustrated in Fig. 9.4.1. Now $\xi_j = \ln p_j(\mathbf{z})$, so that a large $\xi_j(\mathbf{z}_{\text{actual}})$ implies that $p_j(\mathbf{z}_{\text{actual}})$ is large, while a small $\xi_j(\mathbf{z}_{\text{actual}})$ implies that $p_j(\mathbf{z}_{\text{actual}})$ is small. Thus if $\xi_j(\mathbf{z}_{\text{actual}}) \in \mathbf{\Omega}_{\text{tail}, j}$, it follows that $p_j(\mathbf{z}_{\text{actual}})$ is small; i.e., it is

Figure 9.4.1 Tail of distribution

unlikely that the observation $\mathbf{z} = \mathbf{z}_{\text{actual}}$ came from hypothesis $\mathcal{H}_j$. Thus a reasonable decision rule when $M = 2$ is

$$\boxed{\begin{array}{ll} \text{Given } \mathbf{z} = \mathbf{z}_{\text{actual}}, & \\ \quad \text{Choose } \mathcal{H}_1 \text{ if} & \xi_1(\mathbf{z}_{\text{actual}}) \notin \boldsymbol{\Omega}_{\text{tail},1} \\ & \xi_2(\mathbf{z}_{\text{actual}}) \in \boldsymbol{\Omega}_{\text{tail},2} \\ \quad \text{Choose } \mathcal{H}_2 \text{ if} & \xi_1(\mathbf{z}_{\text{actual}}) \in \boldsymbol{\Omega}_{\text{tail},1} \\ & \xi_2(\mathbf{z}_{\text{actual}}) \notin \boldsymbol{\Omega}_{\text{tail},2} \\ \quad \text{No decision if} & \xi_1(\mathbf{z}_{\text{actual}}) \notin \boldsymbol{\Omega}_{\text{tail},1} \\ & \xi_2(\mathbf{z}_{\text{actual}}) \notin \boldsymbol{\Omega}_{\text{tail},2} \\ \quad \text{Reject both hypotheses if} & \xi({}_1\mathbf{z}_{\text{actual}}) \in \boldsymbol{\Omega}_{\text{tail},1} \\ & \xi_2(\mathbf{z}_{\text{actual}}) \in \boldsymbol{\Omega}_{\text{tail},2} \end{array}} \tag{9.4.1}$$

This concept extends in an obvious fashion to the case $M > 2$.

The decision rule (9.4.1) is based on the concept of accepting a hypothesis $\mathcal{H}_1$ only if the observation $\mathbf{z} = \mathbf{z}_{\text{actual}}$ could have resulted from $\mathcal{H}_1$ with *high probability* and $\mathcal{H}_2$ with *low probability*. In many applications this is a much more reasonable rule than forcing a decision to be made in all cases. If an observation $\mathbf{z} = \mathbf{z}_{\text{actual}}$ is inconclusive (could have come from either $\mathcal{H}_1$ or $\mathcal{H}_2$ with high probability), it often is better to acknowledge the fact and not try to make a decision. If it is absolutely necessary to make a decision, the rule (9.4.1) can still be used as an intermediate step so that one at least knows the decision should not be trusted. This point of view can also be applied even if a priori probabilities ρ_j are available.

EXAMPLE: Consider

$$\mathcal{H}_1: \quad z = z_1, \qquad z_1 \text{ is } N(1, 1)$$
$$\mathcal{H}_2: \quad z = z_2, \qquad z_2 \text{ is } N(5, 1)$$

where there are no a priori probabilities. From (9.3.8), the decision functions are

$$2\xi_1(z) = -\ln(2\pi) - (z - 1)^2$$
$$2\xi_2(z) = -\ln(2\pi) - (z - 5)^2$$

It is more convenient to work with the equivalent decision functions

$$\tilde{\xi}_1(z) = (z - 1)^2$$
$$\tilde{\xi}_2(z) = (z - 5)^2$$

A reasonable choice for the tails in (9.4.1) then yields the following decision rule: Given $z = z_{\text{actual}}$,

$$\begin{aligned}
\text{Choose } \mathcal{H}_1 \text{ if } & \tilde{\xi}_1(z_{\text{actual}}) < 9 \\
& \tilde{\xi}_2(z_{\text{actual}}) > 9 \\
\text{Choose } \mathcal{H}_2 \text{ if } & \tilde{\xi}_1(z_{\text{actual}}) > 9 \\
& \tilde{\xi}_2(z_{\text{actual}}) < 9 \\
\text{No decision if } & \tilde{\xi}_1(z_{\text{actual}}) < 9 \\
& \tilde{\xi}_2(z_{\text{actual}}) < 9 \\
\text{No decision if } & \tilde{\xi}_1(z_{\text{actual}}) > 9 \\
& \tilde{\xi}_2(z_{\text{actual}}) > 9
\end{aligned}$$

This rule reduces to

$$\begin{aligned}
\text{Choose } \mathcal{H}_1 \text{ if } & -2 < z_{\text{actual}} < 2 \\
\text{Choose } \mathcal{H}_2 \text{ if } & 4 < z_{\text{actual}} < 8 \\
\text{No decision if } & 2 < z_{\text{actual}} < 4 \\
\text{Reject both hypotheses if } & \phantom{-2 <} z_{\text{actual}} < -2 \\
& \phantom{-2 <} z_{\text{actual}} > 8
\end{aligned}$$

This example is illustrated in Fig. 9.4.2.

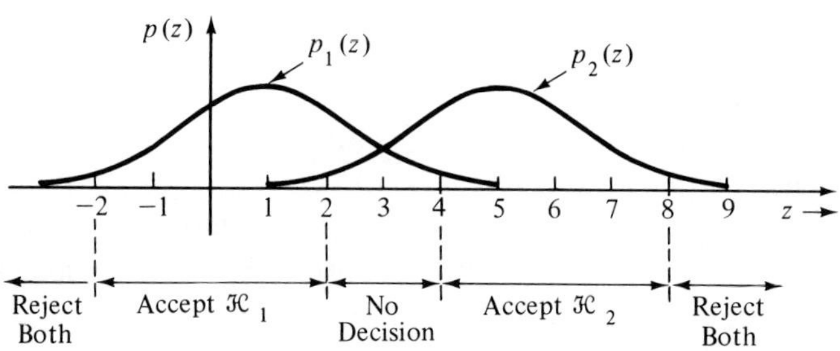

Figure 9.4.2 Illustration of example.

9.4.2 No Alternative Hypothesis

Another Fisher-type model occurs when there is only one hypothesis and it is desired to decide whether or not it is valid. The situation can be expressed as

$\mathcal{H}_1$: $\mathbf{z}$ is random vector with probability density $p_1(\mathbf{z})$
$\mathcal{H}_2$: $\mathcal{H}_1$ is not true

One approach to this problem follows directly from (9.4.1):

$$\boxed{\begin{array}{l}\text{Given } \mathbf{z} = \mathbf{z}_{\text{actual}},\\ \quad \text{Choose } \mathcal{H}_1 \text{ if } \xi_1(\mathbf{z}_{\text{actual}}) \notin \mathbf{\Omega}_{\text{tail},1}\\ \quad \text{Choose } \mathcal{H}_2 \text{ if } \xi_1(\mathbf{z}_{\text{actual}}) \in \mathbf{\Omega}_{\text{tail},1}\end{array}} \tag{9.4.2}$$

For this case one can evaluate one type of probability of error $\alpha_{2,1}$ but not the other $\alpha_{1,2}$. The distance measure concept does not make sense.

9.4.3 Unknown Parameters: General

A third Fisher-type problem occurs when the hypothesized model contains unknown parameters. For example,

$\mathcal{H}_j$: $\mathbf{z}$ is random vector with probability density $p_j(\mathbf{z} : \boldsymbol{\beta}_j)$ which is a known functional form of the parameter vector $\boldsymbol{\beta}_j$ whose value is unknown, $j = 1, \ldots, M$

For such a problem, a reasonable choice of decision function is $\xi_j(\mathbf{z} : \hat{\boldsymbol{\beta}})$, where

$$\begin{aligned}\xi_j(\mathbf{z} : \hat{\boldsymbol{\beta}}) &= \ln\{\max_{\boldsymbol{\beta}_j} p_j(\mathbf{z} : \boldsymbol{\beta}_j)\}\\ &= \ln p_j(\mathbf{z} : \hat{\boldsymbol{\beta}}_j)\end{aligned} \tag{9.4.3}$$

$\hat{\boldsymbol{\beta}}_j$: value of that maximizes $p_j(\mathbf{z} : \boldsymbol{\beta}_j)$ for given $\mathbf{z}$

$\hat{\boldsymbol{\beta}}_j$ is the maximum likelihood estimate of $\boldsymbol{\beta}_j$ made from $\mathbf{z}$ assuming that $\mathcal{H}_j$ is true (see Chapter 11).

If they are available, a priori probabilities, p_j, can be used to help choose a decision rule. However, the Fisher models of Sections 9.4.1 and 9.4.2 can also be studied when unknown parameters are present. The probability density of $\ln\{p_j(\mathbf{z} : \hat{\boldsymbol{\beta}}_j)\}$ is different from that of $\ln\{p_j(\mathbf{z} : \boldsymbol{\beta}_j)\}$, where $\boldsymbol{\beta}_j$ is the true value.

9.4.4 Unknown Parameters: Gaussian

Consider

$\mathcal{H}_j$: $\mathbf{z} = \mathbf{z}_j, \quad \mathbf{z}_j = \mathbf{H}_j\mathbf{x} + \mathbf{v}_j, \quad j = 1, \ldots, M$

$\mathbf{v}_j$ is $N(\mathbf{0}, \mathbf{R}_j)$

$\mathbf{x}$: completely unknown

$\mathbf{x}$: K_1-dimensional vector

$\mathbf{z}$: K_2-dimensional vector

$K_2 \geq K_1$

Assume that $\mathbf{R}_j^{-1}$ exists. Using the Fisher estimation theory of Section 5.2

$$\hat{\mathbf{x}}_j = (\mathbf{H}_j'\mathbf{R}_j^{-1}\mathbf{H}_j)^{-1}\mathbf{H}_j'\mathbf{R}_j^{-1}\mathbf{z}$$

Thus

$$\begin{aligned} 2\xi_j(\mathbf{z} : \hat{\mathbf{x}}_j) &= c_j - [\mathbf{z} - \mathbf{H}_j\hat{\mathbf{x}}_j]'\mathbf{R}_j^{-1}[\mathbf{z} - \mathbf{H}_j\hat{\mathbf{x}}_j] \\ &= c_j - \mathbf{z}'\{\mathbf{R}_j^{-1} - \mathbf{R}_j^{-1}\mathbf{H}_j(\mathbf{H}_j'\mathbf{R}_j^{-1}\mathbf{H}_j)^{-1}\mathbf{H}_j'\mathbf{R}_j^{-1}\}\mathbf{z} \qquad (9.4.4) \\ c_j &= -K_2 \ln(2\pi) - \ln|\mathbf{R}_j| \end{aligned}$$

assuming that $(\mathbf{H}'\mathbf{R}^{-1}\mathbf{H})^{-1}$ exists. Investigation of the probability density of $\xi_j(\mathbf{z} : \hat{\mathbf{x}}_j)$ yields

> When $\mathcal{H}_j$ is true, $-2\xi_j(\mathbf{z} : \hat{\mathbf{x}}_j) + c_j$ is a chi-squared random variable with $K_2 - K_1$ degrees of freedom (9.4.5)

An outline of how to prove (9.4.5) is as follows: Define

$$\boldsymbol{\theta} = \mathbf{R}^{-1} - \mathbf{R}^{-1}\mathbf{H}[\mathbf{H}'\mathbf{R}^{-1}\mathbf{H}]^{-1}\mathbf{H}'\mathbf{R}^{-1}$$

so that

$$\mathbf{z} = \mathbf{H}\mathbf{x} + \mathbf{v}$$

$$\mathbf{z}'\boldsymbol{\theta}\mathbf{z} = \mathbf{v}'\boldsymbol{\theta}\mathbf{v}$$

Linear transformations can be used to show that one can assume that $E\{\mathbf{v}\mathbf{v}'\} = \mathbf{R} = \mathbf{I}$ with no loss in generality. More linear transformations can then be used to show that

$$\mathbf{v}'\boldsymbol{\theta}\mathbf{v} = \tilde{\mathbf{v}}'\tilde{\boldsymbol{\theta}}\tilde{\mathbf{v}}$$

$$E\{\tilde{\mathbf{v}}\tilde{\mathbf{v}}'\} = \mathbf{I}$$

$$\tilde{\boldsymbol{\theta}} = \left(\begin{array}{c|c} \mathbf{I} & 0 \\ \hline 0 & 0 \end{array}\right)$$

$$\mathbf{I}: K_2 - K_1 \text{ unit matrix}$$

The details of the above derivation of (9.4.5) are nontrivial.

EXAMPLE: Consider

$$\mathcal{H}_1: \mathbf{z} = \begin{pmatrix} 1 \\ 1 \end{pmatrix} x + \mathbf{v}$$

$$\mathbf{z} = \begin{pmatrix} z_1 \\ z_2 \end{pmatrix} \qquad \mathbf{v} = \begin{pmatrix} v_1 \\ v_2 \end{pmatrix}$$

$$\mathbf{v} \text{ is } N(\mathbf{0}, \mathbf{I})$$

$$x: \text{completely unknown}$$

$$\mathcal{H}_2: \text{hypothesis } \mathcal{H}_1 \text{ is not true}$$

Then the estimate of x is simply

$$\hat{\mathbf{x}} = \frac{z_1 + z_2}{2}$$

and from (9.4.4)

$$2\xi_1(\mathbf{z}, \hat{\mathbf{x}}) = -2\ln(2\pi) - \left(\frac{z_1 - z_2}{\sqrt{2}}\right)^2$$

where

$$\frac{z_1 - z_2}{\sqrt{2}} \text{ is } N(0, 1)$$

Thus from (9.4.2), a reasonable decision rule is

$$\text{Accept } \mathcal{H}_1 \text{ if } \left|\frac{z_1 - z_2}{\sqrt{2}}\right| \leq 3$$

$$\text{Reject } \mathcal{H}_1 \text{ if } \left|\frac{z_1 - z_2}{\sqrt{2}}\right| > 3$$

9.4.5 Discussion of Fisher

In the Fisher models of Section 9.4.1 and 9.4.2 the Fisher aspects arise because there were no a priori probabilities or no alternative hypotheses. Hence log likelihood functions could still be used as decision functions. However, it is very important to emphasize that *the log likelihood function is* **not** *the only possible decision function.* In Section 9.3.1, the use of the log likelihood function was shown to be optimum in a certain sense for Bayesian models. However, this proof does not apply to Fisher models. For Fisher models, other decision functions may be just as good or better. Unfortunately a complete discussion of other possible decision functions is beyond the scope of this book.

The Fisher models of Sections 9.4.3 and 9.4.4 have hypotheses with completely unknown parameters. This required the introduction of a new concept, the choice of the decision function to be the log likelihood function evaluated at the maximum likelihood estimate of the unknown parameters. From a theoretical point of view, this is a major step. Many good arguments can be developed, but, in this book, the only justification to be given is that it is a "reasonable" thing to do.

In the Gaussian case it is interesting to compare (9.4.4) with the corresponding Bayesian (9.3.8), which can be rewritten as

$$\begin{aligned} 2\xi_j(\mathbf{z}) = &-K_2\ln(2\pi) - \ln|\mathbf{H}_j\boldsymbol{\psi}_j\mathbf{H}'_j + \mathbf{R}_j| \\ &- \mathbf{z}'\{\mathbf{R}_j^{-1} - \mathbf{R}_j^{-1}\mathbf{H}_j[\boldsymbol{\psi}_j^{-1} + \mathbf{H}'_j\mathbf{R}_j^{-1}\mathbf{H}_j]^{-1}\mathbf{H}'_j\mathbf{R}_j^{-1}\}\mathbf{z} \end{aligned} \tag{9.4.6}$$

In Chapter 5 it was argued that a Fisher model estimator could be obtained from a Bayesian model where $E\{\mathbf{x}\mathbf{x}'\} = \boldsymbol{\psi}$ by letting $\boldsymbol{\psi} \to \infty\mathbf{I}$. Comparison of the Bayesian (9.4.6) with (9.4.4) shows that this is apparently true in the present case, as the **z**-dependent term in (9.4.6) becomes the **z**-dependent term in (9.4.4) for $\boldsymbol{\psi} = \infty\mathbf{I}$. One of the "log terms" in (9.4.6) "blows up," but this is relatively minor. However, a basic difference is that the **z**-dependent term in (9.4.6) is chi-squared with K_2 degrees of freedom no matter how large $\boldsymbol{\psi}_j$ becomes, while the *z*-dependent term in (9.4.4) is chi-squared with K_2 −

K_1 degrees of freedom. This illustrates again the fact that a Fisher model involving an unknown vector $\mathbf{x}$ is a fundamentally different model than a Bayesian model involving a random vector $\mathbf{x}$ with very large covariance matrix.

9.5 Weighted Least Squares

In estimation theory, weighted least squares was discussed as a way of estimating without choosing a model for the uncertainties. The same sort of arguments apply in hypothesis testing.

The basic idea is as follows: Assume that $\mathbf{z}$ is observed and there are M hypotheses:

$\mathcal{H}_j$: $\mathbf{z}$ is approximately equal to $\mathbf{f}_j(\mathbf{x})$ for some value of $\mathbf{x}$, where $\mathbf{f}_j$ is a known function of $\mathbf{x}$, $j = 1, \ldots, M$

The choice is made by determining which $\mathbf{f}_j$ can provide the closest fit to the observed $\mathbf{z} = \mathbf{z}_{\text{actual}}$. A reasonable measure of closeness of fit is

$$J_j(\mathbf{x}) = [\mathbf{z} - \mathbf{f}_j(\mathbf{x})]'\mathbf{R}^{-1}[\mathbf{z} - \mathbf{f}_j(\mathbf{x})]$$

where $\mathbf{R}$ is some chosen positive definite weighting matrix. (Here the symbol J does not denote divergence.) Let $\hat{\mathbf{x}}_j$ denote the value of $\mathbf{x}$ which minimizes $J_j(\mathbf{x})$ for $\mathbf{z} = \mathbf{z}_{\text{actual}}$. Then a reasonable decision rule for deciding between $\mathcal{H}_1$ and $\mathcal{H}_2$ is

$$\boxed{\text{Choose } \mathcal{H}_k \text{ if } j = k \text{ yields minimum } J_j(\hat{\mathbf{x}}_j),\ j = 1, \ldots, M} \qquad (9.5.1)$$

In the special case where $\mathbf{f}_j(\mathbf{x}) = \mathbf{H}_j\mathbf{x}$ it follows that

$$J_j(\hat{\mathbf{x}}_j) = \mathbf{z}'[\mathbf{R}_j^{-1} - \mathbf{R}_j^{-1}\mathbf{H}_j(\mathbf{H}_j'\mathbf{R}_j^{-1}\mathbf{H}_j)^{-1}\mathbf{H}_j'\mathbf{R}_j^{-1}]\mathbf{z}$$

so there is a close relationship between the $J_j(\hat{\mathbf{x}}_j)$ and the Fisher $\xi_j(\mathbf{z} : \hat{\mathbf{x}}_j)$ of (9.4.4).

Hypothesis testing based on (9.5.1) could be called *minimum residual* hypothesis testing. $J_j(\hat{\mathbf{x}}_j)$ is "residual left over" after $\mathbf{f}_j(\mathbf{x})$ is fitted to $\mathbf{z}_{\text{actual}}$.

As in estimation theory, the weighted-least-squares approach yields no measure of performance, as no models for the uncertainty are assumed. Thus probability of errors and distance measures play no role.

9.6 Unknown-but-Bounded

Hypothesis testing for an unknown-but-bounded model is conceptually simpler than for a stochastic model. The hypotheses are

$$\mathcal{H}_j\colon\ \mathbf{z} \in \mathbf{\Omega}_{z,j}, \qquad j = 1, \ldots, M$$

where the $\mathbf{\Omega}_{z,j}$ are specified sets.

9.6.1 Set-Theoretic Decision Making

Assume that $M = 2$. Assume that an observation $\mathbf{z} = \mathbf{z}_{\text{actual}}$ is made. The decision rule to decide whether $\mathcal{H}_1$ or $\mathcal{H}_2$ is true is as follows:

$$
\boxed{\begin{aligned}
&\text{Given } \mathbf{z} = \mathbf{z}_{\text{actual}}, \\
&\qquad \text{Choose } \mathcal{H}_1 \text{ if } \mathbf{z}_{\text{actual}} \in \mathbf{\Omega}_{z,1} \\
&\qquad\qquad\qquad\qquad \mathbf{z}_{\text{actual}} \notin \mathbf{\Omega}_{z,2} \\
&\qquad \text{Choose } \mathcal{H}_2 \text{ if } \mathbf{z}_{\text{actual}} \notin \mathbf{\Omega}_{z,1} \\
&\qquad\qquad\qquad\qquad \mathbf{z}_{\text{actual}} \in \mathbf{\Omega}_{z,2} \\
&\text{Reject both } \mathcal{H}_1 \text{ and } \mathcal{H}_2 \text{ if } \mathbf{z}_{\text{actual}} \notin \mathbf{\Omega}_{z,1} \\
&\qquad\qquad\qquad\qquad \mathbf{z}_{\text{actual}} \notin \mathbf{\Omega}_{z,2} \\
&\qquad \text{No decision if } \mathbf{z}_{\text{actual}} \in \mathbf{\Omega}_{z,1} \\
&\qquad\qquad\qquad\qquad \mathbf{z}_{\text{actual}} \in \mathbf{\Omega}_{z,2}
\end{aligned}} \tag{9.6.1}
$$

Extension of the hypothesis-testing logic to M hypothesis

$$\mathcal{H}_j\colon\ \mathbf{z} \in \mathbf{\Omega}_{z,j}, \qquad j = 1, \ldots, M$$

is straightforward.

The basic logic behind (9.6.1) and the $M > 2$ case is simple. If the observation $\mathbf{z}$ could not have occurred assuming that $\mathcal{H}_j$ is true, the hypothesis $\mathcal{H}_j$ is rejected. No single hypothesis is accepted until all other alternative hypotheses are rejected. This type of logic makes the testing quite independent of the value of M. For example, if $M = 1$ so there is no alternative hypothesis (as in Section 9.4.2), the same logic is used. The hypothesis is assumed to be true until it is rejected.

The test (9.6.1) is related in spirit to the Fisher test (9.4.1). However, as in estimation theory, the performance of unknown-but bounded model hypothesis testing differs radically from Bayesian and Fisher models. If a decision is made for the unknown-but-bounded model, it is made with no error. However, no decisions may be made.

An important and interesting relationship exists between the above decision making and the estimation concepts of Chapter 5, Section 5.6. Consider

$$
\begin{aligned}
\mathcal{H}_j\colon\ \mathbf{z} &= \mathbf{H}_j\mathbf{x} + \mathbf{v} \\
\mathbf{x} &\in \mathbf{\Omega}_{x,j} \\
\mathbf{v} &\in \mathbf{\Omega}_{v,j}
\end{aligned}
$$

Then

$$
\begin{aligned}
&\mathcal{H}_j\colon\ \mathbf{z} \in \mathbf{\Omega}_{z,j} \\
&\quad \mathbf{\Omega}_{z,j} = \text{vector sum of } \mathbf{\Omega}_{\mathbf{H}_j\mathbf{x}} \text{ and } \mathbf{\Omega}_{v,j} \\
&\quad \mathbf{\Omega}_{H_jx} = \{\mathbf{z}\colon\ \mathbf{H}_j\mathbf{x} = \mathbf{z},\ \mathbf{x} \in \mathbf{\Omega}_{x,j}\}
\end{aligned}
$$

Now define as in Chapter 5 (see Section 5.6.1)

$$\mathbf{\Omega}_{\text{est},j} = \mathbf{\Omega}_{x,j} \cap \mathbf{\Omega}_{x,j|z_{\text{actual}}}$$

$$\mathbf{\Omega}_{x,j|z} = \{\mathbf{x}\colon \mathbf{z} - \mathbf{H}_j\mathbf{x} \in \mathbf{\Omega}_{v,j}\}$$

so that $\mathbf{\Omega}_{\text{est},j}$ is the estimate set which, given the observation $\mathbf{z} = \mathbf{z}_{\text{actual}}$, must contain $\mathbf{x}$ if hypothesis $\mathcal{H}_j$ is true. Now consider the following two statements:

$$\begin{aligned} &1.\ \text{If } \mathbf{z}_{\text{actual}} \notin \mathbf{\Omega}_{z,j}, \text{ then } \mathbf{\Omega}_{\text{est},j} \text{ does not exist (is empty)} \\ &2.\ \text{If } \mathbf{\Omega}_{\text{est},j} \text{ does not exist, then } \mathbf{z}_{\text{actual}} \notin \mathbf{\Omega}_{z,j} \end{aligned} \tag{9.6.2}$$

These statements are "reasonable" and appear to be true in general, although no formal proof has been obtained. Henceforth they are assumed to be true in general. The conclusion that follows from these statements is that the hypothesis-testing problem can be replaced by an estimation problem. In particular (9.6.1) can be replaced by

$$\boxed{\begin{aligned} \text{Choose } \mathcal{H}_1 \text{ if } \quad & \mathbf{\Omega}_{\text{est},1} \text{ exists} \\ & \mathbf{\Omega}_{\text{est},2} \text{ does not exist} \\ \text{Choose } \mathcal{H}_2 \text{ if } \ & \mathbf{\Omega}_{\text{est},1} \text{ does not exist} \\ & \mathbf{\Omega}_{\text{est},2} \text{ exists} \\ \text{Reject both } \mathcal{H}_1 \text{ and } \mathcal{H}_2 \text{ if } \ & \mathbf{\Omega}_{\text{est},1} \text{ does not exist} \\ & \mathbf{\Omega}_{\text{est},2} \text{ does not exist} \\ \text{No decision if } \ & \mathbf{\Omega}_{\text{est},1} \text{ exists} \\ & \mathbf{\Omega}_{\text{est},2} \text{ exists} \end{aligned}} \tag{9.6.3}$$

EXAMPLE: Consider

$$\mathcal{H}_1\colon z = x + v$$

$$|x| \leq 2 \qquad |v| \leq 1$$

$$\mathcal{H}_2\colon \mathcal{H}_1 \text{ is not true}$$

Then using the logic of (9.6.1) the decision rule is

$$\text{Choose } \mathcal{H}_1 \text{ if } |z| \leq 3$$

$$\text{Reject } \mathcal{H}_1 \text{ if } |z| > 3$$

Using the estimate set concept

$$\mathbf{\Omega}_{\text{est},1} = \{x\colon |x| \leq 2, |z - x| \leq 1\}$$

and a decision rule such as (9.6.3) is

$$\text{Choose } \mathcal{H}_1 \text{ if } \mathbf{\Omega}_{\text{est},1} \text{ exists}$$

$$\text{Reject } \mathcal{H}_1 \text{ if } \mathbf{\Omega}_{\text{est},1} \text{ does not exist}$$

Both rules yield the same decision.

9.6.2 Bounding Ellipsoids: Decision Making

Consider

$$\mathcal{H}_j\colon \mathbf{z} \in \mathbf{\Omega}_{z,j}$$
$$\mathbf{\Omega}_{z,j} = \{\mathbf{z}\colon [\mathbf{z} - \mathbf{m}_j]'\mathbf{\Gamma}_j^{-1}[\mathbf{z} - \mathbf{m}_j] \leq 1\}, \qquad j = 1, \ldots, M$$

The decisions are based on the decision functions

$$\gamma_j(\mathbf{z}) = 1 - [\mathbf{z} - \mathbf{m}_j]'\mathbf{\Gamma}^{-1}[\mathbf{z} - \mathbf{m}_j], \qquad j = 1, \ldots, M$$

For example, if $M = 2$ and $\mathbf{z} = \mathbf{z}_{\text{actual}}$ is observed, the decision rule is

$$\begin{aligned}
&\text{Choose } \mathcal{H}_1 \text{ if } \begin{cases} \gamma_1(\mathbf{z}_{\text{actual}}) \geq 0 \\ \gamma_2(\mathbf{z}_{\text{actual}}) < 0 \end{cases} \\
&\text{Choose } \mathcal{H}_2 \text{ if } \begin{cases} \gamma_1(\mathbf{z}_{\text{actual}}) < 0 \\ \gamma_2(\mathbf{z}_{\text{actual}}) \geq 0 \end{cases} \\
&\text{Reject } \mathcal{H}_1 \text{ and } \mathcal{H}_2 \text{ if } \begin{cases} \gamma_1(\mathbf{z}_{\text{actual}}) < 0 \\ \gamma_2(\mathbf{z}_{\text{actual}}) < 0 \end{cases} \\
&\text{No decision if } \begin{cases} \gamma_1(\mathbf{z}_{\text{actual}}) \geq 0 \\ \gamma_2(\mathbf{z}_{\text{actual}}) \geq 0 \end{cases}
\end{aligned} \tag{9.6.4}$$

Note the resemblance between $\gamma_j(\mathbf{z})$ of (9.6.4) and the Gaussian log likelihood function $\xi_j(\mathbf{z})$ of (9.3.8). They both have the same $\mathbf{z}$ dependence, yet the decision rules applied to them are different.

Consider

$$\mathcal{H}_j\colon \mathbf{z} = \mathbf{H}_j\mathbf{x} + \mathbf{v}, \qquad j = 1, \ldots, M$$
$$\mathbf{x} \in \mathbf{\Omega}_{j,x} = \{\mathbf{x} : \mathbf{x}'\boldsymbol{\psi}_j^{-1}\mathbf{x} \leq 1\}$$
$$\mathbf{v} \in \mathbf{\Omega}_{j,v} = \{\mathbf{v} : \mathbf{v}'\mathbf{R}_j^{-1}\mathbf{v} \leq 1\}$$

This is equivalent to

$$\begin{aligned}
&\mathcal{H}_j\colon \mathbf{z} \in \mathbf{\Omega}_{z,j}, \qquad j = 1, \ldots, M \\
&\mathbf{\Omega}_{z,j}\colon \text{ vector sum of } \mathbf{\Omega}_{H,x} \text{ and } \mathbf{\Omega}_{v_j} \\
&\mathbf{\Omega}_{H,x} = \{\mathbf{z} : \mathbf{z}'[\mathbf{H}_j\boldsymbol{\psi}_j\mathbf{H}_j']^{-1}\mathbf{z} \leq 1\} \\
&\mathbf{\Omega}_{v_j} = \{\mathbf{z} : \mathbf{z}'\mathbf{R}_j^{-1}\mathbf{z} \leq 1\}
\end{aligned} \tag{9.6.5}$$

Now $\mathbf{\Omega}_{z,j}$ is not an ellipsoid but following the ideas of Section 4.3 there exists a bounding ellipsoid $\mathbf{\Omega}_{bj}$:

$$\mathbf{\Omega}_{bj} \supset \mathbf{\Omega}_{z,j}$$
$$\mathbf{\Omega}_{bj} = \{\mathbf{z} : \mathbf{z}'[\beta^{-1}\mathbf{H}_j\boldsymbol{\psi}_j\mathbf{H}_j' + (1 - \beta)^{-1}\mathbf{R}_j]^{-1}\mathbf{z} \leq 1\}$$
$$1 \geq \beta \geq 0$$

Thus as an approximation, the hypothesis (9.6.5) can be replaced by

$$\mathcal{H}_j\colon \mathbf{z} \in \mathbf{\Omega}_{bj} \tag{9.6.6}$$

and the decision functions are

$$\gamma_j(\mathbf{z}) = 1 - \mathbf{z}'[\beta^{-1}\mathbf{H}_j\boldsymbol{\psi}_j\mathbf{H}_j' + (1-\beta)^{-1}\mathbf{R}_j]^{-1}\mathbf{z} \tag{9.6.7}$$

which are used just as in (9.6.4). If $\gamma_j(\mathbf{z}) < 0$ for any admissible β, $0 \leq \beta \leq 1$, $\mathcal{H}_j$ is rejected. Rejection occurs if $\gamma_j(\mathbf{z}) < 0$ holds for any value of β, $0 \leq \beta \leq 1$, as $\boldsymbol{\Omega}_{b,j}$ is a bounding ellipsoid for all β. If $\boldsymbol{\Omega}_{b,j}$ does not exist for some β, then $\boldsymbol{\Omega}_{z,j}$ does not exist. Thus in theory the best decision function is (9.6.7), where β is the value that maximizes $\mathbf{z}'[\beta^{-1}\mathbf{H}_j\boldsymbol{\psi}_j\mathbf{H}_j' + (1-\beta)^{-1}\mathbf{R}_j]^{-1}\mathbf{z}$ for the given $\mathbf{z} = \mathbf{z}_{\text{actual}}$. In practice, of course, some fixed value of β may be used instead.

Consider the estimation problem of using $\mathbf{z}$ to find a bounding ellipsoid estimate set $\boldsymbol{\Omega}_{\hat{x}}$ which contains $\mathbf{x}$. Consider (9.6.5) and assume that $\mathcal{H}_j$ is true. Then from (5.6.3)

$$\begin{aligned}
\boldsymbol{\Omega}_{\hat{x}} &= \{\mathbf{x} : (\mathbf{x} - \hat{\mathbf{x}})'\boldsymbol{\Sigma}_j^{-1}(\mathbf{x} - \hat{\mathbf{x}}) \leq 1\} \\
\hat{\mathbf{x}} &= [(1-\rho)\boldsymbol{\psi}_j^{-1} + \rho\mathbf{H}_j'\mathbf{R}_j^{-1}\mathbf{H}_j]^{-1}[\rho\mathbf{H}_j'\mathbf{R}_j^{-1}\mathbf{z}] \\
\boldsymbol{\Sigma} &= (1-\delta^2)[(1-\rho)\boldsymbol{\psi}_j^{-1} + \rho\mathbf{H}_j'\mathbf{R}_j^{-1}\mathbf{H}_j]^{-1} \\
\delta^2 &= \mathbf{z}'[(1-\rho)^{-1}\mathbf{H}_j\boldsymbol{\psi}_j\mathbf{H}_j' + \rho^{-1}\mathbf{R}_j]^{-1}\mathbf{z} \\
1 &\geq \rho \geq 0
\end{aligned} \tag{9.6.8}$$

Now $\boldsymbol{\Omega}_{\hat{x}}$ does not exist if $1 - \delta^2 < 0$ for any admissible ρ_j, $0 \leq \rho \leq 1$. Thus comparison of δ^2 of (9.6.8) with $\gamma_j(\mathbf{z})$ of (9.6.7) shows that the statement (9.6.2) is true for bounding ellipsoids.

9.6.3 Set-Theoretic Distance Measure†

Assume that $M = 2$. Let D denote the *distance* between the two hypotheses, where D is to play the same role for unknown-but-bounded models as the divergence J and Bhattacharyya distance B do for stochastic models. One reasonable way to define D is

$$\begin{aligned}
D &= -\ln \frac{V_{\text{int}}}{(V_1 V_2)^{1/2}} \\
V_j&: \text{ volume of } \boldsymbol{\Omega}_j, \qquad j = 1, 2 \\
V_{\text{int}}&: \text{ volume of } \boldsymbol{\Omega}_1 \cap \boldsymbol{\Omega}_2
\end{aligned} \tag{9.6.9}$$

Equation (9.6.9) is reasonable in the following sense: If $V_{\text{int}} = 0$, one can always choose the correct hypothesis and $V_{\text{int}} = 0$ implies that $D = \infty$. If $\boldsymbol{\Omega}_1$ and $\boldsymbol{\Omega}_2$ are identical sets, one can never make a decision and $\boldsymbol{\Omega}_1$ identical to $\boldsymbol{\Omega}_2$ implies that $V_1 = V_2 = V_{\text{int}}$, which implies that $D = 0$. The definition of (9.6.9) is, of course, not the only possibility.

9.6.4 Bounding Ellipsoid: Distance Measures†

The distance measure D of (9.6.9) is studied for

$$\begin{aligned}
\mathcal{H}_j&: \mathbf{z} \in \boldsymbol{\Omega}_j \qquad j = 1, 2 \\
\boldsymbol{\Omega}_j &= \{\mathbf{z} : [\mathbf{z} - \mathbf{m}_j]'\boldsymbol{\Gamma}_j^{-1}[\mathbf{z} - \mathbf{m}_j] \leq 1\}
\end{aligned}$$

A bounding ellipsoid is used to replace $\mathbf{\Omega}_1 \cap \mathbf{\Omega}_2$ so D becomes

$$D = -\ln \frac{V_{\text{int},b}}{\sqrt{V_1 V_2}}$$

$$V_j\text{: volume of } \mathbf{\Omega}_j, \qquad j = 1, 2$$

$$V_{\text{int},b}\text{: volume of } \mathbf{\Omega}_{\text{int},b}$$

where $\mathbf{\Omega}_{\text{int},b}$ is a bounding ellipsoid for $\mathbf{\Omega}_1 \cap \mathbf{\Omega}_2$ given by (see Exercise 5.13)

$$\begin{aligned}
\mathbf{\Omega}_{\text{int},b} &= \{\mathbf{z} : [\mathbf{z} - \mathbf{m}]'\mathbf{\Gamma}_b^{-1}[\mathbf{z} - \mathbf{m}] \leq 1\} \\
\mathbf{m} &= [\rho\mathbf{\Gamma}_1^{-1} + (1 - \rho)\mathbf{\Gamma}_2^{-1}]^{-1}[\rho\mathbf{\Gamma}_1^{-1}\mathbf{m}_1 + (1 - \rho)\mathbf{\Gamma}_2^{-1}\mathbf{m}_2] \\
\mathbf{\Gamma}_b &= (1 - \delta^2)[\rho\mathbf{\Gamma}_1^{-1} + (1 - \rho)\mathbf{\Gamma}_2^{-1}]^{-1} \\
\delta^2 &= [\mathbf{m}_1 - \mathbf{m}_2]'[\rho^{-1}\mathbf{\Gamma}_1 + (1 - \rho)^{-1}\mathbf{\Gamma}_2]^{-1}[\mathbf{m}_1 - \mathbf{m}_2] \\
1 &\geq \rho \geq 0
\end{aligned}$$

The volume of an ellipsoid defined by $\mathbf{\Gamma}$ is proportional to $|\mathbf{\Gamma}|^{1/2}$. Thus after some manipulation, D becomes

$$\begin{aligned}
2D = {} & 2\ln|\rho^{-1}\mathbf{\Gamma}_1 + (1 - \rho)^{-1}\mathbf{\Gamma}_2| \\
& - \ln|\mathbf{\Gamma}_1| - \ln|\mathbf{\Gamma}_2| \\
& - K\ln\{1 - [\mathbf{m}_1 - \mathbf{m}_2]'[\rho^{-1}\mathbf{\Gamma}_1 + (1 - \rho)^{-1}\mathbf{\Gamma}_2]^{-1}[\mathbf{m}_1 - \mathbf{m}_2]\}, \\
& \qquad\qquad 1 \geq \rho \geq 0
\end{aligned}$$

This unknown-but-bounded distance D bears a close structural resemblance to the Gaussian Bhattacharyya distance B of (9.3.17).

9.7 Discussion

For Bayesian and Fisher models, theorems on the optimality of the use of log likelihood functions as decision functions are available. However, many types of decision functions are possible. Log likelihood decision functions are used in practice because they work and because often nothing better is available. There exists a lot of freedom in the choice of decision rule. The choice depends on the designer's judgment on which types of error are most important. The Bayesian and Fisher models can be confusing because of all the different symbols and formulas. However, in the Gaussian case, the basic decision mechanism is closely related to that of weighted least squares, and weighted least squares provides a very simple interpretation of how the decisions are made. In other words, the basic principle is, *Given the observation* $\mathbf{z} = \mathbf{z}_{\text{actual}}$, *choose the hypothesis (model) which is closest to* $\mathbf{z}_{\text{actual}}$.

The nature of unknown-but-bounded models automatically specifies the exact nature of the decision functions and rules. As in Part II there is is a very strong structural resemblance between the equations resulting from the Gaussian models and the unknown-but-bounded ellipsoidal models.

However, also as in Part II, the behavior of the resulting tests is radically different. This is as expected, as the models are fundamentally different.

Historical Notes and References: Chapter 9

Section 9.3, 9.4

These basic ideas are discussed in many places in many ways. The approach used here is patterned on the Mathematical Statistics references of Book List 2: Cramer [1] and Wilks [1] were actually used. The author was introduced to the use of the Bhattacharyya distance by discussions with T. Kailath, see Kailath [1]; Bhattacharryya [1]. Many different distance measures are compared in Kullbeck [1].

Section 9.6

To the author's knowledge, hypothesis testing for unknown-but-bounded models is a new and untried technique.

Exercises

9.1. Consider the Bayesian model

$$\mathcal{H}_j\colon \mathbf{z} = \mathbf{z}_j, \qquad \mathbf{z}_j \text{ is } N(\mathbf{m}_j, \mathbf{I})$$

$$\mathbf{m}_1 = \begin{pmatrix} 2 \\ 0 \end{pmatrix} \qquad \mathbf{m}_2 = \begin{pmatrix} 1 \\ 1 \end{pmatrix}$$

$$\rho_1 = \rho_1 = \tfrac{1}{2}$$

Assume that the decision rule (9.3.2) is to be used.

a. Evaluate the total probability of error, P_e.

b. Evaluate the Bhattacharyya distance B and compute an upper bound for P_e using (9.3.16).

Comment: To do part (a) you will have to use tables of Gaussian probabilities, which are not in this book. Such tables are widely available.

9.2. Consider

$$\mathcal{H}_j\colon \mathbf{z} = \mathbf{z}_j, \qquad j = 1, 2$$

$$\mathbf{z}_j \text{ has probability density } p_j(\mathbf{z})$$

$$\rho_j\colon \text{a priori probability that } \mathcal{H}_j \text{ is true}$$

Define

$$\alpha_{j,k} \text{ probability that } \mathcal{H}_j \text{ is chosen when } \mathcal{H}_k \text{ is true}$$

Consider the cost or risk function

$$c = \gamma_1 \rho_2 \alpha_{1,2} + \gamma_2 \rho_1 \alpha_{2,1}$$

where the $\gamma_j > 0$, $j = 1, 2$, are chosen to "weight" the different types of errors. Specify a decision function and a decision rule which minimizes c.

9.3. Consider the Bayesian model

$$\mathcal{H}_j\colon \mathbf{z} = \mathbf{z}_j, \qquad j = 1, 2$$

$$\mathbf{z}_j \text{ has probability density } p_j(\mathbf{z})$$

$$\rho_j\colon \text{a priori probability that } \mathcal{H}_j \text{ is true}$$

Let P_e denote the total probability of error.

a. Prove that

$$P_e \leq \rho_1^s \rho_2^{1-s} \int [p_1(\mathbf{z})]^s [p_2(\mathbf{z})]^{1-s}\, d\mathbf{z}$$

for any s, $0 \leq s \leq 1$.

b. Assume that $\mathbf{z}$ is a scalar and z_j is $N(m_j, \mathbf{\Gamma}_j)$, $j = 1, 2$. Evaluate

$$\int [p_1(z)]^s [p_2(z)]^{1-s}\, dz$$

9.4. Derive (9.3.17).

9.5. Consider

$$\mathbf{z}_j \text{ is } N(\mathbf{m}_j, \mathbf{\Gamma}_j)$$

$$\mathbf{m}_1 = \begin{pmatrix}1\\0\end{pmatrix} \qquad \mathbf{m}_2 = \begin{pmatrix}0\\1\end{pmatrix}$$

$$\mathbf{\Gamma}_1 = \begin{pmatrix}\alpha & 0\\0 & 1\end{pmatrix} \qquad \mathbf{\Gamma}_2 = \begin{pmatrix}\alpha & 0\\0 & 2\end{pmatrix}$$

Evaluate the divergence J and the Bhattacharyya distance B. What happens to these distances as $\alpha \longrightarrow 0$? Does this make sense? *Comment:* Consider the models with $\alpha = 0$. Given an observation, ask yourself how you would decide which model applies.

9.6. Consider the case where there is only one hypothesis:

$$\mathcal{H}_1\colon \quad z = z_1$$

$$z_1\colon \text{random variable with probability density } p_1(z)$$

$$p_1(z) = \begin{cases} e^{-z}, & z \geq 0 \\ 0, & z < 0 \end{cases}$$

It is desired to use the observation to decide if the hypothesis is valid. Specify a decision function and decision level so that $\alpha = 0.1$, where

$$\alpha\colon \text{probability that } \mathcal{H}_1 \text{ is rejected when } \mathcal{H}_1 \text{ is actually true}$$

9.7. Consider a model with no a priori probabilities:

$$\mathcal{H}_j\colon \mathbf{z} = \mathbf{H}_j x + \mathbf{v}, \qquad j = 1, 2$$

$$\mathbf{H}_1 = \begin{pmatrix}1\\0\end{pmatrix} \qquad \mathbf{H}_2 = \begin{pmatrix}0\\1\end{pmatrix} \qquad \mathbf{z} = \begin{pmatrix}z_1\\z_2\end{pmatrix}$$

$$\mathbf{v} \text{ is } N(\mathbf{0}, \mathbf{I})$$

Consider the following two cases (models for x):
a. x is $N(0, 100)$; x, $\mathbf{v}$ independent (Bayesian).
b. x is completely unknown (Fisher).

Give explicit decision functions and rules for these two cases. Show that both cases yield "almost" the same decisions.

9.8. Consider

$$\mathcal{H}_j: \quad \mathbf{z} = \mathbf{H}_j x + \mathbf{v}, \qquad j = 1, 2$$

$$x: \text{completely unknown}$$

$$\mathbf{H}_1 = \begin{pmatrix} 1 \\ 0 \end{pmatrix} \quad \mathbf{H}_2 = \begin{pmatrix} 0 \\ 1 \end{pmatrix} \quad \mathbf{z} = \begin{pmatrix} z_1 \\ z_2 \end{pmatrix}$$

a. Assume that $\mathbf{v}$ is $N(\mathbf{0}, \mathbf{I})$ and that there are no a priori probabilities on which the hypothesis is true. Consider the two-dimensional space with coordinates z_1 and z_2. Divide this space into two regions so that

If $\mathbf{z}$ is in region 1, $\mathcal{H}_1$ is chosen
If $\mathbf{z}$ is in region 2, $\mathcal{H}_2$ is chosen

b. Assume that $\mathbf{v} \in \mathbf{\Omega}_v = \{\mathbf{v} : \mathbf{v}'\mathbf{v} \leq 1\}$. Divide the two-dimensional z_1, z_2 space into four regions so that

If z is in region 1, $\mathcal{H}_1$ is true
If z is in region 2, $\mathcal{H}_2$ is true
If z is in region 3, no decision is possible
If z is in region 4, both hypotheses are wrong

9.9. Consider

$$\mathbf{z} = \mathbf{H}\mathbf{x} + \mathbf{v}$$

$$\mathbf{v} \text{ is } N(\mathbf{0}, \mathbf{R})$$

Assume that there are two hypotheses,

$$\mathcal{H}_1: \mathbf{x} \in \mathbf{\Omega}_{x_2}$$

$$\mathcal{H}_2: \mathbf{x} \in \mathbf{\Omega}_{x_2}$$

where the sets $\mathbf{\Omega}_{x_1}$ and $\mathbf{\Omega}_{x_2}$ are disjoint. There is no a priori probability on which hypothesis is true. Develop an explicit decision function (or functions) and decision rule to use $\mathbf{z}$ to make a choice between the hypotheses. *Comment:* There are several "acceptable" answers as well as many "unacceptable" ones.

10

Hypothesis Testing: Linear Dynamic Systems

The basic ideas of hypothesis testing given in Chapter 9 are now extended to the case where $\mathbf{z}(t)$, $0 \leq t \leq T$, or $\mathbf{z}(n)$, $n = 1, \ldots, N$, is observed. New ideas arising from the presence of dynamics are also introduced.

Discrete-time and continuous-time Gaussian models are discussed in Sections 10.1 and 10.2. Weighted-least-squares and unknown-but-bounded models are considered in Sections 10.3 and 10.4. Sequential hypothesis testing is discussed in Section 10.5. Finite memory operation and event detection are discussed in Sections 10.6 and 10.7. Tests based on the whiteness of the feedback residuals are mentioned in Section 10.8. Applications are discussed in Section 10.9.

This chapter does not contain many new ideas. Little hypothesis-testing theory itself is discussed, as all the ideas of Chapter 9 on the forming of hypotheses and decision rules still apply. The major role of this chapter is simply to specify methods of computing decision functions for linear dynamic systems. It turns out that the desired decision functions can be computed by simple modification of the filters of Chapters 6 and 7. Thus the reader should not be worried by the larger number of complicated-appearing equations in this chapter. The equations have simple interpretations, and the complication is mostly bookkeeping. The length of this chapter is due to handling of special cases and worrying about a variety of relatively "minor" details.

10.1 Discrete-Time Gaussian Process

Assume that $\mathbf{z}(n\Delta)$, $n = 1, \ldots N$, is observed and that there are M hypotheses

$$\begin{aligned}&\mathcal{H}_j\colon \mathbf{z}(n\Delta) = \mathbf{z}_j(n), \qquad n = 1, \ldots, N, \\ &\mathbf{z}_j \text{ is discrete-time Gaussian process,} \qquad j = 1, \ldots, M\end{aligned} \tag{10.1.1}$$

Given the observations, a choice on which hypothesis is probably true is made by first processing the actual observations $\mathbf{z}_{\text{actual}}(n)$, $n = 1, \ldots, N$, to get a decision function and then using a decision rule (see Fig. 9.1). By direct analogy with Sections 9.3 and 9.4, the desired decision functions are the log likelihood functions, $\xi_j(N\Delta)$:

$$\xi_j(N\Delta)\text{: log likelihood function for } \mathbf{z}_N$$

$$\mathbf{z}_N = \begin{bmatrix} \mathbf{z}(\Delta) \\ \cdot \\ \cdot \\ \cdot \\ \mathbf{z}(N\Delta) \end{bmatrix}$$

Once $\xi_j(N\Delta)$ has been obtained, the decision rules follow exactly as in Sections 9.3 and 9.4. Therefore the main problem to be discussed here is now to compute $\xi_j(N\Delta)$ in some efficient manner [that does not require the brute force approach of forming $\mathbf{z}_N$ and using an equation from Section 9.3 such as (9.3.8)]. It turns out that

The log likelihood function $\xi_j(N\Delta)$ can be computed recursively essentially by just *squaring* and *integrating* the residuals of the feedback term of the optimum filter of (6.3.1).	(10.1.2)

This is a highly useful result, as it means that all the important concepts are already known. This pleasant state of affairs is now proved and discussed.

10.1.1 Log Likelihood Functions

Define

$$\mathbf{z}_N = \begin{bmatrix} \mathbf{z}(\Delta) \\ \cdot \\ \cdot \\ \cdot \\ \mathbf{z}(N\Delta) \end{bmatrix}$$

so that the hypotheses (10.1.1) can be rewritten as

$$\mathcal{H}_j\colon \mathbf{z}_N = \mathbf{z}_{N,j}, \qquad \mathbf{z}_{N,j} \text{ is a random vector with probability density } p_j(\mathbf{z}_N),\ j = 1, \ldots, M$$

The log likelihood function $\xi_j(N\Delta)$ is defined as

$$\xi_j(N\Delta) = \ln\{p_j(\mathbf{z}_N)\}$$

By Bayes' rule (see Appendix D.6)

$$p_j(\mathbf{z}_N) = p_j(\mathbf{z}_{N-1}, \mathbf{z}(N\Delta)) = p_j(\mathbf{z}_{N-1})p_j(\mathbf{z}(N\Delta) \mid \mathbf{z}_{N-1})$$

where $p_j(\mathbf{z}(N\Delta) \mid \mathbf{z}_{N-1})$ is the conditional probability density of $\mathbf{z}(N\Delta)$ given $\mathbf{z}_{N-1}$, assuming that $\mathcal{H}_j$ is true. Thus

$$\boxed{\xi_j(N\Delta) = \xi_j(N\Delta - \Delta) + \ln\{p_j(\mathbf{z}(N\Delta) \mid \mathbf{z}_{N-1})\}, \qquad \xi_j(0) = 0} \tag{10.1.3}$$

The simple steps leading to (10.1.3) actually constitute the heart of the desired derivation. To get a more useful form, more symbols are introduced and manipulated and the Gaussian assumption is employed but (10.1.3) is the crucial concept. Note that (10.1.3) is valid without a Gaussian assumption. General formulas for $p(\mathbf{z}(N\Delta) \mid \mathbf{z}_{N-1})$ will be derived in Chapter 13; see, for example (13.2.8).

Define

$$\begin{aligned}
\hat{\mathbf{z}}_j(N\Delta \mid N\Delta - \Delta) &= E\{\mathbf{z}(N\Delta) \mid \mathbf{z}_{N-1}\}: \qquad \mathcal{H}_j \text{ is true} \\
\boldsymbol{\delta}_j(N\Delta) &= \mathbf{z}(N\Delta) - \hat{\mathbf{z}}_j(N\Delta \mid N\Delta - \Delta) \\
\boldsymbol{\Sigma}_{z,j}(N\Delta \mid N\Delta - \Delta) &= E\{\boldsymbol{\delta}_j(N\Delta)\boldsymbol{\delta}'_j(N\Delta)\}
\end{aligned}$$

where the subscript j means $\hat{\mathbf{z}}_j(N\Delta \mid N\Delta - \Delta)$ is the conditional expectation of $\mathbf{z}(N\Delta)$ given $\mathbf{z}_{N-1}$ assuming that hypothesis $\mathcal{H}_j$ is true, etc. Using the Gaussian assumption, (see Appendix E),

$$\begin{aligned}
p_j(\mathbf{z}(N\Delta) \mid \mathbf{z}_{N-1}) = {} & [(2\pi)^K \mid \boldsymbol{\Sigma}_{z,j}(N\Delta \mid N\Delta - \Delta)|]^{-1/2} \\
& \times \exp\{-\tfrac{1}{2}[\boldsymbol{\delta}'_j(N\Delta)\boldsymbol{\Sigma}_{z,j}^{-1}(N\Delta \mid N\Delta - \Delta)\boldsymbol{\delta}_j(N\Delta)]\}
\end{aligned}$$

where K is the dimension of $\mathbf{z}(N\Delta)$. Thus substituting, into (10.1.3) gives,

$$\begin{aligned}
2\xi_j(N\Delta) = {} & 2\xi(N\Delta - \Delta) - \ln\{|\boldsymbol{\Sigma}_{z,j}(N\Delta \mid N\Delta - \Delta)|\} \\
& - K\ln(2\pi) - \boldsymbol{\delta}'_j(N\Delta)\boldsymbol{\Sigma}_{z,j}^{-1}(N\Delta \mid N\Delta - \Delta)\boldsymbol{\delta}_j(N\Delta) \qquad (10.1.4) \\
\xi_j(0) = {} & 0
\end{aligned}$$

It is often convenient to view the hypothesis (10.1.1) as

$$\begin{aligned}
\mathcal{H}_j\colon \mathbf{z}(n\Delta) = \mathbf{y}_j(n\Delta) + \mathbf{v}_j(n\Delta), \qquad & n = 1, \ldots, N \\
& j = 1, \ldots, M
\end{aligned}$$

$\mathbf{y}_j$: Gaussian process

$\mathbf{v}_j(n\Delta)$: zero mean white stochastic Gaussian process (10.1.5)

$$E\{\mathbf{v}(n_1\Delta)\mathbf{v}'(n_2\Delta)\} = \begin{cases} \Delta^{-1}\mathbf{R}_j(n_1\Delta), & n_1 = n_2 \\ \mathbf{0}, & n_1 \neq n_2 \end{cases}$$

$\mathbf{y}_j, \mathbf{v}_j$: independent

This model is as general as before, as one can set $\mathbf{R}_j = \mathbf{0}$ if desired. For the model (10.1.5), the ξ_j of (10.1.4) become

$$
\boxed{
\begin{aligned}
2\xi_j(N\Delta) &= \xi_{j,\text{bias}}(N\Delta) + \xi_{j,\text{observation}}(N\Delta) \\
\xi_{j,\text{bias}}(N\Delta) &= -\sum_{n=1}^{N} \ln\{|\boldsymbol{\Sigma}_{y,j}(n\Delta \mid n\Delta - \Delta) + \Delta^{-1}\mathbf{R}_j(n\Delta)|\} \\
&\quad - NK\ln\{2\pi\} \\
\xi_{j,\text{observation}}(N\Delta) &= -\sum_{n=1}^{N} \boldsymbol{\delta}'_j(n\Delta)[\boldsymbol{\Sigma}_{y,j}(n\Delta \mid n\Delta - \Delta) \\
&\quad + \Delta^{-1}\mathbf{R}_j(n\Delta)]^{-1}\boldsymbol{\delta}_j(n\Delta) \\
\hat{\mathbf{y}}_j(n\Delta \mid n\Delta - \Delta) &= E\{\mathbf{y}_j(n\Delta) \mid \mathbf{z}_{N-1}\}: \qquad \mathcal{H}_j \text{ is true} \\
\boldsymbol{\Sigma}_{y,j}(N\Delta \mid N\Delta - \Delta) &= E\{[\mathbf{y}_j(N\Delta) - \hat{\mathbf{y}}_j(N\Delta \mid N\Delta - \Delta)] \\
&\quad \times [\mathbf{y}_j(n\Delta) - \hat{\mathbf{y}}_j(N\Delta \mid Nn - \Delta)]'\} \\
\boldsymbol{\delta}_j(N\Delta) &= \mathbf{z}(N\Delta) - \hat{\mathbf{y}}_j(N\Delta \mid N\Delta - \Delta)
\end{aligned}
}
\tag{10.1.6}
$$

The bias component $\xi_{j,\text{bias}}(N\Delta)$ is deterministic and can be calculated in advance. In many applications it can be considered to be part of the decision rule. In some applications it can be ignored entirely and only the observation-dependent component $\xi_{j,\text{observation}}(N\Delta)$ need be considered. Equation (10.1.6) is illustrated in Fig. 10.1.1.

The $\mathbf{y}_j(n\Delta)$ of (10.1.5) can be an arbitrary Gaussian process. However, specialization to state space models enables the use of the theory of Chapter 6 to actually implement a filter to get the conditional expectation $\hat{\mathbf{y}}_j(n\Delta \mid n\Delta - \Delta)$. To illustrate, consider (10.1.5), where

$$
\begin{aligned}
\mathbf{y}_j(n) &= \tilde{\mathbf{y}}_j(n) + \mathbf{m}_j(n) \\
E\{\tilde{\mathbf{y}}_j(n)\} &= \mathbf{0} \\
\tilde{\mathbf{y}}_j(n) &= \mathbf{H}_j(n)\mathbf{x}_j(n) \\
\mathbf{x}_j(n+1) &= \boldsymbol{\Phi}_j(n)\mathbf{x}_j(n) + \mathbf{G}_j(n)\mathbf{w}_j(n) \\
E\{\mathbf{w}_j(n_1)\mathbf{w}'_j(n_2)\} &= \begin{cases} \mathbf{Q}_j(n_1), & n_1 = n_2 \\ \mathbf{0}, & n_1 \neq n_2 \end{cases}
\end{aligned}
$$

$\mathbf{m}_j(n)$: mean value of $\mathbf{y}_j(n)$

$E\{\mathbf{x}(0)\mathbf{x}'(0)\} = \boldsymbol{\psi}$

$\mathbf{x}(0)\ \mathbf{w}_j(n)$: zero mean, independent, Gaussian

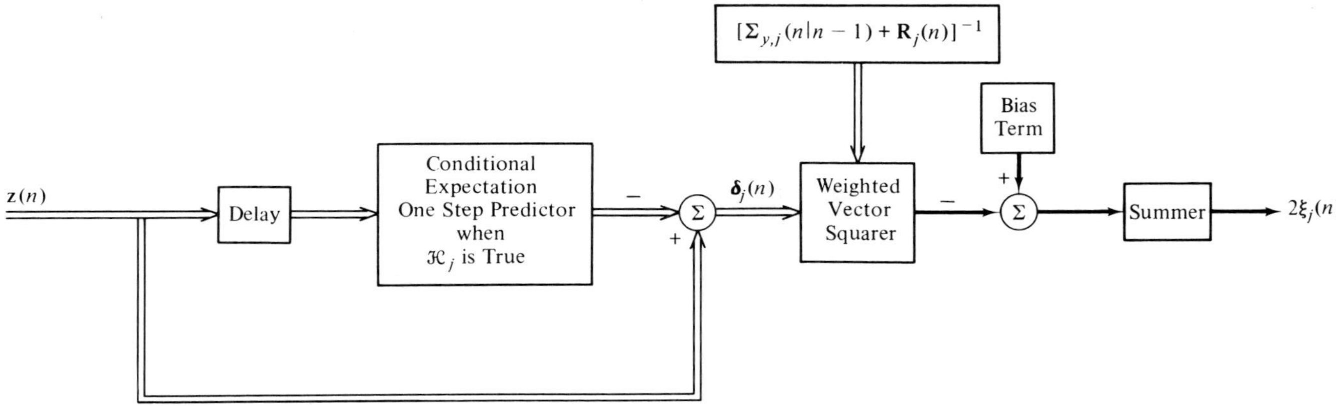

Figure 10.1.1 Log likelihood function: discrete time, Gaussian.

Then $\xi_j(N)$ of (10.1.6) is given by ($\Delta = 1$)

$$
\begin{aligned}
\xi_{j,\text{observation}}(N) &= -\sum_{n=1}^{N} \boldsymbol{\delta}_j'(n)[\mathbf{H}_j(n)\boldsymbol{\Sigma}_j(n|n-1) \\
&\qquad \times \mathbf{H}_j'(n) + \mathbf{R}_j(n)]^{-1}\boldsymbol{\delta}_j(n) \\
\boldsymbol{\delta}_j(n+1) &= \mathbf{z}(n+1) - \mathbf{H}_j(n+1) \\
&\qquad \times \boldsymbol{\Phi}_j(n)\hat{\mathbf{x}}_j(n|n) - \mathbf{m}_j(n+1) \\
\xi_{j,\text{bias}}(N) &= -NK\ln(2\pi) \\
&\qquad - \sum_{n=1}^{N} \ln|\mathbf{H}_j(n)\boldsymbol{\Sigma}_j(n|n-1)\mathbf{H}_j'(n) + \mathbf{R}_j(n)|
\end{aligned}
\tag{10.1.7}
$$

$\hat{\mathbf{x}}_j(n|n)$, $\boldsymbol{\Sigma}_j(n+1|n)$: given by (6.3.1) with modiffcation for $\mathbf{m}_j(n)$

This system is illustrated in block diagram form in Fig. 10.1.2. Comparison of Fig. 10.1.2 with Fig. 6.3.3 shows that only the "minor" addition of the "squarer" and "summer" are needed to obtain the observation-dependent component of the log likelihood function.

EXAMPLE: Assume that you have to model some scalar, stationary, zero mean process $z(n)$, $n = 1, \ldots$. Assume that after much thought and soul searching, you decide that a good model is

$$
\begin{aligned}
x(n+1) &= (0.5)x(n) + w(n) \\
z(n) &= x(n) + v(n) \\
E\{v^2(n)\} &= 4 \\
E\{w^2(n)\} &= 1 \\
E\{x^2(n)\} &= \tfrac{4}{3}, \qquad \text{all } n
\end{aligned}
$$

$x(0)$, $v(n)$, $w(n)$: white, zero mean, independent, Gaussian

Now suppose that you are lucky enough to have some actual observations $z_{\text{actual}}(n)$, $n = 1, \ldots, N$. Unless you have complete faith in your modeling ability, you will want to verify your model by checking the hypothesis:

$\mathcal{H}_1$: model is valid
$\mathcal{H}_2$: model is not valid

Thus you could evaluate $\xi_1(N)$, the log likelihood function for $\mathcal{H}_1$. From (10.1.6) and (10.1.7),

$$
\begin{aligned}
\xi_1(N) = &-\sum_{n=1}^{N} \ln|\Sigma(n|n-1) + 4| - N\ln(2\pi) \\
&-\sum_{n=1}^{N} \frac{[z(n) - (0.5)\hat{x}(n-1|n-1)]^2}{\Sigma(n|n-1) + 4}
\end{aligned}
$$

where from (6.3.1),

$$
\begin{aligned}
\hat{x}(n+1|n+1) &= (0.5)\hat{x}(n|n) + K(n+1)[z(n+1) - (0.5)\hat{x}(n|n)] \\
\Sigma(n|n) &= \Sigma(n|n-1) - \frac{\Sigma^2(n|n-1)}{4 + \Sigma(n|n-1)}
\end{aligned}
$$

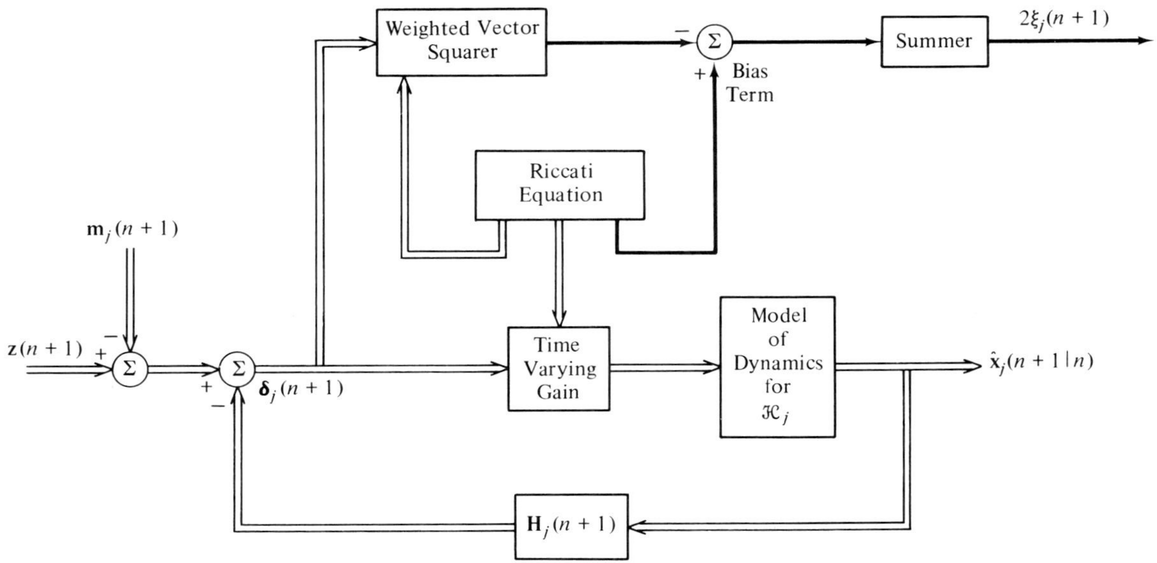

Figure 10.1.2 Log likelihood function: discrete time, Gaussian, driven dynamic system model for hypotheses $\mathcal{H}_j$.

$$\Sigma(n|n-1) = (0.25)\Sigma(n-1|n-1) + 1$$

$$K(n) = \frac{\Sigma(n|n)}{4}$$

As N increases,

$$\Sigma(N|N) \longrightarrow \Sigma_\infty$$

Thus if N is large, the *transient* in $\Sigma(n|n)$ can be ignored, and an approximate $\xi_1(N)$ is given by

$$\xi_1(N) = -N\ln[2\pi(\Sigma_\infty + 4)] - \sum_{n=1}^{N} \frac{[z(n) - (0.5)\hat{x}(n-1|n-1)]^2}{(0.25)\Sigma_\infty + 5}$$

$$\hat{x}(n+1|n+1) = 0.5\hat{x}(n|n) + \frac{\Sigma_\infty}{4}[z(n+1) - 0.5\hat{x}(n|n)]$$

If the model is valid, then $\xi_1(N)$ is Gaussian for large N with specified mean and variances (see Section 10.1.3). Hence the model's validity can be checked by comparing the $\xi_1(N)$ computed for the actual observations $z_{\text{actual}}(1) \cdots z_{\text{actual}}(N)$ with the Gaussian distribution to see if they are consistent. See Section 9.4.2.

For $M = 2$, it is sometimes more convenient to work with the log likelihood ratio defined by

$$\lambda(N\Delta) = \lambda(\mathbf{z}_N)$$

$$\lambda(N\Delta) = \ln\left\{\frac{p_1(\mathbf{z}_N)}{p_2(\mathbf{z}_N)}\right\}.$$

$$\lambda(N\Delta) = \xi_1(N\Delta) - \xi_2(N\Delta)$$

It follows directly that for the model (10.1.5)

$$\begin{aligned}
2\lambda(N\Delta) &= \lambda_{\text{bias}}(N\Delta) + \lambda_{\text{observation}}(N\Delta) \\
\lambda_{\text{bias}}(N\Delta) &= -\sum_{n=1}^{N} \ln\left\{\frac{|\mathbf{\Sigma}_{y,1}(n\Delta|n\Delta - \Delta) + \Delta^{-1}\mathbf{R}_1(n\Delta)|}{|\mathbf{\Sigma}_{y,2}(n\Delta|n\Delta - \Delta) + \Delta^{-1}\mathbf{R}_2(n\Delta)|}\right\} \\
\lambda_{\text{observation}}(N\Delta) &= -\sum_{n=1}^{N} \boldsymbol{\delta}_1'(n\Delta)[\mathbf{\Sigma}_{y,1}(n\Delta|n\Delta - \Delta) \\
&\quad + \Delta^{-1}\mathbf{R}_1(n\Delta)]^{-1}\boldsymbol{\delta}_1(n\Delta) \\
&\quad + \sum_{n=1}^{N} \boldsymbol{\delta}_2'(n\Delta)[\mathbf{\Sigma}_{y,2}(n\Delta|n\Delta - \Delta) \\
&\quad + \Delta^{-1}\mathbf{R}_2(n\Delta)]^{-1}\boldsymbol{\delta}_2(n\Delta)
\end{aligned} \tag{10.1.8}$$

10.1.2 Equal Correlation Functions

An important special case occurs when the $\mathbf{y}_j$ of (10.1.5) have equal correlation functions and the hypotheses differ only in the mean values. This leads to the *matched filter* concept of communication theory.

Consider

$$
\begin{aligned}
&\mathcal{H}_j\colon\ \mathbf{z}(n\Delta) = \mathbf{y}_j(n\Delta) + \mathbf{v}(n\Delta), \qquad j = 1, \ldots, M \\
&\qquad \mathbf{y}_j(n\Delta) = \mathbf{m}_j(n\Delta) + \tilde{\mathbf{y}}(n\Delta) \\
&\mathbf{m}_j(n\Delta)\colon \text{deterministic (mean value)}, \qquad j = 1, \ldots, M \\
&\tilde{\mathbf{y}}(n\Delta)\colon \text{ zero mean Gaussian process} \\
&\mathbf{v}(n\Delta)\colon \text{ zero mean white Gaussian process} \\
&\qquad E\{\mathbf{v}(n_1\Delta)\mathbf{v}'(n_2\Delta)\} = \begin{cases} \Delta^{-1}\mathbf{R}(n\Delta), & n_1 = n_2 \\ 0, & n_1 \neq n_2 \end{cases} \\
&\tilde{\mathbf{y}},\ \mathbf{v}\colon \text{independent}
\end{aligned}
\tag{10.1.9}
$$

Thus the hypotheses differ only with respect to the deterministic process, i.e., their mean values. The zero mean Gaussian processes, $\tilde{\mathbf{y}}(n\Delta) + \mathbf{v}(n\Delta)$, are the same for all hypotheses.

Define the matrix $\mathfrak{F}_n[\,]$ as the linear operation on $\tilde{\mathbf{y}}(m\Delta) + \mathbf{v}(m\Delta)$, $m = 1, \ldots, n-1$, which yields $\hat{\tilde{\mathbf{y}}}(n\Delta \mid n\Delta - \Delta)$ so that

$$
\begin{aligned}
\hat{\mathbf{y}}_j(n\Delta \mid n\Delta - \Delta) &= \mathbf{m}_j(n\Delta) + \mathfrak{F}_n[\mathbf{z}(m\Delta) - \mathbf{m}_j(m\Delta),\ m = 1, \ldots, n-1] \\
\boldsymbol{\delta}_j(n\Delta) &= \mathbf{z}(n\Delta) - \mathfrak{F}_n[\mathbf{z}(m\Delta),\ m = 1, \ldots, n-1] \\
&\quad - \{\mathbf{m}_j(n\Delta) - \mathfrak{F}_n[\mathbf{m}_j(m\Delta),\ m = 1, \ldots, n-1]\}
\end{aligned}
$$

Note that $\mathfrak{F}_N$ is independent of which hypothesis is true. Then manipulation of (10.1.6) yields for $\mathcal{H}_j$ of (10.1.9)

$$
\boxed{
\begin{aligned}
&2\xi_j(N\Delta) = 2\tilde{\xi}_j(N\Delta) + \gamma(N\Delta) \\
&\tilde{\xi}_j(N\Delta) = \sum_{n=1}^{N} [\mathbf{z}(n\Delta) - \mathfrak{F}_n[\mathbf{z}(m\Delta),\ m = 1, \ldots, n-1]]' \\
&\qquad \times [\boldsymbol{\Sigma}_y(n\Delta \mid n\Delta - \Delta) + \Delta^{-1}\mathbf{R}(n\Delta)]^{-1}\mathbf{d}_j(n\Delta) \\
&\mathbf{d}_j(n\Delta) = \mathbf{m}_j(n\Delta) - \mathfrak{F}_n[\mathbf{m}_j(m\Delta),\ m = 1, \ldots, n-1] \\
&\gamma(n\Delta)\colon \text{terms involving } \mathbf{z}(n\Delta) + \text{terms involving } \mathbf{m}_j(n\Delta)
\end{aligned}
}
\tag{10.1.10}
$$

Now $\gamma(n\Delta)$ does not contain any terms involving $\mathbf{z}(n\Delta)$ that depend on the hypothesis, i.e., on the $\mathbf{m}_j(n\Delta)$. Hence insofar as hypothesis testing is concerned, $\gamma(n\Delta)$ can be ignored and only the modified likelihood function $\tilde{\xi}_j(N\Delta)$ need be implemented. Generation of $\tilde{\xi}_j(N\Delta)$ of (10.1.10) is illustrated in Fig. 10.1.3. Figures 10.1.1 and 10.1.3 are different systems, but, insofar as decision making is concerned, they are equivalent if (10.1.9) holds in the sense that by choosing appropriate decision rules the same performance can be obtained. Remember that explicit equations for $\mathfrak{F}_n$ can be obtained by using dynamic system models and using the equations of Chapter 6.

The special case where $\tilde{\mathbf{y}}(n\Delta) = \mathbf{0}$ yields

$$
\tilde{\xi}_j(N\Delta) = \sum_{n=1}^{N} \Delta\mathbf{z}'(n\Delta)\mathbf{R}^{-1}(n\Delta)\mathbf{m}_j(n\Delta) \tag{10.1.11}
$$

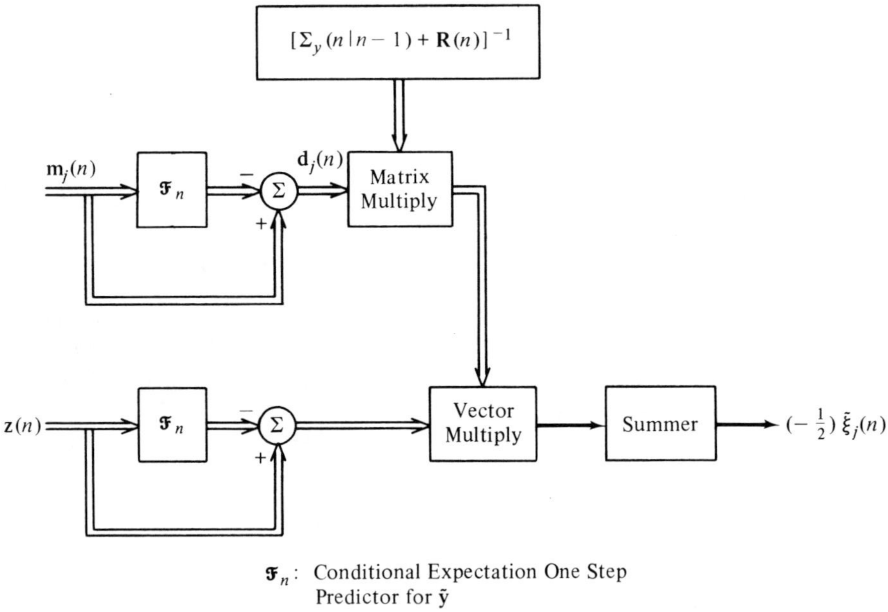

Figure 10.1.3 Modified log likelihood function: equal covariance functions, discrete time, Gaussian.

It is (10.1.11) that motivates the use of the term *matched filter*. (See the corresponding continuous-time result in Section 10.2.2.)

These results were derived as a special case of the general results of Section 10.1.1. However, they can also be derived by starting with the corresponding model of Section 9.3.3. which involved equal covariance matrices.

EXAMPLE: Assume that you observe

$$z(n) = \sin(\omega n) + v(n), \qquad n = 1, \ldots, N$$

$$E\{v^2(n)\} = R$$

$$v(n): \text{ zero mean white Gaussian}$$

Assume that you want to decide whether the value of ω is $\omega_1 = 1, \omega_2 = 2$, or $\omega_3 = 3$. You could do this by computing

$$\tilde{\xi}_j(N) = \sum_{n=1}^{N} z(n)\sin(\omega_j n), \qquad j = 1, 2, 3$$

and choosing the value of ω_j which yields the largest $\xi_j(N)$.

10.1.3 Probability Density of $\xi_j(N\Delta)$

The probability density of $\xi_j(N\Delta)$ of (10.1.6) can be calculated when $\mathcal{H}_j$ is true. In the development of the Bayesian theory of Section 6.3, it was shown that

$$\boldsymbol{\delta}_j(n\Delta) = \mathbf{z}(n\Delta) - \hat{\mathbf{z}}_j(n\Delta \mid \mathbf{z}_{n-1}) = \mathbf{z}(n\Delta) - \hat{\mathbf{y}}_j(n\Delta \mid \mathbf{z}_{n-1})$$

is a zero mean stochastic white process with

$$E\{\boldsymbol{\delta}_j(n_1\Delta)\boldsymbol{\delta}_j'(n_2\Delta)\} = \begin{cases} \boldsymbol{\Sigma}_{y_j}(n_1\Delta \mid n_1\Delta - \Delta) + \Delta^{-1}\mathbf{R}_j(n_1\Delta), & n_1 = n_2 \\ \mathbf{0}, & n_1 \neq n_2 \end{cases}$$

if $\mathcal{H}_j$ is true. Thus using the ideas of Section (9.3.4), it follows that if **z** is a K-dimensional vector in (10.1.6)

$$\boxed{\begin{array}{l} \text{When } \mathcal{H}_j \text{ is true,} \\ \quad -\xi_{j,\text{observation}}(N\Delta) \text{ is a chi-squared random variable} \\ \quad \qquad \text{with } NK \text{ degrees of freedom} \\ \text{For large values of } NK, \\ \quad -\xi_{j,\text{observation}}(N\Delta) \text{ is approximately } N(NK, 2NK) \end{array}} \tag{10.1.12}$$

A better feel for (10.1.12) may be obtained by noting that for large NK

$$-\left(\frac{1}{NK}\right)\xi_{j,\text{observation}}(N\Delta) \text{ is approximately } N[1, 2(KN)^{-1}]$$

Thus as the number of observations N increases, the variable $-(1/NK)\xi_{j,\text{observation}}(N\Delta)$ tends to converge to the constant value 1 if $\mathcal{H}_j$ is true.

EXAMPLE: Consider the example of Section 10.1.1. If the model actually corresponds to the observations, then for large N

$$\frac{1}{N}\sum_{n=1}^{N}\frac{[z(n) - (0.5)\hat{\mathbf{x}}(n-1 \mid n-1)]^2}{0.25\boldsymbol{\Sigma}_\infty + 5} \text{ becomes } N[1, 2/N]$$

In the case of equal correlation functions, the modified log likelihood function $\tilde{\xi}_j(N\Delta)$ of (10.1.10) is Gaussian no matter which hypothesis is true. After a lot of manipulation it follows that

$$\boxed{\begin{array}{l} \text{When } \mathcal{H}_k \text{ is true} \\ \tilde{\xi}_j(N\Delta) = N(\mu_{kj}(N\Delta), \mu_{jj}(N\Delta)) \\ \mu_{kj}(N\Delta) = \displaystyle\sum_{n=1}^{N} \mathbf{d}_k(n)[\boldsymbol{\Sigma}_y(n\Delta \mid n\Delta - \Delta) + \Delta^{-1}\mathbf{R}(n\Delta)]^{-1}\mathbf{d}_j(n) \end{array}} \tag{10.1.13}$$

Hence decision rules and the probability of various types of error can easily be developed.

These results are an extension of the ideas of Section 9.3.4.

10.1.4 Distance Measures†

Manipulations like those of Section 10.1.1 yield difference equations for the two distance measures introduced in Section 9.3.5.

Consider the hypothesis (10.1.1). Define

$J(N\Delta)$: divergence between $z_1(n\Delta)$ and $z_2(n\Delta)$,
$$n = 1, \ldots, N$$
$\mathbf{B}(N\Delta)$: Bhattacharyya distance between $z_1(n\Delta)$ and $z_2(n\Delta)$,
$$n = 1, \ldots, N$$

To simplify the notation, only scalar $z(n\Delta)$ are considered. Generalization to the case of vector $\mathbf{z}$ is straightforward.

To express the results in a nice form, a lot of definitions are needed. Define

$$\mathbf{z}_{N,j} = \begin{bmatrix} z_j(\Delta) \\ \cdot \\ \cdot \\ \cdot \\ z_j(N\Delta) \end{bmatrix} \qquad \mathbf{m}_{N,j} = \begin{bmatrix} m_j(\Delta) \\ \cdot \\ \cdot \\ \cdot \\ m_j(N\Delta) \end{bmatrix}$$
$$m_j(n\Delta) = E\{z_j(n\Delta)\}$$

Define matrices (linear operators) $\mathfrak{F}_{N,1}$, $\mathfrak{F}_{N,2}$, and $\mathfrak{F}_{N,1+2}$ to be such that

$$E\{z_j(N\Delta) \mid \mathbf{z}_{N-1,j}\} = \mathfrak{F}_{N,j}[\mathbf{z}_{N-1,j} - \mathbf{m}_{N-1,j}] + m_j(N\Delta), \qquad j = 1, 2$$

$$E\{z_j(N) + z_2(N) \mid \mathbf{z}_{N-1,1} + \mathbf{z}_{N-1,2}\} = \mathfrak{F}_{N,1+2}[\mathbf{z}_{N-1,1} + \mathbf{z}_{N-1,2} - \mathbf{m}_{N-1,1} - \mathbf{m}_{N-1,2}] + m_1(N\Delta) + m_2(N\Delta)$$

Thus $\mathfrak{F}_{N,j}$ can be viewed as the linear system which operates on the zero mean component of $z(n\Delta)$, $n = 1, \ldots, N-1$, to give the zero mean component of the conditional expectation of $z(N\Delta)$ for hypotheses $\mathcal{H}_j$, $j = 1, 2$. $\mathfrak{F}_{N,1+2}$ has a similar interpretation for the process $z_{1+2}(n\Delta) = z_1(n\Delta) + z_2(n\Delta)$, z_1 and z_2 independent. Define

$\Sigma_{z,1|2}(N\Delta \mid N\Delta - \Delta)$: one-step prediction error variance when $z_1(N\Delta)$ is predicted under the assumption that $\mathcal{H}_2$ is true

$$\Sigma_{z,1|2}(N\Delta \mid N\Delta - \Delta) = E\{[z_1(N\Delta) - m_1(N\Delta) - \mathfrak{F}_{N,2}(\mathbf{z}_{N-1,1} - \mathbf{m}_{N-1,1})]^2\}$$

$\Sigma_{z,2}(N\Delta \mid N\Delta - \Delta)$: one-step prediction error variance when $z_2(N\Delta)$ is predicted from $z_2(n\Delta)$, $\quad n = 1, \ldots, N-1$

$d_2(N\Delta \mid N\Delta - \Delta)$: one-step prediction error mean value when $m_1(n\Delta) - m_2(n\Delta)$ is "passed through" $\mathfrak{F}_{N,2}$

$$d_2(N\Delta \mid N\Delta - \Delta) = m_1(N\Delta) - m_2(N\Delta) - \mathfrak{F}_{N,2}[\mathbf{m}_{N-1,1} - \mathbf{m}_{N-1,2}]$$

with similar definitions for $\Sigma_{z,2|1}(N\Delta \mid N\Delta - \Delta)$, $\Sigma_{z,1}(N\Delta \mid N\Delta - \Delta)$, and

$d_1(N\Delta\,|\,N\Delta - \Delta)$. Define

$\Sigma_{z,1+2}(N\Delta\,|\,N\Delta - \Delta)$: one-step prediction error variance when $z_1(N\Delta) + z_2(N\Delta)$ is predicted from $z_1(n\Delta) + z_2(n\Delta)$, $n = 1, \ldots, N-1$

$d_{1+2}(N\Delta\,|\,N\Delta - \Delta)$: one-step prediction error mean value when $m_1(n\Delta) - m_2(n\Delta)$ is passed through $\mathfrak{F}_{N,1+2}$

$$d_{1+2}(N\Delta\,|\,N\Delta - \Delta) = m_1(N\Delta) - m_2(N\Delta) - \mathfrak{F}_{N,1+2}[\mathbf{m}_{N-1,1} - \mathbf{m}_{N-1,2}]$$

Then it can be shown that for (10.1.1) the divergence and Bhattacharyya distance are given by

$$\begin{aligned} 2[J(N\Delta) - J(N\Delta - \Delta)] &= -2 \\ &\quad + \frac{\Sigma_{z,1|2}(N\Delta\,|\,N\Delta - \Delta) + d_2^2(N\Delta)}{\Sigma_{z,2}(N\Delta\,|\,N\Delta - \Delta)} \\ &\quad + \frac{\Sigma_{z,2|1}(N\Delta\,|\,N\Delta - \Delta) + d_1^2(N\Delta)}{\Sigma_{z,1}(N\Delta\,|\,N\Delta - \Delta)} \\ J(0) &= 0 \end{aligned} \tag{10.1.14}$$

$$\begin{aligned} 4[B(N\Delta) - B(N\Delta - \Delta] &= 2\ln\{(\tfrac{1}{2})\Sigma_{z,1+2}(N\Delta\,|\,N\Delta - \Delta)\} \\ &\quad - \ln\{\Sigma_{z,1}(N\Delta\,|\,N\Delta - \Delta)\} \\ &\quad - \ln\{\Sigma_{z,2}(N\Delta\,|\,N\Delta - \Delta)\} \\ &\quad + \frac{d_{1+2}^2(N\Delta)}{\Sigma_{z,1+2}(N\Delta\,|\,N\Delta - \Delta)} \\ B(0) &= 0 \end{aligned} \tag{10.1.15}$$

These results are valid for arbitrary processes, but explicit equations for the Σ_z and d can be obtained by using dynamic system models and the results of Chapter 6.

Consider the special case

$$\begin{aligned} \mathcal{H}_j\colon z(n\Delta) &= z_j(n\Delta) \\ z_j(n\Delta) &= \tilde{y}(n\Delta) + m_j(n\Delta) + v(n\Delta), \qquad j = 1, 2 \end{aligned}$$

so that the $z(n\Delta)$ differ only in mean value. Then $\mathfrak{F}_{N,1+2} = \mathfrak{F}_{N,1} = \mathfrak{F}_{N,2} \triangleq \mathfrak{F}_N$ and

$$\begin{aligned} \frac{8}{\Delta}[B(N\Delta) - B(N\Delta - \Delta)] &= \frac{1}{\Delta}[J(N\Delta - J(N\Delta - \Delta)] \\ &= \frac{d^2(N\Delta)}{\Delta\Sigma_y(N\Delta\,|\,N\Delta - \Delta) + R(N\Delta)} \\ d(N\Delta) &= m_1(N\Delta) - m_2(N\Delta) - \mathfrak{F}_N[\mathbf{m}_{N-1,1} - \mathbf{m}_{N-1,2}] \\ \Sigma_y(N\Delta\,|\,N\Delta - \Delta) &= E\{\tilde{y}(N\Delta) - \mathfrak{F}_N\tilde{\mathbf{z}}_{N-1}\}^2 \end{aligned}$$

where $\tilde{\mathbf{z}}_{N-1}$ is the zero mean component of $\mathbf{z}_{N-1}$. In the very special case where the $y_j(n\Delta)$ are deterministic $[\tilde{y}(n\Delta) = 0]$,

$$\frac{8}{\Delta}[B(N\Delta) - B(N\Delta - \Delta)] = \frac{1}{\Delta}[J(N\Delta) - J(N\Delta - \Delta)]$$

$$= \frac{[m_1(N\Delta) - m_2(N\Delta)]^2}{R(n\Delta)}$$

The details of the proofs of (10.1.14) and (10.1.15) are straightforward but tedious and dull. The divergence result (10.1.14) follows most directly by substituting (10.1.8) into (9.3.14), although it can be derived directly from (9.3.17). The Bhattacharyya distance result (10.1.15) can be derived from (9.3.17) by using the following basic relationships from Appendix E.2:

$$\begin{aligned}
\mathbf{z}_N &= \begin{bmatrix} z(1) \\ \cdot \\ \cdot \\ \cdot \\ z(N) \end{bmatrix} \qquad E\{\mathbf{z}_N\} = \mathbf{0} \\
\mathbf{\Gamma}_N &= E\{\mathbf{z}_N\mathbf{z}_N'\} \\
\mathbf{\Gamma}_N &= \left[\begin{array}{c|c} \Gamma(N) & \boldsymbol{\gamma}'(N-1) \\ \hline \boldsymbol{\gamma}(N-1) & \mathbf{\Gamma}_{N-1} \end{array}\right] \\
E\{z(N)\,|\,\mathbf{z}_{N-1}\} &= \mathfrak{F}_N\mathbf{z}_{N-1} \\
\mathfrak{F}_N &= \boldsymbol{\gamma}'(N-1)\mathbf{\Gamma}_{N-1}^{-1} \\
|\mathbf{\Gamma}_N| &= \Sigma_z(N\Delta\,|\,N\Delta - \Delta)\,|\mathbf{\Gamma}_{N-1}| \\
\Sigma_z(N\Delta\,|\,N\Delta - \Delta) &= E\{z(N) - \mathfrak{F}_N\mathbf{z}_{N-1})^2\}
\end{aligned} \tag{10.1.16}$$

10.1.5 Discussion of Discrete Time

The discussions have concentrated on formulas for the log likelihood function $\xi_j(N\Delta)$ for each hypothesis. The actual hypothesis testing is to be done using decision rules as discussed in Chapter 9, and there are many ways to proceed. Both the Bayesian and Fisher model concepts of Sections 9.3 and 9.4 can be used.

One Fisher model of Chapter 9 involved the presence of unknown parameters (see Section (9.4.3). This concept also extends to the present case provided that the problem makes sense. Assume that the $\mathbf{y}$ are generated by a dynamic system model with input $\mathbf{w}$. Then decision functions can be obtained for certain types of completely unknown $\mathbf{w}(n\Delta)$ by using the Fisher estimation concepts of Chapter 6.

The basic recursion (10.1.3) is valid in general, but the rest of the development assumed Gaussian processes. When the hypotheses involve non-Gaussian processes, the discussions of Section 9.3.7 apply equally in the present case; i.e., the formulas derived for Gaussian hypotheses are often very effective for non-Gaussian processes.

The discussion of Section 9.3.8 on the design and analysis of nonoptimum decision functions also extends in a straightforward fashion to the present case.

10.2 Continuous-Time Gaussian Processes

Assume that $\mathbf{z}(t)$, $0 \leq t \leq T$, is observed and that there are M hypotheses

$$\mathcal{H}_j\colon \mathbf{z}(t) = \mathbf{z}_j(t), \qquad 0 \leq t \leq T \qquad \mathbf{z}_j(t) \text{ is continuous-time Gaussian process}, \qquad j = 1, \ldots, M$$

The corresponding log likelihood functions are desired to use as decision functions. Application of the discrete to continuous time limit procedure to the result of Section 10.1 leads to the basic result in a simple fashion. Unfortunately the "great god Mathematics" is not kind to us mortal engineers and some "problems" arise as a consequence of the use of continuous-time white stochastic process models. These problems are not fundamental but they lead to a variety of "fine points" which have to be discussed.

The actual continuous-time hypotheses to be considered are

$$\begin{aligned}
\mathcal{H}_j\colon\ & \mathbf{z}(t) = \mathbf{z}_j(t) \\
& \mathbf{z}_j(t) = \mathbf{y}_j(t) + \mathbf{v}(t), \qquad 0 \leq t \leq T \\
& \qquad\qquad\qquad\qquad\qquad j = 1, \ldots, M \\
& \mathbf{y}_j(t)\colon \text{Gaussian process with no white component} \\
& \mathbf{m}_j(t) = E\{\mathbf{y}_j(t)\} \\
& \mathbf{v}(t)\colon \text{zero mean white stochastic Gaussian process} \\
& E\{\mathbf{v}(t)\mathbf{v}'(\tau)\} = \delta(t - \tau)\mathbf{R}(t) \\
& \quad \mathbf{R}^{-1}(t)\colon \text{exists} \\
& \mathbf{v}(t), \mathbf{y}_j(t)\colon \text{independent processes}
\end{aligned} \tag{10.2.1}$$

It is assumed that as $n \longrightarrow \infty$, $\Delta \longrightarrow 0$, $n\Delta \longrightarrow t$, the discrete-time model (10.1.5) becomes the continuous-time model (10.2.1). These continuous-time hypothesis models are not quite as general as those of (10.1.5), as (10.2.1) assumes that the white processes all have the same $\mathbf{R}(t)$ and that $\mathbf{R}^{-1}(t)$ exists. Thus the corresponding discrete-time model has $\mathbf{R}_j(n\Delta) = \mathbf{R}(n\Delta)$, where $\mathbf{R}^{-1}(n\Delta)$ exists.

10.2.1 Log Likelihood Function: Direct Limit

The discrete to continuous time limit is now applied to $\xi_1(N\Delta)$. Equation (10.1.6) expresses $\xi_j(N\Delta)$ as

$$2\xi_j(N\Delta) = \xi_{j,\text{bias}}(N\Delta) + \xi_{j,\text{observation}}(N\Delta)$$

Consider first $\xi_{j,\text{bias}}(N\Delta)$. Using (10.1.6), $\xi_{j,\text{bias}}(N\Delta)$ can be expressed as

$$\xi_{j,\text{bias}}(N\Delta) = \tilde{\xi}_{j,\text{bias}}(N\Delta) + \alpha(N\Delta)$$

$$\tilde{\xi}_{j,\text{bias}}(N\Delta) = -\sum_{n=1}^{N} \ln|\mathbf{I} + \Delta\mathbf{R}^{-1}(n\Delta)\boldsymbol{\Sigma}_{y,j}(n\Delta\,|\,n\Delta - \Delta)|$$

$$\alpha(N\Delta) = -NK\ln(2\pi) - \sum_{n=1}^{N} \ln|\Delta^{-1}\mathbf{R}(n\Delta)|$$

Using the fact that for small Δ

$$|\mathbf{I} + \Delta\mathbf{A}| \approx 1 + \Delta\text{tr}(\mathbf{A})$$

$$\ln[1 + \Delta\text{tr}(\mathbf{A})] \approx \Delta\text{tr}(\mathbf{A})$$

yields in the limit as $N \longrightarrow \infty$, $\Delta \longrightarrow 0$, $N\Delta \longrightarrow T$,

$$\begin{aligned}\tilde{\xi}_{j,\text{bias}}(T) &= -\int_0^T \text{tr}\{\mathbf{R}^{-1}(t)\boldsymbol{\Sigma}_{y,j}(t\,|\,t)\}\,dt \\ \alpha(T) &= -\infty\end{aligned} \tag{10.2.2}$$

where $\boldsymbol{\Sigma}_{y,j}(t\,|\,t)$ is the continuous-time version of $\boldsymbol{\Sigma}_{y,j}(n\Delta\,|\,n\Delta - \Delta)$. The fact that $\alpha(N\Delta) \longrightarrow -\infty$ is disconcerting until one recalls that the log likelihood functions are desired so a decision can be made on which hypothesis $\mathcal{H}_j$, $j = 1, \ldots, M$, is true. Since the model (10.2.1) is such that $\alpha(N\Delta)$ is the same for all hypotheses, its "blow up" is not important, as $\alpha(N\Delta)$ can be ignored for hypothesis testing. Therefore, in the continuous-time case, consideration will be given to a *modified* log likelihood function $\tilde{\xi}_j(N\Delta)$ given by

$$2\tilde{\xi}_j(T) = \tilde{\xi}_{j,\text{bias}}(T) + \xi_{j,\text{observation}}(T) \tag{10.2.3}$$

Equation (10.2.3) is not the same as the modified log likelihood function of Section 10.1.2.

Now consider the observation term of (10.1.6). If the fact that $\boldsymbol{\delta}_j(n\Delta)$ contains a white noise term is ignored, then the limit $N \longrightarrow \infty$, $\Delta \longrightarrow 0$, $N\Delta \longrightarrow T$, gives

$$\xi_{j,\text{observation}}(T) = -\int_0^T \boldsymbol{\delta}_j'(t)[\mathbf{R}(t)]^{-1}\boldsymbol{\delta}_j(t)\,dt \tag{10.2.4}$$

Thus the desired limit of (10.1.6) is a modified log likelihood function:

$$\begin{aligned}
2\tilde{\xi}_j(T) &= -\int_0^T \boldsymbol{\delta}_j'(t)\mathbf{R}^{-1}(t)\boldsymbol{\delta}_j(t)\,dt + \text{bias term} \\
\text{Bias} &= -\int_0^T \text{tr}[\mathbf{R}^{-1}(t)\boldsymbol{\Sigma}_{y,j}(t\,|\,t)]\,dt \\
\hat{\mathbf{y}}_j(t\,|\,t)&: \text{ conditional expectation of } \mathbf{y}(t) \text{ given } \mathbf{z}(\tau),\ 0 \le \tau \le t, \text{ assuming that } \mathcal{H}_j \text{ is true} \\
\boldsymbol{\Sigma}_{y,j}(t\,|\,t) &= E\{[\mathbf{y}_j(t) - \hat{\mathbf{y}}_j(t\,|\,t)][\mathbf{y}_j(t) - \hat{\mathbf{y}}_j(t\,|\,t)]'\} \\
\boldsymbol{\delta}_j(t) &= \mathbf{z}(t) - \hat{\mathbf{y}}_j(t\,|\,t)
\end{aligned} \tag{10.2.5}$$

As in the discrete-time case, explicit equations for $\hat{\mathbf{y}}$ and $\boldsymbol{\Sigma}_y$ can be obtained by using dynamic system models and the results of Chapter 7. The implementation of (10.2.5) is illustrated in Fig. 10.2.1.

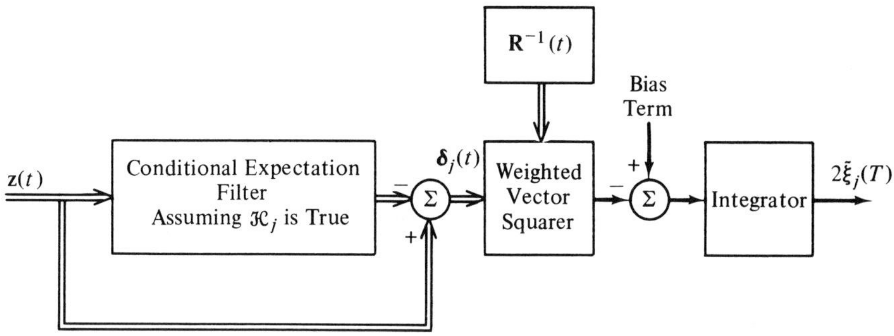

Figure 10.2.1 Log likelihood function: continuous time, Gaussian.

From a mathematical point of view, the validity of (10.2.5) is open to question. $\xi_j(T)$ is obtained by a nonlinear operation on $\mathbf{z}(t)$, and $\mathbf{z}(t)$ has a white component $\mathbf{v}(t)$. As discussed in Section 4.2.4, strange mathematical things can happen with white continuous-time stochastic process. It turns out that such strange things do occur with $\xi_j(T)$. If the indicated limit process is done in a "mathematically correct" fashion by considering $\Delta^2 v^2(n\Delta)$ terms, the result is (10.2.5) without the deterministic (bias) term. However, if the nature of the limit is defined differently, then (10.2.5) can be derived as being correct even with the $\Delta^2 v^2(n\Delta)$ terms. Following the discussions of Section 4.2, (10.2.5) corresponds to the use of Stratanovich calculus, while (10.2.5) without the bias term corresponds to Ito calculus.

10.2.2 Equal Correlation Functions

Consider

$$
\begin{aligned}
\mathcal{H}_j:\ & \mathbf{z}(t) = \mathbf{y}_j(t) + \mathbf{v}(t) \\
& \mathbf{y}_j(t) = \mathbf{m}_j(t) + \tilde{\mathbf{y}}(t) \\
& \mathbf{m}_j(t)\text{: deterministic (mean value),} \quad j = 1, \ldots, M \\
& \tilde{\mathbf{y}}(t)\text{: zero mean Gaussian process, no white component} \\
& \mathbf{v}(t)\text{: zero mean white Gaussian process} \\
& E\{\mathbf{v}(t)\mathbf{v}'(\tau)\} = \delta(t-\tau)\mathbf{R}(t)
\end{aligned}
\tag{10.2.6}
$$

This corresponds to the discrete-time $\mathcal{H}_j$ of (10.1.9).

Define $\mathfrak{F}_t$ as the linear operator (filter) which is such that when any $\mathcal{H}_j$ is true,

$$\hat{\mathbf{y}}_j(t \mid t) = \mathbf{m}_j(t) + \mathfrak{F}_t[\mathbf{z}(\tau) - \mathbf{m}_j(\tau), 0 \leq \tau \leq t]$$

Then application of the discrete to continuous time limit, $n \longrightarrow \infty$, $\Delta \longrightarrow 0$, $n\Delta \longrightarrow t$, to (10.1.10) yields

$$\begin{aligned} \tilde{\xi}_j(T) &= \int_0^T [\mathbf{z}(t) - \mathfrak{F}_t[\mathbf{z}(\tau), 0 \leq \tau \leq t]]'\mathbf{R}^{-1}(t)\mathbf{d}_j(t)\, dt \\ \mathbf{d}_j(t) &= \mathbf{m}_j(t) - \mathfrak{F}_t[\mathbf{m}_j(\tau), 0 \leq \tau \leq t] \end{aligned} \tag{10.2.7}$$

In the special case when

$$\tilde{\mathbf{y}}(t) = \mathbf{0}$$

(10.2.7) becomes [see also (10.1.11)]

$$\tilde{\xi}_j(T) = \int_0^T \mathbf{z}'(t)\mathbf{R}^{-1}(t)\mathbf{m}_j(t)\, dt \tag{10.2.8}$$

In the very special continuous-time case where $z(t)$ is a scalar and $R(t) = R$,

$$\tilde{\xi}_j(T) = \frac{1}{R}\int_0^T z(t)m_j(t)\, dt \tag{10.2.9}$$

The system of (10.2.9) is sometimes called a *correlator-detector*, as it is a decision function which can be used to decide whether a signal $m_j(t)$ is present by *correlating* (i.e., multiplying and integrating) $m_j(t)$ with the observations $z(t)$. The system is also sometimes called a *matched filter* because it can sometimes be implemented by putting $z(t)$ through a linear system called a matched filter. Consider a linear system with input $z(t)$, $0 \leq t < T$, output $x(T)$, and weighting pattern $W(t, T)$, so that

$$x(T) = \int_0^T W(t, T)z(t)\, dt$$

If the system can be designed so that

$$W(t, T) = m_j(t), \qquad 0 \leq t \leq T$$

then the system is "matched" to $m_j(t)$; i.e., it is a matched filter.

The correlator-detector and matched filter interpretations can be extended to the vector case and discrete-time systems if desired.

10.2.3 Probability Density of $\xi_j(T)$†

Consider (10.1.12). In the limit $N \longrightarrow \infty$, $\Delta \longrightarrow 0$, $N\Delta \longrightarrow T$, $\xi_j(T)$ is Gaussian with infinite mean and variance. The modified $\tilde{\xi}_j(T)$ of (10.2.5) also has infinite mean and variance. This ridiculous state of affairs is one of the consequences of using a continous-time white stochastic process model. $\tilde{\xi}_j(T)$ of (10.2.5) contains a $z^2(t)$ term, which means that it has a $v^2(t)$ term and $E\{v^2(t)\} = \infty$.

This problem can be partially alleviated by writing (10.2.5) as

$$\begin{aligned}
2\tilde{\xi}_j(T) &= \tilde{\tilde{\xi}}_j(T) + \beta(T) - \int_0^T \operatorname{tr}[\mathbf{R}^{-1}(t)\boldsymbol{\Sigma}_{y,j}(t\,|\,t)]dt \\
\tilde{\tilde{\xi}}_j(T) &= \int_0^T 2\mathbf{z}'(t)\mathbf{R}^{-1}(t)\hat{\mathbf{y}}_j(t\,|\,t)\,dt \\
&\quad - \int_0^T \hat{\mathbf{y}}'_j(t\,|\,t)\mathbf{R}^{-1}(t)\hat{\mathbf{y}}_j(t\,|\,t)\,dt \\
\beta(T) &= -\int_0^T \mathbf{z}'(t)\mathbf{R}^{-1}(t)\mathbf{z}(t)\,dt
\end{aligned} \tag{10.2.10}$$

Note first that $\beta(T)$ is causing the infinite mean and variance problems and second that $\beta(T)$ is independent of which hypothesis is true. Hence the modified-modified log likelihood function $\tilde{\tilde{\xi}}_j(T)$ of (10.2.10) can be used to make decisions in place of $\tilde{\xi}_j(T)$, and $\tilde{\tilde{\xi}}_j(T)$ is much better behaved (mathematically). Unfortunately there is still a problem, as $\mathbf{z}'(t)\mathbf{R}^{-1}(t)\hat{\mathbf{y}}_j(t\,|\,t)$ contains $\mathbf{v}'(t)\mathbf{R}^{-1}(t)\hat{\mathbf{y}}_j(t\,|\,t)$. The limiting operation was $\hat{\mathbf{y}}_j(n\Delta\,|\,n\Delta - \Delta) \longrightarrow \hat{\mathbf{y}}_j(t\,|\,t)$, so $\hat{\mathbf{y}}_j(t\,|\,t)$ does not (formally) include the end point and

$$E\{\mathbf{v}'(t)\mathbf{R}^{-1}(t)\hat{\mathbf{y}}_j(t\,|\,t)\} = 0$$

However, if $\hat{\mathbf{y}}(t\,|\,t)$ does include the end point, a different result is obtained. This apparent dilemma is closely related to the problem associated with (10.2.5). See also discussions in Section 4.2.3.

In the special case of equal correlation functions, there are no problems. It follows directly from (10.1.13) that for the modified log likelihood function $\tilde{\xi}_j(T)$ of (10.2.7)

$$\boxed{\begin{aligned}
&\text{When } \mathcal{H}_k \text{ is true,} \\
&\quad \tilde{\xi}_j(T) \text{ is } N(\mu_{jk}(T), \mu_{jj}(T)) \\
&\quad \mu_{jk}(T) = \int_0^T \mathbf{d}'_k(t)\mathbf{R}^{-1}(t)\mathbf{d}_j(t)\,dt
\end{aligned}} \tag{10.2.11}$$

EXAMPLE: Consider

$$\begin{aligned}
\mathcal{H}_j\colon z(t) &= z_j(t), \qquad j = 1, 2 \\
z_1(t) &= t + v(t) \\
z_2(t) &= 1 + v(t) \\
E\{v(t)v(\tau)\} &= \delta(t - \tau) \\
&v(t)\text{: zero mean Gaussian}
\end{aligned}$$

From (10.2.11), or more particularly (10.2.9),

$$\begin{aligned}
\tilde{\xi}_1(T) &= \int_0^T t z(t)\,dt \\
\tilde{\xi}_2(T) &= \int_0^T z(t)\,dt
\end{aligned}$$

Thus when $\mathcal{H}_1$ is true,

$$\tilde{\xi}_1(T) \text{ is } N\left(\frac{T^3}{3}, \frac{T^3}{3}\right)$$

$$\tilde{\xi}_2(T) \text{ is } N\left(\frac{T^2}{2}, T\right)$$

When $\mathcal{H}_2$ is true,

$$\tilde{\xi}_1(T) \text{ is } N\left(\frac{T^2}{2}, \frac{T^3}{3}\right)$$

$$\tilde{\xi}_2(T) \text{ is } N(T, T)$$

10.2.4 One Solution to White Model Problems†

Some "problems" associated with the white model process were discussed in Sections 10.2.1 and 10.2.3. One "answer" is now discussed.

Consider the functions $\xi_j^*(T), j = 1, \ldots, M$, generated by adding an extra linear system to Fig. 10.2.1 to give the system of Fig. 10.2.2, where

$$\xi_j^*(T) = -\int_0^T \boldsymbol{\delta}_j^{*\prime}(t)\mathbf{R}^{-1}(t)\boldsymbol{\delta}_j^*(t)\, dt$$

$$\frac{d}{dt}\boldsymbol{\delta}_j^*(t) = \alpha\boldsymbol{\delta}_j^*(t) + \boldsymbol{\delta}_j(t)$$

$$\alpha < 0$$

Now, $\boldsymbol{\delta}_j^*(T)$ is a well-behaved Gaussian process with no white component, as the extra linear system's integration action "smooths out" the "whiteness" of $\boldsymbol{\delta}_j(t)$. Hence the $\xi_j^*(T), j = 1, \ldots, M$, are well-defined random processes whose probability densities, etc., exist. Therefore the "problems" go away if the $\xi_j^*(T)$ are used as decision function. The use of the $\xi_j^*(T)$ can be justified in an engineering sense if the extra linear system's time constant ($|\alpha|^{-1}$) is very small relative to the "time constants" associated with the hypothesis dynamics. In terms of bandwidth arguments, the bandwidth of the extra linear system should be large compared to the bandwidths of hypothesized processes. Figure 10.2.2 is, of course, only an example of one way to proceed.

10.2.5 Distance Measures†

Application of the discrete to continuous time limit to the discrete-time formulas (10.1.14) and (10.1.15) for $J(N\Delta)$ and $\beta(N\Delta)$ yield the desired formulas for the corresponding $J(T)$ and $\beta(T)$ distances.

Consider the scalar process

$$z_1(t) = y_1(t) + v(t)$$

$$z_2(t) = y_2(t) + v(t)$$

defined as in (10.2.1). Define by direct analogy with the definition of Section

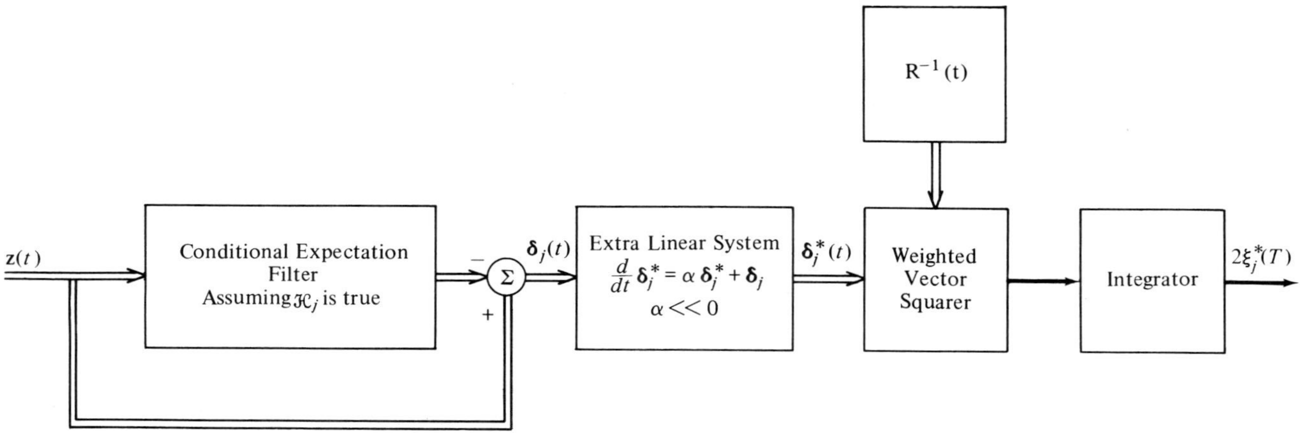

Figure 10.2.2 Modified log likelihood functions.

10.1.4

$\left.\begin{array}{l}\mathfrak{F}_{T,1}\\ \mathfrak{F}_{T,2}\\ \mathfrak{F}_{T,1+2}\end{array}\right\}$ linear operation which generates conditional expectations for nonzero mean components

$\left.\begin{array}{l}\Sigma_{y,1|2}(T|T)\\ \Sigma_{y,2|1}(T|T)\\ \Sigma_{y,2}(T|T)\\ \Sigma_{y,1}(T|T)\\ \Sigma_{y,1+2}(T|T)\end{array}\right\}$ error variance associated with estimating $y(T)$ from $z(t)$, $0 \leq t \leq T$

$\left.\begin{array}{l}d_1(T)\\ d_2(T)\\ d_{1+2}(T)\end{array}\right\}$ mean value errors associated with passing $m_1(t) - m_2(t)$, $0 \leq t \leq T$ through $\mathfrak{F}_{T,1}$, $\mathfrak{F}_{T,2}$, and $\mathfrak{F}_{T,1+2}$

Then application of the discrete to continuous time limit to (10.1.14) and (10.1.15) yields for the $\mathfrak{IC}_j$ of (10.2.1)

$$\boxed{\begin{aligned} 2R(T)\frac{dJ(T)}{dT} &= \Sigma_{y,1|2}(T|T) + \Sigma_{y,2|1}(T|T) \\ &\quad - \Sigma_{y,1}(T|T) - \Sigma_{y,2}(T|T) \\ &\quad + d_1^2(T) + d_2^2(T) \\ J(0) &= 0 \end{aligned}} \tag{10.2.12}$$

$$\boxed{\begin{aligned} 4R(T)\frac{dB(T)}{dT} &= \Sigma_{y,1+2}(T|T) - \Sigma_{y,1}(T|T) \\ &\quad - \Sigma_{y,2}(T|T) + \tfrac{1}{2}d_{1+2}^2(T) \\ B(0) &= 0 \end{aligned}} \tag{10.2.13}$$

In the special case where the $y_j(t)$ have the same covariance functions,

$$\begin{aligned} 8R(T)\frac{dB(T)}{dt} &= R(T)\frac{dJ(T)}{dT} = d^2(T) \\ d(T) &= m_1(T) - m_2(T) - \mathfrak{F}_T[m_1(t) - m_2(t), \\ &\qquad 0 \leq t \leq T] \end{aligned} \tag{10.2.14}$$

Once again more explicit equations can be obtained by using the dynamic system models and results of Chapter 7.

10.2.6 Discussion of Continuous Time

The discrete to continuous time limit operation is basically a simple operation, but some confusion arises as a consequence of using a white process model. The mathematics yields a well-defined answer in the limit.

The confusion lies in the fact that different ways of taking the limit can yield different answers. (Problems do not exist in the case of equal correlation functions.) However, it is the author's opinion that the differencc between the various answers [such as (10.2.5) with or without the bias term] are usually only of academic interest. If one wants to worry about the difference, it is also the author's opinion that (10.2.5) is closest "to the real world" as it is written.

Since the confusion arises because of the white observation process $\mathbf{v}(t)$, one is tempted to redo the analysis for the case where $\mathbf{v}(t) = \mathbf{0}$ $[\mathbf{R}(t) = \mathbf{0}]$. Unfortunately this does not really help, as it leads to the *singular detection* problem, which means that, mathematically, perfect decisions can be made from arbitrarily short time intervals of observations. This phenomenon follows from the preceding equations, but the details are not given.

10.3 Weighted Least Squares: Minimum Residual Decision Making

Assume that $\mathbf{z}(t), 0 \leq t \leq T$, is observed and that there are M hypotheses:

$\mathcal{H}_j$: $\mathbf{z}(t)$ can be approximated by the output of one of M linear dynamical systems

$$\frac{d}{dt}\mathbf{x}(t) = \mathbf{F}_j(t)\mathbf{x}(t) + \mathbf{G}_j(t)\mathbf{w}(t), \qquad j = 1, \ldots, M \tag{10.3.1}$$

$$\mathbf{z}(t) = \mathbf{H}_j(t)\mathbf{x}(t)$$

The choice of $\mathcal{H}_j$ is to be determined by the linear dynamical system that can generate an output "closest" to the observed $\mathbf{z}(t) = \mathbf{z}_{\text{actual}}(t), 0 \leq t \leq T$, assuming that $\mathbf{x}(0)$ and $\mathbf{w}(t), 0 \leq t \leq T$, are free to be chosen in some manner. The corresponding static case concept was discussed in Section 9.5. The crux of the following analysis is the evaluation of residuals which determine how close a model can fit the observations.

Consider first the special case where, given $\mathbf{z}(t), 0 \leq t \leq T$, it is desired to find the $\mathbf{x} = \mathbf{x}(0)$ which minimizes

$$2J_j(T) = \int_0^T [\mathbf{z}(t) - \mathbf{H}_j(t)\mathbf{x}(0)]'\mathbf{R}_j^{-1}(t)[\mathbf{z}(t) - \mathbf{H}_j(t)\mathbf{x}(0)]\,dt$$

and then find

$$2\hat{J}_j(T) = \int_0^T [\mathbf{z}(t) - \mathbf{H}_j(t)\hat{\mathbf{x}}_j(0|T)]'\mathbf{R}^{-1}(t)[\mathbf{z}(t) - \mathbf{H}_j(t)\hat{\mathbf{x}}(0|T)]\,dt \tag{10.3.2}$$

where $\hat{\mathbf{x}}_j(0|T)$ is the $\mathbf{x} = \mathbf{x}(0)$ which minimizes $J_j(T)$ for the given $\mathbf{z}(t)$, $0 \leq t \leq T$. From Chapter 5 (extended to continuous time)

$$\hat{\mathbf{x}}_j(0|T) = \left[\int_0^T \mathbf{H}_j'(t)\mathbf{R}_j^{-1}(t)\mathbf{H}_j(t)\,dt\right]^{-1} \int_0^T \mathbf{H}_j'(t)\mathbf{R}_j^{-1}(t)\mathbf{z}(t)\,dt$$

Differentiation of $\hat{J}(T)$ with respect to T gives

$$\frac{d}{dT}2\hat{J}_j(T) = [\mathbf{z}(T) - \mathbf{H}_j(T)\hat{\mathbf{x}}_j(0\,|\,T)]'\mathbf{R}_j^{-1}(T)[\mathbf{z}(T) - \mathbf{H}_j(T)\hat{\mathbf{x}}_j(0\,|\,T)] + \theta_j(T)$$

where

$$\theta_j(T) = -\int_0^T \left[\frac{\partial \hat{\mathbf{x}}_j(0\,|\,T)}{\partial T}\right]' \mathbf{H}_j'(t)\mathbf{R}_j^{-1}(t)[\mathbf{z}(t) - \mathbf{H}_j(t)\hat{\mathbf{x}}_j(0\,|\,T)]\,dt$$

or

$$\theta_j(T) = -\left[\frac{\partial \hat{\mathbf{x}}_j(0\,|\,T)}{\partial T}\right]' \left\{\int_0^T \mathbf{H}_j'(t)\mathbf{R}_j^{-1}(t)\mathbf{z}(t)\,dt - \left[\int_0^T \mathbf{H}_j'(t)\mathbf{R}_j^{-1}(t)\mathbf{H}_j(t)\,dt\right]\hat{\mathbf{x}}_j(0\,|\,T)\right\}$$

or substituting for $\hat{\mathbf{x}}_j(0\,|\,T)$,

$$\theta_j(T) = 0$$

Thus

$$2\hat{J}_j(T) = \int_0^T [\mathbf{z}(t) - \mathbf{H}_j(t)\hat{\mathbf{x}}_j(0\,|\,t)]'\mathbf{R}_j^{-1}(t)[\mathbf{z}(t) - \mathbf{H}_j(t)\hat{\mathbf{x}}_j(0\,|\,t)]\,dt \qquad (10.3.3)$$

Now, (10.3.2) uses $\hat{\mathbf{x}}_j(0\,|\,T)$ while (10.3.3) uses $\hat{\mathbf{x}}(0\,|\,t)$, and this difference can be very important. For example, to use (10.3.2) one must

1. Wait until time T.
2. Calculate $\hat{\mathbf{x}}_j(0\,|\,T)$.
3. Evaluate (10.3.2).

On the other hand, $\hat{\mathbf{x}}_j(0\,|\,t)$ can be generated by a dynamic system, so (10.3.3) can be implemented in real time and gives $\hat{J}_j(T)$ continually for increasing T.

Now consider the following more general version of the same problem. Given $\mathbf{z}(t)$, $0 \le t \le T$, and the model

$$\begin{aligned} \frac{d}{dt}\mathbf{x}(t) &= \mathbf{F}_j(t)\mathbf{x}(t) + \mathbf{G}_j(t)\mathbf{w}(t) \\ \boldsymbol{\epsilon}(t) &= \mathbf{z}(t) - \mathbf{H}_j(t)\mathbf{x}(t) \end{aligned} \qquad (10.3.4)$$

Find the $\mathbf{x}(0)$ and $\mathbf{w}(t)$, $0 \le t \le T$, which minimize

$$2J_j(T) = \mathbf{x}'(0)\boldsymbol{\psi}_j^{-1}\mathbf{x}(0) + \int_0^T [\boldsymbol{\epsilon}'(t)\mathbf{R}_j^{-1}(t)\boldsymbol{\epsilon}(t) + \mathbf{w}'(t)\mathbf{Q}_j^{-1}(t)\mathbf{w}(t)]\,dt$$

Let $\hat{\mathbf{x}}_j(0\,|\,T)$ denote the minimizing initial condition and $\hat{\mathbf{w}}_j(t\,|\,T)$ denote the minimizing input, with $\hat{\mathbf{x}}_j(t\,|\,T)$ denoting the resulting trajectory. Define

$$\begin{aligned} 2\hat{J}_j(T) &= \hat{\mathbf{x}}_j'(0\,|\,T)\boldsymbol{\psi}_j^{-1}\hat{\mathbf{x}}_j(0\,|\,T) + \int_0^T [\hat{\boldsymbol{\epsilon}}_j'(t\,|\,T)\mathbf{R}_j^{-1}(t)\hat{\boldsymbol{\epsilon}}_j(t\,|\,T) \\ &\quad + \hat{\mathbf{w}}_j'(t\,|\,T)\mathbf{Q}_j^{-1}(t)\hat{\mathbf{w}}_j(t\,|\,T)]\,dt \\ \hat{\boldsymbol{\epsilon}}_j(t\,|\,T) &= \mathbf{z}(t) - \mathbf{H}_j(t)\hat{\mathbf{x}}_j(t\,|\,T) \end{aligned} \qquad (10.3.5)$$

It can be proved that

$$\boxed{\begin{aligned} 2\hat{J}_j(T) &= \int_0^T \hat{\boldsymbol{\epsilon}}_j'(t\,|\,t)\mathbf{R}_j^{-1}(t)\hat{\boldsymbol{\epsilon}}_j(t\,|\,t)\,dt \\ \hat{\boldsymbol{\epsilon}}_j(t\,|\,t) &= \mathbf{z}(t) - \mathbf{H}_j(t)\hat{\mathbf{x}}_j(t\,|\,t) \end{aligned}} \tag{10.3.6}$$

Equation (10.3.6) is a more general version of (10.3.3), and all the comments on the value of (10.3.3) apply even more so to (10.3.6). For example, to evaluate $\hat{J}(T)$ using (10.3.5), one must calculate $\hat{\mathbf{x}}(t\,|\,T)$, i.e., the smoothed estimate, while (10.3.6) requires only $\hat{\mathbf{x}}(t\,|\,t)$. The actual proof of (10.3.6) is tedious, requires a good knowledge of the calculus of variations, and is not given.

Now return to the basic problem of deciding between the $\mathcal{H}_j$ of (10.3.1). Weighted-least-squares or minimum residual decision making is done by evaluating $\hat{J}_j(T), j = 1, \ldots, M$, using (10.3.6) and using the rule

$$\text{Choose the } \mathcal{H}_j \text{ which yields minimum value of } \hat{J}_j(T), \qquad j = 1, \ldots, M \tag{10.3.7}$$

A system to generate $\hat{J}_j(T)$ is illustrated in Fig. 10.3.1. As expected, the minimum residual decision function $\hat{J}_j(T)$ is closely related to the log likelihood

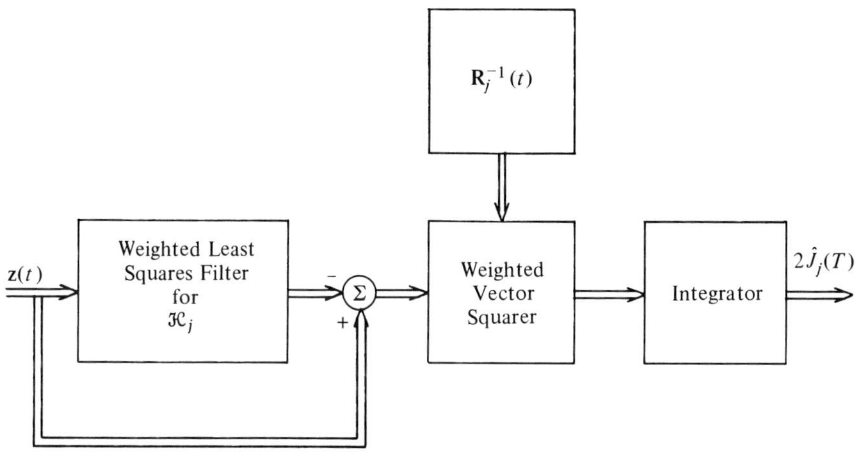

Figure 10.3.1 Generation of minimum residual decision function [see (10.3.6)].

function $\xi_j(T)$ for Gaussian processes. In fact, if one is worried about the white process dilemmas associated with $\xi_j(T)$, the weighted-least-squares logic provides a convenient conceptual method of sidestepping them.

Only the continuous-time case was discussed, but similar results and conclusions are obtained in the discrete-time case.

10.4 Unknown-but-Bounded

As in Chapter 9, hypothesis testing and decision making for unknown-but bounded dynamic models is a simple concept which is evident (or at least almost evident) by inspection.

Consider the discrete-time hypotheses

$$\begin{aligned} \mathcal{H}_j\colon \mathbf{z}(n) &= \mathbf{z}_j(n), \qquad n = 1, \ldots, N \\ \mathbf{z}_j(n) &= \mathbf{y}_j(n) + \mathbf{v}_j(n) \qquad J = 1, \ldots, M \\ &\mathbf{y}_j(n), \mathbf{v}_j(n)\colon \text{unknown-but-bounded processes} \end{aligned}$$

Define

$\mathbf{\Omega}_{\text{est}, j}(N \mid N)$: estimate set which contains the actual $\mathbf{y}(N)$ as calculated from the actual observations $\mathbf{z}_{\text{actual}}(1) \cdots \mathbf{z}_{\text{actual}}(N)$ assuming that $\mathcal{H}_j$ is true

Then the decision rule, by direct analogy with (9.6.3), is ($M = 2$)

$$\boxed{\begin{array}{ll} \text{Choose } \mathcal{H}_1 \text{ if} & \mathbf{\Omega}_{\text{est},1}(N \mid N) \text{ exists} \\ & \mathbf{\Omega}_{\text{est},2}(N \mid N) \text{ does not exist} \\ \text{Choose } \mathcal{H}_2 \text{ if} & \mathbf{\Omega}_{\text{est},1}(N \mid N) \text{ does not exist} \\ & \mathbf{\Omega}_{\text{est},2}(N \mid N) \text{ exists} \\ \text{Reject both } \mathcal{H}_1 \text{ and } \mathcal{H}_2 \text{ if} & \\ & \mathbf{\Omega}_{\text{est},1}(N \mid N) \text{ does not exist} \\ & \mathbf{\Omega}_{\text{est},2}(N \mid N) \text{ does not exist} \\ \text{No decision if} & \mathbf{\Omega}_{\text{est},1}(N \mid N) \text{ exists} \\ & \mathbf{\Omega}_{\text{est},2}(N \mid N) \text{ exists} \end{array}} \tag{10.4.1}$$

This rule generalizes to $M > 2$, and the same concept also applies for continuous-time models.

For ellipsoidal models, the estimate sets can be replaced by bounding ellipsoids as given in Chapters 6 and 7. For example, consider (6.6.4). The bounding ellipsoid will fail to exist if $\delta^2(N + 1) > 1$. In some applications the special cases of (6.6.6) or (6.6.7) (for discrete time) can be used to reduce "on-line" computation requirements. If (6.6.6) is used, $\boldsymbol{\delta}^2(N + 1)$ is computed for use as a decision function but is not used to update the $\mathbf{\Sigma}$ equations. If (6.6.7) is used, $q(N + 1)$ is used, as the decision function. It is clear that (6.6.7) always provides a better (or at least not worse) decision function than (6.6.6). Similar discussions apply to (7.5.2), (7.5.3) and (7.5.4).

The distance measure D definition of (9.6.9) should extend to discrete- and continuous-time systems in a straightforward fashion. However, to the author's knowledge, the details have never been worked out.

When comparing unknown-but-bounded and Gaussian model hypothesis testing, there are two important points. First, for unknown-but-bounded models, the filters of Chapters 6 and 7 provide the decision function directly, while for Gaussian models it is necessary to add a squarer and integrator. Second, for unknown-but-bounded models, no decision using $\mathbf{z}(1) \cdots \mathbf{z}(N)$ may be possible, but alternatively a perfect decision may be possible using only $\mathbf{z}(1) \cdots \mathbf{z}(n)$ for $n < N$. In this sense, unknown-but-bounded models automatically yield a *sequential* hypothesis-testing procedure, to be discussed in Section 10.5.

10.5 Sequential Hypothesis Testing

Consider the use of log likelihoods as decision functions for hypotheses involving stochastic models. Following the ideas of Chapter 9, one method to make a decision (in discrete time) is

1. Observe $\mathbf{z}(1) \cdots \mathbf{z}(N)$.
2. Calculate $\xi_j(N), j = 1, \ldots, M$.
3. Make decision by choosing the "most likely" hypothesis.

However, a more general decision making approach is

1. Observe $\mathbf{z}(1)$.
2. Calculate $\xi_j(1), j = 1, \ldots, M$.
3. (a) If one hypothesis is "clearly" true, choose it and stop observing.
   (b) Otherwise make another observation.
4. Calculate $\xi_j(2), j = 1, \ldots, M$, etc.

This second approach to decision making is called sequential hypothesis testing. At each observation time, one either chooses a hypothesis or decides to wait for more observations.

The philosophy of sequential hypothesis testing is related to the Fisher concepts of Section 9.4.1. To illustrate one explicit sequential hypothesis-testing technique, define as in Section 9.4.1

$$\mathbf{\Omega}_{\text{tail},j}(n) = \{\xi : \xi < d_j(n)\} \tag{10.5.1}$$

The $d_j(n)$ are chosen so that if $\mathcal{H}_j$ is true, the probability of $\xi_j(n) \in \mathbf{\Omega}_{\text{tail},j}(n)$ is very small. Then one sequential testing rule is as follows. Assume that $\mathbf{z}(1) \cdots \mathbf{z}(n)$ has been observed. Then

$$\text{Choose } \mathcal{H}_k \text{ and stop observing if} \begin{cases} \xi_k(n) \notin \mathbf{\Omega}_{\text{tail},k}(n) \\ \xi_j(n) \in \mathbf{\Omega}_{\text{tail},j}(n), \quad \text{all } j \neq k \\ \qquad\qquad j = 1, \ldots, M \end{cases} \tag{10.5.2}$$

Make no decision and wait for next observation $\mathbf{z}(n+1)$ otherwise

There are, of course, many other possible variations of the basic idea. The logic illustrated by (10.5.2) is called a *stopping rule*.

Sequential hypothesis testing for stochastic models requires the evaluation of the log likelihood decision functions for increasing time. Thus the dynamic systems of Figs. 10.1.1–10.2.1 are very useful, as they evaluate $\xi_j(n)$, $n = 1, \ldots, N$; $\xi_j(t)$, $0 \le t \le T$; etc.—not just $\xi_j(N)$ and $\xi_j(T)$.

EXAMPLE: Consider the example of Section 10.2.3

$$\begin{aligned} \mathcal{H}_j&: z(t) = z_j(t), \qquad j = 1, 2 \\ z_1(t) &= t + v(t) \\ z_2(t) &= 1 + v(t) \\ E\{v(t)v(\tau)\} &= \delta(t - \tau) \\ v(t)&: \text{ zero mean Gaussian} \end{aligned}$$

Consider the following decision functions:

$$\tilde{\xi}_1(T) = \int_0^T tz(t)\, dt$$

$$\tilde{\xi}_2(T) = \int_0^T z(t)\, dt$$

Then

$$\tilde{\xi}_2(T) \text{ is } N\left(\frac{T^3}{3}, \frac{T^3}{3}\right) \qquad \text{when } \mathcal{H}_1 \text{ is true}$$

$$\tilde{\xi}_2(T) \text{ is } N(T, T) \qquad \text{when } \mathcal{H}_2 \text{ is true}$$

Define

$$\gamma_1(t) = \frac{\tilde{\xi}_1(t) - (t^3/3)}{(t^3/3)^{1/2}}$$

$$\gamma_2(t) = \frac{\tilde{\xi}_2(t) - t}{t^{1/2}}$$

so that

$$\gamma_1(t) \text{ is } N(0, 1) \qquad \text{when } \mathcal{H}_1 \text{ is true}$$

$$\gamma_2(t) \text{ is } N(0, 1) \qquad \text{when } \mathcal{H}_2 \text{ is true}$$

Then a reasonable sequential hypothesis testing decision rule is continue observing until

$$|\gamma_1(t)| > 3$$
$$|\gamma_2(t)| < 3$$

or until

$$|\gamma_1(t)| < 3$$
$$|\gamma_2(t)| > 3$$

Note that these sequential testing discussions have concerned stochastic models and log likelihood decision functions. For unknown-but-bounded models, special discussions are not needed, as unknown-but-bounded hypothesis testing is inherently a sequential technique. Observations are continued until a decision is made.

10.6 Finite Memory

The log likelihood generators of Section 10.1 and weighted-least-squares logic of Section 10.3 can easily be converted to finite memory operation.

In discrete time, finite memory equations can be obtained from (10.1.6) by simply replacing the $\sum_{n=1}^{N}$ by $\sum_{n=N-m}^{N}$. This can be implemented as a difference equation by noting that an equation such as

$$\xi(N) = \sum_{n=N-m}^{N} \theta(n)$$

can be implemented as

$$\xi(N+1) = \xi(N) + \theta(N+1) - \theta(N-m)$$

In continuous time, a delay line and a substractor are simply inserted before the integrator in Fig. 10.2.1.

A finite memory decision function may be implemented with or without a corresponding finite memory estimator to get the $\hat{\mathbf{y}}$. Such choices depend on the particular problem considered.

Sequential hypothesis testing can also be done with a finite memory system.

10.7 Event Detection

In many applications, an observed time process suddenly or gradually changes its character at some instant of time because of some change in the system being observed. Detection of such changes is called event detection. Assume that $\mathbf{z}(n)$, $n = 1, \ldots,$ is observed and that the event (change) takes place (or begins to take place) at time N_0 so that

$$\mathbf{z}(n) = \mathbf{z}_1(n), \qquad n = 1, \ldots, N_0 - 1$$
$$\mathbf{z}(n) = \mathbf{z}_2(n), \qquad n = N_0, \ldots$$

where $\mathbf{z}_1(n)$ and $\mathbf{z}_2(n)$ are processes with different properties.

10.7.1 Unknown-but-Bounded

Event detection for an unknown-but-bounded model is very simple. One simply implements a filter based on the model for $\mathbf{z}_1(n)$ and uses it to process $\mathbf{z}(n)$. If the estimate set $\mathbf{\Omega}_{\hat{x}}(N|N)$ fails to exist for some N, it is clear that the event took place at some earlier time. A priori knowledge of the nature of $\mathbf{z}_2(n)$ and/or the approximate time the event is to occur are of little help.

10.7.2 Bayesian

In a completely Bayesian formulation, $\mathbf{z}_1(n)$ and $\mathbf{z}_2(n)$ have stochastic models, and there exists a probability structure on the time, N_0, that the event occurs. Such Bayesian models involving probabilistic structures on event time N_0 are interesting to analyze, but the details are left as a non-trivial "exercise for the reader."

10.7.3 Completely Unknown Event Time

Consider a case where the event time N_0 is completely unknown while $\mathbf{z}_1(n)$ and $\mathbf{z}_2(n)$ are completely specified by stochastic models. Thus N_0 is Fisher while $\mathbf{z}_1(n)$ and $\mathbf{z}_2(n)$ are Bayesian.

A straightforward approach to testing $\mathbf{z}(1) \cdots \mathbf{z}(N)$ evokes $N + 1$ hypotheses

$$\begin{aligned} &\mathcal{H}_0\text{: event has not occurred,} && n = 1, \ldots, N \\ &\mathcal{H}_j\text{: event occurred at time } N - j, && j = 0, \ldots, N - 1 \end{aligned} \tag{10.7.1}$$

This approach can be implemented by calculating N log likelihood functions $\xi_j(N), j = 1, \ldots, N$: $\xi_j(N)$ is the log likelihood function for

$$\mathbf{z}(n) = \begin{cases} \mathbf{z}_1(n), & n = 1, \ldots, j \\ \mathbf{z}_2(n), & n = j + 1, \ldots, N \end{cases} \tag{10.7.2}$$

This straightforward solution has obvious practical shortcomings when N is increasing.

A variation on the concept of (10.7.1) and (10.7.2) is to use a "sliding window" of M hypotheses. To illustrate the idea, assume that $N > M$ and consider the M hypotheses:

$$\begin{aligned} &\mathcal{H}_j\text{: event occurred at time } N - j, && j = 0, \ldots, M - 2 \\ &\mathcal{H}_0\text{: event has not occurred,} && n = 1, \ldots, N \end{aligned} \tag{10.7.3}$$

The decision is made using the corresponding M log likelihood functions $\xi_j(N), j = 0, \ldots, M - 2$, and $\xi_0(N)$. If $\mathcal{H}_0$ is accepted at time N, the next observation $\mathbf{z}(N + 1)$ is taken and the hypotheses are changed to

$$\begin{aligned} &\mathcal{H}_j\text{: event occurred at time } N + 1 - j, && j = 0, \ldots, M - 2 \\ &\mathcal{H}_0\text{: event has not occurred,} && n = 1, \ldots, N + 1 \end{aligned} \tag{10.7.4}$$

and the process is repeated.

The special sliding window case where $M = 2$ can be viewed as a predict and check scheme. The observations $\mathbf{z}(1) \cdots \mathbf{z}(N - 1)$ are used to calculate $\hat{\mathbf{z}}(N|N - 1)$, a prediction of the next observation, assuming that the event has not occurred. Then the actual observation $\mathbf{z}(N)$ is checked against the prediction $\hat{\mathbf{z}}(N|N - 1)$ by using fancy hypothesis tests or by simply checking the size of the difference.

The sliding window of M hypotheses (10.7.3) and (10.7.4) implicitly assumed that the log likelihoods $\xi_j(N)$ are based on all available observations $\mathbf{z}(1) \cdots \mathbf{z}(N)$. In some cases, it is better to use finite memory decision functions, with or without finite memory filters.

Extension of these sliding window, etc., concepts to continuous-time models requires quantization of possible event times into a finite number of possibilities.

10.7.4 No Alternative Hypothesis

Consider a case where $\mathbf{z}_1(n)$ is modeled as a stochastic process but both the event time and the new process $\mathbf{z}_2(n)$ are completely unknown. A reasonable approach uses $\xi_1(N, N - m)$, a finite memory log likelihood function calculated from $\mathbf{z}(N) \cdots \mathbf{z}(N - m)$ based on the model for $\mathbf{z}_1(n)$. $\xi_1(N, N - m)$ is a stochastic process with specified probability distributions assuming that the event has not occurred. If a value is calculated which lies in the tail of this distribution, it is assumed that the event has occurred. A simpler version is possible when $\mathbf{z}_1(n)$, $n = 1, \ldots,$ and $\xi_1(N, N - m)$, $N = 1, \ldots,$ are stationary stochastic processes. In such a case, one merely observes $\xi_1(N, N - m)$ for increasing N and "looks for" jumps or trends which are inconsistent with past behavior.

In the present case, where there is no alternative hypothesis, there is very little special justification for the use of the log likelihood function as a decision function. It is reasonable, but other decision functions may be just as good or better. In fact, for event detection, a variation on the ideas of Section 10.8 on testing the whiteness of the feedback residuals is a very nice concept.

10.7.5 Discussion of Event Detection

For unknown-but-bounded models, event detection is conceptually straightforward, but for stochastic (weighted-least-squares) models there are many ways to proceed. The "mathematically optimum" approach is usually not practical, so engineering judgment is needed to choose a reasonable scheme which hopefully will do the job.

Only the case of detecting a single event was considered. There are, of course, applications where a sequence of events can occur. Such problems simply require still more engineering judgment.

Event detection and sequential hypothesis testing are similar in that a decision (choose hypothesis or detect event) is attempted for each new observation. However, event detection and sequential testing involve fundamentally different concepts and should not be confused. In sequential testing, the process being observed $\mathbf{z}(n)$, $n = 1, \ldots,$ has the same structure for all time. In event detection, the observed process changes its structure (character) at some time. For stochastic models, a sequential testing system and an event detection system may bear little resemblance. However, for unknown-but-bounded models, the two types of systems are identical.

10.8 Testing Whiteness of Residuals

The discussions of this chapter have so far concentrated on the computation and use of the log likelihood function as a decision function. However, as has been emphasized earlier (see Section 9.4), the log likelihood function is not necessarily an optimum decision function for various Fisher-type formulations. One alternative approach is now briefly discussed.

To be explicit, assume that the problem is Fisher with no alternative hypothesis, so that

$$\mathcal{H}_1: \mathbf{z}(n): \text{specified Gaussian process}$$

$$\mathcal{H}_2: \mathcal{H}_1 \text{ is not true}$$

Define

$$\hat{\mathbf{z}}_1(n|n-1): \text{prediction of } \mathbf{z}(n) \text{ using } \mathbf{z}(1) \cdots \mathbf{z}(n-1) \text{ and the assumption that } \mathcal{H}_1 \text{ is true}$$

$$\boldsymbol{\delta}_{z,1}(n) = \mathbf{z}(n) - \hat{\mathbf{z}}_1(n|n-1)$$

From Section 6.3.3,

If $\mathcal{H}_1$ is true, then $\boldsymbol{\delta}_{z,1}(n)$ is a white stochastic process

This is true independent of the Gaussian assumption. Therefore instead of using the log likelihood function, one can decide whether $\mathcal{H}_1$ is true by deciding whether or not $\boldsymbol{\delta}_{z,1}(n)$ is a white process. There are, of course, many ways to test whether or not a given process is white, and a complete discussion will not be attempted here. However, one obvious approach is to use $\alpha(N)$ as a decision function, where

$$\alpha(N) = \sum_{n=1}^{N} \boldsymbol{\delta}'_{z,1}(n)\boldsymbol{\delta}_{z,1}(n-1)$$

If $\mathcal{H}_1$ is true, $\alpha(N)$ is a zero mean random variable with computable variance. If $\mathcal{H}_1$ is not true, then one would expect that the $\boldsymbol{\delta}_{z,1}(n)$ would not be white and hence $|\alpha(N)|$ would become very large. There are many variations of this idea.

10.9 Application of Hypothesis-Testing Theory

A few of the practical aspects of hypothesis testing are now discussed. The discussions are restricted to stochastic (Bayesian, Fisher) models. This section is closely related to Chapter 8, which discussed applications of the estimation theory for linear models. Almost all the ideas of Chapter 8 also apply to hypothesis testing and will not be repeated. The following is concerned primarily with new aspects introduced by hypothesis testing.

10.9.1 Development of Models

Hypothesis testing can introduce the need for multiple models for the multiple hypotheses and, if appropriate, a priori probabilities. The one modeling aspect of hypothesis testing that has no estimation counterpart is the problem of specifying the hypotheses to be considered. Often this is a critical step which influences both performance and the difficulty of implementation.

To illustrate the modeling problem associated with specifying the hypotheses, consider a situation where one is making observations on some physical process which is being controlled by a law based on a mathematical model. The basic problem is to decide whether the mathematical model is valid. The simplest approach is to view the overall model as a one hypothesis problem (no alternative hypothesis). However, usually one has some idea of "how the model will go bad," so it may be possible to specify K possible "bad models," in which case the overall model can become one with $M = K + 1$ hypotheses. A formulation with $K + 1$ hypotheses may yield better performance but will definitely be more difficult to implement. As with all modeling problems, there is no simple answer. The author merely wants to emphasize the importance of giving careful consideration to the choice of hypotheses.

10.9.2 Choice of Decision Functions

In many cases, the "ideal" decision functions are based on the log likelihood functions. This is especially true for Bayesian models and Fisher models involving specified alternative hypotheses. For non-Gaussian models, a good practical approach is often to use a decision function based on a Gaussian assumption.

In the case of Fisher models with no alternative hypotheses, the log likelihood function no longer dominates and other decision functions can prove equally or more effective. For example, a decision function for testing the "whiteness of the residuals" has a lot of appeal. There is no general rule of approach except to use engineering judgment. In some cases, one has a "feeling" for what the alternatives will be but uses a no-alternative formulation because the "feeling" cannot be translated in a mathematically formulated hypothesis. This feeling, however, may prove helpful in choosing a decision function.

For linear Gaussian models the log likelihood function is closely tied to the optimum filter corresponding to the hypothesis. The filter part of the system is by far the most complicated and difficult part to implement. Therefore an obvious approach to simplify the design is to replace the optimum filter by a simpler and easier-to-implement nonoptimum filter. The subject of designing simpler, practical filters was discussed extensively in Section 8.2 and so is not repeated.

As discussed in Section 8.2, the filter's memory span can be an important design consideration. However, for stochastic model hypotheses, one also has to worry about the memory span of the integrator (which integrates the squared feedback error terms of, say, Fig. 10.1.1). An integrator remembers back to time zero, but in practice it may be acceptable and/or desirable to replace the integrator by some "long time constant" system which slowly forgets the past.

For stochastic models, the feedback error terms are squared, weighted, and then summed or integrated. In practice, sensitivity to the weights should not be expected, so approximations can usually be used.

Note that it is not essential to base the design of a practical decision function on the design of a practical filter. However, the author recommends this point of view, as it provides the most insight (at least to the author). The "picture" of a filter as a feedback system around a model of the dynamics, with decision function determined by the size of the feedback error correction, can help guide a lot of engineering decisions.

10.9.3 Decision Rules

For stochastic models, one must choose decision rules as well as decision functions (for unknown-but-bounded models, the decision rule is automatic).

The main conceptual problem is to decide what types of error are most important. In practice the difficult problem may be to determine the probability density of the decision function under the various hypotheses. The probability density formulas derived under a Gaussian assumption may become good approximations for non-Gaussian hypotheses, especially for long observation time spans (because of the central limit theorem).

In practice a good, albeit brute force, approach to choosing a decision rule for given decision functions is to simply simulate the models and decision functions on a computer and then adjust the decision rule until "satisfactory" performance is obtained. An even better approach is, of course, to use actual samples of real rather than simulated data.

10.9.4 Good Models Not Available

If the models for the hypotheses are not very good, there are various ways to proceed. The discussions of Section 8.2.3 apply. One general approach worth mentioning is to "train" the decision-making system. For example, consider a stochastic model and assume that the form of the decision function is specified and that the problem is to choose a decision rule. Assume that real data are being processed with some fixed decision rule and assume that one can decide which hypothesis was true (at some later time or from other information). Then one knows when the rule gave wrong answers, so it can be adjusted to give better performance. Obviously the decision function as well as the decision rule could be adjusted. Sometimes this is called a "teacher" (the knowledge of which hypothesis was actually true), which is "training" the decision maker until it "learns." The terminology "supervised and unsupervised learning" is sometimes used to denote how "smart" the "teacher" is. Another jargon is to simply call it an "adaptive" hypothesis-testing scheme.

10.9.5 System, Signal, Measurement, etc., Design

As discussed in Section 8.3, the equations for the error covariance matrix of an estimate provide a powerful vehicle for many problems involving the choice of sensors and the design of "signals." The equations for the divergence and the Bhattacharyya distance play an analogous role in system and signal design for hypothesis testing using stochastic models. These distance measures do not provide (of themselves) an absolute measure of performance, but relative performance is often all that is needed; i.e., it is desired to choose the sensor system or signal that yields the maximum distance.

10.10 Discussion

An important point of this chapter is the ease with which a filter can be "converted" into a decision function for hypothesis testing. For Bayesian and Fisher Gaussian models and weighted-least-squares concepts, only a squarer and integrator (summer) have to be added to the filter. For an unknown-but-bounded model, the filter itself automatically acts like a decision maker.

The discussions centered on models of the form $\mathbf{z}(n) = \mathbf{y}(n) + \mathbf{v}(n)$, where $\mathbf{y}(n)$ was viewed as a "signal" and $\mathbf{v}(n)$ was a white process viewed as observation noise. However, all the discussions of Chapter 3 on using white-process-driven dynamic systems as building blocks for more complex models apply in the present context. Many very complicated models can be converted into the forms studied here.

A point made in the introduction to this chapter will be reemphasized. The chapter has concentrated on the development of formulas for decision functions. The use and interpretation of these formulas is as in Chapter 9. Thus for actual applications involving dynamic systems, one must combine Chapters 9 and 10 (and use the filter equations of Chapters 6 and 7). The only really new hypothesis-testing ideas developed in this chapter are in Sections 10.5 and 10.7 on sequential hypothesis testing and event detection.

Historical Notes and References; Chapter 10

Sections 10.1, 10.2

The development of the discrete-time likelihood function and the corresponding continuous-time results are taken from Schweppe [4], which resulted from discussions with R. Price. More sophisticated developments can be found in Duncan [1]; Kailath [3], [4]. The distance measure results are taken from Schweppe [7], [8]. Most of the basic continuous-time results are also available in the Communications Book List 5, but the reader may find it difficult to correlate this book with the communication theory developments, as the approaches are different. Van Trees [1], [2], and Snyder [1] help "bridge the gap".

Section 10.3

The proof of (10.3.3) was done with L. Gardner and arose from discussions with R. Kalman.

Section 10.5

Wald [1] contains related material.

Section 10.8

See Book List 2 on Mathematical Statistics.

Section 10.9

Discussions on communication signal design are found in Schweppe [7], [8].

Exercises

10.1. Consider

$$\mathcal{H}_j\colon\ z(n) = x_j + c(n), \qquad j = 1, 2$$

$$x_1 \text{ is } N(m_1, \mathbf{\Gamma}_1)$$

$$x_2 \text{ is } N(m_2, \mathbf{\Gamma}_2)$$

$$c(n+1) = \mathbf{\Phi}_c c(n) + v(n)$$

$$E\{v(n_1)v(n_2)\} = \begin{cases} R_c, & n_1 = n_2 \\ 0, & n_1 \neq n_2 \end{cases}$$

$c(0)$ is $N(0, \psi_c)$

$v(n)$: Gaussian

$c(0)$, $v(n)$, x_j: independent

a. Write a difference equation for the log likelihood function $\xi_1(N)$.
b. What is the probability density of $\xi_1(N)$ for large N assuming that $\mathcal{H}_1$ is true?

10.2. Consider

$$z(t) = y_j(t) + v(t)$$
$$v(t)\text{: zero mean Gaussian}$$
$$E\{v(t)v(\tau)\} = \delta(t - \tau)R$$

a. Assume that

$$\mathcal{H}_j\colon y_j(t) = \sin(\omega_j t), \qquad j = 1, \ldots, M$$

where the ω_j are given. Design a hypothesis-testing system to decide which frequency is present. *Comment:* If you use matched filters, you will have designed a simplified Doppler filter bank.

b. Redo part (a) assuming that

$$\mathcal{H}_j\colon\ y_j(t) = \sin(\omega_j t) + x(t), \qquad j = 1, \ldots, M$$
$$\frac{d}{dt}x(t) = Fx(t) + w(t)$$
$$x(0) \text{ is } N(0, \psi)$$
$$w(t)\text{: zero mean normal}$$
$$E\{w(t)w(\tau)\} = \delta(t - \tau)Q$$

10.3. Derive (10.1.11) directly from (9.3.9).

10.4. Consider a sequence of observations $z(n)$, $n = 1, \ldots$. Define

$$J(N) = \sum_{n=1}^{N} [z(n) - ah(n)]^2$$
$$h(n),\ n = 1, \ldots : \text{is specified}$$
$$\hat{a}(N)\text{: value of } a \text{ which minimizes } J(N)$$
$$\hat{J}(N) = \sum_{n=1}^{N} [z(n) - \hat{a}(N)h(n)]^2$$

Derive a difference equation for $\hat{J}(N)$ of the form

$$\hat{J}(N + 1) = \hat{J}(N) + \text{specified function of } z(N + 1) \text{ and } \hat{a}(N)$$

10.5. Consider

$$x(n + 1) = \Phi x(n) + w(n), \qquad n = 1, \ldots$$
$$x(0) = 0$$
$$z(n) = Hx(n)$$
$$|w(n)|^2 \leq Q$$

Assume that it is known that either $H = 1$ or $H = 2$. Write the equations for a system which will process $x(n)$, $n = 1, \ldots, N$, to try to decide the value of H.

10.6. Consider

$$x(n+1) = \Phi x(n) + w(n)$$
$$x(0) = 0$$
$$z(n) = x(n) + i(n)$$

Assume that $i(n)$ is a process which represents the effect of an erratic sensor [observing $x(n)$] which usually makes perfect measurements but which sometimes introduces a very large error. Assume that the only information available about $i(n)$ is

$$i(n) = 0, \qquad \text{almost all values of } n$$
$$i(n)\text{: very large for a few values of } n$$

No a priori knowledge of the times n when $i(n) \neq 0$ is available. A $z(n)$ when $i(n) \neq 0$ is called a bad data point. Design a system which will attempt to detect these bad data points for the following two cases.

a. $E\{w(n)\} = 0$

$$E\{w(n_1)w(n_2)\} = \begin{cases} Q, & n_1 = n_2 \\ 0, & n_1 \neq n_2 \end{cases}$$

b. $|w(n)|^2 \leq Q$

10.7. Consider the zero mean Gaussian model

$$z(n) = x(n) + v(n)$$
$$x(n+1) = \Phi x(n) + w(n)$$
$$E\{w(n_1)w(n_2)\} = \begin{cases} Q, & n_1 = n_2 \\ 0, & n_1 \neq n_2 \end{cases}$$
$$x(n)\text{: stationary process}$$
$$E\{x^2(n)\} = \Gamma$$
$$E\{v(n_1)v(n_2)\} = \begin{cases} R, & n_1 = n_2 \\ 0, & n_1 \neq n_2 \end{cases}$$
$$w(n),\ v(n)\text{: independent}$$

Assume that it is desired to decide whether a sequence of observations $z_{\text{actual}}(n)$, $n = 1, \ldots, N$, are consistent with this model. One way to do this is to use a decision function

$$d(N) = \sum_{n=1}^{N} \frac{z^2(n)}{\Gamma + R}$$

If the model is valid, then $d(N)$ has a mean value of N, and a decision could be based on the size of $d(N) - N$. This decision function is not equivalent to $\xi(N)$, the log likelihood function, but $d(N)$ is obviously much simpler to implement. Explain in words why the use of $\xi(N)$ should provide "better" performance.

10.8. Consider

$$\mathbf{x}(n+1) = \mathbf{\Phi}\mathbf{x}(n) + \mathbf{G}w(n)$$

$$\mathbf{z}(n) = \mathbf{x}(n) + \mathbf{v}(n)$$

$$\mathbf{\Phi} = \begin{bmatrix} \Phi_{11} & 1 \\ 0 & 1 \end{bmatrix} \qquad \mathbf{G} = \begin{bmatrix} 0 \\ 1 \end{bmatrix} \qquad \mathbf{x}(n) = \begin{bmatrix} x_1(n) \\ x_2(n) \end{bmatrix}$$

$$\mathbf{x}(0) = \mathbf{0}$$

$$w(n),\ n = 1, \ldots : \text{completely unknown}$$

$$E\{\mathbf{v}(n_1)\mathbf{v}'(n_2)\} = \begin{cases} \mathbf{I}, & n_1 = n_2 \\ \mathbf{0}, & n_1 \neq n_2 \end{cases}$$

$$n = 1, \ldots, N$$

Assume that there are two hypotheses

$$\mathcal{H}_1 : \Phi_{11} = 0.5, \qquad n = 1, \ldots, N$$

$$\mathcal{H}_2 : \Phi_{11} = 0.25, \qquad n = 1, \ldots, N$$

Develop a hypothesis-testing system which can be used to process $\mathbf{z}(n)$, $n = 1, \ldots, N$, to make a decision as to which hypothesis is true. Define the decision function (or functions) explicitly. Describe in words how you would choose the decision rule. *Comment:* Two ways to approach this problem are (1) combine formulas from Sections 10.1.1. and 6.4.2, and (2) think about the nature of the model and combine ideas of Section 3.2 with formulas from Section 10.1.1.

10.9. Prove (10.3.6). *Hint:* Start with the results used to solve Exercise 7.8 of Chapter 7. Write the adjoint variable $\mathbf{p}(t)$ as $\mathbf{p}(t \mid T)$ to indicate dependence on terminal time T. Using (10.3.5),

$$\frac{d}{dT}\hat{J}(T) = \tfrac{1}{2}\hat{\boldsymbol{\epsilon}}'(T \mid T)\mathbf{R}^{-1}(T)\hat{\boldsymbol{\epsilon}}(T \mid T) + \text{other terms}$$

After a lot of work you can show that the "other terms" are zero. *Comment:* The details are tedious and are not recommended except to those who enjoy such manipulations. A simpler proof is probably possible using *Hamilton-Jacobi theory.*

PART IV

Nonlinear Models

After many chapters of linear models, we finally acknowledge the existence of nonlinear models. Since the "real world" is nonlinear, the reader may feel that it is about time. However, it turns out that many of the workable nonlinear techniques actually make use of linear model results and concepts, so the experiences of our earlier journeys are definitely of great value.

Our travels through nonlinear models consist of initial flights through the rarefied air of theoretical developments. These flights are smooth and quite scenic. However, to reach our destination of a "usable" set of equations, we have to come down to earth and proceed through a maze of very bumpy roads. There is simply no best single way to go.

Chapters 11 and 12 are both devoted to a nonlinear static model which is a nonlinear version of the model of Chapter 5. Chapter 13 is devoted to estimation for nonlinear discrete-time dynamic systems and hence is related to Chapter 6.

The details of discrete to continuous time limit process whereby the discrete-time results of Chapter 13 are extended to continuous time are not presented. The limiting procedure is straightforward but tedious, and one has to worry about the definitions of the limit (see Section 4.2). From an application point of view, the land of continuous-time nonlinear models is covered with dense fog, and an attempt to provide enough candlepower to see the road would require a whole new book.

11

Estimation: Static Nonlinear Systems

The state estimation problem model for a nonlinear, time-invariant system with no dynamics is

$$\begin{aligned} \mathbf{z} &= \mathbf{h}(\mathbf{x}, \mathbf{v}) \\ \mathbf{z}&: K_2 \text{ vector; observation} \\ \mathbf{x}&: K_1 \text{ vector; uncertain state whose value is to be estimated from } \mathbf{z} \\ \mathbf{v}&: K_2 \text{ vector; uncertain errors in observation (noise)} \\ \mathbf{h}(\mathbf{x}, \mathbf{v})&: K_2 \text{ vector; nonlinear function of } \mathbf{x} \text{ and } \mathbf{v} \end{aligned} \tag{11.1}$$

The model is often specialized to the case where

$$\mathbf{z} = \mathbf{h}(\mathbf{x}) + \mathbf{v} \tag{11.2}$$

The problem of interest is to determine some estimator (function) which can be used to convert a particular observation, $\mathbf{z}_{\text{actual}}$, into an estimate $\hat{\mathbf{x}}$ of the true value of the state $\mathbf{x}$.

The nonlinear model (11.1) or (11.2) is just a generalization of the $\mathbf{z} = \mathbf{H}\mathbf{x} + \mathbf{v}$ model of Chapter 5. It is possible to read the present chapter without first reading Chapter 5, but this is not recommended unless the reader already has some background in estimation theory.

Consideration of the ideas underlying the static case provides a solid basis for understanding the dynamic cases of Chapter 13. However, a nonlinear static model $\mathbf{z} = \mathbf{h}(\mathbf{x}, \mathbf{v})$ is encountered directly in a wide variety of applications. Some of these areas of application are briefly summarized in Fig. 11.1. In the statistics and system identification examples of Fig. 11.1, the

Area of Application	State $\mathbf{x}$	Observation $\mathbf{z}$
Surveying	Location of Towns on Grid	Distance between Towns
"Loran type" Navigation	Location of Ship on Ocean	"Times" of Arrival of Signals from Known Locations
Electric Power Network Monitoring	Voltage (Magnitude and Angle) at all Buses (Nodes) of Network	Voltage Magnitudes and Line Flows (Watts and Vars)
Statistics	Mean and Covariance of Random Vector	Samples of Random Vector
System Identification	Parameters Which Specify Linear Dynamic System (Time Constants, etc.)	Output of System

Figure 11.1. Examples of application areas for static $\mathbf{z} = \mathbf{h}(\mathbf{x}, \mathbf{v})$ model.

state vector $\mathbf{x}$ of (11.1) is really an uncertain parameter vector of some more general model with other "natural" states. In such cases, the term *parameter estimation* is more appropriate than state estimation. In some applications, such as the statistics example of Fig. 11.1, the $\mathbf{z} = \mathbf{h}(\mathbf{x}, \mathbf{v})$ model form is not natural. However, for the sake of presentation, this chapter emphasizes the *state estimation* problem arising from the static model $\mathbf{z} = \mathbf{h}(\mathbf{x}, \mathbf{v})$.

Sections 11.1–11.5 discuss theory, in particular, Bayesian, Fisher, relation of Bayesian to Fisher, weighted least squares, linear estimators, and unknown-but-bounded. Section 11.6 summarizes the theoretical concepts. Sections 11.7–11.9 discuss methods of implementation, such as iterative techniques, parameterization, and changes of variables. Stochastic approximation is briefly considered in Section 11.10.

Discussions on error analyses to determine how well an estimator performs are deferred until Chapter 12.

11.1 Bayesian

In a Bayesian model, both $\mathbf{v}$ and $\mathbf{x}$ are modeled as random vectors. Let $p(\mathbf{x})$ and $p(\mathbf{v})$ be the probability densities of $\mathbf{x}$ and $\mathbf{v}$. Assume that $\mathbf{x}$ and $\mathbf{v}$ are independent and that the probability densities, $p(\mathbf{x})$ and $p(\mathbf{v})$, are known.

11.1.1 General Estimators

Bayesian estimation theory is based on $p(\mathbf{x}\,|\,\mathbf{z})$, the conditional probability density of the state $\mathbf{x}$ given the observation $\mathbf{z}$. The function $p(\mathbf{x}\,|\,\mathbf{z})$ is sometimes called the a posteriori probability density, while $p(\mathbf{x})$ is called the a priori probability density. Using Bayes' rule,

$$p(\mathbf{x}\,|\,\mathbf{z}) = \frac{p(\mathbf{z}\,|\,\mathbf{x})p(\mathbf{x})}{p(\mathbf{z})} \tag{11.1.1}$$

For $\mathbf{z} = \mathbf{h}(\mathbf{x}) + \mathbf{v}$, $p(\mathbf{z}\,|\,\mathbf{x})$ can be obtained directly from $p(\mathbf{v})$ as

$$p(\mathbf{z}\,|\,\mathbf{x}) = p(\mathbf{v})\,|_{\mathbf{v}=\mathbf{z}-\mathbf{h}(\mathbf{x})}$$

Given $p(\mathbf{v})$ and $p(\mathbf{x})$, it is, in principle, possible to calculate $p(\mathbf{x}\,|\,\mathbf{z})$.

Given an observation, $\mathbf{z} = \mathbf{z}_{\text{actual}}$, $p(\mathbf{x}\,|\,\mathbf{z}_{\text{actual}})$ viewed as a function $\mathbf{x}$ can be used to define an estimate of $\mathbf{x}$ in several ways. Let $\hat{\mathbf{x}}$ denote some estimate of $\mathbf{x}$. Consider, first, the case where x is a scalar. Four different estimators are

1. *Conditional expectation:*

$$\hat{x} = \int_{-\infty}^{\infty} x\, p(x\,|\,\mathbf{z})\, dx$$

   This estimator is sometimes called (a) the mean of $p(x\,|\,\mathbf{z})$, (b) the conditional expectation of x given $\mathbf{z}$, or (c) the regression of x on $\mathbf{z}$. In this book it is called the conditional expectation estimator. It is optimum for many types of criteria and classes of probability distributions. (See Exercise 11.9.)

2. *Most probable:*

$$\hat{x} = \text{value of } x \text{ which maximizes } p(x\,|\,\mathbf{z})$$

   This estimator is sometimes called (a) the most probable, (b) the maximum a posteriori probability (MAP), or (c) the mode. In this book it is called the most probable estimator. The term *mode* actually refers to any local maximum of $p(x\,|\,\mathbf{z})$, while $\hat{x}$ is a global maximum (which still may not be unique).

3. *Median:*

$$\hat{x} = \text{value of } x \text{ determined by the condition}$$

$$\int_{-\infty}^{\hat{x}} p(x\,|\,\mathbf{z})\, dx = \int_{\hat{x}}^{\infty} p(x\,|\,\mathbf{z})\, dx$$

   This estimator is called the median. It may not be unique.

4. *Min-max:*

$$\hat{x} = \text{value of } x \text{ which minimizes the maximum possible error}$$

   This estimator is called the minimax. It may not exist and may not be unique.

The differences between these four estimators is illustrated in Fig. 11.1.1 in terms of a particular (and pathological) $p(x|\mathbf{z})$ for some $\mathbf{z} = \mathbf{z}_{\text{actual}}$. Note that the most probable estimator is most liable to give "unreasonable" answers when applied to complex (and pathological) problems.

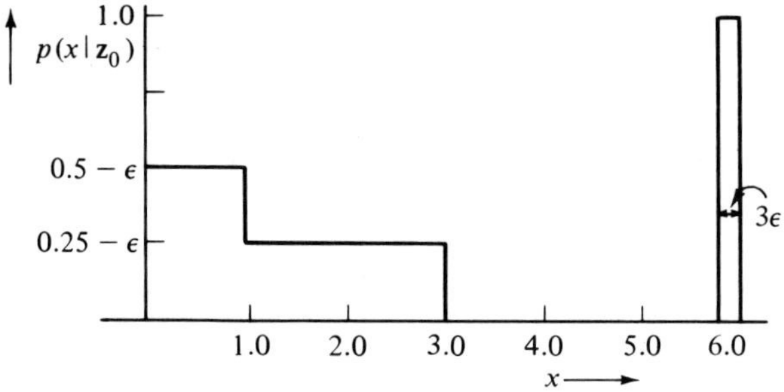

For small ϵ,

$$\hat{x} = \int_{-\infty}^{\infty} xp(x|\mathbf{z}_{\text{actual}})\,dx \approx 1.25 \qquad \text{(Conditional expectation)}$$

$$\hat{x} = \max_x p(x|\mathbf{z}_{\text{actual}}) = 6.0 \qquad \text{(Most probable)}$$

$$\left.\begin{aligned} &\hat{x}: \int_{-\infty}^{\hat{x}} p(x|\mathbf{z}_{\text{actual}})\,dx = \int_{\hat{x}}^{\infty} p(x|\mathbf{z}_{\text{actual}})\,dx \\ &\hat{x} \approx 1 \end{aligned}\right\} \qquad \text{(Median)}$$

$$\hat{x} = 3.0 \qquad \text{(Minimax)}$$

Figure 11.1.1 Comparison of four Bayesian estimators.

When $\mathbf{x}$ is a K_1 vector, the four estimators generalize as follows:

Conditional expectation: $\hat{\mathbf{x}} = \int_{-\infty}^{\infty} \cdots \int_{-\infty}^{\infty} \mathbf{x}\, p\,(\mathbf{x}|\mathbf{z})\,dx_1 \cdots dx_{K_1}$ (11.1.2)

Most probable (type 1): $\hat{\mathbf{x}}$ = value of $\mathbf{x}$ which maximizes $p(\mathbf{x}|\mathbf{z})$ (11.1.3)

Most probable (type 2): $\hat{x}_k$ = value of x_k which maximizes $p(x_k|\mathbf{z})$ (11.1.4)

Median: $\hat{x}_k$ = value of x_k determined by the condition

$$\int_{-\infty}^{\hat{x}_k} p(x_k|\mathbf{z})\,dx_k = \int_{\hat{x}_k}^{\infty} p(x_k|\mathbf{z})\,dx_k \qquad (11.1.5)$$

Min-max: $\hat{x}_k$ = value of x_k which minimizes the maximum possible error in kth component (11.1.6)

where for (11.1.4) to (11.1.6) $\hat{\mathbf{x}}' = [\hat{x}_1, \ldots, \hat{x}_{K_1}]$.

Note that there are two most probable estimators and that they can give different answers.

The conditional expectation $\hat{\mathbf{x}}$ of (11.1.2) is an unbiased estimate; i.e., $E\{\hat{\mathbf{x}}\} = E\{\mathbf{x}\}$. In general, the other three methods yield biased estimates.

Substitution of (11.1.1) into (11.1.2)–(11.1.6) shows that $p(\mathbf{z})$ has no effect on the most probable, median, and minimax estimators and enters the conditional expectation estimator as a scale factor. Hence when evaluating $p(\mathbf{x}\,|\,\mathbf{z})$ from (11.1.1), the denominator, $p(\mathbf{z})$, is often of no importance.

For the Gaussian case where

$$\begin{aligned}&\mathbf{z} = \mathbf{h}(\mathbf{x}) + \mathbf{v}\\ &\mathbf{x} \text{ is } N(\mathbf{0}, \boldsymbol{\psi})\\ &\mathbf{v} \text{ is } N(\mathbf{0}, \mathbf{R})\\ &\mathbf{x}, \mathbf{v}\text{: independent}\end{aligned}$$

it follows that

$$p(\mathbf{z}\,|\,\mathbf{x}) = [(2\pi)^{K_2}|\,R\,|]^{-1/2} \exp\{-(\tfrac{1}{2})[\mathbf{z} - \mathbf{h}(\mathbf{x})]'\mathbf{R}^{-1}[\mathbf{z} - \mathbf{h}(\mathbf{x})]\} \tag{11.1.7}$$

so that from (11.1.1)

$$\begin{aligned}p(\mathbf{x}\,|\,\mathbf{z}) = \left[\frac{1}{p(\mathbf{z})}\right][(2\pi)^{K_1+K_2}\,|\,R\,|\,|\,\boldsymbol{\psi}\,|]^{-1/2} \exp\Big[-\Big(\frac{1}{2}\Big)\{&\mathbf{x}'\boldsymbol{\psi}^{-1}\mathbf{x}\\ &+ [\mathbf{z} - \mathbf{h}(\mathbf{x})]'\mathbf{R}^{-1}[\mathbf{z} - \mathbf{h}(\mathbf{x})]\}\Big]\end{aligned} \tag{11.1.8}$$

In (11.1.8), $p(\mathbf{z})$ has not been evaluated, but as noted previously, $p(\mathbf{z})$ can often be ignored in the calculation of an estimate.

In general the preceding four estimators yield different estimates. However, for Gaussian $\mathbf{x}$ and $\mathbf{v}$ and a linear model $\mathbf{z} = \mathbf{H}\mathbf{x} + \mathbf{v}$, the conditional expectation, most probable, and median estimators all reduce to the estimator of Section 5.1. When $\mathbf{x}$ and $\mathbf{v}$ are not Gaussian, the four estimators in general yield different values even for a linear model.

EXAMPLE: Consider the scalar Bayesian model

$$\begin{gathered}z = Hx + v\\ x \text{ is } N(0, \psi) \qquad v \text{ is } N(0, R)\\ x \text{ and } v\text{: independent}\end{gathered}$$

Then z is $N(0, H^2\psi + R)$, so

$$p(z) = [2\pi(H^2\psi + R)]^{-1/2} \exp[-(\tfrac{1}{2})\{z^2[H^2\psi + R]^{-1}\}]$$

and from (11.1.8)

$$p(x\,|\,z) = \left(\frac{2\pi R\psi}{H^2\psi + R}\right)^{-1/2} \exp\left[\left(-\frac{1}{2}\right)\left\{\frac{x^2}{\psi} + \frac{(z - Hx)^2}{R} - \frac{z^2}{H^2\psi + R}\right\}\right]$$

After manipulation

$$\begin{aligned}p(x\,|\,z) = {}&[2\pi(H^2R^{-1} + \psi^{-1})^{-1}]^{-1/2}\\ &\times \exp[(-\tfrac{1}{2})\{[x - (H^2R^{-1} + \psi^{-1})^{-1}HR^{-1}z]^2[H^2R^{-1} + \psi^{-1}]^{-1}\}]\end{aligned}$$

Thus for both

$$\hat{x} = \int x\, p(x \mid z)\, dx$$

and

$$\hat{x} = \text{value of } x \text{ which maximizes } p(x \mid z)$$

it follows that

$$\hat{x} = [H^2R^{-1} + \psi^{-1}]^{-1}HR^{-1}z$$

which is the estimator of Section 5.1, see (5.1.1).

11.1.2 Linear Estimators

The estimators of (11.1.2)–(11.1.6) are all, in general, nonlinear functions. The design of estimators which are restricted to be linear functions is now discussed.

Consider $\mathbf{z} = \mathbf{h}(\mathbf{x}, \mathbf{v})$ and define

$$\begin{aligned}
\mathbf{m}_z &= E\{\mathbf{z}\} \\
\mathbf{m}_x &= E\{\mathbf{x}\} \\
\mathbf{\Gamma}_{zz} &= E\{(\mathbf{z} - \mathbf{m}_z)(\mathbf{z} - \mathbf{m}_z)'\} \\
\mathbf{\Gamma}_{xz} &= E\{(\mathbf{x} - \mathbf{m}_x)(\mathbf{z} - \mathbf{m}_z)'\}
\end{aligned}$$

Assume that a linear estimator of the form

$$\hat{\mathbf{x}} = \mathbf{W}\mathbf{z} + \mathbf{W}_0$$

is desired. It is reasonable to choose $\mathbf{W}$ and $\mathbf{W}_0$ to minimize $E\{(\mathbf{x} - \hat{\mathbf{x}})(\mathbf{x} - \hat{\mathbf{x}})'\}$. The result is that, given the Bayesian model $\mathbf{z} = \mathbf{h}(\mathbf{x}, \mathbf{v})$, the best linear estimator is given by

$$\boxed{\hat{\mathbf{x}} = \mathbf{m}_x + \mathbf{\Gamma}_{xz}\mathbf{\Gamma}_{zz}^{-1}[\mathbf{z} - \mathbf{m}_z]} \qquad (11.1.9)$$

The argument leading to (11.1.9) for the nonlinear model is essentially the same argument used to get (5.1.2) for the linear model (see Section 5.1.2). Note however, that there is a major difference in the usefulness of the equation. In (5.1.2), the estimator is expressed directly in terms of $\mathbf{H}$, $\mathbf{R}$, etc., which are specified directly by the model. In (11.1.9) the estimator is specified in terms of means and covariance matrices which can be very difficult to calculate from the $\mathbf{z} = \mathbf{h}(\mathbf{x}, \mathbf{v})$ model.

Equation (11.1.9) specifies the best linear estimator. This does not mean that it yields a good estimate. In many applications one of the general nonlinear estimators of Section 11.1.1 (conditional expectation, most probable, etc.) will yield a much better estimate. In fact, in many cases a simple ad hoc estimator developed by just "looking at" the structure of $\mathbf{h}$ will do much better than the best linear estimator of (11.1.9).

EXAMPLE: Consider

$$z_1 = x^2 + v_1$$
$$z_2 = x + v_2$$
$$x \text{ is } N(0, \psi)$$
$$\mathbf{v} = \begin{pmatrix} v_1 \\ v_2 \end{pmatrix} \qquad \mathbf{v} \text{ is } N(0, \mathbf{R}) \qquad \mathbf{R} = \begin{pmatrix} R_{11} & 0 \\ 0 & R_{22} \end{pmatrix}$$
$$x, \mathbf{v}\text{: independent}$$

For this model

$$\mathbf{m}_z = \begin{bmatrix} \psi \\ 0 \end{bmatrix} \qquad m_x = 0$$
$$\mathbf{\Gamma}_{xx} = \psi$$
$$\mathbf{\Gamma}_{xz} = E\{x[x^2 + v_1 - \psi, x + v_2]\} = [0 \quad \psi]$$
$$\mathbf{\Gamma}_{zz} = E\left\{\begin{pmatrix} x^2 + v_1 - \psi \\ x + v_2 \end{pmatrix}(x^2 + v_1 - \psi \quad x + v_2)\right\}$$
$$= \begin{pmatrix} \mathbf{\Gamma}_{zz,11} & 0 \\ 0 & \psi + R_{22} \end{pmatrix}$$

Thus from (11.1.9)

$$\hat{x} = \psi(\psi + R_{22})^{-1}z_2$$

which means that the best linear estimator completely ignores the observation z_1 (this conclusion can also be deduced directly from the model). Thus the best linear estimator will perform very poorly (relative to a nonlinear estimator) if, for example, $R_{22} \gg R_{11}$.

11.1.3 Linear-Type Nonlinear Estimators†

The $\hat{\mathbf{x}}$ of (11.1.9) is optimized with respect to $\mathbf{W}$ and $\mathbf{W}_0$ but can still be very poor relative to one of the estimators discussed in Section 11.1.1. However, (11.1.9) is easy to implement. One way to maintain most of the simplicity of (11.1.9) and still improve performance is to use *linear-type* nonlinear estimators.

For simplicity, assume that x and z are scalars. In place of $\hat{x} = Wz + W_0$, write

$$\hat{x} = W_1 z + W_2 z^2 + \cdots + W_q z^q + W_0$$

and then find the W_i and W_0 which minimize

$$E\{[x - (W_1 z + \cdots + W_q z^q + W_0)]^2\}$$

Obviously this approach extends to the vector case in various ways. Furthermore, it is not necessary to use powers; i.e., $\hat{x}$ such as

$$\hat{x} = \sum_{i=1}^{q} W_i \theta_i(z) + W_0 \tag{11.1.10}$$

could be used where the $\theta_i(z)$ are functions chosen by engineering judgment.

The estimator of (11.1.10) is not a linear estimator. However, implementation is of the same order of difficulty if q is small and the method of design (i.e., determination of the W_i) is conceptually the same as for the true linear estimator (11.1.9). Hence (11.1.10) is called a linear-type nonlinear estimator. The big problem associated with these estimators is obvious; how does one choose the $\theta_i(z)$ of (11.1.10) so that a small q yields a good estimate? This question often seems to have no answer.

11.1.4 Discussion of Bayesian

In Section 11.1.1 four estimators based on different criteria (conditional expectation, most probable, median, and min-max) were discussed. The choice of which of these estimators to use depends on whether one is an idealist or a realist:

1. Idealistic answer: Use the estimator whose criterion best matches the needs of the problem.
2. Realistic answer: Use the estimator which is easiest to implement.

The realistic answer is motivated by many factors. One factor is the fact that in many cases (especially when extended to dynamic systems, as in Chapter 13) none of the estimates can be calculated exactly and the general theory is merely a guide to the choice of a "reasonable" estimation technique. A second factor is the fact that for many (unimodal) $p(\mathbf{x}|\mathbf{z})$ all the estimators yield roughly the same estimate. In the subsequent developments of this and following chapters it will be seen that the most probable estimator (type 1 or type 2) is often the easiest to compute or approximate and hence is also the "most popular" when actual implementation is attempted. On the other hand, the conditional expectation estimate minimizes many properties of the estimate error so the conditional expectation is the "most popular" when theoretical studies are being done (see Exercise 11.9).

In Section 11.1.2, linear estimators for nonlinear models were discussed. If the necessary covariance matrices can be obtained, the linear estimator is easy to implement. However, as illustrated by the example in Section 11.1.2, its performance may be very poor (relative to even some simple nonlinear estimator). Therefore the "realistic approach" of choosing the $\hat{\mathbf{x}}$ which is easiest to implement should not be carried too far. (In other words, even the realist should have some ideals.)

11.2 Fisher

In a Fisher model $\mathbf{z} = \mathbf{h}(\mathbf{x}, \mathbf{v})$, the error, $\mathbf{v}$, is assumed to be a random vector with known probability density $p(\mathbf{v})$, while the state to be estimated, $\mathbf{x}$, is a completely unknown vector.

Fisher estimation theory for nonlinear problems is based on the likelihood function

$$p(\mathbf{z} : \mathbf{x}) = \text{probability density of } \mathbf{z} \text{ for specified } \mathbf{x}$$

Viewed as a function of $\mathbf{z}$ and $\mathbf{x}$, the likelihood function $p(\mathbf{z} : \mathbf{x})$ is the same as the conditional probability density $p(\mathbf{z}|\mathbf{x})$ of the Bayesian model. The special notation $p(\mathbf{z} : \mathbf{x})$ and terminology is used as in a Fisher formulation; $\mathbf{x}$ is not a random variable and Bayes rule equations such as (11.1.1) do not hold as $p(\mathbf{x})$ is not defined.

11.2.1 Maximum Likelihood Estimator

The most common and, in general, "powerful" estimate for Fisher models is the maximum likelihood estimate. If $\hat{\mathbf{x}}$ is the maximum likelihood estimate, then

$$\hat{\mathbf{x}} : \text{value of } \mathbf{x} \text{ which maximizes } p(\mathbf{z} : \mathbf{x}) \text{ for } \mathbf{z} = \mathbf{z}_{\text{actual}}$$

Maximum likelihood estimators have been investigated extensively in the statistical literature and many nice properties have been found (see also Section 11.2.4). $\hat{\mathbf{x}}$, of course, may not be unique.

For the Gaussian case where $\mathbf{z} = \mathbf{h}(\mathbf{x}) + \mathbf{v}$ and $\mathbf{v}$ is $N(\mathbf{0}, \mathbf{R})$, the likelihood function is

$$p(\mathbf{z} : \mathbf{x}) = [(2\pi)^{K_2} |\mathbf{R}|]^{-1/2} \exp\{-(\tfrac{1}{2})\{[\mathbf{z} - \mathbf{h}(\mathbf{x})]'\mathbf{R}^{-1}[\mathbf{z} - \mathbf{h}(\mathbf{x})]\}\}$$

Thus the maximum likelihood estimate $\hat{\mathbf{x}}$ minimizes

$$J(\mathbf{x}) = [\mathbf{z} - \mathbf{h}(\mathbf{x})]'\mathbf{R}^{-1}[\mathbf{z} - \mathbf{h}(\mathbf{x})] \tag{11.2.1}$$

As expected, the likelihood function $p(\mathbf{z} : \mathbf{x})$ is identical in form to the conditional density $p(\mathbf{z}|\mathbf{x})$.

It should be noted that, in general, the maximum likelihood estimator is biased; i.e.,

$$E\{\hat{\mathbf{x}}\} \neq \mathbf{x} \tag{11.2.2}$$

EXAMPLE: Consider

$$z = x^{1/2} + v$$

$$x\text{: unknown}$$

$$v \text{ is } N(0, R)$$

Let $\hat{x}$ be the maximum likelihood estimator. Then from (11.2.1)

$$\hat{x} = z^2$$

It then follows that

$$\begin{aligned} E\{\hat{x}\} &= E\{x + 2xv + v^2\} \\ &= x + R \end{aligned}$$

so $\hat{x}$ is biased with a bias error

$$b = x - E\{\hat{x}\} = R$$

In the case of a linear Gaussian model, $\mathbf{z} = \mathbf{H}\mathbf{x} + \mathbf{v}$, $\mathbf{v}$ is $N(\mathbf{0}, \mathbf{R})$, the maximum likelihood $\mathbf{x}$ is as developed in Section 5.2. (See Exercise 5.9.)

As indicated in the introduction to this chapter, the model form $\mathbf{z} = \mathbf{h}(\mathbf{x}, \mathbf{v})$ is not always the natural form. However, the maximum likelihood estimator is based on the likelihood function $p(\mathbf{z} : \mathbf{x})$, which is not restricted to $\mathbf{z} = \mathbf{h}(\mathbf{x}, \mathbf{v})$ types of models.

EXAMPLE: Consider

$$\mathbf{z}_N = \begin{bmatrix} z(1) \\ \cdot \\ \cdot \\ \cdot \\ z(N) \end{bmatrix}.$$

$z(n)$ is $N(m, \mathbf{\Gamma})$

$z(n)$, $n = 1, \ldots, N$: independent

$m, \mathbf{\Gamma}$; unknown.

Thus $\mathbf{x}$ the unknown state is

$$\mathbf{x} = \begin{bmatrix} m \\ \mathbf{\Gamma} \end{bmatrix}$$

For this case

$$p(\mathbf{z}_N : m, \mathbf{\Gamma}) = (2\pi\mathbf{\Gamma})^{-N/2} \exp\{-(\tfrac{1}{2})J(\mathbf{x})\}$$

$$J(\mathbf{x}) = \sum_{n=1}^{N} \frac{(z(n) - m)^2}{\mathbf{\Gamma}}$$

To find $\hat{\mathbf{x}}$ (i.e., $\hat{m}$ and $\hat{\mathbf{\Gamma}}$), one solves

$$\frac{\partial p(\mathbf{z} : \mathbf{x})}{\partial \mathbf{x}} = \mathbf{0}$$

or

$$\frac{\partial p(\mathbf{z} : m, \mathbf{\Gamma})}{\partial m} = -2 \sum_{n=1}^{N} \frac{z(n) - m}{\mathbf{\Gamma}} = 0$$

$$\frac{\partial p(\mathbf{z} : m, \mathbf{\Gamma})}{\partial \mathbf{\Gamma}} = (2\pi\mathbf{\Gamma})^{-N/2} \exp\left\{-\left(\frac{1}{2}\right)J(x)\right\}\left[-\frac{N}{2}\mathbf{\Gamma}^{-1} + \frac{\sum_{n=1}^{N}(z(n) - m)^2}{2\mathbf{\Gamma}^2}\right]$$
$$= 0$$

The result is:

$$\hat{m} = \frac{1}{N}\sum_{n=1}^{N} z(n)$$

$$\hat{\mathbf{\Gamma}} = \frac{1}{N}\sum_{n=1}^{N} (z(n) - \hat{m})^2$$

For this result $\hat{m}$ is unbiased,

$$E\{\hat{m}\} = m$$

but $\hat{\mathbf{\Gamma}}$ is biased,

$$E\{\hat{\mathbf{\Gamma}}\} = \frac{1}{N} \sum_{n=1}^{N} E\left\{z(n) - \frac{1}{N} \sum_{n=1}^{N} z(m)\right\}^2$$

$$= \left(\frac{N-1}{N}\right) \mathbf{\Gamma}$$

11.2.2 Other Estimators†

The maximum likelihood estimate is by no means the only way of developing estimators for nonlinear Fisher models. Many other types are discussed in the literature.

In Section 11.1.2, a linear estimate was developed for a nonlinear Bayesian model by essentially repeating the linear Bayesian model theory of Chapter 5. Therefore it is tempting to try to develop a similar linear estimator theory for Fisher nonlinear models. In general, however, this cannot be done, as the linear Fisher model theory of Chapter 5 was based on the concept of forcing the estimator to be unbiased and then finding the particular linear unbiased estimator which minimized the error covariance matrix. For a nonlinear model, there usually exists no linear estimator which is unbiased and without such a condition, there is no "reasonable" way to proceed when $\mathbf{x}$ is completely unknown.

11.2.3 Interval Estimators†

The maximum likelihood estimate yields a point estimate, i.e., an estimate $\hat{\mathbf{x}}$ of $\mathbf{x}$ is calculated from $\mathbf{z}$. An *interval estimate* involves a somewhat different point of view. To illustrate, assume that x is a scalar. Then instead of computing $\hat{\mathbf{x}}$ to be the point estimate of x, one computes an $x_{\min}$ and $x_{\max}$ which define an interval

$$x_{\min} \leq x \leq x_{\max}$$

where $x_{\min}$ and $x_{\max}$ are specified so that the probability of the true but unknown x lying within the interval is some specified quantity. The interval is sometimes called a confidence interval.

EXAMPLE: Consider

$$z = x + v$$

$$v \text{ is } N(0, R)$$

$$x\text{: unknown}$$

Given an observation z_{actual}, define

$$x_{\min} = z_{\text{actual}} - 3(R)^{1/2}$$

$$x_{\max} = z_{\text{actual}} + 3(R)^{1/2}$$

Then one can say that x lies in the interval $x_{\min} \le x \le x_{\max}$ with probability ρ:

$$\rho = \frac{1}{(2\pi R)^{1/2}} \int_{-3R^{1/2}}^{3R^{1/2}} \exp\left(\frac{v^2}{2R}\right) dv$$

Interval estimators and confidence intervals are important concepts, as they emphasize that often it is more important to know with "some certainty" that x lies in some interval than to simply have some estimate of x. The interval estimate philosophy is not emphasized in this book (for Bayesian and Fisher models). Instead the error analysis approach is used where the point estimate $\hat{x}$ and its error properties are studied. The two concepts are usually equivalent in practice.

11.2.4 Consistent, Efficient, and Sufficient Estimators†

Some important general ideas associated with Fisher estimation theory are the concepts of consistent, efficient, and sufficient estimators. These ideas are now briefly discussed.

Consider a Fisher model where

$$\mathbf{z}_N = \begin{bmatrix} \mathbf{z}(1) \\ \cdot \\ \cdot \\ \cdot \\ \mathbf{z}(N) \end{bmatrix} \tag{11.2.3}$$

$$\mathbf{z}(n) = \mathbf{h}(\mathbf{x}, \mathbf{v}(n)), \qquad n = 1, \ldots N$$

Let $\hat{\mathbf{x}}_N$ denote an estimate of $\mathbf{x}$ computed from $\mathbf{z}_N$. Then $\hat{\mathbf{x}}_N$ is said to be a *consistent estimate* if $\hat{\mathbf{x}}_N$ converges to $\mathbf{x}$ as N goes to infinity.

Consider $\mathbf{z} = \mathbf{h}(\mathbf{x}, \mathbf{v})$ and let $\hat{\mathbf{x}}$ be some unbiased estimate of $\mathbf{x}$. Then $\hat{\mathbf{x}}$ is an *efficient estimate* if there exists no other unbiased estimate which has a smaller error covariance matrix. Consider (11.2.3). The estimator which yields $\hat{\mathbf{x}}_N$ is *asymptotically efficient* if $\hat{\mathbf{x}}_N$ becomes efficient as N goes to infinity. Efficiency can also be defined relative to the lower bound (Cramer-Rao) of the estimate error variance to be discussed in Section 12.3.

Consider $\mathbf{z} = \mathbf{h}(\mathbf{x}, \mathbf{v})$ and let $\hat{\mathbf{x}}$ be some estimate. In vague, general terms, $\hat{\mathbf{x}}$ is a *sufficient statistic* if $\hat{\mathbf{x}}$ contains "all the information" in $\mathbf{z}$ about $\mathbf{x}$. This definition can be made more precise.

The maximum likelihood estimate of Section 11.2.1 is a "good" estimate because it can be shown that if an efficient estimator exists, the maximum likelihood estimate will be efficient, and if a sufficient statistic exists, the maximum likelihood estimate will be a function of it. It can also be shown that under fairly general conditions, the maximum likelihood estimate is asymptotically efficient.

For the linear Fisher model, $\mathbf{z} = \mathbf{H}\mathbf{x} + \mathbf{v}$, $\mathbf{v}$ is $N(\mathbf{0}, \mathbf{R})$, the estimator of Chapter 5 yields consistent and efficient estimates.

11.2.5 Discussion of Fisher

The maximum likelihood estimator is definitely a reasonable approach and has many nice properties. As a general rule of thumb that is almost universally applicable, one can say that a maximum likelihood estimator is the "best" one can do for nonlinear Fisher models. However, for most nonlinear models the maximum likelihood estimate will be biased, and it does not, in general, minimize any criteria such as the estimate error covariance matrix.

11.3 Relationship Between Fisher and Bayesian†

Fisher and Bayesian models both conceptually use the observation $\mathbf{z}$ to calculate some function of $\mathbf{x}: p(\mathbf{x} \mid \mathbf{z})$ if Bayesian, $p(\mathbf{z}:\mathbf{x})$ if Fisher. The estimate $\hat{\mathbf{x}}$ is then obtained by some further operation. In Section 5.3 the Bayesian to Fisher relationship was discussed for linear models. This argument is now generalized to nonlinear models.

Bayesian estimation is based on

$$p(\mathbf{x} \mid \mathbf{z}) = \frac{p(\mathbf{z} \mid \mathbf{x})p(\mathbf{x})}{p(\mathbf{z})}$$

However, as noted in Section 11.1, $p(\mathbf{z})$ does not affect certain estimators of $\mathbf{x}$. In the following, a Bayesian estimator is considered to be only a function of $p(\mathbf{z} \mid \mathbf{x})p(\mathbf{x})$.

Consider first how to go "from Bayesian to Fisher." If the a priori distribution $p(\mathbf{x})$ is "spread out" so that it is flat relative to $p(\mathbf{z} \mid \mathbf{x})$ (viewed as a function of $\mathbf{x}$), then the Bayesian estimator is effectively based only on $p(\mathbf{z} \mid \mathbf{x})$. As noted in Section 11.2, $p(\mathbf{z}:\mathbf{x})$ and $p(\mathbf{z} \mid \mathbf{x})$ are the same functions of $\mathbf{x}$. Thus a Bayesian model estimator becomes closer and closer to a Fisher model estimator as the a priori distribution $p(\mathbf{x})$ "spreads out." The "Fisher to Bayesian route" is the same procedure used in the linear case, i.e., introduction of an extra measurement.

As emphasized in earlier chapters, the ability to make a Bayesian model estimator "look like" a Fisher model estimator or vice versa does not mean that the models are equivalent. Fisher and Bayesian models involve fundamentally different assumptions.

11.4 Weighted Least Squares

A Bayesian model assumes that both $\mathbf{x}$ and $\mathbf{v}$ are random vectors, while a Fisher model assumes a probabilistic structure only for $\mathbf{v}$. These probabilistic assumptions yielded functions of $\mathbf{x}$ given $\mathbf{z}$, $p(\mathbf{x} \mid \mathbf{z})$, or $p(\mathbf{z}:\mathbf{x})$, which are used as a basis for calculating an estimate $\hat{\mathbf{x}}$.

The need for probabilistic assumptions can be removed entirely simply by using engineering judgment to choose some function of $\mathbf{x}$ and $\mathbf{z}$ which, given $\mathbf{z}$, can be used to calculate an estimate $\hat{\mathbf{x}}$. Let $J(\mathbf{x}:\mathbf{z})$ denote such a function. The $\hat{\mathbf{x}}$ usually obtained from $J(\mathbf{x}:\mathbf{z})$ is

$$\hat{\mathbf{x}} = \text{value of } \mathbf{x} \text{ which minimizes } J(\mathbf{x}:\mathbf{z}) \tag{11.4.1}$$

The $\hat{\mathbf{x}}$ of (11.4.1) is most analogous to the most probable Bayesian estimator and the maximum likelihood Fisher estimator. Obviously one could also obtain estimators other than (11.4.1) from some $J(\mathbf{x}:\mathbf{z})$, just as was done with $p(\mathbf{x}\,|\,\mathbf{z})$.

For $\mathbf{z} = \mathbf{h}(\mathbf{x}) + \mathbf{v}$, a common $J(\mathbf{x}:\mathbf{z})$ is the weighted-least-squares criterion for which

$$J(\mathbf{x}:\mathbf{z}) = [\mathbf{z} - \mathbf{h}(\mathbf{x})]'\mathbf{R}^{-1}[\mathbf{z} - \mathbf{h}(\mathbf{x})] \tag{11.4.2}$$

where $\mathbf{R}$ is a positive definite matrix. Comparison of (11.4.2) with the Fisher case for Gaussian random vectors, (11.2.1), shows that the same $\hat{\mathbf{x}}$ will result if the engineering judgment behind the choice of $\mathbf{R}$ in (11.4.2) yielded the same values as those in (11.2.1).

In the special linear model case, (11.4.2) specializes to that considered in Section 5.4.

11.5 Unknown-but-Bounded

For unknown-but-bounded models, the basic concepts of Section 5.6 on the linear model also apply to the nonlinear case. Consider

$$\mathbf{z} = \mathbf{h}(\mathbf{x}) + \mathbf{v}$$
$$\mathbf{x} \in \Omega_x$$
$$\mathbf{v} \in \Omega_v$$

Define

$$\Omega_{x|z} = \{\mathbf{x}:\mathbf{z} - \mathbf{h}(\mathbf{x}) \in \Omega_v\}$$

Then the estimate set Ω_{est} is given by

$$\Omega_{\text{est}} = \Omega_x \cap \Omega_{x|z}$$

The difficult part of the problem is, of course, to actually calculate the estimate set or some properties thereof. To illustrate, consider the special case where

$$\Omega_v = \{\mathbf{v}:\mathbf{v}'\mathbf{R}^{-1}\mathbf{v} \leq 1\}$$
$$\Omega_x = \{\mathbf{x}:\mathbf{x}'\boldsymbol{\psi}^{-1}\mathbf{x} \leq 1\}$$

In this case

$$\Omega_{x|z} = \{\mathbf{x}:[\mathbf{z} - \mathbf{h}(\mathbf{x})]'\mathbf{R}^{-1}[\mathbf{z} - \mathbf{h}(\mathbf{x})] \leq 1\}$$

The set $\Omega_{x|z}$ is not an ellipsoid and in general there is no easy way to bound it by an ellipsoid. The set $\Omega_{x|z}$ may not be convex and may not be connected.

Its structure and shape depend on the nonlinear function $\mathbf{h}(\mathbf{x})$. The set Ω_{est} is

$$\Omega_{\text{est}} = \{\mathbf{x} : [\mathbf{z} - \mathbf{h}(\mathbf{x})]'\mathbf{R}^{-1}[\mathbf{z} - \mathbf{h}(\mathbf{x})] \leq 1 \text{ and } \mathbf{x}'\boldsymbol{\psi}^{-1}\mathbf{x} \leq 1\}$$

It can be bounded by Ω_b:

$$\Omega_b \supset \Omega_{\text{est}}$$

$$\Omega_b = \{\mathbf{x} : J(\mathbf{x} : \mathbf{z}) \leq 1\}$$

$$J(\mathbf{x} : \mathbf{z}) = \rho[\mathbf{z} - \mathbf{h}(\mathbf{x})]'\mathbf{R}^{-1}[\mathbf{z} - \mathbf{h}(\mathbf{x})] + (1 - \rho)\mathbf{x}'\boldsymbol{\psi}^{-1}\mathbf{x}$$

any ρ such that

$$0 \leq \rho \leq 1$$

These Ω_{est} and Ω_b are, in general, not ellipsoids and may not be convex or connected.

Now consider the ellipsoidal model with $\boldsymbol{\psi}^{-1} = \mathbf{0}$ so that

$$\Omega_{\text{est}} = \Omega_{x \mid z} = \{\mathbf{x} : J(\mathbf{x} : \mathbf{z}) \leq 1\}$$

$$J(\mathbf{x} : \mathbf{z}) = [\mathbf{z} - \mathbf{h}(\mathbf{x})]'\mathbf{R}^{-1}[\mathbf{z} - \mathbf{h}(\mathbf{x})]$$

Let $\hat{\mathbf{x}}$ denote a value of $\mathbf{x}$ which minimizes $J(\mathbf{x} : \mathbf{z})$. Then if

$$J(\mathbf{x} : \mathbf{z})\,|_{\mathbf{x}=\hat{\mathbf{x}}} \leq 1$$

it follows that

$$\hat{\mathbf{x}} \in \Omega_{\text{est}}$$

If only one $\hat{\mathbf{x}}$ exists, it is reasonable to view $\hat{\mathbf{x}}$ as the center of Ω_{est}, and if desired, to treat $\hat{\mathbf{x}}$ as an "estimate" of $\mathbf{x}$.

11.6 Summary of Types of Estimators

Given an observation $\mathbf{z}$, the following four functions of $\mathbf{x}$ arose in the estimation theory of the preceding sections:

$p(\mathbf{x} \mid \mathbf{z})$	Bayesian
$p(\mathbf{z} : \mathbf{x})$	Fisher
$J(\mathbf{x} : \mathbf{z})$	Weighted least squares
$J(\mathbf{x} : \mathbf{z})$	Unknown-but-bounded, where $\Omega_{\text{est}} = \{\mathbf{x} : J(\mathbf{x} : \mathbf{z}) \leq 1\}$

These functions of $\mathbf{x}$ (for $\mathbf{z} = \mathbf{z}_{\text{actual}}$) can all be viewed as "measuring" the uncertainty remaining in the true value of $\mathbf{x}$ after $\mathbf{z} = \mathbf{z}_{\text{actual}}$ has been observed. An estimate $\hat{\mathbf{x}}$ is obtained by performing some further operation on this function of $\mathbf{x}$.

Many different types of point estimators have been given in this chapter and Chapter 5 on the linear model. An attempt is now made to summarize some of these ideas and the associated terminology. This is intended only to be a guide, and many (important) details have been omitted. (see also Fig. 5.12.1)

The basic definitions are repeated for convenience.

Linear model:

$\mathbf{z} = \mathbf{H}\mathbf{x} + \mathbf{v}$

Nonlinear model:

$\mathbf{z} = \mathbf{h}(\mathbf{x}, \mathbf{v})$

Bayesian formulation:

$\mathbf{x}, \mathbf{v}$: independent random vectors

Gaussian case:

$\mathbf{x}, \mathbf{v}$: have normal distributions

Conditional density:

$p(\mathbf{x} \mid \mathbf{z})$

Fisher formulation:

$\mathbf{x}$: unknown parameter vector

$\mathbf{v}$: random vector

Gaussian case:

$\mathbf{v}$ has normal distribution

Likelihood function:

$p(\mathbf{z} : \mathbf{x})$

Weighted-least-squares formulation:

No models for $\mathbf{x}$ and $\mathbf{v}$; estimator defined in terms of minimizing some function, $J(\mathbf{x} : \mathbf{z})$

Unknown-but-bounded formulation:

$\mathbf{x}$ and $\mathbf{v}$: constrained to lie in sets $\mathbf{\Omega}_x$ and $\mathbf{\Omega}_v$

Linear estimator:

$\hat{\mathbf{x}} = \mathbf{W}\mathbf{z} + \mathbf{W}_0$, where $\mathbf{W}$ and $\mathbf{W}_0$ are specified a priori

Nonlinear estimator:

$\hat{\mathbf{x}} = \mathbf{g}(\mathbf{z})$, where the function $\mathbf{g}(\cdot)$ is specified a priori; often $\mathbf{g}(\mathbf{z})$ is not specified explicitly; i.e., $\hat{\mathbf{x}}$ may be specified by $\hat{\mathbf{x}}$ maximizes $\mathbf{f}(\mathbf{z}, \mathbf{x})$, where the function $\mathbf{f}(\cdot)$ is specified a priori

Unbiased estimator:

$E\{\hat{\mathbf{x}}\} = \mathbf{x}$ Fisher, for any $\mathbf{x}$

$E\{\hat{\mathbf{x}}\} = E\{\mathbf{x}\}$ Bayesian

Biased estimator:

$E\{\hat{\mathbf{x}}\} \neq \mathbf{x}$ Fisher

$E\{\hat{\mathbf{x}}\} \neq E\{\mathbf{x}\}$ Bayesian

Covariance matrix of estimate error; Bayesian or Fisher:

If $\hat{\mathbf{x}}$ is unbiased, $\mathbf{\Sigma} = E\{[\mathbf{x} - \hat{\mathbf{x}}][\mathbf{x} - \hat{\mathbf{x}}]'\}$

Estimate set: unknown-but-bounded

$$\Omega_{\text{est}} = \{\mathbf{x} : J(\mathbf{x} : \mathbf{z}) \leq 1\}$$

Some of the important types of estimators obtained from a Bayesian formulation are compared in Fig. 11.6.1. Sixteen cases are listed. Cases 1, 5, 6, 9, and 13 all yield the same estimator; i.e., if

$$E\{\mathbf{v}\mathbf{v}'\} = \mathbf{R},\ E\{\mathbf{x}\mathbf{x}'\} = \boldsymbol{\psi}$$

and if $\mathbf{v}$ and $\mathbf{x}$ are zero mean then

$$\hat{\mathbf{x}} = \boldsymbol{\Sigma}\mathbf{H}'\mathbf{R}^{-1}\mathbf{z}$$

$$\boldsymbol{\Sigma} = [\mathbf{H}'\mathbf{R}^{-1}\mathbf{H} + \boldsymbol{\psi}^{-1}]^{-1}$$

The cases which are labeled nonlinear and/or biased mean that, in general, the estimator is nonlinear and/or biased, although there may exist special cases.

Some of the important types of estimators obtained from a Fisher formulation are compared in Fig. 11.6.2. Six cases are listed. Cases 1, 5, and 6 all yield the same estimator; i.e., if

$$E\{\mathbf{v}\mathbf{v}'\} = \mathbf{R}$$

and if $\mathbf{v}$ is zero mean, then

$$\hat{\mathbf{x}} = \boldsymbol{\Sigma}\mathbf{H}'\mathbf{R}^{-1}\mathbf{z}$$

$$\boldsymbol{\Sigma} = [\mathbf{H}'\mathbf{R}^{-1}\mathbf{H}]^{-1}$$

As in Fig. 11.6.1, cases labeled nonlinear or biased may not be in special cases.

In a weighted-least-squares formulation, only one estimate was defined, the value of $\mathbf{x}$ which minimizes $J(\mathbf{x} : \mathbf{z})$. Nothing can be said about the properties of $\hat{\mathbf{x}}$ unless some model for $\mathbf{x}$ and $\mathbf{v}$ is assumed.

In an unknown-but-bounded formulation, the estimate is best considered to be the whole set Ω_{est}. If a point estimator is desired, the center of Ω_{est} is a reasonable choice. There are many ways to define a center of a set, but the value of $\mathbf{x}$ which minimizes $J(\mathbf{x} : \mathbf{z})$ is a reasonable choice.

11.7 Iterative Methods for Finding a Minimum

Consider the three estimators most probable Bayesian; maximum likelihood Fisher; and weighted-least-squares. In all three cases, the estimator $\hat{\mathbf{x}}$ is obtained as the value of $\mathbf{x}$ which maximizes (or minimizes) some function of $\mathbf{x}$ (assuming that $\mathbf{z} = \mathbf{z}_{\text{actual}}$ is given). For unknown-but-bounded models, the center of the estimate set can also sometimes be found by minimizing some function of $\mathbf{x}$. Hence it is appropriate to deviate briefly from estimation theory per se and consider the problem of finding the value of $\mathbf{x}$ which minimizes some $J(\mathbf{x})$. A standard approach to such a problem is the iterative solution.

Estimator	Definition	Case	Model: Linear	Model: Nonlinear	Model: Gaussian	Model: Non-Gaussian	Properties of Estimator: Linear	Properties of Estimator: Nonlinear	Properties of Estimator: Unbiased	Properties of Estimator: Biased
Conditional Expectation of $\mathbf{x}$ given $\mathbf{z}$	$\hat{\mathbf{x}} = \int \mathbf{x} p(\mathbf{x} \mid \mathbf{z})\, d\mathbf{x}$	1	X		X		X		X	
		2	X			X		X	X	
		3		X	X			X	X	
		4		X		X		X	X	
Linear, Minimum Variance	Linear estimator with min $\mathbf{\Sigma}$	5	X		X		X		X	
		6	X			X	X		X	
		7		X	X		X		X	
		8		X		X	X		X	
Most Probable	$\mathbf{x}$ maximizes $p(\mathbf{x} \mid \mathbf{z})$ or x_i maximizes $p(x_i \mid \mathbf{z})$ $i = 1 \ldots$	9	X		X		X		X	
		10	X			X		X		X
		11		X	X			X		X
		12		X		X		X		X
Median	$\hat{x}_i$ $i = 1 \ldots$ is such that $\int_{-\infty}^{\hat{x}_i} p(x_i \mid \mathbf{z})\, dx_i = \int_{\hat{x}_i}^{\infty} p(x_i \mid \mathbf{z})\, dx_i$	13	X		X		X		X	
		14	X			X		X		X
		15		X	X			X		X
		16		X		X		X		X

Figure 11.6.1 Types of Bayesian estimators.

Estimator	Definition	Model					Properties of Estimator			
		Case	Linear	Non-linear	Gaussian	Non-Gaussian	Linear	Non-linear	Un-biased	Biased
Maximum Likelihood	$\hat{\mathbf{x}}$ maximizes $p(\mathbf{z} : \mathbf{x})$	1	X		X		X		X	
		2	X			X		X		X
		3		X	X			X		X
		4		X		X		X		X
Linear, Minimum Variance, Unbiased	Unbiased linear estimator with min $\mathbf{\Sigma}$	5	X		X		X		X	
		6	X			X	X		X	
		7		X			not exist			

Figure 11.6.2. Types of Fisher estimators.

Let $\hat{\mathbf{x}}$ denote the value of $\mathbf{x}$ which minimizes $J(\mathbf{x})$. It is assumed that $\hat{\mathbf{x}}$ exists. Then $\hat{\mathbf{x}}$ has the property that

$$\mathbf{J}^{(1)}(\mathbf{x})|_{\mathbf{x}=\hat{\mathbf{x}}} = \mathbf{J}^{(1)}(\hat{\mathbf{x}}) = \mathbf{0}$$
$$\mathbf{J}^{(1)}(\mathbf{x}) = \frac{\partial J(\mathbf{x})}{\partial \mathbf{x}} = \left[\frac{\partial J(\mathbf{x})}{\partial x_1} \cdots \frac{\partial J(\mathbf{x})}{\partial x_{K_1}}\right] \tag{11.7.1}$$

$\mathbf{J}^{(1)}(\mathbf{x})$ is the *gradient* of $J(\mathbf{x})$. Note that $\mathbf{J}^{(1)}(\mathbf{x})$ is defined to be a row vector. Assume for the moment that $J(\mathbf{x})$ has only one extremum, a minimum, so that there is only one unique $\hat{\mathbf{x}}$ which yields (11.7.1).

The basic iterative technique consists of solving the difference equation

$$\boxed{\tilde{\mathbf{x}}_{i+1} = \tilde{\mathbf{x}}_i - \mathbf{B}(i)\mathbf{J}^{(1)\prime}(\tilde{\mathbf{x}}_i), \qquad i = 1, 2, \ldots} \tag{11.7.2}$$

for specified $\mathbf{B}(i)$ until $\tilde{\mathbf{x}}_i$ converges; i.e., until (hopefully)

$$\tilde{\mathbf{x}}_{i+1} - \tilde{\mathbf{x}}_i \longrightarrow \mathbf{0}$$

The K_1 by K_1 matrix $\mathbf{B}(i)$ is called the gain matrix. If convergence occurs, it follows from (11.7.1) and (11.7.2) that

$$\tilde{\mathbf{x}}_i = \hat{\mathbf{x}}$$

for any gain matrix $\mathbf{B}(i)$ provided that it is of full rank.

The basic idea behind (11.7.2) is illustrated in Fig. 11.7.1 for a scalar

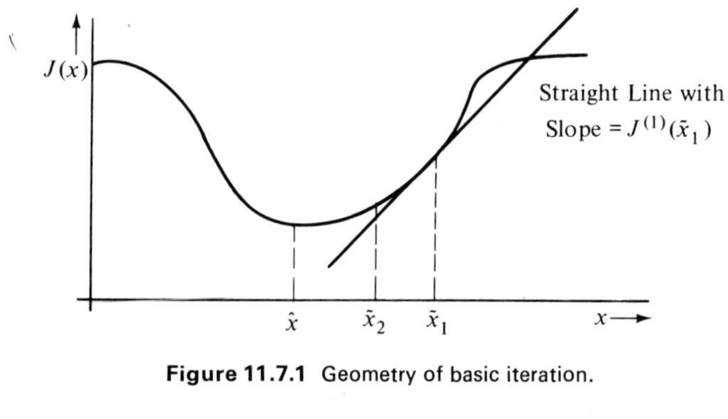

Figure 11.7.1 Geometry of basic iteration.

x. Assume that a guess $\tilde{x}_1$ has been made on the value $\hat{x}$ which minimizes $J(x)$. If $\tilde{x}_1$ is as shown in Fig. 11.7.1, the gradient $J^{(1)}(\tilde{x}_1)$ (or slope or derivative) is positive. Hence if the gain $B(1)$ is positive, the $\tilde{x}_2$ given by (11.7.2) will be smaller than $\tilde{x}_1$, i.e., *toward* $\hat{x}$, and might be as illustrated in Fig. 11.7.1. Thus the gradient tells which "direction" to go to choose $\tilde{x}_2$. The gain $B(i)$ controls "how far" to go in each step. If $B(i)$ is too large, $\tilde{x}_{i+1}$ may "overshoot" and go into *oscillations* about $\hat{x}$. If $B(i)$ is too small, $\tilde{x}_i$ will approach $\hat{x}$ very slowly and may not even get there. It is clear that

$$B(i) > 0$$

is a necessary condition, at least for large i.

A somewhat different geometrical picture results if (11.7.2) is viewed as an iteration to find the $\mathbf{x}$ which makes $\mathbf{J}^{(1)}(\mathbf{x}) = \mathbf{0}$ rather than the $\mathbf{x}$ which minimizes $J(\mathbf{x})$.

A still different geometrical interpretation of (11.7.2) is illustrated in Fig. 11.7.2 for the case of a two-dimensional $\mathbf{x}$. Figure 11.7.2 shows the contour lines where

$$\text{contour line} = \{\mathbf{x} : J(\mathbf{x}) \text{ is constant}\}$$

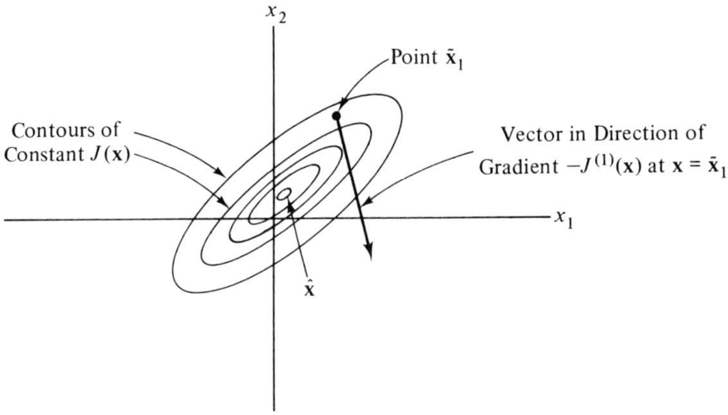

Figure 11.7.2 Contour lines and gradient vector.

and illustrates the nature of the gradient, $\mathbf{J}^{(1)}(\mathbf{x})$. Figure 11.7.3 illustrates two possible iterative paths that might result when using a gain

$$\mathbf{B}(i) = b\,\mathbf{I}$$

where b is a positive constant. As seen in Fig. 11.7.3, an overly small value

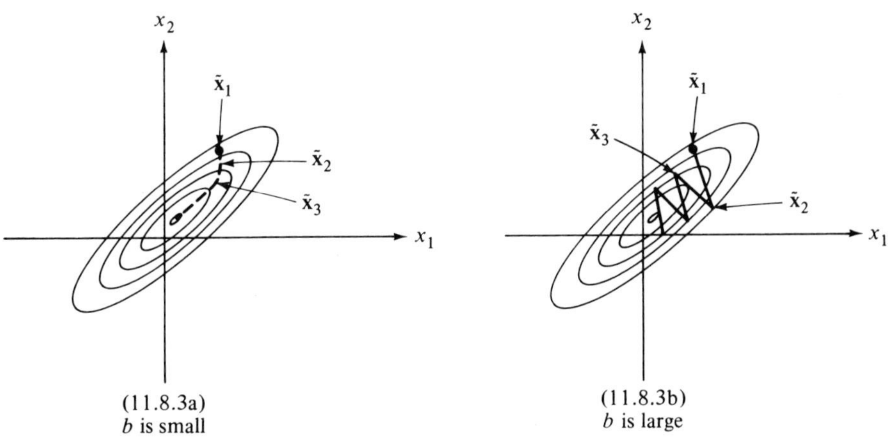

Figure 11.7.3 Iterative path for $\mathbf{B}\ (i) = b\mathbf{I}$.

of b causes slow convergence, while an overly large value causes oscillations.

Iterative solutions are conceptually very simple. The gain matrix $\mathbf{B}(i)$ must be full rank if $\tilde{\mathbf{x}}_{i+1} = \tilde{\mathbf{x}}_i$ is to imply $\mathbf{J}^{(1)}(\mathbf{x}_i) = \mathbf{0}$. On the basis of Fig. 11.7.1 it is reasonable to want $\mathbf{B}(i) \geq \mathbf{0}$. However, there is a lot of freedom in the choice of $\mathbf{B}(i)$, and the choice of gain sequence is a big problem. The gain is important, as it effects the number of iterations needed for convergence. With a bad gain, convergence may not even occur. Several methods of choosing a gain sequence are now discussed.

To illustrate the general equations, the case where

$$J(\mathbf{x}) = (\tfrac{1}{2})[\mathbf{z} - \mathbf{h}(\mathbf{x})]'\mathbf{R}^{-1}[\mathbf{z} - \mathbf{h}(\mathbf{x})] \tag{11.7.3}$$

is used as an example. Equation (11.7.3) can correspond to a Fisher maximum likelihood estimator or to a weighted-least-squares formulation. A factor of $(\frac{1}{2})$ is included in (11.7.3) for convenience. For (11.7.3)

$$\mathbf{J}^{(1)'}(\mathbf{x}) = -\mathbf{H}^{(1)'}(\mathbf{x})\mathbf{R}^{-1}[\mathbf{z} - \mathbf{h}(\mathbf{x})]$$

$$\mathbf{H}^{(1)}(\mathbf{x}) = \frac{\partial \mathbf{h}(\mathbf{x})}{\partial \mathbf{x}} = \begin{bmatrix} \dfrac{\partial h_1(\mathbf{x})}{\partial x_1} & \cdots & \dfrac{\partial h_1(\mathbf{x})}{\partial x_{K_1}} \\ \vdots & & \vdots \\ \dfrac{\partial h_{K_2}(\mathbf{x})}{\partial x_1} & \cdots & \dfrac{\partial h_{K_2}(\mathbf{x})}{\partial x_{K_1}} \end{bmatrix} \tag{11.7.4}$$

where $\mathbf{H}^{(1)}(\mathbf{x})$ is often called the *Jacobian* matrix. Then (11.7.2) becomes

$$\tilde{\mathbf{x}}_{i+1} = \tilde{\mathbf{x}}_i + \mathbf{B}(i)\mathbf{H}^{(1)'}(\tilde{\mathbf{x}}_i)\mathbf{R}^{-1}[\mathbf{z} - \mathbf{h}(\tilde{\mathbf{x}}_i)] \tag{11.7.5}$$

11.7.1 Newton-Raphson†

The Newton-Raphson iteration (gain sequence) can be derived as follows. Assume that some guess $\tilde{\mathbf{x}}_i$ of $\hat{\mathbf{x}}$ is available. Define

$$\mathbf{x} = \tilde{\mathbf{x}}_i + \boldsymbol{\delta}_i$$

A Taylor series expansion of $J(\mathbf{x})$ yields

$$J(\mathbf{x}) = J(\tilde{\mathbf{x}}_i) + \mathbf{J}^{(1)}(\tilde{\mathbf{x}}_i)\boldsymbol{\delta} + (\tfrac{1}{2})\boldsymbol{\delta}'\mathbf{J}^{(2)}(\tilde{\mathbf{x}}_i)\boldsymbol{\delta} + \cdots$$

where $\mathbf{J}^{(2)}(\mathbf{x})$ is often called the *Hessian* matrix (11.7.6)

$$\mathbf{J}^{(2)}(\tilde{\mathbf{x}}) = \left.\frac{\partial \mathbf{J}^{(1)}(\mathbf{x})}{\partial \mathbf{x}}\right|_{\mathbf{x}=\tilde{\mathbf{x}}}$$

The problem of finding $\hat{\mathbf{x}}$ can be expressed as the problem of finding the $\boldsymbol{\delta}$ which makes the derivative of $J(\mathbf{x})$ with respect to $\boldsymbol{\delta}$ equal to $\mathbf{0}$, i.e., to find the $\boldsymbol{\delta}$ which satisfies

$$\mathbf{0} = \mathbf{J}^{(1)'}(\tilde{\mathbf{x}}_i) + \mathbf{J}^{(2)}(\hat{\mathbf{x}}_i)\boldsymbol{\delta} + \cdots \tag{11.7.7}$$

This is a nonlinear equation in $\boldsymbol{\delta}$ because of the higher-order terms. However, if $\tilde{\mathbf{x}}_i$ is close to $\hat{\mathbf{x}}$, the $\boldsymbol{\delta}$ which satisfies (11.7.7) should be small and it is reasonable to linearize (11.7.7) by dropping the higher-order terms. Let $\tilde{\boldsymbol{\delta}}_i$ denote the value which satisfies

$$\mathbf{0} = \mathbf{J}^{(1)'}(\tilde{\mathbf{x}}_i) + \mathbf{J}^{(2)}(\tilde{\mathbf{x}}_i)\boldsymbol{\delta}$$

Define

$$\tilde{\mathbf{x}}_{i+1} = \tilde{\mathbf{x}}_i + \tilde{\boldsymbol{\delta}}_i$$

Then

$$\boxed{\begin{aligned}\tilde{\mathbf{x}}_{i+1} &= \tilde{\mathbf{x}}_i - \mathbf{B}(i)\mathbf{J}^{(1)'}(\tilde{\mathbf{x}}_i)\\ \mathbf{B}(i) &= [\mathbf{J}^{(2)}(\tilde{\mathbf{x}}_i)]^{-1}\end{aligned}} \tag{11.7.8}$$

In the special case of (11.7.3)

$$\mathbf{J}^{(2)}(\mathbf{x}) = \mathbf{H}^{(1)'}(\mathbf{x})\mathbf{R}^{-1}\mathbf{H}^{(1)}(\mathbf{x}) - \frac{\partial}{\partial \mathbf{x}}[\mathbf{H}^{(1)'}(\mathbf{x})\mathbf{R}^{-1}\mathbf{z}] \tag{11.7.9}$$

One variation on the basic Newton-Raphson is to use a constant gain, $\mathbf{B}(i) = \mathbf{B}$, where

$$\mathbf{B} = [\mathbf{J}^{(2)}(\tilde{\mathbf{x}}_1)]^{-1}$$

where $\tilde{\mathbf{x}}_1$ is the initial guess of $\hat{\mathbf{x}}$. Another variation is to use a *piecewise constant* gain. For example,

$$\begin{aligned}\mathbf{B}(i) &= \mathbf{B}_1, \qquad i = 1, \ldots, 4\\ \mathbf{B}(i) &= \mathbf{B}_4, \qquad i = 4, \ldots, 8 \qquad \text{etc.}\end{aligned}$$

where

$$\mathbf{B}_k = [\mathbf{J}^{(2)}(\tilde{\mathbf{x}}_k)]^{-1}$$

There are many other possible variations. In some applications, such variations can save a lot of computer time.

One potential disadvantage of the Newton-Raphson method should be noted. The $\mathbf{B}(i)$ of (11.7.8) need not have the positive definite property which was deemed desirable for convergence.

It should be emphasized that in practice (11.7.8) is often not solved as written. Instead one solves the system of linear equations

$$[\mathbf{J}^{(2)}(\tilde{\mathbf{x}}_i)][\tilde{\mathbf{x}}_{i=1} - \tilde{\mathbf{x}}_i] = \mathbf{J}^{(1)'}(\tilde{\mathbf{x}}_i) \tag{11.7.10}$$

for $\tilde{\mathbf{x}}_{i+1} - \tilde{\mathbf{x}}_i$. See the discussion of Section 5.11.

11.7.2 Linearization of Model†

The Newton-Raphson gain is obtained by expanding $J(\mathbf{x})$ about some $\tilde{\mathbf{x}}_i$ and then *linearizing* the equations by dropping higher-order terms. In the case of (11.7.3), where

$$J(\mathbf{x}) = (\tfrac{1}{2})[\mathbf{z} - \mathbf{h}(\mathbf{x})]'\mathbf{R}^{-1}[\mathbf{z} - \mathbf{h}(\mathbf{x})]$$

an alternative approach is to expand $\mathbf{h}(\mathbf{x})$ about some $\tilde{\mathbf{x}}_i$ and then linearize, i.e., linearize the model.

A Taylor expansion of $\mathbf{h}(\mathbf{x})$ yields

$$\mathbf{h}(\mathbf{x}) = \mathbf{h}(\tilde{\mathbf{x}}_i) + \mathbf{H}^{(1)}(\tilde{\mathbf{x}}_i)\boldsymbol{\delta} + \cdots$$
$$\mathbf{x} = \tilde{\mathbf{x}}_i + \boldsymbol{\delta}$$

Dropping the higher-order terms and substituting yields

$$\mathbf{J}(\mathbf{x}) = [\mathbf{z} - \mathbf{h}(\tilde{\mathbf{x}}_i) - \mathbf{H}^{(1)}(\tilde{\mathbf{x}}_i)\boldsymbol{\delta}]'\mathbf{R}^{-1}[\mathbf{z} - \mathbf{h}(\tilde{\mathbf{x}}_i) - \mathbf{H}^{(1)}(\tilde{\mathbf{x}}_i)\boldsymbol{\delta}]$$

Let $\tilde{\boldsymbol{\delta}}_i$ denote the value of $\boldsymbol{\delta}$ which minimizes this $J(\mathbf{x})$ and define

$$\tilde{\mathbf{x}}_{i+1} = \tilde{\mathbf{x}}_i + \tilde{\boldsymbol{\delta}}_i$$

Then (see Section 5.4 for methods to find $\tilde{\boldsymbol{\delta}}_i$)

$$\boxed{\begin{aligned} \tilde{\mathbf{x}}_{i+1} &= \tilde{\mathbf{x}}_i + \mathbf{B}(\mathbf{i})\mathbf{H}^{(1)'}(\mathbf{x}_i)\mathbf{R}^{-1}[\mathbf{z} - \mathbf{h}(\mathbf{x}_i)] \\ \mathbf{B}(i) &= [\mathbf{H}^{(1)'}(\tilde{\mathbf{x}}_i)\mathbf{R}^{-1}\mathbf{H}^{(1)}(\tilde{\mathbf{x}}_i)]^{-1} \end{aligned}} \tag{11.7.11}$$

This gain is like the Newton-Raphson gain of (11.7.8) and (11.7.9) except that a term of (11.7.9) has been dropped.

As with the Newton-Raphson, many simplifications and variations are possible. For example, a constant (or piecewise constant) gain $\mathbf{B}$ can be used where $\mathbf{B}$ is evaluated for some $\mathbf{x}$. Also implementation in the form of (11.7.10) should be considered.

Note that $\mathbf{B}(i)$ of (11.7.11) is always positive definite (assuming that the inverse exists).

11.7.3 Simple Gains†

A very simple gain is

$$\mathbf{B}(i) = b(i)\,\mathbf{I} \tag{11.7.12}$$

where the scalar $b(i)$ are chosen in some clever fashion. They may be constant. The effect of using (11.7.12) was illustrated in Fig. 11.7.3.

In the special case of (11.7.3) a simple gain is

$$\begin{aligned} \mathbf{B}_i &= b\mathbf{D}_i^{-1} \\ \mathbf{D}_i &= \text{diagonal}\{\mathbf{H}^{(1)'}(\tilde{\mathbf{x}}_i)\mathbf{R}^{-1}\mathbf{H}^{(1)}(\tilde{\mathbf{x}}_i)\} \end{aligned} \tag{11.7.13}$$

where *diagonal* means $\mathbf{D}$ is $\mathbf{H}^{(1)'}\mathbf{R}^{-1}\mathbf{H}^{(1)}$ with all off-diagonal terms set to zero. This is classified as a simple gain, as the matrix inversion is easy to perform.

11.7.4 Gradient Methods†

A simple gradient method (also called steepest descent) is as follows:

1. Evaluate $\mathbf{J}^{(1)}(\tilde{\mathbf{x}}_1)$ for the initial guess $\tilde{\mathbf{x}}_1$.

2. By a separate iteration or search procedure, find the value of b such that $J(\tilde{\mathbf{x}}_2)$ is minimum for

$$\tilde{\mathbf{x}}_2 = \tilde{\mathbf{x}}_1 - b\mathbf{J}^{(1)}(\tilde{\mathbf{x}}_1)$$

3. Evaluate $\mathbf{J}^{(1)}(\tilde{\mathbf{x}}_2)$ and repeat.

This gradient method basically introduces a separate suboptimization problem to find the best $\mathbf{B}(i)$ of the form

$$\mathbf{B}(i) = b(i)\,\mathbf{I}$$

In most cases this reduces the total number of iterations needed and thus is very advantageous if $\mathbf{J}^{(1)}(\tilde{\mathbf{x}}_i)$ is difficult to compute. Of course, it may be difficult to solve the suboptimization problem to find the $b(i)$.

A disadvantage of the simple gradient method is that the search always is in the direction of the gradient and this can cause problems in the presence of "valleys" or "ridges" in the $J(\mathbf{x})$ surface. Such problems lead to the use of methods which search in "all" directions. There are many possibilities depending on the nature of the logic behind the choice of direction. Three relatively simple and related methods are (in order of increasing complexity), 1) conjugate directions, 2) conjugate gradient, and 3) Fletcher-Powell (this terminology is not always followed). The Fletcher-Powell iteration has proven to be very successful in a wide variety of state estimation problems. A Fletcher-Powell iteration "starts out" like a simple gradient search and then "adapts itself" to become more like a Newton-Raphson.

A detailed presentation of all these concepts is beyond the scope of this book. They are mentioned here primarily to point out their existence.

11.7.5 Analysis of Iteration†

Several possible gains have been stated. Exact analysis of how they perform is very difficult, but some equations are now manipulated in an attempt to provide some insight. To be specific, discussion is confined to the $J(\mathbf{x})$ of (11.7.3), so the iteration is of the form of (11.7.5).

As before, let $\hat{\mathbf{x}}$ denote the $\mathbf{x}$ which minimizes $J(\mathbf{x})$ so that

$$\mathbf{J}^{(1)\prime}(\hat{\mathbf{x}}) = \mathbf{H}^{(1)\prime}(\hat{\mathbf{x}})\mathbf{R}^{-1}[\mathbf{z} - \mathbf{h}(\hat{\mathbf{x}})] = \mathbf{0} \qquad (11.7.14)$$

Define

$$\begin{aligned} \boldsymbol{\delta}_i &= \hat{\mathbf{x}} - \tilde{\mathbf{x}}_i \\ \mathbf{H}^{(1)}(\tilde{\mathbf{x}}_i) &= \mathbf{H}^{(1)}(\hat{\mathbf{x}}) + \boldsymbol{\Delta}_i \end{aligned} \qquad (11.7.15)$$

Then from (11.7.5)

$$\boldsymbol{\delta}_{i+1} = \boldsymbol{\delta}_i + \mathbf{B}_i[\mathbf{H}^{(1)\prime}(\hat{\mathbf{x}}) + \boldsymbol{\Delta}_i']\mathbf{R}^{-1}[\mathbf{z} - \mathbf{h}(\hat{\mathbf{x}}) + \mathbf{H}^{(1)\prime}(\hat{\mathbf{x}})\boldsymbol{\delta}_i + \mathbf{g}_i]$$

where $\mathbf{g}_i$ is a function of higher-order terms in $\boldsymbol{\delta}_i$. Using (11.7.14),

$$\begin{aligned} \boldsymbol{\delta}_{i+1} &= \mathbf{A}_i\boldsymbol{\delta}_i + \mathbf{a}_i \\ \mathbf{A}_i &= \mathbf{I} - \mathbf{B}_i\mathbf{H}^{(1)\prime}(\hat{\mathbf{x}})\mathbf{R}^{-1}\mathbf{H}^{(1)}(\hat{\mathbf{x}}) \\ \mathbf{a}_i &= \mathbf{B}_i\{\boldsymbol{\Delta}_i'\mathbf{R}^{-1}[\mathbf{z} - \mathbf{h}(\hat{\mathbf{x}}) + \mathbf{H}^{(1)\prime}(\hat{\mathbf{x}})\boldsymbol{\delta}_i + \mathbf{g}_i] + \mathbf{H}^{(1)\prime}(\hat{\mathbf{x}})\mathbf{R}^{-1}\mathbf{g}_i\} \end{aligned} \qquad (11.7.16)$$

Now if $\boldsymbol{\delta}_i \longrightarrow \mathbf{0}$, $\tilde{\mathbf{x}}_i$ converges to $\hat{\mathbf{x}}$. If $\boldsymbol{\delta}_i \longrightarrow \mathbf{0}$, then $\boldsymbol{\Delta}_i \longrightarrow \mathbf{0}$, $\mathbf{g}_i \longrightarrow \mathbf{0}$, and $\mathbf{a}_i \longrightarrow \mathbf{0}$. Thus it is reasonable to expect the matrix $\mathbf{A}_i$ to be an important quantity which helps determine how the $\boldsymbol{\delta}_i$ behave. For example, if the $\mathbf{A}_i$, $i = 1, \ldots,$ have eigenvalues with magnitude greater than 1, convergence cannot be expected. If all the eigenvalues of $\mathbf{A}_i$ have magnitudes much less than 1, rapid convergence can be expected (although not guaranteed). Study of the eigenvalues of $\mathbf{A}_i$ cannot prove convergence in a true mathematical sense but does provide a good engineering tool. The $\mathbf{A}_i$ of (11.7.16) is of the form

$$\mathbf{A}_i = \mathbf{I} - \mathbf{B}_i\mathbf{S}(\hat{\mathbf{x}})$$
$$\mathbf{S}(\hat{\mathbf{x}}) = \mathbf{H}^{(1)\prime}(\hat{\mathbf{x}})\mathbf{R}^{-1}\mathbf{H}^{(1)}(\hat{\mathbf{x}})$$

$\mathbf{S}(\hat{\mathbf{x}})$ is a positive definite matrix. Assume that the gain $\mathbf{B}_i$ is a positive definite matrix. Then it can be shown that (see Exercise 11.10)

$\mathbf{B}_i \leq 2\mathbf{S}^{-1}(\hat{\mathbf{x}})$ guarantees that the eigenvalues of $\mathbf{A}_i$ are real and less than 1 in magnitude (11.7.17)

The optimum gains (in the present limited sense) are

$$\mathbf{B}_i = \mathbf{S}^{-1}(\hat{\mathbf{x}})$$

as this yields zero eigenvalues for $\mathbf{A}_i$.

Now (11.7.17) is not overly useful as written, as $\hat{\mathbf{x}}$ is not known a priori. However, in keeping with the heuristic level of arguments being used, the following statement is not unreasonable:

Define	
$\mathbf{S}(\tilde{\mathbf{x}}_i) = \mathbf{H}^{(1)\prime}(\tilde{\mathbf{x}}_i)\mathbf{R}^{-1}\mathbf{H}^{(1)}(\tilde{\mathbf{x}}_i)$	
Then if $\mathbf{B}_i \leq 2\mathbf{S}^{-1}(\tilde{\mathbf{x}}_i)$ the iteration (11.7.5) probably converges and an optimum gain is	(11.7.18)
$\mathbf{B}_i = \mathbf{S}^{-1}(\tilde{\mathbf{x}}_i)$	

This statement is admittedly vague, but the author has found it to be useful.

11.7.6 Relation to Linear Models†

Consider the linear model problem

$$J(\mathbf{x}) = [\mathbf{z} - \mathbf{H}\mathbf{x}]'\mathbf{R}^{-1}[\mathbf{z} - \mathbf{H}\mathbf{x}]$$

Let $\hat{\mathbf{x}}$ minimize $J(\mathbf{x})$. From Chapter 5,

$$\hat{\mathbf{x}} = [\mathbf{H}'\mathbf{R}^{-1}\mathbf{H}]^{-1}\mathbf{H}'\mathbf{R}^{-1}\mathbf{z}$$

However, another way to solve for $\hat{\mathbf{x}}$ is by an iteration:

$$\mathbf{x}_{i+1} = \mathbf{x}_i + \mathbf{B}_i\mathbf{H}'\mathbf{R}^{-1}[\mathbf{z} - \mathbf{H}\mathbf{x}_i] \tag{11.7.19}$$

It follows from the ideas of Section 11.7.5 that

$$\tilde{\mathbf{x}}_i \longrightarrow \hat{\mathbf{x}} \qquad \text{if } \mathbf{0} < \mathbf{B}_i < 2[\mathbf{H}'\mathbf{R}^{-1}\mathbf{H}]^{-1}$$

The iteration (11.7.19) converges in one step for any $\mathbf{x}_0$ if

$$\mathbf{B}_0 = (\mathbf{H}'\mathbf{R}^{-1}\mathbf{H})^{-1}$$

EXAMPLE: Consider the Fisher linear model

$$\mathbf{z} = \begin{bmatrix} z_1 \\ z_2 \end{bmatrix} = \mathbf{H}x + \mathbf{v}$$

$$\mathbf{H} = \begin{bmatrix} 1 \\ 1 \end{bmatrix} \qquad \mathbf{v} \text{ is } N(\mathbf{0}, \mathbf{R})$$

$$\mathbf{R} = \begin{bmatrix} 1 & 0 \\ 0 & 4 \end{bmatrix}$$

The maximum likelihood estimate x minimizes

$$J(x) = (\mathbf{z} - \mathbf{H}x)'\mathbf{R}^{-1}(\mathbf{z} - \mathbf{H}x)$$

and is given directly by

$$\hat{x} = \tfrac{4}{5}z_1 + \tfrac{1}{5}z_2$$

However, assume that an iterative solution is attempted:

$$\tilde{x}_{i+1} = \tilde{x}_i - B_i \frac{\partial J(x)}{\partial x}\bigg|_{x=\tilde{x}_i}$$

$$\tilde{x}_0 = \text{initial guess of } \hat{x}$$

or

$$\tilde{x}_{i+1} = \tilde{x}_i + B_i \left[\frac{4z_1 + z_2}{4} - \frac{5}{4}\tilde{x}_i\right]$$

so

$$\tilde{x}_{i+1} = \left[1 - \frac{5}{4}B_i\right]\tilde{x}_i + B_i\left[\frac{4z_1 + z_2}{4}\right]$$

The best $B_i = (\mathbf{H}'\mathbf{R}^{-1}\mathbf{H})^{-1} = \frac{4}{5}$ and yields $\hat{x}$ in one iteration for any $\tilde{x}_0$:

$$\tilde{x}_1 = \frac{4z_1 + z_2}{5} = \hat{x}$$

However, any constant $B_i = B$, $0 < B < 2(\frac{4}{5})$, yields

$$\tilde{x}_{i+1} = \Phi\tilde{x}_i + B\left[\frac{4z_1 + z_2}{4}\right]$$

with $|\Phi| < 1$, so $\tilde{x}_i$ will converge to a constant $\tilde{x}_\infty$, where

$$\tilde{x}_\infty = \left(1 - \frac{5}{4}B\right)\tilde{x}_\infty + B\left[\frac{4z_1 + z_2}{4}\right]$$

so

$$\tilde{x}_\infty = \frac{4z_1 + z_2}{5} = \hat{x} \qquad \text{for any } x_0$$

If $B \geq 2(\frac{4}{5})$, $\tilde{x}_i$ does not converge.

11.7.7 Stopping Rules†

Thus far only the basic iteration process has been considered. A related question is, What stopping rule should be used? In other words, how does one decide that $\tilde{\mathbf{x}}_{i+1} - \tilde{\mathbf{x}}_i$ is small enough to conclude that $\tilde{\mathbf{x}}_i \approx \hat{\mathbf{x}}$? Three obvious stopping rules are as follows:

1. Stop when all components of $\tilde{\mathbf{x}}_{i+1} - \tilde{\mathbf{x}}_i$ are "small enough."
2. Stop when $J(\tilde{\mathbf{x}}_{i+1}) - J(\tilde{\mathbf{x}}_i)$ is "small enough."
3. Stop when all components of $\mathbf{J}^{(1)}(\tilde{\mathbf{x}}_i)$ are "small enough."

The definition of "small enough" is, of course, a matter of engineering judgment.

When deciding what is small enough, it is important to remember that $\hat{\mathbf{x}} \neq \mathbf{x}_{\text{true}}$, so there is no reason to iterate until $\tilde{\mathbf{x}}_i = \hat{\mathbf{x}}$. As long as $\tilde{\mathbf{x}}_i$ is "relatively close" to $\hat{\mathbf{x}}$, it is probably as good an estimate of $\mathbf{x}_{\text{true}}$ as $\hat{\mathbf{x}}$ itself. This statement is admittedly vague, but the underlying concept is important.

In the special case when

$$\begin{aligned} &\mathbf{z} = \mathbf{h}(\mathbf{x}) + \mathbf{v} \qquad && \mathbf{x}:\; K_1 \text{ vector} \\ &\mathbf{v} \text{ is } N(\mathbf{0}, \mathbf{R}) && \mathbf{v}:\; K_2 \text{ vector} \end{aligned}$$

a different type of stopping rule can sometimes be used. In Section 9.4.4, it was discussed that for a Fisher model where $J(\mathbf{x}) = [\mathbf{z} - \mathbf{Hx}]'\mathbf{R}^{-1}[\mathbf{z} - \mathbf{Hx}]$ and $\hat{\mathbf{x}}$ minimizes $J(\hat{\mathbf{x}})$, then $J(\mathbf{x})$ is a chi-squared random variable with $K_2 - K_1$ degrees of freedom. If the error $\mathbf{x}_{\text{true}} - \hat{\mathbf{x}}$ is small enough, then this same property also applies approximately for the nonlinear model. Thus a stopping rule can be based directly on the size of $J(\tilde{\mathbf{x}}_i)$ itself.

11.7.8 Discussion of Iterative Solutions

The basic iteration (11.7.2) is a powerful tool, but there is no guarantee that $\tilde{\mathbf{x}}_i$ will converge to $\hat{\mathbf{x}}$. As the simplest counterexample, consider a $J(x)$ which has two local minima (at $\hat{x}$ and x_0) as in Fig. 11.7.4. If the initial guess $\tilde{x}_1$ is close to x_0, $\tilde{x}_1$ may converge to x_0 rather than $\hat{x}$. Even if $\tilde{x}_1$ is close to $\hat{x}$, the gain sequence $B(i)$ may be such that $\tilde{x}_i$ converges to x_0. The basic iteration works well in many problems, but it is not perfect and the reader must never assume its success without further study or actual implementation.

The best gain sequence is the one which converges with the least amount of expended computer time. A simple gain sequence such as $\mathbf{B}(i) = b\,\mathbf{I}$ reduces the computation time per iteration but may require more iterations (if it converges at all) and hence more overall computation time than a more sophisticated approach. Unfortunately there is no general answer to the question of which gain sequence is best.

The analysis of Section 11.7.5 provides only a guide on how to choose gains. There is still no absolute guarantee of convergence.

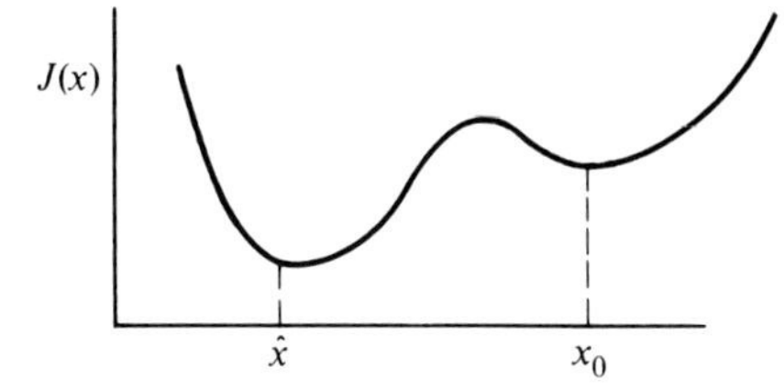

Figure 11.7.4 A case where iterations may not converge to true minimum.

11.8 Parameterization†

Iterative solutions as in Section 11.7 are one way to obtain an estimate from a function of $\mathbf{x}$ such as $p(\mathbf{x}|\mathbf{z})$, $p(\mathbf{z}:\mathbf{x})$, or $J(\mathbf{x}:\mathbf{z})$. Such iterative solutions are local in nature and yield only a single vector estimate $\hat{\mathbf{x}}$ with no understanding of the overall uncertainty in $\hat{\mathbf{x}}$, i.e., in the shape of $p(\mathbf{x}|\mathbf{z})$, $p(\mathbf{z}:\mathbf{x})$, or $J(\mathbf{x}:\mathbf{z})$. A more global approach is to approximate the given function of $\mathbf{x}$ by some collection of numbers which *characterize* the overall function. This is called parameterization as a function of $\mathbf{x}$ in being parameterized by a finite set of numbers. Three approaches are considered: expansions, quantizations, and moments.

Consider a Bayesian formulation where the estimator is based on the conditional probability density, $p(\mathbf{x}|\mathbf{z})$. One convenient parameterization of $p(\mathbf{x}|\mathbf{z})$ is to expand $p(\mathbf{x}|\mathbf{z})$ in terms of functions of $\mathbf{x}$ and assume that

$$p(\mathbf{x}|\mathbf{z}) \approx \sum_{m=1}^{M} a_m(\mathbf{z})\theta_m(\mathbf{x}) \tag{11.8.1}$$

The $\theta_m(\mathbf{x})$, $m = 1, \ldots, M$, are predetermined functions of $\mathbf{x}$. The a_m, $m = 1, \ldots, M$, are the parameters which characterize $p(\mathbf{x}|\mathbf{z})$, where the a_m depend on the observation $\mathbf{z}$. One reasonable way to obtain the $a_m(\mathbf{z})$ is as the values which minimize

$$\int_{-\infty}^{\infty} \cdots \int_{-\infty}^{\infty} \left[p(\mathbf{x}|\mathbf{z}) - \sum_{m=1}^{M} a_m(\mathbf{z})\theta_m(\mathbf{x}) \right]^2 dx_1 \cdots dx_{K_1} \tag{11.8.2}$$

If one chooses the $\theta_m(\mathbf{x})$ to be orthonormal so that

$$\int_{-\infty}^{\infty} \cdots \int_{-\infty}^{\infty} \theta_{m_1}(\mathbf{x})\theta_{m_2}(\mathbf{x})\, dx_1 \cdots dx_{K_1} = \begin{cases} 1, & m_1 = m_2 \\ 0, & m_1 \neq m_2 \end{cases}$$

then the $a_m(\mathbf{z})$ which minimize (11.8.2) are given by

$$a_m(\mathbf{z}) = \int \cdots \int p(\mathbf{x}|\mathbf{z})\theta_m(\mathbf{x})\, dx_1 \cdots dx_{K_1}$$

Using (11.8.1), the conditional expectation estimator

$$\hat{\mathbf{x}} = \int \cdots \int p(\mathbf{x}|\mathbf{z})\mathbf{x}\, dx_1 \cdots dx_{K_1}$$

becomes

$$\hat{\mathbf{x}} = \sum_{m=1}^{M} \xi_m a_m(\mathbf{z})$$

$$\xi_m = \int \cdots \int \mathbf{x}\theta_m(\mathbf{x})\, dx_1 \cdots dx_{K_1}$$

The big problem is how to choose the $\theta_m(\mathbf{x})$, $m = 1, \ldots, M$, so that the $a_m(\mathbf{z})$ and ξ_m can be easily evaluated and so that a small M yields a good estimate.

Consider a Bayesian formulation. "Quantize" $\mathbf{x}$ into M values $\boldsymbol{\chi}_1 \cdots \boldsymbol{\chi}_M$ and assume that $p(\mathbf{x} \mid \mathbf{z})$ is "represented" by

$$p(\boldsymbol{\chi}_m \mid \mathbf{z}), \qquad m = 1, \ldots, M$$

This can be viewed as a special case of (11.8.1). The conditional expectation estimator can be approximated by

$$\hat{\mathbf{x}} = \frac{\sum_{m=1}^{M} p(\boldsymbol{\chi}_m \mid \mathbf{z})\boldsymbol{\chi}_m}{\sum_{m=1}^{M} p(\boldsymbol{\chi}_m \mid \mathbf{z})} \tag{11.8.3}$$

The most probable estimator can be approximated by

$$\hat{\mathbf{x}} = \boldsymbol{\chi}_m, \qquad \text{where } \boldsymbol{\chi}_m \text{ is the value which yields the maximum } p(\boldsymbol{\chi}_m \mid \mathbf{z}),\ m = 1, \ldots, M \tag{11.8.4}$$

The big problem here is that the needed value of M can become overly large. Even for scalar x, M may have to be too large to be practical.

A third method of parameterization or characterizing some function of $\mathbf{x}$ is in terms of its moments; i.e., for a Bayesian formulation

$$\int \cdots \int \mathbf{x}^m p(\mathbf{x} \mid \mathbf{z})\, dx, \qquad m = 1, \ldots, M$$

where $\mathbf{x}^1 = \mathbf{x}$, $\mathbf{x}^2 = \mathbf{x}\mathbf{x}'$, etc. Such an approach automatically yields the conditional expectation estimator as the first moment. The big problem here is the fact that it is often difficult to calculate the moments.

The above parameterization concepts extend to Fisher and weighted-least-squares formulations in a relatively straightforward fashion.

In an unknown-but-bounded formulation where

$$\boldsymbol{\Omega}_{\text{est}} = \{\mathbf{x} : J(\mathbf{x} : \mathbf{z}) \leq 1\}$$

the parameterization problem is to find some more convenient function $\tilde{J}(\mathbf{x} : \mathbf{z})$ such that

$$J(\mathbf{x} : \mathbf{z}) \geq \tilde{J}(\mathbf{x} : \mathbf{z}), \qquad \text{all } \mathbf{x} \tag{11.8.5}$$

Possible $\tilde{J}(\mathbf{x} : \mathbf{z})$ are

$$\tilde{J}(\mathbf{x} : \mathbf{z}) = \sum_{m=1}^{M} a_m(\mathbf{z})\theta_m(\mathbf{x})$$

$$\tilde{J}(\mathbf{x} : \mathbf{z}) = [\mathbf{x} - \hat{\mathbf{x}}(\mathbf{z})]'\boldsymbol{\Sigma}^{-1}(\mathbf{z})[\mathbf{x} - \hat{\mathbf{x}}(\mathbf{z})]$$

The big problem is, of course, to assure that (11.8.5) is satisfied at least for "most" $\mathbf{x}$.

All the preceding parameterization techniques are conceptually straightforward and are simply standard methods of characterizing a function by a finite set of parameters. However, all the techniques have big problems associated with them, and in many cases none of them are practical. Nevertheless they should be considered, as they yield a global characterization rather than simply a single point estimate $\hat{\mathbf{x}}$.

11.9 Change of Variables

A recurring theme of this book is the importance of the choice of the model. Many aspects of model choosing for linear problems discussed in Chapter 8 carry over to nonlinear problems and are not repeated here. However, the practical implementation problems of nonlinear estimation introduce some additional possibilities—in particular, the use of a change of variables.

Assume that in terms of the "natural" state vector $\mathbf{x}$ the model is "very" nonlinear. There may exist another state vector $\mathbf{y}$ related by the one-to-one nonlinear transformation

$$\mathbf{x} = \boldsymbol{\theta}(\mathbf{y}) \tag{11.9.1}$$

for which the model is "more" linear in the sense that

$$\mathbf{z} = \mathbf{h}(\mathbf{x}) + \mathbf{v}$$

$$\mathbf{z} = \mathbf{h}[\boldsymbol{\theta}(\mathbf{y})] + \mathbf{v}$$

$$\mathbf{z} \approx \mathbf{H}\mathbf{y} + \mathbf{v}$$

In such a case, it would be reasonable to first calculate $\hat{\mathbf{y}}$ and get $\hat{\mathbf{x}}$ by

$$\hat{\mathbf{x}} = \boldsymbol{\theta}[\hat{\mathbf{y}}] \tag{11.9.2}$$

Use of (11.9.2) need not be optimal in any mathematical sense, but it is reasonable and that is usually the most important point.

Another approach is to try to choose a nonlinear transform of the observations

$$\tilde{\mathbf{z}} = \boldsymbol{\lambda}(\mathbf{z})$$

so that

$$\tilde{\mathbf{z}} = \boldsymbol{\lambda}[\mathbf{h}(\mathbf{x}) + \mathbf{v}] \tag{11.9.3}$$

can (hopefully) be viewed as approximately

$$\tilde{\mathbf{z}} = \boldsymbol{\lambda}[\mathbf{h}(\mathbf{x})] + \tilde{\mathbf{v}} \tag{11.9.4}$$

where $\tilde{\mathbf{v}}$ is some new uncertain vector and

$$\boldsymbol{\lambda}[\mathbf{h}(\mathbf{x})] \approx \mathbf{H}\mathbf{x}$$

Sometimes there is enough freedom in the choice of uncertainty models to justify the use of (11.9.4) instead of (11.9.3).

Changes of variables can be helpful for iterative and parametric solutions even if they do not linearize the problem. For example, a change of variables may change the "shape" of $J(\mathbf{x})$ into a more suitable form.

Unfortunately there seems to be no general way to choose the "right" change of variables. Success depends on the particular problem and the engineer's insight. However, the possibility of a change of variables should always be considered.

11.10 Stochastic Approximation†

Stochastic approximation is a term applied to a general class of iterative procedures for finding the minimum (or zero) of a function from "noisy" observations of the function. Stochastic approximation algorithms are related to some of the ideas of (Section 11.7).

The power and beauty of stochastic approximation theory is that it provides simple, easy to implement *gain sequences* which guarantee convergence without depending (explicitly) on knowledge of the function to be minimized or the noise properties. Unfortunately, convergence is usually extremely slow. This is to be expected, as "good performance" cannot be expected if no (or very little) knowledge of the nature of the problem is built into the algorithm. In other words, the strength of stochastic approximation (simplicity, little a priori knowledge) is also its weakness. Because of this weakness, the theory of stochastic approximation is not presented here. However, it may have value in specialized types of applications.

11.11 Discussion

From a conceptual point of view, the nonlinear model $\mathbf{z} = \mathbf{h}(\mathbf{x}, \mathbf{v})$ is not much harder than the linear model $\mathbf{z} = \mathbf{H}\mathbf{x} + \mathbf{v}$ of Chapter 5. The differences between the linear and nonlinear cases arise when one tries to convert the concepts into an estimator which can be implemented on a digital computer. For a linear model, the estimator is given by matrix manipulations (such as $\hat{\mathbf{x}} = [\mathbf{H}'\mathbf{R}^{-1}\mathbf{H}]^{-1}\mathbf{H}'\mathbf{R}^{-1}\mathbf{z}$), which can be "easily" implemented. For the nonlinear model, the estimator is rarely specified in such a convenient closed form, and the main problem is usually the specification of some technique which can be implemented and which yields acceptable answers. This situation is well illustrated by the Bayesian formulation where the conditional expectation estimator is "optimum" but the most probable estimator is often used, as it is easier to implement.

Iterative techniques to find the minimum or maximum of some function can often be used to implement the most probable (Bayesian), maximum likelihood (Fisher), weighted-least-squares, and center of set (unknown-but-bounded) estimators. Iterative solutions are the most common general techniques used. Their major disadvantage is their local nature; they may not converge or may converge to the wrong answer (a local instead of global minimum). Even after a successful convergence, they yield a point estimate, not a global picture. Parameterization techniques such as representing $p(\mathbf{x} \mid \mathbf{z})$ by $p(\boldsymbol{\chi}_m \mid \mathbf{z})$, $m = 1, \ldots, M$, are used less frequently because of the problem of choosing a practical parameterization. Parameterization has the advantage that it is global; i.e., it enables consideration of the whole functions $p(\mathbf{x} \mid \mathbf{z})$, $p(\mathbf{z} : \mathbf{x})$, or $J(\mathbf{x} : \mathbf{z})$.

The global parameterization and local iterative techniques can sometimes be effectively combined. For example, a most probable Bayesian estimator might be obtained in two steps:

1. Find $\boldsymbol{\chi}_m$, the value which yields the maximum $p(\boldsymbol{\chi}_m \mid \mathbf{z})$, $m = 1, \ldots, M$.
2. Iterate to find the actual maximum of $p(\mathbf{x} \mid \mathbf{z})$ using $\boldsymbol{\chi}_m$ as the starting point.

With such an approach, crude parameterization can be used to get in the "right ball park" so that the iterations will converge rapidly to the right answer.

There are, of course, other ways besides parameterization to get a ball park answer to be used to start an iteration. For example, if

$$\begin{bmatrix} \mathbf{z}_1 \\ \mathbf{z}_2 \end{bmatrix} = \begin{bmatrix} \mathbf{h}_1(\mathbf{x}) \\ \mathbf{h}_2(\mathbf{x}) \end{bmatrix} + \mathbf{v}$$

and $\mathbf{z}_1$ is the same dimension as $\mathbf{x}$, a ball park answer can sometimes be obtained by solving

$$\mathbf{z}_1 = \mathbf{h}_1(\mathbf{x})$$

for $\mathbf{x}$ in some fashion.

The use of changes of variables can be very effective. The reader should not be afraid of making approximations which change a nonlinear model into a simpler or more convenient form. After all, the "optimum" estimate for an "exact" model can rarely be exactly implemented.

Historical Notes and References: Chapter 11

Section 11.1

The basic ideas are available many places in many forms in Book Lists 4 and 5 on Stochastic Systems and Communications.

Section 11.2

The basic ideas are taken directly from Mathematical Statistics, where much more extensive discussions can be found, see Book List 2. Wilks [1] and Cramer [1] were used. See also Barankin [1]; Ferguson [1]; Lelam [1]; Sakrison [2].

Section 11.7

The discussions on iterative methods only touch the surface of a large field. One of many possible references is Wilde and Beightler [1]. See also Book List 3 on Deterministic Systems. Masiello and Schweppe [1] and Schweppe [11] discuss methods of choosing simple, easy to compute gains for a particular high dimensional problem associated with electric power systems.

Section 11.10

Albert and Gardner [1] is a good general reference. A few of many applications are in Gardner [1] and Sakrison [1].

Exercises

For more details on optimality criteria as in 11.9, see Benedict and Sondhi [1]; Brown [1]; Sherman [1]; Zakai [1].

Exercises

11.1. Consider the linear Bayesian model

$$z = x + v$$

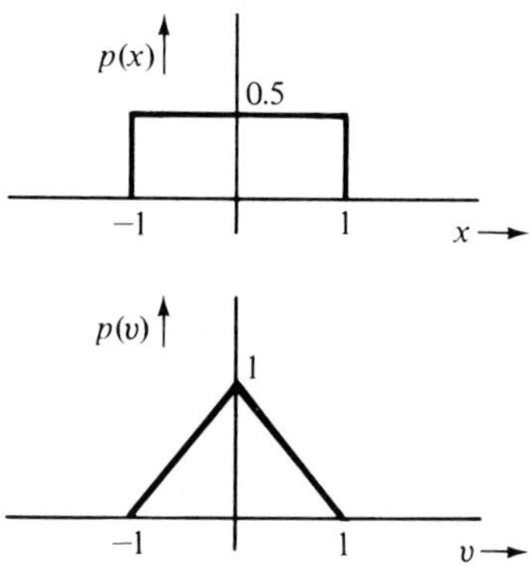

Figure P11.1

a. Plot $p(z \mid x)$ versus x for $z = 0.5$.
b. Assume that an observation $z = z_{\text{actual}}$, $z_{\text{actual}} = 0.5$ is made. What is the value for the estimate $\hat{x}$ for the following estimators?

(1) Conditional expectation.

(2) Most probable.

(3) Median.

(4) Minimax.

11.2. Consider the Bayesian model

$$z_1 = x^2 + v_1$$
$$z_2 = x + v_2$$
$$x \text{ is } N(0, \psi)$$
$$\mathbf{v} = \begin{bmatrix} v_1 \\ v_2 \end{bmatrix} \qquad \mathbf{v} \text{ is } N(\mathbf{0}, \mathbf{R}) \qquad \mathbf{R} = \begin{bmatrix} R_{11} & 0 \\ 0 & R_{22} \end{bmatrix}$$
$$x, \mathbf{v}: \text{independent}$$

a. Let $\hat{x}$ denote the most probable estimate of x. $\hat{x}$ is given by the solution of a cubic equation. What is the equation?

b. A "reasonable" estimator which is easier to implement is

$$\hat{x} = W_1(z_1)^{1/2} + W_2 z_2 + W_0$$

Discuss how you would calculate "optimum" values for W_1, W_2, and W_0.

11.3. Consider a random variable z which can take on values $z = 0, 1, 2, \ldots$. Let $p(z)$ denote the probability function:

$$p(z) = \frac{\mu^z e^{-\mu}}{z!}$$

z has a *Poisson distribution.* (Note that since z assumes only discrete values, $p(z)$ is not a probability density.) Let $z(1). \ldots z(N)$ denote N independent observations of z. Let $\hat{\mu}_N$ denote the maximum likelihood estimate of μ computed from $z(1) \cdots z(N)$. Show that

$$\hat{\mu}_N = \frac{1}{N} \sum_{n=1}^{N} z(n)$$

11.4. Consider the observations

$$z(t) = h(x, t) + v(t), \qquad 0 \le t \le T$$

x: unknown

$v(t)$: continuous-time white Gaussian stochastic process

$$E\{v(t)v(\tau)\} = \delta(t - \tau)R$$

Let $\hat{x}$ denote a maximum-likelihood-type estimate of x given $z(t)$, $0 \le t \le T$.

a. Does the likelihood function exist?

b. Give an argument which enables $\hat{x}$ to be expressed as the value of x which minimizes some explicit function of h, R, and $z(t)$, $0 \le t \le T$. What is this function?

Hint: Formulate the discrete-time version

$$z(n\Delta) = h(x, n\Delta) + v(n\Delta)$$

and use the discrete to continuous time limit.

11.5. Consider the unknown-but-bounded model

$$\mathbf{z} = \mathbf{h}(\mathbf{x}) + \mathbf{v}$$

$$\mathbf{h}(\mathbf{x}) = \begin{bmatrix} \sqrt{x_1^2 + x_2^2} \\ x_1 + x_2 \end{bmatrix}$$

$$v_1^2 + v_2^2 \leq 1$$

$$x_1, x_2: \text{ unknown}$$

Assume that the particular observation

$$\mathbf{z}_{\text{actual}} = \begin{bmatrix} 4 \\ 1 \end{bmatrix}$$

is obtained.

a. Sketch $\boldsymbol{\Omega}_{\text{est}}$.

b. Assume that a point estimate $\hat{\mathbf{x}}$ is to be used where $\hat{\mathbf{x}}$ is the value(s) of $\mathbf{x}$ which minimizes

$$[\mathbf{z} - \mathbf{h}(\mathbf{x})]'[\mathbf{z} - \mathbf{h}(\mathbf{x})]$$

Indicate on the sketch of part (a) where the value(s) of $\hat{\mathbf{x}}$ lie. Do not actually solve for $\hat{\mathbf{x}}$.

11.6. Consider the Fisher model

$$\mathbf{z} = \begin{bmatrix} \sqrt{x_1^2 + x_2^2} \\ 3\sqrt{x_1^2 + x_2^2} \\ \tan^{-1}\left(\dfrac{x_1}{x_2}\right) \end{bmatrix} + \mathbf{v}$$

where x_1 and x_2 are completely unknown and $\mathbf{v}$ is $N(0, \mathbf{R})$. It is desired to use $\mathbf{z}$ to obtain an estimate of $\mathbf{x}$.

a. Assume that a maximum likelihood estimator is desired and that an iterative solution of the form

$$\hat{\mathbf{x}}_{i+1} = \hat{\mathbf{x}}_i - \mathbf{B}(i)\,\mathbf{f}(\mathbf{z}, \hat{\mathbf{x}}_i)$$

$$\mathbf{B}(i): \text{ gain matrix}$$

is to be used. What is a good choice for $\mathbf{f}(\mathbf{z}, \hat{\mathbf{x}}_i)$?
Discuss briefly how you might choose a gain sequence $\mathbf{B}(i)$.

b. Assume that you were willing to accept a reasonable but not a maximum likelihood estimator. Discuss one way you might use a change of variable to obtain a new estimator. Assume that $\mathbf{v}$ is small.

Hint: Consider a *polar coordinate* system.

11.7. Consider

$$z(t) = \sin(\omega t) + v(t), \qquad 0 \leq t \leq T$$

frequency ω: unknown

$v(t)$: continuous time white stochastic process

$$E\{v(t)v(\tau)\} = \delta(t - \tau)R$$

In radars, a standard way to estimate ω is the *Doppler filter bank*, which can

be represented by

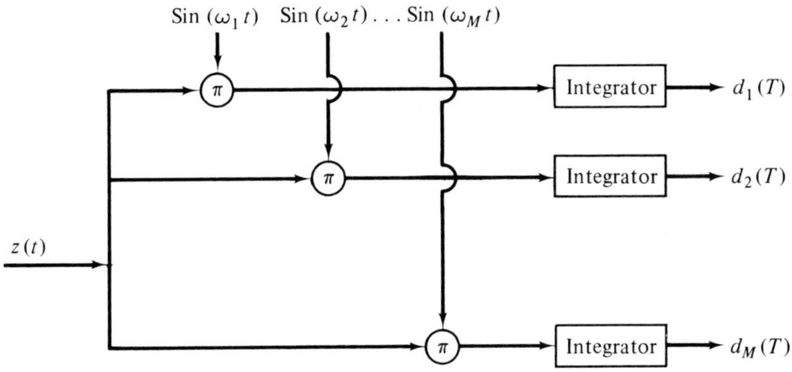

Figure P11.7

where $\hat{\omega}$ is the ω_j which yields the largest $d_j(T)$. Justify the use of Doppler filter banks in terms of maximum likelihood estimation. *Comment:* In a "real" problem,

$$z(t) = A \sin(\omega t + \theta) + v(t)$$

where the amplitude A and phase angle θ are also unknown. Thus actual radar Doppler filter banks are somewhat different.

11.8. Consider

$$z_1(n) = Hx(n) + v(n)$$

$$z_2(n) = x(n)$$

$$\tilde{\mathbf{x}}(n) = \begin{bmatrix} x(n) \\ v(n) \end{bmatrix} \qquad z(n) = \begin{bmatrix} z_1(n) \\ z_2(n) \end{bmatrix}$$

$$E[\tilde{\mathbf{x}}(n_1)\tilde{\mathbf{x}}'(n_2)] = \begin{cases} R, & n_1 = n_2 \\ \mathbf{0}, & n_1 \neq n_2 \end{cases}$$

$$\mathbf{R} = \begin{bmatrix} 1 & \rho \\ \rho & 1 \end{bmatrix}$$

$\tilde{\mathbf{x}}(n)$: Gaussian process, zero mean

The value of H is completely unknown, and it is desired to use $\mathbf{z}(n), n = 1, \ldots, N$, to estimate the value of H.

a. A "reasonable" approach is to use least-squares concepts to yield an estimate $\hat{\hat{H}}(N)$ defined as the value of H which minimizes

$$J = \sum_{n=1}^{N} [z_1(n) - Hz_2(n)]^2$$

What is $\hat{\hat{H}}(N)$? This $\hat{\hat{H}}(N)$ is a biased estimate. What is the bias? *Hint:*

To evaluate the bias, note that

$$E\left\{\frac{xv}{x^2}\right\} = \int \frac{xp(x)}{x^2}\left[\int vp(v\,|\,x)\,dv\right]dx$$

and use formulas of Appendix *E*.

b. A "more reasonable" approach is to use a maximum likelihood estimate $\hat{H}(N)$. What is $\hat{H}(N)$? Is $\hat{H}(N)$ biased? *Hint:* Define $\mathbf{z}(n) = \tilde{\mathbf{H}}\tilde{\mathbf{x}}(n)$, evaluate the log likelihood function, and then minimize with respect to H.

11.9. Consider the scalar random variable x_1 and the random vector $\mathbf{x}_2$. Let $g(\mathbf{x}_2)$ be some arbitrary function of $\mathbf{x}_2$. Let d denote a measure of the *distance* between x_1 and $g(\mathbf{x}_2)$ defined by

$$d = E\{l[x_1 - g(\mathbf{x}_2)]\}$$

$$l(\cdot)\text{: specified function}$$

The particular function $g(\mathbf{x}_2)$ which minimizes d can be viewed as the optimum Bayesian estimator (relative to l) of x_1 that can be calculated using $\mathbf{x}_2$.

a. Consider $l(y) = y^2$. Show that the minimum of $d = E\{[x_1 - g(\mathbf{x}_2)]^2\}$ occurs for $g(\mathbf{x}_2) = E\{x_1\,|\,\mathbf{x}_2\}$. *Hint:* Write

$$\begin{aligned} d &= E\{[x_1 - E\{x_1\,|\,\mathbf{x}_2\} + E\{x_1\,|\,\mathbf{x}_2\} - g(\mathbf{x}_2)]^2\} \\ &= d_{\text{cross}} + \tilde{d} \end{aligned}$$

$$d_{\text{cross}} = E\{[x_1\,|\,\mathbf{x}_2\}][E\{x_1\,|\,\mathbf{x}_2\} - g(\mathbf{x}_2)]\}$$

and then use (D.21) in Appendix D to get $d_{\text{cross}} = 0$.

b. Consider the $l(\cdot)$, which are symmetric, nonnegative, convex functions. From (D.13)

$$d = E\{E\{l[x_1 - g(\mathbf{x}_2)]\,|\,\mathbf{x}_2\}\}$$

or

$$d = \int\left[\int l[x_1 - g(\mathbf{x}_2)]\,p[x_1\,|\,x_2]\,dx_1\right]d\mathbf{x}_2$$

Assume that $p[x_1\,|\,x_2]$ is a symmetric function about $E\{x_1\,|\,x_2\}$. Convince yourself from the geometry of $l(\cdot)$ and $p(x_1\,|\,\mathbf{x}_2)$ that $g(\mathbf{x}_2) = E\{x_1\,|\,\mathbf{x}_2\}$ minimizes d.

c. Now assume that x_1 and $\mathbf{x}_2$ are Gaussian. Use (E.9) and (E.10) in Appendix E to show that

$$d = \int q(x_2)\,dx_2$$

$$q(\mathbf{x}_2) = [2\pi\mathbf{\Gamma}_{1|2}]^{-1/2}\int l[x_1 - g(\mathbf{x}_2)]\exp\left\{-\frac{(x_1 - E\{x_1\,|\,\mathbf{x}_2\})^2}{2\mathbf{\Gamma}_{1|2}}\right\}dx_1$$

Define $\hat{c}$ to be the value of c which minimizes

$$\int l(y)\exp\left\{-\frac{[y - c]^2}{2\mathbf{\Gamma}_{1|2}}\right\}dy$$

Show that for "any" $l(\cdot)$, $d = E\{l[x_1 - g(\mathbf{x}_2)]\}$ is minimized by $g(\mathbf{x}_2)$

$= E\{x_1 \mid \mathbf{x}_2\} + \hat{c}$. *Hint:* Make a change of variables

$$y = x_1 - g(\mathbf{x}_2)$$

so

$$q(\mathbf{x}_2) = [2\pi\boldsymbol{\Gamma}_{1|2}]^{-1/2} \int l(y) \exp\left\{-\frac{[y + g(\mathbf{x}_2) - E\{x_1 \mid \mathbf{x}_2\}]^2}{2\boldsymbol{\Gamma}_{1|2}}\right\} dy$$

Comment: The results for parts *a*, *b*, and *c* can be generalized to the case where x_1 is a vector.

11.10. Prove (11.7.17). *Hint:* Let λ_k, $k = 1, \ldots,$ denote the eigenvalues of $\mathbf{A}_i$. Then for any invertible matrix $\mathbf{C}$

$$|\mathbf{I}(\lambda - 1) + (\mathbf{C}\mathbf{B}_i\mathbf{C}')(\mathbf{C}\mathbf{S}^{-1}(\hat{\mathbf{x}})\mathbf{C}')^{-1}| = 0$$

If $\mathbf{C}$ is chosen so that

$$\mathbf{C}\mathbf{S}^{-1}(\hat{\mathbf{x}})\mathbf{C}' = \mathbf{D}$$

$$\mathbf{C}\mathbf{B}\mathbf{C}' = \mathbf{I}$$

$$\mathbf{D} = \begin{bmatrix} D_{11} & & \mathbf{0} \\ & D_{22} & \\ \mathbf{0} & & \ddots \end{bmatrix} \qquad \mathbf{D}_{kk} > 0$$

then

$$\lambda_k = 1 - D_{kk}^{-1}$$

12

Performance of Estimators: Static Nonlinear Systems

In Chapter 11, estimation techniques were discussed for the static case models

$$\mathbf{z} = \mathbf{h}(\mathbf{x}, \mathbf{v}) \tag{12.1}$$

$$\mathbf{z} = \mathbf{h}(\mathbf{x}) + \mathbf{v} \tag{12.2}$$

where $\mathbf{z}$ is the observation (a K_2 vector), $\mathbf{x}$ is the uncertain state to be estimated (a K_1 vector), $\mathbf{v}$ is the uncertainty in the observation, and $\mathbf{h}$ is a known vector function. The performance of estimators is now discussed for a Fisher formulation where $\mathbf{x}$ is unknown and $\mathbf{v}$ is a random vector.

The discussions are restricted to Fisher models for simplicity. They can be extended to Bayesian models in various ways. One approach is to use the concepts of Section 11.3 and "view" a Bayesian model as a Fisher model with an extra observation.

Linearized error analysis for arbitrary estimators is discussed in Section 12.1 and similar results are given in Section 12.2 for Gaussian $\mathbf{v}$ for maximum likelihood estimators. Lower bounds on the estimate error covariance matrix are discussed in Sections 12.3 and 12.5. Ambiguity functions are discussed in Section 12.4. Monte Carlo methods are briefly considered in Section 12.6. A final discussion is given in Section 12.7.

12.1 Linearized Error Analysis (Sensitivity Analysis)

Assume that an estimate $\hat{\mathbf{x}}$ is obtained by a nonlinear operation on $\mathbf{z}$ as follows:

$$\hat{\mathbf{x}} = \mathbf{w}(\mathbf{z}) \tag{12.1.1}$$

where $\mathbf{w}$ is some K_1 vector nonlinear function of the K_2 vector $\mathbf{z}$. This $\mathbf{w}$ could be one of the nonlinear estimators of Chapter 11, but for the present it can be any $\mathbf{w}$ (whose necessary derivatives exist). Only (12.2) when $\mathbf{z} = \mathbf{h}(\mathbf{x}) + \mathbf{v}$ is considered. Assume that

$$E\{\mathbf{v}\} = \mathbf{0} \qquad E\{\mathbf{v}\mathbf{v}'\} = \mathbf{R}$$

A Gaussian assumption is not used. Let $\mathbf{x}_t = \mathbf{x}_{\text{true}}$ denote the true but unknown state.

Let $\mathbf{b}(\mathbf{x}_t)$ denote the bias error in $\hat{\mathbf{x}}$ so

$$\mathbf{x}_t - E\{\hat{\mathbf{x}}\} = \mathbf{b}(\mathbf{x}_t) \tag{12.1.2}$$

where the expectation is over $\mathbf{v}$. Substituting (12.2) into (12.1.1) gives

$$\hat{\mathbf{x}} = \mathbf{w}[\mathbf{h}(\mathbf{x}_t) + \mathbf{v}]$$

Expanding $\mathbf{w}$ in a Taylor series expansion about $\mathbf{h}(\mathbf{x}_t)$ gives

$$\begin{aligned} \hat{\mathbf{x}} &= \mathbf{w}[\mathbf{h}(\mathbf{x}_t)] + \mathbf{W}^{(1)}[\mathbf{h}(\mathbf{x}_t)]\mathbf{v} + \cdots \\ \mathbf{W}^{(1)}[\mathbf{h}(\mathbf{x}_t)] &= \left.\frac{\partial \mathbf{w}(\mathbf{z})}{\partial \mathbf{z}}\right|_{\mathbf{z}=\mathbf{h}(\mathbf{x}_t)} \end{aligned} \tag{12.1.3}$$

Now assume that the uncertainty $\mathbf{v}$ is small enough so that (in some unspecified average sense) the higher-order terms in (12.1.3) can be neglected. Then from (12.1.3) and (12.1.2) and the fact that $\mathbf{v}$ is zero mean

$$\mathbf{w}[\mathbf{h}(\mathbf{x}_t)] = \mathbf{x}_t - \mathbf{b}(\mathbf{x}_t)$$

so that (12.1.3) becomes, after neglecting the higher-order terms,

$$\mathbf{x}_t - \hat{\mathbf{x}} = -\mathbf{W}^{(1)}[\mathbf{h}(\mathbf{x}_t)]\mathbf{v} + \mathbf{b}(\mathbf{x}_t)$$

Thus for the estimator (12.1.1), the estimate error is

$$\boxed{\begin{aligned} E\{(\mathbf{x}_t - \hat{\mathbf{x}})(\mathbf{x}_t - \hat{\mathbf{x}})'\} &= \mathbf{\Sigma} + \mathbf{b}(\mathbf{x}_t)\mathbf{b}'(\mathbf{x}_t) \\ \mathbf{\Sigma} &= \mathbf{W}^{(1)}[\mathbf{h}(\mathbf{x}_t)]\mathbf{R}\mathbf{W}^{(1)\prime}[\mathbf{h}(\mathbf{x}_t)] \end{aligned}} \tag{12.1.4}$$

Now (12.1.4) is not directly usable, as the partial derivatives are evaluated at the true and hence unknown value $\mathbf{x}_t$. However, two ways to proceed are

1. If an estimate $\hat{\mathbf{x}} = \mathbf{w}(\mathbf{z}_{\text{actual}})$ for some observation $\mathbf{z} = \mathbf{z}_{\text{actual}}$ has been made, assume that

$$\mathbf{W}^{(1)}[\mathbf{h}(\mathbf{x}_t)] \approx \mathbf{W}^{(1)}[\mathbf{h}(\hat{\mathbf{x}})]$$

and then use

$$\Sigma = \mathbf{W}^{(1)}[\mathbf{h}(\hat{\mathbf{x}})]\mathbf{R}\mathbf{W}^{(1)\prime}[\mathbf{h}(\hat{\mathbf{x}})]$$

as an error analysis formula for the given $\hat{\mathbf{x}}$.

2. Choose some value $\mathbf{x}_0$ and interpret

$$\Sigma = \mathbf{W}^{(1)}[\mathbf{h}(\mathbf{x}_0)]\mathbf{R}\mathbf{W}^{(1)\prime}[\mathbf{h}(\mathbf{x}_0)]$$

as the covariance matrix of the error that would result if $\mathbf{x}_t$ had the value $\mathbf{x}_0$.

Thus for small $\mathbf{v}$, (12.1.4) can be used to get a "feel" for Σ, the covariance matrix of the errors.

EXAMPLE: Consider

$$z = (x)^{1/2} + v$$

$$x\text{: unknown}$$

$$E\{v\} = 0 \qquad E\{v^2\} = R$$

A "reasonable" nonlinear estimator is

$$\hat{x} = w(z) = z^2$$

For this estimator

$$\begin{aligned} W^{(1)}[h(x_t)] &= 2z\,|_{z=h(x_t)} \\ &= 2h[x_t] \\ &= 2x_t^{1/2} \end{aligned}$$

so from (12.1.4)

$$\Sigma = 4Rx_t$$

This example is discussed further in Section 12.2.

In the special case of a linear model $\mathbf{z} = \mathbf{H}\mathbf{x} + \mathbf{v}$ and a linear estimator $\hat{\mathbf{x}} = \mathbf{W}\mathbf{z}$, then

$$\begin{aligned} \mathbf{W}^{(1)}[\mathbf{h}(\mathbf{x}_t)] &= \mathbf{W} \\ \mathbf{b} &= [\mathbf{W}\mathbf{H} - \mathbf{I}]\mathbf{x}_t \\ \Sigma &= \mathbf{W}\mathbf{R}\mathbf{W}' \end{aligned}$$

which naturally corresponds to the result of Section 5.9. Thus (12.1.4) can be simply interpreted as the linear model result of Section 5.9 where the $\mathbf{W}$ matrix of the linear estimator is replaced by the matrix of partial derivatives of the nonlinear estimator.

It must be emphasized the nonlinear Fisher estimates are usually biased and that no general way to calculate the bias error, $\mathbf{b}(\mathbf{x}_t)$, seems to exist other than brute force. This is unfortunate, as bias errors can dominate, i.e., be larger than the error variances. This problem is often ignored because the bias errors are hard to compute, but such justification obviously is rather hard to defend. In the linear model of Chapter 5, all biases were zero unless modeling errors were present.

Because of the bias error problem, many authors call the analysis leading to $\mathbf{\Sigma}$ of (12.1.4) a sensitivity analysis instead of a linearized error analysis. There is good justification for such terminology, as $\mathbf{\Sigma}$ really measures the sensitivity of $\hat{\mathbf{x}}$ to the uncertainty $\mathbf{v}$ instead of measuring the total error in $\hat{\mathbf{x}}$.

The above analysis assumes the estimator is specified *explicitly* as is (12.1.1) where the function $\mathbf{w}(\cdot)$ is given. Many of the estimators of Chapter 11 were specified *implicitly* as for example, $\hat{\mathbf{x}}$ minimizes $J(\mathbf{x})$. In such cases $\mathbf{w}(\cdot)$ is not known. Section 12.2 considers such situations.

12.2 Linear Error Analysis: Maximum Likelihood Estimator

A linearized error analysis like that of Section 12.1 is now done for maximum likelihood estimators. Only the special case

$$\begin{gathered}\mathbf{z} = \mathbf{h}(\mathbf{x}) + \mathbf{v} \\ \mathbf{v} \text{ is } N(\mathbf{0}, \mathbf{R})\end{gathered} \tag{12.2.1}$$

is considered here. More general models are discussed in Section 12.3.

If the model were linear ($\mathbf{z} = \mathbf{Hx} + \mathbf{v}$) as in Chapter 5, then $\mathbf{\Sigma}$, the estimate error covariance matrix, would be $\mathbf{\Sigma} = [\mathbf{HR}^{-1}\mathbf{H}]^{-1}$. For the nonlinear model (12.2.1) it turns out [as should be expected by analogy with Section (12.1)] that an approximate $\mathbf{\Sigma}$ is of the form $[\mathbf{H}^{(1)}\mathbf{R}^{-1}\mathbf{H}^{(1)}]^{-1}$, where $\mathbf{H}^{(1)}$ is a matrix of partial derivatives of $\mathbf{h}(\mathbf{x})$. The necessary details are now given.

The maximum likelihood estimate $\hat{\mathbf{x}}$ is the value of $\hat{\mathbf{x}}$ which maximizes the likelihood function $p(\mathbf{z}: \mathbf{x})$ or equivalently $\ln p(\mathbf{z}: \mathbf{x})$. For the assumed model (12.2.1)

$$\begin{gathered}\ln p(\mathbf{z}: \mathbf{x}) = c - [\mathbf{z} - \mathbf{h}(\mathbf{x})]'\mathbf{R}^{-1}[\mathbf{z} - \mathbf{h}(\mathbf{x})] \\ c\text{: constant}\end{gathered}$$

so $\hat{\mathbf{x}}$ is specified by

$$\mathbf{H}^{(1)}(\hat{\mathbf{x}})\mathbf{R}^{-1}[\mathbf{z} - \mathbf{h}(\hat{\mathbf{x}})] = \mathbf{0} \tag{12.2.2}$$

Define

$$\boldsymbol{\delta}_x = \mathbf{x}_t - \hat{\mathbf{x}}$$

Assume for the moment that $\mathbf{x}$ is a scalar. Then a Taylor series about $\hat{x} = x_t$ of (12.2.2) yields

$$\begin{aligned}\mathbf{H}^{(1)}(x_t)\mathbf{R}^{-1}[\mathbf{z} - \mathbf{h}(x_t)] + \{\mathbf{H}^{(2)}(x_t)\mathbf{R}^{-1}[\mathbf{z} - \mathbf{h}(x_t)] \\ - \mathbf{H}^{(1)}(x_t)\mathbf{R}^{-1}\mathbf{H}^{(1)}(x_t)\}\delta_x + \cdots = 0\end{aligned}$$

Assume that δ_x is small enough so the higher-order terms can be neglected. Then using $\mathbf{z} = \mathbf{h}(x_t) + \mathbf{v}$ gives

$$[\mathbf{H}^{(1)}(x_t)\mathbf{R}^{-1}\mathbf{H}^{(1)}(x_t) + \mathbf{H}^{(2)}(x_t)\mathbf{R}^{-1}\mathbf{v}]\,\delta_x = \mathbf{H}^{(1)}(x_t)\mathbf{R}^{-1}\mathbf{v}$$

Since $E\{\mathbf{v}\delta_x\} \neq 0$, it follows that $E\{\delta_x\} \neq 0$, so x is biased (in general). However, to proceed further with the error analysis, the $\mathbf{H}^{(2)}(x_t)$ term will now be ignored. This corresponds to assuming that $\mathbf{H}^{(1)}(\hat{\mathbf{x}}) \approx \mathbf{H}^{(1)}(x_t)$. With this assumption

$$E\{\delta_x^2\} = [\mathbf{H}^{(1)}(x_t)\mathbf{R}^{-1}\mathbf{H}^{(1)}(x_t)]^{-1}$$

In the case of vector $\mathbf{x}$, the corresponding approximation for the error covariance matrix $\mathbf{\Sigma}$ is

$$\begin{aligned} \mathbf{\Sigma} &= E[\boldsymbol{\delta}_x\,\boldsymbol{\delta}_x]' = E\{(\mathbf{x}_t - \mathbf{x})(\mathbf{x}_t - \mathbf{x})'\} \\ \mathbf{\Sigma} &\approx [\mathbf{H}^{(1)}[\mathbf{x}_t]\mathbf{R}^{-1}\mathbf{H}^{(1)}[\mathbf{x}_t]]^{-1} \end{aligned} \tag{12.2.3}$$

The argument leading to (12.2.3) implies that $\hat{\mathbf{x}}$ is unbiased, $E\{\boldsymbol{\delta}_x\} = \mathbf{0}$, but in practice this is usually not true. Thus (12.2.3), like (12.1.4), should be viewed as a sensitivity analysis formula rather than a complete error analysis. Bias errors can dominate.

The fact that (12.2.3) involves the true but unknown $\mathbf{x}_t$ is handled just as was done with (12.1.4).

EXAMPLE: Consider

$$z = (x)^{1/2} + v$$
$$v \text{ is } N(0, R)$$

This is the same model as the example of Section 12.1 except that v is now also assumed to be Gaussian. For this case

$$\hat{x} = z^2$$

is the maximum likelihood estimate, so that error analysis of (12.1.4) could be applied. However, using (12.2.3),

$$H^{(1)}(x_t) = \frac{1}{2(x_t)^{1/2}}$$
$$\Sigma = 4Rx_i$$

which is what (12.1.4) also gave. This example was purposely chosen, as an exact error analysis is also possible as follows:

$$\hat{x} = x_t + 2vx_t^{1/2} + v^2$$

so

$$x_t - E\{\hat{x}\} = b(x_t) = -R \qquad \text{bias eror}$$

Using Appendix E.4,

$$E\{(x_t - \hat{x})^2\} = 4Rx_t + 3R^2$$

Thus if Σ is the exact error variance,

$$\Sigma = E\{(x_t - x - b)^2\} = E\{(x_t - x)^2\} - b^2$$
$$= 4Rx_t + 2R^2$$

Therefore it is seen that the linear error analysis of either (12.1.4) or (12.2.3) yields a "good" answer if $x_t \gg R$ but is "bad" if $R \gg x_t$.

Note that the Gaussian assumption was used only to yield (12.2.2). Hence the error analysis formula (12.2.3) applies for non-Gaussian $\mathbf{v}$ if $\hat{\mathbf{x}}$ is chosen (by engineering judgment because it is reasonable) to be the value of $\mathbf{x}$ which minimizes

$$J(\mathbf{x}) = [\mathbf{z} - \mathbf{h}(\mathbf{x})]'\mathbf{R}^{-1}[\mathbf{z} - \mathbf{h}(\mathbf{x})]$$

12.3 Lower Bound on Estimate Error Variance†

The error analyses of Section 12.1 and 12.2 were based on linearization and engineering type arguments for specific models. A general mathematical theory which yields a lower bound on the estimate error covariance for more general models is now developed. The result (12.2.3) follows as a special case of the general theory. The bound to be discussed is often called the *Cramer-Rao lower bound* or the *information inequality*.

12.3.1 General Case†

Assume that an observation $\mathbf{z}$ is made where $\mathbf{z}$ is a random vector with probability density specified by an unknown parameter vector $\mathbf{x}$. Let $p(\mathbf{z}: \mathbf{x})$ denote the likelihood function. The model relating $\mathbf{z}$ to $\mathbf{x}$ might be as in (12.1) and (12.2) but could be of another form if desired.

For simplicity the case of scalar x is considered first. Let x_t denote the true but unknown value of x. Consider some estimator $\hat{x} = w(\mathbf{z})$ and let $b(x)$ denote the bias error of $\hat{x}$ so that

$$b(x_t) = x_t - E\{\hat{x}\}$$

Now $E\{\hat{x}\}$ denotes expectation over $\mathbf{z}$ for fixed x_t and can be expressed in terms of the likelihood function as

$$E\{\hat{x}\} = \int \cdots \int \hat{x}p(\mathbf{z}: x_t)\, dz_1 \cdots dz_{K_2}$$

or, in simpler notation,

$$E\{\hat{x}\} = \int \hat{x}p(\mathbf{z}: x_t)\, d\mathbf{z}$$

Thus

$$\int (x_t - \hat{x})p(\mathbf{z}: x_t)\, d\mathbf{z} = b(x_t)$$

Taking the partial derivative of both sides with respect to x_t yields

$$\int (x_t - \hat{x})\left[\frac{\partial}{\partial x} p(\mathbf{z}: x_t)\right] d\mathbf{z} = -1 + b^{(1)}(x_t)$$

$$b^{(1)}(x_t) = \left.\frac{\partial b(x)}{\partial x}\right|_{x=x_t}$$

$$\frac{\partial}{\partial x} p(\mathbf{z}: x_t) = \left.\frac{\partial p(\mathbf{z}: x)}{\partial x}\right|_{x=x_t}$$

which can be rewritten as

$$\begin{aligned} \int (x_t - \hat{x})\left[\frac{\partial}{\partial x} \ln p(\mathbf{z}: x_t)\right] p(\mathbf{z}: x_t)\, d\mathbf{z} &= -1 + b^{(1)}(x_t) \\ \frac{\partial}{\partial x} \ln p(\mathbf{z}: x_t) &= \left.\frac{\partial \ln p(\mathbf{z}: x)}{\partial x}\right|_{x=x_t} \end{aligned} \tag{12.3.1}$$

The Schwartz inequality states that for "any" $f_1(\mathbf{z}), f_2(\mathbf{z})$, and $p(\mathbf{z}) \geq 0$,

$$\left[\int f_1(\mathbf{z}) f_2(\mathbf{z}) p(\mathbf{z})\, d\mathbf{z}\right]^2 \leq \left[\int f_1^2(\mathbf{z}) p(\mathbf{z})\, d\mathbf{z}\right]\left[\int f_2^2(\mathbf{z}) p(\mathbf{z})\, d\mathbf{z}\right]$$

Using the Schwartz inequality on (12.3.1) gives

$$\left\{\int (x_t - \hat{x})^2 p(\mathbf{z}: x_t)\, d\mathbf{z}\right\}\left\{\int \left[\frac{\partial}{\partial x} \ln p(\mathbf{z}: x_t)\right]^2 p(\mathbf{z}: x_t)\, d\mathbf{z}\right\} \geq [-1 + b^{(1)}(x_t)]^2$$

Since

$$E\{(x_t - \hat{x})^2\} = \int (x_t - \hat{x})^2 p(\mathbf{z}: x_t)\, dz$$

it follows that a lower bound on the estimate error variance for any estimate $\hat{x}$ has been obtained; i.e.,

$$E\{(x_t - \hat{x})^2\} \geq \frac{[-1 + b^{(1)}(x_t)]^2}{\int [(\partial/\partial x) \ln p(\mathbf{z}: x_t)]^2 p(\mathbf{z}: x_t)\, d\mathbf{z}}, \tag{12.3.2}$$

where $\hat{x}$ is any estimate

A different way of writing (12.3.2) is

$$E\{(x_t - \hat{x})^2\} \geq \frac{[-1 + b^{(1)}(x_t)]^2}{E\{[(\partial/\partial x) \ln p(\mathbf{z}: x_t)]^2\}}$$

If the necessary derivatives exist, it follows that

$$\frac{\partial}{\partial x} \int p(\mathbf{z}: x)\, d\mathbf{z} = \int \left[\frac{\partial}{\partial x} \ln p(\mathbf{z}: x)\right] p(\mathbf{z}: x)\, d\mathbf{z} = 0$$

and thus

$$\frac{\partial^2}{\partial x^2}\int p(\mathbf{z}: x)\,d\mathbf{z} = \int \frac{\partial^2}{\partial x}[\ln p(\mathbf{z}: x)]\,p(\mathbf{z}: x)\,d\mathbf{z} + \int \left[\frac{\partial}{\partial x}\ln p(\mathbf{z}: x)\right]^2 p(\mathbf{z}: x)\,d\mathbf{z} = \mathbf{0} \tag{12.3.3}$$

Define

$$\frac{\partial^2}{\partial x^2}[\ln p(\mathbf{z}: x_t)] = \frac{\partial^2}{\partial x^2}[\ln p(\mathbf{z}: x)]\bigg|_{x=x_t}$$

Then (12.3.2) can be rewritten in an equivalent form:

$$E\{(x_t - \hat{x})^2\} \geq \frac{[-1 + b^{(1)}(x_t)]^2}{-\int (\partial^2/\partial x^2)[\ln p(\mathbf{z}: x_t)]\,p(\mathbf{z}: x_t)\,d\mathbf{z}}, \quad \text{any } \hat{x} \tag{12.3.4}$$

The case where $\mathbf{x}$ is a vector is now outlined. For simplicity assume that $\hat{\mathbf{x}}$ is unbiased so that

$$\int [\hat{\mathbf{x}} - \mathbf{x}_t]\,p(\mathbf{z}: \mathbf{x}_t)\,d\mathbf{z} = \mathbf{0}$$

Taking the partial with respect to $\mathbf{x}_t$ gives

$$\int [\hat{\mathbf{x}} - \mathbf{x}_t]\left[\frac{\partial p(\mathbf{z}: \mathbf{x}_t)}{\partial \mathbf{x}}\right]' d\mathbf{z} = \mathbf{I}$$

or

$$\int [\hat{\mathbf{x}} - \mathbf{x}_t]\left[\frac{\partial \ln p(\mathbf{z}: \mathbf{x}_t)}{\partial \mathbf{x}}\right]' p(\mathbf{z}: \mathbf{x}_t)\,d\mathbf{z} = \mathbf{I} \tag{12.3.5}$$

Now multiply (12.3.5) by two arbitrary vectors $\mathbf{y}_1$ and $\mathbf{y}_2$ to give

$$\int \{\mathbf{y}_1'(\hat{\mathbf{x}} - \mathbf{x}_t)\}\left\{\left[\frac{\partial \ln p(\mathbf{z}: \mathbf{x}_t)}{\partial \mathbf{x}}\right]'\mathbf{y}_2\right\} p(\mathbf{z}: \mathbf{x}_t)\,d\mathbf{z} = \mathbf{y}_1'\mathbf{y}_2$$

The Schwartz inequality gives

$$\begin{aligned} &(\mathbf{y}_1'\mathbf{\Sigma}\mathbf{y}_1)(\mathbf{y}_2'\mathbf{B}\mathbf{y}_2) \geq (\mathbf{y}_1'\mathbf{y}_2)^2 \\ &\mathbf{\Sigma} = E\{(\hat{\mathbf{x}} - \mathbf{x}_t)(\hat{\mathbf{x}} - \mathbf{x}_t)'\} \\ &\mathbf{B} = \int \left\{\left[\frac{\partial \ln p(\mathbf{z}: \mathbf{x}_t)}{\partial \mathbf{x}}\right]\left[\frac{\partial \ln p(\mathbf{z}: \mathbf{x}_t)}{\partial \mathbf{x}}\right]'\right\} p(\mathbf{z}: \mathbf{x}_t)\,d\mathbf{z} \end{aligned} \tag{12.3.6}$$

Since $\mathbf{y}_1$ and $\mathbf{y}_2$ are arbitrary, they can be chosen such that (assuming the inverse exists)

$$\mathbf{y}_2 = \mathbf{B}^{-1}\mathbf{y}_1$$

so that (12.3.6) becomes

$$\mathbf{y}_1'\mathbf{\Sigma}\mathbf{y}_1 \geq \mathbf{y}_1'\mathbf{B}^{-1}\mathbf{y}_1$$

But since $\mathbf{y}_1$ is still arbitrary, it can be concluded that

$$\boldsymbol{\Sigma} \geq \mathbf{B}^{-1} \tag{12.3.7}$$

which is the matrix, unbiased version of (12.3.2). The matrix analogy to (12.3.4) is

$$\mathbf{B} = \int \left\{ \frac{\partial}{\partial x} \left[\frac{\partial}{\partial x} \ln p(\mathbf{z}: \mathbf{x}_t) \right] \right\} p(\mathbf{z}: \mathbf{x}_t)\, d\mathbf{z} \tag{12.3.8}$$

12.3.2 Gaussian Case†

Now consider the model

$$\begin{gathered} \mathbf{z} = \mathbf{h}(\mathbf{x}) + \mathbf{v} \\ \mathbf{v} \text{ is } N(\mathbf{O}, \mathbf{R}) \end{gathered} \tag{12.3.9}$$

Then

$$\ln p(\mathbf{z}: \mathbf{x}) = c - [\mathbf{z} - \mathbf{h}(\mathbf{x})]'\mathbf{R}^{-1}[\mathbf{z} - \mathbf{h}(x)]$$
$$c\text{: constant}$$

If x is a scalar,

$$\frac{\partial^2}{\partial x^2} \ln p(\mathbf{z}: \mathbf{x}_t) = -\mathbf{H}^{(1)}(x_t)\mathbf{R}^{-1}\mathbf{H}^{(1)}(x_t) + \mathbf{H}^{(2)}(x_t)\mathbf{R}^{-1}[\mathbf{z} - \mathbf{h}(x_t)]$$

so

$$\int \frac{\partial^2}{\partial x^2} [\ln p(\mathbf{z}: x_t)] p(\mathbf{z}: x_t)\, dz = E\left\{ \frac{\partial^2}{\partial x} [\ln p(\mathbf{z}: x_t)] \right\} = -\mathbf{H}^{(1)}(x_t)\mathbf{R}^{-1}\mathbf{H}^{(1)}(x_t)$$

Repeating the argument for vector $\mathbf{x}$ gives the result that for the model (12.3.9), the lower bound (12.3.7) using (12.3.8) is

$$\boxed{\begin{gathered} \boldsymbol{\Sigma} = E\{\mathbf{x}_t - \hat{\mathbf{x}})(\mathbf{x}_t - \hat{\mathbf{x}})'\} \\ \boldsymbol{\Sigma} \geq [\mathbf{H}^{(1)\prime}(\mathbf{x}_t)\mathbf{R}^{-1}\mathbf{H}^{(1)}(\mathbf{x}_t)]^{-1} \\ \hat{\mathbf{x}}\text{: an unbiased estimate} \end{gathered}} \tag{12.3.10}$$

Comparison of (12.3.10) with (12.2.3) shows that the linearized error analysis for the maximum likelihood estimate done in Section 12.2 has actually yielded a lower bound on the estimate error covariance matrix for unbiased estimators. The example of Section 12.2 illustrated this point.

A lower bound such as (12.3.10) has some applications. For example, if the lower bound is greater than some specified performance level, the system obviously will not work. However, it would be "nicer" in general to have an upper bound. An upper bound was not presented for the simple reason that the author does not know of a practical, general approach.

12.4 Ambiguity Functions

The ambiguity function is both a conceptual and practical aid in error analysis. The term ambiguity function is usually employed in radar problems where the components of $\mathbf{x}$ denote range and range rate (and amplitude and phase) of a moving point reflector. The concept is simply used here in a more general setting. The exact definition of a radar ambiguity function may differ in details from the present definition, but the basic ideas are the same.

Assume that the model is

$$\mathbf{z} = \mathbf{h}(\mathbf{x}, \mathbf{v})$$
$$\mathbf{x}\text{: completely unknown}$$
$$p(\mathbf{z}: \mathbf{x})\text{: likelihood function}$$

Then if $\hat{\mathbf{x}}$ is the maximum likelihood estimate, $\hat{\mathbf{x}}$ is the value of $\mathbf{x}$ which maximizes $\ln p(\mathbf{z}: \mathbf{x})$. Now the log likelihood function $\ln p(\mathbf{z}: \mathbf{x})$, viewed as a function of $\mathbf{z}$, is random variable, as $\mathbf{z}$ is a random vector. The ambiguity function is denoted by $\gamma(\mathbf{x}_t, \mathbf{x})$ and is defined as the average of the log likelihood function,

$$\begin{aligned} \gamma(\mathbf{x}_t: \mathbf{x}) &= \int \ln[p(\mathbf{z}: \mathbf{x})]\, p(\mathbf{z}: \mathbf{x}_t)\, d\mathbf{z} \\ &= E\{\ln p(z: \mathbf{x})\} \end{aligned} \tag{12.4.1}$$
$$\mathbf{x}_t\text{: true value of } \mathbf{x}$$

In the case where $\mathbf{z} = \mathbf{h}(\mathbf{x}) + \mathbf{v}$

$$p(\mathbf{z}: \mathbf{x}) = p[\mathbf{h}(\mathbf{x}_t) + \mathbf{v}: \mathbf{x}]$$

and the expectation is over $\mathbf{v}$. In the case where $\mathbf{z} = \mathbf{h}(\mathbf{x}) + \mathbf{v}$ and $\mathbf{v}$ is $N(\mathbf{O}, \mathbf{R})$, then if $\mathbf{z}$ is a K_2 vector

$$2 \ln p(\mathbf{z}: \mathbf{x}) = -\ln[(2\pi)^{K_2} |\mathbf{R}|] - [\mathbf{z} - \mathbf{h}(\mathbf{x})]'\mathbf{R}^{-1}[\mathbf{z} - \mathbf{h}(\mathbf{x})]$$

Since $E\{\mathbf{v}'\mathbf{R}^{-1}\mathbf{v}\} = K_2$,

$$2\gamma(\mathbf{x}_t, \mathbf{x}) = -\ln[(2\pi)^{K_2} |\mathbf{R}|] - K_2 - [\mathbf{h}(\mathbf{x}_t) - \mathbf{h}(\mathbf{x})]'\mathbf{R}^{-1}[\mathbf{h}(\mathbf{x}_t) - \mathbf{h}(\mathbf{x})] \tag{12.4.2}$$

For a linear model where $\mathbf{z} = \mathbf{H}\mathbf{x} + \mathbf{v}$ and $\mathbf{v}$ is $N(\mathbf{O}, \mathbf{R})$

$$2\gamma(\mathbf{x}_t, \mathbf{x}) = -\ln[(2\pi)^{K_2} |\mathbf{R}|] - K_2 - [\mathbf{x}_t - \mathbf{x}]'\mathbf{H}'\mathbf{R}^{-1}\mathbf{H}[\mathbf{x}_t - \mathbf{x}] \tag{12.4.3}$$

The shape of the ambiguity function $\gamma(\mathbf{x}_t, \mathbf{x})$ as a function of $\mathbf{x}$ can give an overall feel for the estimator performance to be expected. To see why, assume that x is a scalar and consider the two cases illustrated in Fig. 12.4.1 and 12.4.2. For Fig. 12.4.1, the model is assumed to be such that typical $\ln[p(\mathbf{z}: x)]$ versus x curves for different observation values $\mathbf{z}_a$, $\mathbf{z}_b$, and $\mathbf{z}_c$ are as in Fig. 12.4.1(a)–(c). In this case the ambiguity function $\gamma(x_t, x) =$

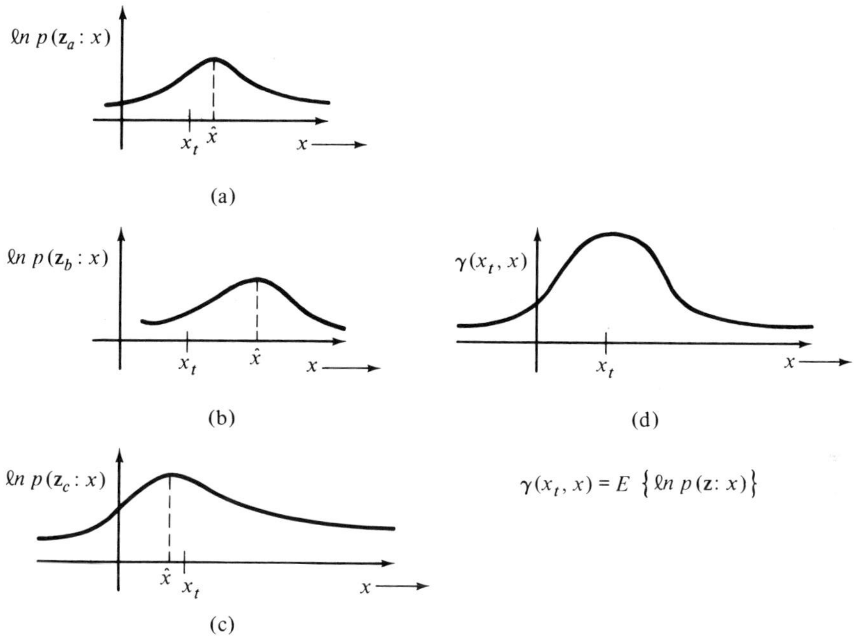

Figure 12.4.1 Possible ln $p(\mathbf{z}:x)$ and corresponding ambiguity functions.

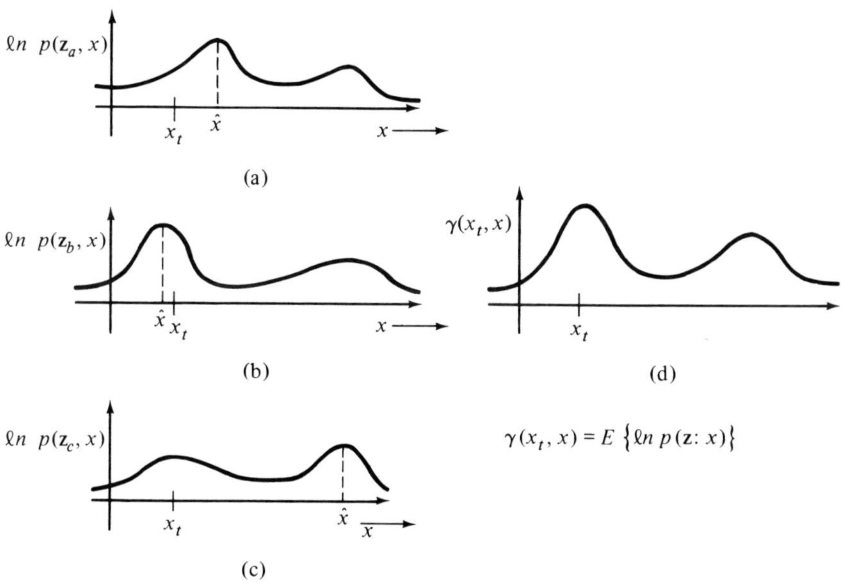

Figure 12.4.2 Possible ln $p(\mathbf{z}:x)$ and corresponding ambiguity functions.

$E\{\ln[p(\mathbf{z}: x)]\}$ might be as in Fig. 12.4.1(d). Figure 12.4.2 is the same as Fig. 12.4.1 except that the model is assumed to be different in the sense that $\ln p(\mathbf{z}: x)$ has two peaks so the corresponding ambiguity function, $\gamma(x_t, x)$ also has two peaks. This multipeak situation can cause many problems. If an iterative technique as in Section 11.7 is used to find $\hat{\mathbf{x}}$, the iteration may converge to a local rather than a global peak, so in Fig. 12.4.2(a) and (b) the indicated $\hat{x}$ may be found. Even if the global maximum is found, the effect of observation error may make the "wrong" peak of $\ln p(\mathbf{z}: \mathbf{x})$ a maximum, as illustrated in Fig. 12.4.2(c). The ambiguity function of Fig. 12.4.2(d) warns the engineer that multipeak problems for $\ln p(\mathbf{z}: \mathbf{x})$ are to be expected. The term ambiguity function is used because the extra peaks give rise to *ambiguities.*

In the case of Fig. 12.4.1, one could conclude that the local error analysis of Sections 12.1–12.3 might be good approximations. In the case of Fig. 12.4.2, a local error analysis can be meaningless except, of course, that the Cramer-Rao inequality still provides a lower bound.

The shape of the ambiguity function "right around" x_t is also a useful feature. Consider the two ambiguity functions of Fig. 12.4.3. Intuitively one might expect to be able to estimate x_t better in case (a) where the ambiguity function has a "sharp" peak than in case (b) where the ambiguity function has a "flat" peak. Thus institution can be made more rigorous by considering the

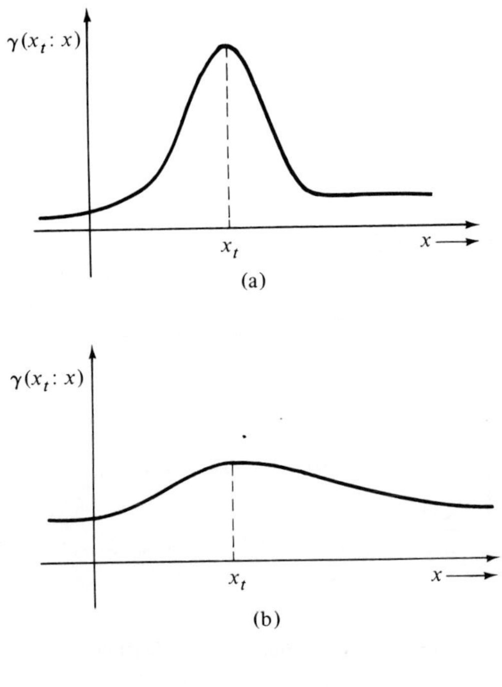

Figure 12.4.3 Two ambiguity functions.

inequality of (12.3.4). From (12.4.1)

$$\int \frac{\partial^2}{\partial x^2}[\ln p(\mathbf{z}: x)]\, p(\mathbf{z}: x_t)\, d\mathbf{z} = \frac{\partial^2}{\partial x^2}\gamma(x_t, x)$$

so (12.3.4) can be rewritten as

$$\boxed{E\{(x_t - \hat{x})^2\} \geq \frac{[-1 + b^{(1)}(x_t)]^2}{-(\partial^2/\partial x^2)\gamma(x_t, x)\,|_{x=x_t}}} \tag{12.4.4}$$

Thus the lower bound on the estimate error covariance matrix depends on curvature of the ambiguity function $\gamma(x_t, x)$ at $x = x_t$.

EXAMPLE: Consider

$$z_1 = \sin x + v_1$$

$$z_2 = x + v_2$$

$$\mathbf{v} = \begin{bmatrix} v_1 \\ v_2 \end{bmatrix} \qquad \mathbf{v} \text{ is } N(\mathbf{0}, \mathbf{R})$$

$$\mathbf{R} = \begin{bmatrix} R_{11} & 0 \\ 0 & R_{22} \end{bmatrix}$$

Then

$$2\gamma[x_t, x] = c - \frac{[\sin x_t - \sin x]^2}{R_{11}} - \frac{(x_t - x)^2}{R_{22}}$$

c: constant

Consider two cases:

Case 1: $R_{22} = \infty$; i.e., z_2 effectively not present.

Case 2: $R_{11} \approx R_{22}$.

In case 1, where z_2 is not available, it is clear from the model that no matter how small R_{11} is, one can never hope to estimate x_{actual}, as $\hat{x} + 2\pi m$, $m = 1, \ldots,$ are all equally good estimates. In case 2, the measurement z_2 can be used to resolve the ambiguity. The corresponding ambiguity functions are given in Fig. 12.4.4. The curvature of $\gamma(x_t : x)$ at x_t is

$$\left.\frac{\partial^2\gamma(x_t : x)}{\partial x^2}\right|_{x=x_t} = -\frac{\cos^2 x_t}{R_{11}} - \frac{1}{R_{22}}$$

so from (12.4.4) and neglecting biases (which are actually present)

$$\mathbf{\Sigma} \geq \left[\frac{\cos^2 x_t}{R_{11}} + \frac{1}{R_{22}}\right]^{-1}$$

Assume that $x_t = 0$ and $R_{11} \ll R_{22}$. Then the observation z_2 has little effect on the lower bound (or on a linearized local error analysis). However, the observation z_2 may still be absolutely essential to resolve the ambiguities.

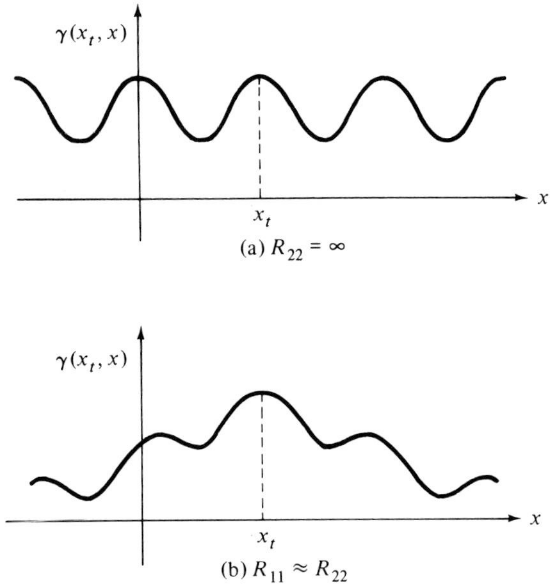

Figure 12.4.4 Ambiguity function $\gamma(x_t, x) = c - \dfrac{(\sin x_t - \sin x)^2}{R_{11}} - \dfrac{(x_t - x)^2}{R_{22}}$.

The discussions thus far have centered on the case of scalar x. The concept obviously generalizes to the use of vector $\mathbf{x}$. Equally obviously, the ambiguity function becomes difficult to work with when the dimension of $\mathbf{x}$ is high.

Since $\gamma(\mathbf{x}_t, \mathbf{x})$, the mean value of the log likelihood function, is a useful concept, it is tempting to carry the basic idea further and evaluate the higher moments of the log likelihood function. This may be practical in some cases, but the subject is not pursued here.

As a final point, the close relationship between the ambiguity function and the divergence distance measure of Part III on hypothesis testing should be noted. Consider two hypotheses:

$$\mathcal{H}_j : \mathbf{z} = \mathbf{z}_j$$
$$\mathbf{z}_j = \mathbf{h}(\mathbf{x}_j, \mathbf{v}), \qquad j = 1, 2$$
$$\mathbf{x}_j\text{: specified}$$

Then the likelihood function $p(\mathbf{z} : \mathbf{x}_j)$ can be interpreted as

$$p(\mathbf{z} : \mathbf{x}_j)\text{: probability density of } \mathbf{z} \text{ when } \mathcal{H}_j \text{ is true,} \qquad j = 1, 2$$

From (9.3.12), the divergence J between $\mathcal{H}_1$ and $\mathcal{H}_2$ is

$$J = \int [p(\mathbf{z} : \mathbf{x}_1) - p(\mathbf{z} : \mathbf{x}_2)] \ln\left\{\frac{p(\mathbf{z} : \mathbf{x}_1)}{p(\mathbf{z} : \mathbf{x}_2)}\right\} dz \qquad (12.4.5)$$

The relationship between J and $\gamma(\mathbf{x}_j : \mathbf{x})$ follows directly by comparing (12.4.5) with (12.4.1).

12.5 A Better Lower Bound†

The inequality of Section 12.3 yields a lower bound on the error variance (or mean square error). The ambiguity function arguments of Section 12.4 show that this bound may be very poor. However, there exists another lower bound which can yield a better (higher) lower bound and which can be related in general terms to the ambiguity function. This lower bound is often called a *Barankin bound.* For simplicity, only scalar x and unbiased estimators $\hat{x}$ are considered.

Since $\hat{x}$ is assumed to be unbiased, it follows by definition that for all values of x

$$\int \hat{x} p(\mathbf{z} : x)\, d\mathbf{z} = x$$

Choose K values of x, x_k, $k = 1, \ldots, K$. Then for arbitrary values of w_k, $k = 1, \ldots, K$, it follows that

$$\int \hat{x}\left[\sum_{k=1}^{K} w_k p(\mathbf{z} : x_k)\right] d\mathbf{z} = \sum_{k=1}^{K} w_k x_k$$

Let x_t denote the true value of x. Subtracting $\sum_{k=1}^{K} w_k x_t$ from both sides yields, after some manipulation,

$$\int (\hat{x} - x_t)\left[\frac{\sum_{k=1}^{K} w_k p(\mathbf{z} : x_k)}{p(\mathbf{z} : x_t)}\right] p(\mathbf{z} : x_t)\, d\mathbf{z} = \sum_{k=1}^{K} w_k (x_k - x_t)$$

Application of the Schwartz inequality yields

$$\left\{\int (\hat{x} - x_t)^2 p(\mathbf{z} : x_t)\, d\mathbf{z}\right\}\left\{\int \left[\frac{\sum_{k=1}^{K} w_k p(\mathbf{z} : x_k)}{p(\mathbf{z} : x_t)}\right]^2 p(\mathbf{z} : x_t)\, d\mathbf{z}\right\} \geq \left[\sum_{k=1}^{K} w_k (x_k - x_t)\right]^2$$

which can be rewritten as

$$E\{(\hat{x} - x_t)^2\} \geq \frac{\left[\sum_{k=1}^{K} w_k (x_k - x_t)\right]^2}{\int \left[\sum_{k=1}^{K} w_k p(\mathbf{z} : x_k)/p(\mathbf{z} : x_t)\right]^2 p(\mathbf{z} : x_t)\, d\mathbf{z}} \tag{12.5.1}$$

Since (12.5.1) holds for arbitrary w_k and x_k, $k = 1, \ldots, K$, it follows that the "tightest" bound is found by maximizing the right-hand side over w_k and x_k. The result of this maximization is the Barankin bound. Use of a larger value of K improves the bound.

To use (12.5.1), it is necessary to pick the "best" x_k and w_k, $k = 1, \ldots, K$ (and a value for K), in some fashion. Brute force computer search is possible but not intellectually satisfying. One reasonable approach is to choose the x_k, $k = 1, \ldots, K$, and then do an optimization over the w_k in some fashion. This can be effective, especially in the Gaussian case. The choice of x_k values can be a problem but placing the x_k at the local maximum of the ambiguity function of Section 12.4 is not unreasonable.

The lower bound of (12.3.2) for unbiased estimates can be derived as the limit of the special case of (12.5.1) when $\epsilon \longrightarrow 0$, where

$$K = 2 \qquad x_1 = x_t \qquad x_2 = x_t + \epsilon$$
$$w_1 = -\frac{1}{\epsilon} \qquad w_2 = \frac{1}{\epsilon}$$

12.6 Monte Carlo Methods

At this point, the reader who has really understood what has been said in this chapter is probably disappointed. Several different approaches were presented, but they have many shortcomings, such as requiring small errors and ignoring potentially very important phenomena such as biases. Ambiguity functions provide a general approach, but they become very unwieldy for large dimensional $\mathbf{x}$. One way around these problems is to use Monte Carlo techniques.

The Monte Carlo approach to error analysis is as follows:

1. Use a random number generator to obtain typical samples of $\mathbf{v}$, say, $\mathbf{v}_j, j = 1, \ldots, M$.
2. Pick a value of $\mathbf{x}$ (say, $\mathbf{x}_{\text{true}}$) and then compute the resulting $\mathbf{z}$, say,
$$\mathbf{z}_j = \mathbf{h}(\mathbf{x}_{\text{true}}, \mathbf{v}_j), \qquad j = 1, \ldots, M$$
3. Process each $\mathbf{z}_j$ through the estimator to get the resulting estimates, $\hat{\mathbf{x}}_j, j = 1, \ldots, M$.
4. Look at the errors $\mathbf{x}_{\text{true}} - \hat{\mathbf{x}}_j, j = 1, \ldots, M$, and decide how well the estimator works.

This approach is simplicity itself and automatically includes the case of large errors and bias errors. One big disadvantage is the need to first implement the estimator itself. Also, the required computation time can be excessive, as the number of samples M may have to be large. To be truly "scientific," an M of several thousand might be needed but $M = 20$ or 30 is often "safe" (although the author has often used $M = 2$–5 and "crossed his fingers").

12.7 Discussion

It should be restated that the preceding discussions on a Fisher formulation can be extended to a Bayesian formulation if desired.

By straightforward, albeit somewhat hand-waving, linearization arguments for small uncertainty, the error covariance matrix $\mathbf{\Sigma}$ for an arbitrary estimator, $\hat{\mathbf{x}} = \mathbf{w}(\mathbf{z})$, $\mathbf{z} = \mathbf{h}(\mathbf{x}) + \mathbf{v}$ was shown to be

$$\mathbf{\Sigma} \approx \mathbf{W}^{(1)}\mathbf{R}\mathbf{W}^{(1)\prime} \tag{12.7.1}$$

while for the maximum likelihood estimator for Gaussian $\mathbf{v}$,

$$\mathbf{\Sigma} \approx [\mathbf{H}^{(1)\prime}\mathbf{R}^{-1}\mathbf{H}^{(1)}]^{-1} \tag{12.7.2}$$

These two formulas are useful tools. However, it is important to emphasize two of their shortcomings. First, they are local analyses which ignore the global picture such as the shape of the ambiguity function. Second, even if the local picture is sufficient, (12.7.1) and (12.7.2) give only the error variances. In many applications, bias errors are more important.

Equation (12.7.2) can be evaluated directly from the model ($\mathbf{h}$ and $\mathbf{R}$) without implementing the estimator. This makes it an extremely useful tool in doing initial "feasibility studies" to see whether a given $\mathbf{z} = \mathbf{h}(\mathbf{x}) + \mathbf{v}$ situation can yield acceptable results. Equation (12.7.2) often provides a good "feel" for the problem even if the errors are not Gaussian. In many applications the shortcomings of (12.7.2) are ignored under the philosophy that "some sort of answer is better than no answer at all." Such a philosophy can of course, also lead to trouble.

The lower bound inequality of Section 12.3 provides a more mathematically precise approach to error analysis. For Gaussian errors (large or small) it enables (12.7.2) to be given an exact interpretation in terms of a lower bound for unbiased estimators. Unfortunately there is, in general, no guarantee that the lower bound given by the inequality can be achieved. Furthermore, although a bound including bias can be formulated, the evaluation of the bias errors remains very difficult. The lower bound of Section 12.5 provides a method to try to improve the lower bound of Section 12.3 but the needed calculations can easily become prohibitive.

The ambiguity function is a useful concept for visualizing some of the global aspects of the problem. It also provides another way of viewing the lower bound of Section 12.3.

Monte Carlo techniques are the only completely general approach to error analysis.

Historical Notes and References: Chapter 12

Section 12.3

The lower bound is often called the Cramer-Rao lower bound or the information inequality and is sometimes viewed as part of the theory of efficient estimation. See Book List 2, Mathematical Statistics. Cramer [1] and Wilkes [1] were used. This bound has been employed in a wide variety of situations, of which Sklar and Schweppe [1] and Schweppe [6] are examples of application to radar problems.

Section 12.4

The term ambiguity function seems to have originated in the radar context in Woodward [1].

Section 12.5

Barankin [1] is the original reference. The author used Hoeffesteder and McAulay [1]. Another type of generalization of the Cramer-Rao bound is discussed in Bhattacharyya [2].

Exercises

12.1. Consider a Fisher model (see also Exercise 11.2)

$$z_1 = x^2 + v_1$$

$$z_2 = x + v_2$$

v_1, v_2: zero mean independent Gaussian

$$E\{v_1^2\} = 1$$

$$E\{v_2^2\} = 4$$

x: completely unknown

Let x_t denote the true value of x.

a. Evaluate a lower bound on the error variance for any estimator and sketch its behavior as a function of x_t.

b. A maximum likelihood estimate of x requires the solution of a cubic equation. Therefore it might be desired to use one of the following estimators instead:

$$\hat{x} = \frac{1}{2}[z_1 + z_2^{1/2}] \qquad \text{or} \qquad \hat{x} = \frac{4z_1 + z_2^{1/2}}{5}$$

Perform a linearized error analysis on these two estimators to determine the error variance assuming that the observation uncertainties are small. Sketch their behavior as a function of x_t.

c. Discuss how you might attempt to calculate the bias errors for the maximum likelihood estimator and for the estimators of part (b). Do not attempt to actually evaluate the bias errors.

12.2. Consider

$$z = \sqrt{x}(1 + v)$$

x: completely unknown

v is $N(0, R)$

It is desired to estimate x using the observation z.

a. A reasonable estimate to use is $\hat{x} = z^2$. Evaluate the bias error and the error variance for this estimate. To evaluate the error variance, you may use either exact or linearized techniques, but state which approach you are following.

b. Let $\hat{x}$ denote the maximum likelihood estimate. What is $\hat{x}$? Repeat the error analysis of part a for this $\hat{x}$.

12.3. Consider

$$\mathbf{z} = \mathbf{h}(x) + \mathbf{v}$$

$$\mathbf{v} \text{ is } N[\mathbf{0}, \mathbf{R}(x)]$$

$$x\text{: completely unknown}$$

Let $\hat{x}$ denote the maximum likelihood estimate. Calculate a lower bound (Cramer-Rao) on Σ as in Section 12.3 assuming that x is unbiased.

12.4. Consider the observations

$$z(t) = h(x, t) + v(t), \qquad 0 \leq t \leq T$$

$$x\text{: unknown}$$

$$v(t)\text{: continuous-time white stochastic process}$$

$$E\{v(t)v(\tau)\} = \delta(t - \tau)R$$

a. Derive a lower bound for the variance of the error of any estimator $\hat{x}$.

b. An ambiguity function as defined in these notes does not exist. However, a closely related function can be used to yield the same type of information. What is this function?

Hint: Use the discrete to continuous time limit. See Exercise 11.4.

12.5. Consider the Fisher model

$$z_1 = (x_1^2 + x_2^2)^{1/2} + v_1$$

$$z_2 = x_1 + x_2 + v_2$$

$$\mathbf{v} = \begin{bmatrix} v_1 \\ v_2 \end{bmatrix} \qquad \mathbf{v} \text{ is } N(\mathbf{0}, \mathbf{I})$$

Let $\gamma(\mathbf{x}_t, \mathbf{x})$ denote the ambiguity function.

a. Sketch on an $x_1 - x_2$ plane, contours of *constant ambiguity*,

$$\gamma(\mathbf{x}_t, \mathbf{x}) = c \text{ for } \mathbf{x}_t = \begin{bmatrix} 1 \\ 1 \end{bmatrix}.$$

b. Let $\hat{\mathbf{x}}$ denote a maximum likelihood estimate of $\mathbf{x}$. Will $\hat{\mathbf{x}}$ be unique?

c. Visualize in your mind the effect a third observation $z_3 = x_1 + v_3$ will have on the ambiguity function and $\hat{x}$.

13

Estimation: Discrete-Time Nonlinear Dynamic Systems

Some of the ideas of Chapters 11 and 12 are now extended to nonlinear discrete-time dynamic systems.

Two nonlinear models are emphasized; the no-dynamics system

$$\mathbf{z}(n) = \mathbf{h}(\mathbf{x}, n) + \mathbf{v}(n), \qquad n = 1, \ldots, N \tag{13.1}$$

and the driven dynamic system

$$\begin{aligned} \mathbf{x}(n+1) &= \boldsymbol{\phi}[\mathbf{x}(n), n] + \mathbf{G}(n)\mathbf{w}(n) \\ \mathbf{z}(n) &= \mathbf{h}[\mathbf{x}(n), n] + \mathbf{v}(n), \qquad n = 1, \ldots, N \end{aligned} \tag{13.2}$$

$\boldsymbol{\phi}$: K_1 vector nonlinear function of the K_1 vector state $\mathbf{x}$ and time n
$\mathbf{h}$: K_2 vector nonlinear function of the K_1 vector state $\mathbf{x}$ and time n
$\mathbf{G}$: K_1 by K_3 matrix
$\mathbf{w}, \mathbf{v}$: uncertain processes
$\mathbf{x}(0)$: uncertain initial state

However, much of the analysis to follow is also valid for more general formulations such as

$$\begin{aligned} \mathbf{x}(n+1) &= \boldsymbol{\phi}[\mathbf{x}(n), n] + \mathbf{G}[\mathbf{x}(n), n]\mathbf{w}(n) \\ \mathbf{z}(n) &= \mathbf{h}[\mathbf{x}(n), n] + \mathbf{v}(n) \end{aligned} \tag{13.3}$$

or

$$\begin{aligned} \mathbf{x}(n+1) &= \boldsymbol{\phi}[\mathbf{x}(n), n, \mathbf{w}(n)] \\ \mathbf{z}(n) &= \mathbf{h}[\mathbf{x}(n), n, \mathbf{v}(n)] \end{aligned} \tag{13.4}$$

In many applications there are also known inputs $\mathbf{u}(n)$, so the model of interest might be

$$\mathbf{x}(n+1) = \phi[\mathbf{x}(n), n] + \mathbf{G}[\mathbf{x}(n), n]\mathbf{w}(n) + \mathbf{B}(n)\mathbf{u}(n)$$

or, still more generally,

$$\mathbf{x}(n+1) = \boldsymbol{\phi}[\mathbf{x}(n), \mathbf{u}(n), n] + \mathbf{G}[\mathbf{x}(n), n]\mathbf{w}(n)$$

Note, however, that since $\mathbf{u}(n)$ is a known function of time, $\boldsymbol{\phi}[\mathbf{x}(n), \mathbf{u}(n), n]$ can be viewed simply as $\boldsymbol{\phi}[\mathbf{x}(n), n]$. The $\mathbf{w}(n)$ and $\mathbf{v}(n)$ are modeled as discrete-time white processes.

When a stochastic (Bayesian) model is used for the uncertainty, it is assumed that $\mathbf{v}(n)$, $\mathbf{w}(n)$, and $\mathbf{x}(0)$ are random vectors with

$$\begin{aligned} \mathbf{v}(n_1), \mathbf{v}(n_2)&: \text{ independent} \qquad n_1 \neq n_2 \\ \mathbf{w}(n_1), \mathbf{w}(n_2)&: \text{ independent} \qquad n_1 \neq n_2 \\ \mathbf{v}(n_1), \mathbf{w}(n_2), \mathbf{x}(0)&: \text{ independent, all } n_1 \text{ and } n_2 \end{aligned}$$

The definition of a white process used in Section 3.1.2 involved "uncorrelated in time" processes rather than "independent in time" processes, so the $\mathbf{w}(n)$ and $\mathbf{v}(n)$ are special white processes.

For probabilistic (Bayesian) models, the general theory is primarily concerned with the behavior of the probability densities and is mostly only an exercise in manipulating Bayes' rule. The only really difficult part is keeping the notation straight. The notation $p[\mathbf{v}(n)]$, $p[\mathbf{w}(n)]$, $p[\mathbf{x}(n)]$, etc., is used to denote the probability densities of $\mathbf{v}(n)$, $\mathbf{w}(n)$, and $\mathbf{x}(n)$, etc. It would be more precise to use the notation $p[\mathbf{x}, n]$ instead of $p[\mathbf{x}(n)]$, as it is the probability density of $\mathbf{x}$ at time n that is of interest, but the notation $p[\mathbf{x}(n)]$ seems to yield the "simplest"-appearing equations.

This chapter is organized somewhat like Chapter 11; first theory, then practice. The first group of sections presents the theory. Section 13.1 discusses the propagation of uncertainty for both Bayesian and unknown-but-bounded models. Section 13.2 discusses Bayesian filtering and smoothing. Sections 13.3–13.7 discuss Fisher, weighted-least-squares, finite memory estimation, and unknown-but-bounded models and performance analysis. Section 13.8 summarizes the theoretical development. The second group of sections discusses schemes for obtaining practical results. Approximate propagation of uncertainty is discussed in Section 13.9. Recursive filters are considered in Section 13.10. Although this chapter is on discrete time models, the continuous time case is briefly discussed in Section 13.11.

13.1 Propagation of Uncertainty

Given uncertainty models for $\mathbf{x}(0)$ and $\mathbf{w}(n)$, $n = 1, \ldots$, it is desired to find the uncertainty model for the $\mathbf{x}(n)$ resulting from (13.4) (when no observations are made).

13.1.1 Bayesian

Consider $\mathbf{x}(n+1) = \boldsymbol{\phi}[\mathbf{x}(n), \mathbf{w}(n), n]$. Assume that $\mathbf{x}(0)$ and $\mathbf{w}(n)$, $n = 1, 2, \ldots$, are modeled as independent random vectors with known probability densities. It is desired to evaluate $p[\mathbf{x}(n+1)]$ in terms of $p[\mathbf{x}(n)]$.

By definition of marginal distributions

$$p[\mathbf{x}(n+1)] = \int p[\mathbf{x}(n+1), \mathbf{x}(n)]\, d\mathbf{x}(n)$$

Using Bayes' rule

$$\boxed{p[\mathbf{x}(n+1)] = \int p[\mathbf{x}(n+1) \,|\, \mathbf{x}(n)]\, p[\mathbf{x}(n)]\, d\mathbf{x}(n)} \tag{13.1.1}$$

This is the desired result, as it expresses $p[\mathbf{x}(n+1)]$ in terms of $p[\mathbf{x}(n)]$ and other computable functions. If $\mathbf{x}(n+1) = \boldsymbol{\phi}[\mathbf{x}(n), n] + \mathbf{G}(n)\mathbf{w}(n)$ and $\mathbf{G}(n)$ is a nonsingular matrix, it follows from the independence of the $\mathbf{w}(n)$, $n = 1, \ldots$, that (see Appendix D.4)

$$p[\mathbf{x}(n+1) \,|\, \mathbf{x}(n)] = \frac{1}{|\mathbf{G}(n)|}\, p[\mathbf{w}(n)] \Big|_{\mathbf{w}(n) = \mathbf{G}^{-1}(n)\{\mathbf{x}(n+1) - \boldsymbol{\phi}[\mathbf{x}(n), n]\}} \tag{13.1.2}$$

where $p[\mathbf{w}(n)]$ is the probability density of $\mathbf{w}(n)$. For more general models, similar expressions can be written in various ways.

Equation (13.1.1) is a recursive functional equation. $p[\mathbf{x}(n)]$ is a scalar function of the K_1 vector $\mathbf{x}$. Except for special cases, this equation is not suitable for direct implementation on a computer. For example, to numerically calculate $p[\mathbf{x}(n+1)]$, it is necessary to have already calculated and stored the value of $p[\mathbf{x}(n)]$ for all possible values of $\mathbf{x}$ at time n.

In terms of state space concepts, (13.1.1) shows that $p[\mathbf{x}(n)]$ can be viewed as the "state" of the system. Since $p[\mathbf{x}(n)]$ is a function, this state can be considered to be an infinite dimensional vector.

A Gaussian assumption on $\mathbf{x}(0)$ and $\mathbf{w}(n)$ enables more explicit equations to be written but, in general, does not alleviate the basic difficulty associated with solving functional equations. Of course, if the nonlinear model becomes linear, the results of Section 4.1.2 can be derived from (13.1.1).

EXAMPLE: Consider the linear, zero mean, Gaussian scalar model

$$x(n+1) = \Phi x(n) + w(n)$$

$$E\{x^2(0)\} = \Gamma(0)$$

$$E\{w(n_1)w(n_2)\} = \begin{cases} Q, & n_1 = n_2 \\ 0, & n_1 \neq n_2 \end{cases}$$

Since $x(0)$ and $w(0)$ are Gaussian,

$$p[x(0)] = [2\pi\Gamma(0)]^{-1/2} \exp\left\{\frac{x^2(0)}{2\Gamma(0)}\right\}$$

$$p[x(1)\,|\,x(0)] = [2\pi Q]^{-1/2} \exp\left\{\frac{[x(1) - \Phi x(0)]^2}{2Q}\right\}$$

Thus from (13.1.1)

$$p[x(1)] = [4\pi^2\Gamma(0)Q]^{-1/2} \int \exp\left\{\frac{x^2(0)}{2\Gamma(0)} + \frac{[x(1) - \Phi x(0)]^2}{2Q}\right\} dx(0)$$

or, after a lot of rearrangement,

$$p[x(1)] = [4\pi^2\Gamma(0)Q]^{-1/2} \int \exp\left\{\frac{Q + \Phi^2\Gamma(0)}{2\Gamma(0)Q}\left[x(0) - \frac{x(1)\Phi\Gamma(0)}{Q + \Phi^2\Gamma(0)}\right]^2 + \frac{x^2(1)}{2[Q + \Phi^2\Gamma(0)]}\right\} dx(0)$$

Since

$$\int \exp[(x - a)^2/2\Gamma]\, dx = (2\pi\Gamma)^{1/2},$$

$$p[x(1)] = [2\pi(Q + \Phi^2\Gamma(0))]^{-1/2} \exp\left\{\frac{x^2(1)}{2[Q + \Phi^2\Gamma(0)]}\right\}$$

Thus $x(1)$ is zero mean Gaussian with

$$E\{x^2(1)\} = \Gamma(1) = \Phi^2\Gamma(0) + Q$$

The same argument can be carried out for general time n. Naturally the result is the same as obtained in Section 4.1.2, i.e.,

$$\Gamma(n+1) = \Phi^2\Gamma(n) + Q$$

13.1.2 Unknown-but-Bounded

Consider $\mathbf{x}(n+1) = \boldsymbol{\phi}[\mathbf{x}(n), n] + \mathbf{G}(n)\mathbf{w}(n)$. Assume that unknown-but-bounded models are used for the $\mathbf{x}(0)$ and $\mathbf{w}(n)$:

$$\mathbf{x}(0) \in \boldsymbol{\Omega}_x(0)$$

$$\mathbf{w}(n) \in \boldsymbol{\Omega}_w(n), \qquad n = 1, 2, \ldots$$

It is desired to express $\boldsymbol{\Omega}_x(n+1)$, the set in K_1 space that contains $\mathbf{x}(n+1)$, in terms of $\boldsymbol{\Omega}_\mathbf{x}(n)$.

Define

$$\boldsymbol{\Omega}_{Gw}(n) = \{\mathbf{x} : \mathbf{x} = \mathbf{G}(n)\mathbf{w}, \mathbf{w} \in \boldsymbol{\Omega}_\mathbf{w}(n)\}$$

$$\tilde{\boldsymbol{\Omega}}_x(n+1\,|\,n) = \{\mathbf{x} : \mathbf{x} = \boldsymbol{\phi}[\mathbf{y}, n], \mathbf{y} \in \boldsymbol{\Omega}_\mathbf{x}(n)\}$$

Then

$$\boxed{\mathbf{\Omega}_x(n+1) = \text{vector sum of } \mathbf{\Omega}_{Gw}(n) \text{ and } \tilde{\mathbf{\Omega}}_x(n+1\,|\,n)} \tag{13.1.3}$$

This is the desired result. Note that it is essentially the same as (4.3.5) for the linear model. The logic is exactly the same.

Many sets in K_1 space can be defined in terms of some scalar function $J(\mathbf{x})$; for example,

$$\mathbf{\Omega}_x(n) = \{\mathbf{x} : J(\mathbf{x}, n) \leq 1\} \tag{13.1.4}$$

If this is possible, (13.1.3) can be viewed as a recursive functional equation relating $J(\mathbf{x}, n+1)$ to $J(\mathbf{x}, n)$. For a Bayesian model, $p[\mathbf{x}(n)]$ could be considered the state of the system. In a similar fashion $J(\mathbf{x}, n)$ can be considered the state for the unknown-but-bounded model. Instead of (13.1.4), it might be possible to represent $\mathbf{\Omega}_x(n)$ by its support function $s(\mathbf{n})$. With such a representation, the support function can be considered the state of the system. Note, however, that not all sets can be represented by their support function; for example, consider nonconvex sets.

Both unknown-but-bounded models and Bayesian models yield recursive functional equations. They are of different types, but in general neither is "harder" or "easier" than the other to calculate. In practice, approximate solutions are needed in both cases.

13.2 Bayesian Estimation

Consider

$$\mathbf{x}(n+1) = \boldsymbol{\phi}[\mathbf{x}(n), \mathbf{w}(n), n]$$
$$\mathbf{z}(n) = \mathbf{h}[\mathbf{x}(n), \mathbf{v}(n), n]$$
$$\mathbf{w}(n_1), \mathbf{w}(n_2), \mathbf{v}(n_1), \mathbf{v}(n_2), \mathbf{v}(0)\text{: independent}$$

Define

$$\mathbf{z}_N = \begin{bmatrix} \mathbf{z}(1) \\ \cdot \\ \cdot \\ \cdot \\ \mathbf{z}(N) \end{bmatrix}$$

13.2.1 No Dynamics

Consider the no-dynamics special case where $\mathbf{z}(n) = \mathbf{h}[\mathbf{x}, \mathbf{v}(n), n]$. Since

$$p[\mathbf{x}\,|\,\mathbf{z}_{N+1}] = p[\mathbf{x}\,|\,\mathbf{z}_N, \mathbf{z}(N+1)]$$

Bayes' rule gives

$$p[\mathbf{x}\,|\,\mathbf{z}_{N+1}] = \frac{p[\mathbf{x}, \mathbf{z}_N, \mathbf{z}(N+1)]}{p[\mathbf{z}_N, \mathbf{z}(N+1)]}$$

or after more applications of Bayes' rule,

$$p[\mathbf{x} \mid \mathbf{z}_{N+1}] = \frac{p[\mathbf{z}(N+1) \mid \mathbf{x}, \mathbf{z}_N]\, p[\mathbf{x} \mid \mathbf{z}_N]}{p[\mathbf{z}(N+1) \mid \mathbf{z}_N]}$$

From the independence of the $\mathbf{v}(n)$,

$$p[\mathbf{z}(N+1) \mid \mathbf{x}, \mathbf{z}_N] = p[\mathbf{z}(N+1) \mid \mathbf{x}] \tag{13.2.1}$$

so that

$$p[\mathbf{x} \mid \mathbf{z}_{N+1}] = \frac{p[\mathbf{z}(N+1) \mid \mathbf{x}]\, p[\mathbf{x} \mid \mathbf{z}_N]}{p[\mathbf{z}(N+1) \mid \mathbf{z}_N]} \tag{13.2.2}$$

Now

$$\int p[\mathbf{x} \mid \mathbf{z}_{N+1}]\, d\mathbf{x} = 1$$

and thus the denominator of (13.2.1) is the integral of the numerator. Hence

$$\boxed{\begin{aligned} p[\mathbf{x} \mid \mathbf{z}_{N+1}] &= \frac{p[\mathbf{z}(N+1) \mid \mathbf{x}]\, p[\mathbf{x} \mid \mathbf{z}_N]}{\int p[\mathbf{z}(N+1) \mid \mathbf{x}]\, p[\mathbf{x} \mid \mathbf{z}_N]\, d\mathbf{x}} \\ p[\mathbf{x} \mid \mathbf{z}_0] &= p[\mathbf{x}(0)] \end{aligned}} \tag{13.2.3}$$

This is the desired answer, as it expresses $p[\mathbf{x} \mid \mathbf{z}_{N+1}]$ in terms of $p[\mathbf{x} \mid \mathbf{z}_N]$ and a computable function $p[\mathbf{z}(N+1) \mid \mathbf{x}]$. In the special case $\mathbf{z}(n) = \mathbf{h}(\mathbf{x}, n) + \mathbf{v}(n)$,

$$p[\mathbf{z}(N+1) \mid \mathbf{x}] = p[\mathbf{v}(N+1)]\,\big|_{\mathbf{v}(N+1) = \mathbf{z}(N+1) - \mathbf{h}(\mathbf{x}, N+1)} \tag{13.2.4}$$

Once $p(\mathbf{x} \mid \mathbf{z}_{N+1})$ is available, any of the estimates (conditional expectation, most probable, median, minimax) of Chapter 11 can be computed.

13.2.2 Filtering

The complete driven dynamic system model appears to be more complicated than the no-dynamics model, but all the arguments of Section 13.2.1 leading to (13.2.3) still apply. Thus by direct analogy with (13.2.3),

$$p[\mathbf{x}(N+1) \mid \mathbf{z}_{N+1}] = \frac{p[\mathbf{z}(N+1) \mid \mathbf{x}(N+1)]\, p[\mathbf{x}(N+1) \mid \mathbf{z}_N]}{\int p[\mathbf{z}(N+1) \mid \mathbf{x}(N+1)]\, p[\mathbf{x}(N+1) \mid \mathbf{z}_N]\, d\mathbf{x}(N+1)} \tag{13.2.5}$$

To see why this is true, note that the derivation of (13.2.3) used the special properties of the no-dynamics model only to get (13.2.1) and that particular argument also applies to the driven dynamic system model.

Rewrite $p[\mathbf{x}(N+1) \mid \mathbf{z}_N]$ as

$$p[\mathbf{x}(N+1) \mid \mathbf{z}_N] = \int p[\mathbf{x}(N+1), \mathbf{x}(N) \mid \mathbf{z}_N]\, d\mathbf{x}(N)$$

Using Bayes' rule,

$$p[\mathbf{x}(N+1)\,|\,\mathbf{z}_N] = \int p[\mathbf{x}(N+1)\,|\,\mathbf{x}(N), \mathbf{z}_N]p[\mathbf{x}(N)\,|\,\mathbf{z}_N]\,d\mathbf{x}(N) \tag{13.2.6}$$

Using the independence of the $\mathbf{v}(n)$,

$$p[\mathbf{x}(N+1)\,|\,\mathbf{x}(N), \mathbf{z}_N] = p[\mathbf{x}(N+1)\,|\,\mathbf{x}(N)] \tag{13.2.7}$$

Substitution of (13.2.7) into (13.2.6) and then (13.2.6) into (13.2.5) gives

$$\boxed{\begin{aligned} &p[\mathbf{x}(N+1)\,|\,\mathbf{z}_{N+1}] \\ &\quad = \frac{p[\mathbf{z}(N+1)\,|\,\mathbf{x}(N+1)] \int p[\mathbf{x}(N+1)\,|\,\mathbf{x}(N)]p[\mathbf{x}(N)\,|\,\mathbf{z}_N]\,d\mathbf{x}(N)}{\iint p[\mathbf{z}(N+1)|\,\mathbf{x}(N+1)]p[\mathbf{x}(N+1)|\,\mathbf{x}(N)]p[\mathbf{x}(N)|\,\mathbf{z}_N]d\mathbf{x}(N)d\mathbf{x}(N+1)} \\ &p[\mathbf{x}(0)\,|\,\mathbf{z}_0] \\ &\quad = p[\mathbf{x}(0)] \end{aligned}} \tag{13.2.8}$$

where the denominator is the integral of the numerator. This is the desired answer, as it expresses $p[\mathbf{x}(N+1)\,|\,\mathbf{z}_{N+1}]$ in terms of $p[\mathbf{x}(N)\,|\,\mathbf{z}_N]$ and functions $p[\mathbf{z}(N+1)\,|\,\mathbf{x}(N+1)]$ and $p[\mathbf{x}(N+1)\,|\,\mathbf{x}(N)]$, which can be calculated, for example, as in (13.1.2) and (13.2.4). Given $p[\mathbf{x}(N+1)|\,\mathbf{z}_{N+1}]$, any of the estimates of Chapter 11 can be computed (conceptually).

As in Section 13.1.1, it must be emphasized that (13.2.8) is a functional equation. In general, it cannot be solved directly on a computer. For the general nonlinear model, a Gaussian assumption is little help. For the linear Gaussian model, the results of Chapter 6 can, of course, be derived as a special case of (13.2.8) provided that one has the fortitude.

13.2.3 Smoothing: General Case

Equations are now derived for the conditional probability density $p(\mathbf{x}_N\,|\,\mathbf{z}_N)$, where

$$\mathbf{x}_N = \begin{bmatrix} \mathbf{x}(1) \\ \cdot \\ \cdot \\ \cdot \\ \mathbf{x}(N) \end{bmatrix}$$

$$\mathbf{z}_N = \begin{bmatrix} \mathbf{z}(1) \\ \cdot \\ \cdot \\ \cdot \\ \mathbf{z}(N) \end{bmatrix}$$

and where N is fixed. These results are therefore applicable to the smoothing problem, i.e., to obtaining an estimate of all the states $\mathbf{x}(n)$, $n = 1, \ldots, N$, given all the measurements $\mathbf{z}(1) \ldots \mathbf{z}(N)$. However, unlike the filtering case, a Gaussian assumption does "help" reduce the computational problem.

For the driven dynamic system model, Bayes' rule gives

$$p[\mathbf{x}_N \mid \mathbf{z}_N] = \frac{p[\mathbf{z}_N \mid \mathbf{x}_N] p[\mathbf{x}_N]}{p[\mathbf{z}_N]} \tag{13.2.9}$$

Consider first $p[\mathbf{z}_N \mid \mathbf{x}_N]$. Bayes' rule gives

$$p[\mathbf{z}_N \mid \mathbf{x}_N] = p[\mathbf{z}(N) \mid \mathbf{x}_{N-1}, \mathbf{x}(N), \mathbf{z}_{N-1}] p[\mathbf{z}_{N-1} \mid \mathbf{x}_{N-1}, \mathbf{x}(N)]$$

Using the model and the independence of the $\mathbf{v}(n)$ and $\mathbf{w}(n)$,

$$p[\mathbf{z}_N \mid \mathbf{x}_N] = p[\mathbf{z}(N) \mid \mathbf{x}(N)] p[\mathbf{z}_{N-1} \mid \mathbf{x}_{N-1}]$$

so that

$$p[\mathbf{z}_N \mid \mathbf{x}_N] = \prod_{n=1}^{N} p[\mathbf{z}(n) \mid \mathbf{x}(n)] \tag{13.2.10}$$

Now consider $p[\mathbf{x}_N]$. Using Bayes' rule,

$$p[\mathbf{x}_N] = p[\mathbf{x}(N) \mid \mathbf{x}_{N-1}] p[\mathbf{x}_{N-1}]$$

From the model

$$p[\mathbf{x}(N) \mid \mathbf{x}_{N-1}] = p[\mathbf{x}(N) \mid \mathbf{x}(N-1)]$$

so

$$p[\mathbf{x}_N] = \prod_{n=1}^{N} p[\mathbf{x}(n) \mid \mathbf{x}(n-1)] p[\mathbf{x}(0)] \tag{13.2.11}$$

Substituting (13.2.10) and (13.2.11) into (13.2.9) gives

$$\boxed{p[\mathbf{x}_N \mid \mathbf{z}_N] = \frac{\left\{\prod_{n=1}^{N} p[\mathbf{z}(n) \mid \mathbf{x}(n)]\right\}\left\{\prod_{n=1}^{N} p[\mathbf{x}(n) \mid \mathbf{x}(n-1)]\right\} p[\mathbf{x}(0)]}{p[\mathbf{z}_N]}} \tag{13.2.12}$$

This is the desired result. The quantities $p(\mathbf{z}(n) \mid \mathbf{x}(n)]$ and $p[\mathbf{x}(n+1) \mid \mathbf{x}(n)]$ can be calculated, for example, as in (13.2.4) and (13.1.2). Similarly, $p(\mathbf{z}_N)$ is the integral of the numerator.

13.2.4 Smoothing: Gaussian Case

Now consider the special Gaussian case where

$$\mathbf{x}(n+1) = \boldsymbol{\phi}[\mathbf{x}(n), n] + \mathbf{G}(n)\mathbf{w}(n)$$
$$\mathbf{z}(n) = \mathbf{h}[\mathbf{x}(n), n] + \mathbf{v}(n)$$
$$\mathbf{v}(n) \text{ is } N[\mathbf{0}, \mathbf{R}(n)]$$
$$\mathbf{w}(n) \text{ is } N[\mathbf{0}, \mathbf{Q}(n)]$$
$$\mathbf{x}(0) \text{ is } N[\mathbf{0}, \boldsymbol{\psi}]$$
$$\mathbf{G}(n)\text{: nonsingular}$$

Use of the Gaussian structure and model on 13.2.12 yields

$$p[\mathbf{x}_N \,|\, \mathbf{z}_N] = \frac{\alpha_2}{p[\mathbf{z}_N]} \exp\left\{\frac{1}{2}\left[\sum_{n=1}^{N} [\mathbf{z}(n) - \mathbf{h}[\mathbf{x}(n), n]]'\mathbf{R}^{-1}(n)[\mathbf{z}(n) - \mathbf{h}[\mathbf{x}(n), n]] + \mathbf{x}'(0)\boldsymbol{\psi}^{-1}\mathbf{x}(0) + \sum_{n=1}^{N} [\mathbf{x}(n) - \boldsymbol{\phi}[\mathbf{x}(n-1), n-1]]' \times [\mathbf{G}(n)\mathbf{Q}(n)\mathbf{G}'(n)]^{-1}[\mathbf{x}(n) - \boldsymbol{\phi}[\mathbf{x}(n-1), n-1]]\right]\right\} \quad (13.2.13)$$

where α_2 is a constant.

Now $p[\mathbf{z}_N]$ is very difficult to calculate. However, as discussed in Chapter 11, many estimates, $\hat{\mathbf{x}}_{N|N}$ of $\mathbf{x}_N$ can be obtained from $p[\mathbf{x}_N \,|\, \mathbf{z}_N]$ without evaluating $p[\mathbf{z}_N]$. For example, one most probable estimate $\hat{\mathbf{x}}_{N|N}$ is the value of $\mathbf{x}_N$ which maximizes $p[\mathbf{x}_N \,|\, \mathbf{z}_N]$. This $\hat{\mathbf{x}}_{N|N}$ is given by

Let $\hat{\mathbf{x}}(0 \,|\, N)$, $\hat{\mathbf{x}}(n \,|\, N)$, and $\hat{\mathbf{w}}(n \,|\, N)$, $n = 1, \ldots, N$, denote the particular $\mathbf{x}$ and $\mathbf{w}$ which minimize $J(\mathbf{z}_N : \mathbf{x}_N)$:

$$J(\mathbf{z}_N : \mathbf{x}_N) = \sum_{n=1}^{N} [\mathbf{z}(n) - \mathbf{h}(\mathbf{x}(n), n]]'\mathbf{R}^{-1}(n)[\mathbf{z}(n) - \mathbf{h}[\mathbf{x}(n), n]] + \sum_{n=1}^{N} \mathbf{w}'(n)\mathbf{Q}^{-1}(n)\mathbf{w}(n) + \mathbf{x}'(0)\boldsymbol{\psi}^{-1}\mathbf{x}(0)$$

subject to the constraint

$$\hat{\mathbf{x}}(n+1 \,|\, N) = \boldsymbol{\phi}[\hat{\mathbf{x}}(n \,|\, N), n] + \mathbf{G}(n)\hat{\mathbf{w}}(n \,|\, N), \qquad n = 1, \ldots, N \quad (13.2.14)$$

Then

$$\hat{\mathbf{x}}_{N|N} = \begin{bmatrix} \hat{\mathbf{x}}(1 \,|\, N) \\ \cdot \\ \cdot \\ \cdot \\ \hat{\mathbf{x}}(N \,|\, N) \end{bmatrix}$$

The model leading to (13.2.14) assumed $\mathbf{G}(n)$, $\mathbf{R}(n)$, and $\mathbf{Q}(n)$ to be specified functions of time. However, the estimator for a more general model involving $\mathbf{G}[\mathbf{x}(n), n]$, $\mathbf{R}[\mathbf{x}(n), n]$, and $\mathbf{Q}[\mathbf{x}(n), n]$ follows directly simply by substituting $\mathbf{G}(n) = \mathbf{G}[\mathbf{x}(n), n]$, etc., into (13.2.14). The estimator of (13.2.14) can be derived without the assumption that $\mathbf{G}(n)$ is nonsingular.

It should be restated that the $\hat{\mathbf{x}}_{N|N}$ of (13.2.14) is the most probable $\mathbf{x}_N$. It does not necessarily yield the most probable value of $\mathbf{x}(N)$.

13.3 Fisher

As in Chapter 6, Fisher models of several types are possible depending on whether $\mathbf{w}(n)$ and/or $\mathbf{x}(0)$ are assumed to be completely unknown, while $\mathbf{v}(n)$ is modeled as a sequence of independent random variables. For such

Fisher models, likelihood functions rather than conditional probability densities are of interest, and it is possible to develop many equations similar to the preceding Bayesian equations. However, the details will not be given, as one can obtain the formulas by the "trick" of using the Bayesian equation and "flattening out" the corresponding probability densities $p[\mathbf{w}(n)]$ and $p[\mathbf{x}(0)]$. As emphasized repeatedly in this book, such a trick does not make the Bayesian and Fisher models equal; it merely removes the need to go through all the details.

13.4 Weighted Least Squares

In the weighted-least-squares approach, a model for the uncertainty is not used; a criterion for the estimator is merely chosen. One logical extension of the linear model ideas of Section 6.5 is as follows:

Choose the $\mathbf{x}(0)$ and $\mathbf{w}(n)$, $n = 0, \ldots, N$, which minimize $J(\mathbf{z}_N : \mathbf{x}_N)$ of (13.2.14) subject to the constant

$$\mathbf{x}(n+1) = \boldsymbol{\phi}(\mathbf{x}(n), n) + \mathbf{G}(n)\mathbf{w}(n) \tag{13.4.1}$$

where $\mathbf{Q}$, $\mathbf{R}$, and $\boldsymbol{\psi}$ are chosen by engineering judgment

The state estimates are $\hat{\mathbf{x}}(n)$, $n = 1, \ldots, N$, the $\mathbf{x}$ resulting from minimizing $\mathbf{x}(0)$ and $\mathbf{w}(n)$, $n = 1, \ldots, N$.

This weighted-least-squares criterion corresponds to the Bayesian smoothing solution for Gaussian processes of Section 13.2.4.

13.5 Finite Memory†

As discussed in Section 6.7, a finite memory estimator is one that estimates the state using only the past m observations. A recursive equation for the corresponding conditional probability density is now derived.

Consider a no-dynamics model $\mathbf{z}(n) = \mathbf{h}[\mathbf{x}, \mathbf{v}(n), n]$, $n = 1, \ldots, \mathbf{v}(n)$ independent. Define

$$\mathbf{z}_{N,N-m} = \begin{bmatrix} \mathbf{z}(N) \\ \cdot \\ \cdot \\ \cdot \\ \mathbf{z}(N-m) \end{bmatrix}$$

$$\mathbf{z}_N = \begin{bmatrix} \mathbf{z}(N) \\ \cdot \\ \cdot \\ \cdot \\ \mathbf{z}(1) \end{bmatrix}$$

By Bayes' rule

$$p[\mathbf{x} \mid \mathbf{z}_N] = \frac{p[\mathbf{x}, \mathbf{z}_{N,N-m}, \mathbf{z}_{N-m-1}]}{p[\mathbf{z}_N]}$$

Using Bayes' rule a few more times and using the fact that

$$p[\mathbf{z}_{N,N-m} \mid \mathbf{x}, \mathbf{z}_{N-m-1}] = p[\mathbf{z}_{N,N-m} \mid \mathbf{x}]$$

yields, after manipulation,

$$p[\mathbf{x} \mid \mathbf{z}_{N,N-m}] = \frac{p[\mathbf{z}_N]\,p[\mathbf{x}]}{p[\mathbf{z}_{N,N-m}]\,p[\mathbf{z}_{N-m-1}]}\,\frac{p[\mathbf{x} \mid \mathbf{z}_N]}{p[\mathbf{x} \mid \mathbf{z}_{N-m-1}]} \tag{13.5.1}$$

A similar expression can be written for $p[\mathbf{x} \mid \mathbf{z}_{N+1,N+1-m}]$. Substitution of (13.2.2) into the equation for $p[\mathbf{x} \mid \mathbf{z}_{N+1,N+1-m}]$ and using (13.5.1) gives, after more manipulation,

$$\begin{aligned} p[\mathbf{x} \mid \mathbf{z}_{N+1,N+1-m}] &= \alpha\frac{p[\mathbf{z}(N+1) \mid \mathbf{x}]}{p[\mathbf{z}(N-m) \mid \mathbf{x}]}\,p[\mathbf{x} \mid \mathbf{z}_{N,N-m}] \\ \alpha &= \frac{p[\mathbf{z}_{N+1,N+1-m}]}{p[\mathbf{z}_{N,N-m}]} \end{aligned} \tag{13.5.2}$$

α can be expressed as an integral of the rest of the right-hand side of (13.5.2).

Equation (13.5.2) is the desired recursion which is valid for $N > m$. The same startup problem exists that was found in the linear model of Section 6.7. Equation (13.5.2) is a functional recursion, so the computational problems of the growing memory filters of Section 13.2 still exist.

A driven dynamic system model such as (13.2) can also be analyzed in the same general way, but the equations are much more complicated.

13.6 Unknown-but-Bounded†

Consider the case where unknown-but-bounded models are used for the uncertainty in $\mathbf{x}(0)$, $\mathbf{v}(n)$, and $\mathbf{w}(n)$. The general set logic given in Section 6.6.1 applies directly to the nonlinear model. Hence it is not repeated.

As discussed in Section 13.1.2, sets can be expressed in terms of functions. Hence the unknown-but-bounded filtering solution yields functional recursive equations. As with their Bayesian model compatriots, exact solutions are difficult.

13.7 Performance Analysis

As stated many times in this book, knowledge of the error associated with an estimate is equally as important as the estimate itself. For example, the discussion in Part II on Bayesian linear models emphasized the role of the error covariance matrix $\mathbf{\Sigma}$.

For the nonlinear dynamics system of interest here, it is possible to write general formulas. For Bayesian models, equations for the conditional error covariance matrix (see Appendix E) can be written, while for Fisher models, the ideas of Chapter 12 can be extended to dynamic system models. Unfortunately, these general equations are usually very difficult to actually solve. The only way the author has ever obtained a performance analysis (for nonlinear dynamic systems) he was happy with was by Monte Carlo techniques, which are an obvious generalization of Section 12.6.

13.8 Discussion of Theoretical Development

The basic ideas of the preceding sections are really quite simple. The Bayesian model equations resulted from nothing more than repeated, albeit tedious, use of Bayes' rule, while unknown-but-bounded concepts are identical to those of the linear model.

The key point has been stated before but should be reemphasized. The concepts are easy but, in most cases, result in functional equations which are in general "impossible" to implement exactly. One exception to the general rule of functional equations occurred in Section 13.2.4 for the special case of

1. Bayesian smoothing solution.
2. Most probable estimator.
3. Gaussian distribution.

This special case led to a minimization problem which is still difficult but which is usually an improvement over a functional equation. For example, an estimate $\hat{\mathbf{x}}_N$ or $\hat{\mathbf{x}}(N)$ can be obtained by the use of the calculus of variations (or optimal control theory) to "reduce" the problem to a two-point boundary value problem. This special case may not correspond to the minimization of any particular error criterion, but it is not unreasonable.

The above leads one to ask two questions:

1. Why were the general equations derived?
2. What does one do in practice?

There are three answers to the first question. First, some people find it fun to derive general equations. Second, the equations can be solved analytically or approximately by numerical techniques in some cases. Third, and most important, the equations do provide some insight into how to proceed in practice. Partial answers to the second question are given in the rest of the chapter.

13.9 Approximate Propagation of Uncertainty

Exact equations for the propagation of uncertainty were developed in Section 13.1. Unfortunately the equations are usually not suitable for solution. The use of linearization techniques to get approximate simple solutions is now discussed. For simplicity, only Bayesian models of the following form are considered:

$$\mathbf{x}(n+1) = \boldsymbol{\phi}[\mathbf{x}(n), n] + \mathbf{G}[\mathbf{x}(n)]\mathbf{w}(n)$$

$$E\{\mathbf{x}(0)\} = \mathbf{m}_x(0)$$

$$E\{[\mathbf{x}(0) - \mathbf{m}_x(0)][\mathbf{x}(0) - \mathbf{m}_x(0)]'\} = \boldsymbol{\psi}$$

$$E\{\mathbf{w}(n_1)\mathbf{w}'(n_2)\} = \begin{cases} \mathbf{Q}, & n_1 = n_2 \\ \mathbf{0}, & n_1 \neq n_2 \end{cases}$$

$\mathbf{x}(0) - \mathbf{m}_x(0)$ and $\mathbf{w}(n)$: zero mean Gaussian independent

Recall that a special case of $\boldsymbol{\phi}[\mathbf{x}(n), n]$ is

$$\boldsymbol{\phi}[\mathbf{x}(n), n] = \boldsymbol{\phi}[\mathbf{x}(n)] + \mathbf{B}[\mathbf{x}(n)]\mathbf{u}(n)$$

$\mathbf{u}(n)$: known input

13.9.1 Linearization

Define $\mathbf{x}_{\text{nom}}(n)$ to be the nominal solution when there is no uncertainty; i.e.,

$$\mathbf{x}_{\text{nom}}(n+1) = \boldsymbol{\phi}[\mathbf{x}_{\text{nom}}(n), n]$$

$$\mathbf{x}_{\text{nom}}(0) = \mathbf{m}_x(0)$$

Define

$$\mathbf{d}_x(n) = \mathbf{x}(n) - \mathbf{x}_{\text{nom}}(n)$$

A Taylor series expansion of $\boldsymbol{\phi}(\mathbf{x})$ yields

$$\boldsymbol{\phi}[\mathbf{x}(n), n] = \boldsymbol{\phi}[\mathbf{x}_{\text{nom}}(n), n] + \boldsymbol{\Phi}^{(1)}[\mathbf{x}_{\text{nom}}(n)]\,\mathbf{d}_x(n) + \text{high-order terms}$$

$$\boldsymbol{\Phi}^{(1)}[\mathbf{x}_{\text{nom}}(n)] = \left.\frac{\partial \boldsymbol{\phi}(\mathbf{x})}{\partial \mathbf{x}}\right|_{\mathbf{x}=\mathbf{x}_{\text{nom}}(n)}$$

If the higher-order terms in $\mathbf{d}_x$ are dropped,

$$\mathbf{d}_x(n+1) = \boldsymbol{\Phi}^{(1)}[\mathbf{x}_{\text{nom}}(n)]\mathbf{d}_x(n) + \mathbf{G}[\mathbf{x}_{\text{nom}}(n)]\mathbf{w}(n) \tag{13.9.1}$$

Thus if the system is such that the actual solution $\mathbf{x}(n)$ stays close to the nominal solution $\mathbf{x}_{\text{nom}}(n)$, then it follows from (13.9.1) and Section 4.1.2 that

$$\boxed{\begin{aligned} &\mathbf{x}(n) \text{ is } N[\mathbf{x}_{\text{nom}}(n), \mathbf{\Gamma}(n)] \\ &\mathbf{x}_{\text{nom}}(n+1) = \boldsymbol{\phi}[\mathbf{x}_{\text{nom}}(n), n] \\ &\qquad \mathbf{\Gamma}(n) = E\{\mathbf{x}(n) - \mathbf{x}_{\text{nom}}(n)][\mathbf{x}(n) - \mathbf{x}_{\text{nom}}(n)]'\} \\ &\quad \mathbf{\Gamma}(n+1) = \mathbf{\Phi}^{(1)}[\mathbf{x}_{\text{nom}}(n)]\mathbf{\Gamma}(n)\mathbf{\Phi}^{(1)'}[\mathbf{x}_{\text{nom}}(n)] \\ &\qquad\qquad + \mathbf{G}[\mathbf{x}_{\text{nom}}(n)]\mathbf{Q}\mathbf{G}'[\mathbf{x}_{\text{nom}}(n)] \\ &\qquad \mathbf{\Gamma}(0) = \boldsymbol{\psi} \end{aligned}} \tag{13.9.2}$$

The key assumption underlying (13.9.2) is that $\mathbf{x}(n) - \mathbf{x}_{\text{nom}}(n)$ is small. A necessary condition for this is that the noise $\mathbf{w}(n)$ is small (or rather $\mathbf{GQG}'$ is small) in some sense. However, this is not sufficient and care must always be exercised in the use of (13.9.2).

EXAMPLE: Consider

$$x(n+1) = \frac{x^2(n)}{4} + w(n) + 1$$

$$x(0) = 2 \qquad E\{w(n)\} = 0$$

$$E\{w(n_1)w(n_2)\} = \begin{cases} Q, & n_1 = n_2 \\ 0, & n_1 \neq n_2 \end{cases}$$

Then $x_{\text{nom}}(n)$ satisfies

$$x_{\text{nom}}(n+1) = \frac{[\mathbf{x}_{\text{nom}}(n)]^2}{4} + 1$$

$$x_{\text{nom}}(0) = 2$$

so

$$x_{\text{nom}}(n) = 2, \qquad \text{all } n$$

Now

$$\Phi^{(1)}[x_{\text{nom}}(n)] = \frac{\partial}{\partial x}\left(\frac{x^2}{4}\right)\bigg|_{x=2} = 1$$

so from (13.9.2)

$$\Gamma(n+1) = \Gamma(n) + Q$$
$$\Gamma(0) = 0$$

Thus the linearized result of (13.9.2) says that $x(n)$ is $N(2, \Gamma(n))$, where $\Gamma(n)$ is growing without bound. Hence no matter how small the value of the noise $w(n)$ (how small Q is), it is clear that the linearized equation (13.9.1) becomes invalid for large time as the assumption that $x_{\text{nom}}(n)$ is close to $x(n)$ becomes invalid.

13.9.2 Second-Order Terms†

Equation (13.9.2) results from truncating a Taylor series expansion of $\boldsymbol{\phi}(\mathbf{x})$ and keeping only the linear terms. The effect of keeping the second-order term is now considered. To keep the notation from becoming awkward, only the case of scalar $x(n)$ is considered and $\mathbf{G}[x(n), n]$ is assumed to be G. Otherwise the model is as in Section 13.9.1.

If the second-order term in the expansion of $\phi(x)$ is kept, (13.9.1) becomes, for scalar x,

$$
\begin{aligned}
d_x(n+1) &= \Phi^{(1)}[x_{\text{nom}}(n)]\, d_x(n) + \Phi^{(2)}[x_{\text{nom}}(n)]\, d_x^2(n) + Gw(n) \\
\Phi^{(2)}[x_{\text{nom}}(n)] &= \left.\frac{\partial^2\phi(x)}{\partial x}\right|_{x=x_{\text{nom}}(n)}
\end{aligned}
\tag{13.9.3}
$$

To further simplify the notation, $\Phi^{(1)}$ and $\Phi^{(2)}$ will be written instead of $\Phi^{(1)}[x_{\text{nom}}(n)]$ and $\Phi^{(2)}[x_{\text{nom}}(n)]$. Define

$$
\begin{aligned}
m_d(n) &= E\{d_x(n)\} \\
\Gamma(n) &= E\{[d_x(n) - m_d(n)]^2\} \\
E\{d_x^2(n)\} &= \Gamma(n) + m_d^2(n)
\end{aligned}
$$

Then from (13.9.3)

$$
\begin{aligned}
m_d(n+1) &= [\Phi^{(1)}]m_d(n) + [\Phi^{(2)}][\Gamma(n) + m_d^2(n)] \\
m_d(0) &= 0
\end{aligned}
\tag{13.9.4}
$$

Now rewrite (13.9.3) as

$$
\begin{aligned}
d_x(n+1) - m_d(n+1) = {} & [\Phi^{(1)}][d_x(n) - m_d(n)] + [\Phi^{(2)}]\{[d_x(n) - m_d(n)]^2 \\
& + 2m_d(n)[d_x(n) - m_d(n)] - \Gamma(n)\} + \mathbf{G}w(n)
\end{aligned}
$$

To obtain a useful equation, it is necessary to evaluate $E\{[d_x(n)]^3\}$ and $E\{[d_x(n)]^4\}$. This would be easy *if* $d_x(n)$ were Gaussian, but $d_x(n)$ is not Gaussian becuase (13.9.3) is a nonlinear equation. However, in the interest of obtaining an equation, it is now *assumed* that $d_x(n)$ is Gaussian. Thus from Appendix E.4

$$
\begin{aligned}
E\{[d_x(n) - m_d(n)]^3\} &= 0 \\
E\{[d_x(n) - m_d(n)]^4\} &= 2\Gamma(n)
\end{aligned}
$$

Thus

$$
\begin{aligned}
\Gamma(n+1) = {} & [\Phi^{(1)}]^2\Gamma(n) + 4[\Phi^{(1)}\Phi^{(2)}]m_d(n)\Gamma(n) \\
& + [\Phi^{(2)}]^2\{2\Gamma(n) - \Gamma^2(n) + 4m_d^2(n)\Gamma(n)\} \\
& + G^2Q \\
\Gamma(0) = {} & \Gamma
\end{aligned}
\tag{13.9.5}
$$

Thus when the second-order term in the expansion of $\Phi(x)$ is included,

$$\boxed{\begin{array}{c} x(n) \text{ is } N[x_{\text{nom}}(n) + m_d(n), \Gamma(n)] \\ m_d(n)\text{: given by (13.9.4)} \\ \Gamma(n)\text{: given by (13.9.5)} \end{array}} \tag{13.9.6}$$

EXAMPLE: Consider the example in Section 13.9.1 so

$$x(n+1) = \frac{x^2(n)}{4} + w(n) + 1$$

$$x(0) = 2 \qquad E\{w(n)\} = 0$$

$$E\{w(n_1)w(n_2)\} = \begin{cases} Q, & n_1 = n_2 \\ 0, & n_1 \neq n_2 \end{cases}$$

$$x_{\text{nom}}(n) = 2, \qquad \text{all } n$$

$$\Phi^{(1)}[x_{\text{nom}}(n)] = 1$$

$$\Phi^{(2)}[x_{\text{nom}}(n)] = \tfrac{1}{2}$$

Then from (13.9.6)

$$m_d(n+1) = m_d(n) + \tfrac{1}{2}[\Gamma(n) + m_d^2(n)]$$

$$\Gamma(n+1) = \Gamma(n)\left[\tfrac{3}{2} + 2m_d(n) + m_d^2(n) - \frac{\Gamma(n)}{4}\right] + Q$$

$$\Gamma(0) = \psi$$

$$m_d(0) = 0$$

Thus inclusion of the second-order term yields an $x(n)$ whose mean value $(2 + m_d(n))$ also increases without bound.

13.10 Recursive Filters

Two fundamental rules of engineering are

1. When in doubt, use feedback.
2. If faced with a nonlinear problem, use linearization.

Application of these rules to the filtering problem for nonlinear models leads in various ways to computational algorithms, which sometimes work very well.

For simplicity of discussion, only Bayesian models are considered.

13.10.1 Feedback Tracking System

Consider the first rule: *When in doubt use feedback.* The filtering problem for the driven dynamic system of (13.3) can be stated in general terms as follows: Given the model

$$\mathbf{x}(n+1) = \boldsymbol{\phi}[\mathbf{x}(n), n] + \mathbf{G}[\mathbf{x}(n), n]\mathbf{w}(n)$$
$$\mathbf{z}(n) = \mathbf{h}[\mathbf{x}(n), n] + \mathbf{v}(n)$$

it is desired to build a system which uses the observations $\mathbf{z}(n)$ to track (or follow) the system state $\mathbf{x}(n)$. A reasonable "solution" to this problem is shown in Fig. 13.10.1, where $\hat{\mathbf{x}}(n|n)$ is the output which is "tracking" $\mathbf{x}(n)$.

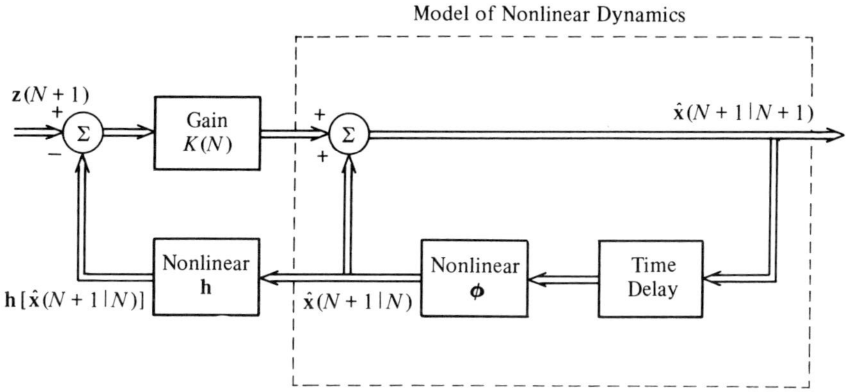

Figure 13.10.1 Feedback tracking system.

The logic behind Fig. 13.10.1 can be summarized as follows:

1. Build a model of the nonlinear dynamic system.
2. Use the model output to predict what the next observation will be.
3. Use the difference between the model prediction and the actual observation to "correct" the model.

This type of argument is sometimes called the *model reference* approach. Recall that the model form includes the case of known inputs where $\boldsymbol{\phi}[\mathbf{x}(n), n] = \boldsymbol{\phi}[\mathbf{x}(n)] + \mathbf{B}[\mathbf{x}(n)]\mathbf{u}(n)$.

Figure 13.10.1 is a "conventional" engineering solution. The difficult part of implementation is to choose the gains $\mathbf{K}(n)$ which determine how and how much of the errors, $\mathbf{z}(n+1) - \mathbf{h}[\hat{\mathbf{x}}(n+1), n+1]$, are "feedback" to drive the model. If the gains are too low, the tracking system will be "sluggish," so that, if the initial model state $\hat{\mathbf{x}}(0|0)$ is not "good," the tracking system will not "lock on" to the true system state $\mathbf{x}(n)$. If the gains are too high, the tracking system may overshoot, oscillate, and fail to track $\mathbf{x}(n)$ very closely. See the analogous discussion for linear models in Section 8.2.2.

13.10.2 Linearization About Nominal

The second rule—*Given a nonlinear problem, try to linearize it*—can be applied in different ways. The following way is closely related to the ideas of Section 13.9.1.

Choose some nominal trajectory, $\mathbf{x}_{\text{nom}}(n)$, $n = 1, \ldots,$ through state space where

$$\mathbf{x}_{\text{nom}}(n+1) = \boldsymbol{\phi}[\mathbf{x}_{\text{nom}}(n), n] \qquad \mathbf{x}_{\text{nom}}(0)\text{: specified}$$

Define

$$\mathbf{d}_x(n) = \mathbf{x}(n) - \mathbf{x}_{\text{nom}}(n)$$

$$\boldsymbol{\Phi}^{(1)}[\mathbf{x}_{\text{nom}}(n), n] = \left.\frac{\partial \boldsymbol{\phi}(\mathbf{x}, n)}{\partial \mathbf{x}}\right|_{\mathbf{x}=x_{\text{nom}}(n)}$$

$$\mathbf{H}^{(1)}[\mathbf{x}_{\text{nom}}(n), n] = \left.\frac{\partial \mathbf{h}(\mathbf{x}, n)}{\partial \mathbf{x}}\right|_{\mathbf{x}=x_{\text{nom}}(n)}$$

$$\mathbf{d}_z(n) = \mathbf{z}(n) - \mathbf{h}[\mathbf{x}_{\text{nom}}(n), n]$$

Then Taylor series expansions of the nonlinear model yield

$$\begin{aligned} \mathbf{d}_x(N+1) &= \boldsymbol{\Phi}^{(1)}[\mathbf{x}_{\text{nom}}(N), N]\mathbf{d}_x(N) + \mathbf{G}[\mathbf{x}_{\text{nom}}(N), N]\mathbf{w}(N) + \cdots \\ \mathbf{d}_z(N) &= \mathbf{H}^{(1)}[\mathbf{x}_{\text{nom}}(N), N]\mathbf{d}_x(N) + \mathbf{v}(N) + \cdots \end{aligned} \tag{13.10.1}$$

If the higher-order terms (in $\mathbf{d}_x(N)$) are dropped, (13.10.1) becomes linear and the theory of Chapter 6 can be used to obtain $\hat{\mathbf{d}}_x(N+1)(N+1)$, etc. If the usual Bayesian model for $\mathbf{x}(0)$, $\mathbf{v}(n)$, and $\mathbf{w}(n)$ is used, combining (6.2.5) and Section 6.3.7 with (13.10.1) leads to the following equations:

$$\begin{aligned} \hat{\mathbf{x}}(N+1|N+1) &= \hat{\mathbf{d}}_x(N+1|N+1) + \mathbf{x}_{\text{nom}}(N+1) \\ \hat{\mathbf{d}}_x(N+1|N+1) &= \boldsymbol{\Phi}^{(1)}[\mathbf{x}_{\text{nom}}(N), N]\,\hat{\mathbf{d}}_x(N|N) \\ &\quad + \mathbf{K}(N+1)\{\mathbf{z}(N+1) - \mathbf{h}[\mathbf{x}_{\text{nom}}(N+1), N+1] \\ &\quad - \mathbf{H}^{(1)}[\mathbf{x}_{\text{nom}}(N), N]\boldsymbol{\Phi}^{(1)}[\mathbf{x}_{\text{nom}}(N), N]\,\hat{\mathbf{d}}_x(N|N)\} \\ \mathbf{x}_{\text{nom}}(N+1) &= \boldsymbol{\phi}[\mathbf{x}_{\text{nom}}(N), N] \\ \mathbf{K}(N+1) &= \boldsymbol{\Sigma}(N+1|N+1)\mathbf{H}^{(1)}[\mathbf{x}_{\text{nom}}(N+1), N+1] \\ &\quad \times \mathbf{R}^{-1}(N+1) \\ \boldsymbol{\Sigma}(N+1|N+1) &= \{\mathbf{H}^{(1)}[\mathbf{x}_{\text{nom}}(N+1), N+1]\mathbf{R}^{-1}(N+1) \\ &\quad \times \mathbf{H}^{(1)}[\mathbf{x}_{\text{nom}}(N+1), N+1] + \boldsymbol{\Sigma}^{-1}(N+1|N)\}^{-1} \\ \boldsymbol{\Sigma}(N+1|N) &= \boldsymbol{\Phi}^{(1)}[\mathbf{x}_{\text{nom}}(N), N]\boldsymbol{\Sigma}(N|N)\boldsymbol{\Phi}^{(1)\prime}[\mathbf{x}_{\text{nom}}(N), N] \\ &\quad + \mathbf{G}[\mathbf{x}_{\text{nom}}(N), N]\mathbf{Q}(N)\mathbf{G}'[\mathbf{x}_{\text{nom}}(N), N] \\ \boldsymbol{\Sigma}(0|0) &= \boldsymbol{\psi} \end{aligned} \tag{13.10.2}$$

The implementation of (13.10.2) is illustrated in Fig. 13.10.2. The feedback loop generating the nominal trajectory $\mathbf{x}_{\text{nom}}(n)$ is "running free"

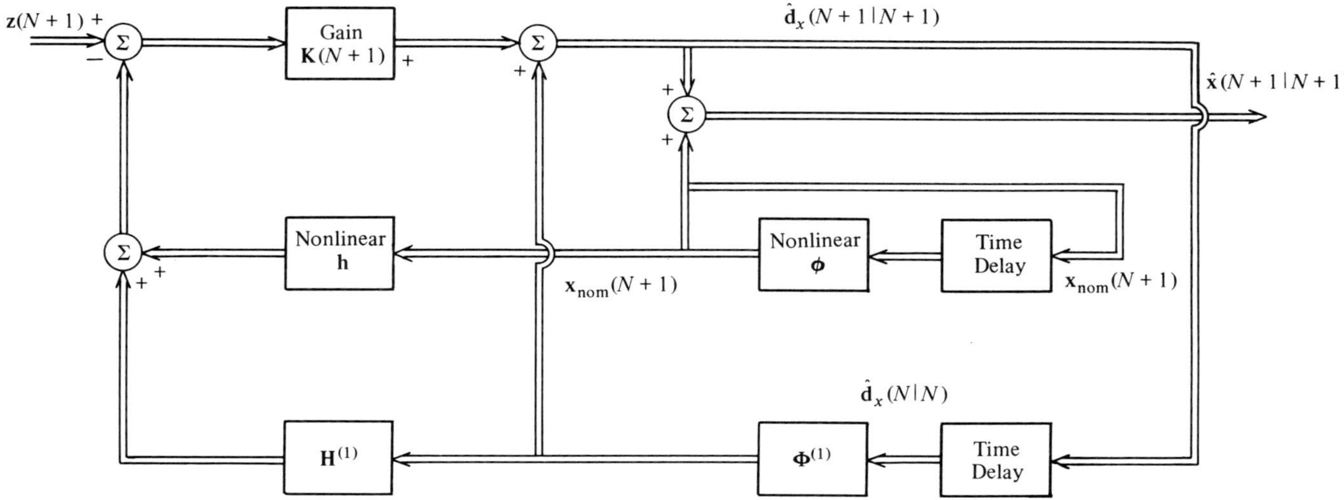

$\mathbf{K}(N+1)$ determined from $\mathbf{H}^{(1)}[\mathbf{x}_{nom}(N)]$, $\boldsymbol{\Phi}^{(1)}[\mathbf{x}_{nom}(N)]$, $\mathbf{Q}$, $\mathbf{R}$, $\boldsymbol{\psi}$, $\mathbf{G}[\mathbf{x}_{nom}(N)]$

Figure 13.10.2 Estimator for linearization about $\mathbf{x}_{nom}(n)$.

and is not effected by $\mathbf{z}(n)$. Since the gains $\mathbf{K}(n+1)$ depend only on $\mathbf{x}_{\text{nom}}(n)$, they can be precomputed. Note that Fig. 13.10.2 is similar to but not the same as Fig. 13.10.1.

The disadvantage of linearization about a nominal trajectory is that $\hat{\mathbf{x}}(n) - \mathbf{x}_{\text{nom}}(n)$ can easily become large even if $\hat{\mathbf{x}}(0) - \mathbf{x}_{\text{nom}}(0)$ is small. If this occurs, the filter equations become meaningless.

13.10.3 Linearization About Old Estimate

A simple and obvious modification of (13.10.2) and Fig. 13.10.2 is to replace

$$\mathbf{x}_{\text{nom}}(N) \quad \text{by} \quad \hat{\mathbf{x}}(N\,|\,N)$$
$$\mathbf{x}_{\text{nom}}(N+1) \quad \text{by} \quad \hat{\mathbf{x}}(N+1\,|\,N) = \boldsymbol{\phi}[\hat{\mathbf{x}}(N\,|\,N), N]$$

so that the linearization is done about the old state estimate $\hat{\mathbf{x}}(N\,|\,N)$. Now in (13.10.2)

$$\hat{\mathbf{d}}_x(N\,|\,N) = \hat{\mathbf{x}}(N\,|\,N) - \mathbf{x}_{\text{nom}}(N)$$

Thus by setting $\mathbf{x}_{\text{nom}}(N) = \hat{\mathbf{x}}(N\,|\,N)$, (13.10.2) becomes

$$\begin{aligned}
\hat{\mathbf{x}}(N+1\,|\,N+1) &= \hat{\mathbf{x}}(N+1\,|\,N) + \mathbf{K}(N+1)\{\mathbf{z}(N+1) \\
&\qquad - \mathbf{h}[\hat{\mathbf{x}}(N+1\,|\,N), N+1]\} \\
\hat{\mathbf{x}}(N+1\,|\,N) &= \boldsymbol{\phi}[\hat{\mathbf{x}}(N\,|\,N), N] \\
\mathbf{K}(N+1) &= \boldsymbol{\Sigma}(N+1\,|\,N+1) \\
&\qquad \times \mathbf{H}^{(1)'}[\hat{\mathbf{x}}(N+1\,|\,N), N+1]\mathbf{R}^{-1}(N+1) \\
\boldsymbol{\Sigma}(N+1\,|\,N+1) &= \{\mathbf{H}^{(1)'}[\hat{\mathbf{x}}(N+1\,|\,N), N+1] \\
&\qquad \times \mathbf{R}^{-1}(N+1)\mathbf{H}^{(1)}[\hat{\mathbf{x}}(N+1\,|\,N), N+1] \\
&\qquad + \boldsymbol{\Sigma}^{-1}(N+1\,|\,N)\}^{-1} \\
\boldsymbol{\Sigma}(N+1\,|\,N) &= \boldsymbol{\Phi}^{(1)}[\hat{\mathbf{x}}(N\,|\,N), N]\boldsymbol{\Sigma}(N\,|\,N)\boldsymbol{\Phi}^{(1)'}[\hat{\mathbf{x}}(N\,|\,N), N] \\
&\qquad + \mathbf{G}[\hat{\mathbf{x}}(N\,|\,N), N]\mathbf{Q}(N)\mathbf{G}'[\hat{\mathbf{x}}(N\,|\,N), N] \\
\boldsymbol{\Sigma}(0\,|\,0) &= \boldsymbol{\psi} \\
\hat{\mathbf{x}}(0\,|\,0) &= \mathbf{0}
\end{aligned} \tag{13.10.3}$$

This filter is of the form of Fig. 13.10.1 where the gains $\mathbf{K}(N+1)$ are calculated using the linear theory of Chapter 6. To be very precise, the form of Fig. 13.10.1 corresponding to (13.10.3) should be modified to indicate the explicit dependence of $\mathbf{K}(N+1)$ on $\hat{\mathbf{x}}(N\,|\,N)$.

Although (13.10.3) is the result of only a "minor modification" of (13.10.2), the two filters can exhibit completely different behaviors. One obvious disadvantage to (13.10.3) is that the gains depend on the estimate

$\hat{\mathbf{x}}(n|n)$ and thus cannot be precomputed. The full equation has to be solved in real time.

EXAMPLE: Consider

$$x(n+1) = \left[\frac{x^2(n)}{4} + 1\right] + w(n)$$

$$z(n) = x^3(n) + v(n)$$

$$x(0) = 2$$

$$E\{w(n_1)w(n_2)\} = \begin{cases} Q, & n_1 = n_2 \\ 0, & n_1 \neq n_2 \end{cases}$$

$$E\{v(n_1)v(n_2)\} = \begin{cases} R, & n_1 = n_2 \\ 0, & n_1 \neq n_2 \end{cases}$$

Consider first the use of a nominal solution and (13.10.2). Let

$$x_{\text{nom}}(n) = 2$$

$$\Phi^{(1)}[x_{\text{nom}}(n)] = \frac{\partial}{\partial x}\left[\frac{x^2}{4} + 1\right]\Bigg|_{x=2} = 1$$

$$H^{(1)}[x_{\text{nom}}(n)] = \frac{\partial}{\partial x}[x^3]\Bigg|_{x=2} = 12$$

Then from (13.10.2)

$$\hat{x}(N+1|N+1) = \hat{d}_x(N+1|N+1) + 2$$

$$\hat{d}_x(N+1|N+1) = \hat{d}_x(N|N) + K(N+1)\{z(N+1) - 8 - 12\hat{d}_x(N|N)\}$$

$$K(N+1) = 12\Sigma(N+1|N+1)R^{-1}$$

$$\Sigma(N+1|N+1) = [144R^{-1} + (\Sigma(N|N) + Q)^{-1}]^{-1}$$

This is a nice simple result which yields a $\Sigma(N|N)$ which becomes constant for large N. However, since by the examples in Sections 13.9.1 and 13.9.2, the difference $x(n) - x_{\text{nom}}(n)$ will not stay small, so for large N, the above equations become meaningless even for very small Q and R. This situation can be improved by using (13.10.3) instead of (13.10.2) to give

$$\hat{x}(N+1|N+1) = \hat{x}(N+1|N) + K(N+1)\{z(N+1) - [\hat{x}(N+1|N)]^3\}$$

$$\hat{x}(N+1|N) = \frac{\hat{x}^2(N|N)}{4} + 1$$

$$K(N+1) = \Sigma(N+1|N+1)3[\hat{x}(N+1|N)]^2R^{-1}$$

$$\Sigma(N+1|N+1) = \left\{\frac{9[\hat{x}(N+1|N)]^4}{R} + \left[\frac{[\hat{x}(N+1|N)]^2}{4}\Sigma(N|N) + Q\right]^{-1}\right\}^{-1}$$

The equations are now more complex, but they should work "reasonably well" provided that Q and R are small enough.

The corresponding equations for the linear model are (6.3.1), (6.2.5), and (6.2.7). Obviously (13.10.3) can be written in many different forms just as in the linear case of Chapter 6.

In application of (13.10.3), a trick that is sometimes useful is to increase **Q** by some amount. The idea is that the increased **Q** accounts in some way for the extra errors resulting from the linearization. The problem is, of course, to choose the "right" value for **Q**.

13.10.4 Linearization About New or Smoothed Estimate†

The filter (13.10.3) is straightforward to implement because $\hat{\mathbf{x}}(N+1|N+1)$ is expressed as an explicit function of $\hat{\mathbf{x}}(N|N)$ and $\mathbf{z}(N+1)$. However, from an idealist point of view, it can be argued that there are more accurate approaches.

One possibility is to replace

$$\mathbf{H}^{(1)}[\hat{\mathbf{x}}(N+1|N), N+1] \quad \text{by} \quad \mathbf{H}^{(1)}[\hat{\mathbf{x}}(N+1|N+1), N+1]$$

i.e., when linearizing $\mathbf{h}[\mathbf{x}(N+1), N+1]$ to linearize about $\hat{\mathbf{x}}(N+1|N+1)$ instead of $\hat{\mathbf{x}}(N+1|N) = \boldsymbol{\phi}[\hat{\mathbf{x}}(N|N), N]$. If this is done in (13.10.3), the result is a set of equations which implicitly define $\hat{\mathbf{x}}(N+1|N+1)$ but which still have to be solved, usually in some iterative fashion.

An extension of this idea is to replace

$$\boldsymbol{\Phi}^{(1)}[\hat{\mathbf{x}}(N|N), N] \quad \text{by} \quad \boldsymbol{\Phi}^{(1)}[\hat{\mathbf{x}}(N|N+1), N]$$

where $\hat{\mathbf{x}}(N|N+1)$ is calculated using the linear model smoothing theory of Chapter 6. This again converts (13.10.3) into an implicit rather than explicit set of equations for $\hat{\mathbf{x}}(N+1|N+1)$.

13.10.5 Fisher-Bayesian Type of Argument: Second-Order Terms†

The argument used in Section 6.2.2 to derive the optimum filter for a linear model was as follows:

1. Assume that $\hat{\mathbf{x}}(N|N)$ and its statistical properties are given.
2. Calculate $\hat{\mathbf{x}}(N+1|N)$ and its statistics.
3. View $\hat{\mathbf{x}}(N+1|N)$ as a Fisher measurement on $\mathbf{x}(N+1)$.
4. Combine the *two* observations $\hat{\mathbf{x}}(N+1|N)$ and $\mathbf{z}(N+1)$ to get $\hat{\mathbf{x}}(N+1|N+1)$.

This logic is now applied to the nonlinear case to yield the filter of (13.10.3) plus *second-order terms.* Techniques related to those of Section 13.9.2 are used.

For simplicity consider the case where x is a scalar and the dynamic model is

$$x(n+1) = \phi[x(n)] + G(n)w(n)$$
$$z(n) = h[x(n)] + v(n)$$

Let $\hat{x}(n|m)$ be an estimate of $x(n)$ calculated from $z(1) \cdots z(m)$. Define

$$\delta_x(n|m) = x(n) - \hat{x}(n|m)$$

Let $\Sigma(n|m)$ denote the variance of $\delta_x(n|m)$:

$$\Sigma(n|m) = E\{[\delta_x(n|m) - E\{\delta_x(n|m)\}]^2\}$$

Consider

$$\hat{x}(N+1|N) = \phi[\hat{x}(N|N)]$$

A Taylor series expansion about $\hat{x}(N|N)$ yields

$$\hat{x}(N+1|N) = x(N+1) - \delta_x(N+1|N)$$

$$\begin{aligned}\delta_x(N+1|N) = \Phi^{(1)}[\hat{x}(N|N)]\delta_x(N|N) + \Phi^{(2)}[\hat{x}(N|N)]\delta_x^2(N|N) \\ + G(N)w(N) + \cdots\end{aligned} \tag{13.10.4}$$

Consider

$$z(N+1) = h[x(N+1)] + v(N+1)$$

A Taylor series expansion about $\hat{x}(N+1|N)$ gives

$$\begin{aligned}z^*(N+1) &= H^{(1)}[\hat{x}(N+1|N)]x(N+1) + v^*(N+1) \\ v^*(N+1) &= v(N+1) + H^{(2)}[\hat{x}(N+1|N)] \\ &\quad + [\delta_x(N+1|N)]^2 + \cdots\end{aligned} \tag{13.10.5}$$

where

$$\begin{aligned}z^*(N+1) = z(N+1) - h[\hat{x}(N+1|N)] \\ + H^{(1)}[\hat{x}(N+1|N)][\hat{x}(N+1|N)]\end{aligned}$$

Now $\hat{x}(N+1|N)$ and $z^*(N+1)$ can be considered to be observations, as they can be calculated from $z(N+1)$ and $\hat{x}(N|N)$. Since (13.10.4) and (13.10.5) are of the form

$$\hat{x}(N+1|N) = x(N+1) + \text{error}$$

$$z^*(N+1) = [H^{(1)}]x(N+1) + \text{error}$$

the argument of Section 6.2 can be employed to get an equation for $\hat{x}(N+1|N+1)$ if the mean and variance of the errors, $\delta_x(N+1|N)$ and $v^*(N+1)$, can be evaluated.

The only easy way to obtain the statistical properties of $\delta_x(N+1|N)$ and $v^*(N+1)$ is to keep making assumptions until an answer is obtained. Consider first $\delta_x(N+1|N)$. Assume that higher-order terms in (13.10.4) for $\delta_x(N|N)$ above δ^2 can be dropped and that

$$\delta_x(N|N) \text{ is } N(0, \Sigma(N|N))$$

The zero mean Gaussian assumption yields

$$E\{\delta_x^3(N|N)\} = 0$$

$$E\{\delta_x^4(N|N)\} = 2\Sigma(N|N)$$

Then from (13.10.4),

$$E\{\delta_x(N+1\,|\,N)\} \approx \Phi^{(2)}[\hat{x}(N\,|\,N)]\Sigma(N\,|\,N)$$

$$\begin{aligned}\Sigma(N+1\,|\,N) = {} & [\Phi^{(1)}[\hat{x}(N\,|\,N)]]^2\Sigma(N\,|\,N) + G^2(N)Q(N) \\ & + [\Phi^{(2)}[\hat{x}(N\,|\,N]]^2[2\Sigma(N\,|\,N) - \Sigma^2(N\,|\,N)]\end{aligned} \tag{13.10.6}$$

where, when taking the expectation, $\hat{x}(N|N)$ in $\Phi^{(1)}$ and $\Phi^{(2)}$ is viewed as a constant. Now consider $v^*(N+1)$. Assume that higher-order terms in (13.10.5) for $\delta_x(N+1\,|\,N)$ can be dropped and that

$$\delta_x(N+1\,|\,N) \text{ is Gaussian}$$

Then

$$\begin{aligned}E\{v^*(N+1)\} &= H^{(2)}[\hat{x}(N+1\,|\,N)]\{E^2[\delta_x(N+1\,|\,N)] \\ &\quad + \Sigma(N+1\,|\,N)\} \\ R^*(N+1) &= R(N+1) + 2[H^{(2)}[\hat{x}(N+1\,|\,N)]]^2\Sigma(N+1\,|\,N)\end{aligned} \tag{13.10.7}$$

where $R^*(N+1)$ is the variance of $v^*(N+1)$. In taking the expectation, $\hat{x}(N+1\,|\,N)$ in $H^{(2)}$ is again viewed as a constant.

The preceding approximate equations for the mean and variance of $\delta_x(N+1\,|\,N)$ and $v^*(N+1)$ can now be combined with the argument of Section 6.2. In Section 6.2 the errors were zero mean, but the nonzero mean values can be handled by subtraction. The result is

$$\begin{aligned}\hat{x}(N+1\,|\,N+1) = {} & \Sigma(N+1\,|\,N+1)\{\Sigma^{-1}(N+1\,|\,N)[\hat{x}(N+1\,|\,N) \\ & - E\{\delta_x(N+1\,|\,N)\}] + H^{(1)}[\hat{x}(N+1\,|\,N)] \\ & \times R^{*-1}(N+1)[z^*(N+1) - E\{v^*(N+1)\}]\}\end{aligned} \tag{13.10.8}$$

$$\begin{aligned}\Sigma^{-1}(N+1\,|\,N+1) = {} & \Sigma^{-1}(N+1\,|\,N) \\ & + [H^{(1)}[\hat{x}(N+1\,|\,N)]]^2R^{*-1}(N+1)\end{aligned}$$

If all terms involving the second partial derivatives, $H^{(2)}$ and $\Phi^{(2)}$, are dropped, manipulation shows that (13.10.6), (13.10.7), and (13.10.8) reduce to the simple linearized filter of (13.10.3). This should be no surprise.

The analysis was done for scalar x. The case of vector $\mathbf{x}$ follows in a similar fashion except for the notational complications involved in handling the $\mathbf{H}^{(2)}$ and $\mathbf{\Phi}^{(2)}$.

13.10.6 Empirical Gains†

An altogether different approach to obtaining the gains $\mathbf{K}(N+1)$ of Fig. 13.10.1 is now considered wherein the gains are determined at least partially by the tracking errors, $\mathbf{z}(n) - \mathbf{h}[\hat{\mathbf{x}}(n\,|\,n), n]$.

Assume that x is a scalar. From (13.10.3)

$$K(N+1) = \Sigma(N+1\,|\,N+1)H^{(1)}[\hat{x}(N+1\,|\,N), N]R^{-1}(N+1)$$

Using the linear model theory of Chapter 6, $\Sigma(N+1 \mid N+1)$ can be viewed as

$$\Sigma(N+1 \mid N+1) = E[\delta_x(N+1 \mid N+1)]^2$$
$$\delta_x(N+1 \mid N+1) = x(N+1) - \hat{x}(N+1 \mid N+1)$$

One example of the empirical gain concept is to estimate $\Sigma(N+1 \mid N+1)$ from the estimates and observations themselves rather than compute it from (13.10.3). For a linearized model, the tracking errors are

$$\begin{aligned} z(N+1) &- h[\hat{x}(N+1 \mid N+1), N+1] \\ &= H^{(1)}[\hat{x}(N+1 \mid N), N+1]\delta_x(N+1 \mid N+1) + v(N+1) \end{aligned}$$

Taking the expected values over v and $\delta_x(N+1 \mid N+1)$,

$$\begin{aligned} E\{[z(N+1) &- h[\hat{x}(N+1 \mid N+1), N+1]]^2\} \\ &= R(N+1) + [H^{(1)}[\hat{x}(N+1 \mid N+1), N+1]]^2\Sigma(N+1 \mid N+1) \end{aligned}$$

Assume the problem is such that

$$\left.\begin{array}{l} R(n) \\ H^{(1)}[\hat{x}(n+1 \mid n), n+1] \\ \Sigma(n+1 \mid n+1) \end{array}\right\} \text{are approximately constant,} \qquad n = N-m, \ldots, N$$

Then it is reasonable to estimate $\Sigma(N+1 \mid N+1)$ by

$$\begin{aligned} \Sigma(N+1 \mid N+1) &\approx \frac{1}{m} \sum_{n=N-m}^{N} [H^{(1)}[\hat{x}(n \mid n), n]]^{-2} \\ &\times \{[z(n) - h[\hat{x}(n \mid n), n]]^2 - R(n)\} \end{aligned} \tag{13.10.9}$$

From a statistical point of view, the reader may rightly question the meaningfulness of (13.10.9). However, in some applications it may work. Extension of the concept of (13.10.9) to vector $\mathbf{x}$ is not necessarily trivial unless $\mathbf{z}$ is the same dimension as $\mathbf{x}$. However, the reader with ingenuity can figure out several reasonable approaches [for example, try lumping, say, $\mathbf{z}(N)$, $\mathbf{z}(N-1)$, and $\mathbf{z}(N-2)$ into one vector].

There are many other variations on the basic idea of using the tracking errors themselves to control the size of the gains. One reasonable variation is to choose a set of, say, three gains $\mathbf{K}_1$, $\mathbf{K}_2$, and $\mathbf{K}_3$ along with a rule which picks the gain depending on the behavior of the tracking errors. This approach could be viewed as a problem in hypothesis testing. For example, an "event" may occur which causes the state $\mathbf{x}(n+1)$ to "jump." This event is detected by looking at the tracking errors. After detection of the event, the gain is increased until a good track is reestablished. (See the discussion in Chapter 10 on event detection.)

13.10.7 Sampled Data

Consider a continuous-time dynamic system model where discrete-time (sampled data) measurements are made and modeled by an instantaneous sampler:

$$\frac{d}{dt}\mathbf{x}(t) = \mathbf{f}[\mathbf{x}(t), t] + \mathbf{G}[\mathbf{x}(t)]\mathbf{w}(t)$$

$$\mathbf{z}(n\Delta) = \mathbf{h}[\mathbf{x}(n\Delta), n\Delta] + \mathbf{v}(n\Delta)$$

$$\mathbf{x}(n\Delta) = \mathbf{x}(t)\big|_{t=n\Delta}$$

In the corresponding linear model case discussion of Section 6.15 an averaging sampler and other types of observation error modeling were discussed, but since these discussions also apply in general terms to the nonlinear case, they are not repeated.

Consider first the problem of solving the deterministic equation

$$\frac{d}{dt}\mathbf{x}(t) = \mathbf{f}[\mathbf{x}(t), t]$$

$$\mathbf{x}(0)\text{: specified}$$

In general, closed-form solutions are not possible, and it is necessary to use numerical integration techniques and a digital computer. Let τ denote the *integration step size.* Two basic approaches to numerical integration are the explicit and the implicit techniques. In an explicit technique $\mathbf{x}(t+\tau)$ is expressed as

$$\mathbf{x}(t+\tau) = \boldsymbol{\theta}[\mathbf{x}(t), t, \tau]$$

For small τ, one possible choice is

$$\mathbf{x}(t+\tau) = \mathbf{x}(t) + \tau\mathbf{f}[\mathbf{x}(t), t]$$

However, much more sophisticated approaches such as *Runge Kutta* are usually used. In an implicit technique $\mathbf{x}(t+\tau)$ is expressed implicitly as

$$\mathbf{x}(t+\tau) = \boldsymbol{\theta}[\mathbf{x}(t), \mathbf{x}(t+\tau), t, \tau]$$

For small τ, one possible choice is

$$\mathbf{x}(t+\tau) = \mathbf{x}(t) + \tau\frac{\mathbf{f}[\mathbf{x}(t+\tau), t+\tau] + \mathbf{f}[\mathbf{x}(t), t]}{2}$$

With an implicit technique it is necessary to solve an algebraic system of equations for $\mathbf{x}(t+\tau)$ at each time step, and this is usually done by iteration. This disadvantage must be weighed against the fact that implicit numerical integration schemes are usually more accurate for the same τ and all other factors being equal (which they never are).

Numerical integration concepts provide an obvious way of developing the discrete-time fomulas needed for sampled data situations. Consider, first, explicit integration. If Δ (time between observations) and τ (integration step size) are equal, then

$$\hat{\mathbf{x}}(n\Delta + \Delta \mid n\Delta) = \boldsymbol{\phi}[\hat{\mathbf{x}}(n\Delta \mid n\Delta), n\Delta] = \boldsymbol{\theta}[\mathbf{x}(t), t, \tau] \Big|_{\substack{t = n\Delta \\ \tau = \Delta \\ \mathbf{x}(t) = \hat{\mathbf{x}}(n\Delta \mid n\Delta)}}$$

If $\Delta = M\tau$, where M is an integer, then $\boldsymbol{\phi}[(\hat{\mathbf{x}}n\Delta \mid n\Delta), n\Delta]$ is determined numerically by M integration steps. Now consider implicit numerical integration. If $\Delta = \tau$, one possibility is

$$\hat{\mathbf{x}}(n\Delta + \Delta \mid n\Delta) = \boldsymbol{\theta}[\mathbf{x}(t), \mathbf{x}(t + \tau), t, \tau] \Big|_{\substack{t = n\Delta \\ \tau = \Delta \\ \mathbf{x}(t) = \hat{\mathbf{x}}(n\Delta \mid n\Delta) \\ \mathbf{x}(t+\tau) = \hat{\mathbf{x}}(n\Delta + \Delta \mid n\Delta)}}$$

However, a different approach is

$$\hat{\mathbf{x}}(n\Delta + \Delta \mid n\Delta) = \boldsymbol{\theta}[\mathbf{x}(t), \mathbf{x}(t + \tau), t, \tau] \Big|_{\substack{t = n\Delta \\ \tau = \Delta \\ \mathbf{x}(t) = \hat{\mathbf{x}}(n\Delta \mid n\Delta) \\ \mathbf{x}(t+\tau) = \hat{\mathbf{x}}(n\Delta + \Delta \mid n\Delta + \Delta)}}$$

where $\hat{\mathbf{x}}(n\Delta + \Delta \mid n\Delta + \Delta)$ is defined by, say, (13.10.3). This last approach yields an implicit equation for $\hat{\mathbf{x}}(n\Delta + \Delta \mid n\Delta + \Delta)$, which in general has to be solved by iterative techniques. It is clear that there are many ways to proceed, and unfortunately the best approach is not obvious.

The problem of how to convert the effects of the uncertainty term $\mathbf{G}[\mathbf{x}(t)]\mathbf{w}(t)$ into an equivalent discrete-time effect is also one with many possibilities, none of which is best. For "small" $\mathbf{w}(t)$, linearization concepts are the logical approach.

13.10.8 Discussion of Recursive Filters

The preceding discussions can be summarized as follows:

Sections 13.10.1–13.10.3:	Basic ideas
Sections 13.10.4–13.10.6:	Extensions
Section 13.10.7:	Sampled data

The basic ideas of Sections 13.10.1–13.10.3 can work very well in many cases. The extensions of Sections 13.10.4–13.10.6 complicate the filter, and the question of whether such extensions yield "enough" improvement to justify the complications depends entirely on the explicit problem of interest. However, the basic rule "*Use the simplest filter that works satisfactorily*" indicates that the basic ideas of Sections 13.10.1–13.10.3 should always be tried first and that the ideas of Sections 13.10.4–13.10.6 should be considered only as a last resort (and they may not work either).

No attempt has been made to discuss all possible extensions of the basic ideas. The reader can develop many other possibilities simply by combining the ideas of Sections 13.10.4–13.10.6 in various ways. An almost uncountable number of possibilities result in the sampled data case by combining the ideas of Sections 13.10.4–13.10.6 with Section 13.10.7.

In Section 6.13, the term *divergence of Kalman Filter* was discussed in the context of modeling errors. This term is also applied to recursive filters for nonlinear models to describe situations when the recursion does not work as well as the designer had hoped. This is a fairly common occurrence.

The discussions have been restricted to Bayesian models. Extension to Fisher and weighted-least-squares is straightforward. Some of the concepts can also be applied to unknown-but-bounded models, but care is needed, as the unknown-but-bounded model is fundamentally different. The preceding recursive filters are designed to generate point estimates, while the unknown-but-bounded concept is based on global sets.

The point estimate character of the preceding feedback filters also points out a major deficiency. No error or performance analysis is available unless the errors are actually small so that $\mathbf{\Sigma}(N|N)$ can be viewed as an error covariance matrix. It should be emphasized that *acceptable tracking performance may be obtained using some* $\mathbf{\Sigma}(N|N)$ *to get the gains even if* $\mathbf{\Sigma}(N|N)$ *does not correspond to the error covariance matrix.* In general there seems to be only one reliable way to determine how well the filter will perform. Do a Monte Carlo trial on a digital computer by simulating the system, adding random errors, programming the filter, and seeing what happens.

13.11 Continuous Time

As mentioned in the introduction to Part IV, the author has chosen not to present the details of applying the discrete-to-continuous-time limit process to the discrete-time results of this chapter. These details were included in an earlier manuscript version of this book but were removed as it was not obvious to the author how these results could be applied in practice. However, a few general comments are still appropriate.

Consider a Bayesian model and the functional equations for $p[\mathbf{x}(n)]$ of Section 13.1.1. If $\mathbf{x}(n)$, and $\mathbf{w}(n)$ are assumed to be Gaussian, the discrete to continuous limit can be carried out to yield the *Chapman-Kolmogorov* partial differential equation for the probability density of $\mathbf{x}(t)$. This result can then be combined with the functional equations of Section 13.2.2 for $p[\mathbf{x}(n)|\mathbf{z}_n]$ (and a Gaussion $\mathbf{v}(n)$) to yield a partial differential equation for the conditional probability density of $\mathbf{x}(t)$ given $\mathbf{z}(\tau)$ $0 \leq \tau \leq t$. The details of this limiting process are tricky and should not be attempted without help.

The partial differential equation for the condition probability density, of $\mathbf{x}(t)$ has the form of the Chapman-Kolmogorov equation plus a driving term which depends on $\mathbf{z}(t)$. Unfortunately the form of this driving term depends on how the limit is defined; and whether Itô or Stratonovich calculus is used (see Section 4.2). The author prefers the Stratonovich result but the whole question appears to be of only academic interest.

The discrete-to-continuous-time limit can also be applied to the recursive filters of Section 13.10. Then results can have value in applications but the details are relatively straightforward and so are left as the famous "exercise for the reader."

13.12 Discussion

The discussions of this chapter have ignored the question of the stability of nonlinear stochastic models. Stochastic stability can be an important and very difficult question, which is outside the chosen range of this book.

It is obvious that a simple answer does not exist to the question asked in Section 13.8: What does one do in practice? The development of successful estimators for nonlinear models is more of an art than a science. (The question of whether engineering is an art or a science will not be discussed.)

To further confuse the issue, the ideas of Section 13.10 by no means exhaust the range of possible ways to try to obtain a practical filter. The "change of variables" concept of Section 11.9 is very important in all nonlinear estimator problems. The parameterization ideas of Section 11.8 can be extended in various ways to the dynamic situation.

Historical Notes and References: Chapter 13

Mendel and Gieseking [1] is a bibliography of over 900 references, many of which are related to this chapter. Some of these referencs are discussed below, but many are not.

Section 13.2, 13.3, 13.4

See Book List 4 on Stochastic Systems. See also discussion of Section 13.10.

Section 13.5

A similar development is given in Jazwinski [2].

Section 13.10

The basic idea of handling nonlinear models by combining linearization with the linear model theory has been discussed in an almost uncountable number of papers. The resulting filters are often called an *extended Kalman filter*. The second order term discussions of Section 13.10.5 are the result of discussions with L. Gardner.

Some of the many papers on the basic linearization, second order terms, and other approximations are Bass, Noran, and Schwartz [1]; Bellman, Kagiwada and Sridhar [1]; Bryson and Frazier [1]; Cox [1], [2], [3]; Jazwinski [1]; Kushner [3]; Larson and Peschon [1]; Mowerey [1]; Ohap and Stubberad [1]; Schmidt [1]; Schwartz and Stear [1]; Smith [1]; Snyder [1]; Stridhar and

Detchmendy [1]. Some of these references are actually on the corresponding continuous time problem. See also Book List 4 on Stochastic Systems.

Section 13.11

Many of the references for Section 13.10, also consider the continuous time case. Papers on the general partial differential equations include Duncan [1]; Fisher and Stear [1]; Kushner [1], [2]; Stratonovich [1], [2], [3], [4]. As discussed in the text, these papers do not always give the same answer.

Section 13.12

For discussions on stochastic stability, see Kushner [4].

Exercises

13.7 is taken from Magill [1].

Exercises

Exercises 13.1–13.3 are all based on the following Bayesian model:

$$x(n+1) = x^3(n) + w(n)$$

$$z(n) = x^2(n) + v(n)$$

$x(0)$ is $N(0, \psi)$

$w(n)$ is $N(0, Q)$

$v(n)$ is $N(0, R)$

$x(0), w(n), v(n)$: independent

$w(n_1), w(n_2)$: independent, $n_1 \neq n_2$

$v(n_1), v(n_2)$: independent, $n_1 \neq n_2$

13.1. Write a functional equation expressing $p[x(N+1)|\mathbf{z}_{N+1}]$ in terms of $p[x(N)|\mathbf{z}_N]$. Be as explicit as possible but do not attempt to evaluate any integrals.

13.2. Write the equations which define $\hat{\mathbf{x}}_{N|N}$, the most probable estimate of

$$\mathbf{x}_N = \begin{bmatrix} x(1) \\ \cdot \\ \cdot \\ \cdot \\ x(N) \end{bmatrix}$$

given $\mathbf{z}(1) \cdots \mathbf{z}(N)$. Do not attempt an explicit solution.

13.3. Write the equations for two feedback tracking system filters based on the linearization concepts of Sections 13.10.2 and 13.10.3.

13.4. Consider

$$x(n+1) = \mathbf{\Phi}x(n) + w(n)$$

$$z(n) = x(n) + v(n)$$

and the following two types of models for the uncertainty in $x(0)$, $w(n)$ and $v(n)$.

Bayesian: $x(0)$, $w(n)$, $v(n)$: independent

$w(n)$, $v(n)$: white processes (independent)

$$p[x(0)] = \begin{cases} \dfrac{1}{2a}, & |x(0)| \leq a \\ 0, & \text{otherwise} \end{cases}$$

$$p[w(n)] = \begin{cases} \dfrac{1}{2b}, & |w(n)| \leq b \\ 0, & \text{otherwise} \end{cases}$$

$$p[v(n)] = \begin{cases} \dfrac{1}{2c}, & |v(n)| \leq c \\ 0, & \text{otherwise} \end{cases}$$

Unknown-but-bounded:

$$\begin{aligned} x(0) &\in \mathbf{\Omega}_x(0) & \mathbf{\Omega}_x(0) &= \{x: |x| \leq a\} \\ w(n) &\in \mathbf{\Omega}_w(n) & \mathbf{\Omega}_w(n) &= \{w: |w| \leq b\} \\ v(n) &\in \mathbf{\Omega}_v(n) & \mathbf{\Omega}_v(n) &= \{v: |v| \leq c\} \end{aligned}$$

Define

$$\mathbf{x}_N = \begin{bmatrix} x(0) \\ \cdot \\ \cdot \\ \cdot \\ x(N) \end{bmatrix} \qquad \mathbf{z}_N = \begin{bmatrix} z(1) \\ \cdot \\ \cdot \\ \cdot \\ z(N) \end{bmatrix}$$

Define

Bayesian: $\hat{\mathbf{x}}_N$: value of $\mathbf{x}_N$ which maximizes $p[\mathbf{x}_N | \mathbf{z}_N]$ of (13.2.12)

Unknown-but-bounded: $\mathbf{\Omega}_{\text{est}}(N)$: minimum set containing $\mathbf{x}_N$ that can be calculated from $\mathbf{z}_N$

Show that $\hat{\mathbf{x}}_N$ is not unique and in fact that any $\mathbf{x}_N \in \mathbf{\Omega}_{\text{est}}(N)$ is a possible $\hat{\mathbf{x}}_N$. *Comment:* This example can be generalized to provide a tie between unknown-but-bounded models and most probable estimates for Bayesian models.

13.5. Consider

$$x(n+1) = \mathbf{\Phi}x(n) + w(n)$$

$$z(n) = x(n) + v(n)$$

$x(0)$, $w(n)$, $v(n)$: independent

$w(n)$, $v(n)$: white

$w(n)$ is $N(0, Q)$

$v(n)$ is $N(0, R)$

$$p[x(0)] = \begin{cases} \dfrac{1}{2a}, & |x(0)| \leq a \\ 0, & \text{otherwise} \end{cases}$$

Derive an equation for $\hat{x}(N|N)$, the conditional expectation of $x(N)$ given $\mathbf{z}_N$,

$$\mathbf{z}_N = \begin{bmatrix} z(1) \\ \cdot \\ \cdot \\ \cdot \\ z(N) \end{bmatrix}$$

Hint:

$$p[x(N)\,|\,\mathbf{z}_N] = \int p[x(N)\,|\,\mathbf{z}_N,\, x(0)]p[x(0)]\, dx(0)$$

$p[x(N)\,|\,\mathbf{z}_N,\, x(0)]$: Gaussian density which can be evaluated as a function of $x(0)$ using Chapter 6 equations

Comment: This example can be generalized to a general theory for handling non-Gaussian initial conditions.

13.6. Consider the following dynamical system:

$$x(n + 1) = a[1 + w(n)]x(n), \qquad a > 1$$

$$z(n) = b[1 + v(n)]x(n), \qquad b > 1$$

$w(n)$ and $v(n)$ are white Gaussian random variables with zero mean and covariances:

$$E[w(n_1)w(n_2)] = \begin{cases} 2 \log a, & n_1 = n_2 \\ 0, & n_1 \neq n_2 \end{cases}$$

$$E[v(n_1)v(n_2)] = \begin{cases} 2 \log b, & n_1 = n_2 \\ 0, & n_1 \neq n_2 \end{cases}$$

$w(n)$, $v(n)$, and $x(0)$ are uncorrelated. $x(0)$ is a zero mean Gaussian random variable with covariance ψ. It is desired to find an estimate of $x(N)$ given measurements $z(1) \cdots z(N)$.

a. Develop a system which uses the observations $z(n)$ to track the system state $x(n)$ as in Section 13.10.2.
b. By taking logarithms of the equations a linear system can be obtained. Assume that the noises are small, so that

$$\log [1 + w(n)] \approx w(n) - \frac{w^2(n)}{2}$$

$$\log [1 + v(n)] \approx v(n) - \frac{v^2(n)}{2}$$

Find an estimator which gives a good estimate of $\log x(n)$ given $z(1) \cdots z(n)$ and then deduce a reasonable estimator for finding $\hat{x}(n|n)$.

13.7. Consider the M different models

$$\mathbf{x}(n + 1) = \mathbf{\Phi}_j(n)\mathbf{x}(n) + \mathbf{G}_j(n)\mathbf{w}(n)$$

$$\mathbf{z}(n) = \mathbf{H}_j(n)\mathbf{x}(n) + \mathbf{v}(n)$$

$$\mathbf{x}(0),\ \mathbf{v}(n),\ \mathbf{w}(n)\text{: independent}$$
$$\mathbf{x}(0) \text{ is } N(\mathbf{0}, \boldsymbol{\psi}_j)$$
$$\mathbf{v}(n) \text{ is } N(\mathbf{0}, \mathbf{R}_j(n))$$
$$\mathbf{w}(n) \text{ is } N(\mathbf{0}, \mathbf{Q}_j(n))$$
$$j = 1, \ldots, M$$

For a given sequence of observations $\mathbf{z}(1) \cdots \mathbf{z}(N)$, one of these M models is always applicable. Define

ρ_j: a priori probability that $\mathbf{z}(1) \cdots$ is generated by jth model, $j = 1, \ldots, M$

$$\sum_{j=1}^{M} \rho_j = 1$$

The a priori probabilities ρ_j are the probabilities before any observations are made. Define the random variable

e_j: event that jth model is true

$$p[e_j] = \rho_j$$

$p[\mathbf{z}_N \mid e_j]$ = probability of $\mathbf{z}_N$ given that the jth model is true

Define

$\hat{\mathbf{x}}(N \mid N)$: conditional expectation of $\mathbf{x}(N)$ given $\mathbf{z}_N$

$\hat{\mathbf{x}}_j(N \mid N)$: conditional expectation of $\mathbf{x}(N)$ given $\mathbf{z}_N$ and assuming that e_j has occurred

$\xi_j(N)$: log likelihood function assuming that event e_j is true

$$p(\mathbf{z}_N \mid e_j) = e^{\xi_j(N)}$$

$\hat{\mathbf{x}}_j(N \mid N)$ and $\xi_j(N)$ can be easily calculated using the equations of Chapters 6 and 10. Show that

$$\hat{\mathbf{x}}(N \mid N) = \sum_{j=1}^{M} c_j(N)\hat{\mathbf{x}}_j(N \mid N)$$

$$c_j(N) = \frac{\rho_j e^{\xi_j(N)}}{\sum_{k=1}^{M} \rho_k e^{\xi_k(N)}}$$

Hint:

$$p[\mathbf{x}(N) \mid \mathbf{z}_N] = \sum_{j=1}^{M} p[\mathbf{x}(N), e_j \mid \mathbf{z}_N]$$
$$= \sum_{j=1}^{M} p[\mathbf{x}(N) \mid \mathbf{z}_N, e_j] c_j(N)$$
$$c_j(N) = p[e_j \mid \mathbf{z}_N]$$
$$= \frac{p[e_j] p[\mathbf{z}_N \mid e_j]}{\sum_{k=1}^{M} p[\mathbf{z}_N \mid e_k] p[e_k]}$$

PART V

Survey Discussions

The first four parts of our journey were highly organized tours which followed well-defined paths (static to discrete-time to continuous-time models). The fifth and last part of our trip is completely different, as each chapter is an independent discussion of a separate subject. System identification, a subject which has been mentioned several times in earlier chapters, is discussed in Chapter 14. The discussions draw extensively on ideas and concepts from many of the earlier chapters. The subject of stochastic control is considered in Chapter 15; these discussions use results of earlier chapters only in a few places. Chapters 14 and 15 are called surveys for the lack of a better term.

14

System Identification

System identification has been mentioned at various points in the book and can arise in a very wide range of diverse physical (and social, economic, etc.) problems. Basically, as discussed in Chapter 2, system identification arises when mathematical models for the process of interest cannot be completely specified and the model has to be "identified" using actual observations of the output and possibly input. Four examples are summarized in a highly overly simplified form in Figure 14.1.

A mathematical model can be viewed in terms of two aspects:

1. Structure.
2. Numerical values (parameter values).

The problem of identifying a model from observations of the system's behavior can be divided into three steps:

1. Hypothesize reasonable structure.
2. Estimate numerical values for hypothesized structure.
3. Test validity of the model (structure plus estimated values).

Often these three steps are repeated many times as the third step (test validity) may reject the model structure hypothesized in the first step. The discussions to follow consider the last two steps. The problem of hypothesizing structures which "look good" is left up to the engineer. Most emphasis is given to the

Problem Area	Nature of Problem	Problem to be Solved
Electric Power Load Forecasting	Electric power demand (megawatts) is nonstationary stochastic process which also depends on weather	Find low order mathematical model describing the process
Inertial Guidance System	Performance and methods of "updating" depend on "gyro drift rates"	Estimate the drift rates
Electric Power Generating Plants	Dynamics of plant are represented by high order, nonlinear set of equations with unknown parameters	Find a low order simple model which approximates plant dynamics about some nominal operating point
Radar Detection in Presence of Clutter	Radar return includes effects of very many randomly oriented (and moving) reflectors	View the clutter as a random channel and identify a model for the channel

Figure 14.1. Examples of system identification.

second step of estimating the numerical values for a given structure as the testing of step 3 is closely related to the ideas of Part III of this book.

System identification problems are often divided into two types:

1. Off Line: Observations are recorded and system identification is done in *nonreal time*, usually by reprocessing the observations many times.
2. On Line: The system identification is to be done in real time, usually by a recursive algorithm.

However the dividing line between on and off line techniques is fuzzy. The on-line case usually results in what is often called an adaptive system (see the discussion in Chapter 2). All four examples of Fig. 14.1 can yield off-line or on-line problems depending on the particular nature of the situation.

Model structures are discussed in Section 14.1. Section 14.2 considers the state augmentation approach to system identification. Maximum likelihood techniques are discussed in Sections 14.3, 14.4, and 14.5. Model reference, correlation, and instrument variable concepts are discussed in Section 14.6, 14.7, and 14.8. Unknown-but-bounded models are considered in Section 14.9 .The choice of system input is discussed in Section 14.10. The model reduction problem is considered in Section 14.11. Nonlinear structures are discussed in Section 14.12. A brief final discussion is given in Section 14.13.

14.1 Model Structures

A basic principle of system identification is that you cannot identify a completely unknown system. In general it is necessary to put some restrictions on the class of systems in which lies the system to be identified. Consider models of the following structure:

$$\mathbf{x}(n+1) = \mathbf{\Phi x}(n) + \mathbf{Gw}(n) + \mathbf{Bu}(n)$$
$$\mathbf{z}(n) = \mathbf{Hx}(n) + \mathbf{v}(n)$$

$\mathbf{x}(n)$: state (K_1 vector)
$\mathbf{z}(n)$: observation (K_2 vector) (14.1.1)
$\mathbf{v}(n)$: observation white uncertain process (K_2 vector)
$\mathbf{w}(n)$: input white uncertain process (K_3 vector)
$\mathbf{u}(n)$: input deterministic process which is known (K_4 vector)

For stochastic models

$$\mathbf{w}(n),\ \mathbf{v}(n),\ \mathbf{x}(0)\text{: zero mean uncorrelated}$$

$$E\{\mathbf{w}(n_1)\mathbf{w}'(n_2)\} = \begin{cases} \mathbf{Q}, & n_1 = n_2 \\ \mathbf{0}, & n_1 \neq n_2 \end{cases}$$

$$E\{\mathbf{v}(n_1)\mathbf{v}'(n_2)\} = \begin{cases} \mathbf{R}, & n_1 = n_2 \\ \mathbf{0}, & n_1 \neq n_2 \end{cases}$$

$$E\{\mathbf{x}(0)\mathbf{x}'(0)\} = \boldsymbol{\psi}$$

For unknown-but-bounded–set-theoretic models

$$\mathbf{w}(n) \in \boldsymbol{\Omega}_w$$
$$\boldsymbol{\Omega}_w = \{\mathbf{w}: \mathbf{w}'\mathbf{Q}^{-1}\mathbf{w} \leq 1\}$$
$$\mathbf{v}(n) \in \boldsymbol{\Omega}_v$$
$$\boldsymbol{\Omega}_v = \{\mathbf{v}: \mathbf{v}'\mathbf{R}^{-1}\mathbf{v} \leq 1\}$$
$$\mathbf{x}(0) \in \boldsymbol{\Omega}_x(0)$$
$$\boldsymbol{\Omega}_x(0) = \{\mathbf{x}: \mathbf{x}'\boldsymbol{\psi}^{-1}\mathbf{x} \leq 1\}$$

Some of the values of $\boldsymbol{\Phi}$, $\mathbf{B}$, $\mathbf{G}$, $\mathbf{H}$, $\mathbf{R}$, $\mathbf{Q}$, are assumed completely unknown. Define

$\boldsymbol{\alpha}$: vector containing all the unknown parameters in $\boldsymbol{\Phi}$, $\mathbf{B}$, $\mathbf{G}$, $\mathbf{H}$, $\mathbf{R}$, $\mathbf{Q}$,

If the degree of the model (K_1 the dimension of $\mathbf{x}$) is specified,

1. The hypothesized structure is (14.1.1) for fixed K_1.
2. The numerical values to be estimated are $\boldsymbol{\alpha}$.

If the model degree (K_1) is not known, then each possible value of K_1 is considered to be a different structure.

The above formulation is obviously not the most general possible. Therefore before proceeding with discussions on estimation and testing, various aspects of the structure formulation are discussed. The discussions are confined to discrete time systems but almost all of the concepts carry through to continuous time.

14.1.1 Linearity

The model structure of (14.1.1) is linear. Some of the general identification techniques to be discussed can also (conceptually at least) be applied to nonlinear structures. The linear structure discussion conveys the basic ideas. Section 14.12 discusses the nonlinear case directly.

14.1.2 Parameterization

The model (14.1.1) views the uncertainty in the system itself in terms of unknown parameters; the values of $\boldsymbol{\Phi}$, $\mathbf{B}$, etc. and the model degree. In almost all applications, only certain parameters of the matrices are unknown. For example, $\boldsymbol{\Phi}$ may be assumed to be of the form

$$\mathbf{\Phi} = \begin{bmatrix} 0 & 1 & 0 & \cdots & 0 \\ 0 & 0 & 1 & \cdots & 0 \\ \cdot & \cdot & \cdot & \cdot & \cdot \\ \cdot & \cdot & \cdot & \cdot & \cdot \\ \cdot & \cdot & \cdot & \cdot & \cdot \\ \alpha_1 & \alpha_2 & \alpha_3 & \cdots & \alpha_{K_1} \end{bmatrix}$$

where the values of $\alpha_1 \cdots \alpha_{K_1}$ are unknown (the value of K_1 may also be unknown). Similar modeling is used for the **B**, **G**, **H**, **Q**, and **R** matrices. Thus $\boldsymbol{\alpha}$, the vector of all the unknown parameters of $\mathbf{\Phi}$, **B**, etc., may be of low dimension.

The model (14.1.1) assumes that $\boldsymbol{\alpha}$ is completely unknown (Fisher). In certain applications, a priori knowledge such as "$\boldsymbol{\alpha}$ is a random vector with specified probability density" (Bayesian) or "$\boldsymbol{\alpha}$ belongs to some set" (unknown but bounded) may be available. Discussion is restricted to the case of unknown $\boldsymbol{\alpha}$ to simplify the discussions. As discussed in Sections 5.3 and 11.3, a priori knowledge can be treated as an extra measurement.

Note that there is no conceptual need for the unknown parameter $\boldsymbol{\alpha}$ to correspond directly to explicit parameters in $\mathbf{\Phi}$, **B**, etc. For example, one might choose $\boldsymbol{\alpha}$ to contain the eigenvalues of $\mathbf{\Phi}$ rather than explicit elements of the $\mathbf{\Phi}$ matrix. In fact, an important part of the system identification problem is to choose an $\boldsymbol{\alpha}$ of the most appropriate form. This problem of choosing $\boldsymbol{\alpha}$ is not discussed much in this book, but its importance must be emphasized.

14.1.3 Time Invariance

The model (14.1.1) is a time-invariant system. Time-varying structures such as

$$\mathbf{x}(n+1) = \mathbf{\Phi}(n)\mathbf{x}(n) + \mathbf{G}(n)\mathbf{w}(n) + \mathbf{B}(n)\mathbf{u}(n)$$

are conceivable. Assume that the model degree is specified. Let $\boldsymbol{\alpha}$ denote the vector of uncertain values of the $\mathbf{\Phi}(n)$, etc., $n = 1, \ldots,$ matrices. There are two cases of interest:

1. $\boldsymbol{\alpha}$ is constant.
2. $\boldsymbol{\alpha}(n)$ is time-varying.

Examples of these two cases are

$$\mathbf{\Phi}(n) = \begin{bmatrix} 0 & 1 & 0 \\ 0 & 0 & 1 \\ \alpha_1 & n\alpha_2 & n^2\alpha_2 \end{bmatrix}$$

$$\mathbf{\Phi}(n) = \begin{bmatrix} 0 & 1 & 0 \\ 0 & 0 & 1 \\ \alpha_1(n) & \alpha_2(n) & \alpha_3(n) \end{bmatrix}$$

The case where $\boldsymbol{\alpha}$ is a constant is relatively minor extension of (14.1.1).

When $\boldsymbol{\alpha}(n)$ is time-varying, the problem is more complicated. A completely unknown $\boldsymbol{\alpha}(n)$, $n = 1, \ldots$, can sometimes be considered, or $\boldsymbol{\alpha}(n)$ can be modeled as some stochastic or unknown-but-bounded process. A case of more practical importance occurs when $\boldsymbol{\alpha}(n)$ is "slowly time-varying." In such a case, $\boldsymbol{\alpha}(n)$ can be assumed to be constant over short time intervals. With luck, this time interval will be long enough to identify the parameter values using the model (14.1.1).

The case of time-varying $\boldsymbol{\alpha}(n)$ will be briefly discussed again in later sections.

14.1.4 Perfect Knowledge of Input u(n)

There are two input processes in (14.1.1), $\mathbf{w}(n)$ and $\mathbf{u}(n)$. It is assumed that the input $\mathbf{u}(n)$ is perfectly known. In some applications, the input $\mathbf{u}(n)$ is not known a priori but is observed; for example,

$$\tilde{\mathbf{z}}(n) = \mathbf{u}(n) + \tilde{\mathbf{v}}(n)$$

Such a situation can be put in the form of (14.1.1) using the ideas of Section 3.2 by associating $\mathbf{u}(n)$ with $\tilde{\mathbf{z}}(n)$ and incorporating $\tilde{\mathbf{v}}(n)$ into $\mathbf{w}(n)$.

14.1.5 White Uncertainty

The model (14.1.1) assumes that the input and observation uncertainties $\mathbf{w}(n)$ and $\mathbf{v}(n)$ are white processes. However, the ideas of Section 3.2. can be used to express more general cases in the form of (14.1.1). Similarly, nonzero mean stochastic processes or non-origin-centered set processes can be handled using the ideas of Section 3.2.

14.1.6 Identifiability

Given (14.1.1), it is not always possible to identify all the unknown system parameter values from knowledge of $\mathbf{z}(n)$ and $\mathbf{u}(n)$, $n = 1, \ldots$.

Consider Fig. 14.1.1. It is obvious that parameters associated with $\boldsymbol{\Phi}_1$ and $\boldsymbol{\Phi}_4$ cannot be identified. Similarly, parameters associated with $\boldsymbol{\Phi}_3$ can be identified only if the initial conditions $\mathbf{x}_3(0)$ allow it. If $\mathbf{x}_3(0) = \mathbf{0}$, values associated with $\boldsymbol{\Phi}_3$ cannot be identified. In terms of the controllability and observability concepts discussed in Appendix *B*,

System $\boldsymbol{\Phi}_1$: not observable

System $\boldsymbol{\Phi}_3$: not controllable

System $\boldsymbol{\Phi}_4$: not observable, not controllable

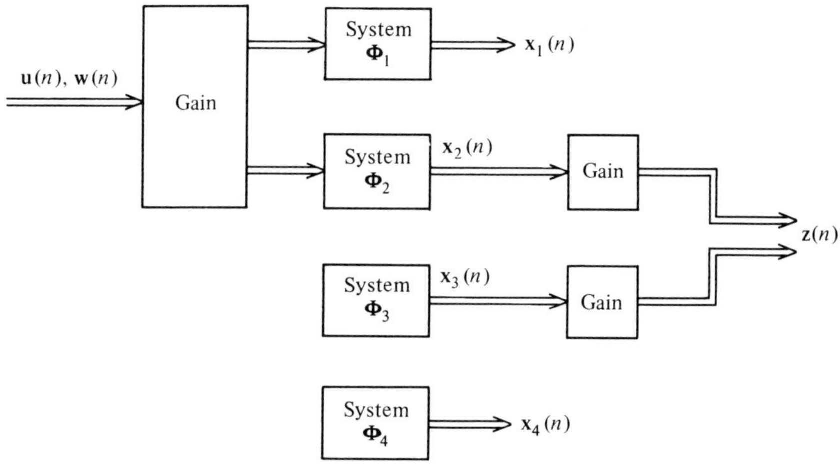

Figure 14.1.1

Now consider a scalar version of (14.1.1):

$$\begin{aligned} x(n+1) &= \Phi x(n) + Bu(n) \\ z(n) &= Hx(n) + v(n) \\ x(0) &= 0 \end{aligned} \tag{14.1.2}$$

Then

$$z(N) = HB \sum_{n=0}^{N-1} \Phi^{N-n-1} u(n) + v(N), \qquad N = 1, \ldots$$

Assume that values of H and B are both unknown. Obviously individual values for H and B cannot be identified from knowledge of the input $u(n)$ and output $z(n)$ [even if Φ is known and $v(n) = 0$]. Only values for the product HB can be identified.

The discussions associated with Fig. 14.1.1 and the simple example of (14.1.2) point out that careful definitions are needed when one asks whether or not a system can be identified. As one example, consider Fig. 14.1.1 and assume that Φ_4 is unknown. The system is unidentifiable in the sense that Φ_4 cannot be identified. However, if identification is defined in terms of being able to predict the system's output response to some input, there is no need to know Φ_4, so the system may be considered to be identifiable. As another similar example, consider (14.1.2). If H and B are unknown, the system is unidentifiable in the sense that individual values of H and B cannot be identified. However, if identification is defined in terms of the ability to predict a system's input-output characteristics, the system (14.1.2) is identifiable, as the product HB can be identified and only HB affects the input-output characteristics. One way to view (14.1.2) with unknown H and B is to say that the system is identifiable (can identify its input-output behavior), but the identification is not unique (many possible combinations of values for H and B are possible).

It would now be ideal to introduce an exact definition which specifies when a system is or is not identifiable. However, the factors already discussed, the presence of the uncertainties $\mathbf{w}$ and $\mathbf{v}$ (which may be stochastic or unknown but bounded), and the possibility of modeling errors makes a general definition very difficult. Thus a definition is not given. It seems best to view each case separately in terms of the actual physical problem of interest.

14.1.7 Autoregressive and Moving Average Models

In Section 3.4.2, autoregressive, moving average, and mixed autoregressive-moving average models were defined where, for example, with $z(n) = y(n)$,

Autoregressive:

$$y(n+1) = \Phi_1 y(n) + \Phi_2 y(n-1) + w(n)$$

Moving Average:

$$y(n) = G_1 w(n) + G_2 w(n-1)$$

Mixed Autoregressive-Moving Average:

$$y(n+1) = \Phi_1 y(n) + \Phi_2 y(n-1) + G_1 w(n) + G_2 w(n-1)$$

Such models can be put in the general form of (14.1.1) but for system identification it is often more desirable to work with the "natural" form directly.

It will be seen that autoregressive and moving average models can yield "nice" results while a mixed model does not. Thus, it should be noted that the state space model

$$\begin{aligned} x(n+1) &= \Phi x(n) + w(n) \\ z(n) &= x(n) + v(n) \end{aligned}$$

corresponds to the mixed autoregressive-moving average model

$$\begin{aligned} y(n+1) &= \Phi y(n) + w(n) + v(n+1) + \Phi v(n) \\ z(n) &= y(n) \end{aligned}$$

14.1.8 Structural Errors

Up to now, it has been tacitly assumed that the model structure (14.1.1) actually fits the real world for some value of the unknown paramcter $\boldsymbol{\alpha}$ (and degree of model). This is usually a naive assumption. For example, the actual system may be a nonlinear system with "dead bands", "saturation", and other such phenomena present. In such a case, a structure like (14.1.1) might be used because an exact structure is too complicated. Another example is a linear system of the form of (14.1.1) for some large value of K_1 (degree of model) when "practical consideration" causes one to want to identify an approximate model of low degree; i.e., in (14.1.1) K_1 is fixed at some small value. In both examples one encounters structural errors. At the beginning of this chapter, system identification was divided into three steps: 1) hypothesize structure, 2) estimate values, 3) test validity. The presence of possible structural errors emphasizes the need for the third step of testing.

When structural errors are expected, one philosophy of approach is to assume that $\mathbf{Q}$ the size of the uncertain process $\mathbf{w}(n)$ driving the system is unknown, even if the real system is known to have no uncertain inputs. The idea is that the identified value of $\mathbf{Q}$ will account for the structural errors in some sense.

EXAMPLE: Assume that the actual observations $z_{\text{actual}}(n)$ are

$$z_{\text{actual}}(n) = \sin(\omega n), \qquad n = 1, \ldots, N$$

First assume that the structure of the model to be identified is hypothesized to be

$$x(n+1) = \Phi x(n)$$

where Φ is unknown so that

$$x(n) = \Phi^n x(0)$$

An estimate of Φ could be obtained but it is clear that the resulting model does not constitute a good fit to the observations so the identification actually yields a meaningless model. A different approach would be to hypothesize the structure

$$\begin{aligned} x(n+1) &= \Phi x(n) + w(n) \\ Q \quad &: \text{"size" of } w(n) \\ Q, \Phi \quad &: \text{both unknown} \end{aligned}$$

In this case the "identified" value of Q would be large. Thus even though the identified model will still not behave like the observations, one is "told" by the identified model that there is a lot of uncertainty in the model. The best procedure would, of course, be to hypothesize a second order structure which could have a sinusoidal output.

14.1.9 Discussion of Structure

The structure (14.1.1) is definitely not the most general possible model, but it covers a very wide range of cases of interest. It should be evident that a crucial part of any system identification problem lies in making the vector $\boldsymbol{\alpha}$ of unknown parameters as small as possible. Care must also be taken to ensure that the unknown parameter $\boldsymbol{\alpha}$ can actually be estimated (in some sense).

There is no simple set of rules which enables one to perform the crucial initial stage of hypothesizing the model structure. This is the same basic type of problem discussed in Section 8.1 and the same rules apply. One must understand the physical process as well as possible and then exercise good engineering judgment.

14.2 State Augmentation: Nonlinear Estimation Theory

System identification by state augmentation was briefly mentioned in Chapter 2. This point of view is now considered again.

Consider the model (14.1.1). Assume that the model degree is known. Let $\boldsymbol{\alpha}$ denote the vector of all unknown system parameters (unknown values of $\boldsymbol{\Phi}$, $\mathbf{B}$, etc.). Then by using an augmented state vector $\tilde{\mathbf{x}}(n)$, the problem can be modeled as

$$\begin{aligned} \tilde{\mathbf{x}}(n) &= \begin{bmatrix} \mathbf{x}(n) \\ \boldsymbol{\alpha}(n) \end{bmatrix} \\ \boldsymbol{\alpha}(n+1) &= \boldsymbol{\alpha}(n) \\ \tilde{\mathbf{x}}(n+1) &= \boldsymbol{\phi}[\tilde{\mathbf{x}}(n), \mathbf{w}(n), \mathbf{u}(n)] \\ \mathbf{z}(n) &= \mathbf{h}[\tilde{\mathbf{x}}(n), \mathbf{v}(n)] \end{aligned} \tag{14.2.1}$$

EXAMPLE: Consider

$$x(n+1) = \Phi x(n) + w(n) + u(n)$$
$$z(n) = Hx(n) + v(n)$$
$$\Phi, H\text{: unknown}$$

Define

$$\boldsymbol{\alpha} = \begin{bmatrix} \alpha_1 \\ \alpha_2 \end{bmatrix} = \begin{bmatrix} \Phi \\ H \end{bmatrix}$$

Then the nonlinear model (14.2.1) is

$$\begin{aligned} x(n+1) &= \alpha_1(n)\, x(n) + w(n) + u(n) \\ \alpha_1(n+1) &= \alpha_1(n) \\ \alpha_2(n+1) &= \alpha_2(n) \\ z(n) &= \alpha_2(n)x(n) + v(n) \\ \tilde{\mathbf{x}}(n) &= \begin{bmatrix} x(n) \\ \alpha_1(n) \\ \alpha_2(n) \end{bmatrix} \end{aligned}$$

This state augmentation approach can be extended easily to the case of time-varying system parameters $\boldsymbol{\alpha}(n)$ which are modeled as stochastic or unknown-but-bounded processes. For example, if $\boldsymbol{\alpha}(n)$ is modeled as a stochastic process, (14.2.1) might be replaced by

$$\tilde{\mathbf{x}}(n) = \begin{bmatrix} \mathbf{x}(n) \\ \mathbf{y}(n) \end{bmatrix}$$

$$\mathbf{y}(n+1) = \boldsymbol{\theta}\mathbf{y}(n) + \mathbf{D}\mathbf{w}_y(n)$$

$$\boldsymbol{\alpha}(n) = \mathbf{L}\mathbf{y}(n)$$

where $\boldsymbol{\theta}$, $\mathbf{D}$, and $\mathbf{L}$ are specified matrices and $\mathbf{w}_y(n)$ is a white stochastic process with specified covariance matrix.

The problem of using the $\mathbf{z}(1) \cdots$ [and knowledge of the $\mathbf{u}(n)$, $n = 1, \ldots,$] to identify $\boldsymbol{\alpha}$ [or $\boldsymbol{\alpha}(n)$] can now be viewed as a nonlinear state estimation problem, and the discussions and concepts of Part IV can be applied. Thus in one sense, no further discussion of system identification is required, as the system identification problem has been transformed into a problem which has already been discussed extensively.

The discussions are not terminated at this point for the simple reason that Part IV did not provide any "best" one way to solve a nonlinear state estimation problem. A major conclusion of Part IV was that the best way to proceed depends heavily on the explicit nature of the problem. System identification leads to very special types of nonlinear estimation problems, so specialized discussions are needed.

In the sections to follow the state augmentation approach is not emphasized, as the author feels that it is much more appropriate to approach the system identification problem directly. However, there are special cases where state augmentation works very well.

EXAMPLE: Consider

$$x_1(n+1) = \Phi x_1(n) + w(n) + \mathrm{bu}(n)$$

$$z(n) = x_1(n) + v(n)$$

$$u(n); \quad n = 1 \ldots : \text{known input}$$

$$b \text{ is only unknown } (\alpha = b)$$

Define

$$\mathbf{x}(n+1) = \begin{bmatrix} x_1(n) \\ x_2(n) \end{bmatrix}$$

$$x_2(n+1) = x_2(n) = b$$

so

$$\mathbf{x}(n+1) = \begin{bmatrix} \Phi & u(n) \\ 0 & 1 \end{bmatrix} \mathbf{x}(n) + \begin{bmatrix} w(n) \\ 0 \end{bmatrix}$$

The desired estimator follows from Chapter 6.

However in such special cases, the techniques of the following sections also work very well.

14.3 Maximum Likelihood: Gaussian Model, General Case

Consider a stochastic uncertainty version of (14.1.1) so $\mathbf{z}(n) = 1 \ldots$ is a stochastic process which is completely specified except for the values of the $\boldsymbol{\alpha}$ vector. Such a problem formulation leads directly to the use of maximum likelihood concepts to try to find an estimate of $\boldsymbol{\alpha}$ (see Section 11.2). To be more specific, define

$$\mathbf{z}_N = \begin{bmatrix} \mathbf{z}(1) \\ \cdot \\ \cdot \\ \cdot \\ \mathbf{z}(N) \end{bmatrix}$$

$p(\mathbf{z}_N : \boldsymbol{\alpha})$: probability density of $\mathbf{z}_N$ for given $\boldsymbol{\alpha}$
$\xi(N : \boldsymbol{\alpha}) = \ln p(\mathbf{z}_N : \boldsymbol{\alpha})$: log likelihood function
$\hat{\boldsymbol{\alpha}}(N) =$ value of $\boldsymbol{\alpha}$ which maximizes $\xi(N : \boldsymbol{\alpha})$ for particular $\mathbf{z}_N$

Thus $\hat{\boldsymbol{\alpha}}(N)$ is the maximum likelihood estimate of the unknown system parameter $\boldsymbol{\alpha}$ given the observations $\mathbf{z}(1) \cdots \mathbf{z}(N)$. Note that it is being tacitly assumed that there are no structural errors; i.e., there actually exists some unknown $\boldsymbol{\alpha}$ for which the model corresponds exactly to $\mathbf{z}(n)$, $n = 1 \cdots$.

From (14.1.1) the assumed model is

$$\begin{aligned} \mathbf{x}(n+1) &= \boldsymbol{\Phi}\mathbf{x}(n) + \mathbf{G}\mathbf{w}(n) + \mathbf{B}\mathbf{u}(n) \\ \mathbf{z}(n) &= \mathbf{H}\mathbf{x}(n) + \mathbf{v}(n) \end{aligned}$$

$$\begin{aligned} &\mathbf{x}: K_1 \text{ vector} \\ &\mathbf{z}: K_2 \text{ vector} \\ &\mathbf{w}: K_3 \text{ vector} \\ &\mathbf{u}: K_4 \text{ vector} \end{aligned}$$

$$\mathbf{x}(0), \mathbf{w}(n), \mathbf{v}(n): \text{zero mean independent Gaussian} \tag{14.3.1}$$

$$E\{\mathbf{x}(0)\mathbf{x}'(0)\} = \boldsymbol{\psi}$$

$$E\{\mathbf{w}(n_1)\mathbf{w}'(n_2)\} = \begin{cases} \mathbf{Q} & n_1 = n_2 \\ \mathbf{0} & n_1 \neq n_2 \end{cases}$$

$$E\{\mathbf{v}(n_1)\mathbf{v}'(n_2)\} = \begin{cases} \mathbf{R} & n_1 = n_2 \\ \mathbf{0} & n_1 \neq n_2 \end{cases}$$

$\boldsymbol{\alpha}$: vector of unknown parameters (elements) of $\boldsymbol{\Phi}$, $\mathbf{B}$, $\mathbf{G}$, $\mathbf{H}$, $\mathbf{Q}$, $\mathbf{R}$

The degree of model (dimension of $\mathbf{x}$) is for the moment assumed to be known (see Section 14.3.5).

The equation for the log likelihood function $\xi(N : \boldsymbol{\alpha})$ can be written in different ways. The "filtering form" and "smoothing form" will be discussed.

14.3.1 Filtering Form of Likelihood Function

The filtering form of the equation for $\xi(n:\boldsymbol{\alpha})$ has already been derived in Section 10.1.1 but for ease of reference, the derivation is outlined here (see Section 10.1.1 for more details).

Using Bayes' rule

$$\xi(N:\boldsymbol{\alpha}) = \xi(N-1:\boldsymbol{\alpha}) + \ln p\{\mathbf{z}(N) \mid \mathbf{z}_{N-1}:\boldsymbol{\alpha}\}$$

Define

$\hat{\mathbf{z}}(n \mid n-1:\boldsymbol{\alpha})$: conditional expectation of $\mathbf{z}(n)$ given $\mathbf{z}(1) \cdots \mathbf{z}(n-1)$ for specified $\boldsymbol{\alpha}$

$\boldsymbol{\delta}_z(n:\boldsymbol{\alpha}) = \mathbf{z}(n) - \hat{\mathbf{z}}(n \mid n-1:\boldsymbol{\alpha})$

$\boldsymbol{\Sigma}_z(n \mid n-1:\boldsymbol{\alpha}) = E\{\boldsymbol{\delta}_z(n:\boldsymbol{\alpha})\boldsymbol{\delta}_z'(n:\boldsymbol{\alpha})\}$, where the expectation is taken assuming that $\mathbf{z}(1) \cdots$ is obtained from system with specified $\boldsymbol{\alpha}$.

Using the Gaussian assumption

$$\begin{aligned}
2\xi(N:\boldsymbol{\alpha}) &= \xi_{\text{bias}}(N:\boldsymbol{\alpha}) + \xi_{\text{observation}}(N:\boldsymbol{\alpha}) \\
\xi_{\text{bias}}(N:\boldsymbol{\alpha}) &= -NK_2 \ln(2\pi) - \sum_{n=1}^{N} \ln|\boldsymbol{\Sigma}_z(n \mid n-1:\boldsymbol{\alpha})| \\
\xi_{\text{observation}}(N:\boldsymbol{\alpha}) &= -\sum_{n=1}^{N} \boldsymbol{\delta}_z'(n:\boldsymbol{\alpha})\boldsymbol{\Sigma}_z^{-1}(n \mid n-1:\boldsymbol{\alpha})\boldsymbol{\delta}_z(n:\boldsymbol{\alpha})
\end{aligned} \tag{14.3.2}$$

The equations for $\hat{\mathbf{z}}$ and $\boldsymbol{\Sigma}_z$ are derived in Chapter 6. Expressing (6.2.7) and (6.3.1) in the present notation and including the effect of the input $\mathbf{Bu}(n)$ gives

$$\begin{aligned}
\hat{\mathbf{x}}(n+1 \mid n+1:\boldsymbol{\alpha}) &= \boldsymbol{\Phi}\hat{\mathbf{x}}(n \mid n:\boldsymbol{\alpha}) + \mathbf{Bu}(n) \\
&\quad + \boldsymbol{\Sigma}(n+1 \mid n:\boldsymbol{\alpha})\mathbf{H}'\boldsymbol{\Sigma}_z^{-1}(n+1 \mid n:\boldsymbol{\alpha}) \\
&\quad \times \boldsymbol{\delta}_z(n+1:\boldsymbol{\alpha}) \\
\boldsymbol{\delta}_z(n+1:\boldsymbol{\alpha}) &= \mathbf{z}(n+1) - \hat{\mathbf{z}}(n+1 \mid n:\boldsymbol{\alpha}) \\
\hat{\mathbf{z}}(n+1 \mid n:\boldsymbol{\alpha}) &= \mathbf{H}\boldsymbol{\Phi}\hat{\mathbf{x}}(n \mid n:\boldsymbol{\alpha}) + \mathbf{Bu}(n) \\
\boldsymbol{\Sigma}(n+1 \mid n+1:\alpha) &= \boldsymbol{\Sigma}(n+1 \mid n:\boldsymbol{\alpha}) \\
&\quad - \boldsymbol{\Sigma}(n+1 \mid n:\alpha)\mathbf{H}'\boldsymbol{\Sigma}_z^{-1}(n+1 \mid n:\boldsymbol{\alpha}) \\
&\quad \times \mathbf{H}\boldsymbol{\Sigma}(n+1 \mid n:\boldsymbol{\alpha}) \\
\boldsymbol{\Sigma}(n+1 \mid n:\boldsymbol{\alpha}) &= \boldsymbol{\Phi}\boldsymbol{\Sigma}(n \mid n:\boldsymbol{\alpha})\boldsymbol{\Phi}' + \mathbf{GQG}' \\
\boldsymbol{\Sigma}_z(n+1 \mid n:\boldsymbol{\alpha}) &= \mathbf{R} + \mathbf{H}\boldsymbol{\Sigma}(n+1 \mid n:\boldsymbol{\alpha})\mathbf{H}' \\
\boldsymbol{\Sigma}(0 \mid 0) &= \boldsymbol{\psi} \\
\hat{\mathbf{x}}(0 \mid 0) &= \mathbf{0}
\end{aligned} \tag{14.3.2 (continued)}$$

The dependence of the $\mathbf{\Phi}$, $\mathbf{B}$, $\mathbf{H}$, $\mathbf{Q}$ and $\mathbf{R}$ matrices on $\boldsymbol{\alpha}$ is not explicitly indicated.

14.3.2 Smoothing Form for Likelihood Function

Equation (14.3.2) is called the filtering form because it uses the filter estimate $\hat{\mathbf{x}}(n \mid n : \boldsymbol{\alpha})$. The smoothing form for $\xi(N : \boldsymbol{\alpha})$ involves the smoothed estimate $\hat{\mathbf{x}}(n \mid N : \boldsymbol{\alpha})$.

The smoothing form for the log likelihood function $\xi(N : \boldsymbol{\alpha})$ is given by

$$
\begin{aligned}
2\xi(N:\boldsymbol{\alpha}) &= \xi_{\text{bias}}(N:\boldsymbol{\alpha}) + \xi_{\text{observation}}(N:\boldsymbol{\alpha}) \\
\xi_{\text{bias}}(N:\boldsymbol{\alpha})&: \text{as in (14.3.2)} \\
\xi_{\text{observation}}(N:\boldsymbol{\alpha}) &= -\sum_{n=1}^{N} [\mathbf{z}(n) - \mathbf{H}\hat{\mathbf{x}}(n \mid N:\boldsymbol{\alpha})]'\mathbf{R}^{-1}[\mathbf{z}(n) \\
&\quad - \mathbf{H}\hat{\mathbf{x}}(n \mid N:\boldsymbol{\alpha})] - \sum_{n=1}^{N} [\hat{\mathbf{x}}(n \mid N:\boldsymbol{\alpha}) \\
&\quad - \mathbf{\Phi}\hat{\mathbf{x}}(n-1 \mid N:\boldsymbol{\alpha}) - \mathbf{B}\mathbf{u}(n-1)]' \\
&\quad \times [\mathbf{G}\mathbf{Q}\mathbf{G}']^{-1}[\hat{\mathbf{x}}(n \mid N:\boldsymbol{\alpha}) \\
&\quad - \mathbf{\Phi}\hat{\mathbf{x}}(n-1 \mid N:\boldsymbol{\alpha}) - \mathbf{B}\mathbf{u}(n-1)] \\
&\quad - \hat{\mathbf{x}}'(0 \mid N:\boldsymbol{\alpha})\boldsymbol{\psi}^{-1}\hat{\mathbf{x}}(0 \mid N:\boldsymbol{\alpha}) \\
\hat{\mathbf{x}}(n \mid N:\boldsymbol{\alpha})&: \text{conditional expectation of } \mathbf{x}(n) \text{ given} \\
&\quad \mathbf{z}_N \text{ for specified } \boldsymbol{\alpha}
\end{aligned}
\tag{14.3.3}
$$

Equations (14.3.2) and (14.3.3) are just alternate ways of writing $\xi(N: \boldsymbol{\alpha})$. Equations for the smoothed estimate $\hat{\mathbf{x}}(n \mid N)$ were discussed in Chapter 6. The possible problem of a nonexisting $[\mathbf{GQG}']^{-1}$ will be discussed shortly.

Equation (14.3.3) could be derived by showing that the $\xi_{\text{observation}}(N : \boldsymbol{\alpha})$ terms in (14.3.2) and (14.3.3) are equal. Such equality was actually stated without formal proof in Section 10.3 (see (10.3.5)(10.3.6) for continuous time version). However a different type of a derivation of (14.3.3) will be outlined here. As a first step, consider $p[\mathbf{z}_N, \mathbf{x}_N]$ where

$$
\mathbf{x}_N = \begin{bmatrix} \mathbf{x}(0) \\ \cdot \\ \cdot \\ \cdot \\ \mathbf{x}(N) \end{bmatrix}
$$

By Bayes' rule

$$
p[\mathbf{z}_N, \mathbf{x}_N] = p[\mathbf{x}_N \mid \mathbf{z}_N]p[\mathbf{z}_N]
$$

Then using (13.2.12) and specializing as in (13.2.13) to the linear Gaussian case gives

$$p[\mathbf{z}_N, \mathbf{x}_N] = c \exp\Big\{-\Big(\frac{1}{2}\Big)\Big[\sum_{n=1}^{N} [\mathbf{z}(n) - \mathbf{H}\mathbf{x}(n)]'\mathbf{R}^{-1}[\mathbf{z}(n) - \mathbf{H}\mathbf{x}(n)]$$
$$+ \sum_{n=1}^{N} [\mathbf{x}(n) - \mathbf{\Phi}\mathbf{x}(n-1) - \mathbf{B}\mathbf{u}(n-1)]'[\mathbf{GQG}']^{-1}$$
$$\times [\mathbf{x}(n) - \mathbf{\Phi}\mathbf{x}(n-1) - \mathbf{B}\mathbf{u}(n-1)] + \mathbf{x}'(0)\boldsymbol{\psi}^{-1}\mathbf{x}(0)\Big]\Big\}$$
$$c \text{ is constant which depends on } \mathbf{R}, \mathbf{Q}, \boldsymbol{\psi} \tag{14.3.4}$$

assuming again that $[\mathbf{GQG}']^{-1}$ exists. The second step is to combine equations of Appendix E with tedious manipulation to prove that for any zero mean Gaussian random vectors $\mathbf{z}$ and $\mathbf{x}$,

$$E\left\{\begin{bmatrix}\mathbf{z}\\ \mathbf{x}\end{bmatrix}[\mathbf{z}' \quad \mathbf{x}']\right\} = \mathbf{\Gamma} = \begin{bmatrix}\mathbf{\Gamma}_{zz} & \mathbf{\Gamma}_{zx}\\ \mathbf{\Gamma}_{xz} & \mathbf{\Gamma}_{xx}\end{bmatrix}$$
$$\mathbf{z} : K_5 \text{ vector}$$
$$\mathbf{x} : K_6 \text{ vector}$$

that

$$p[\mathbf{z}] = \left[\frac{|\mathbf{\Gamma}|2\pi^{K_6}}{|\mathbf{\Gamma}_{zz}|}\right]^{1/2} p[\mathbf{z}, \mathbf{x}]\Big|_{\mathbf{x}=\hat{\mathbf{x}}(\mathbf{z})} \tag{14.3.5}$$
$$\hat{\mathbf{x}}(\mathbf{z}): \text{ conditional expectation of } \mathbf{x} \text{ given } \mathbf{z}$$

Applying (14.3.5) to (14.3.4) and taking logarithms gives

$$\xi(N:\boldsymbol{\alpha}) = \text{bias term} + \xi_{\text{observation}}(N:\boldsymbol{\alpha}) \tag{14.3.6}$$

where $\xi_{\text{observation}}(N:\boldsymbol{\alpha})$ is as in (14.3.3). A lot of clever and tedious manipulation of determinates of partitioned matrices shows that the bias term of (14.3.6) is equal to $\xi_{\text{bias}}(N:\boldsymbol{\alpha})$ of (14.3.2).

Now consider the question of the possible non-existence of $[\mathbf{GQG}']^{-1}$. Equation (14.3.3) can be rewritten so that the problem goes away, but it is desired to maintain the given form, as it is convenient for certain applications to be discussed later. Thus, one of various possible approximate approaches to using (14.3.3) in its present form is discussed. Consider a simple example of two-dimensional state $\mathbf{x}(n)$, and a scalar input $w_1(n)$, where

$$G = \begin{bmatrix}G_1\\ G_2\end{bmatrix}$$
$$E[w_1^2(n)] = Q_1$$

Clearly $[\mathbf{GQ}_1\mathbf{G}']^{-1}$ does not exist. Define

$$\tilde{\mathbf{G}}\tilde{\mathbf{w}}(n) = \begin{bmatrix}G_1 & 0\\ G_2 & 1\end{bmatrix}\begin{bmatrix}w_1(n)\\ w_2(n)\end{bmatrix}$$
$$E\left\{\begin{bmatrix}w_1(n)\\ w_2(n)\end{bmatrix}[w_1(n)w_2(n)]\right\} = \begin{bmatrix}Q_1 & 0\\ 0 & Q_2\end{bmatrix} = \tilde{\mathbf{Q}}$$

As $Q_2 \longrightarrow 0$, $\tilde{\mathbf{G}}\,\tilde{\mathbf{w}}(n) \longrightarrow \mathbf{G}w_1(n)$. Now as long as $Q_2 \neq 0$, $[\tilde{\mathbf{G}}\tilde{\mathbf{Q}}\tilde{\mathbf{G}}']^{-1}$ exists; in fact,

$$[\tilde{\mathbf{G}}\tilde{\mathbf{Q}}\tilde{\mathbf{G}}']^{-1} = \frac{1}{G_1^2}\begin{bmatrix} Q_1^{-1} + G_2^2 Q_2^{-1} & G_1 G_2 Q_2^{-1} \\ G_1 G_2 Q_2^{-1} & G_1^2 Q_2^{-1} \end{bmatrix}$$

Thus, one approach is to replace the non-existent $[\mathbf{GQG}']^{-1}$ with $[\tilde{\mathbf{G}}\tilde{\mathbf{Q}}\tilde{\mathbf{G}}']^{-1}$, where Q_2 is chosen to be very small. This idea generalizes directly to a K_1 − dimensional state and a K_3 − dimensional input uncertainty.

14.3.3 Iterative Solutions: Off Line

Equation (14.3.2) or (14.3.3) provides a method of evaluating $\xi(N:\boldsymbol{\alpha})$ for a specified value of $\boldsymbol{\alpha}$. There remains the difficult problem of finding the value $\hat{\boldsymbol{\alpha}}(N)$ which maximizes $\xi(N:\boldsymbol{\alpha})$. The basic approach is to develop an iterative sequence $\boldsymbol{\alpha}_i$, $i = 1 \cdots$ for fixed N which converges to the maximum likelihood estimate $\hat{\boldsymbol{\alpha}}(N)$. Such an iterative procedure can be used for off-line (nonreal time) identification where the observations $\mathbf{z}(1) \cdots \mathbf{z}(N)$ are stored and can be continually reprocessed.

Consider Section 11.7 where the problem was to find a value of $\mathbf{x}$ which minimizes $J(\mathbf{x})$. In the present case, the problem is to find a value of $\boldsymbol{\alpha}$ which minimizes $-\xi(N:\boldsymbol{\alpha})$. Hence a logical approach is to choose some initial guess $\boldsymbol{\alpha}_0$ and then perform an iteration as in Section 11.7; i.e.,

$$\boldsymbol{\alpha}_{i+1} = \boldsymbol{\alpha}_i + \mathbf{B}(i)[\boldsymbol{\xi}^{(1)}(N:\boldsymbol{\alpha}_i)]'$$

$$\boldsymbol{\xi}^{(1)}(N:\boldsymbol{\alpha}_i) = \left.\frac{\partial \xi(N:\boldsymbol{\alpha})}{\partial \boldsymbol{\alpha}}\right|_{\boldsymbol{\alpha}=\boldsymbol{\alpha}_i} \tag{14.3.7}$$

$\mathbf{B}(i)$: gain matrix

which hopefully converges $\boldsymbol{\alpha}_i \longrightarrow \hat{\boldsymbol{\alpha}}(N)$. (14.3.7) looks very simple but in practice the evaluation of the gradient $\xi^{(1)}(N:\boldsymbol{\alpha})$ of the log likelihood function and acceptable gains $\mathbf{B}(l)$ can be very tedious. (Note the $\mathbf{B}(i)$ of (14.3.7) is obviously not the $\mathbf{B}$ of the model(14.3.1).) The degree of difficulty of course depends on the specific problem of interest.

If the dimension of $\boldsymbol{\alpha}$ (number of unknown parameters) is small, one other path is to simply evaluate $\xi(N:\boldsymbol{\alpha})$ for many different values of $\boldsymbol{\alpha}$ which span the range of possible $\boldsymbol{\alpha}$. This approach is related to the parameterization concepts of Section 11.8.

Thus far the discussions apply equally to either form for $\xi(N:\boldsymbol{\alpha})$ (14.3.2) or (14.3.3) but the filtering form (14.3.2) would probably be used as it is easier to implement. However, the smoothing form of (14.3.3) leads to a special type of iterative solution. Consider $\xi_{\text{observation}}(N:\boldsymbol{\alpha})$ of (14.3.3). Assume for the moment that $\hat{\mathbf{x}}(n \mid N:\boldsymbol{\alpha})$ $n = 1 \ldots N$ is specified and $\mathbf{Q}$, $\mathbf{G}$, and $\mathbf{R}$ are known so $\boldsymbol{\alpha}$ contains only elements of $\boldsymbol{\Phi}$, $\mathbf{B}$, and $\mathbf{H}$. Then the equation

$$\frac{\partial \xi_{\text{observation}}(N : \boldsymbol{\alpha})}{\partial \boldsymbol{\alpha}} = \mathbf{0} \qquad \text{for fixed } \hat{\mathbf{x}}(n \mid N : \boldsymbol{\alpha}) \tag{14.3.8}$$

yields a set of equations which are linear in $\boldsymbol{\alpha}$ and hence can be solved explicitly. Thus when **G**, **Q**, and **R** are known, the following iteration is a possible way of finding a value of $\boldsymbol{\alpha}$ which minimizes $-\xi_{\text{observation}}(N : \boldsymbol{\alpha})$.

1. Make initial guess $\boldsymbol{\alpha}_0$ and compute $\hat{\mathbf{x}}(n \mid N : \boldsymbol{\alpha}_0)$ $n = 1 \cdots N$ (14.3.9)
2. Solve the linear equations (14.3.8) using $\hat{\mathbf{x}}(n \mid N : \boldsymbol{\alpha}_0)$ to get a new value $\boldsymbol{\alpha}_1$.
3. Compute $\hat{\mathbf{x}}(n \mid N : \boldsymbol{\alpha}_1)$.

etc.

This yields a sequence $\boldsymbol{\alpha}_i$ with a corresponding $-\xi_{\text{observation}}(N : \boldsymbol{\alpha}_i)$ which never increases. This iteration can be easier to implement than (14.3.7), but it has several shortcomings. First, **G**, **Q**, and **R** are assumed to be known. Second, it yields the $\boldsymbol{\alpha}$ which minimizes $-\xi_{\text{observation}}(N : \boldsymbol{\alpha})$ and this is not necessarily the maximum likelihood estimate which minimizes $-\xi(N : \boldsymbol{\alpha})$. From a practical point of view, these shortcomings are not necessarily crucial. If **G**, **Q** and **R** contain unknowns, ad hoc techniques for estimating them may suffice, and the effect of the neglected $\xi_{\text{bias}}(N : \boldsymbol{\alpha})$ term may not be important. If so, the iteration (14.3.9) can be viewed as a way to obtain an approximation to the true maximum likelihood estimate. In special cases it is possible to modify the basic logic underlying (14.3.9) to obtain a true maximum likelihood estimate.

14.3.4 Recursive Feedback Solutions: On Line

In some applications, on-line real-time operation is desired where estimates $\hat{\boldsymbol{\alpha}}(N)$ are to be obtained for increasing N in real time. An iteration as in Section 14.3.3 would suffice if the computer was big and fast enough but usually some sort of recursive logic which does not require storage and reprocessing of $\mathbf{z}(1) \ldots \mathbf{z}(N)$ is required. In special cases this is possible (see Section 14.4) but in general the resulting estimate is some (hopefully close) approximation to the true maximum likelihood estimate. K. Åström (see bibliography) has called the problem of obtaining a satisfactory recursive algorithm a "fiddler's paradise". The following discussions are intended to help stimulate the fiddler's imagination.

One logical approach is to try to turn an iteration into a recursion by applying the logic of Section 13.10 to the present problem. The state augmentation ideas of Section 14.2 may be helpful. The feedback tracking formulation illustrated by Figure 13.10.1 can be a good point of view to employ.

The choice of gains is of course difficult but may be possible if approximate values of $\boldsymbol{\alpha}$ are known a priori.

If the unknown parameters $\boldsymbol{\alpha}$ are modeled as constants as in (14.3.1), the iterative solutions of Section 14.3.3 will often yield an estimate $\hat{\boldsymbol{\alpha}}(N)$ which for large N is independent of N; i.e. $\hat{\boldsymbol{\alpha}}(N) \approx \hat{\boldsymbol{\alpha}}(N+1)$ for large N. If the same behavior is desired for a recursive feedback solution, the algorithm has to be chosen so that $\hat{\boldsymbol{\alpha}}(N) \longrightarrow$ constant for large N. However in some applications such convergence to a final answer is not desired. For example, assume the unknown system parameters are slowly time varying, i.e. $\boldsymbol{\alpha} = \boldsymbol{\alpha}(n)$. Then the algorithm should be chosen so that $\hat{\boldsymbol{\alpha}}(N)$ tracks the actual $\boldsymbol{\alpha}(N)$. Thus in terms of state estimation, the identifier should have limited or finite memory. There are various ways to do this, see corresponding discussions in Sections 8.2.3 and 10.6.

The iteration of (14.3.9) based on the smoothing form (14.3.3) leads to an interesting class of recursive algorithms. Equation (14.3.3) requires knowledge of $\hat{\mathbf{x}}(n \mid N)$ but in many cases $\hat{\mathbf{x}}(n \mid n+m)$ is a good approximation to $\hat{\mathbf{x}}(n \mid N)$ for sufficiently large m. The $\hat{\mathbf{x}}(n \mid n+m)$ can be generated recursively using a "fixed lag" smoothing estimator. Such thinking leads to the type of recursive algorithm indicated in Figure 14.3.1.

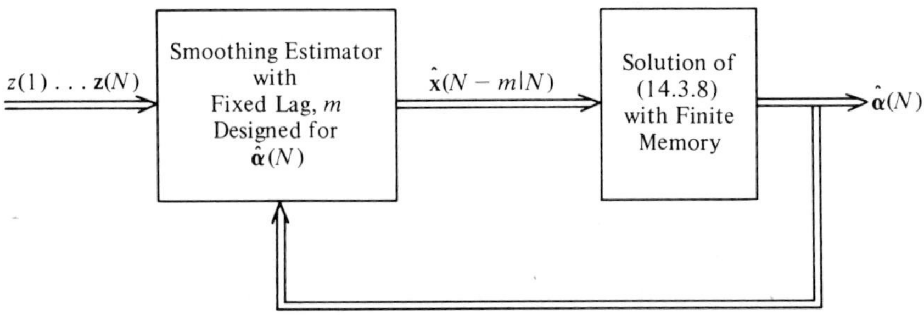

Figure 14.3.1 Possible structure for recursive algorithm.

There are many more possibilities. May your fiddling bring forth beautiful music!

14.3.5 Determining Degree of Model

In Sections 14.3.1–14.3.4 the degree of the model (dimension K_1 of $\mathbf{x}$) is assumed to be specified. If the model degree is not known, separate structures for "all possible" model degrees can be hypothesized and then tested. For example, if the model degree K_1 can be 1, 2, or 3, one tests the three hypotheses

$$\mathcal{H}_j: K_1 = j \qquad j = 1, \ldots, 3$$

to see which is best.

One way to do this is as follows. The values of the log likelihood functions for the maximizing value of $\boldsymbol{\alpha}$ are evaluated for various possible model degrees. The degree which yields the maximum value is then chosen. To be more explicit, define:

$\xi(N:\boldsymbol{\alpha}, K_1)$: log likelihood for model of degree K_1

$\hat{\boldsymbol{\alpha}}(N, K_1)$: maximum likelihood estimate of $\boldsymbol{\alpha}$ given $\mathbf{z}(1) \cdots \mathbf{z}(N)$ and K_1 degree model

$\hat{K}_1$: value of K_1 which yields maximum $\xi[N:\hat{\boldsymbol{\alpha}}(N, K_1), K_1]$

Then $\hat{K}_1$ is the model degree to be used. This technique is closely related in concept to the ideas of Section 6.16.4 on adaptive curve fitting, where the degree of a curve-fitting polynomial was determined from the observations.

14.3.6 Structural Errors: Goodness of Fit

In practice, the assumed model structure is usually not perfect; i.e., there exist no values of $\boldsymbol{\alpha}$ and model degree such that (14.3.1) corresponds to the actual observed $\mathbf{z}(1) \cdots$. Therefore it is necessary to check the "goodness of fit" of the model to the observations.

The needed basic idea comes from the hypothesis testing concepts of Chapter 9—in particular from a combination of the ideas of Sections 9.4.2 and 9.4.3. Assume that the model degree is fixed. Define two hypotheses:

$\mathcal{H}_1$: $\mathbf{z}(1) \cdots \mathbf{z}(N)$ is generated from (14.3.1) for some value of $\boldsymbol{\alpha}$

$\mathcal{H}_2$: $\mathcal{H}_1$ is not true

One way to decide between these two hypotheses is to relate the actual value of $\xi[N:\hat{\boldsymbol{\alpha}}(N)]$ to its probability density evaluated assuming that $\mathcal{H}_1$ is true. If the actual value lies in the tail of the distribution, $\mathcal{H}_1$ is rejected. In other words, $\xi[N:\hat{\boldsymbol{\alpha}}(N)]$ is a measure of how close the structure (14.3.1) can fit the actual observations. If this fit is not good enough, the structure is not valid. Other tests are of course possible. One alternate approach is to test the whiteness of the residuals under $\mathcal{H}_1$ (see Section 10.8). This whiteness test has proven to be very powerful in practice.

14.3.7 Discussion

The maximum likelihood concept provides a viable, general, and very powerful approach to estimating the numerical values of a linear system with Gaussian processes. The parameters to be identified may specify the system's structure ($\boldsymbol{\Phi}$, $\mathbf{B}$, $\mathbf{G}$, $\mathbf{H}$ matrices) or the uncertainties ($\mathbf{Q}$, $\mathbf{R}$ matrices) or combinations thereof. The degree of model can be determined from the observations. The goodness of the fit can be checked to determine if structure modeling errors are important.

An obvious question is, How crucial is the Gaussian assumption? A detailed mathematical answer will not be given. However as a general rule of thumb: *If many observations are available, the Gaussian assumption is not crucial*. In other words, if the number of observations N is large, all the preceding formulas can be used with "confidence" even if the Gaussian assumption is not satisfied. This is a theme that has been repeated many times in this book.

There is, of course, one drawback in the above development. The general equations are quite (very) complicated and no one best way of calculating $\hat{\boldsymbol{\alpha}}(N)$ was given. In the next section, various special cases which yield simpler equations and explicit formulae for the maximum likelihood $\hat{\boldsymbol{\alpha}}(N)$ are discussed.

14.4 Maximum Likelihood: Gaussian Model, Special Cases

Some explicit special cases of maximum likelihood are now considered.

14.4.1 Perfect Observations of Complete State

Consider the special case of (14.3.1) where

$$\mathbf{x}(n+1) = \mathbf{\Phi x}(n) + \mathbf{Gw}(n) + \mathbf{Bu}(n)$$

$$\mathbf{z}(n) = \mathbf{x}(n)$$

$$\mathbf{x}(0) = 0$$

$$\mathbf{G} = \mathbf{I}$$

$$\left.\begin{aligned} \mathbf{Q}&: \text{unknown} \\ \mathbf{\Phi}, \mathbf{B}&: \text{partially unknown} \end{aligned}\right\} \text{so } \boldsymbol{\alpha} \text{ contains } \mathbf{Q} \text{ and elements of } \mathbf{B}, \mathbf{\Phi}$$

Since $\mathbf{z}(n) = \mathbf{x}(n)$, it follows that

$$\hat{\mathbf{z}}(n \mid n-1 : \boldsymbol{\alpha}) = \mathbf{\Phi z}(n-1) + \mathbf{Bu}(n-1)$$

$$\boldsymbol{\delta}_z(n : \boldsymbol{\alpha}) = \mathbf{z}(n) - \mathbf{\Phi z}(n-1) - \mathbf{Bu}(n-1)$$

$$\mathbf{\Sigma}_z(n \mid n-1 : \boldsymbol{\alpha}) = \mathbf{Q}$$

Thus from (14.3.2)

$$2\xi(N : \boldsymbol{\alpha}) = -NK_2 \ln(2\pi) - N \ln|\mathbf{Q}| - \sum_{n=1}^{N} \boldsymbol{\delta}_z'(n : \boldsymbol{\alpha})\mathbf{Q}^{-1}\boldsymbol{\delta}_z(n : \boldsymbol{\alpha})$$

or equivalently

$$2\xi(N : \boldsymbol{\alpha}) = -NK_2 \ln(2\pi) - N \ln|\mathbf{Q}| - \operatorname{tr}\left\{\mathbf{Q}^{-1}\left[\sum_{n=1}^{N} \boldsymbol{\delta}_z(n : \boldsymbol{\alpha})\boldsymbol{\delta}_z'(n : \boldsymbol{\alpha})\right]\right\}$$

The estimate $\hat{\mathbf{Q}}$ must satisfy

$$\frac{\partial \xi(N:\boldsymbol{\alpha})}{\partial \mathbf{Q}} = \mathbf{0}$$

Thus using the formulas of Appendix C

$$\hat{\mathbf{Q}} = \frac{1}{N} \sum_{n=1}^{N} \boldsymbol{\delta}_z(n:\boldsymbol{\alpha})\, \boldsymbol{\delta}_z'(n:\boldsymbol{\alpha}), \qquad \text{for } \boldsymbol{\alpha} = \hat{\boldsymbol{\alpha}}$$

Substitution back into the equation for $\xi(N:\boldsymbol{\alpha})$ leads to the following desired result, which specifies the system parameter estimates $\hat{\mathbf{Q}}$, $\hat{\boldsymbol{\Phi}}$, and $\hat{\mathbf{B}}$ for the case of perfect observations of the complete state:

$$\boxed{\begin{aligned} \hat{\mathbf{Q}} &= \frac{1}{N} \sum_{n=1}^{N} [\mathbf{z}(n) - \hat{\boldsymbol{\Phi}}\mathbf{z}(n-1) - \hat{\mathbf{B}}\mathbf{u}(n-1)] \\ &\quad \times [\mathbf{z}(n) - \hat{\boldsymbol{\Phi}}\mathbf{z}(n-1) - \hat{\mathbf{B}}\mathbf{u}(n-1)]' \\ &\text{where } \hat{\boldsymbol{\Phi}} \text{ and } \hat{\mathbf{B}} \text{ minimize} \\ \xi &= \left| \sum_{n=1}^{N} [\mathbf{z}(n) - \boldsymbol{\Phi}\mathbf{z}(n-1) - \mathbf{B}\mathbf{u}(n-1)] \right. \\ &\quad \left. \times [\mathbf{z}(n) - \boldsymbol{\Phi}\mathbf{z}(n-1) - \mathbf{B}\mathbf{u}(n-1)]' \right| \end{aligned}} \tag{14.4.1}$$

The result (14.4.1) was derived assuming that $\mathbf{Q}$ was unknown. If the value of $\mathbf{Q}$ is known, $\boldsymbol{\alpha}$ contains only elements of $\boldsymbol{\Phi}$ and $\mathbf{B}$ and the result is

$$\begin{aligned} &\hat{\boldsymbol{\Phi}} \text{ and } \hat{\mathbf{B}} \text{ minimize} \\ \xi &= \sum_{n=1}^{N} [\mathbf{z}(n) - \boldsymbol{\Phi}\mathbf{z}(n-1) - \mathbf{B}\mathbf{u}(n-1)]'\mathbf{Q}^{-1}[\mathbf{z}(n) \\ &\quad - \boldsymbol{\Phi}\mathbf{z}(n-1) - \mathbf{B}\mathbf{u}(n-1)] \end{aligned} \tag{14.4.2}$$

Setting the partial derivative of ξ of (14.4.2) with respect to the unknown values, $\boldsymbol{\alpha}$ of $\boldsymbol{\Phi}$ and $\mathbf{B}$ equal to zero results in a system of linear equations, specifying $\hat{\boldsymbol{\alpha}}$ and hence $\hat{\boldsymbol{\Phi}}$ and $\hat{\mathbf{B}}$. These equations for $\hat{\boldsymbol{\alpha}}$ can be solved recursively in time if desired.

EXAMPLE: Consider

$$\begin{aligned} x(n+1) &= \Phi x(n) + w(n) + u(n) \\ x(0) &= 0 \\ z(n) &= x(n) \qquad n = 0, 1, \ldots N \\ \boldsymbol{\alpha} &= \begin{bmatrix} \Phi \\ Q \end{bmatrix} \end{aligned}$$

From (14.4.1)

$$\xi = \sum_{n=1}^{N} [z(n) - \Phi z(n-1) - u(n-1)]^2$$

$$\frac{\partial \xi}{\partial \Phi} = -2 \sum_{n=1}^{N} z(n-1)[z(n) - \Phi z(n-1) - u(n-1)] = 0$$

so

$$\hat{\Phi} = \frac{\sum_{n=1}^{N} z(n-1)[z(n) - u(n-1)]}{\sum_{n=1}^{N} z^2(n-1)}$$

$$\hat{Q} = \frac{1}{N}\sum_{n=1}^{N} (z(n) - \hat{\Phi}z(n-1) - u(n))^2$$

This example is also a special case of (14.4.3) to be derived in Section 14.4.2.

14.4.2 Autoregressive Model

Consider the special case of (14.3.1) where

$$y(n) = \Phi_1 y(n-1) + \Phi_2 y(n-2) \dots + \Phi_{K_1} y(n-K_1) + w(n-1) + bu(n)$$

$$z(n) = y(n)$$

$w(n)$: zero mean white Gaussian process

$$E\{w(n_1)w(n_2)\} = \begin{cases} Q & n_1 = n_2 \\ 0 & n_1 \neq n_2 \end{cases}$$

In a more concise notation

$$y(n) = \boldsymbol{\theta}'\mathbf{y}_{n,K_1} + w(n-1) + bu(n)$$

$$\boldsymbol{\theta} = \begin{bmatrix} \Phi_1 \\ \cdot \\ \cdot \\ \cdot \\ \Phi_{K_1} \end{bmatrix} \qquad \mathbf{y}_{n,K_1} = \begin{bmatrix} y(n-1) \\ \cdot \\ \cdot \\ \cdot \\ y(n-K_1) \end{bmatrix}$$

To simplify discussion of initial conditions, assume that

$$u(0) = 0$$

$$y(0) = y(-1) = y(-2) = \cdots = 0$$

$$z(0) = z(-1) = z(-2) = \cdots = 0$$

Assume that Q, $\boldsymbol{\theta}$, and b are unknown so

$$\boldsymbol{\alpha} = \begin{bmatrix} b \\ \boldsymbol{\theta} \\ Q \end{bmatrix}$$

Since $w(n)$ is a white process it follows that

$$\hat{z}(n \mid n-1) = \boldsymbol{\theta}'\mathbf{z}_{n,K_1} + bu(n-1)$$

$$\mathbf{z}_{n,K_1} = \begin{bmatrix} z(n-1) \\ \cdot \\ \cdot \\ \cdot \\ z(n-K_1) \end{bmatrix}$$

$$\boldsymbol{\Sigma}_z(n \mid n-1 : \boldsymbol{\alpha}) = Q$$

and hence from (14.3.2)

$$2\xi(N:\alpha) = -NK_2 \ln(2\pi) - N \ln|Q| - \sum_{n=1}^{N} [z(n) - \boldsymbol{\theta}'\mathbf{z}_{n,K_1} - bu(n-1)]^2 Q^{-1}$$

Therefore the estimates $\hat{\boldsymbol{\theta}}$, $\hat{b}$ and $\hat{Q}$ satisfy

$$\frac{\partial\xi(N:\boldsymbol{\alpha})}{\partial b} = \sum_{n=1}^{N} [z(n) - \boldsymbol{\theta}'\mathbf{z}_{n,K_1} - bu(n-1)]u(n-1) = 0$$

$$\frac{\partial\xi(N:\boldsymbol{\alpha})}{\partial\boldsymbol{\theta}} = \sum_{n=1}^{N} [z(n) - \boldsymbol{\theta}'\mathbf{z}_{n,K_1} - bu(n-1)]\mathbf{z}_{n,K_1} = \mathbf{0}$$

$$\frac{2\xi(N:\boldsymbol{\alpha})}{\partial Q} = \frac{N}{Q} - \frac{\sum_{n=1}^{N} [z(n) - \boldsymbol{\theta}'\mathbf{z}_{n,K_1} - bu(n-1)]^2}{Q^2}$$

so that

$$\begin{bmatrix} \hat{b} \\ \hat{\boldsymbol{\theta}} \end{bmatrix} = \mathbf{A}^{-1}\mathbf{a}$$

$$\mathbf{A} = \begin{bmatrix} \sum_{n=1}^{N} u^2(n-1) & \sum_{n=1}^{N} u(n-1)\mathbf{z}'_{n,K_1} \\ \sum_{n=1}^{N} \mathbf{z}_{n,K_1}u(n-1) & \sum_{n=1}^{N} \mathbf{z}_{n,K_1}\mathbf{z}'_{n_1K_1} \end{bmatrix} \tag{14.4.3}$$

$$\mathbf{a} = \begin{bmatrix} \sum_{n=1}^{N} z(n)u(n-1) \\ \sum_{n=1}^{N} \mathbf{z}_{n,K_1}z(n) \end{bmatrix}$$

$$\hat{Q} = \frac{1}{N} \sum_{n=1}^{N} [z(n) - \hat{\boldsymbol{\theta}}'\mathbf{z}_{n,K_1} - \hat{b}u(n-1)]^2$$

providing that $\mathbf{A}^{-1}$ exists. (See example of Section 14.4.1.)

If on-line operation is desired, a recursive relation between $\mathbf{A}^{-1}(N+1)$ [the $\mathbf{A}$ of (14.4.3) for $N+1$ observation] and $\mathbf{A}^{-1}(N)$ can be developed using the matrix identity in Appendix A, (A.3).

Equation (14.4.3) applies to a simple autoregressive model with scalar input and output. It is relatively straightforward to repeat the derivation for the more general model:

$$\mathbf{y}(n) = \boldsymbol{\Phi}_1\mathbf{y}(n-1) + \boldsymbol{\Phi}_2\mathbf{y}(n-2) + \cdots + \boldsymbol{\Phi}_{K_1}\mathbf{y}(n-K_1) + \mathbf{w}(n-1)$$
$$+ \mathbf{B}_1\mathbf{u}(n) + \mathbf{B}_2\mathbf{u}(n-1) + \cdots \mathbf{B}_{K_2}\mathbf{u}(n-K_2-1)$$

$$\mathbf{z}(n) = \mathbf{y}(n)$$

However, the ideas do not extend readily to mixed autoregressive moving average models such as

$$y(n) = \boldsymbol{\Phi}_1 y(n-1) + \boldsymbol{\Phi}_2 y(n-2) + \cdots + \boldsymbol{\Phi}_{K_1} y(n-K_1) + \mathbf{G}_1 w(n-1)$$
$$+ \mathbf{G}_2 w(n-2) + \cdots + \mathbf{G}_{K_2} w(n-K_2) + \mathbf{B}u(n)$$

14.4.3 No Input Uncertainty

Consider the special case of (14.3.1) where

$$
\begin{aligned}
\mathbf{x}(n+1) &= \mathbf{\Phi x}(n) + \mathbf{Bu}(n) \\
\mathbf{z}(n) &= \mathbf{Hx}(n) + \mathbf{v}(n) \\
\mathbf{x}(0) &= 0
\end{aligned}
$$

$$
\left.\begin{aligned}
\mathbf{R}&\text{: unknown} \\
\mathbf{H}, \mathbf{\Phi}, \mathbf{B}&\text{: partially unknown}
\end{aligned}\right\} \text{so } \boldsymbol{\alpha} \text{ contains } \mathbf{R} \text{ and elements of } \mathbf{H}, \mathbf{\Phi}, \mathbf{B}.
$$

Since there is no input or initial condition uncertainty, it follows that

$$
\begin{aligned}
\hat{\mathbf{z}}(n \mid n-1) &= \mathbf{H}\hat{\mathbf{x}}(n) \\
\hat{\mathbf{x}}(n+1) &= \mathbf{H}\,\hat{\mathbf{x}}(n) + \mathbf{Bu}(n) \\
\hat{\mathbf{x}}(0) &= \mathbf{0} \\
\boldsymbol{\delta}_z(n:\boldsymbol{\alpha}) &= \mathbf{z}(n) - \mathbf{H}\hat{\mathbf{x}}(n) \\
\mathbf{\Sigma}_z(n \mid n-1:\boldsymbol{\alpha}) &= \mathbf{R}
\end{aligned}
$$

Thus (14.3.2) becomes

$$
\xi(N:\boldsymbol{\alpha}) = -NK_2 \ln(2\pi) - N \ln|\mathbf{R}| - \operatorname{tr}\left\{\mathbf{R}^{-1} \sum_{n=1}^{N} \boldsymbol{\delta}_z(n:\boldsymbol{\alpha})\boldsymbol{\delta}_z'(n:\boldsymbol{\alpha})\right\}
$$

Following the steps used in Section 14.4.1, yields

$$
\boxed{
\begin{aligned}
&\hat{\mathbf{R}} = \frac{1}{N} \sum_{n=1}^{N} [\mathbf{z}(n) - \hat{\mathbf{H}}\hat{\mathbf{x}}(n)][\mathbf{z}(n) - \hat{\mathbf{H}}\hat{\mathbf{x}}(n)]' \\
&\hat{\mathbf{x}}(n+1) = \hat{\mathbf{\Phi}}\hat{\mathbf{x}}(n) + \hat{\mathbf{B}}\mathbf{u}(n) \\
&\hat{\mathbf{x}}(0) = 0 \\
&\text{where } \hat{\mathbf{\Phi}}, \hat{\mathbf{B}} \text{ and } \hat{\mathbf{H}} \text{ are values which minimize} \\
&\xi = \left| \sum_{n=1}^{N} [\mathbf{z}(n) - \hat{\mathbf{H}}\hat{\mathbf{x}}(n)][\mathbf{z}(n) - \hat{\mathbf{H}}\hat{\mathbf{x}}(n)]' \right|
\end{aligned}
}
\tag{14.4.4}
$$

Like (14.4.1) the result (14.4.4) is not overly surprising. Physically, it merely says to find the system parameter values ($\hat{\mathbf{\Phi}}$, $\hat{\mathbf{B}}$, $\hat{\mathbf{H}}$) which yield an output $\hat{\mathbf{H}}\hat{\mathbf{x}}(n)$ which is closest to $\mathbf{z}(n)$ when the input is $\mathbf{u}(\mathbf{n})$. (See also Section 14.5.1.)

Equation (14.4.4) assumes that $\mathbf{R}$ is unknown. In the case where $\mathbf{R}$ is known, (14.4.4) becomes

$$
\begin{gathered}
\hat{\mathbf{\Phi}}, \hat{\mathbf{B}}, \text{ and } \hat{\mathbf{H}} \text{ minimize} \\
\xi = \sum_{n=1}^{N} [\mathbf{z}(n) - \hat{\mathbf{H}}\hat{\mathbf{x}}(n)]'\mathbf{R}^{-1}[\mathbf{z}(n) - \hat{\mathbf{H}}\hat{\mathbf{x}}(n)]
\end{gathered}
\tag{14.4.5}
$$

14.4.4 Moving Average Model: No Input Uncertainty†

Consider the following moving average model with no input uncertainty

$$y(n) = \sum_{m=0}^{M} a_m u(n-m)$$

$$z(n) = y(n) + v(n)$$

$$\boldsymbol{\alpha} = \begin{bmatrix} a_0 \\ \cdot \\ \cdot \\ \cdot \\ a_M \\ R \end{bmatrix}$$

This model is analogous to that of Section 14.4.3 except that a different dynamic system model is considered.

The desired results follow by noting that the derivation of the log likelihood function $\xi(N:\boldsymbol{\alpha})$ in Section 14.3 did not assume a state space representation for the dynamic system. Thus the basic form of $\xi(N:\boldsymbol{\alpha})$ of (14.3.2) still applies. For the present case

$$\hat{z}(n \mid n-1) = \sum_{m=0}^{M} a_m u(n-m)$$

$$\delta_z(n:\boldsymbol{\alpha}) = z(n) - \sum_{m=0}^{M} a_m u(n-m)$$

$$\boldsymbol{\Sigma}_z(n \mid n-1:\boldsymbol{\alpha}) = R$$

$$2\xi(N:\boldsymbol{\alpha}) = -N\ln(2\pi) - N\ln R - \sum_{n=1}^{N} \frac{\boldsymbol{\delta}_z^2(n:\boldsymbol{\alpha})}{R}$$

Thus

$$\frac{\partial\xi(N:\boldsymbol{\alpha})}{\partial a_m} = \sum_{n=1}^{N} u(n-m)\left[z(n) - \sum_{m=0}^{M} a_m u(n-m)\right]$$

This yields a system of $M+1$ linear equations in the a_m $m = 0 \cdots M$ which can be easily solved. A simple equation for the estimate of R also results.

14.4.5 Time Invariant Estimator

In all of the preceding special cases, the estimate $\hat{\mathbf{z}}(n \mid n-1)$ was obvious by inspection and the estimation theory of Chapter 6 did not have to be used. A special case involving the estimators of Chapter 6 is now considered.

The basic model (14.3.1) is a time invariant system; i.e., $\boldsymbol{\Phi}$, $\mathbf{G}$, $\mathbf{B}$, $\mathbf{H}$, $\mathbf{Q}$ and $\mathbf{R}$ are constant. However the corresponding filter (estimator) may be time varying because of the finite observation interval (and the ideas of Section 14.3

generalize easily to time varying $\mathbf{\Phi}$, etc.). The special case to be considered now assumes that:

1. The basic model is such that the filter becomes a time invariant system for long observation intervals. (See Section 6.3.5.)
2. The observation interval is long enough so that the filter can be considered to be time invariant for the whole time interval.

Note that there is no assumption that $\mathbf{x}(n)$ is a stationary process.

Under the above assumptions, (14.3.2) can be written as follows (see (6.3.6))

$$
\begin{aligned}
2\xi(N:\boldsymbol{\alpha}) &= \xi_{\text{bias}}(N:\boldsymbol{\alpha}) + \xi_{\text{observation}}(N:\boldsymbol{\alpha}) \\
\xi_{\text{bias}}(N:\boldsymbol{\alpha}) &= -N\ln[(2\pi)^{K_2}\,|\mathbf{\Sigma}_z|] \\
\xi_{\text{observation}}(N:\boldsymbol{\alpha}) &= -\sum_{n=1}^{N}\boldsymbol{\delta}_z'(n:\boldsymbol{\alpha})\mathbf{\Sigma}_z^{-1}\boldsymbol{\delta}_z(n:\boldsymbol{\alpha}) \\
\hat{\mathbf{x}}(n+2\,|\,n+1:\boldsymbol{\alpha}) &= \mathbf{\Phi}\hat{\mathbf{x}}(n+1\,|\,n:\boldsymbol{\alpha}) + \mathbf{B}\mathbf{u}(n+1) + \tilde{\mathbf{K}}_\infty\boldsymbol{\delta}_z(n+1:\boldsymbol{\alpha}) \\
\boldsymbol{\delta}_z(n+1:\boldsymbol{\alpha}) &= \mathbf{z}(n+1) - \hat{\mathbf{z}}(n+1\,|\,n:\boldsymbol{\alpha}) \qquad (14.4.6)\\
\hat{\mathbf{z}}(n+1\,|\,n:\boldsymbol{\alpha}) &= \mathbf{H}\hat{\mathbf{x}}(n+1\,|\,n:\boldsymbol{\alpha}) \\
\mathbf{\Sigma}_\infty &= \mathbf{\Sigma}_\infty(1\,|\,0) - \mathbf{\Sigma}_\infty(1\,|\,0)\mathbf{H}'\mathbf{\Sigma}_z^{-1}\mathbf{H}\mathbf{\Sigma}_\infty(1\,|\,0) \\
\mathbf{\Sigma}_\infty(1\,|\,0) &= \mathbf{\Phi}\mathbf{\Sigma}_\infty\mathbf{\Phi}' + \mathbf{G}\mathbf{Q}\mathbf{G}' \\
\mathbf{\Sigma}_z &= \mathbf{R} + \mathbf{H}\mathbf{\Sigma}_\infty(1\,|\,0)\mathbf{H}' \\
\tilde{\mathbf{K}}_\infty &= \mathbf{\Phi}\mathbf{\Sigma}_\infty\mathbf{H}'\mathbf{R}^{-1}
\end{aligned}
$$

Note that the equations are written in terms of $\hat{\mathbf{x}}(n+1\,|\,n)$, rather than $\hat{\mathbf{x}}(n\,|\,n)$. The vector of unknown parameters $\boldsymbol{\alpha}$ as defined in Section 14.1 contains the unknown values of the elements of $\mathbf{\Phi}$, $\mathbf{H}$, $\mathbf{B}$, $\mathbf{R}$, $\mathbf{G}$, and $\mathbf{Q}$. However, from (14.4.6) it is seen that for this special case $\xi(N:\boldsymbol{\alpha})$ depends only on $\mathbf{\Phi}$, $\mathbf{H}$, $\mathbf{B}$, $\mathbf{\Sigma}_z$, $\tilde{\mathbf{K}}_\infty$. Therefore define

$$\tilde{\boldsymbol{\alpha}} = \text{unknown parameters in } \mathbf{\Phi}, \mathbf{H}, \mathbf{B}, \tilde{\mathbf{K}}_\infty$$

The estimation problem can now be stated as follows:

$$\text{Find the values of } \tilde{\boldsymbol{\alpha}}, \mathbf{\Sigma}_z, \text{ which maximize } \xi(N) \qquad (14.4.7)$$

where all the elements of $\tilde{\mathbf{K}}_\infty$ and $\mathbf{\Sigma}_z$ are assumed to be unknown. The maximization of $\xi(N)$ of (14.4.6) with respect to $\mathbf{\Sigma}_z$ can now be performed directly to yield

$$\hat{\mathbf{\Sigma}}_z(\hat{\tilde{\boldsymbol{\alpha}}}) = \frac{1}{N}\left[\sum_{n=1}^{N}\boldsymbol{\delta}_z(n:\hat{\tilde{\boldsymbol{\alpha}}})\boldsymbol{\delta}_z'(n:\hat{\tilde{\boldsymbol{\alpha}}})\right] \qquad (14.4.8)$$

where $\hat{\tilde{\boldsymbol{\alpha}}}$ minimizes

$$\tilde{\xi}(N:\tilde{\boldsymbol{\alpha}}) = \ln\left|\sum_{n=1}^{N}\boldsymbol{\delta}_z(n:\tilde{\boldsymbol{\alpha}})\boldsymbol{\delta}_z'(n:\tilde{\boldsymbol{\alpha}})\right|$$

After manipulation (see Appendix C)

$$\frac{\partial \tilde{\xi}(N:\tilde{\boldsymbol{\alpha}})}{\partial \tilde{\boldsymbol{\alpha}}} = -\sum_{n=1}^{N} \boldsymbol{\delta}'_z(n:\tilde{\boldsymbol{\alpha}})\hat{\boldsymbol{\Sigma}}_z^{-1}(\tilde{\boldsymbol{\alpha}})\frac{\partial \hat{\mathbf{z}}(n\,|\,n-1:\tilde{\boldsymbol{\alpha}})}{\partial \tilde{\boldsymbol{\alpha}}}$$

where difference equations for evaluating $(\partial/\partial\tilde{\boldsymbol{\alpha}})\hat{\mathbf{z}}(n\,|\,n-1:\tilde{\boldsymbol{\alpha}})$ follow easily from (14.4.6). A suggested iterative algorithm is (see Sec. 11.7)

$$\tilde{\boldsymbol{\alpha}}_{i+1} = \tilde{\boldsymbol{\alpha}}_i - \mathbf{B}(i+1)\frac{\partial \tilde{\xi}'(N:\tilde{\boldsymbol{\alpha}})}{\partial \tilde{\boldsymbol{\alpha}}}\bigg|_{\tilde{\boldsymbol{\alpha}}=\tilde{\boldsymbol{\alpha}}_i}$$
$$\mathbf{B}(i+1) = \left\{\sum_{n=1}^{N}\left[\frac{\partial \hat{\mathbf{z}}(n\,|\,n-1:\tilde{\boldsymbol{\alpha}})}{\partial \tilde{\boldsymbol{\alpha}}}\right]' \hat{\boldsymbol{\Sigma}}_z^{-1}(\tilde{\boldsymbol{\alpha}})\frac{\partial \hat{\mathbf{z}}(n\,|\,n-1:\tilde{\boldsymbol{\alpha}})}{\partial \tilde{\boldsymbol{\alpha}}}\right\}^{-1}_{\tilde{\boldsymbol{\alpha}}=\tilde{\boldsymbol{\alpha}}_i} \tag{14.4.9}$$

In a general sense, a change of variables has been made so that the values of all the elements of $\tilde{\mathbf{K}}_\infty$ and $\boldsymbol{\Sigma}_z$ are being estimated instead of the values of the unknown parameters in $\mathbf{R}$ and $\mathbf{GQG}'$. Let k_5 and k_6 denote the number of unknown parameters in $\mathbf{R}$, $\mathbf{GQG}'$ and in $\boldsymbol{\Sigma}_z$, $\tilde{\mathbf{K}}_\infty$ respectively. When $k_5 > k_6$, then $\mathbf{R}$, $\mathbf{GQG}'$ are not completely identifiable (see Section 14.1.6). In such a case, consideration of $\boldsymbol{\Sigma}_z$ and $\tilde{\mathbf{K}}_\infty$ as the unknowns has many advantages. When $k_5 = k_6$ it may be possible to use $\hat{\hat{\boldsymbol{\alpha}}}$ and $\hat{\boldsymbol{\Sigma}}_z$ to solve for unique estimates of the unknowns in $\mathbf{R}$ and $\mathbf{GQG}'$. If $k_5 < k_6$, then the "change of variables" obviously introduces extra unknowns and its validity is questionable in a strict mathematical sense.

14.4.6 Discussion

The preceding discussions are not intended to tabulate all the special cases for which "nice" results can be obtained. Various extended versions of the special cases also yield simple results. On the other hand, certain apparently minor variations on the special cases can radically change the nature of the problem.

In some special cases, "nice" algorithms also result when the minimization of $\xi(N:\boldsymbol{\alpha})$ is viewed as the minimization of $\xi(N:\boldsymbol{\alpha}, \hat{\mathbf{x}}(n\,|\,n), n = 1, \ldots, N)$ or $\xi(N:\boldsymbol{\alpha}, \hat{\mathbf{x}}(n\,|\,N), n = 1, \ldots, N)$ subject to the constraint that $\hat{\mathbf{x}}(n\,|\,n)$, or $\hat{x}(n\,|\,N)$ satisfy the necessary filtering or smoothing equations and the constraints are handled by introducing Lagrange multipliers (or costate variables or adjoint variables). The details of this approach are closely related to optimal control theory concepts.

14.5 Maximum Likelihood: Error Analysis†

In Sections 14.3 and 14.4, the evaluation of $\hat{\boldsymbol{\alpha}}(N)$ the maximum likelihood estimate of the unknown model parameters $\boldsymbol{\alpha}$ from the observations $\mathbf{z}(1) \cdots \mathbf{z}(N)$, was discussed. The corresponding error analysis question is now considered; i.e., the difference between $\hat{\boldsymbol{\alpha}}(N)$ and the true value of $\boldsymbol{\alpha}$ is considered. The discussions apply to the model of (14.3.1) where the observations are assumed to be exactly represented by (14.3.1) for some value of $\boldsymbol{\alpha}$ and specified model degree. The errors in the estimate $\hat{\boldsymbol{\alpha}}(N)$ arise only from the uncertainty in $\mathbf{x}(0)$, $\mathbf{w}(n)$, and $\mathbf{v}(n)$. It is important not to confuse the following error analyses with the testing step of the three step (hypothesize—estimate—test) approach to system identification.

In Chapter 12, error analysis for Fisher models of the form

$$\mathbf{z} = \tilde{\mathbf{h}}(\mathbf{x}, \mathbf{v})$$

$\mathbf{x}$: completely unknown

$\mathbf{v}$: random vector

was discussed. The system identification problem of (14.3.1) can be viewed as

$$\mathbf{z}_N = \tilde{\mathbf{h}}_N[\boldsymbol{\alpha} : \tilde{\mathbf{v}}_N]$$

$$\mathbf{z}_N = \begin{bmatrix} \mathbf{z}(1) \\ \cdot \\ \cdot \\ \cdot \\ \mathbf{z}(N) \end{bmatrix}$$

$\tilde{\mathbf{v}}_N$: contains $\mathbf{x}(0)$, $\mathbf{w}(n)$, $\mathbf{v}(n)$, $\qquad n = 1, \ldots, N$

$\boldsymbol{\alpha}$: completely unknown

Hence the concepts of Chapter 12 can be applied directly in the present case.

In Chapter 12, a key concept is the ambiguity function (see Section 12.4) which is the average value of the log likelihood function

$$\gamma(\mathbf{x}_t, \mathbf{x}) = \int \{\ln p(\mathbf{z} : \mathbf{x})\} p(\mathbf{z} : \mathbf{x}_t)\, d\mathbf{z}$$

The ambiguity function provides a global error analysis (i.e., an indication of whether multiple peak log likelihood functions are to be expected). Consideration of its curvature,

$$\left. \frac{\partial^2 \gamma(\mathbf{x}_t, \mathbf{x})}{\partial \mathbf{x}^2} \right|_{\mathbf{x}=\mathbf{x}_t}$$

enables one to evaluate a lower bound on the parameter estimate error covariance matrix (where the lower bound corresponds to that obtained from a linearized error analysis). Since such uses are discussed in Section 12.4, only the system identification ambiguity function itself is discussed here.

In system identification, the vector of unknowns is $\boldsymbol{\alpha}$. Thus after including the N dependence in the notation the ambiguity function is $\gamma(N : \boldsymbol{\alpha}_t, \boldsymbol{\alpha})$

$$\gamma(N : \boldsymbol{\alpha}_t, \boldsymbol{\alpha}) = \int \{\ln[p(\mathbf{z}_N : \boldsymbol{\alpha})]\} p(\mathbf{z}_N : \boldsymbol{\alpha}_t)\, d\mathbf{z}_N$$

or in terms of $\xi(N : \boldsymbol{\alpha})$

$$\gamma(N : \boldsymbol{\alpha}_t, \boldsymbol{\alpha}) = \int \xi(N : \boldsymbol{\alpha}) p(\mathbf{z}_N : \boldsymbol{\alpha}_t)\, d\mathbf{z}_N \tag{14.5.1}$$

or in still simpler notation

$$\gamma(N : \boldsymbol{\alpha}_t, \boldsymbol{\alpha}) = E\{\xi(N : \boldsymbol{\alpha})\}$$

As noted in Section 12.4, the ambiguity function is closely related to the divergence J of hypothesis testing. Thus for the Gaussian case of (14.3.1) the equation for $\gamma(N : \boldsymbol{\alpha}_.; \boldsymbol{\alpha})$ is closely related to (10.1.14). Define (by analogy with Section 10.1.4)

$\mathfrak{F}_{N,\alpha}$: linear system which operates on the zero mean component of $\mathbf{z}(n)$, $\mathbf{n} = 1, \ldots, N-1$, to give the zero mean component of the conditional expectation of $\mathbf{z}(N)$ assuming that $\mathbf{z}(1) \cdots \mathbf{z}(N)$ is generated by (14.3.1) with given value of $\boldsymbol{\alpha}$

$\mathbf{m}_\alpha(n)$: mean value of $\mathbf{z}(n)$ as obtained from (14.3.1) with given value of $\boldsymbol{\alpha}$; $\mathbf{m}_\alpha(N)$ depends on $\mathbf{u}(n)$, $n = 1, \ldots, N-1$

$$\mathbf{z}_N = \begin{bmatrix} \mathbf{z}(1) \\ \cdot \\ \cdot \\ \cdot \\ \mathbf{z}(N) \end{bmatrix}$$

$$\mathbf{m}_{N,\alpha} = \begin{bmatrix} \mathbf{m}_\alpha(1) \\ \cdot \\ \cdot \\ \cdot \\ \mathbf{m}_\alpha(N) \end{bmatrix}$$

$$\tilde{\boldsymbol{\delta}}_z(N \mid N-1) = \mathbf{z}(N) - \mathbf{m}_{\alpha_t}(N) - \mathfrak{F}_{N,\alpha}[\mathbf{z}_{N-1} - \mathbf{m}_{N-1,\alpha_t}]$$

$$\mathbf{d}(N \mid N-1) = \mathbf{m}_{\alpha_t}(N) - \mathbf{m}_\alpha(N) - \mathfrak{F}_{N,\alpha}[\mathbf{m}_{N-1,\alpha_t} - \mathbf{m}_{N-1,\alpha}]$$

Then in (14.3.2)

$$\hat{\mathbf{z}}(N \mid N-1 : \boldsymbol{\alpha}) = \mathbf{m}_\alpha(N) + \mathfrak{F}_{N,\alpha}[\mathbf{z}_{N-1} - \mathbf{m}_{N-1,\alpha}]$$

$$\begin{aligned} \boldsymbol{\delta}_z(N : \boldsymbol{\alpha}) &= \mathbf{z}(N) - \hat{\mathbf{z}}(N \mid N-1 : \boldsymbol{\alpha}) \\ &= \tilde{\boldsymbol{\delta}}_z(N \mid N-1) + \mathbf{d}(N \mid N-1) \end{aligned}$$

Define

$$\boldsymbol{\Sigma}_{z:\alpha_t|\alpha}(N \mid N-1) = E\{\tilde{\boldsymbol{\delta}}_x(N \mid N-1)\tilde{\boldsymbol{\delta}}_z'(N \mid N-1)\}$$

Then from (14.3.2) and (14.5.1) the ambiguity function is given by

$$\boxed{\begin{aligned} 2\gamma(N : \boldsymbol{\alpha}_t, \boldsymbol{\alpha}) &= -NK_2 \ln(2\pi) - \sum_{n=1}^{N} \ln|\boldsymbol{\Sigma}_z(n \mid n-1 : \boldsymbol{\alpha})| \\ &\quad - \sum_{n=1}^{N} \operatorname{tr}\{\boldsymbol{\Sigma}_z^{-1}(n \mid n-1 : \boldsymbol{\alpha}) \\ &\quad \times E\{\boldsymbol{\delta}_z(n : \boldsymbol{\alpha})\boldsymbol{\delta}_z'(n : \boldsymbol{\alpha})\}\} \\ E\{\boldsymbol{\delta}_z(n : \boldsymbol{\alpha})\boldsymbol{\delta}_z'(n : \alpha)\} &= \mathbf{d}(n \mid n-1)\mathbf{d}'(n \mid n-1) \\ &\quad + \boldsymbol{\Sigma}_{z:\alpha_t|\alpha}(n \mid n-1) \end{aligned}} \tag{14.5.2}$$

The $\boldsymbol{\Sigma}_z(n \mid n-1 : \boldsymbol{\alpha})$ of (14.5.2) and (14.3.2) can be viewed as

$$\boldsymbol{\Sigma}_z(n \mid n-1 : \boldsymbol{\alpha}) = \boldsymbol{\Sigma}_{z:\boldsymbol{\alpha}_t|\boldsymbol{\alpha}}(n \mid n-1) \qquad \text{for } \boldsymbol{\alpha}_t = \boldsymbol{\alpha}$$

Equation (14.5.2), like (14.3.2), is a complicated-looking equation because it applies to a very general model and the vector of unknown system parameters $\boldsymbol{\alpha}$ has not been specified. However, explicit equations for the $\boldsymbol{\Sigma}_z$ and $\mathbf{d}_\alpha$ can be written down in state space form using the equations of Chapters 4 and 6 and (14.5.2) can be numerically evaluated for specified $\boldsymbol{\alpha}$ and $\boldsymbol{\alpha}_t$ in a straightforward fashion. Of course, in special cases such as those of Section 14.4 much simpler equations result.

EXAMPLE: Consider the model (see also Section 14.4.1)

$$x(n+1) = \Phi x(n) + w(n)$$

$$z(n) = x(n)$$

$$E\{w(n_1)w(n_2)\} = \begin{cases} Q, & n_1 = n_2 \\ 0, & n_1 \neq n_2 \end{cases}$$

$$E\{x^2(0)\} = \psi$$

$$w(n), x(0)\text{: zero mean Gaussian}$$

Assume that only the value of Φ is unknown, so

$$\alpha = \Phi$$

Let Φ_t denote the true value of Φ and $x_t(n)$ denote the true state. Define

$$\Gamma_t(n) = E\{x_t^2(n)\}$$

so

$$\Gamma_t(n+1) = \Phi_t^2 \Gamma_t(n) + Q$$

$$E\{x_t(n+1)x_t(n)\} = \Phi_t \Gamma_t(n)$$

Now

$$\hat{z}(n \mid n-1 \mid \Phi) = \Phi z(n-1)$$

$$= \Phi x_t(n-1)$$

Since the model is zero mean $[u(n) = 0]$, $d_\alpha = 0$ and

$$\delta_z(n : \Phi) = x_t(n) - \Phi x_t(n-1)$$

$$\Sigma_{z:\Phi_t|\Phi}(n \mid n-1) = E\{\delta_z^2(n : \Phi)\}$$

$$= \Gamma_t(n) - 2\Phi\Phi_t\Gamma_t(n-1) + \Phi^2\Gamma_t(n-1)$$

Since

$$\Sigma_z(n \mid n-1 : \Phi) = Q$$

(14.5.2) becomes

$$2\gamma(N : \Phi_t, \Phi) = -N\ln(2\pi) - N\ln Q - \sum_{n=1}^{N} \frac{\Gamma_t(n) - 2\Phi\Phi_t\Gamma_t(n-1) + \Phi^2\Gamma_t(n-1)}{Q}$$

or

$$2\gamma(N:\Phi_t,\Phi) = -N\ln(2\pi Q) - \sum_{n=1}^{N} \frac{(\Phi_t - \Phi)^2\Gamma_t(n-1) + Q}{Q}$$

From (12.4.4), for any unbiased estimate $\hat{\Phi}$ of Φ_t, a lower bound on the estimate error variance is given by

$$E\{(\Phi_t - \hat{\Phi})^2\} \geq \frac{1}{-\partial^2\gamma(N:\Phi_t,\Phi)/\partial\Phi^2}\bigg|_{\Phi=\Phi_t}$$

so

$$E\{(\Phi_t - \hat{\Phi})^2\} \geq \frac{Q}{\sum_{n=1}^{N}\Gamma_t(n-1)}$$

EXAMPLE: Consider a model with no input uncertainty (see also Section 14.4.3):

$$x(n+1) = \Phi x(n) + u(n)$$

$$z(n) = x(n) + v(n)$$

$$x(0) = 0 \qquad u(n)\text{: known}$$

$$E\{v(n_1)v(n_2)\} = \begin{cases} R, & n_1 = n_2 \\ 0, & n_1 \neq n_2 \end{cases}$$

$$v(n)\text{:} \quad \text{zero mean Gaussian}$$

Assume that Φ is the only unknown, so $\boldsymbol{\alpha} = \Phi$. Then from (14.5.2)

$$2\gamma(N:\Phi_t,\Phi) = -N\ln(2\pi R) - \sum_{n=1}^{N} \frac{[x_t(n) - \hat{x}(n)]^2 + R}{R}$$

$$x_t(n) = \sum_{m=1}^{n} \Phi_t^{n-m}u(m) \qquad \hat{x}(n) = \sum_{m=1}^{n} \Phi^{n-m}u(m)$$

The vector versions of (12.3.2), (12.3.3) and (12.4.4) imply that

$$\frac{\partial\gamma(N:\boldsymbol{\alpha}_t,\boldsymbol{\alpha})}{\partial\boldsymbol{\alpha}^2}\bigg|_{\boldsymbol{\alpha}=\boldsymbol{\alpha}_t} = -E\left\{\left[\frac{\partial\xi(N:\boldsymbol{\alpha})}{\partial\boldsymbol{\alpha}}\right]'\left[\frac{\partial\xi(N:\boldsymbol{\alpha})}{\partial\boldsymbol{\alpha}}\right]\right\}\bigg|_{\boldsymbol{\alpha}=\boldsymbol{\alpha}_t}$$

In the special case where the filter is time invariant (as in Section 14.4.5), it follows that $\mathbf{q}(n+1:\boldsymbol{\alpha})$ is a stationary process where

$$\mathbf{q}(n+1:\boldsymbol{\alpha}) = \frac{\partial\xi(n+1:\boldsymbol{\alpha})}{\partial\boldsymbol{\alpha}} - \frac{\partial\xi(n:\boldsymbol{\alpha})}{\partial\boldsymbol{\alpha}}$$

Thus, for large N

$$\frac{\partial\gamma(N:\boldsymbol{\alpha}_t,\boldsymbol{\alpha})}{\partial\boldsymbol{\alpha}^2}\bigg|_{\boldsymbol{\alpha}=\boldsymbol{\alpha}_t} \approx -\sum_{n=1}^{N}\mathbf{q}(n:\boldsymbol{\alpha}_t)\mathbf{q}'(n:\boldsymbol{\alpha}_t) \tag{14.5.3}$$

If $\boldsymbol{\alpha}$ is replaced by $\hat{\boldsymbol{\alpha}}$, (14.5.3) may be much easier to implement than (14.5.2), if the error analysis is to be done after the observations are made. It can be shown that $\mathbf{q}(n:\boldsymbol{\alpha}_t)$ is a zero mean process.

14.6 Model Reference: Weighted Least Squares

Maximum likelihood theory is based on an explicit hypothesized structure (14.3.1) involving stochastic processes (Gaussian). A different approach to system identification which relies more on physical intuition is the model reference—weighted least squares concept. With this approach the hypothesized structure does not involve stochastic models.

Consider Fig. 14.6.1. Figure 14.6.1(a) is the real world where $\mathbf{u}(n)$ and

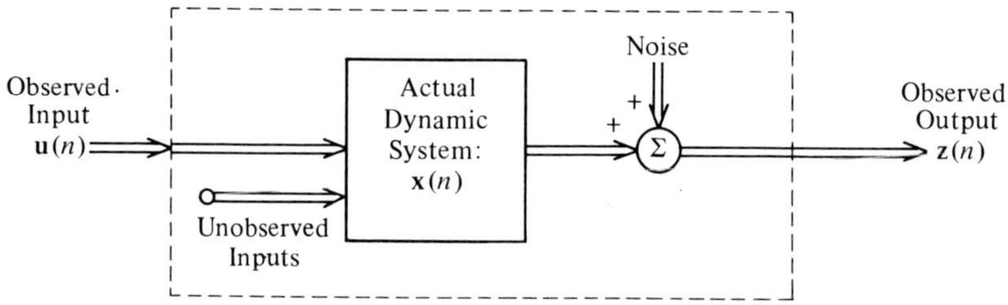

(a) Real World

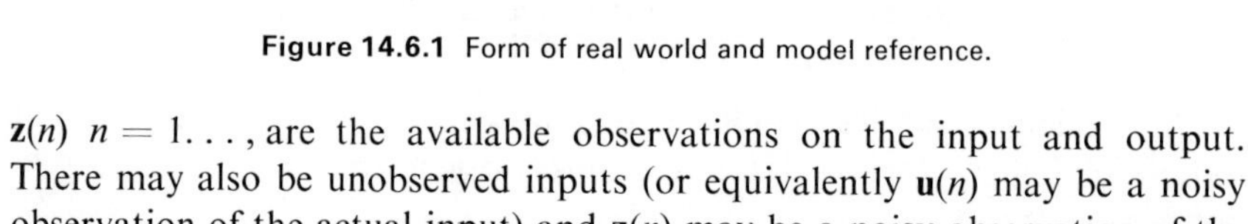

(b) Model Reference

Figure 14.6.1 Form of real world and model reference.

$\mathbf{z}(n)$ $n = 1 \ldots$, are the available observations on the input and output. There may also be unobserved inputs (or equivalently $\mathbf{u}(n)$ may be a noisy observation of the actual input) and $\mathbf{z}(n)$ may be a noisy observation of the output. To simplify the discussion, the actual dynamic system is assumed to be known to have a state vector $\mathbf{x}(n)$ of known dimension. Figure 14.6.1(b) is the *model reference* which is hypothesized to have the structure

$$\mathbf{x}_{\text{model}}(n+1) = \mathbf{\Phi x}_{\text{model}}(n) + \mathbf{B u}_{\text{model}}(n)$$

$$\mathbf{z}_{\text{model}}(n) = \mathbf{H x}_{\text{model}}(n) \tag{14.6.1}$$

$$\boldsymbol{\alpha}: \text{ unknown parameters in } \mathbf{\Phi}, \mathbf{B}, \mathbf{H}$$

The model reference approach to system identification is

> Choose $\boldsymbol{\alpha}$, $\mathbf{u}_{\text{model}}(n)$ and $\mathbf{z}_{\text{model}}(n)$ $n = 1 \ldots$ such that (14.6.1) is satisfied and
>
> $$\mathbf{u}(n) - \mathbf{u}_{\text{model}}(n) \; n = 1, \ldots,$$
>
> and
>
> $$\mathbf{z}(n) - \mathbf{z}_{\text{model}}(n) \; n = 1, \ldots,$$
>
> are as small as possible in some sense.

This is a very reasonable approach. The problem is that there are many different ways to define a criterion which determines how close $\mathbf{u}_{\text{model}}(n)$ and $\mathbf{z}_{\text{model}}(n)$ are to $\mathbf{u}(n)$ and $\mathbf{z}(n)$, $n = 1, \ldots, N$. Three cases are now discussed.

Consider the special case where it is assumed that in the real world of Figure 14.6.1(a), the observation $\mathbf{z}(n)$ is felt to be very "close" to $\mathbf{x}(n)$, the state of the actual system. Then a reasonable model reference approach is

> Set $\mathbf{z}_{\text{model}}(n) = \mathbf{x}_{\text{model}}(n) = \mathbf{z}(n)$. Choose $\boldsymbol{\alpha}$ to minimize
>
> $$J(\boldsymbol{\alpha}) = \sum_{n=0}^{N-1} [\mathbf{B}\mathbf{u}_{\text{model}}(n) - \mathbf{B}\mathbf{u}(n)]'\mathbf{Q}^{-1}[\mathbf{B}\mathbf{u}_{\text{model}}(n) - \mathbf{B}\mathbf{u}(n)]$$
>
> $\mathbf{Q}$: positive definite matrix chosen by engineering judgement

Using (14.6.1) and the condition $\mathbf{x}_{\text{model}}(n) = \mathbf{z}(n)$, the criteria $J(\boldsymbol{\alpha})$ can be written

$$J(\boldsymbol{\alpha}) = \sum_{n=1}^{N} [\mathbf{z}(n) - \boldsymbol{\Phi}\mathbf{z}(n-1) - \mathbf{B}\mathbf{u}(n-1)]'\mathbf{Q}^{-1}[\mathbf{z}(n) - \boldsymbol{\Phi}\mathbf{z}(n-1) - \mathbf{B}\mathbf{u}(n-1)]$$

This case is closely related to the maximum likelihood special case of Section 14.4.1; see (14.4.2).

Consider a different special case where it is assumed that in the real world of Figure 14.6.1(a), there are no unobserved inputs so $\mathbf{u}(n)$ is the true input. Also assume initial conditions are known perfectly. Then a reasonable model reference approach is

> Set $\mathbf{u}_{\text{model}}(n) = \mathbf{u}(n)$. Choose $\boldsymbol{\alpha}$ to minimize
>
> $$J(\boldsymbol{\alpha}) = \sum_{n=1}^{N} [\mathbf{z}(n) - \mathbf{H}\mathbf{x}_{\text{model}}(n)]'\mathbf{R}^{-1}[\mathbf{z}(n) - \mathbf{H}\mathbf{x}_{\text{model}}(n)]$$
>
> $\mathbf{R}$: positive definite matrix chosen by engineering judgment

This case is closely related to the special maximum likelihood case of Section 14.4.3; see (14.4.5).

Now consider the general case. A reasonable model reference—weighted least squares approach is

Choose $\boldsymbol{\alpha}$ to minimize $\hat{J}(\boldsymbol{\alpha})$

$$\hat{J}(\boldsymbol{\alpha}) = J(\boldsymbol{\alpha} : \hat{\mathbf{x}}_{\text{model}}(0 \mid N), \hat{\mathbf{w}}(n \mid N), n = 1 \ldots N)$$

$$\begin{aligned} J(\boldsymbol{\alpha} : \mathbf{x}_{\text{model}}(0), \mathbf{w}(n), n = 1 \ldots N) = \sum_{n=1}^{N} [\mathbf{z}(n) - \mathbf{H}\mathbf{x}_{\text{model}}(n)]'\mathbf{R}^{-1}[\mathbf{z}(n) \\ - \mathbf{H}\mathbf{x}_{\text{model}}(n)] + \sum_{n=1}^{N} \mathbf{w}'(n)\mathbf{Q}^{-1}\mathbf{w}(n) \\ + \mathbf{x}'_{\text{model}}(0)\boldsymbol{\psi}\mathbf{x}_{\text{model}}(0) \end{aligned} \tag{14.6.2}$$

$\mathbf{Q}$, $\mathbf{R}$, $\boldsymbol{\psi}$ chosen by engineering judgement

$\hat{\mathbf{x}}_{\text{model}}(0 \mid N)$, $\hat{\mathbf{w}}(n \mid N)$: values of $\mathbf{x}_{\text{model}}(0)$ and $\mathbf{w}(n)$ which minimize $J(\boldsymbol{\alpha} : \mathbf{x}_{\text{model}}(0), \mathbf{w}(n)\ n = 1, \ldots, N)$ subject to constraint

$$\mathbf{x}_{\text{model}}(n + 1) = \boldsymbol{\Phi}\mathbf{x}_{\text{model}}(n) + \mathbf{B}\mathbf{u}_{\text{model}}(n)$$

$$\mathbf{B}\mathbf{u}_{\text{model}}(n) = \mathbf{B}\mathbf{u}(n) + \mathbf{w}(n)$$

This general formulation is closely related to the smoothing form of the maximum likelihood approach as discussed in Section 14.3.2 as $\hat{J}(\boldsymbol{\alpha})$ is the same form as $\xi_{\text{observation}}(N\colon \mathbf{a})$ of (14.3.3).

The model reference—weighted least squares concept is very appealing because of its intuitive basis. A recurring theme in this book is that "mathematical niceties" are usually much less important than such intuitive type thinking. However there are dangers associated with the intuitive model reference concept as it can lead to "bad results". For example, model reference—weighted least squares thinking can yield estimates that have large biases which do not decrease with N, the number of observations. This is the main reason maximum likelihood concepts are emphasized in this chapter (although they can also yield biased estimates).

EXAMPLE: Consider the case where the real world is

$$z_1(n) = Hx(n) + v(n) \qquad n = 1 \ldots N$$

$$z_2(n) = x(n)$$

$$\tilde{\mathbf{x}}(n) = \begin{bmatrix} x(n) \\ v(n) \end{bmatrix}$$

$$E\{\tilde{\mathbf{x}}(n_1)\tilde{\mathbf{x}}(n_2)\} = \begin{cases} \boldsymbol{\Gamma}, & n_1 = n_2 \\ 0, & n_1 \neq n_2 \end{cases}$$

$$\boldsymbol{\Gamma} = \begin{bmatrix} 1 & \rho \\ \rho & 1 \end{bmatrix}$$

$\tilde{\mathbf{x}}(n)$: zero mean white Gaussian

H: unknown ($\boldsymbol{\alpha} = H$)

A reasonable model reference approach is to define the estimate $\hat{H}$ as the value of H which minimizes

$$J = \sum_{n=1}^{N} [z_1(n) - Hz_2(n)]^2$$

This gives

$$\hat{H} = \left[\sum_{n=1}^{N} z_2^2(n)\right]^{-1} \sum_{n=1}^{N} z_1(n)z_2(n)$$

An error analysis for this $\hat{H}$ can be done by substituting for $z_1(n)$ and $z_2(n)$ to yield

$$\hat{H} = \left[\sum_{n=1}^{N} x^2(n)\right]^{-1} \sum_{n=1}^{N} [Hx^2(n) + x(n)v(n)]$$

so

$$H - \hat{H} = -\left[\sum_{n=1}^{N} x^2(n)\right]^{-1} \sum_{n=1}^{N} x(n)v(n)$$

Since $E\{x(n)v(n)\} = \rho$, $\hat{H}$ does not converge to H as N increases. In other words, $\hat{H}$ is biased. A maximum likelihood formulation alleviates this problem. (See Exercise 11.8 in Chapter 11.)

Thus although model reference—weighted least squares concepts can yield equations very similar to those of maximum likelihood, such is not always the case.

The model reference concepts discussed here did not include any method of estimating the "size" of the observation and/or input uncertainties. In practice this is often an important aspect of the problem.

An important weakness of the model reference—weighted least squares approach lies in the areas of determining the degree of the model and in checking the "goodness of fit". The discussions have so far tacitly assumed the degree of the system is specified. One cannot determine the degree of the model by minimizing $\hat{J}(\hat{\boldsymbol{\alpha}})$ for different model degrees as $\hat{J}(\hat{\boldsymbol{\alpha}})$ goes to zero (usually) as the model degree increases. In the Gaussian—maximum likelihood case, the $\xi(N:\boldsymbol{\alpha})$ of (14.3.2), the $\xi_{\text{bias}}(N:\boldsymbol{\alpha})$ terms enable one to minimize $\xi(N:\hat{\boldsymbol{\alpha}})$ with respect to model degree. Similarly, the question of whether $\hat{J}(\hat{\boldsymbol{\alpha}})$ is small enough to say that the "fit is good" is difficult to answer so the presence of structural errors presents problems. If one is using a model reference–weighted-least-squares philosophy, these problems of model degree and goodness of fit (structural errors) can be approached by "looking at" the residuals (difference between actual observations and model output) to see if they are small enough. This is a valid engineering approach but it is nice to have resort to the general Gaussian-maximum likelihood formulae. It is also nice to be able to use the error analysis-ambiguity function results of Section 14.5.

14.7 Correlation Methods

The model reference concept of Section 14.6 did not involve the use of probabilistic models. Correlation methods are a different type of intuitive approach which returns to the use of hypothesized probabilistic models. Thus (14.3.1) is the hypothesized structure except that the Gaussian assumption is not needed. Correlation methods are most reasonable when $\mathbf{x}(n)$ is stationary and when there is no known input. Thus, only the following special case of (14.3.1) is considered

$$\begin{aligned} \mathbf{x}(n+1) &= \mathbf{\Phi x}(n) + \mathbf{w}(n) \\ \mathbf{z}(n) &= \mathbf{Hx}(n) + \mathbf{v}(n) \\ \mathbf{x}(n)&: \text{ stationary} \end{aligned}$$

The analysis of Section 4.1.2 yields the following relationships for the covariance matrices $\mathbf{\Gamma}(n)$ and correlation function $\mathcal{P}(n_1, n_2)$ of $\mathbf{x}(n)$.

$$\begin{aligned} \mathbf{\Gamma}_x(n) &= E[\mathbf{x}(n)\mathbf{x}'(n)] = \mathbf{\Gamma}_x \qquad \text{all } n \\ \mathbf{\Gamma}_x &= \mathbf{\Phi\Gamma}_x\mathbf{\Phi}' + \mathbf{Q} \\ \mathcal{P}_x(n+m, n) &= E[\mathbf{x}(n+m)\mathbf{x}'(n)] = \mathcal{P}_x(m) \qquad \text{all } n \\ \mathcal{P}_x(0) &= \mathbf{\Gamma}_x \\ \mathcal{P}_x(m+1) &= \mathbf{\Phi}\mathcal{P}_x(m) \\ \mathbf{\Gamma}_z &= E[\mathbf{z}(n)\mathbf{z}'(n)] \qquad \text{all } n \\ \mathbf{\Gamma}_z &= \mathbf{H\Gamma}_x\mathbf{H}' + \mathbf{R} \\ \mathcal{P}_z(m) &= E[\mathbf{z}(n+m)\mathbf{z}'(n)] \qquad \text{all } n \\ \mathcal{P}_z(m) &= \mathbf{H}\mathcal{P}_x(m)\mathbf{H}' \end{aligned} \tag{14.7.1}$$

Let $\boldsymbol{\alpha}$ denote the vector of unknowns in $\mathbf{\Phi}$, $\mathbf{Q}$, $\mathbf{R}$, and $\mathbf{H}$. Then it is possible to write $\mathbf{\Gamma}_z$ and $\mathcal{P}_z(m)$ of (14.7.1) as $\mathbf{\Gamma}_z(\boldsymbol{\alpha})$ and $\mathcal{P}_z(m:\boldsymbol{\alpha})$.

Since $\mathbf{z}(n)$ is a stationary process, it is reasonable to use $\mathbf{z}(n)$ $n = 1 \cdots N$ to compute estimates $\hat{\mathbf{\Gamma}}_z$ of $\mathbf{\Gamma}_z$ and $\hat{\mathcal{P}}_z(m)$ of $\mathcal{P}_z(m)$ as follows:

$$\begin{aligned} \hat{\mathbf{\Gamma}}_z &= \frac{1}{N}\sum_{n=1}^{N} \mathbf{z}(n)\mathbf{z}'(n) \\ \hat{\mathcal{P}}_z(m) &= \frac{1}{N-m}\sum_{n=m}^{N} \mathbf{z}(n)\mathbf{z}'(n-m) \end{aligned} \tag{14.7.2}$$

The basic idea underlying correlation methods of system identification is to choose the value of $\boldsymbol{\alpha}$ which makes $\mathbf{\Gamma}_z(\boldsymbol{\alpha})$ and $\mathcal{P}_z(m:\boldsymbol{\alpha})$ as close to the estimates $\hat{\mathbf{\Gamma}}_z$ and $\hat{\mathcal{P}}_z(m)$ as possible, $m = 1, 2, \ldots M$. This is a reasonable approach, but there is an obvious problem. What criterion should be used to measure how close $\mathbf{\Gamma}_z(\boldsymbol{\alpha})$ and $\mathcal{P}_z(m:\boldsymbol{\alpha})$ are to $\hat{\mathbf{\Gamma}}_z$ and $\hat{\mathcal{P}}_z(m)$? There is also the related problem of what value of M should be used.

One approach which is sometimes very effective is to choose M such that the equations

$$\hat{\mathbf{\Gamma}}_z = \mathbf{\Gamma}_z(\boldsymbol{\alpha})$$
$$\hat{\mathcal{P}}_z(m) = \mathcal{P}_z(m:\boldsymbol{\alpha}) \qquad m = 1, \dots M$$

yield a unique value of $\boldsymbol{\alpha}$.

EXAMPLE:

Consider the scalar case

$$x(n+1) = \Phi x(n) + w(n)$$
$$z(n) = x(n)$$
$$\boldsymbol{\alpha} = \begin{bmatrix} \Phi \\ Q \end{bmatrix}$$

Then from (14.7.1)

$$\Gamma_z(\boldsymbol{\alpha}) = \Phi^2 \Gamma_z(\boldsymbol{\alpha}) + Q$$
$$\mathcal{P}_z(1:\boldsymbol{\alpha}) = \Phi \Gamma_z(\boldsymbol{\alpha})$$

Use (14.7.2) to define $\hat{\Gamma}_z$, $\hat{\mathcal{P}}_z(1)$, Setting $\Gamma_z(\boldsymbol{\alpha}) = \hat{\Gamma}_z$ and $\mathcal{P}_z(1:\boldsymbol{\alpha}) = \hat{\mathcal{P}}_z(1)$ yields

$$\hat{\Phi} = \frac{\hat{\mathcal{P}}_z(1)}{\hat{\Gamma}_z} = \frac{N \sum_{n=2}^{N} z(n)z(n-1)}{(N-1)\sum_{n=1}^{N} z^2(n)}$$

$$\hat{Q} = \hat{\Gamma}_z - \frac{\hat{\mathcal{P}}_z^2(1)}{\hat{\Gamma}_z}$$

For large N, manipulation shows that

$$\hat{Q} = \frac{1}{N}\sum_{n=1}^{N} (z(n) - \hat{\Phi} z(n-1))^2$$

Thus, in this case, the correlation method yields maximum likelihood results (see Example of Section 14.4.1).

To illustrate the difficulties that can arise, consider the following model of perfect observations of the complete state

$$\mathbf{x}(n+1) = \mathbf{\Phi}\mathbf{x}(n) + \mathbf{w}(n)$$
$$\mathbf{z}(n) = \mathbf{x}(n) \quad n = 1 \dots N$$
$$\mathbf{Q}\text{: known}$$
$$\boldsymbol{\alpha}\text{: contains elements of } \mathbf{\Phi}$$

Define $\hat{\mathbf{\Gamma}}_z$ and $\hat{\mathcal{P}}_z(1)$ as in (14.7.2). Assume there is no value of $\boldsymbol{\alpha}$ (i.e., $\mathbf{\Phi}$) which enables the following equations to be solved.

$$\hat{\mathbf{\Gamma}}_z = \mathbf{\Phi}\hat{\mathbf{\Gamma}}_z\mathbf{\Phi}' + \mathbf{Q}$$
$$\hat{\mathcal{P}}_z(1) = \mathbf{\Phi}\hat{\mathbf{\Gamma}}_z \qquad (14.7.3)$$

Thus, it is necessary to define some scalar criteria which measures how close a given $\boldsymbol{\alpha}$ (i.e., $\boldsymbol{\Phi}$) comes to satisfying (14.7.3). The optimum criterion follows from (14.4.2) which yields for large N that $\hat{\boldsymbol{\alpha}}$ (i.e., $\hat{\boldsymbol{\Phi}}$) minimizes

$$\frac{1}{N}\xi = \mathrm{tr}\{\mathbf{Q}^{-1}[\hat{\boldsymbol{\Gamma}}_z - \boldsymbol{\Phi}\hat{\boldsymbol{\wp}}_z(1) - \hat{\boldsymbol{\wp}}_z(1)\boldsymbol{\Phi}' + \boldsymbol{\Phi}\hat{\boldsymbol{\Gamma}}_z\boldsymbol{\Phi}']\}$$

This criterion is not obvious without maximum likelihood theory.

Correlation techniques do not provide a natural way to "test validity" in the three steps, hypothesize-estimate-test, of system identification. This is a major disadvantage.

14.8 Instrument Variables

Another approach to system identification is the use of "instrument variables". This approach uses an hypothesized structure containing stochastic processes. The concept is easiest to understand when the hypothesized model is expressed in the autoregressive—moving average form (see Section 14.1.7). In order to simplify the notation only the following special case of (14.3.1) is considered.

$$\begin{aligned} y(n) &= \boldsymbol{\Phi}_1 y(n-1) + \boldsymbol{\Phi}_2 y(n-2) + \epsilon(n) \\ z(n) &= y(n) \\ &\epsilon(n)\text{: moving average process} \\ \epsilon(n) &= G_1 w(n) + G_2 w(n-1) \end{aligned}$$

The values of $\boldsymbol{\Phi}_1$, $\boldsymbol{\Phi}_2$, G_1 and G_2 are unknown (which means Q can be assumed to be known). However it is assumed that it is desired to only estimate $\boldsymbol{\Phi}_1$ and $\boldsymbol{\Phi}_2$.

Let $i_1(n)$ and $i_2(n)$ $n = 1\ldots$ denote two time functions, the "instrument variables". Consider the following equations

$$\begin{aligned} A_1(0) &= \boldsymbol{\Phi}_1 A_1(1) + \boldsymbol{\Phi}_2 A_1(2) + B_1 \\ A_2(0) &= \boldsymbol{\Phi}_1 A_2(1) + \boldsymbol{\Phi}_2 A_2(2) + B_2 \\ A_j(k) &= \sum_{n=1}^{N} i_j(n) z(n-k) \\ B_j &= \sum_{n=1}^{N} i_j(n)\epsilon(n) \end{aligned} \tag{14.8.1}$$

In (14.8.1), the $A_j(k)$ are calculable quantities while the B_j are unknowns. Now suppose that the instrument variables $i_j(n)$ are chosen such that

1. $i_j(n)$ are correlated with $z(n)$
2. $i_j(n)$ are uncorrelated with $\epsilon(n)$

(14.8.2)

If the conditions (14.8.2) are satisfied, it is reasonable to expect that for large N, the B_j terms in (14.8.1) can be ignored. Thus estimates $\hat{\boldsymbol{\Phi}}_1$ and $\hat{\boldsymbol{\Phi}}_2$ can be defined by

$$\begin{aligned} A_1(0) &= \hat{\Phi}_1 A_1(1) + \hat{\Phi}_2 A_1(2) \\ A_2(0) &= \hat{\Phi}_1 A_2(1) + \hat{\Phi}_2 A_2(2) \end{aligned} \tag{14.8.3}$$

The obvious problem associated with this approach is the choice of the instrument variables such that (14.8.2) is satisfied. One simple approach to the present case is to choose

$$i_1(n) = z(n-2)$$
$$i_2(n) = z(n-3)$$

EXAMPLE: Consider

$$x(n) = \Phi x(n-1) + w(n-1)$$
$$z(n) = y(n)$$

Let $i(n)$ be an instrument variable. Then

$$\hat{\Phi} = \frac{\sum_{n=1}^{N} i(n)z(n)}{\sum_{n=1}^{N} i(n)z(n-1)}$$

The conditions (14.8.2) are satisfied by choosing

$$i(n) = z(n-1)$$

which also yields the maximum likelihood estimate.

Another method of obtaining the instrument variables is to use a different set of observations (made by a different sensor).

Instrument variables techniques do not provide a natural way to "test validity" in the basic three steps, hypothesize-estimate-test, of system identification. This is a major disadvantage.

14.9 Unknown-But-Bounded†

Consider a system identification problem involving unknown-but-bounded modeling of $\mathbf{x}(0)$, $\mathbf{w}(n)$ and $\mathbf{v}(n)$ in (14.1.1) so

$$\begin{aligned} \mathbf{x}(n+1) &= \mathbf{\Phi}\mathbf{x}(n) + \mathbf{G}\mathbf{w}(n) + \mathbf{B}\mathbf{u}(n) \\ \mathbf{z}(n) &= \mathbf{H}\mathbf{x}(n) + \mathbf{v}(n) \\ \mathbf{w}(n) \in \Omega_w &= \{\mathbf{w} : \mathbf{w}'\mathbf{Q}^{-1}\mathbf{u} \leq 1\} \\ \mathbf{v}(n) \in \Omega_v &= \{\mathbf{v} : \mathbf{v}'\mathbf{R}^{-1}\mathbf{v} \leq 1\} \\ \mathbf{x}(0) \in \Omega_x(0) &= \{\mathbf{x} : \mathbf{x}'\boldsymbol{\psi}^{-1}\mathbf{x} \leq 1\} \end{aligned} \tag{14.9.1}$$

Let $\boldsymbol{\alpha}$ denote the vector of unknown parameters in $\mathbf{\Phi}$, $\mathbf{B}$, $\mathbf{G}$, $\mathbf{H}$, $\mathbf{Q}$, $\mathbf{R}$. This unknown-but-bounded system identification problem has not been investigated (to the author's knowledge) in any detail. However the following is a "reasonable approach" which is based on the hypothesis testing ideas of Chapter 9.

Let $\mathbf{z}_N$ be the vector formed from N observations. Define the hypothesis

$$\mathcal{H}_\alpha: \ \mathbf{z}_N \text{ came from (14.9.1) for specified value of } \boldsymbol{\alpha}$$

Define

$$\Omega_{\tilde{\alpha}}: \text{ set of all } \boldsymbol{\alpha} \text{ such that hypothesis } \mathcal{H}_\alpha \text{ is true}$$

This set $\Omega_{\tilde{\alpha}}$ constitutes the *identification set* as the model (14.9.1) for any $\boldsymbol{\alpha} \in \Omega_{\tilde{\alpha}}$ can be viewed as an identified system. Hence as in the unknown-but-bounded estimation theory of the earlier chapters, the estimate is a set of possible answers.

If $\mathbf{R}$ and/or $\mathbf{Q}$ are included in $\boldsymbol{\alpha}$, some special properties result. Assume for simplicity that $w(n)$ and $v(n)$ and hence Q and R are scalars. It is felt to be intuitively obvious that if R_0 and Q_0 are values which lie in $\Omega_{\tilde{\alpha}}$, then any $R > R_0$ and $Q > Q_0$ also lies in $\Omega_{\tilde{\alpha}}$. Hence a reasonable approach to unknown-but-bounded system identification is to try to choose $\boldsymbol{\Phi}$, $\mathbf{B}$, $\mathbf{H}$ so that R_0 and Q_0 are minimized in some sense. In some cases this approach may yield a unique answer.

The bounding ellipsoid algorithms provide a vehicle for actually obtaining answers. For example, from the ideas of Section 10.4, it follows that a given $\boldsymbol{\alpha}$ cannot lie in $\Omega_{\tilde{\alpha}}$ if it fails to yield an estimate set for any n, $n = 1 \cdots N$. This occurs if $\delta^2(n) > 1$ for any $n = 1 \cdots N$ where $\delta^2(n)$ is defined in (6.6.4).

14.10 Choice of Input†

The basic model (14.1.1) contains a known input $\mathbf{u}(n)$, $n = 0, \ldots$. The choice of this input can have a major effect on how well the system identification can be performed. For example, assume that $\mathbf{x}(0) = 0$ and that $\mathbf{w}(n) = 0$. Obviously $\mathbf{u}(n)$ cannot be zero if any system identification is to be accomplished. It is also felt to be relatively obvious that a $\mathbf{u}(n)$, $n = 0, \ldots,$ with large rapid variations should provide a better system identification than a low-amplitude, constant $\mathbf{u}(n)$. Therefore an interesting question is, *What is the best choice of* $\mathbf{u}(n)$, $n = 0, \ldots,$ *when the possible* $\mathbf{u}(n)$ *are constrained in some fashion*?

In general this question is very difficult to answer, as the best $\mathbf{u}(n)$ depends on both the structure of $\boldsymbol{\Phi}$, $\mathbf{B}$, etc., and their true but unknown values. For example, if $\boldsymbol{\Phi}$ corresponds to a discrete-time oscillator of frequency ω, a good $\mathbf{u}(n)$ would have "a lot of energy at" the frequency ω. However, if ω is not known, it is difficult to design the $\mathbf{u}(n)$ before the system is identified.

If values of the system parameters are known approximately a priori, an approach to designing $\mathbf{u}(n)$ can be formulated. Let $\boldsymbol{\alpha}_{\text{nom}}$ denote the nominal value of the unknown $\boldsymbol{\alpha}$. Assume that $\mathbf{x}(0)$, $\mathbf{w}(n)$, and $\mathbf{v}(n)$ are stochastic processes. Then using the Cramer-Rao inequality it is conceptually possible to evaluate a lower bound on the error variance of a maximum likelihood estimate (see Section 14.5) where this bound depends on $\mathbf{u}(n)$, $n = 0, \ldots, N - 1$, and on $\boldsymbol{\alpha}_{\text{nom}}$. The determination of the $\mathbf{u}(n)$, n $= 0, \ldots,$ which minimizes this bound can be viewed as signal design problem which is related to the discussions of Sections 8.3 and 10.9.5. Unfortunately, the optimum input obtained this way may be sensitive to the assumed nominal values, $\boldsymbol{\alpha}_{\text{nom}}$, and thus be of limited practical importance.

14.11 Model Reduction†

The following discussion constitutes a change in direction toward a problem which is not really system identification but which can be approached using the ideas of system identification.

Consider a system of equations

$$\begin{aligned} \mathbf{x}(n+1) &= \boldsymbol{\Phi}\mathbf{x}(n) + \mathbf{B}u(n) \\ y(n) &= \mathbf{H}\mathbf{x}(n) \\ \mathbf{x}(n)&: K_1 \text{ vector} \\ u(n), y(n)&: \text{scalars} \\ \boldsymbol{\Phi}, \mathbf{B}, \mathbf{H}&: \text{given} \end{aligned} \tag{14.9.1}$$

If K_1 is large, it may be desired to reduce the dimension of the model and develop an approximate model such as

$$\begin{aligned} \tilde{\mathbf{x}}(n+1) &= \tilde{\boldsymbol{\Phi}}\tilde{\mathbf{x}}(n) + \tilde{\mathbf{B}}u(n) \\ \tilde{y}(n) &= \tilde{\mathbf{H}}\tilde{\mathbf{x}}(n) \\ \tilde{\mathbf{x}}(n)&: \tilde{K}_1 \text{ vector} \\ K_1 &\gg \tilde{K}_1 \end{aligned} \tag{14.9.2}$$

where hopefully

$$\tilde{y}(n) \approx y(n)$$

for some wide class of $u(n)$ This is called the model reduction problem.

This model reduction problem is obviously related in many respects to the system identification problem, and there are various ways to proceed. To illustrate, one particular approach related to Section 14.9 is briefly discussed. Assume that the reduced model is to be valid for M different inputs $u_m(n)$, $n = 0, \ldots, N - 1$, $m = 1, \ldots, M$. Let $y_m(n)$ denote the corresponding output from the exact model (14.9.1). Assume that instead of

(14.9.2) a reduced model of the following form is desired:

$$\begin{aligned} \tilde{\mathbf{x}}(n+1) &= \tilde{\mathbf{\Phi}}\tilde{\mathbf{x}}(n) + \tilde{\mathbf{B}}u(n) + \mathbf{G}w(n) \\ \tilde{y}(n) &= \tilde{\mathbf{H}}\mathbf{x}(n) \\ w(n)&: \text{ white unknown-but-bounded process} \\ |w(n)|^2 &\leq Q \end{aligned} \tag{14.9.3}$$

The model reduction is accomplished by finding the (or some) value of $\tilde{\mathbf{\Phi}}$, $\tilde{\mathbf{B}}$, $\tilde{\mathbf{G}}$, and $\tilde{\mathbf{H}}$ with a minimum value of Q such that (14.9.3) is "consistent with" the $y_m(n)$, $n = 1, \ldots, N$, $m = 1, \ldots, M$. In this approach, a deterministic high-dimensional model (14.9.1) is reduced (replaced) by a low-dimensional model with an uncertain (unknown-but-bounded) input. The reduced model generates a set which always contains the true model's output.

There are obviously many possible extensions to the above ideas.

14.12 Nonlinear Structures

Throughout this chapter, the assumed structure of the model to be identified has been linear. In some applications a nonlinear structure may be desired. For example, instead of (14.1.1), one might assume a structure such as

$$\begin{aligned} \mathbf{x}(n+1) &= \boldsymbol{\phi}[\mathbf{x}(n), \mathbf{w}(n), \mathbf{u}(n)] \\ \mathbf{z}(n) &= \mathbf{h}[\mathbf{x}(n), \mathbf{v}(n)] \\ \boldsymbol{\alpha}&: \text{ vector of unknown parameter values} \end{aligned}$$

Three ways to tackle such a problem are

1. Use equations of Section 13.2 to get a likelihood function which is then maximized over $\boldsymbol{\alpha}$.
2. Combine the linearized recursive filters of Section 13.10 with the maximum likelihood results of Sections 14.3 and 14.4.
3. Extend the model reference ideas of Section 14.6 to nonlinear structures in any reasonable way.

The first approach is general but impractical except in very special cases. The second approach can be very effective provided that the linearization is valid (small errors). The effectiveness of the third approach of course depends on the "reasonableness" of the engineer. As expected, the second and third approaches can yield the same results.

As noted earlier in this chapter, a linear structure may yield acceptable results even if the basic system is nonlinear (but not "too" nonlinear). If $\mathbf{Q}$ is identified, its "size" may effectively compensate for the unmodeled nonlinearities.

14.13 Discussion

Only system identification techniques related to the concepts of earlier chapters were discussed. However, the techniques that were discussed are useful in an extremely wide variety of problems.

The basic three step procedure of system identification discussed here is:

1. Hypothesize mathematical structure with unknown parameters $\boldsymbol{\alpha}$.
2. Estimate numerical values of $\boldsymbol{\alpha}$.
3. Test validity of model.

where many "iterations" with new hypothesized structures may be required. The first step *hypothesize* is up to the reader. The discussions of this chapter have concentrated on the second step *estimate*. However, the third step *test* is equally important and must never be ignored.

In Sections 14.2–14.8, five different approaches (state augmentation, maximum likelihood, model reference, correlation, and instrument variables) to the problem of estimating the unknown parameters ($\boldsymbol{\alpha}$) were discussed. All five are applicable to stochastic type input and observation uncertainties. Figure 14.3.1 summarizes the author's opinions on their usefulness. When tackling a system identification problem, the author feels the following approach is widely applicable.

1. Make a Gaussian assumption and formulate a maximum likelihood

State Augmentation	General approach but not tailored to specific needs of system identification. Not recommended except as possible method of tracking small time variations in system parameter values.
Maximum Likelihood: Gaussian case	Most powerful general approach. Disadvantage is possible computation requirements. Gaussian assumption can usually be ignored.
Model Reference: Weighted Least Squares	Conceptually simple. Choice of criterion can be difficult. Yields equations similar to those of maximum likelihood.
Correlation Methods	Conceptually simple. Choice of criterion can be difficult. Can yield easy to implement equations.
Instrument Variables	Yields easy to implement equations. Choice of instrument variables can be difficult.

Figure 14.13.1 Comparison of parameter estimation techniques for system identification.

approach. Implement it if possible (even if the Gaussian assumption is not valid).

2. If maximum likelihood is implemented, consider presenting it to others (in reports and presentations) in terms of model reference concepts. This makes the approach more understandable to those without specialized training.

3. If the computation requirements of maximum likelihood are unacceptable, consider correlation or instrument variable concepts or some simple model reference criterion which is hopefully related to the maximum likelihood equations.

Historical Notes and References: Chapter 14

System identification has been approached in many ways by many authors, and no attempt will be made to provide even a partially complete discussion of past work. However, the work of K. Åström and his associates on maximum likelihood techniques deserves special mention as they pioneered theoretical developments and, even more important, actual applications.

Many of the ideas of this chapter arose from discussions with F. Galiana, R. Moore, and R. Masiello. See Moore and Schweppe [1]; Galiana and Schweppe [1]; DeVille and Schweppe [1] for application to electric power system problems. The discussions of Section 14.3.3 on using the "smoothing form" were motivated by Bar-Shalom [1]. There is a close tie between system identification and econometric theory as in Wonnacott and Wonnacott [1]. The excellent survey paper by Åström and Eykhoff [1] contains an extensive bibliography.

Exercises

14.1. Do Exercise 11.8.

14.2. Consider (14.4.3). Let $\hat{b}(N)$ and $\hat{\boldsymbol{\theta}}(N)$ denote the estimates with the dependence on the number of observations N explicitly indicated. Write a set of nonlinear difference equations which express $\hat{b}(N+1)$, $\hat{\boldsymbol{\theta}}(N+1)$ in terms of $\hat{b}(N)$, $\hat{\boldsymbol{\theta}}(N)$, $z(N+1)$ and other variables which also satisfy difference equations.

14.3. Consider the Bayesian, Gaussian model

$$x(n+1) = \Phi x(n) + Bu(n)$$
$$z(n) = Hx(n) + v(n)$$
$$x(0) = 0$$

$$u(n)\text{: known}$$

$$E\{v(n)\} = 0$$

$$E\{v(n_1)v(n_2)\} = \begin{cases} R & n_1 = n_2 \\ 0 & n_1 \neq n_2 \end{cases}$$

Assume Φ and R are known. Write explicit equations for the maximum likelihood estimate of

a. H assuming that B is known

b. B assuming that H is known.

14.4. Consider the zero mean Bayesian model

$$x(n+1) = \Phi x(n) + w(n) \qquad n = 1 \ldots N$$

$$z(n) = x(n) + v(n)$$

$$x(0) = 0$$

$$E\{w(n_1)w(n_2)\} = \begin{cases} Q & n_1 = n_2 \\ 0 & n_1 \neq n_2 \end{cases}$$

$$E\{v(n_1)v(n_2)\} = \begin{cases} R & n_1 = n_2 \\ 0 & n_1 \neq n_2 \end{cases}$$

Assume N is large.

a. Consider a model reference approach where $\alpha = \Phi$ and $\hat{\Phi}$ minimizes

$$J = \sum_{n=1}^{N} [z(n) - \Phi z(n-1)]^2$$

Convince yourself that $\hat{\Phi}$ is biased; i.e.

$$E\{\hat{\Phi}\} \neq \Phi_t$$

$$\Phi_t\text{: true value of } \Phi$$

Hint: Since

$$z(n) = \Phi_t z(n-1) + w(n-1) + v(n) - \Phi_t v(n-1)$$

then

$$\Phi_t - \hat{\Phi} = \frac{\Phi_t \sum_{n=1}^{N} v^2(n-1)}{\sum_{n=1}^{N} z^2(n-1)} + \text{other terms}$$

b. Consider an instrument variable approach. Choose $i(n) = z(n-2)$. Is the resulting $\hat{\Phi}$ biased?

c. Consider a correlation approach. Define

$$\hat{\mathcal{P}}_z(m) = \frac{1}{N-m} \sum_{n=m}^{N} z(n)z(n-m)$$

Assume R and Q are known and write $\hat{\Phi}$ in terms of $\mathcal{P}_z(0)$.

d. Consider a correlation approach where only R is known (Q, Φ unknown). Write $\hat{\Phi}$ in terms of $\hat{\mathcal{P}}_z(1)$, $\hat{\mathcal{P}}_z(0)$. Is $\hat{\Phi}$ biased?

e. Consider a correlation approach where R, Q and Φ are unknown. Write $\hat{\Phi}$ in terms of $\hat{\mathcal{P}}_z(2)$, $\hat{\mathcal{P}}_z(1)$. Compare with instrument variable estimate.

f. Write the equations which must be solved to get a maximum likelihood approach. Do for both filtering and smoothing forms of likelihood function. Do not attempt to solve.

g. Consider a state augmentation approach where R and Q are known while Φ is unknown. Write the equations for feedback tracking solution using the ideas of Section 13.10.3.

14.6. Consider the model of Section 14.4.2 with $u(n) = 0$. This model can be rewritten in different notation as

$$z(n) = \mathbf{H}(n)\mathbf{x} + v(n)$$

$$\mathbf{x} = \boldsymbol{\theta} \qquad v(n) = w(n-1)$$

$$\mathbf{H}(n) = \mathbf{z}'_{n,K_1}$$

Application of the linear estimation theory (Fisher) of Part II to these equations yields

$$\hat{\mathbf{x}} = \boldsymbol{\Sigma}(n)Q^{-1}\sum_{n-1}^{N}\mathbf{H}(n)z(n)$$

$$\boldsymbol{\Sigma}(n) = Q\left[\sum_{n=1}^{N}\mathbf{H}'(n)\mathbf{H}(n)\right]$$

Show that these estimates are the same as (14.4.3).

14.7. Consider the Bayesian, Gaussian zero mean model

$$\begin{bmatrix} x_1(n+1) \\ x_2(n+1) \end{bmatrix} = \begin{bmatrix} 0 & 0 \\ 1 & 0 \end{bmatrix}\begin{bmatrix} x_1(n) \\ x_2(n) \end{bmatrix} + \begin{bmatrix} u(n) \\ 0 \end{bmatrix}$$

$$z(n) = [H_1 \quad H_2]\begin{bmatrix} x_1(n) \\ x_2(n) \end{bmatrix} + v(n)$$

$$\mathbf{x}(0) = 0$$

$$u(n)\text{: known input}$$

$$E\{v(n_1)v(n_2)\} = \begin{cases} R & n_1 = n_2 \\ 0 & n_1 \neq n_2 \end{cases}$$

What are the equations for the maximum likelihood estimate of H_1 and H_2? *Hint:* Use the theory of Section 14.4.3 or express the model as a moving average and use Section 14.4.4.

14.8. Consider the unknown-but-bounded model

$$x(n+1) = \Phi x(n) + w(n)$$

$$z(n) = x(n)$$

$$|w(n)| \leq Q^{1/2}$$

Assume that Φ and Q are unknown. What equations would you use to obtain an estimate of the values of Φ and Q?

15

Stochastic Control

The basic models to be considered are Bayesian of the form

$$
\begin{aligned}
&\mathbf{x}(n+1) = \boldsymbol{\phi}[\mathbf{x}(n)] + \mathbf{G}\mathbf{w}(n) + \mathbf{B}\mathbf{u}(n) \\
&\mathbf{z}(n) = \mathbf{h}[\mathbf{x}(n)] + \mathbf{v}(n) \\
&\mathbf{w}(n)\text{: zero mean white Gaussian process} \\
&E\{\mathbf{w}(n_1)\mathbf{w}'(n_2)\} = \begin{cases} \mathbf{Q}, & n_1 = n_2 \\ 0, & n_1 \neq n_2 \end{cases} \\
&\mathbf{v}(n)\text{: zero mean white Gaussian process} \\
&E\{\mathbf{v}(n_1)\mathbf{v}'(n_2)\} = \begin{cases} \mathbf{R}, & n_1 = n_2 \\ \mathbf{0}, & n_1 \neq n_2 \end{cases} \\
&\mathbf{x}(0)\text{: Gaussian vector} \\
&E\{\mathbf{x}(0)\} = \mathbf{m} \qquad E\{\mathbf{x}(0)\mathbf{x}'(0)\} = \boldsymbol{\psi} \\
&\mathbf{w}(n), \mathbf{v}(n), \mathbf{x}(0)\text{: independent} \\
&\mathbf{u}(n)\text{: control signal}
\end{aligned} \tag{15.1}
$$

These are not the most general possible models, but they suffice to present the basic concepts. The $\boldsymbol{\phi}[\,]$, $\mathbf{G}$, $\mathbf{B}$, $\mathbf{h}[\,]$, $\mathbf{R}$, and $\mathbf{Q}$ are assumed to be time-invariant only to simplify the notation. The stochastic control problem is to specify the controls (inputs) $\mathbf{u}(n)$, $n = 1, \ldots, N$, so that the state $\mathbf{x}(n)$, $n = 1, \ldots, N$, or some function thereof behaves in some desired fashion subject to possible constraints on the allowable controls. The final time, N, may or may not be fixed. The final time may be infinity.

Stochastic control is a big field in which a wide variety of results of varying usefulness has been obtained. Loosely speaking there are two parts to the solution of a stochastic control problem:

1. Formulation of the stochastic aspects as a well-defined, meaningful deterministic problem.
2. Determination of the desired controls.

An almost equivalent way to view these two steps is

1. Convert the original dynamic system involving uncertainties (stochastic processes) into a deterministic dynamic system whose states are probability densities (or means, covariance matrices, etc.)
2. Use deterministic control design techniques on the new deterministic dynamic system.

This point of view was discussed briefly in Chapter 2. In this chapter, the first step of problem formulation is emphasized. To get "optimum" controls, the second step of solution often employs mathematical optimization techniques such as dynamic programming, the maximum principle, the calculus of variations, and the Kuhn-Tucker theorem. These techniques are not assumed to be part of the readers' background, and it is ridiculous to try to present them here, as many excellent books are available. Hopefully the stochastic control problem formulations to be discussed enable a reader with such mathematical optimization techniques at his disposal to proceed on his own.

There are many ways to formulate a stochastic control problem. General types of formulations are discussed in Section 15.1. Sections 15.2 and 15.3 discuss two general classes of problems, open loop and closed loop. In Section 15.4 and 15.5, an important and useful result, the separation theorem for closed-loop controls, is discussed. Adaptive control and the dual control problems are mentioned in Sections 15.6 and 15.7. Unknown-but-bounded models are discussed very briefly in Section 15.8.

The discussions are presented in terms of discrete-time systems. However, extension of the concepts to continuous-time systems is relatively straightforward. The Gaussian assumption in (15.1) is made primarily for convenience, and many of the ideas to follow apply to non-Gaussian models as well.

Some of the ideas are closely related to general concepts discussed in Chapter 2. It may be appropriate to briefly reread Chapter 2 before proceeding.

15.1 Types of Stochastic Control Problems

Two important features specify the type of problem: the criteria and the open-versus closed-loop question.

15.1.1 Criteria

Discussions center on mathematical criteria; i.e., scalar functions whose magnitude specifies how well a particular control performs. In practice, consideration of a single scalar criteria is rarely sufficient to specify a control and engineering judgment is required to make the necessary tradeoffs between conflicting desires. Scalar criteria are discussed here merely to illustrate what is possible.

Consider for a moment a deterministic problem: $\mathbf{w} = \mathbf{v} = 0$, $\mathbf{x}(0)$ known. A widely used class of criteria is to choose the $\mathbf{u}(n)$, $n = 1, \ldots, N$, to minimize a cost $\tilde{c}$ given by

$$\tilde{c} = \sum_{n=1}^{N} \{c_1[\mathbf{x}(n)] + c_2[\mathbf{u}(n-1)]\} \tag{15.1.1}$$

where the $c_1[\,]$ and $c_2[\,]$ are always nonnegative. This is not the most general criterion but it suffices for present purposes.

Now consider (15.1) with the uncertainties present. A cost $\tilde{c}$ such as (15.1.1) is still reasonable, but $\tilde{c}$ is now a random variable whose probability density $p(\tilde{c})$ is determined, in part, by the chosen controls $\mathbf{u}(1) \cdots \mathbf{u}(N)$. Hence one reasonable criterion is to choose the $\mathbf{u}$ to minimize the cost c given by

$$c = E\{\tilde{c}\} = \int \tilde{c}p(\tilde{c})\, d\tilde{c} \tag{15.1.2}$$

where $\tilde{c}$ is given by (15.1.1). Further discussions center on "average" costs as in (15.1.2), but other criteria are possible. For example, one alternative criterion is to choose the $\mathbf{u}$ to minimize

$$\int_{c_0}^{\infty} p(\tilde{c})\, d\tilde{c}$$

Another criterion is to choose the $\mathbf{u}$ to minimize $\hat{c}$, where $\hat{c}$ is the value of $\tilde{c}$ which maximizes $p(\tilde{c})$.

A somewhat different, but related, criterion occurs when it is desired to make $\mathbf{x}(N)$ for specified N close to some target set. A reasonable criterion for such a desire is to choose $\mathbf{u}$ to *maximize the probability* $\mathbf{x}(N) \in \mathbf{\Omega}_{\text{target}}$, where $\mathbf{\Omega}_{\text{target}}$ is some specified set.

15.1.2 Open-Loop and Closed-Loop Control

Consider (15.1). There are two fundamentally different ways the controls can be specified.

1. Open loop: $\mathbf{u}(n)$, $n = 0, \ldots, N-1$, is a fixed time function completely specified before the control starts.
2. Closed Loop: $\mathbf{u}(n)$ is determined as some specific function of $\mathbf{z}(1), \cdots \mathbf{z}(n)$.

In the open-loop case, $\mathbf{u}(n)$, $n = 0, \ldots, N-1$, is a deterministic time function. In the closed-loop case, $\mathbf{u}(n)$, $n = 0, \ldots, N-1$, is a stochastic process, as it is a function of a stochastic process.

Define

$$\mathbf{z}_n = \begin{bmatrix} \mathbf{z}(1) \\ \cdot \\ \cdot \\ \cdot \\ \mathbf{z}(n) \end{bmatrix}$$

Then a closed-loop control is often indicated by writing

$$\mathbf{u}(n) = \mathbf{u}(n \colon \mathbf{z}_n)$$

to emphasize its dependence on the observations. Note that this is a *causal* control in that the control at time n depends only on the observations up to time n. Another case of interest occurs when the control depends only on $\mathbf{z}(1) \cdots \mathbf{z}(n-1)$, i.e., $\mathbf{u}(n \colon \mathbf{z}_{n-1})$. For simplicity, only $\mathbf{u}(n \colon \mathbf{z}_n)$ is discussed.

A special type of closed-loop control problem occurs when the complete state of the system can be observed perfectly, i.e., when $\mathbf{z}(n) = \mathbf{x}(n)$. In such a case, it follows that one can consider $\mathbf{u}(n) = \mathbf{u}[n \colon \mathbf{x}(n)]$. Of course, even in this special case, the control is a stochastic process.

Let $\mathbf{u}^*(n)$ and $\mathbf{u}^*(n \colon \mathbf{z}_n)$ denote the optimum open-loop controls and closed-loop controls which minimize some cost c. Let c^*_{open} and c^*_{closed} denote the resulting minimum costs. Then

$$c^*_{\text{closed}} \leq c^*_{\text{open}} \tag{15.1.3}$$

as the class of functions $\mathbf{u}(n \colon \mathbf{z}_n)$ of n and $\mathbf{z}_n$ includes, as a special case, time functions $\mathbf{u}(n)$ which do not depend on $\mathbf{z}_n$.

15.1.3 Constraints

Often it is desired to impose constraints which must be satisfied.

Consider first the question of constraints on the allowable controls. For open-loop controls, common constraints are

$$\textit{Peak amplitude}\text{ type:}\quad \begin{aligned} &\mathbf{u}(n) \in \Omega_u(n) \\ &\Omega_u(n)\text{: specified set} \end{aligned} \tag{15.1.4}$$

$$\textit{Energy}\text{ type:}\quad \begin{aligned} &\sum_{n=1}^{N-1} b[\mathbf{u}(n)] \leq \beta \\ &\beta\text{: fixed number} \end{aligned} \tag{15.1.5}$$

For scalar u, common examples are

$$\mathbf{\Omega}_u(n) = \{\mathbf{u}: |u| \leq 1\}$$

$$b[u(n)] = |u(n)| \qquad \text{or} \qquad u^2(n)$$

For closed-loop controls, the stochastic nature of the **u** introduces more possibilities. Constraints such as (15.1.4) and (15.1.5) are still possible but it is easier to handle average constraints such as

$$E\{\mathbf{u}(n: \mathbf{z}_n)\} \in \mathbf{\Omega}_u(n)$$

$$E\left\{\sum_{n=0}^{N-1} b[\mathbf{u}(n: \mathbf{z}_n)]\right\} \leq \beta$$

Now consider constraints on the allowable states $\mathbf{x}(n)$. Desirable constraints are often of the form

$$\begin{gathered} \mathbf{x}(n) \in \mathbf{\Omega}_x(n), \qquad n = 1, \ldots, N \\ \mathbf{\Omega}_x(n)\text{: specified set} \end{gathered} \tag{15.1.6}$$

Such constraints are often difficult to satisfy exactly because of the stochastic nature of the $\mathbf{x}(n)$. So it is tempting to use instead

$$E\{x(n)\} \in \mathbf{\Omega}_x(n), \qquad n = 1, \ldots, N$$

15.1.4 Discussion of Types of Problems

It is clear that the number of possible types of stochastic control problems is limited primarily by one's imagination. The preceding discussions have only skimmed the surface. Perhaps the single most important concept is the difference between open and closed-loop controls. They are fundamentally different.

The Gaussian assumption was not employed in the above discussions.

15.2 Open-Loop Control

In an open-loop control problem, it is desired to determine the control as a function of time before the "action takes place." Hence the observations $\mathbf{z}(n)$ play no role, and the only aspect of the model of interest is the dynamics

$$\mathbf{x}(n + 1) = \boldsymbol{\phi}[\mathbf{x}(n)] + \mathbf{G}\mathbf{w}(n) + \mathbf{B}\mathbf{u}(n) \tag{15.2.1}$$

plus the desired criterion and constraints.

Conceptually one simply uses the concepts of Section 13.1 to evaluate the evolution of the probability density of the state (i.e., $p[\mathbf{x}(n)]$ of (13.1.1)) for arbitrary controls, $\mathbf{u}(n)$, $n = 0, \ldots, N - 1$. The result is a functional equation such as

$$p[\mathbf{x}(n + 1)] = g\{p[\mathbf{x}(n)], \mathbf{u}(n)\} \tag{15.2.2}$$

Conceptually the design of the controls then proceeds exactly as with deterministic systems except that functional equations such as (15.2.2) are viewed as dynamic systems to be controlled. If an "optimum" open-loop control is desired, the appropriate mathematical optimization techniques are then used to find the controls $\mathbf{u}(n)$, $n = 0, \ldots, N-1$, which minimize the cost and satisfy the constraints. This logic also applies to non-Gaussian models.

The difficulty lies in the fact that functional difference equations are difficult dynamic systems to analyze. The resulting deterministic control problems are even harder to actually solve. Hence only special cases and/or approximate solutions are usually attempted. Two such examples are now discussed.

15.2.1 System, Signal, Measurement, etc., Design

In Sections 8.3 and 10.9.5, system, signal, and measurement design problems were discussed for estimation and hypothesis-testing problems, respectively. One problem considered was to design signals $n = 1, \ldots, N$ or $0 \leq t \leq T$ which give the "best" estimates or decisions subject to constraints on the class of allowable signals. The criteria and dynamics are not of the same form being considered in this chapter, but signal design can be viewed as a special case of open-loop control.

The hypothesis-testing signal design concepts were based on the use of distance measures such as the Bhattacharyya distance instead of probability densities. This was done because the deterministic dynamic system for a distance is a set of ordinary difference or differential equations instead of the functional difference or partial differential equations which govern the probability densities. The use of distance measures instead of probability densities can be viewed as a technique to get a more tractable problem.

15.2.2 Linear Systems

The difficulties of working with probability densities go away for linear dynamic systems such as

$$\mathbf{x}(n+1) = \mathbf{\Phi}\mathbf{x}(n) + \mathbf{G}\mathbf{w}(n) + \mathbf{B}\mathbf{u}(n)$$

Because of the Gaussian assumption on $\mathbf{x}(0)$ and $\mathbf{w}(n)$, $\mathbf{x}(n)$ is also Gaussian; i.e.,

$$\begin{aligned} \mathbf{x}(n) &\text{ is } N[\mathbf{m}(n), \mathbf{\Gamma}(n)] \\ \mathbf{m}(n+1) &= \mathbf{\Phi}\mathbf{m}(n) + \mathbf{B}\mathbf{u}(n) \\ \mathbf{m}(0) &= E\{\mathbf{x}(0)\} \\ \mathbf{\Gamma}(n+1) &= \mathbf{\Phi}\mathbf{\Gamma}(n)\mathbf{\Phi}' + \mathbf{G}\mathbf{Q}\mathbf{G}' \\ \mathbf{\Gamma}(0) &= \boldsymbol{\psi} \end{aligned} \tag{15.2.3}$$

Hence the ordinary difference equations for $\mathbf{m}(n)$ and $\mathbf{\Gamma}(n)$ can be viewed as the desired deterministic dynamic system in place of the functional difference equation for the probability density. A further simplification arises because the control $\mathbf{u}(n)$ affects only the mean value $\mathbf{m}(n)$ of $\mathbf{x}(n)$. The covariance matrix, $\mathbf{\Gamma}(n)$, is independent of the control, but for general cost such as

$$c = E \sum_{n=1}^{N} \{c_1[\mathbf{x}(n)] + c_2[\mathbf{u}(n-1)]\}$$

the cost does depend on how $\mathbf{\Gamma}(n)$ behaves.

Let $\mathbf{u}^*(n)$ denote the optimum open-loop control which minimizes c. Then, in general, $\mathbf{u}^*(n)$, $n = 0, \ldots, N-1$, depends on $c_1[\,]$, $c_2[\,]$, $\mathbf{\Phi}$, $\mathbf{B}$, $\mathbf{G}$, $\mathbf{Q}$, $\boldsymbol{\psi}$, and n.

Consider a criterion with a special cost function of the form

$$\begin{aligned} c &= E \sum_{n=1}^{N} \{[\mathbf{x}(n) - \mathbf{d}(n)]'\mathbf{C}_1[\mathbf{x}(n) - \mathbf{d}(n)] + c_2[\mathbf{u}(n-1)]\} \\ &\mathbf{d}(n),\ n = 1, \ldots, N\text{: specified} \\ &\qquad\qquad \mathbf{C}_1\text{: positive definite matrix} \end{aligned} \tag{15.2.4}$$

The $\mathbf{d}(n)$ can be viewed as some desired trajectory which $\mathbf{x}(n)$ is to follow. Expansion of (15.2.4) yields

$$\begin{aligned} c &= \sum_{n=1}^{N} \{\text{tr}[\mathbf{\Gamma}(n)\mathbf{C}_1]\} + \tilde{c} \\ \tilde{c} &= \sum_{n=1}^{N} \{[\mathbf{m}(n) - \mathbf{d}(n)]'\mathbf{C}_1[\mathbf{m}(n) - \mathbf{d}(n)] + c_2[\mathbf{u}(n-1)]\} \end{aligned} \tag{15.2.5}$$

The open-loop control problem can be viewed as involving a deterministic dynamic system

$$\mathbf{m}(n+1) = \mathbf{\Phi m}(n) + \mathbf{Bu}(n)$$

with criterion cost $\tilde{c}$. This deterministic control problem is of a form which has been investigated extensively for various control costs $c_2[\mathbf{u}(n)]$ and constraints on the allowable $\mathbf{u}(n)$, $n = 0, \ldots, N-1$. For the special cost of (15.2.4), the optimum control $\mathbf{u}^*(n)$, $n = 0, \ldots, N-1$, depends on only $\mathbf{\Phi}$, $\mathbf{B}$, $\mathbf{C}_1$, $\mathbf{c}_2[\,]$, $\mathbf{m}$, and $\mathbf{d}(n)$, $n = 1, \ldots, N$.

15.3 Closed-Loop Control: General

A closed-loop stochastic control problem differs from an open-loop problem in a fundamental way. The control is to be a function of the observations. Hence the control is a stochastic instead of a deterministic process. Mathematical optimization is in general much harder, as it is desired to find the best function, i.e., to find the best $\mathbf{u}(n\colon \mathbf{z}_n)$, where $\mathbf{u}$ is a function. For a general cost such as (15.1.1) and (15.1.2) the optimum controls $\mathbf{u}^*(n\colon \mathbf{z}_n)$ also depend on $\phi[\,]$, $\mathbf{G}$, $\mathbf{B}$, $\mathbf{Q}$, $\mathbf{h}[\,]$, $\mathbf{R}$, $\boldsymbol{\psi}$, and $\mathbf{m}$.

Because of their difficulty, solution of closed-loop stochastic control problems is almost always an exercise in approximation. An exception to this rule can occur for linear systems, but discussion of this *separation theorem* is deferred to Section 15.4. Various approximate approaches to the general problem are now discussed. The Gaussian assumption is not needed.

15.3.1 A Reasonable Approach

When mathematics does not easily yield the desired answer, one must rely on judgment. The form of a "reasonable" closed-loop stochastic control system is indicated in Fig. 15.3.1. The measurements $\mathbf{z}(m)$, $m = 1, \ldots, n$,

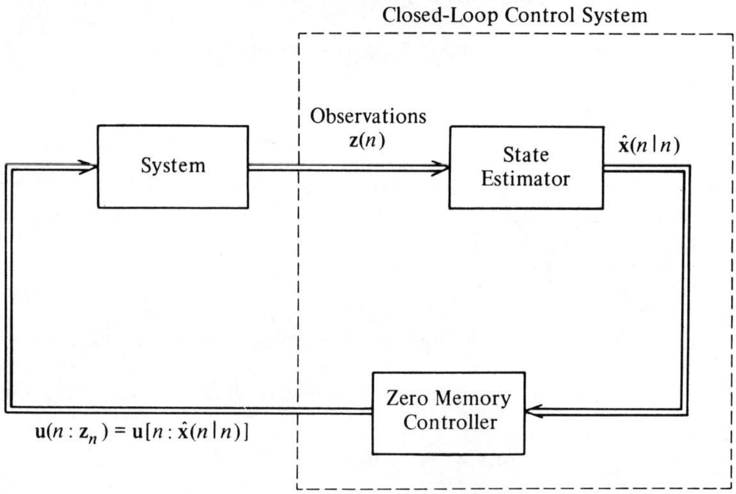

Figure 15.3.1 Form of reasonable closed-loop stochastic control system.

are put through a state estimator to yield $\hat{\mathbf{x}}(n|n)$. This state estimate is then put into a zero memory controller which is such that the control $\mathbf{u}(n\colon \mathbf{z}_n)$ is explicitly only a function of $\hat{\mathbf{x}}(n|n)$, i.e., $\mathbf{u}[n\colon \hat{\mathbf{x}}(n|n])$, where, of course, $\hat{\mathbf{x}}(n|n)$ depends on $\mathbf{z}_n$. There are various ways the control law (i.e., the function $\mathbf{u}[n\colon \hat{\mathbf{x}}(n|n)]$ can be specified.

The control law can be developed assuming that the state estimate $\hat{\mathbf{x}}(n|n)$ is perfect; i.e., $\hat{\mathbf{x}}(n|n) = \mathbf{x}(n)$. This is equivalent to assuming that the observations $\mathbf{z}(n)$ are perfect observations of the whole state; i.e., $\mathbf{z}(n) = \mathbf{x}(n)$. The control law design may or may not include consideration of the effects of the input disturbance, $\mathbf{w}(n)$. The behavior of such a control can then be analyzed taking into account the errors in $\hat{\mathbf{x}}(N|N)$ to see if the performance is satisfactory. If the performance is not satisfactory, an engineering judgment modification may be possible.

If there are *state space constraints* such as (15.1.6), a reasonable approach is to try to choose a control so that

$$\hat{\mathbf{x}}(n|n) \in \tilde{\Omega}_x(n) \qquad n = 1, \ldots, N$$

where $\tilde{\Omega}_x(n)$ is the original set $\Omega_x(n)$ reduced by the "size of the errors" expected in $\hat{\mathbf{x}}(N|N)$.

EXAMPLE: Consider

$$x(n+1) = \Phi x(n) + w(n) + u(n)$$

$$z(n) = x(n) + v(n)$$

$$x(0) = 0$$

Assume that it is desired to choose a closed-loop control $u(n|\mathbf{z}_n)$ so that $x(N) \approx 10$ but $x(n)$ "never" is larger than 15. Let $\hat{x}(n|n)$ be an estimate obtained using the equations of Chapter 6 with $\Sigma(n|n)$ the corresponding error variance. Then a reasonable approach is to choose

$$u(n|\mathbf{z}_n) = K(n)\hat{x}(n|n) + b$$

where b and $K(n)$, $n = 1, \ldots,$ are chosen so that $E\{x(N)\}$, is as close to 10 as possible subject to the constraint that

$$E\{x(n)\} + \alpha\sqrt{\Sigma(n|n)} \leq 15, \qquad \text{all } n$$

where $\alpha = 3$ is a reasonable value.

15.3.2 Fixed Structural Form

A variation on the reasonable approach of Fig. 15.3.1 is to specify a priori the structural form of the control law and then use mathematical optimization techniques to determine the best control law of the specified structure.

A simple example is the best way to illustrate the concept. Assume that a general structure of the form of Fig. 15.3.1 is to be used. Further assume that the control at time n is constrained to be a linear function of $\hat{\mathbf{x}}(n|n)$ such as

$$\mathbf{u}(n\colon \mathbf{z}_n) = \mathbf{K}(n)\hat{\mathbf{x}}(n|n) + \mathbf{b}(n)$$

where $\mathbf{K}(n)$ is a gain matrix. Use of the control law in Fig. 15.3.1 enables (conceptually) the evaluation of the probability density $p[\mathbf{x}(n)]$ (or some related function) of $\mathbf{x}(n)$, $n = 1, \ldots, N$, in terms of the "free" time functions $\mathbf{K}(n)$, $n = 0, \ldots, N-1$. Determination of the optimum $\mathbf{K}(n)$, $\mathbf{b}(n)$, $n = 0, \ldots, N-1$, to give the desired behavior of $p[\mathbf{x}(n)]$ is thus conceptually as done in open-loop control and signal design problems. (See Example of Section 15.3.1.)

15.3.3 Closed-Open-Loop Control

A common approach to closed-loop control is to use a closed-loop control designed using open-loop logic.

Consider a control problem where it is desired to determine $\mathbf{u}(n\colon \mathbf{z}_n)$, $n = 0, \ldots, N - 1$. Consider some time $m < N$. Assume that it is possible to use the observations $\mathbf{z}_m$ up to time m to determine the optimum open-loop control for the remaining times $n = m, \ldots, N$. Let $\mathbf{u}(n\colon \mathbf{z}_m)$, $n = m, \ldots, N$, denote the resulting open-loop controls for the remaining time. Now consider the following logic:

1. At $n = 0$, compute the optimum open-loop control $\mathbf{u}(n)$, $n = 0, \ldots, N - 1$, from a priori information. Apply $\mathbf{u}(0)$.
2. At $n = 1$, compute the optimum open-loop control $\mathbf{u}(n\colon \mathbf{z}_1)$ for $n = 1, \ldots, N - 1$ from $\mathbf{z}(1)$. Apply $\mathbf{u}(1\colon \mathbf{z}_1)$.
3. At $n = m$, compute the optimum open-loop control $\mathbf{u}(n\colon \mathbf{z}_m)$ for $n = m, \ldots, N - 1$ from $\mathbf{z}(1) \cdots \mathbf{z}(m)$. Apply $\mathbf{u}(m\colon \mathbf{z}_m)$.

etc.

The process is continued until the final time—or indefinitely if N is infinite. The result is a closed-loop control that is based on an open-loop logic and is called *closed-open*. It is also called open-loop feedback control.

An optimum open-loop control $\mathbf{u}(n)$, $n = 0, \ldots, N - 1$, can be viewed as the best control that is possible by ignoring all observations $\mathbf{z}(n)$, $n = 1, \ldots, N$. Thus this closed-open-loop logic can be viewed as a control which uses the available observations but ignores the fact that future observations will be made. To illustrate the difference between the "optimum" closed-open-loop control and the truly optimum closed-loop control, consider a case where the covariance matrix of the input and observation uncertainty $\mathbf{w}(n)$ and $\mathbf{v}(n)$ are time-varying, i.e., $\mathbf{Q}(n)$ and $\mathbf{R}(n)$. Let $u^*_{\text{closed}}(n\colon \mathbf{z}_n)$ and $\mathbf{u}^*_{\text{closed-open}}(n\colon \mathbf{z}_n)$ denote the respective optimum controls. Then

$$\mathbf{u}^*_{\text{closed}}(n\colon \mathbf{z}_n) \text{ depends on } \mathbf{z}_n, \boldsymbol{\phi}[\,], \mathbf{G}, \mathbf{B}, \mathbf{h}[\,], \boldsymbol{\psi}, n\colon$$

$$\mathbf{Q}(m), \qquad m = 1, \ldots, N - 1$$

$$\mathbf{R}(m), \qquad m = 1, \ldots, N - 1$$

while

$$\mathbf{u}^*_{\text{closed-open}}(n\colon \mathbf{z}_n) \text{ depends on } \mathbf{z}_n, \boldsymbol{\phi}[\,], \mathbf{G}, \mathbf{B}, \mathbf{h}[\,], \boldsymbol{\psi}, n\colon$$

$$\mathbf{Q}(m), \qquad m = 1, \ldots, N - 1$$

$$\mathbf{R}(m), \qquad m = 1, \ldots, n$$

In some cases the difference between $\mathbf{u}^*_{\text{closed}}(n\colon \mathbf{z}_n)$ and $\mathbf{u}^*_{\text{closed-open}}(n\colon \mathbf{z}_n)$ may be small or even zero, but in other cases the two types of controls may yield radically different performances.

A closed-open-loop control is often restricted to the form of Fig. 15.3.1. Thus the control at time m is an explicit function of $\hat{\mathbf{x}}(m\,|\,m)$ rather than $\mathbf{z}_m$.

EXAMPLE: Consider

$$x(n+1) = \Phi x(n) + w(n) + u(n), \qquad n = 1, \ldots, 20$$

$$z(n) = x(n) + v(n)$$

$$E[x(0)] = m$$

$$E\{(x(0) - m)^2\} = \psi$$

$$E[v(n_1)v(n_2)] = \begin{cases} R(n_1), & n_1 = n_2 \\ 0, & n_1 \neq n_2 \end{cases}$$

Assume that

$$\psi: \text{ very large}$$

$$R(1) \cdots R(10): \text{ very large}$$

$$R(11) \cdots R(20): \text{ very small}$$

Thus the initial conditions are very poorly known and the observations are very noisy until $n = 11$, so that the state is "effectively unknown" for the first 10 time steps. Intuitively a "good" control law would wait until $n > 10$ to apply control (make $u(n) \neq 0$) as any nonzero control used during $n = 1, \ldots, 10$ will be based on "bad information" and will probably be wasted. An optimum closed- open loop control would not have this property, as it would not know that good information will become available after $n > 10$. The optimum closed-loop control would employ this knowledge and hence would probably wait until $n > 10$ to apply control.

15.4 Separation Theorem: General

The reasonable closed-loop control system of Fig. 15.3.1 can be viewed as a separation of the overall closed-loop control system into two parts:

1. State estimator: A dynamic system with memory which converts $\mathbf{z}_n$ into $\hat{\mathbf{x}}(n \mid n)$.
2. Zero memory control: A nondynamic system (no memory) which converts $\hat{\mathbf{x}}(n \mid n)$ into $\mathbf{u}[n: \hat{\mathbf{x}}(n \mid n)]$.

In some cases this separation is optimum as well as reasonable. This important result goes by many names. In this book it is called the *separation theorem.*

There are actually several versions of the separation theorem. Two of them are summarized in Fig. 15.4.1. In this section the first, more general version (I) is discussed. The second more explicit version (II) is considered in Section 15.5, where the matrix $\mathbf{A}(n)$, etc., are defined.

A rigorous derivation of the most general version of the separation theorem is not attempted. However, heuristic discussions are provided to convey some of the basic ideas.

Basic Assumption:
Linear system, Bayesian model, Gaussian
Definitions:
$\mathbf{u}^*(n:\mathbf{z}_n)$ is optimum closed-loop control
Two versions of Separation Theorem:

I. $\mathbf{u}^*(n:\mathbf{z}_n)$ can be obtained as in Figure 15.3.1, i.e. $\mathbf{u}^*(n:\mathbf{z}_n)$ can be written as an explicit function of $\hat{\mathbf{x}}(n\,|\,n)$ rather than $\mathbf{z}_n$.

II. Assume a quadratic criterion is used. Let $\mathbf{u}^*_{\text{det}}(n:\mathbf{x}(n)) = \mathbf{A}(n)\mathbf{x}(n)$ be the optimum control for the corresponding deterministic problem ($\boldsymbol{\psi} = \mathbf{0}$, $\mathbf{Q} = \mathbf{0}$, $\mathbf{R} = \mathbf{0}$). Then $\mathbf{u}^*(n:\mathbf{z}_n) = \mathbf{u}^*_{\text{det}}(n:\hat{\mathbf{x}}(n\,|\,n)) = \mathbf{A}(n)\hat{\mathbf{x}}(n\,|\,n)$

Figure 15.4.1. Two versions of separation theorem.

Consider a Bayesian model of the form of (15.1) where it is desired to find the closed-loop control law $\mathbf{u}^*(n:\mathbf{z}_n)$ which minimizes

$$c = E\left\{\sum_{n=1}^{N} c_1[\mathbf{x}(n)] + c_2[\mathbf{u}(n-1:\mathbf{z}_{n-1})]\right\} \tag{15.4.1}$$

subject to constraints if desired. It is felt to be relatively obvious that instead of having the control at time n depend on $\mathbf{z}_n$, it could depend on $p[\mathbf{x}(n)\,|\,\mathbf{z}_n]$, the conditional probability density of $\mathbf{x}(n)$ given $\mathbf{z}_n$. Thus it is reasonable to

$$\text{Replace} \quad \mathbf{u}^*(n:\mathbf{z}_n) \quad \text{by} \quad \mathbf{u}^*(n:p[\mathbf{x}(n)\,|\,\mathbf{z}_n]) \tag{15.4.2}$$

The Gaussian assumption has not yet been used.

Replacement of consideration of $\mathbf{u}^*(n:\mathbf{z}_n)$ by consideration of $\mathbf{u}^*(n:p[\mathbf{x}(n)\,|\,\mathbf{z}_n])$ seems to be a step backward, as $\mathbf{z}_n$ is a vector while $p[\mathbf{x}(n)\,|\,\mathbf{z}_n]$ is a function. However, consider the case where the nonlinear model (15.1) is replaced by the linear model:

$$\begin{aligned} \mathbf{x}(n+1) &= \boldsymbol{\Phi}\mathbf{x}(n) + \mathbf{G}\mathbf{w}(n) + \mathbf{B}\mathbf{u}(n) \\ \mathbf{z}(n) &= \mathbf{H}\mathbf{x}(n) + \mathbf{v}(n) \end{aligned} \tag{15.4.3}$$

Because of the Gaussian assumption, the conditional probability density $p[\mathbf{x}(n)\,|\,\mathbf{z}_n]$ becomes a Gaussian density with mean $\hat{\mathbf{x}}(n\,|\,n)$ and covariance matrix $\boldsymbol{\Sigma}(n\,|\,n)$, where $\hat{\mathbf{x}}(n\,|\,n)$ and $\boldsymbol{\Sigma}(n\,|\,n)$ can be calculated from the linear estimation theory of Chapter 6. Hence for this special case it is possible to

$$\text{Replace} \quad \mathbf{u}^*[n:p[\mathbf{x}(n)\,|\,\mathbf{z}_n]] \quad \text{by} \quad \mathbf{u}^*[n:\hat{\mathbf{x}}(n\,|\,n)] \tag{15.4.4}$$

It would be more precise to write $\mathbf{u}^*[n:\hat{\mathbf{x}}(n\,|\,n), \boldsymbol{\Sigma}(n\,|\,n)]$ instead of $\mathbf{u}^*[n:\hat{\mathbf{x}}(n\,|\,n]$, but $\boldsymbol{\Sigma}(n\,|\,n)$ does not depend on $\mathbf{z}_n$. This is the desired separation (see I in Fig. 15.4.1), as by combining (15.4.2) and (15.4.4) it is possible to

$$\text{Replace} \quad \mathbf{u}^*(n:\mathbf{z}_n) \quad \text{by} \quad \mathbf{u}^*[n:\hat{\mathbf{x}}(n\,|\,n)] \tag{15.4.5}$$

This heuristic argument has thus "shown" that separation (15.4.5) occurs for a linear Gaussian model such as (15.4.3) for very general criteria such as (15.4.1).

It should be noted that for general criteria such as (15.4.1) the resulting optimum closed-loop control $\mathbf{u}^*[n: \hat{\mathbf{x}}(n|n)]$ may be a nonlinear function of $\hat{\mathbf{x}}(n|n)$. Hence even with the linear model and all the Gaussian assumptions on $\mathbf{x}(0)$, $\mathbf{v}(n)$, and $\mathbf{w}(n)$, the resulting controls and hence $\mathbf{x}(n)$ may not be Gaussian. However, the $p[\mathbf{x}(n)|\mathbf{z}_n]$ is still a Gaussian density because it is assumed that at time n the controls $\mathbf{u}$ applied at time $m = 1, \ldots, n-1$ are known. Hence a more precise notation would be to write $p[\mathbf{x}(n)|\mathbf{z}_n]$ as $p[\mathbf{x}(n)|\mathbf{z}_n, \mathbf{u}_{n-1}]$, where

$$\mathbf{u}_{n-1} = \begin{bmatrix} u(0) \\ u(1:\mathbf{z}_1) \\ \cdot \\ \cdot \\ \cdot \\ u(n-1:\mathbf{z}_{n-1}) \end{bmatrix}$$

Similarly, when considering $\mathbf{u}[n: \hat{\mathbf{x}}(n|n)]$, it should be noted that $\hat{\mathbf{x}}(n|n)$ depends on $\mathbf{u}_{n-1}$ as well as on $\mathbf{z}_n$.

The fact that separation takes place for the linear model does not mean the control problem is solved. In general, it can be very difficult to find the optimum function $\mathbf{u}^*[n: \hat{\mathbf{x}}(n|n)]$. In general, $\mathbf{u}^*[n: \hat{\mathbf{x}}(n|n)]$ depends on $\hat{\mathbf{x}}(n|n)$ and on $\mathbf{\Phi}$, $\mathbf{G}$, $\mathbf{B}$, $\mathbf{H}$, $\mathbf{Q}$, $\mathbf{R}$, $\boldsymbol{\psi}$, $\mathbf{m}$, $c_1[\,]$, and $c_2[\,]$.

The replacement of $\mathbf{u}(n: \mathbf{z}_n)$ by $\mathbf{u}[n: p[\mathbf{x}(n)|\mathbf{z}_n]]$ as in (15.4.2) can be viewed as a type of separation theorem which applies to general nonlinear models without any Gaussian assumptions. This serves to emphasize why optimum closed-loop controls are very difficult to calculate. As discussed in Chapter 13, it is very difficult to calculate $p[\mathbf{x}(n)|\mathbf{z}_n]$. Introduction of the control aspects only makes an "impossible" problem more impossible.

Note that the optimum closed-loop $\mathbf{u}^*[n: \hat{\mathbf{x}}(n|n)]$ can be computed or an optimum closed-open-loop $\mathbf{u}^*[n: \hat{\mathbf{x}}(n|n)]$ as in Section 15.3.3 can be computed. Hence the closed versus closed-open-loop question is "independent" of the separation theorem, as it is being discussed here.

15.5 Separation Theorem: Linear Model; Quadratic Criteria

As noted in Section 15.4, even if separation takes place so that $\mathbf{u}^*(n: \mathbf{z}_n)$ can be replaced by $\mathbf{u}^*[n: \hat{\mathbf{x}}(n|n)]$, it may still be very difficult to calculate $\mathbf{u}^*[n: \hat{\mathbf{x}}(n|n)]$. However, the special case of quadratic criteria and no constraints is an exception which yields version II in Fig. 15.4.1.

The model to be considered is

$$\begin{gathered}\mathbf{x}(n+1)=\mathbf{\Phi}\mathbf{x}(n)+\mathbf{G}\mathbf{w}(n)+\mathbf{B}\mathbf{u}(n)\\ \mathbf{z}(n)=\mathbf{H}\mathbf{x}(n)+\mathbf{v}(n)\\ \mathbf{w}(n),\mathbf{v}(n),\mathbf{x}(0)\text{: independent zero mean Gaussian}\\ E\{\mathbf{w}(n_1)\mathbf{w}'(n_2)\}=\begin{cases}\mathbf{Q}, & n_1=n_2\\ \mathbf{0}, & n_1\neq n_2\end{cases}\\ E\{\mathbf{v}(n_1)\mathbf{v}'(n_2)\}=\begin{cases}\mathbf{R}, & n_1=n_2\\ \mathbf{0}, & n_1\neq n_2\end{cases}\\ E\{\mathbf{x}(0)\mathbf{x}'(0)\}=\boldsymbol{\psi}\end{gathered}\tag{15.5.1}$$

It is desired to find the closed-loop control $\mathbf{u}(n:\mathbf{z}_n)$ which minimizes

$$c=E\sum_{n-1}^{N}\{\mathbf{x}'(n)\mathbf{C}_1\mathbf{x}(n)+\mathbf{u}'(n-1:\mathbf{z}_{n-1})\mathbf{C}_2\mathbf{u}(n-1:\mathbf{z}_{n-1})\}\tag{15.5.2}$$

where $\mathbf{C}_1$ and $\mathbf{C}_2$ are specified positive definite matrices. Extension to a criterion involving $[\mathbf{x}(n)-\mathbf{d}(n)]'\mathbf{C}_1[\mathbf{x}(n)-\mathbf{d}(n)]$ for specified $\mathbf{d}(n)$ is relatively straightforward. The problem is discussed in three steps:

1. Deterministic.
2. Input uncertainty; perfect state observation.
3. Input and observation uncertainty.

It is shown that the control law developed for the deterministic case actually applies to the second two cases involving uncertainty.

15.5.1 Deterministic Case†

Consider the deterministic version of (15.5.1) and (15.5.2) where $\mathbf{x}(0)$ is known and there are no input and observation uncertainties, $\mathbf{w}(n)$ and $\mathbf{v}(n)$ ($\mathbf{Q}=\mathbf{R}=\boldsymbol{\psi}=\mathbf{0}$). In this case the observations $\mathbf{z}$ play no role and the expectation in (15.5.2) can be dropped, as all processes are deterministic.

For the deterministic case, optimal control theory (dynamic programming, maximum principle, etc.) can be used to find the optimum open-loop controls $\mathbf{u}^*(n)$, $n=1,\ldots,N-1$, which are the $\mathbf{u}(n)$, $n=1,\ldots,N-1$, which minimize the cost c of (15.5.2). For deterministic problems, the open- and closed-loop controls are the same; i.e.,

$$\mathbf{u}^*(n)=\mathbf{u}^*(n:\mathbf{x}(n))\tag{15.5.3}$$

For a linear deterministic model with quadratic criteria [such as the deterministic version of (15.5.1) and (15.5.2),] the optimum closed-loop control is a linear function of the state; i.e.,

$$\mathbf{u}^*(n: \mathbf{x}(n)) = \mathbf{A}(n)\mathbf{x}(n) \tag{15.5.4}$$

where the matrix $\mathbf{A}(n)$, $n = 0, \ldots, N-1$, depends on $\mathbf{\Phi}$, $\mathbf{B}$, $\mathbf{x}(0)$, $\mathbf{C}_1(n)$, and $\mathbf{C}_2(n)$ and can be calculated a priori. The actual equations which yield $\mathbf{A}(n)$ are not discussed here, as they are widely available in the literature.

15.5.2 Input Uncertainties; Perfect Observations†

Now consider (15.5.1) and (15.5.2) where there are input uncertainties $\mathbf{w}(n)$ but perfect observations of the full state; i.e.,

$$\mathbf{z}(n) = \mathbf{x}(n), \qquad n = 1, \ldots, N \tag{15.5.5}$$

$\mathbf{x}(0)$ is also assumed to be known. It is desired to find the optimum closed-loop control $\mathbf{u}^*(n : \mathbf{z}_n)$. As discussed earlier, it follows that

$$\mathbf{u}^*(n : \mathbf{z}_n) = \mathbf{u}^*(n: \mathbf{x}(n))$$

It is now argued that the presence of the white input stochastic process $\mathbf{w}(n)$ does not effect the control law and that the deterministic closed-loop control law (15.5.4) still applies; i.e.,

$$\mathbf{u}^*(n: \mathbf{z}_n) = \mathbf{A}(n)\mathbf{x}(n) = \mathbf{A}(n)\mathbf{z}(n) \tag{15.5.6}$$

To see this, substitute (15.5.1) into (15.5.2) to get

$$\begin{aligned} c = E \sum_{n=1}^{N} \{[&\mathbf{\Phi}\mathbf{x}(n-1) + \mathbf{G}\mathbf{w}(n-1) + \mathbf{B}\mathbf{u}(n-1 : \mathbf{z}_{n-1})]'\mathbf{C}_1[\mathbf{\Phi}\mathbf{x}(n-1) \\ &+ \mathbf{G}\mathbf{w}(n-1) + \mathbf{B}\mathbf{u}(n-1 : \mathbf{z}_{n-1})] + \mathbf{u}'(n-1 : \mathbf{z}_{n-1})\mathbf{C}_2\mathbf{u}(n-1 : \mathbf{z}_{n-1})\} \end{aligned}$$

Using the white properties of $\mathbf{w}(n)$,

$$\begin{aligned} c &= \sum_{n-1}^{N} \operatorname{tr}\{\mathbf{G}\mathbf{Q}\mathbf{G}'\mathbf{C}_1\} + E\{\tilde{c}\} \\ \tilde{c} &= \sum_{n=1}^{N} \{\mathbf{\Phi}\mathbf{x}(n-1) + \mathbf{B}\mathbf{u}(n-1 : \mathbf{z}_{n-1})]'\mathbf{C}_1[\mathbf{\Phi}\mathbf{x}(n-1) \\ &\quad + \mathbf{B}\mathbf{u}(n-1 : \mathbf{z}_{n-1})] + \mathbf{u}'(n-1 : \mathbf{z}_{n-1})\mathbf{C}_2\mathbf{u}(n-1 : \mathbf{z}_{n-1})\} \end{aligned} \tag{15.5.7}$$

Now $\tilde{c}$ is a random variable, but it follows from the deterministic case of Section 15.5.1 that $\tilde{c}$ is minimized by using

$$\mathbf{u}(n: \mathbf{z}_n) = \mathbf{A}(n)\mathbf{x}(n) \tag{15.5.8}$$

Since (15.5.8) minimizes $\tilde{c}$, it minimizes $E(\tilde{c})$ and hence c. Therefore (15.5.6) has been verified.

The fact that the control law is the same in the present case as in the deterministic case does not mean the same controls and performance result. In the present case, $\mathbf{x}(n)$ and hence the control is a random variable. The minimum cost $c_{\min}$ can be evaluated by substituting (15.5.6) in (15.5.1) and

(15.5.2) to yield

$$\begin{aligned}\mathbf{x}^*(n+1) &= [\mathbf{\Phi} + \mathbf{A}(n)\mathbf{B}]\mathbf{x}^*(n) + \mathbf{G}\mathbf{w}(n)\\ c_{\min} &= \sum_{n=1}^{N} \operatorname{tr}\{\mathbf{C}_1\mathbf{\Gamma}(n)\} + \sum_{n=0}^{N-1} \operatorname{tr}\{[\mathbf{A}'(n)\mathbf{C}_2\mathbf{A}(n)]\mathbf{\Gamma}(n)\}\\ \mathbf{\Gamma}(n) &= E\{\mathbf{x}^*(n)\mathbf{x}^{*\prime}(n)\}\end{aligned} \tag{15.5.9}$$

where $\mathbf{\Gamma}(n)$ can be evaluated using the ideas of Chapter 4.

15.5.3 Input and Observation Uncertainties†

Now consider the actual model of (15.5.1) and (15.5.2) where $\mathbf{x}(0)$ is uncertain, there are input uncertainties $\mathbf{w}(n)$, and

$$\mathbf{z}(n) = \mathbf{H}\mathbf{x}(n) + \mathbf{v}(n)$$

This problem is solved by reducing it to the form of the perfect observation model of Section 15.5.2.

Define

$$\boldsymbol{\delta}_x(n|n) = \mathbf{x}(n) - \hat{\mathbf{x}}(n|n) \tag{15.5.10}$$

where $\hat{\mathbf{x}}(n|n)$ is the conditional expectation of $\mathbf{x}(n)$ given $\mathbf{z}(m)$, $m = 1, \ldots, n$. $\hat{\mathbf{x}}(n|n)$ also depends on the controls applied, $m = 1, \ldots, n-1$, which are assumed to be known. It follows from Chapter 6 that $\boldsymbol{\delta}_x(n|n)$ is independent of $\mathbf{z}_n$ and hence of $\hat{\mathbf{x}}(n|n)$ so that

$$E\{\boldsymbol{\delta}_x(n|n)\hat{\mathbf{x}}'(n|n)\} = \mathbf{0} \tag{15.5.11}$$

The cost c of (15.5.2) can be rewritten as

$$\begin{aligned}c = E\sum_{n=1}^{N} \{&[\boldsymbol{\delta}_x(n|n) + \hat{\mathbf{x}}(n|n)]'\mathbf{C}_1[\boldsymbol{\delta}_x(n|n) + \hat{\mathbf{x}}(n|n)]\\ &+\mathbf{u}'(n-1:\mathbf{z}_{n-1})\mathbf{C}_2\mathbf{u}(n-1:\mathbf{z}_{n-1})\}\end{aligned}$$

Expanding and using (15.5.11) gives

$$\begin{aligned}c &= \sum_{n=1}^{N} [\operatorname{tr} \mathbf{\Sigma}(n|n)\mathbf{C}_1] + \tilde{c}\\ \tilde{c} &= E\sum_{n=1}^{N} \{\hat{\mathbf{x}}'(n|n)\mathbf{C}_1\hat{\mathbf{x}}(n|n)\\ &\qquad + \mathbf{u}'(n-1:\mathbf{z}_{n-1})\mathbf{C}_2\mathbf{u}(n-1:\mathbf{z}_{n-1})\}\\ E\{\boldsymbol{\delta}_x(n|n)\boldsymbol{\delta}_x'(n|n)\} &= \mathbf{\Sigma}(n|n)\end{aligned} \tag{15.5.12}$$

To determine the optimum $\mathbf{u}(n:\mathbf{z}_n)$, only $\tilde{c}$ has to be considered. The equation for $\hat{\mathbf{x}}(n|n)$ follows from Chapter 6:

$$\begin{aligned}\hat{\mathbf{x}}(n+1|n+1) &= \mathbf{\Phi}\hat{\mathbf{x}}(n|n) + \mathbf{B}\mathbf{u}(n:\mathbf{z}_n) + \tilde{\mathbf{w}}(n)\\ \hat{\mathbf{x}}(0) &= \mathbf{0}\\ \tilde{\mathbf{w}}(n) &= \mathbf{\Sigma}(n+1|n+1)\mathbf{H}'\mathbf{R}^{-1}\boldsymbol{\delta}_z(n+1|n)\\ \boldsymbol{\delta}_z(n+1|n) &= \mathbf{z}(n+1) - \mathbf{H}[\mathbf{\Phi}\hat{\mathbf{x}}(n|n) + \mathbf{B}\mathbf{u}(n:\mathbf{z}_n)]\end{aligned} \tag{15.5.13}$$

From Chapter 6 $\boldsymbol{\delta}_z(n+1 \mid n)$ is a white process, which means that $\tilde{\mathbf{w}}(n)$ is a white process. The determination of the optimum $\mathbf{u}(n\colon \mathbf{z}_n)$ for (15.5.1) and (15.5.2) is equivalent to determining the optimum $\mathbf{u}(n\colon \mathbf{z}_n)$ for (15.5.13) which minimize $\tilde{c}$ of (15.5.12). Since $\tilde{\mathbf{w}}(n)$ is a white process, this new problem is of the same form as that of Section 15.5.2. Therefore for (15.5.12) and (15.5.13) and hence for (15.5.1) and (15.5.2)

$$\mathbf{u}^*(n\colon \mathbf{z}_n) = \mathbf{u}^*[n\colon \hat{\mathbf{x}}(n \mid n)] = \mathbf{A}(n)\hat{\mathbf{x}}(n \mid n) \tag{15.5.14}$$

The minimum cost $c_{\min}$ can be obtained using the same sort of logic which leads to (15.5.9).

The model (15.5.1) employed a Gaussian assumption but the discussions thus far apparently have not used this assumption. Equation (15.5.12) is valid without a Gaussian assumption. However for a non Gaussian model, there may exist some nonlinear estimator which yields a smaller value of $\boldsymbol{\Sigma}(n \mid n)$ than that given by the linear estimator of Chapter 6. If so, the control of (15.5.14) no longer necessarily minimizes the cost c.

15.5.4 Discussion: Linear Model, Quadratic Criterion, Separation Theorem†

The linear Gaussian model of (15.5.1) with quadratic criteria of (15.5.2) has yielded a very nice and useful result. The corresponding deterministic control problem obtained by ignoring all uncertainties can be solved to yield the optimum deterministic closed-loop control law

$$\mathbf{u}^*[n\colon \mathbf{x}(n)] = \mathbf{A}(n)\mathbf{x}(n) \tag{15.5.15}$$

where the gain matrix $\mathbf{A}(n)$ depends on $\boldsymbol{\Phi}$, $\mathbf{G}$, $\mathbf{B}$, $\mathbf{C}_1$, $\mathbf{C}_2$, and $\mathbf{x}(0)$. When the uncertainties are present, the optimum closed-loop control is given by

$$\mathbf{u}^*(n\colon \mathbf{z}_n) = \mathbf{A}(n)\hat{\mathbf{x}}(n \mid n) \tag{15.5.16}$$

where $\mathbf{A}(n)$ in (15.5.15) and (15.5.16) are identical except that the dependence of $\mathbf{A}(n)$ in (15.5.15) on $\mathbf{x}(0)$ is replaced in (15.5.16) by the same dependence on $E\{\mathbf{x}(0)\} = \mathbf{m}$. The uncertainties may have a large effect on the performance, but they do not affect the control law.

In Section 15.3 it was stated that it is often reasonable to use a control system as in Fig. 15.3.1 where the control law relating $\mathbf{u}$ to $\mathbf{x}(n \mid n)$ is determined by ignoring the uncertainties. Thus for the special case of (15.5.1) and (15.5.2), the reasonable is also optimum.

The separation theorem (version II in Fig. 15.4.1) for a linear Gaussian model with quadratic cost criterion is a beautiful result. However it is not perfect. For example the example at the end of Section 15.3.3 where the state is very poorly known for $n = 1, \ldots, 10$ and good information is not obtained until $n = 11, \ldots, 20$ illustrated a weakness of closed-open-loop control. The separation theorem says that the optimum closed-loop and closed-open-loop controls are the same; i.e., nonzero control [as in (15.5.16)] is employed for $n = 1, \ldots, 10$ and hence the optimum control does not have the good property of waiting until $n > 10$, when good information becomes available. This is not a counterexample in the mathematical sense to the separation theorem (version II in Fig. 15.4.1): it merely says that the criterion is such that the *optimum* control (in a mathematical sense) is *not always* a *good* control (in an intuitive sense). It should also be stated that this Section 15.3.3 example is pathological. It is discussed only because of the author's premise that *nothing is perfect.*

15.6 Dual Control†

Consider a nonlinear model such as (15.1). Assume that it is desired to obtain a state estimate, $\hat{\mathbf{x}}(N|N)$. The nature of the input $\mathbf{u}(n)$ or $\mathbf{u}(n: \mathbf{z}_n)$, $n = 0, \ldots, N - 1$, can affect the accuracy of $\hat{\mathbf{x}}(N|N)$. Hence it is possible to formulate a stochastic control problem where the criterion is to find the controls (inputs) which yield the best estimate. Note that for a linear model, the controls (inputs) do not effect the accuracy of the state estimate.

Now consider a nonlinear model where it is desired to find the controls which minimize some general cost such as c of (15.1.2). In general terms, the cost will be reduced if the accuracy of the state estimates is increased. Thus in some applications, the optimum control can be viewed as a compromise between two conflicting desires: the desire to get an accurate state estimate and the desire to minimize the cost. This dual nature sometimes results in the use of the term *dual control problem.*

The dual nature of the control action can become especially evident in the case of a system with no input uncertainty such as

$$\mathbf{x}(n+1) = \boldsymbol{\phi}[\mathbf{x}(n)] + \mathbf{B}\mathbf{u}(n)$$

$$\mathbf{z}(n) = \mathbf{h}[\mathbf{x}(n)] + \mathbf{v}(n)$$

$$\mathbf{x}(0)\text{: uncertain}$$

and of a criterion of minimizing a cost which depends only on $\mathbf{x}(N)$ such as

$$c = E\{c_1[\mathbf{x}(N)]\}$$

subject to a peak amplitude constraint on the control such as

$$\mathbf{u}(n: \mathbf{z}_n) \in \Omega_u$$

If N is sufficiently large, the initial controls $\mathbf{u}(n: \mathbf{z}_n)$, $n = 0, \ldots, N_0$, may be devoted primarily to improving the accuracy of the state estimate $\hat{\mathbf{x}}(N_0|N_0)$, while the later controls $\mathbf{u}(n: \mathbf{z}_n)$, $n = N_0 + 1, \ldots, N - 1$, are devoted to trying to minimize $E\{c_1[\mathbf{x}(N)]\}$.

The dual nature of the control problem is often just a way of interpreting results and viewing the problem. However, a reasonable approach is sometimes to choose some N_0 and divide the design into two parts. Design $\mathbf{u}(n:\mathbf{z}_n)$, $n = 0, \ldots, N_0$, to yield the best state estimate and then design $\mathbf{u}(n:\mathbf{z}_n)$, $n = N_0 + 1, \ldots, N - 1$, to minimize the cost.

15.7 Adaptive Control†

In many control problems, various parameters of the system model may be uncertain. For example, in a linear model such as (15.5.1), the $\mathbf{\Phi}$, $\mathbf{G}$, $\mathbf{B}$, $\mathbf{H}$, $\mathbf{Q}$, and $\mathbf{R}$ matrices may contain uncertain parameters. Let $\boldsymbol{\alpha}$ denote these uncertain parameters and assume that they are constant. These uncertainties can be incorporated into the problem by augmenting the state vector $\mathbf{x}(n)$ by adding new states $\boldsymbol{\alpha}(n)$ specified by (see section 14.2)

$$\boldsymbol{\alpha}(n+1) = \boldsymbol{\alpha}(n)$$

The resulting model with state $\begin{bmatrix} \mathbf{x}(n) \\ \boldsymbol{\alpha}(n) \end{bmatrix}$ is now nonlinear but conceptually of the same form already studied. The resulting control is often called an *adaptive control.*

In such adaptive control problems, the presence of uncertain but constant parameters often emphasizes the dual nature of the problem mentioned in Section 15.6. Assume that a control for $n = 0, \ldots, N - 1$ is desired. The optimum $\mathbf{u}(n:\mathbf{z}_n)$, $n = 0, \ldots, N_0$, $N > N_0$, may be devoted primarily to yielding the best $\hat{\boldsymbol{\alpha}}(N_0 \mid N_0)$, while the $\mathbf{u}(n:\mathbf{z}_n)$, $n = N_0 + 1, \ldots, N - 1$, may concentrate on minimizing the cost. In practice, a reasonable control logic may choose $\mathbf{u}(n)$ or $\mathbf{u}(n:\mathbf{z}_n)$, $n = 0, \ldots, N_0$, solely on the basis of providing the best identification although the discussions of Section 14.10 indicate the difficulties associated with this task. Chapter 14 discusses the problem of estimating $\boldsymbol{\alpha}$ from a more basic point of view.

15.8 Unknown-but-Bounded

This chapter is devoted to stochastic models but it is appropriate to briefly discuss the problems of control when unknown-but-bounded models are used instead.

For an unknown-but-bounded model, the range of possible criteria differs greatly from the ideas of Section 15.1. For example suppose it is desired to make $\mathbf{y}(n) = \mathbf{C}\mathbf{x}(n)$ follow some predetermined trajectory $\mathbf{d}(n)$ $n = 1, \ldots, N$. For a stochastic (Bayesian) model one might try to minimize

$$c = E\left\{\sum_{n=1}^{N} [\mathbf{y}(n) - \mathbf{d}(n)]'\mathbf{C}_1[\mathbf{y}(n) - \mathbf{d}(n)]\right\}$$

However for an unknown-but-bounded model, one might define

$$\Omega_\delta(n) = \{\boldsymbol{\delta}(n) : \boldsymbol{\delta}(n) = \mathbf{y}(n) - \mathbf{d}(n)\}$$

and then try to minimize the maximum "size" of $\Omega_\delta(n)$ $n = 1, \ldots, N$. Alternately one might be given a *target tube* which is the set of allowable deviations $\boldsymbol{\delta}(n) = \mathbf{y}(n) - \mathbf{d}(n)$ and then choose the minimum "energy" or "fuel" control which keeps the actual deviation within the tube. In many applications, the criteria possible with unknown-but-bounded models are actually more natural than those of stochastic models.

As should be expected, the mathematics associated with developing controls for unknown-but-bounded models is not like that of stochastic control. However the basic concepts of open versus closed loop still apply and reasonable closed-loop controls can be developed using the basic ideas of Sections 15.3.1, 15.3.2. Also as should be expected, truly optimum closed-loop controls for unknown-but-bounded models are like those for stochastic models in that they are in general very difficult to compute.

15.9 Discussion

One important point of the preceding discussions is that there are many *different* stochastic control problems. When reading the literature on stochastic control, it is very important to study the problem statement carefully to decide exactly which problem is being considered and exactly what the controls depend on.

Another important point is that although optimum stochastic control problems can be formulated with relative ease, the computation of the optimum control is usually very difficult. Successful controls for stochastic problems usually result from the designer's ability to use his engineering judgment to develop reasonable controls that work satisfactorily. At best, the mathematical optimization theory usually provides only a guide to how to proceed.

It is the author's opinion that for most applications, attempts to tackle the complete optimum stochastic control problem are unrewarding. If the system to be controlled and the sensors which provide the observations are well designed, the uncertainties are often small enough that a state estimator followed by a reasonable zero memory controller (say, designed from a deterministic system) provides satisfactory performance which is close to optimum. If the uncertainties are large, it often appears that even an optimum control cannot yield satisfactory performance.

It should be emphasized in conclusion that this chapter has definitely not discussed all aspects and approaches to stochastic control problems.

Historical Notes and References: Chapter 15

As in Chapter 14 on System Identification, the available literature is huge, and no attempt will be made to provide even a partial overview. Mendel and Gieseking [1] is a bibliography which contains many references to stochastic control. See also Book List 4 on Stochastic Systems.

Section 15.4 and 15.5

Joseph and Tou [1] is one of the first papers on the "separation theorem." Witsenhausen [4] discusses the "separation theorem" in a very general context.

Section 15.6

See Feldbaum [1].

Section 15.8

Glover and Schweppe [1] discusses certain theoretical aspects, while Glover and Schweppe [2] discusses an application to electric power systems.

Epilog

Our travels together are coming to an end. However, before our final parting it seems appropriate to briefly review the journey.

A huge number of complicated-looking equations were presented. Most of them should be treated like the slides or photographs taken on a more conventional journey. They are nice to have, but one should remember only where they are kept and in general what they are about. The details should be forgotten as they can be looked up if needed.

Many basic concepts were also presented. Unlike the explicit equations, the concepts should be remembered and be available on instant recall. A list of the most important concepts would include the following:

1. Methods of uncertainty modeling and the techniques for obtaining white process state space models from natural models.
2. Effect of choice of uncertainty model on analysis and interpretation of results.
3. Analysis of propagation of uncertainty through dynamic systems.
4. Tracking feedback form of filters.
5. Conversion of a filter into a decision function generator.
6. Use of hypothesis testing as a multipurpose tool.

7. Use of error (performance) analysis for estimation and hypothesis testing as a basic tool in the design of measurement and sensor systems.

The importance of engineering judgment could be added to the above list but it deserves special attention. Mathematical optimality received a lot of attention in the developments but in most applications, the mathematics must be combined with judgment both in the choice of models and the implementation of the results. Relatively sophisticated mathematics is an absolutely necessary ingredient in handling many complex problems but it is not sufficient. Unfortunately it is difficult to write a book on judgment so the author can only state and restate and re-restate that it is needed.

Ideally a list of important concepts related to uncertainty not covered in this book should now be given. However, the list would be very long and would end the book on a negative note. Therefore the list will not be given.

Parting is sweet sorrow. Go forth with certainty into a world of uncertainty.

Appendices

These appendices contain background and supplemental material of a mathematical nature. The topics to be covered are matrix algebra, linear difference and differential equations, vector and matrix gradient functions, probability theory, Gaussian (normal) random vectors, stochastic processes, set theory, and ellipsoids.

Appendix A

Matrix Algebra

A.1 Partitioned Matrices

Consider a K by K matrix $\mathbf{A}$ partitioned as follows:

$$\mathbf{A} = \left[\begin{array}{c|c} \mathbf{A}_{11} & \mathbf{A}_{12} \\ \hline \mathbf{A}_{21} & \mathbf{A}_{22} \end{array}\right] \tag{A.1}$$

Assume that $\mathbf{A}^{-1}$ exists and define

$$\mathbf{B} = \mathbf{A}^{-1}$$

where $\mathbf{B}$ is partitioned as in (A.1) to be

$$\mathbf{B} = \left[\begin{array}{c|c} \mathbf{B}_{11} & \mathbf{B}_{12} \\ \hline \mathbf{B}_{21} & \mathbf{B}_{22} \end{array}\right]$$

Then

$$\mathbf{B} = \mathbf{A}^{-1} = \left[\begin{array}{c|c} \mathbf{A}_{11}^{-1} + \mathbf{A}_{11}^{-1}\mathbf{A}_{12}\mathbf{B}_{22}\mathbf{A}_{21}\mathbf{A}_{11}^{-1} & -\mathbf{A}_{11}^{-1}\mathbf{A}_{12}\mathbf{B}_{22} \\ \hline -\mathbf{B}_{22}\mathbf{A}_{21}\mathbf{A}_{11}^{-1} & \mathbf{B}_{22} \end{array}\right] \tag{A.2}$$

where

$$\mathbf{B}_{22}^{-1} = \mathbf{A}_{22} - \mathbf{A}_{21}\mathbf{A}_{11}^{-1}\mathbf{A}_{12}$$

This result can be proved by multiplying (A.1) by (A.2) and seeing that the unit matrix results.

The $\mathbf{A}^{-1} = \mathbf{B}$ of (A.2) can obviously be rewritten in a different form by simply *reordering* the subscripts. Equating terms in these two expressions for $\mathbf{B}$ yields the following very useful matrix identity:

$$\begin{aligned}[\mathbf{A}_{11} - \mathbf{A}_{12}\mathbf{A}_{22}^{-1}\mathbf{A}_{21}]^{-1} = \mathbf{A}_{11}^{-1} &+ \mathbf{A}_{11}^{-1}\mathbf{A}_{12} \\ &\times [\mathbf{A}_{22} - \mathbf{A}_{21}\mathbf{A}_{11}^{-1}\mathbf{A}_{12}]^{-1}\mathbf{A}_{21}\mathbf{A}_{11}^{-1}\end{aligned} \tag{A.3}$$

with the special case

$$[\mathbf{I} + \mathbf{A}]^{-1} = \mathbf{I} - [\mathbf{I} + \mathbf{A}^{-1}]^{-1} \tag{A.4}$$

Two other useful results for partitioned matrices are

$$|\mathbf{A}| = |\mathbf{A}_{11} - \mathbf{A}_{12}\mathbf{A}_{22}^{-1}\mathbf{A}_{21}||\mathbf{A}_{22}| \tag{A.5}$$

$$\operatorname{tr} \mathbf{A} = \operatorname{tr} \mathbf{A}_{11} + \operatorname{tr} \mathbf{A}_{22} \tag{A.6}$$

A.2 Positive Definite Matrices

In this book, the term *positive definite matrix* also implies that the matrix is symmetric. Let $\mathbf{\Gamma}$ be a K by K positive definite matrix. Then

$$\mathbf{x}'\mathbf{\Gamma}\mathbf{x} > 0, \qquad \text{all } \mathbf{x} \neq 0 \tag{A.7}$$

Let λ_k, $k = 1, \ldots, K$, denote the eigenvalues of $\mathbf{\Gamma}$. Then

$$\lambda_k > 0, \qquad k = 1, \ldots, K \tag{A.8}$$

Any symmetric matrix which satisfies (A.7) or (A.8) is positive definite. For a positive semidefinite matrix

$$\mathbf{x}'\mathbf{\Gamma}\mathbf{x} \geq 0, \qquad \text{all } \mathbf{x} \neq 0$$

$$\lambda_k \geq 0, \qquad k = 1, \ldots, K$$

For any K by K positive semidefinite matrix $\mathbf{\Gamma}$, there exists a matrix $\mathbf{C}$ such that

$$\mathbf{C}\mathbf{C}' = \mathbf{I}$$

$$\mathbf{C}\mathbf{\Gamma}\mathbf{C}' = \begin{bmatrix} \lambda_1 & & & \mathbf{0} \\ & \cdot & & \\ & & \cdot & \\ & & & \cdot \\ \mathbf{0} & & & \lambda_K \end{bmatrix}$$

The K by K positive definite matrix $\mathbf{\Gamma}$ has rank K_1 if only K_1 of the λ_k, $k = 1, \ldots, K$, are greater than zero.

In general, it is impossible to say that one matrix is larger than another matrix unless one defines some scalar measure of the matrix's "size." However, there is an important exception to this rule. Let $\mathbf{\Gamma}_1$ and $\mathbf{\Gamma}_2$ be two positive definite matrices. Then, by definition

$$\boxed{\mathbf{\Gamma}_1 - \mathbf{\Gamma}_2 \text{ is positive definite implies that } \mathbf{\Gamma}_1 > \mathbf{\Gamma}_2} \tag{A.9}$$

Equation (A.9) implies that

$$\begin{aligned} &\operatorname{tr} \mathbf{\Gamma}_1 > \operatorname{tr} \mathbf{\Gamma}_2 \\ &|\mathbf{\Gamma}_1| > |\mathbf{\Gamma}_2| \\ &\mathbf{\Gamma}_{1,kk} > \mathbf{\Gamma}_{2,kk}, \qquad \text{all } k \\ &(\mathbf{\Gamma}_{kk} \text{ is } k\text{th main diagonal element of } \mathbf{\Gamma}) \end{aligned}$$

Equation (A.9) also implies that

$$\mathbf{C}\mathbf{\Gamma}_1\mathbf{C}' \geq \mathbf{C}\mathbf{\Gamma}_2\mathbf{C}'$$

for any matrix $\mathbf{C}$. These ideas can be extended to the case of positive semi-definite matrices in an obvious fashion. Let $\mathbf{\Gamma}_{1,ij}$ and $\mathbf{\Gamma}_{2,ij}$ denote the elements of $\mathbf{\Gamma}_1$ and $\mathbf{\Gamma}_2$. It is important to note that $\mathbf{\Gamma}_{1,ij} > \mathbf{\Gamma}_{2,ij}$, all k, j, does *not* imply that $\mathbf{\Gamma}_1 > \mathbf{\Gamma}_2$.

EXAMPLE: Consider

$$\mathbf{\Gamma}_1 = \begin{bmatrix} 5 & 3 \\ 3 & 5 \end{bmatrix} \qquad \mathbf{\Gamma}_2 = \begin{bmatrix} 4 & 1 \\ 1 & 4 \end{bmatrix}$$

$$\mathbf{\Gamma}_{1,kj} > \mathbf{\Gamma}_{2,kj}, \qquad \text{all } k, j$$

But

$$\mathbf{\Gamma}_1 - \mathbf{\Gamma}_2 = \begin{bmatrix} 1 & 2 \\ 2 & 1 \end{bmatrix}$$

is not positive definite.

In a similar fashion, $\mathbf{\Gamma}_1 > \mathbf{\Gamma}_2$ does *not* imply that $\mathbf{\Gamma}_{1,kj} > \mathbf{\Gamma}_{2,kj}$, all k, j.

EXAMPLE: Consider

$$\mathbf{\Gamma}_1 = \begin{bmatrix} 5 & 0 \\ 0 & 5 \end{bmatrix} \qquad \mathbf{\Gamma}_2 = \begin{bmatrix} 2 & 1 \\ 1 & 2 \end{bmatrix}$$

$\mathbf{\Gamma}_1 - \mathbf{\Gamma}_2$ is positive definite but the off-diagonal elements of $\mathbf{\Gamma}_2$ are larger than those of $\mathbf{\Gamma}_1$.

If (A.9) holds, then

$$\mathbf{\Gamma}_1^{-1} < \mathbf{\Gamma}_2^{-1}$$

If $\mathbf{\Gamma}$ is partitioned,

$$\mathbf{\Gamma} = \left[\begin{array}{c:c} \mathbf{\Gamma}_{11} & \mathbf{\Gamma}_{12} \\ \hdashline \mathbf{\Gamma}'_{12} & \mathbf{\Gamma}_{22} \end{array}\right]$$

and

$$\boldsymbol{\theta} = \boldsymbol{\Gamma}^{-1} = \left[\begin{array}{c|c} \boldsymbol{\theta}_{11} & \boldsymbol{\theta}_{12} \\ \hline \boldsymbol{\theta}'_{12} & \boldsymbol{\theta}_{22} \end{array}\right]$$

then from (A.2)

$$\boldsymbol{\theta}_{11} \geq \boldsymbol{\Gamma}_{11}^{-1}$$

Since

$$\operatorname{tr}|\boldsymbol{\Gamma}| = \lambda_1 + \cdots + \lambda_K$$
$$|\boldsymbol{\Gamma}| = \lambda_1 \cdots \lambda_K$$

it follows that for positive definite $\boldsymbol{\Gamma}$

$$\frac{1}{K}\operatorname{tr}[\boldsymbol{\Gamma}] \geq [|\boldsymbol{\Gamma}|]^{1/K}$$

Appendix B

Linear Difference and Differential Equations

Time is considered to be the independent variable, although this is not necessary. The notational convention used is

t	denotes continuous time
$n\Delta$	denotes discrete time

where n is an integer and Δ is the time between events. Thus writing

$\mathbf{x}(t)$	implies that $\mathbf{x}$ is a continuous-time process
$\mathbf{x}(n\Delta)$	implies that $\mathbf{x}$ is a discrete-time process

In a similar fashion

$\mathbf{x}(t)\Big|_{t=n\Delta}$ is the value of the continuous-time process $\mathbf{x}(t)$ at $t = n\Delta$

B.1 Discrete Time

Consider the linear, first-order, vector difference equation

$$\begin{aligned} &\mathbf{x}(n\Delta + \Delta) = \mathbf{\Phi}(n\Delta)\mathbf{x}(n\Delta) + \Delta\mathbf{G}(n\Delta)\mathbf{w}(n\Delta) \\ &\mathbf{x}(0)\text{: initial condition} \end{aligned} \tag{B.1}$$

This difference equation can be solved to give

$$\begin{aligned}\mathbf{x}(N\Delta + \Delta) = {} & [\boldsymbol{\theta}[N\Delta + \Delta, 0]]\mathbf{x}(0) \\ & + \Delta \sum_{m=0}^{N} [\boldsymbol{\theta}[N\Delta + \Delta, m\Delta + \Delta]]\mathbf{G}(m\Delta)\mathbf{w}(m\Delta)\end{aligned} \tag{B.2}$$

where $\boldsymbol{\theta}(N\Delta, m\Delta)$ is the *transition matrix* associated with $\boldsymbol{\Phi}(n\Delta)$. $\boldsymbol{\theta}(n\Delta, m\Delta)$ satisfies the linear difference matrix equation

$$\boldsymbol{\theta}(n\Delta + \Delta, m\Delta) = \boldsymbol{\Phi}(n\Delta)\boldsymbol{\theta}(n\Delta, m\Delta)$$
$$\boldsymbol{\theta}(m\Delta, m\Delta) = \mathbf{I}$$

so that

$$\boldsymbol{\theta}(N\Delta, m\Delta) = \prod_{n=m}^{N-1} \boldsymbol{\Phi}(n\Delta), \qquad N > m$$

$\boldsymbol{\theta}(n\Delta, m\Delta)$ has the property

$$\boldsymbol{\theta}(N\Delta, m\Delta) = \boldsymbol{\theta}(N\Delta, n\Delta)\boldsymbol{\theta}(n\Delta, m\Delta), \qquad N > n > m$$

In the special case where $\boldsymbol{\Phi}(n\Delta) = \boldsymbol{\Phi}$, all n, $\boldsymbol{\theta}(N\Delta, m\Delta)$ depends only on $N - m$ and can be considered to be a matrix function $\boldsymbol{\theta}[n\Delta]$ of one variable $n\Delta$ defined by

$$\boldsymbol{\theta}(n\Delta + \Delta) = \boldsymbol{\Phi}\boldsymbol{\theta}(n\Delta)$$
$$\boldsymbol{\theta}(0) = \mathbf{I}$$

In this special case

$$\mathbf{x}(N\Delta + \Delta) = \boldsymbol{\theta}(N\Delta + \Delta)\mathbf{x}(0) + \Delta \sum_{m=0}^{N} \boldsymbol{\theta}(N\Delta - m\Delta)\mathbf{G}(m\Delta)\mathbf{w}(m\Delta) \tag{B.3}$$

$\boldsymbol{\theta}(n\Delta)$ can be expressed as

$$\boldsymbol{\theta}(n\Delta) = \boldsymbol{\Phi}^n$$

B.2 Continuous Time

Consider the linear, first-order, vector differential equation

$$\begin{gathered}\frac{d}{dt}\mathbf{x}(t) = \mathbf{F}(t)\mathbf{x}(t) + \mathbf{G}(t)\mathbf{w}(t) \\ \mathbf{x}(0)\text{: initial condition.}\end{gathered} \tag{B.4}$$

This differential equation can be solved to give

$$\begin{gathered}\mathbf{x}(t) = \boldsymbol{\theta}(t, t_0)\mathbf{x}(t_0) + \int_{t_0}^{t} \boldsymbol{\theta}(t, \tau)\mathbf{G}(\tau)\mathbf{w}(\tau)\, d\tau \\ t_0 = 0\end{gathered} \tag{B.5}$$

where $\boldsymbol{\theta}(t, t_0)$ is the fundamental or *transition matrix* associated with $\mathbf{F}(t)$.

$\boldsymbol{\theta}(t, t_0)$ satisfies the linear matrix differential equation

$$\frac{d}{dt}\boldsymbol{\theta}(t, t_0) = \mathbf{F}(t)\boldsymbol{\theta}(t, t_0)$$
$$\boldsymbol{\theta}(t_0, t_0) = \mathbf{I} \tag{B.6}$$

$\boldsymbol{\theta}(t, t_0)$ has the properties

$$\boldsymbol{\theta}(t, t_0) = \boldsymbol{\theta}(t, t_1)\boldsymbol{\theta}(t_1, t_0)$$
$$\boldsymbol{\theta}^{-1}(t, t_0) = \boldsymbol{\theta}(t_0, t)$$

In the special case where $\mathbf{F}(t) = \mathbf{F}$, all t, $\boldsymbol{\theta}(t, t_0)$ depends on only $t - t_0$ and can be considered to be a matrix function $\boldsymbol{\theta}(t)$ of one variable, t, defined by

$$\frac{d}{dt}\boldsymbol{\theta}(t) = \mathbf{F}\boldsymbol{\theta}(t)$$
$$\boldsymbol{\theta}(0) = \mathbf{I} \tag{B.7}$$

In this special case

$$\mathbf{x}(t) = \boldsymbol{\theta}(t)\mathbf{x}(0) + \int_0^t \boldsymbol{\theta}(t - \tau)\mathbf{G}(\tau)\mathbf{w}(\tau)\, d\tau$$

or

$$\mathbf{x}(t) = \boldsymbol{\theta}(t)\mathbf{x}(0) + \boldsymbol{\theta}(t) \int_0^t \boldsymbol{\theta}^{-1}(\tau)\mathbf{G}(\tau)\mathbf{w}(\tau)\, d\tau \tag{B.8}$$

$\boldsymbol{\theta}(t)$ can be expressed in terms of an exponential matrix as

$$\boldsymbol{\theta}(t) = \exp\{\mathbf{F}t\}$$

or similarly as the power series

$$\boldsymbol{\theta}(t) = \mathbf{I} + \mathbf{F}t + \left(\frac{1}{2!}\right)\mathbf{F}^2t^2 + \left(\frac{1}{3!}\right)\mathbf{F}^3t^3 + \cdots$$

B.3 Discrete to Continuous Time Limit

Time-invariant equations are considered for the sake of simplicity.

Consider first the time-invariant, continuous-time differential equation

$$\frac{d\mathbf{x}(t)}{dt} = \mathbf{F}\mathbf{x}(t) + \mathbf{G}_C\mathbf{w}_s(t)$$
$$\mathbf{x}(0) = \mathbf{0} \tag{B.9}$$

where the input $\mathbf{w}_s(t)$ is given by

$$\mathbf{w}_s(t) = \mathbf{w}(n\Delta), \qquad n\Delta \le t < n\Delta + \Delta \tag{B.10}$$

where $\mathbf{w}(n\Delta)$ is a discrete-time process. $\mathbf{w}_s(t)$ and $\mathbf{w}(n\Delta)$ are illustrated in Fig. B.1 for a scalar signal. If $\boldsymbol{\theta}_F(t)$ is the fundamental matrix associated with $\mathbf{F}$,

$$\mathbf{x}(t)\Big|_{t=n\Delta+\Delta} = \boldsymbol{\theta}_F(\Delta)\mathbf{x}(t)\Big|_{t=n\Delta} + \boldsymbol{\theta}_F(\Delta)\int_0^\Delta \boldsymbol{\theta}_F^{-1}(t)\mathbf{G}_C\mathbf{w}(n\Delta)\, dt \tag{B.11}$$

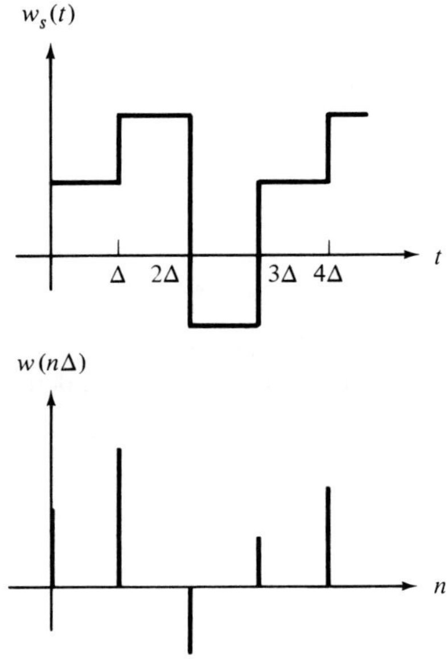

Figure B.1 Relationship between discrete- and continuous-time inputs.

which is of the form of the discrete-time system (B.1) with

$$\mathbf{\Phi} = \boldsymbol{\theta}_F(\Delta)$$

$$\Delta\mathbf{G} = \boldsymbol{\theta}_F(\Delta) \int_0^{\Delta} \boldsymbol{\theta}_F^{-1}(t)\mathbf{G}_C \, dt$$

Hence for an input $\mathbf{w}_s(t)$ such as (B.10), the continuous-time system (B.9) can be replaced by the discrete-time system defined by (B.11) if only $\mathbf{x}(t)|_{t=n\Delta}$ is of interest.

Now assume that Δ is sufficiently small so that

$$\boldsymbol{\theta}_F(\Delta) \approx \mathbf{I} + \Delta\mathbf{F}$$

$$\boldsymbol{\theta}_F^{-1}(t) \approx \mathbf{I} - t\mathbf{F}, \qquad 0 \leq t \leq \Delta$$

Then keeping only terms of order Δ, (B.11) becomes

$$\mathbf{x}(t)\Big|_{t=n\Delta+\Delta} \approx [\mathbf{I} + \Delta\mathbf{F}]\mathbf{x}(t)\Big|_{t=n\Delta} + \Delta\mathbf{G}_C\mathbf{w}(n\Delta)$$

Thus the time-invariant discrete-time system with input $\mathbf{w}(n\Delta)$ and state $\mathbf{x}(n\Delta)$ defined by

$$\begin{aligned} \mathbf{x}(n\Delta + \Delta) &= \mathbf{\Phi}\mathbf{x}(n\Delta) + \mathbf{G}\mathbf{w}(n\Delta) \\ \mathbf{x}(0) &= \mathbf{0} \\ \mathbf{\Phi} &= \mathbf{I} + \Delta\mathbf{F} \\ \mathbf{G} &= \mathbf{G}_C\Delta \end{aligned} \tag{B.12}$$

yields $\mathbf{x}(n\Delta + \Delta)$ approximately equal to $\mathbf{x}(t)|_{t=n\Delta+\Delta}$, where the approximation is good for small Δ. Alternatively, one can say that when Δ is small, the continuous-time process

$$\mathbf{x}_s(t) = \mathbf{x}(n\Delta), \qquad n\Delta \leq t < n\Delta + \Delta$$

is a good approximation to $\mathbf{x}(t)$ in the sense that $\mathbf{x}_s(t) - \mathbf{x}(t)$ is small for all t.

Now assume that a discrete-time system, such as (B.12), has been analyzed and it is desired to obtain "corresponding" results for the continuous-time system

$$\frac{d}{dt}\mathbf{x}(t) = \mathbf{F}\mathbf{x}(t) + \mathbf{G}_C\mathbf{w}(t) \tag{B.13}$$

For small Δ and large n, the discrete-time results for (B.12) are approximately equal to those for (B.13) provided that $\mathbf{w}(t)$ of (B.13) is approximately equal to $\mathbf{w}_s(t)$ of (B.10). In terms of a formal limiting operation, (B.12) gives

$$\Delta^{-1}[\mathbf{x}(n\Delta + \Delta) - \mathbf{x}(n\Delta)] = \mathbf{F}\mathbf{x}(n\Delta) + \mathbf{G}_C\mathbf{w}(n\Delta)$$

so that in the limit as

$$n \longrightarrow \infty$$

$$\Delta \longrightarrow 0$$

$$n\Delta \longrightarrow t$$

(B.12) becomes (B.13) provided that $\mathbf{w}_s(t) \longrightarrow \mathbf{w}(t)$.

The discrete to continuous limit argument does not actually require the use of step functions as in (B.10). It is straightforward to show that the same arguments also hold if $\mathbf{w}_s(t)$ is replaced by a series of ramp functions, etc.

All the preceding arguments can also be carried over to the case of time-varying $\mathbf{\Phi}$, $\mathbf{F}$, and $\mathbf{G}$ provided that the time variations are smooth enough.

Since a discrete to continuous time limit operation is often used in the book, the "recipe" for this operation is now summarized for the general time-varying system case:

Given

$$\mathbf{x}(n\Delta + \Delta) = \mathbf{\Phi}(n\Delta)\mathbf{x}(n\Delta) + \Delta\mathbf{G}(n\Delta)\mathbf{w}(n\Delta)$$

$$\mathbf{x}(0)\text{: initial condition}$$

the use of

$$\mathbf{\Phi}(n\Delta) = \mathbf{I} + \Delta\mathbf{F}(n\Delta)$$

$$n \longrightarrow \infty, \qquad \Delta \longrightarrow 0, \qquad n\Delta \longrightarrow t \tag{B.14}$$

yields

$$\frac{d}{dt}\mathbf{x}(t) = \mathbf{F}(t)\mathbf{x}(t) + \mathbf{G}(t)\mathbf{w}(t)$$

$$\mathbf{x}(0)\text{: initial condition}$$

B.4 Matrix Equations†

Consider the linear matrix differential equation

$$\frac{d}{dt}\mathbf{X}(t) = \mathbf{F}_1(t)\mathbf{X}(t) + \mathbf{X}(t)\mathbf{F}_2(t) + \mathbf{U}(t)$$

$$\mathbf{F}_1, \mathbf{F}_2, \mathbf{X}, \mathbf{U}\text{: matrices}$$

$$\mathbf{X}(0)\text{: initial condition}$$

Such an equation could be solved by rearranging it into vector form, but a direct solution in matrix notation is often more useful.

Define

$$\frac{d}{dt}\boldsymbol{\theta}_1(t, t_0) = \mathbf{F}_1(t)\boldsymbol{\theta}_1(t, t_0)$$

$$\frac{d}{dt}\boldsymbol{\theta}_2(t, t_0) = \boldsymbol{\theta}_2(t, t_0)\mathbf{F}_2(t)$$

$$\boldsymbol{\theta}_1(t, t) = \mathbf{I}$$

$$\boldsymbol{\theta}_2(t, t) = \mathbf{I}$$

Then

$$\mathbf{X}(t) = \boldsymbol{\theta}_1(t, 0)\left\{\mathbf{X}(0) + \int_0^t \boldsymbol{\theta}_1^{-1}(\tau, 0)\mathbf{U}(\tau)\boldsymbol{\theta}_2^{-1}(\tau, 0)\, d\tau\right\}\boldsymbol{\theta}_2(t, 0)$$

This can be proved by differentiation.

B.5 Controllability and Observability†

Consider a linear dynamic system (discrete time, $\Delta = 1$) with input $\mathbf{w}(n)$ and output $\mathbf{y}(n)$ defined by

$$\begin{aligned}\mathbf{x}(n+1) &= \boldsymbol{\Phi}(n)\mathbf{x}(n) + \mathbf{G}(n)\mathbf{w}(n)\\ \mathbf{y}(n) &= \mathbf{H}(n)\mathbf{x}(n)\end{aligned} \tag{B.15}$$

In vague general terms, the concepts of controllability and observability are as follows.

Controllability: The system (B.15) is said to be a *controllable system* if an input $\mathbf{w}(n)$ exists which takes the state from any point to any other point in a specified time.

Observability: The system (B.15) is said to be an *observable system* if it is possible to determine the state from knowledge of the output and input (but no knowledge of initial conditions).

Controllability is determined by $\mathbf{\Phi}(n)$ and $\mathbf{G}(n)$, while observability is determined by $\mathbf{\Phi}(n)$ and $\mathbf{H}(n)$.

More precise definitions of controllability and observability are possible, and explicit conditions on $\mathbf{\Phi}$, $\mathbf{G}$, and $\mathbf{H}$ which guarantee controllability and observability can be derived.

Appendix C

Vector and Matrix Gradient Functions

If $\mathbf{F}(x)$ is a K_1 by K_2 matrix which is a function of the scalar x, then

$$\frac{d}{dx}\mathbf{F}(x) = \begin{bmatrix} \frac{df_{11}(x)}{dx} & \cdots & \frac{df_{1K_2}(x)}{dx} \\ \vdots & & \vdots \\ \frac{df_{K_11}(x)}{dx} & \cdots & \frac{df_{K_1K_2}(x)}{dx} \end{bmatrix}$$

If $f(\mathbf{x})$ is a scalar function of the K-dimensional vector $\mathbf{x}$, then the gradient vector is

$$\frac{\partial f(\mathbf{x})}{\partial \mathbf{x}} = \begin{bmatrix} \frac{\partial f(\mathbf{x})}{\partial x_1} & \cdots & \frac{\partial f(\mathbf{x})}{\partial x_K} \end{bmatrix}$$

where the gradient is a row vector by definition. If $f(\mathbf{X})$ is a scalar function of the K_1 by K_2 matrix $\mathbf{X}$, then the gradient matrix is the K_2 by K_1 matrix

$$\frac{\partial f(\mathbf{X})}{\partial \mathbf{X}} = \begin{bmatrix} \frac{\partial f(\mathbf{X})}{\partial x_{11}} & \cdots & \frac{\partial f(\mathbf{X})}{\partial x_{K_11}} \\ \vdots & & \vdots \\ \frac{\partial f(\mathbf{X})}{\partial x_{1K_2}} & \cdots & \frac{\partial f(\mathbf{X})}{\partial x_{K_1K_2}} \end{bmatrix}$$

If $\mathbf{f}(\mathbf{x})$ is a K_2 vector function of a K_1 vector $\mathbf{x}$, then the Jacobian matrix is the K_2 by K_1 matrix

$$\frac{\partial \mathbf{f}(\mathbf{x})}{\partial \mathbf{x}} = \begin{bmatrix} \dfrac{\partial f_1(\mathbf{x})}{\partial x_1} & \cdots & \dfrac{\partial f_1(\mathbf{x})}{\partial x_{K_1}} \\ \vdots & & \vdots \\ \dfrac{\partial f_{K_2}(\mathbf{x})}{\partial x_1} & \cdots & \dfrac{\partial f_{K_2}(\mathbf{x})}{\partial x_{K_1}} \end{bmatrix}$$

Functions such as $\partial \mathbf{f}(\mathbf{X})/\partial \mathbf{X}$, $\partial \mathbf{F}(\mathbf{x})/\partial \mathbf{x}$, and $\partial \mathbf{F}(\mathbf{X})/\partial \mathbf{X}$ are also well-defined quantities but require *tensor-type* notations which will not be introduced here.

EXAMPLES: If

$$\mathbf{F}(x) = \begin{bmatrix} x^2 & x^3 \\ x & x^4 \end{bmatrix}$$

then

$$\frac{d\mathbf{F}(x)}{dx} = \begin{bmatrix} 2x & 3x^2 \\ 1 & 4x^3 \end{bmatrix}$$

If

$$f(\mathbf{x}) = x_1^2 + x_2^3 + x_1x_2$$

$$\mathbf{x} = \begin{pmatrix} x_1 \\ x_2 \end{pmatrix}$$

then

$$\frac{\partial f(\mathbf{x})}{\partial \mathbf{x}} = [2x_1 + x_2 \quad 3x_2 + x_1]$$

If

$$f(\mathbf{X}) = x_{11}^2x_{21} + x_{21}^3 - x_{11}x_{22}x_{12} + 3x_{21}$$

$$\mathbf{X} = \begin{bmatrix} x_{11} & x_{12} \\ x_{21} & x_{22} \end{bmatrix}$$

then

$$\frac{\partial f(\mathbf{X})}{\partial \mathbf{X}} = \begin{bmatrix} 2x_{11}x_{21} - x_{22}x_{12} & x_{11}^2 + 3x_{21}^2 + 3 \\ -x_{11}x_{22} & -x_{11}x_{12} \end{bmatrix}$$

If

$$\mathbf{f}(\mathbf{x}) = \begin{bmatrix} x_1^2 + x_2^3 \\ x_2 + x_1x_2 \end{bmatrix}$$

$$\mathbf{x} = \begin{pmatrix} x_1 \\ x_2 \end{pmatrix}$$

then

$$\frac{\partial \mathbf{f}(\mathbf{x})}{\partial \mathbf{x}} = \begin{bmatrix} 2x_1 & 3x_2^2 \\ x_2 & 1 + x_1 \end{bmatrix}$$

The trace is a linear operator, so that

$$\frac{d}{dx}\,\mathrm{tr}[\mathbf{F}(x)] = \mathrm{tr}\left[\frac{d\mathbf{F}(x)}{dx}\right]$$

Using this fact, a variety of useful equations can be derived by such as

$$\boxed{\frac{\partial}{\partial \mathbf{X}}\,\mathrm{tr}[\mathbf{AX}] = \mathbf{A}} \tag{C.1}$$

$$\boxed{\frac{\partial}{\partial \mathbf{X}}\,\mathrm{tr}[\mathbf{AX}'] = \mathbf{A}'} \tag{C.2}$$

$$\boxed{\frac{\partial}{\partial \mathbf{X}}\,\mathrm{tr}[\mathbf{AXBX}] = \mathbf{BXA} + \mathbf{AXB}} \tag{C.3}$$

$$\boxed{\frac{\partial}{\partial \mathbf{X}}\,\mathrm{tr}[\mathbf{AXBX}'] = \mathbf{BX}'\mathbf{A} + \mathbf{B}'\mathbf{X}'\mathbf{A}'} \tag{C.4}$$

Many other formulas can be derived from the above by using the facts that

$$\mathrm{tr}[\mathbf{AB}] = \mathrm{tr}[\mathbf{BA}]$$
$$\mathrm{tr}[\mathbf{A}'] = \mathrm{tr}[\mathbf{A}]$$

EXAMPLES: Consider

$$\mathbf{A} = \begin{bmatrix} a_{11} & a_{12} \\ a_{21} & a_{22} \end{bmatrix} \qquad \mathbf{X} = \begin{bmatrix} x_{11} & x_{12} \\ x_{21} & x_{22} \end{bmatrix}$$

Then

$$\mathrm{tr}(\mathbf{AX}) = a_{11}x_{11} + a_{12}x_{21} + a_{21}x_{12} + a_{22}x_{22}$$

so

$$\frac{\partial\,\mathrm{tr}(\mathbf{AX})}{\partial \mathbf{X}} = \begin{bmatrix} a_{11} & a_{12} \\ a_{21} & a_{22} \end{bmatrix} = \mathbf{A}$$

Similarly,

$$\mathrm{tr}[\mathbf{XX}] = x_{11}^2 + 2x_{12}x_{21} + x_{22}^2$$
$$\mathrm{tr}[\mathbf{XX}'] = x_{11}^2 + x_{12}^2 + x_{21}^2 + x_{22}^2$$

so

$$\frac{\partial\,\mathrm{tr}[\mathbf{XX}]}{\partial \mathbf{X}} = 2\begin{bmatrix} x_{11} & x_{12} \\ x_{21} & x_{22} \end{bmatrix} = 2\mathbf{X}$$

$$\frac{\partial\,\mathrm{tr}[\mathbf{XX}']}{\partial \mathbf{X}} = 2\begin{bmatrix} x_{11} & x_{21} \\ x_{12} & x_{22} \end{bmatrix} = 2\mathbf{X}'$$

Using a power series expansion of $e^{\mathbf{X}}$, it can also be shown that

$$\frac{\partial}{\partial \mathbf{X}} \operatorname{tr}[e^{\mathbf{X}}] = e^{\mathbf{X}}$$

A very useful equation results when

$$f(\mathbf{x}) = \mathbf{x}'\mathbf{A}\mathbf{x}$$

Then

$$\boxed{\frac{\partial}{\partial \mathbf{x}} \mathbf{x}'\mathbf{A}\mathbf{x} = \mathbf{x}'\mathbf{A}' + \mathbf{x}'\mathbf{A}} \tag{C.5}$$

This can be viewed as a special case of (C.4) where the matrix $\mathbf{X}$ of (C.4) is the vector $\mathbf{x}$ and $\mathbf{B} = \mathbf{I}$ as

$$\operatorname{tr}[\mathbf{A}\mathbf{x}\mathbf{x}'] = \mathbf{x}'\mathbf{A}\mathbf{x}$$

The determinate is a "nonlinear operator" but simple gradient equations often result. Consider a K by K matrix $\mathbf{X}(\alpha)$ which is a function of the scalar α. Then assuming $\mathbf{X}^{-1}(\alpha)$ exists,

$$\begin{aligned} \frac{\partial}{\partial \alpha} |\mathbf{X}(\alpha)| &= |\mathbf{X}(\alpha)| \operatorname{tr}\left[\mathbf{X}^{-1}(\alpha)\frac{\partial \mathbf{X}(\alpha)}{\partial \alpha}\right] \\ \frac{\partial}{\partial \alpha} \ln |\mathbf{X}(\alpha)| &= \operatorname{tr}\left[\mathbf{X}^{-1}(\alpha)\frac{\partial \mathbf{X}(\alpha)}{\partial \alpha}\right] \end{aligned} \tag{C.6}$$

This leads to formulae such as

$$\begin{aligned} \frac{\partial}{\partial \mathbf{X}} |\mathbf{X}| &= |\mathbf{X}|\mathbf{X}^{-1} \\ \frac{\partial}{\partial \mathbf{X}} \ln |\mathbf{X}| &= \mathbf{X}^{-1} \end{aligned} \tag{C.7}$$

Appendix D

Probability Theory

For simplicity it is always assumed that probability density functions exist, random variables have finite variances, etc.

Consider a K-dimensional random vector $\mathbf{x}$ with elements x_k, $k = 1, \ldots, K$. Let $P(\mathbf{x})$ denote the (cumulative) distribution function and $p(\mathbf{x})$ denote the joint probability density function. Then

$$p(\mathbf{x}) = p(x_1 \cdots x_K) = \frac{\partial^K P(x_1 \cdots x_K)}{\partial x_1 \cdots \partial x_K}$$

$$P(\mathbf{x}) = P(x_1 \cdots x_K) = \int_{-\infty}^{x_K} \cdots \int_{-\infty}^{x_1} p(u_1 \cdots u_K)\, du_1 \cdots du_K$$

Consider a two-dimensional random vector $\mathbf{x}$ where $P(\mathbf{x}) = P(x_1, x_2)$ and $p(\mathbf{x}) = p(x_1, x_2)$ are the distribution and density functions of $\mathbf{x}$, respectively. Let $P(x_j)$ and $p(x_j)$, $j = 1, 2$, be the marginal distributions and densities, respectively. Then

$$P(x_1) = P(x_1, \infty) = \int_{-\infty}^{x_1} \int_{-\infty}^{\infty} p(u, v)\, dv\, du$$

$$p(x_1) = \int_{-\infty}^{\infty} p(x_1, v)\, dv$$

Extension to vectors is straightforward. For $\mathbf{x} = \begin{bmatrix} \mathbf{x}_1 \\ \mathbf{x}_2 \end{bmatrix}$

$$p(\mathbf{x}_1) = \int_{-\infty}^{\infty} \cdots \int_{-\infty}^{\infty} p(x_{1,1}, x_{1,2}, \ldots)\, dx_{1,1}, dx_{1,2}, \ldots$$

or, in simpler notation,

$$\boxed{p(\mathbf{x}_1) = \int p(\mathbf{x}_1, \mathbf{x}_2)\, d\mathbf{x}_2} \tag{D.1}$$

Consider a K_1-dimensional random vector $\mathbf{x}$ with density $p(\mathbf{x}) = p(x_1 \cdots x_{K_1})$. Let $\mathbf{Y}(\mathbf{x})$ denote some K_2 by K_3 matrix function of $\mathbf{x}$. Let $E\{\mathbf{Y}(\mathbf{x})\}$ denote the expectation of $\mathbf{Y}$. Then

$$E\{\mathbf{Y}(\mathbf{x})\} = \begin{bmatrix} E\{y_{11}(\mathbf{x})\} & \cdots & E\{y_{1,K_3}(\mathbf{x})\} \\ \vdots & & \vdots \\ E\{y_{K_2 1}(\mathbf{x})\} & \cdots & E\{y_{K_2,K_3}(\mathbf{x})\} \end{bmatrix}$$

$$\begin{aligned} E\{y_{k,j}(\mathbf{x})\} &= \int_{-\infty}^{\infty} \cdots \int_{-\infty}^{\infty} y_{k,j}(\mathbf{x}) p(\mathbf{x})\, dx_1 \cdots dx_{K_1} \\ &= \int y_{k,j}(\mathbf{x}) p(\mathbf{x})\, d\mathbf{x} \end{aligned}$$

Expectation is a linear operation in the sense that

$$\boxed{E\{\mathbf{H}\mathbf{Y}(\mathbf{x})\} = \mathbf{H}E\{\mathbf{Y}(\mathbf{x})\}} \tag{D.2}$$

$$\boxed{E\{\mathbf{Y}_1(\mathbf{x}) + \mathbf{Y}_2(\mathbf{x})\} = E\{\mathbf{Y}_1(\mathbf{x})\} + E\{\mathbf{Y}_2(\mathbf{x})\}} \tag{D.3}$$

D.1 Means, Covariances, Variances, Standard Deviations, Moments

Let $\mathbf{m}$ denote the mean of the random vector $\mathbf{x}$. Then

$$\mathbf{m} = E\{\mathbf{x}\}$$

Let $\mathbf{\Gamma}$ denote the covariance matrix of $\mathbf{x}$. Then

$$\mathbf{\Gamma} = E\{[\mathbf{x} - \mathbf{m}][\mathbf{x} - \mathbf{m}]'\}$$

$\mathbf{\Gamma}$ is a positive semidefinite, symmetric matrix. The components of $\mathbf{m}$ are the first moments. The components of $\mathbf{\Gamma}$ are the second moments about the mean.

Consider a two-dimensional random vector $\mathbf{x}$ with

$$\mathbf{\Gamma} = \begin{bmatrix} \mathbf{\Gamma}_{11} & \mathbf{\Gamma}_{12} \\ \mathbf{\Gamma}_{12} & \mathbf{\Gamma}_{22} \end{bmatrix}$$

$\mathbf{\Gamma}_{11}$ is the variance of x_1, while $(\mathbf{\Gamma}_{11})^{1/2}$ is the standard deviation of x_1. $\mathbf{\Gamma}_{12}$ is the covariance or cross correlation between x_1 and x_2. The *correlation coefficient* between x_1 and x_2 is

$$\frac{\mathbf{\Gamma}_{12}}{(\mathbf{\Gamma}_{11}\mathbf{\Gamma}_{22})^{1/2}}$$

If $\mathbf{\Gamma}_{12} = 0$, then x_1 and x_2 are uncorrelated. Extension of these ideas to vector $\mathbf{x}_1$ and $\mathbf{x}_2$ is straightforward.

The second moment of a random vector $\mathbf{x}$ is $E\{\mathbf{x}\mathbf{x}'\}$. It follows that

$$\boxed{E\{\mathbf{x}\mathbf{x}'\} = \mathbf{\Gamma} + \mathbf{m}\mathbf{m}'} \tag{D.4}$$

$E\{x^2\}$ is also called the mean square value of x.

D.2 Independence, Uncorrelated, Orthogonal

If the random vectors $\mathbf{x}_1 \cdots \mathbf{x}_K$ are mutually independent, then

$$P(\mathbf{x}) = P(\mathbf{x}_1 \cdots \mathbf{x}_K) = P(\mathbf{x}_1) \cdots P(\mathbf{x}_K)$$

$$p(\mathbf{x}) = p(\mathbf{x}_1 \cdots \mathbf{x}_K) = p(\mathbf{x}_1) \cdots p(\mathbf{x}_K)$$

where $P(\mathbf{x}_j)$, $p(\mathbf{x}_j)$, $j = 1, \ldots, K$, are the marginal distributions and densities. For independent vectors $\mathbf{x}_1$ and $\mathbf{x}_2$

$$E\{\mathbf{Y}_1(\mathbf{x}_1)\mathbf{Y}_2(\mathbf{x}_2)\} = E\{\mathbf{Y}_1(\mathbf{x}_1)\}E\{\mathbf{Y}_2(\mathbf{x}_2)\}$$

Consider a random vector $\mathbf{x} = \begin{bmatrix} \mathbf{x}_1 \\ \mathbf{x}_2 \end{bmatrix}$ with

$$E\{\mathbf{x}\} = \begin{bmatrix} E\{\mathbf{x}_1\} \\ E\{\mathbf{x}_2\} \end{bmatrix} = \begin{bmatrix} \mathbf{m}_1 \\ \mathbf{m}_2 \end{bmatrix} = \mathbf{m}$$

$$E\{[\mathbf{x} - \mathbf{m}][\mathbf{x} - \mathbf{m}]'\} = \mathbf{\Gamma} = \begin{bmatrix} \mathbf{\Gamma}_{11} & \mathbf{\Gamma}_{12} \\ \mathbf{\Gamma}'_{12} & \mathbf{\Gamma}_{22} \end{bmatrix}$$

$\mathbf{x}_1$ and $\mathbf{x}_2$ are uncorrelated if

$$\mathbf{\Gamma}_{12} = \mathbf{0}$$

If $\mathbf{x}_1$ and $\mathbf{x}_2$ are uncorrelated,

$$E\{\mathbf{x}_1\mathbf{x}'_2\} = \mathbf{m}_1\mathbf{m}'_2$$

$\mathbf{x}_1$ and $\mathbf{x}_2$ are orthogonal if

$$E\{\mathbf{x}_1\mathbf{x}'_2\} = \mathbf{0}$$

The following relationships exist between independence, uncorrelated and orthogonal random vectors:

$$\text{independence} \xrightarrow{\text{implies}} \text{uncorrelated}$$

$$\text{uncorrelated plus zero mean} \xrightarrow{\text{implies}} \text{orthogonal}$$

$$\text{orthogonal} \xrightarrow{\text{implies}} \text{uncorrelated}$$

EXAMPLE: Assume that w is a Gaussian random variable with (see Appendix E)

$$E\{w\} = 0$$

$$E\{w^2\} = \psi$$

Then from Appendix E

$$E\{w^3\} = 0$$

Define

$$x_1 = w \qquad x_2 = w^2$$

It is obvious that x_1 and x_2 are not independent. However,

$$E\{x_1 x_2\} = E\{w^3\} = 0$$

so x_1 and x_2 are orthogonal. They are also uncorrelated, as

$$E\{x_1\} = 0$$

$$E\{x_2\} = \psi$$

and

$$E\{x_1(x_2 - \psi)\} = 0$$

D.3 Characteristic Functions†

Consider a random variable x with probability density $p(x)$. The characteristic function $c(\eta)$ of x is defined as

$$c(\eta) = \int_{-\infty}^{\infty} e^{i\eta x} p(x)\, dx = E\{e^{i\eta x}\}$$

where $i = \sqrt{-1}$. If $w(1) \cdots w(n)$ are independent random variables with characteristic functions $c_1(\eta) \cdots c_n(\eta)$, then

$$x = \sum_{m=1}^{n} w(m)$$

has characteristic function $c_x(\eta)$ given by

$$c_x(\eta) = \prod_{m=1}^{n} c_m(\eta)$$

For a vector $\mathbf{x}$ with $p(\mathbf{x})$

$$c(\boldsymbol{\eta}) = \int e^{i\boldsymbol{\eta}'\mathbf{x}} p(\mathbf{x})\, d\mathbf{x}$$

The characteristic function is also called the moment-generating function. Consider a scalar x. Then

$$c(\eta) = \int_{-\infty}^{\infty} e^{i\eta x} p(x)\, dx$$

$$= \int_{-\infty}^{\infty} \left[1 + i\eta x + \frac{(i\eta x)^2}{2} + \cdots\right] p(x)\, dx$$

so that

$$\left.\frac{\partial c(\eta)}{\partial \eta}\right|_{\eta=0} = i \int_{-\infty}^{\infty} x p(x)\, dx$$

$$\left.\frac{\partial^2 c(\eta)}{\partial \eta^2}\right|_{\eta=0} = -\frac{1}{2} \int_{-\infty}^{\infty} x^2 p(x)\, dx \tag{D.5}$$

$$\vdots$$

D.4 Transformation of Variables†

Consider a K-dimensional random vector $\mathbf{x}$ with probability density $p_x(\mathbf{x})$. Consider the K random variables

$$y_k = y_k(\mathbf{x}), \qquad k = 1, \ldots, K$$

Assume that the transformation is one to one so that the inverse transformation can be written as

$$x_k = x_k(\mathbf{y}), \qquad k = 1, \ldots, K$$

where $\mathbf{y}$ has the elements y_k, $k = 1, \ldots, K$. Let $p_y(\mathbf{y})$ denote the probability density of $\mathbf{y}$. Then

$$p_y(\mathbf{y}) = p_x[x_1(\mathbf{y}) \cdots x_K(\mathbf{y})] \left| \frac{\partial \mathbf{x}}{\partial \mathbf{y}} \right|$$

$$\frac{\partial \mathbf{x}}{\partial \mathbf{y}} = \begin{bmatrix} \dfrac{\partial x_1}{\partial y_1} & \cdots & \dfrac{\partial x_1}{\partial y_K} \\ \vdots & & \vdots \\ \dfrac{\partial x_K}{\partial y_1} & \cdots & \dfrac{\partial x_K}{\partial y_K} \end{bmatrix}$$

D.5 Sums of Random Variables

Consider

$$y = x_1 + x_2$$

where x_1 and x_2 are random variables with probability density $p_x(x_1, x_2)$. Let $p_y(y)$ denote the probability density of y. Then

$$p_y(y) = \int_{-\infty}^{\infty} p_x(u, y - u)\, du$$

If x_1 and x_2 are independent so that

$$p_x(x_1, x_2) = p_{x_1}(x_1) p_{x_2}(x_2)$$

then $p_y(y)$ is given by the *convolution*

$$p_y(y) = \int_{-\infty}^{\infty} p_{x_1}(u) p_{x_2}(y - u)\, du$$

Extension to vectors $\mathbf{y}$, $\mathbf{x}_1$, and $\mathbf{x}_2$ and to sums of many random variables is straightforward.

D.6 Conditional Distribution: Bayes' Rule

Assume that $\mathbf{x}$ is a partitioned random vector where $p(\mathbf{x}) = p(\mathbf{x}_1, \mathbf{x}_2)$ is the density and $p(\mathbf{x}_1)$ and $p(\mathbf{x}_2)$ are the marginal densities. Let $p(\mathbf{x}_1 | \mathbf{x}_2)$ denote the conditional probability density of $\mathbf{x}_1$ given $\mathbf{x}_2$. Then the product rule is

$$\boxed{p(\mathbf{x}_1 \mid \mathbf{x}_2) = \frac{p(\mathbf{x}_1, \mathbf{x}_2)}{p(\mathbf{x}_2)}} \tag{D.6}$$

In (D.6) $\mathbf{x}_2$ is sometimes viewed as a given constant value so that $p(\mathbf{x}_1 \mid \mathbf{x}_2)$ is the probability density of $\mathbf{x}_1$ when the random vector $\mathbf{x}_2$ assumes that value. However, $\mathbf{x}_2$ can also be viewed as a random vector so that $p(\mathbf{x}_1 \mid \mathbf{x}_2)$ is a random function, i.e., a function of $\mathbf{x}_1$ whose "shape" depends on the random vector $\mathbf{x}_2$. With either interpretation,

$$\int_{-\infty}^{\infty} p(\mathbf{x}_1 \mid \mathbf{x}_2)\, d\mathbf{x}_1 = 1 \tag{D.7}$$

Using the product rule,

$$\boxed{p(\mathbf{x}_1 \mid \mathbf{x}_2) = \frac{p(\mathbf{x}_2 \mid \mathbf{x}_1) p(\mathbf{x}_1)}{p(\mathbf{x}_2)}} \tag{D.8}$$

Using the definition of marginal distributions and the product rule gives

$$\boxed{\int p(\mathbf{x}_2 \mid \mathbf{x}_1) p(\mathbf{x}_1)\, d\mathbf{x}_1 = \int p(\mathbf{x}_1 \mathbf{x}_2)\, d\mathbf{x}_1 = p(\mathbf{x}_2)} \tag{D.9}$$

Bayes' theorem then follows; i.e.,

$$p(\mathbf{x}_1 \mid \mathbf{x}_2) = \frac{p(\mathbf{x}_2 \mid \mathbf{x}_1) p(\mathbf{x}_1)}{\int p(\mathbf{x}_2 \mid \mathbf{x}_1) p(\mathbf{x}_1)\, d\mathbf{x}_1} \tag{D.10}$$

In this book; the product rule, Bayes' theorem, and other related equations are all called *Bayes' rule.*

D.7 Conditional Expectation

Consider a partitioned $\mathbf{x}$ with $p(\mathbf{x}) = p(\mathbf{x}_1, \mathbf{x}_2)$. Let $\mathbf{y}(\mathbf{x}_1)$ be some function of $\mathbf{x}_1$. Then the conditional expectation of $\mathbf{y}(\mathbf{x}_1)$ given $\mathbf{x}_2$ is

$$E\{\mathbf{y}(\mathbf{x}_1) \mid \mathbf{x}_2\} = \int_{-\infty}^{\infty} \mathbf{y}(\mathbf{x}_1) p(\mathbf{x}_1 \mid \mathbf{x}_2)\, d\mathbf{x}_1 \tag{D.11}$$

The conditional mean of $\mathbf{x}_1$ given $\mathbf{x}_2$ is

$$E\{\mathbf{x}_1 \mid \mathbf{x}_2\} = \int \mathbf{x}_1 p(\mathbf{x}_1 \mid \mathbf{x}_2)\, d\mathbf{x}_1$$

$E\{\mathbf{x}_1 \mid \mathbf{x}_2\}$ is usually called the conditional expectation of $\mathbf{x}_1$ given $\mathbf{x}_2$ rather than the conditional mean. The conditional covariance matrix of $\mathbf{x}_1$ given

$\mathbf{x}_2$ is defined as

$$E\{[\mathbf{x}_1 - E\{\mathbf{x}_1 | \mathbf{x}_2\}][\mathbf{x}_1 - E\{\mathbf{x}_1 | \mathbf{x}_2\}]' | \mathbf{x}_2\} = \int [\mathbf{x}_1 - E\{\mathbf{x}_1 | \mathbf{x}_2\}][\mathbf{x}_1 - E\{\mathbf{x}_1 | \mathbf{x}_2\}]' p(\mathbf{x}_1 | \mathbf{x}_2) \, d\mathbf{x}_1 \tag{D.12}$$

It is very important to emphasize that $E\{\mathbf{x}_1 | \mathbf{x}_2\}$, etc., are functions of $\mathbf{x}_2$. In most cases, $\mathbf{x}_2$ is viewed as a random vector so that the conditional mean, conditional variances, etc., are random quantities.

When the conditional expectation is viewed as a random vector, it has many important properties. Some of these are now listed. Let $\mathbf{Y}(\mathbf{x})$ be some matrix function of $\mathbf{x}$. Then

$$\boxed{E\{E\{\mathbf{Y}(\mathbf{x}_1) | \mathbf{x}_2\}\} = E\{\mathbf{Y}(\mathbf{x}_1)\}} \tag{D.13}$$

$$\boxed{E\{\mathbf{Y}_1(\mathbf{x}_1) + \mathbf{Y}_2(\mathbf{x}_1) | \mathbf{x}_2\} = E\{\mathbf{Y}_1(\mathbf{x}_1) | \mathbf{x}_2\} + E\{\mathbf{Y}_2(\mathbf{x}_1) | \mathbf{x}_2\}} \tag{D.14}$$

$$\boxed{E\{\mathbf{H}\mathbf{Y}(\mathbf{x}_1) | \mathbf{x}_2\} = \mathbf{H}E\{\mathbf{Y}(\mathbf{x}_1) | \mathbf{x}_2\}} \tag{D.15}$$

$$\boxed{E\{\mathbf{Y}_1(\mathbf{x}_2)\mathbf{Y}_2(\mathbf{x}_1) | \mathbf{x}_2\} = \mathbf{Y}_1(\mathbf{x}_2)E\{\mathbf{Y}_2(\mathbf{x}_1) | \mathbf{x}_2\}} \tag{D.16}$$

$$\boxed{E\{\mathbf{Y}(\mathbf{x}_2) | \mathbf{x}_2\} = \mathbf{Y}(\mathbf{x}_2)} \tag{D.17}$$

$$\boxed{E\{\mathbf{H} | \mathbf{x}_2\} = \mathbf{H}} \tag{D.18}$$

$$\boxed{E\{\mathbf{Y}(\mathbf{x}_1) | \mathbf{x}_2\} = E\{\mathbf{Y}(\mathbf{x}_1)\} \qquad \text{if } \mathbf{x}_1, \mathbf{x}_2 \text{ are independent}} \tag{D.19}$$

If $\mathbf{H}^{-1}$ exists, then

$$\boxed{E\{\mathbf{Y}(\mathbf{x}_1) | \mathbf{H}\mathbf{x}_2\} = E\{\mathbf{Y}(\mathbf{x}_1) | \mathbf{x}_2\}} \tag{D.20}$$

The following property is of special importance

$$\boxed{E\{\mathbf{Y}(\mathbf{x}_2)[\mathbf{x}_1 - E\{\mathbf{x}_1 | \mathbf{x}_2\}]\} = \mathbf{0}} \tag{D.21}$$

as it states that the random vector $\mathbf{x}_1 - E\{\mathbf{x}_1 | \mathbf{x}_2\}$ is *orthogonal* to any function of the random vector $\mathbf{x}_2$. Consider the conditional variance of (D.12). Manipulation of (D.12) using (D.21) yields

$$E\{[\mathbf{x}_1 - E\{\mathbf{x}_1 | \mathbf{x}_2\}][\mathbf{x}_1 - E\{\mathbf{x}_1 | \mathbf{x}_2\}] | '\mathbf{x}_2\} = E\{\mathbf{x}_1\mathbf{x}_1' | \mathbf{x}_2\} - E\{[E\{\mathbf{x}_1 | \mathbf{x}_2\}][E\{\mathbf{x}_1 | \mathbf{x}_2\}]'\}$$

Appendix E

Multivariate Gaussian (Normal) Distribution

Consider the K-dimensional random vector $\mathbf{x}$ with probability density $p(\mathbf{x})$. $\mathbf{x}$ is a Gaussian or normal random vector if $p(\mathbf{x})$ is of the form

$$\begin{aligned} p(\mathbf{x}) &= [(2\pi)^K |\mathbf{\Gamma}|]^{-1/2} \exp\{-(\tfrac{1}{2})J(\mathbf{x})\} \\ J(\mathbf{x}) &= (\mathbf{x} - \mathbf{m})'\mathbf{\Gamma}^{-1}(\mathbf{x} - \mathbf{m}) \\ E\{\mathbf{x}\} &= \mathbf{m} \\ E\{(\mathbf{x} - \mathbf{m})(\mathbf{x} - \mathbf{m})'\} &= \mathbf{\Gamma} \end{aligned} \tag{E.1}$$

The K-dimensional vector $\mathbf{m}$ is the mean value vector, while the K by K matrix $\mathbf{\Gamma}$ is the covariance matrix. For scalar x

$$p(x) = \frac{1}{(2\pi\mathbf{\Gamma})^{1/2}} \exp\left\{-\frac{(x - m)^2}{2\mathbf{\Gamma}}\right\}$$

The shape of $p(x)$ is illustrated in Fig. E.1. The terms *Gaussian* and *normal* are both used interchangeably in this book.

The characteristic function $c(\boldsymbol{\eta})$ of a Gaussian $\mathbf{x}$ is

$$c(\boldsymbol{\eta}) = \exp\{-\tfrac{1}{2}[\boldsymbol{\eta}'\mathbf{\Gamma}\boldsymbol{\eta}] + i\boldsymbol{\eta}'\mathbf{m}\} \tag{E.2}$$

If the characteristic function is viewed as the definition of a Gaussian vector, $\mathbf{\Gamma}$ can be a positive semidefinite matrix (i.e., $\mathbf{\Gamma}^{-1}$ need not exist).

The case where $\mathbf{\Gamma}$ is only positive semidefinite corresponds to a *degenerate* Gaussian random vector. Another type of degenerate Gaussian random

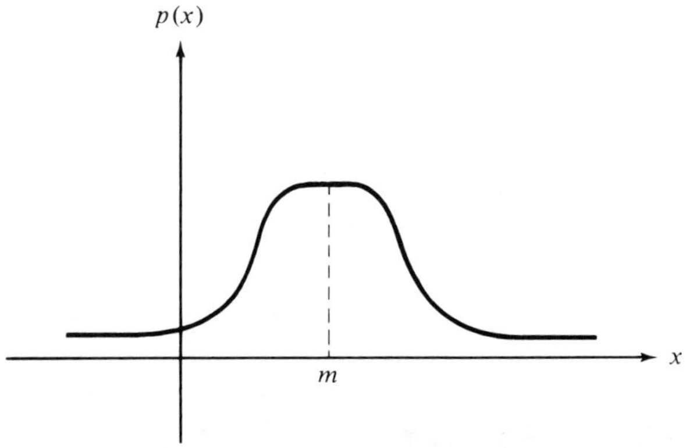

Figure E.1 Gaussian (normal) probability density.

vector occurs when $\mathbf{\Gamma}^{-1}$ is specified but is only positive semidefinite [i.e., $(\mathbf{\Gamma}^{-1})^{-1}$ does not exist].

The fact that $\mathbf{x}$ is a Gaussian (normal) random vector with mean $\mathbf{m}$ and covariance matrix $\mathbf{\Gamma}$ is often indicated by writing

$$\mathbf{x} \text{ is } N(\mathbf{m}, \mathbf{\Gamma})$$

Define the following partitioned vectors and matrices:

$$\mathbf{x} = \begin{bmatrix} \mathbf{x}_1 \\ \hline \mathbf{x}_2 \end{bmatrix} \qquad \mathbf{m} = \begin{bmatrix} \mathbf{m}_1 \\ \hline \mathbf{m}_2 \end{bmatrix}$$
$$\mathbf{\Gamma} = \left[\begin{array}{c|c} \mathbf{\Gamma}_{11} & \mathbf{\Gamma}_{12} \\ \hline \mathbf{\Gamma}'_{12} & \mathbf{\Gamma}_{22} \end{array}\right] \tag{E.3}$$

where $\mathbf{x}_j$, $\mathbf{m}_j$ are K_j-dimensional vectors, $\mathbf{\Gamma}_{jj}$ are K_j by K_j matrices, $j = 1, 2$, and $\mathbf{\Gamma}_{12}$ is a K_1 by K_2 matrix. $K_1 + K_2 = K$. Let $p_j(\mathbf{x}_j)$ denote the probability density of $\mathbf{x}_j$.

Assume that a K_1-dimensional $\mathbf{x}$ is $N(\mathbf{m}, \mathbf{\Gamma})$. Define a K_2-dimensional random vector as a linear transformation of $\mathbf{x}$:

$$\mathbf{y} = \mathbf{H}\mathbf{x} + \mathbf{h}$$

Then

$$\boxed{\mathbf{y} \text{ is } N(\mathbf{H}\mathbf{m} + \mathbf{h}, \mathbf{H}\mathbf{\Gamma}\mathbf{H}')} \tag{E.4}$$

One of many ways to derive (E.4) is to use the characteristic function. Let $c_y(\boldsymbol{\eta})$ denote the characteristic function of $\mathbf{y}$. Then

$$c_y(\boldsymbol{\eta}) = E\{\exp[i\boldsymbol{\eta}'\mathbf{H}\mathbf{x} + i\boldsymbol{\eta}'\mathbf{h}]\}$$

or from (E.2)

$$c_y(\boldsymbol{\eta}) = \exp\{-\tfrac{1}{2}[\boldsymbol{\eta}'\mathbf{H}\boldsymbol{\Gamma}\mathbf{H}'\boldsymbol{\eta}] + i\boldsymbol{\eta}'[\mathbf{H}\mathbf{m} + \mathbf{h}]\}$$

which yields (E.4). If $K_1 \geq K_2$ and $(\mathbf{H}\boldsymbol{\Gamma}\mathbf{H}')^{-1}$ exists, then $p(\mathbf{y})$ is determined as in (E.1) or (E.2). If $[\mathbf{H}\boldsymbol{\Gamma}\mathbf{H}']^{-1}$ does not exist, $p(\mathbf{y})$ is determined in terms of the characteristic function as in (E.2). An important special case of (E.4) is

$$\mathbf{y} = \mathbf{x}_1 + \mathbf{x}_2$$

If $\mathbf{x}$ is $N(\mathbf{m}, \boldsymbol{\Gamma})$, as defined in (E.3),

$$\mathbf{y} \text{ is } N(\mathbf{m}_1 + \mathbf{m}_2, \boldsymbol{\Gamma}_{11} + \boldsymbol{\Gamma}_{12} + \boldsymbol{\Gamma}'_{12} + \boldsymbol{\Gamma}_{22})$$

Extension to sums of many Gaussian random vectors is straightforward.

If $\mathbf{x}_1$ and $\mathbf{x}_2$ are uncorrelated or orthogonal and Gaussian, they are also independent.

Consider a Gaussian $\mathbf{x}$ is partitioned as in (E.3). Then using (E.4),

$$\boxed{\mathbf{x}_j \text{ is } N(\mathbf{m}_j, \boldsymbol{\Gamma}_{jj}), \qquad j = 1, 2} \tag{E.5}$$

This is obviously true if $\mathbf{x}_1$ and $\mathbf{x}_2$ are independent, but it is true in general.

E.1 Conditional Distributions

By Bayes' rule, the conditional probability density $p(\mathbf{x}_1 \mid \mathbf{x}_2)$ is given by

$$p(\mathbf{x}_1 \mid \mathbf{x}_2) = \frac{p(\mathbf{x}_1, \mathbf{x}_2)}{p_2(\mathbf{x}_2)} = \frac{p(\mathbf{x})}{p_2(\mathbf{x}_2)}$$

Substituting (E.1) and (E.5) gives for the Gaussian case

$$p(\mathbf{x}_1 \mid \mathbf{x}_2) = \left[\frac{(2\pi)^{K_1} |\boldsymbol{\Gamma}|}{|\boldsymbol{\Gamma}_{22}|}\right]^{-1/2} \exp\{-(\tfrac{1}{2})J_{1|2}(\mathbf{x})\} \tag{E.6}$$

where

$$J_{1|2}(\mathbf{x}) = J(\mathbf{x}) - J_2(\mathbf{x}_2)$$

Equation (E.6) can be manipulated after a lot of work into the following more useful form:

$$\boxed{\begin{aligned} p(\mathbf{x}_1 \mid \mathbf{x}_2) &= [(2\pi)^{K_1} |\boldsymbol{\Gamma}_{1|2}|]^{-1/2} \exp\{-(\tfrac{1}{2})J_{1|2}(\mathbf{x}_1)\} \\ J_{1|2}(\mathbf{x}_1) &= [\mathbf{x}_1 - \mathbf{m}_{1|2}]'\boldsymbol{\Gamma}_{1|2}^{-1}[\mathbf{x}_1 - \mathbf{m}_{1|2}] \\ \mathbf{m}_{1|2} &= \mathbf{m}_1 + \boldsymbol{\Gamma}_{12}\boldsymbol{\Gamma}_{22}^{-1}[\mathbf{x}_2 - \mathbf{m}_2] \\ \boldsymbol{\Gamma}_{1|2} &= \boldsymbol{\Gamma}_{11} - \boldsymbol{\Gamma}_{12}\boldsymbol{\Gamma}_{22}^{-1}\boldsymbol{\Gamma}'_{12} \end{aligned}} \tag{E.7}$$

If $\mathbf{x}_2$ is considered to be a constant (i.e., not a random vector), then $p(\mathbf{x}_1 \mid \mathbf{x}_2)$ is the probability density of a K_1-dimensional normal vector with mean $\mathbf{m}_{1|2}$ and covariance matrix $\boldsymbol{\Gamma}_{1|2}$.

E.2 Conditional Expectation

Consider the K_1-dimensional random vector $E\{\mathbf{x}_1 | \mathbf{x}_2\}$ which is the conditional expectation of $\mathbf{x}_1$ given $\mathbf{x}_2$. It follows from (E.7) that in the Gaussian case

$$E\{\mathbf{x}_1 | \mathbf{x}_2\} = \mathbf{m}_{1|2} = \mathbf{m}_1 + \mathbf{\Gamma}_{12}\mathbf{\Gamma}_{22}^{-1}(\mathbf{x}_2 - \mathbf{m}_2) \tag{E.8}$$

If $E\{\mathbf{x}_1 | \mathbf{x}_2\}$ is viewed as a random variable, i.e., as a function of the random vector $\mathbf{x}_2$, then for a Gaussian $\mathbf{x}$

$$E\{\mathbf{x}_1 | \mathbf{x}_2\} \quad \text{is} \quad N\{\mathbf{m}_1, \mathbf{\Gamma}_{12}\mathbf{\Gamma}_{22}^{-1}\mathbf{\Gamma}'_{12}\} \tag{E.9}$$

The conditional covariance matrix is defined by (D.12). Substituting (E.7) and (E.8) yields

$$E\{[\mathbf{x}_1 - E\{\mathbf{x}_1 | \mathbf{x}_2\}][\mathbf{x}_1 - E\{\mathbf{x}_1 | \mathbf{x}_2\}]' | \mathbf{x}_2\} = \mathbf{\Gamma}_{1|2} = \mathbf{\Gamma}_{11} - \mathbf{\Gamma}_{12}\mathbf{\Gamma}_{22}^{-1}\mathbf{\Gamma}_{21}$$

Thus in the Gaussian case, the conditional covariance matrix is not a function of $\mathbf{x}_2$; i.e., it is not a random function.

The general properties of the conditional expectation (D.13)–(D.21) also hold in the Gaussian case. Another useful property which can be proved for the Gaussian case is

$$\begin{aligned}&\text{If } \mathbf{x}_2 \text{ and } \mathbf{x}_3 \text{ are independent,}\\ &E\{\mathbf{x}_1 | \mathbf{x}_2, \mathbf{x}_3\} - \mathbf{m}_1 = E\{\mathbf{x}_1 | \mathbf{x}_2\} - \mathbf{m}_1 + E\{\mathbf{x}_1 | \mathbf{x}_3\} - \mathbf{m}_1\end{aligned} \tag{E.10}$$

E.3 Wide Sense and Strict Sense†

Many concepts of random vectors and stochastic processes can be formulated in either a *wide sense* or a *strict sense*. Consider two random vectors $\mathbf{x}$ and $\mathbf{y}$, where

$$\begin{aligned}&\mathbf{x} : \text{not Gaussian}\\ &\mathbf{y} : \text{Gaussian}\\ &E\{\mathbf{x}\} = E\{\mathbf{y}\} = \mathbf{m}\\ &E\{[\mathbf{x} - \mathbf{m}][\mathbf{x} - \mathbf{m}]'\} = E\{[\mathbf{y} - \mathbf{m}][\mathbf{y} - \mathbf{m}]'\} = \mathbf{\Gamma}\end{aligned}$$

Then although $\mathbf{x}$ and $\mathbf{y}$ are different, they have the same means and covariance matrices. Now suppose that some property p_w of $\mathbf{x}$ can be expressed in terms of $\mathbf{m}$ and $\mathbf{\Gamma}$. Let p_s be the corresponding property for the Gaussian $\mathbf{y}$. Then p_w is a wide sense property, while p_s is a strict sense property. A strict sense property is "stronger" than a wide sense property. The following examples should help clarify the strict versus wide sense concept.

Consider

$$\mathbf{x} = \begin{bmatrix} \mathbf{x}_1 \\ \mathbf{x}_2 \end{bmatrix} \qquad \text{non-Gaussian}$$

$$\mathbf{y} = \begin{bmatrix} \mathbf{y}_1 \\ \mathbf{y}_2 \end{bmatrix} \qquad \text{Gaussian}$$

$$E\{\mathbf{x}\} = E\{\mathbf{y}\} = \mathbf{0}$$

$$E\{\mathbf{xx}'\} = E\{\mathbf{yy}'\} = \begin{bmatrix} \mathbf{\Gamma}_{11} & \mathbf{\Gamma}_{12} \\ \mathbf{\Gamma}'_{12} & \mathbf{\Gamma}_{22} \end{bmatrix}$$

Suppose that $\mathbf{\Gamma}_{12} = \mathbf{0}$. Then

1. $\mathbf{y}_1$ and $\mathbf{y}_2$ are independent (a strict sense property).
2. $\mathbf{x}_1$ and $\mathbf{x}_2$ are uncorrelated or orthogonal (a wide sense property).

Both wide sense and strict sense conditional expectations can be defined. The definition of Section D.12 yields a strict sense conditional expectation. For the Gaussian $\mathbf{y}$

$$E_{\text{strict sense}}\{\mathbf{y}_1 \mid \mathbf{y}_2\} = \mathbf{\Gamma}_{12}\mathbf{\Gamma}_{22}^{-1}\mathbf{y}_2$$

For the non-Gaussian $\mathbf{x}$, the wide sense conditional expectation is defined as

$$E_{\text{wide sense}}\{\mathbf{x}_1 \mid \mathbf{x}_2\} = \mathbf{\Gamma}_{12}\mathbf{\Gamma}_{22}^{-1}\mathbf{x}_2$$

Thus

$$\text{Gaussian } \mathbf{y}\text{: } E_{\text{strict sense}}\{\mathbf{y}_1 \mid \mathbf{y}_2\} = E_{\text{wide sense}}\{\mathbf{y}_1 \mid \mathbf{y}_2\}$$

$$\text{Non-Gaussian } \mathbf{x}\text{: } E_{\text{strict sense}}\{\mathbf{x}_1 \mid \mathbf{x}_2\} \neq E_{\text{wide sense}}\{\mathbf{x}_1 \mid \mathbf{x}_2\}$$

More examples of strict and wide sense properties are discussed in Appendix *F*.

E.4 Higher-Order Moments

Consider a scalar Gaussian random variable

$$x \text{ is } N(m, \Gamma)$$

Then

$$E\{x\} = m$$

$$E\{(x - m)^2\} = \Gamma$$

$$E\{(x - m)^3\} = 0$$

$$E\{(x - m)^4\} = 3\Gamma^2$$

or in general

$$E\{(x - m)^{2K+1}\} = 0$$

$$E\{(x - m)^{2K}\} = 1 \times 3 \times \cdots \times (2k - 1)\Gamma^K \qquad \text{(E.11)}$$

$$K = 0, 1, \ldots$$

These results can be derived using the characteristic function by combining (D.5) with (E.2).

E.5 Chi-Squared (χ^2) Random Variables

Consider K independent Gaussian random variables $w(1) \cdots w(k)$:

$$w(k) \text{ is } N(0, 1), \qquad k = 1, \ldots, K$$

A chi-squared random variable with K degrees of freedom is defined as

$$\chi_K^2 = \sum_{k=1}^{K} w^2(k) \tag{E.12}$$

Now consider the Gaussian K-dimensional $\mathbf{x}$ of (E.1). If $\mathbf{\Gamma}$ is a diagonal matrix,

$$J(\mathbf{x}) = (\mathbf{x} - \mathbf{m})'\mathbf{\Gamma}^{-1}(\mathbf{x} - \mathbf{m}) = \sum_{k=1}^{K} \left(\frac{x_k - m_k}{\mathbf{\Gamma}_{kk}}\right)^2$$

Since $(x_k - m_k)/(\mathbf{\Gamma}_{kk})^{1/2}$ is $N(0, 1)$, $J(\mathbf{x})$ is a chi-squared random variable for diagonal $\mathbf{\Gamma}$. However, since any $\mathbf{\Gamma}$ can be diagonalized by the appropriate change of variables, it follows that $J(\mathbf{x})$ is always chi-squared. If $\mathbf{\Gamma}$ does not exist, some care is needed, but the general result is

> If a K-dimensional $\mathbf{x}$ is $N(\mathbf{m}, \mathbf{\Gamma})$, then
>
> $$[\mathbf{x} - \mathbf{m}]'\mathbf{\Gamma}^{-1}[\mathbf{x} - \mathbf{m}] = \chi_{\tilde{K}}^2 \tag{E.13}$$
>
> where the degrees of freedom $\tilde{K}$ equals the rank of $\mathbf{\Gamma}^{-1}$

Consider χ_K^2 as given by (E.12). It follows from (E.11) that

$$E\{\chi_K^2\} = K$$

$$E\{[\chi_K^2 - E\{\chi_K^2\}]^2\} = 2K$$

Tables for the probability density of χ_K^2 for various degrees of freedom are widely available. It can be shown that as K becomes large, χ_K^2 approaches a Gaussian random variable; i.e.,

$$\boxed{\chi_K^2 \longrightarrow N(K, 2K) \text{ as } K \text{ increases}} \tag{E.14}$$

The assumption that $\chi_K^2 = N(K, 2K)$ becomes very good for $K > 30$. It can also be shown that $[2\chi_K^2]^{1/2} \longrightarrow N[\sqrt{2K}, 1]$ as K increases.

E.6 Constant Probability Density Contours

Consider some K-dimensional random vector $\mathbf{x}$ with probability density $p(\mathbf{x})$. The set

$$\Omega = \{\mathbf{x} : p(\mathbf{x}) = \text{constant}\}$$

is a K-dimensional surface defined by all $\mathbf{x}$ with the same probability density. For a Gaussian random vector, it follows from (E.1) that the ellipsoid surface

$$\mathbf{\Omega} = \{\mathbf{x} : [\mathbf{x} - \mathbf{m}]'\mathbf{\Gamma}^{-1}[\mathbf{x} - \mathbf{m}] = \text{constant}\}$$

is such a contour of constant probability density. (Extensive discussions on ellipsoids are given in Appendix G.)

Consider the ellipsoid

$$\mathbf{\Omega}_c = \{\mathbf{x} : [\mathbf{x} - \mathbf{m}]'\mathbf{\Gamma}^{-1}[\mathbf{x} - \mathbf{m}] \leq c^2\}$$

so that $\mathbf{\Omega}_c$ is the set of all $\mathbf{x}$ lying within or on a contour of constant probability density. The probability of obtaining an $\mathbf{x}$ which lies within $\mathbf{\Omega}_c$ is of interest. Define

$$P(c) = \text{probability } \mathbf{x} \text{ lies within } \mathbf{\Omega}_c$$

From Section E.5 if follows that (if $\mathbf{\Gamma}^{-1}$ exists)

$$P(c) = \text{probability } \chi_K^2 \leq c^2$$

Thus $P(c)$ can be easily evaluated. Some representative values are summarized in Fig. E.2. When $c = 1$, $\mathbf{\Omega}_c$ is often called the *one-sigma* ellipsoid as for scalars, a common notation is $\sigma^2 = E\{(x - m)^2\}$. Similarly, for $c = 2$, $\mathbf{\Omega}_c$ is the *two-sigma* ellipsoid, etc. It is important to note from Fig. E.2 that as the dimension K increases, the probability of $\mathbf{x}$ lying within a one-sigma ($c = 1$) ellipsoid becomes very small.

K: dimension of $\mathbf{x}$	$P(c) = .1$	$P(c) = .5$	$P(c) = .9$	$P(c) = .95$
1	—	$c^2 = .46$	$c^2 = 2.71$	$c^2 = 3.84$
2	$c^2 = .02$	1.39	4.61	5.99
3	.115	2.37	6.25	7.81
5	.554	4.35	9.24	11.07
10	2.56	9.34	15.99	18.31

$\mathbf{x}$ is $N(\mathbf{m}, \mathbf{\Gamma})$ ($\mathbf{\Gamma}^{-1}$ exists)
$P(c)$ is probability $\mathbf{x}$ lying in $\mathbf{\Omega}_c$
$\mathbf{\Omega}_c = \{\mathbf{x}: [\mathbf{x} - \mathbf{m}]'\mathbf{\Gamma}^{-1}[\mathbf{x} - \mathbf{m}] \leq c^2\}$

Figure E.2. Probability of Gaussian $\mathbf{x}$ lying within constant probability contour.

Appendix F

Stochastic Processes

In this book, a stochastic process is merely viewed as a family of random variables which is indexed by some parameter which is associated with time. Such a definition, of course, is almost all-encompassing. A discrete-time stochastic process is a sequence of random vectors $\mathbf{x}(n)$, $n = 1, \ldots,$ where the indexing parameter is taking on discrete values (integer values). In a continuous-time stochastic process, the "integer time n" is replaced by "continuous time t."

Consider a scalar discrete-time stochastic process $x(n)$, $n = 1, \ldots$. The properties of the process from $n = 1$ to $n = N$ can be investigated by applying the concepts of Appendices D and E to an N-dimensional vector formed from the $x(n)$, $n = 1, \ldots, N$. This point of view emphasizes that in order to completely define a discrete-time stochastic process $x(n)$, $n = 1$, $\ldots$, it is necessary to specify its complete joint probability density function:

$$p[x(1), x(2), \ldots]$$

Consider a vector discrete-time stochastic process $\mathbf{x}(n)$, $n = 1, \ldots,$ with mean value $\mathbf{m}(n)$. The correlation function $\mathcal{P}(n_1, n_2)$ is defined as

$$\mathcal{P}(n_1, n_2) = E\{[\mathbf{x}(n_1) - \mathbf{m}(n_1)][\mathbf{x}(n_2) - \mathbf{m}(n_2)]'\}$$

Specification of $\mathbf{m}(n)$ and $\mathcal{P}(n_1, n_2)$ for all n_1 and n_2 completely specifies all the first and second moments. If $\mathbf{\Gamma}(n)$ is the covariance matrix of $\mathbf{x}(n)$, then

$$\mathbf{\Gamma}(n) = \mathcal{P}(n, n)$$

Consider a process whose probability densities (such as $p[\mathbf{x}(n)]$) do not change with time and whose joint probability densities (such as $p[\mathbf{x}(n_1), \mathbf{x}(n_2)]$) depend only on time difference $(n_1 - n_2)$. Such a process is called stationary in a strict sense (see Section E.3). A wide sense stationary process is one with constant mean and a correlation function which depends only on time differences. Thus a discrete-time process $\mathbf{x}(n)$ is wide sense stationary if

$$E\{\mathbf{x}(n)\} = \mathbf{m}$$

$$E\{[\mathbf{x}(n_1) - \mathbf{m}][\mathbf{x}(n_2) - \mathbf{m}]'\} = \boldsymbol{\mathcal{P}}(n_1 - n_2)$$

Note that the condition "$p[\mathbf{x}(n)]$ is independent of n" is not sufficient to guarantee that $\mathbf{x}(n)$ is strict sense stationary and the condition "$\boldsymbol{\mathcal{P}}(n, n) = \mathbf{\Gamma}(n)$ is independent of n" does not guarantee that $\mathbf{x}(n)$ is wide sense stationary.

The term *white stochastic process* is used in this book to denote a zero mean process with no correlation in time. However, the literature contains many other names and definitions. Sometimes a white process is assumed to be "independent in time." A discrete-time process $\mathbf{w}(n)$, $n = 1, \ldots,$ is a process of *mutually independent random vectors* if all $\mathbf{w}(n)$ are mutually independent. Such a process is sometimes called a strict sense, discrete-time, white process. A zero mean discrete-time process $\mathbf{w}(n)$, $n = 1, \ldots,$ is a process with uncorrelated or orthogonal random variables if

$$E\{\mathbf{w}(n_1)\mathbf{w}'(n_2)\} = \begin{cases} \mathbf{Q}(n_1), & n_1 = n_2 \\ \mathbf{0}, & n_1 \neq n_2 \end{cases}$$

Such a process is sometimes called a wide sense white process. For continuous time it is often mathematically more convenient to work with processes with independent increments (strict sense property) or uncorrelated or orthogonal increments (wide sense property). A continuous-time process $\mathbf{b}(t)$ has independent increments if the differences

$$\mathbf{b}(t_2) - \mathbf{b}(t_1), \cdots, \mathbf{b}(t_n) - \mathbf{b}(t_{n-1})$$

are independent for all t, $t_1 < t_2 < \cdots < t_n$. $\mathbf{b}(t)$ is a process with uncorrelated or orthogonal increments if

$$E\{[\mathbf{b}(t_2) - \mathbf{b}(t_1)][\mathbf{b}(t_n) - \mathbf{b}(t_{n-1})]'\} = \mathbf{0}$$

for all t, $t_1 < t_2 < \cdots t_n$. The time derivative of a zero mean process with uncorrelated or orthogonal increments is a continuous-time white process; i.e., for scalars

$$\frac{d}{dt} b(t) = w(t)$$

Brownian motion (also called the Weiner process) is a Gaussian process with independent increments.

Markov processes are stochastic processes that have the *Markov property*. Let $\mathbf{x}(n)$, $n = 1, \ldots,$ denote some sequence of random vectors.

The sequence is said to have the Markov property (strict sense) if

$$p[\mathbf{x}(n) \mid \mathbf{x}(1) \cdots \mathbf{x}(n-1)] = p[\mathbf{x}(n) \mid \mathbf{x}(n-1)]$$
$$E\{\mathbf{x}(n) \mid \mathbf{x}(1) \cdots \mathbf{x}(n-1)\} = E\{\mathbf{x}(n) \mid \mathbf{x}(n-1)\}$$

For example, assume that $w(1) \cdots w(n)$ are a sequence of mutually independent random variables. Then the following $x(n)$ are Markov processes:

$$x(n) = w(n)$$
$$x(n) = \sum_{k=1}^{n} w(k) = x(n-1) + w(n)$$
$$x(n) = g[x(n-1)] + w(n)$$

In Section E.3, the wide sense conditional expectation was defined. A sequence $\mathbf{x}(n)$, $n = 1, \ldots,$ is a wide sense Markov process if

$$E_{\text{wide sense}}\{\mathbf{x}(n) \mid \mathbf{x}(1) \cdots \mathbf{x}(n-1)\} = E_{\text{wide sense}}\{\mathbf{x}(n) \mid \mathbf{x}(n-1)\}$$

Let $\mathbf{x}(t)$, $0 \leq t \leq T$, denote some continuous-time stochastic process. This process is said to have the Markov property (strict sense) if

$$p\{\mathbf{x}(t+\tau) \mid \mathbf{x}(s), 0 \leq s \leq t\} = p\{\mathbf{x}(t+\tau) \mid \mathbf{x}(t)\}$$
$$E\{\mathbf{x}(t+\tau) \mid \mathbf{x}(s), 0 \leq s \leq t\} = E\{\mathbf{x}(t+\tau) \mid \mathbf{x}(t)\}$$
$$\tau > 0$$

A continuous-time wide sense Markov process is defined in terms of the wide sense conditional expectation.

Appendix G

Set Theory

The set of all $\mathbf{x}$ which satisfy the relationship $g(\mathbf{x}) \leq 0$ is expressed as

$$\Omega = \{\mathbf{x} : g(\mathbf{x}) \leq 0\}$$

The set of all $\mathbf{x}$ lying on some surface is expressed as

$$\Omega = \{\mathbf{x} : g(\mathbf{x}) = 0\}$$

The fact that $\mathbf{x}$ is an element of Ω is expressed by $\mathbf{x} \in \Omega$. If Ω is the union of Ω_1 and Ω_2,

$$\begin{aligned} \Omega &= \Omega_1 \cup \Omega_2 \\ &= \{\mathbf{x} : \mathbf{x} \in \Omega_1 \text{ or } \mathbf{x} \in \Omega_2\} \end{aligned}$$

If Ω is the intersection of Ω_1 and Ω_2,

$$\begin{aligned} \Omega &= \Omega_1 \cap \Omega_2 \\ &= \{\mathbf{x} : \mathbf{x} \in \Omega_1 \text{ and } \mathbf{x} \in \Omega_2\} \end{aligned}$$

The relationship, $\Omega_2 \supset \Omega_1$ means that every $\mathbf{x}$ in Ω_1 is also contained in Ω_2 so that Ω_2 is *greater than* Ω_1 and Ω_1 is a subset of Ω_2.

G.1 Closed Convex Sets and Support Functions

Loosely speaking, a closed set is one which also contains the points on its boundary. For example,

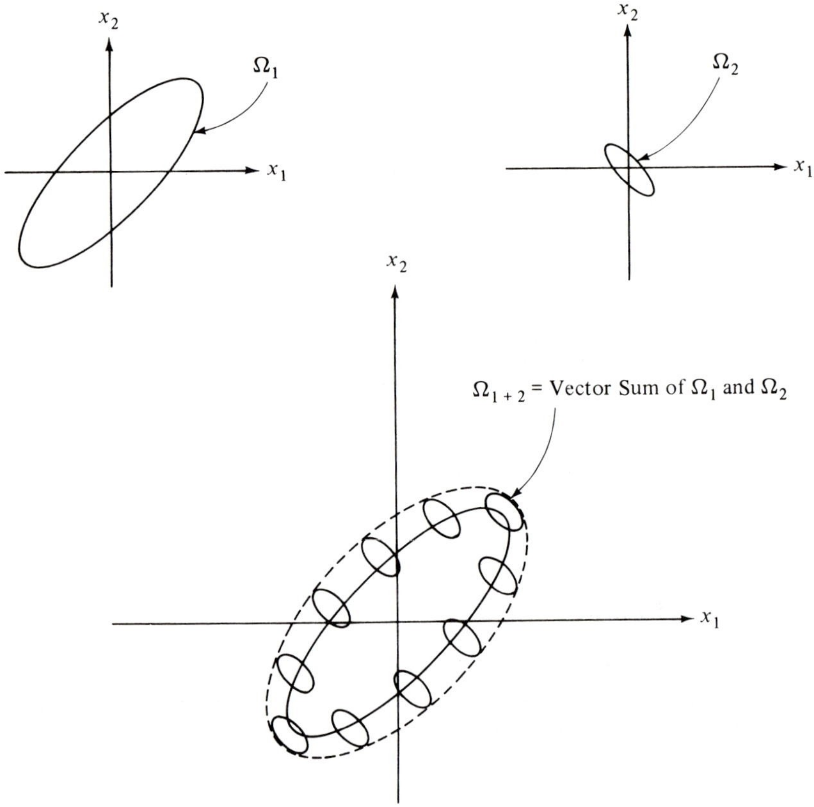

Figure G.3 Vector sum of sets.

This property is felt to be obvious using the geometrical picture furnished by Fig. G.2.

G.3 Linear Transformations, Projections, and Erections

Consider the set $\mathbf{\Omega}_y$ of K_2-dimensional vectors $\mathbf{y}$ of the form

$$\mathbf{y} = \mathbf{Hx} + \mathbf{h}$$

where $\mathbf{x}$ is a K_1-dimensional vector which belongs to the set

$$\mathbf{\Omega}_x = \{\mathbf{x} : g(\mathbf{x}) \leq \mathbf{0}\}$$

Then $\mathbf{\Omega}_y$ can be considered the linear transformation of $\mathbf{\Omega}_x$ given by

$$\mathbf{\Omega}_y = \{\mathbf{y} : \mathbf{y} = \mathbf{Hx} + \mathbf{h},\ g(\mathbf{x}) \leq \mathbf{0}\}$$

Consider some K-dimensional set $\mathbf{\Omega}_x$ where

$$\mathbf{\Omega}_x = \{\mathbf{x} : \mathbf{g}(\mathbf{x}) \leq \mathbf{0}\}$$

$$\mathbf{x} = \begin{bmatrix} \mathbf{x}_1 \\ \mathbf{x}_2 \end{bmatrix}$$

$\mathbf{x}_1 : K_1$-dimensional
$\mathbf{x}_2 : K_2$-dimensional

Let $\mathbf{\Omega}_{x_1}$ denote the projection of $\mathbf{\Omega}_x$ "down on to" the K_1-dimensional subspace of $\mathbf{x}_1$. $\mathbf{\Omega}_{x_1}$ is defined as

$$\mathbf{\Omega}_{x1} = \{\mathbf{x}_1 : \mathbf{g}(\mathbf{x}_1, \mathbf{x}_2) \leq \mathbf{0} \text{ for any } \mathbf{x}_2\} \tag{G.8}$$

This definition is illustrated in Fig. G.4. A projection can be viewed as the

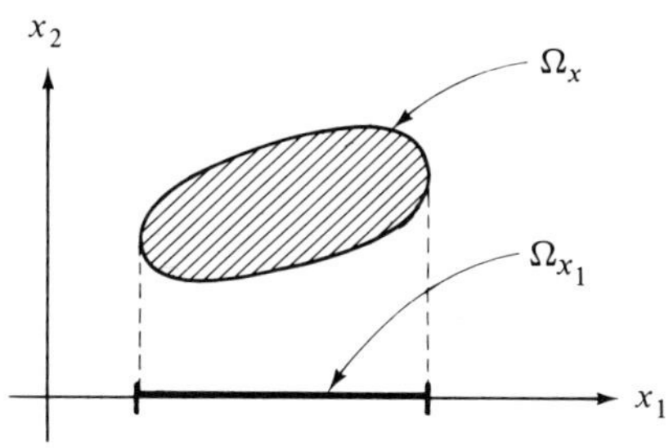

Figure G.4 Example of projection.

Ω_{x_1} : Projection of 2 Dimensional Ω_x Down Onto 1 Dimensional Space of x_1

result of a special linear transformation

$$\mathbf{y} = \mathbf{H}\mathbf{x}, \qquad \mathbf{H} = [\mathbf{I} \quad \mathbf{0}] \tag{G.9}$$

so that $\mathbf{y} = \mathbf{x}_1$

Consider some K-dimensional space with vectors $\mathbf{x}$:

$$\mathbf{x} = \begin{bmatrix} \mathbf{x}_1 \\ \mathbf{x}_2 \end{bmatrix}$$

$\mathbf{x}_1 : K_1$-dimensional
$\mathbf{x}_2 : K_2$-dimensional

Consider a K_1-dimensional set $\mathbf{\Omega}_{x_1}$:

$$\mathbf{\Omega}_{x_1} = \{\mathbf{x}_1 : \mathbf{g}_1(\mathbf{x}_1) \leq \mathbf{0}\}$$

The erection "on top of" $\mathbf{\Omega}_{x_1}$ is the K-dimensional set $\mathbf{\Omega}_x$ defined by

$$\mathbf{\Omega}_x = \{\mathbf{x} : \mathbf{x}_1 \in \mathbf{\Omega}_{x_1}, \text{all } \mathbf{x}_2\}$$

This definition is illustrated in Fig. G.5.

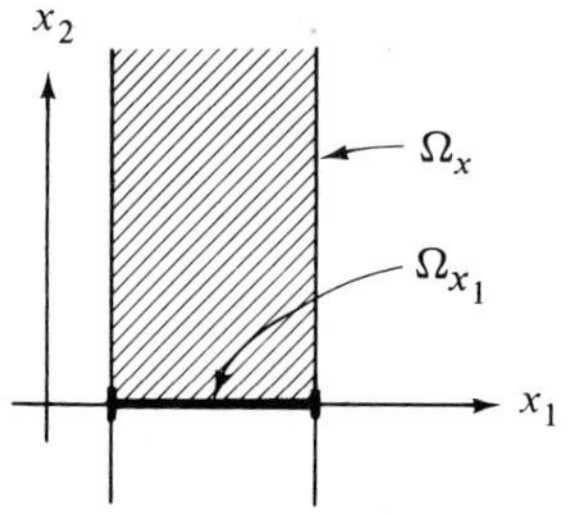

Ω_x : Erection of Two Dimensional Set on Top of One Dimensional Ω_{x_1}

Figure G.5 Example of erection.

It follows from the definitions that

$$\text{Projection of erection of } \mathbf{\Omega}_{x_1} = \mathbf{\Omega}_{x_1}$$

$$\text{Erection of projection of } \mathbf{\Omega}_x \neq \mathbf{\Omega}_x$$

Appendix H

Ellipsoids

The closed convex set called an ellipsoid is of particular interest. An ellipsoid $\boldsymbol{\Omega}$ with center at $\mathbf{m}$ can be defined by

$$\boldsymbol{\Omega} = \{\mathbf{x} : [\mathbf{x} - \mathbf{m}]'\boldsymbol{\Gamma}^{-1}[\mathbf{x} - \mathbf{m}] \leq 1\} \tag{H.1}$$

where $\boldsymbol{\Gamma}$ is a positive definite matrix. The size, shape, and location of an ellipsoid are determined by $\mathbf{m}$ and $\boldsymbol{\Gamma}$. The center of the ellipsoid is determined by $\mathbf{m}$. The orientation (direction of axis) of $\boldsymbol{\Omega}$ is determined by the eigenvectors of $\boldsymbol{\Gamma}$, and the lengths of the semimajor axes of $\boldsymbol{\Omega}$ are determined by eigenvalues of $\boldsymbol{\Gamma}$. This is illustrated in Fig. H.1.

The ellipsoid set is closely related in mathematical structure to the probability density of a Gaussian random vector. This is easily seen by comparing (H.1) with (E.1). See also Appendix E.6.

H.1 Support Functions

Let $s(\boldsymbol{\eta})$ denote the support function of the ellipsoid of (H.1). To find $\mathbf{x}_{max}(\boldsymbol{\eta})$ of (G.4), introduce a Lagrange multiplier p and solve the set of equations

$$\frac{\partial}{\partial \mathbf{x}}\left\{\mathbf{x}'\boldsymbol{\eta} + \frac{p}{2}[(\mathbf{x} - \mathbf{m})'\boldsymbol{\Gamma}^{-1}(\mathbf{x} - \mathbf{m}) - 1]\right\} = \mathbf{0}$$

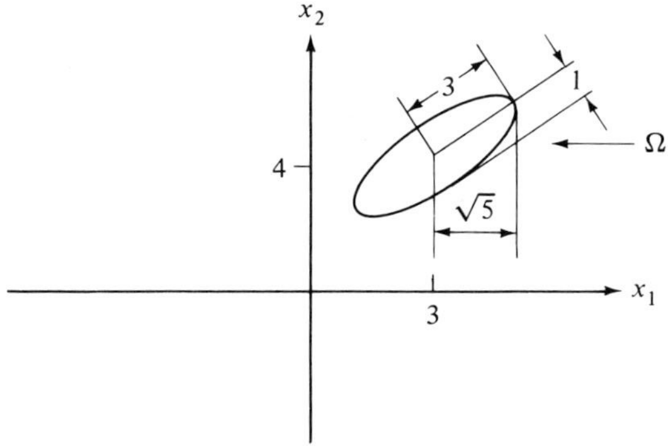

$$\Omega = \left\{ \mathbf{x} \colon [\mathbf{x} - \mathbf{m}]' \Gamma^{-1} [\mathbf{x} - \mathbf{m}] \leqslant 1 \right\}$$

$$\mathbf{m} = \begin{bmatrix} 3 \\ 4 \end{bmatrix} \qquad \mathbf{\Gamma} = \begin{bmatrix} 5 & 4 \\ 4 & 5 \end{bmatrix} \qquad \begin{matrix} \lambda\text{: Eigenvalues of } \mathbf{\Gamma} \\ \lambda = 9, 1 \end{matrix}$$

Figure H.1 Ellipsoid.

to get

$$\mathbf{x}_{\max}(\boldsymbol{\eta}) = \mathbf{m} - \frac{1}{p}\mathbf{\Gamma}\boldsymbol{\eta}$$

$$p = \pm\sqrt{\boldsymbol{\eta}'\mathbf{\Gamma}\boldsymbol{\eta}}$$

Thus the support function for $\mathbf{\Omega}$ of (H.1) is

$$\boxed{s(\boldsymbol{\eta}) = \boldsymbol{\eta}'\mathbf{m} + \sqrt{\boldsymbol{\eta}'\mathbf{\Gamma}\boldsymbol{\eta}}} \tag{H.2}$$

Comparison of $s(\boldsymbol{\eta})$ of (H.2) with the characteristic function $c(\boldsymbol{\eta})$ of (E.2) for a Gaussian $\mathbf{x}$ shows many similarities. The support function (H.2) can be used to define a degenerate ellipsoid, i.e., when $\mathbf{\Gamma}^{-1}$ does not exist.

The vector sum of two ellipsoids with $\mathbf{\Gamma}_1$ and $\mathbf{\Gamma}_2$ and with centers $\mathbf{m}_1$ and $\mathbf{m}_2$ is from (G.6)

$$s_{1+2}(\boldsymbol{\eta}) = [\mathbf{m}_1 + \mathbf{m}_2]'\boldsymbol{\eta} + \sqrt{\boldsymbol{\eta}'\mathbf{\Gamma}_1\boldsymbol{\eta}} + \sqrt{\boldsymbol{\eta}'\mathbf{\Gamma}_2\boldsymbol{\eta}} \tag{H.3}$$

Thus the vector sum of two ellipsoids is usually not an ellipsoid. Note that this property differs from the result of adding two Gaussian random vectors which results in a Gaussian random vector.

Consider two ellipsoids, $\mathbf{\Omega}_1$ and $\mathbf{\Omega}_2$, with common centers (say, the origin) defined by $\mathbf{\Gamma}_1$ and $\mathbf{\Gamma}_2$. Assume that $\mathbf{\Gamma}_1 > \mathbf{\Gamma}_2$ so $\mathbf{\Gamma}_1 - \mathbf{\Gamma}_2$ is positive definite (see Appendix A.2). It follows from (H.2) that

$$s_1(\boldsymbol{\eta}) > s_2(\boldsymbol{\eta}), \qquad \text{for all } \boldsymbol{\eta}$$

which from (G.7) means that $\mathbf{\Omega}_1 \supset \mathbf{\Omega}_2$. Thus if the ellipsoids have common centers, the condition $\mathbf{\Gamma}_1 > \mathbf{\Gamma}_2$ means that the ellipsoid defined by $\mathbf{\Gamma}_1$ completely contains the ellipsoid defined by $\boldsymbol{\Gamma}_2$. Note that this property is "obvious." The support function argument was used only as an example.

H.2 Linear Transformations, Projections, and Erections

Consider

$$\begin{aligned}&\boldsymbol{\Omega}_x = \{\mathbf{x} : [\mathbf{x} - \mathbf{m}]'\boldsymbol{\Gamma}^{-1}[\mathbf{x} - \mathbf{m}] \leq 1\}\\ &\mathbf{x} : K_1\text{-dimensional}\\ &\boldsymbol{\Omega}_y = \{\mathbf{y} : \mathbf{y} = \mathbf{H}\mathbf{x} + \mathbf{h}, \mathbf{x} \in \boldsymbol{\Omega}_x\}\\ &\mathbf{y} : K_2\text{-dimensional}\end{aligned}$$

Then

$$\boxed{\boldsymbol{\Omega}_y = \{\mathbf{y} : [\mathbf{y} - \mathbf{h} - \mathbf{H}\mathbf{m}]'[\mathbf{H}\boldsymbol{\Gamma}\mathbf{H}']^{-1}[\mathbf{y} - \mathbf{h} - \mathbf{H}\mathbf{m}] \leq 1\}} \tag{H.4}$$

This result is similar to that of Appendix E on Gaussian random variables. One of various ways to derive (H.4) is by use of support functions. Let $s_y(\boldsymbol{\eta})$ denote the support function of $\boldsymbol{\Omega}_y$:

$$\begin{aligned} s_y(\boldsymbol{\eta}) &= \max_{y \in \boldsymbol{\Omega}_y} \boldsymbol{\eta}'\mathbf{y}\\ &= \max_{x \in \boldsymbol{\Omega}_x} \boldsymbol{\eta}'\mathbf{H}\mathbf{x} + \boldsymbol{\eta}'\mathbf{h}\\ &= \boldsymbol{\eta}'[\mathbf{H}\mathbf{m} + \mathbf{h}] + \sqrt{\boldsymbol{\eta}'\mathbf{H}\boldsymbol{\Gamma}\mathbf{H}'\boldsymbol{\eta}}\end{aligned}$$

which yields (H.4). If $[\mathbf{H}\boldsymbol{\Gamma}\mathbf{H}']^{-1}$ does not exist, $\boldsymbol{\Omega}_y$ can be considered to be defined in terms of $s_y(\boldsymbol{\eta})$ as in (G.3) instead of as in (H.4).

Partition $\mathbf{x}$, $\mathbf{m}$, and $\boldsymbol{\Gamma}$ of (H.1) as

$$\mathbf{x} = \begin{bmatrix}\mathbf{x}_1\\ \mathbf{x}_2\end{bmatrix} \qquad \boldsymbol{\Gamma} = \begin{bmatrix}\boldsymbol{\Gamma}_{11} & \boldsymbol{\Gamma}_{12}\\ \boldsymbol{\Gamma}_{12} & \boldsymbol{\Gamma}_{22}\end{bmatrix} \qquad \mathbf{m} = \begin{bmatrix}\mathbf{m}_1\\ \mathbf{m}_2\end{bmatrix} \tag{H.5}$$

The set $\boldsymbol{\Omega}_{x_1}$ defined by

$$\boldsymbol{\Omega}_{x_1} = \left\{\mathbf{x}_1 : \begin{bmatrix}\mathbf{x}_1 - \mathbf{m}_1\\ \mathbf{x}_2 - \mathbf{m}_2\end{bmatrix}' \boldsymbol{\Gamma}^{-1} \begin{bmatrix}\mathbf{x}_1 - \mathbf{m}_1\\ \mathbf{x}_2 - \mathbf{m}_2\end{bmatrix} \leq 1, \text{ any } \mathbf{x}_2\right\}$$

is the projection of $\boldsymbol{\Omega}$ onto the subspace of the $\mathbf{x}_1$. To get a more useful expression, consider the linear transformation

$$\begin{aligned}\mathbf{x}_1 &= \mathbf{H}\mathbf{x}\\ \mathbf{H} &= [\mathbf{I} \quad \mathbf{0}]\end{aligned}$$

Then from (H.4)

$$\boldsymbol{\Omega}_{x_1} = \{\mathbf{x}_1 : [\mathbf{x}_1 - \mathbf{m}_1]'\boldsymbol{\Gamma}_{11}^{-1}[\mathbf{x}_1 - \mathbf{m}_1] \leq 1\} \tag{H.6}$$

Equation (H.6) shows that the projection of an ellipsoid is related in mathematical structure to the marginal distribution of a Gaussian random vector, as discussed in Appendix E.

Consider the partitioning of (H.5). The set

$$\boldsymbol{\Omega}_x = \{\mathbf{x} : [\mathbf{x}_1 - \mathbf{m}_1]'\boldsymbol{\Gamma}_{11}^{-1}[\mathbf{x}_1 - \mathbf{m}_1] \leq 1, \text{ all } \mathbf{x}_2\}$$

is the erection on top of $\boldsymbol{\Omega}_{x_1}$:

$$\boldsymbol{\Omega}_{x_1} = \{\mathbf{x}_1 : [\mathbf{x}_1 - \mathbf{m}_1]'\boldsymbol{\Gamma}_{11}^{-1}[\mathbf{x}_1 - \mathbf{m}_1] < 1\}$$

The support function of $\boldsymbol{\Omega}_x$ is

$$s(\boldsymbol{\eta}_1) = \boldsymbol{\eta}_1'\mathbf{m}_1 + \sqrt{\boldsymbol{\eta}_1\boldsymbol{\Gamma}_{11}\boldsymbol{\eta}_1}$$

where $\boldsymbol{\eta}_1$ has the dimension of $\mathbf{x}_1$. Then from (G.3)

$$\boldsymbol{\Omega}_x = \{\mathbf{x} : \mathbf{x}_1'\boldsymbol{\eta}_1 < s(\boldsymbol{\eta}_1), \text{ for all } \boldsymbol{\eta}_1, \boldsymbol{\eta}_1'\boldsymbol{\eta}_1 = 1\}$$

Appendix I

Historical Notes and References: Appendices

Most of this background material is widely available in many places in many ways. Therefore only very selective discussions are given.

Appendix A

Bellman [1]; Bodewig [1]; Franklin [1]; Gantmacher [1]; Householder [1]; are some of an almost uncountable number of good references. The matrix identity (A.3) is found in many places. Bodewig [1] outlines the history of (A.3) and lists some of the many men who discovered and rediscovered the identity starting in 1916. (A.3) will probably be rediscovered again sometime in the future. Bellman and Bechenbach [1] contains many interesting and useful inequalities involving positive definite (and other) matrices.

Appendix C

The trace and determinate matrix gradients results were taken from Athans and Schweppe [1], which uses a matrix algebra discussed in Bodewig [1]. Not all authors use the same definitions. For example, the gradient vector $\partial f(x)/\partial \mathbf{x}$ is sometimes defined as a column vector, etc.

Appendices D and E

See Book Lists 1, 2, and 3, on Probability and Stochastic Processes, Mathematical Statistics and Stochastic Systems.

Appendices G and H

The material on support functions and vector sums of sets is not widely used. Bishop and Phelps [1] are related references. The actual reference used was Witsenhausen [1], [2], [3]. See also Schlaepfer [1].

Bibliography

The Historical Notes and References at the end of each chapter often refer to a basic book list rather than one book in particular. This was done, because the "best" book usually depends on the reader's background, interests, and needs. Each of the following five basic book lists contains books which cover a spectrum of backgrounds and interests.

Book List 1: Probability and Stochastic Processes

Doob [1]; Dubes [1]; Dynkin [1]; Feller [1]; and Parzen [1].

Book List 2: Mathematical Statistics

Anderson [1]; Cramer [1]; Grenander and Rosenblatt [1]; Freund [1]; Kullbeck [1]; and Wilks [1].

Book List 3: Deterministic Systems

Anderson and Moore [1]; Athans and Falb [1]; Brockett [1]; Bryson and Ho [1]; Dertouzos, Athans, Spann, and Mason [1]; DeRusso, Roy, and Close [1]; Ogata [1]; Timothy and Bona [1]; Sage [1]; and Zadeh and Desoer [1].

Book List 4: Stochastic Systems

Åström [1]; Aoki [1]; Bryson and Ho [1]; Bucy and Joseph [1]; Davenport and Root [1]; Jazwinski [3]; Kushner [4], [5]; Laning and Battin [1]; Lavi [1]; Lewis and Odell [1]; Liebelt [1]; Meditch [3]; Melsa and Sage [1]; Nahi [1]; Sage [1]; Sage and Melsa [1]; and Wong [1].

Book List 5: Communications

Helstrom [1]; Middleton [1]; Snyder [1]; Van Trees [1]; and Wiener [1].

Albert, A., and L. Gardner [1], *Stochastic Approximation and Nonlinear Regression.* Cambridge, Mass.: M.I.T. Press, 1967.

Anderson, Brian D. O., and John B. Moore [1], *Linear Optimal Control.* Englewood Cliffs, N. J.: Prentice Hall, 1971.

Anderson, T. [1], *An Introduction to Multivariate Statistical Analysis.* New York: Wiley, 1958.

Aoki, M. [1], *Optimization of Stochastic Systems.* New York: Academic Press, 1970.

Åström, K. [1], *Introduction to Stochastic Control Theory.* New York: Academic Press, 1970.

Åström, K., and P. Eykhoff [1], "System Identification, A Survey," in IFAC Symposium on Identification and Process-Parameter Estimation, Prague, June 1970. See also Report 7006, Lund Institute of Technology, Lund, Sweden, Sept. 1970.

Athans, M., and F. Falb [1], *Optimal Control.* New York: McGraw-Hill, 1966.

Athans, M. and F. Schweppe [1], *Gradient Matrices and Matrix Calculations*, TN-1965–53. Lincoln Lab., M.I.T., Lexington, Mass. 1965.

Athans, M., and F. Schweppe [2], "Optimal Waveform Design via Control Theoretic Concepts," *Inform. Control*, 10, No. 4, April 1967, 355–377.

Athans, M., and E. Tse [1], "A Direct Derivation of the Optimal Linear Filter Using the Maximum Principle," *IEEE Trans. Autom. Control, AC-12*, Dec. 1967, 690–698.

Bar-Shalom, Y. [1], "Optimal Simultaneous State Estimation and Parameter Identification in Linear Discrete Systems," *IEEE Trans. Autom. Control, AC-17*, June 1972, 308–319.

Barankin, E. [1], "Locally Best Unbiased Estimates," *Ann. Math. Stat., 20*, 1949, 477–501.

Bass, R., V. Norum, and L. Schwartz [1], "Optimal Multichannel Nonlinear Filtering," *J. Math. Anal. Appl., 16*, 1, 1966.

Bechenbach, E., and R. Bellman [1], *Inequalities.* Berlin: Springer, 1961.

Bellman, R. [1], *Introduction to Matrix Analyis.* New York: McGraw-Hill, 1960.

Bellman, R., H. Kagurada, R. Kalaba, and R. Sridhar [1], "Invariant Imbedding and Nonlinear Filtering," *J. Astronautical Sci. XII*, No. 3, May—June 1966.

Benedict, T., and M. Sondhi [1], "On a Property of Wiener Filters," *Proc. IRE, 45*, July 1957, 1021–1022.

Bertsekas, D., and I. Rhodes [1], "Recursive State Estimation for a Set-Membership Description of Uncertainty," *IEEE Trans. Autom. Control, AC-16*, April 1971, 117–128.

Bhattacharyya, A. [1], "On a Measure of Divergence Between Two Statistical Populations Defined by Their Probability Densities," *Bull. Math. Soc., 35*, 1947, 99–109.

Bhattacharyya, A. [2], "On Some Analogues of the Amount of Information and Their Use in Statistical Estimation," *Sankhyn, 8,* 1946, part 1, 1–14; part 3, 201–217; part 4, 315–328.

Bishop, E., and Phelps [1], "The Support Functionals of a Convex Set," *Proc. of Symp. in Pure Math.*, 7, AMS, 1963.

Blum, M. [1], "Fixed Memory Least Squares Filters Using Recursive Methods," *IRE Trans. Inform. Theory, IT-3,* 1957, 178–182.

Blum, M. [2], "A Stagewise Parameter Estimation Procedure for Correlation Data," *Numer. Math., 3,* 1961, 202–208.

Bodewig, E., [1], *Matrix Calculus.* New York: Wiley-Interscience, 1956.

Brockett, R. [1], *Finite Dimensional Linear Systems.* New York: Wiley, 1970.

Brown, J. [1], "Asymmetric Non-Mean-Square Error Criteria," *IRE Trans. Autom. Control, AC-7,* Jan. 1962, 64–66.

Bryson, A., and M. Frazier [1], "Smoothing in Linear and Nonlinear Systems," in *Proceedings of the Optimal System Synthesis Symposium, Wright Field, Ohio,* ASD-TDR-63-119, Feb. 1963, 353–364.

Bryson, A., and Y. Ho [1], *Applied Optimal Control,* Waltham, Mass,: Ginn/Blaisdell, 1969.

Bryson, A., and D. Johansen [1] "Linear Filtering for Time Varying Systems Using Measurements Containing Colored Noise," *IEEE Trans. Autom. Control, AC-10,* No. 1, Jan. 1965, 4–11.

Bucy, R., and P. Joseph [1], *Filtering for Stochastic Processes with Applications to Guidance.* New York: Wiley, 1968.

Cox, H. [1], "On the Estimation of State Variables and Parameters for Noisy Dynamic Systems," *IEEE Trans. Autom. Control, AC-9,* Jan. 1964, 5–12.

Cox, H. [2], "Recursive Nonlinear Filtering," *Proc. Natl. Electron. Conf., XXI,* 1965, 770–775.

Cox, H. [3], "Sequential Minimax Estimation," *IEEE Trans. Autom. Control, AC-11,* No. 2, April 1966, 323–324.

Cramer, H. [1], *Mathematical Methods of Statistics.* Princeton, N. J.: Princeton University Press, 1946.

Crittendon, R., E. Dubinsky, D. Labell, and F. Schweppe [1], "A Number of Results from Studies of Digital Filtering and the Convergence of Iterative Techniques," 22G-0061, Lincoln Lab., M.I.T., Lexington, Mass., June 1961.

Davenport, W., and W. Root [1], *An Introduction to the Theory of Random Signals and Noise.* New York: McGraw-Hill, 1958.

Dertouzos, M., M. Athans, R. Spann, and S. Mason [1], *Systems, Networks, and Computation: Basic Concepts.* New York: McGraw-Hill, 1972.

DeRusso, P., K. Roy, and C. Close [1], *State Variables for Engineers.* New York: Wiley, 1965.

DeVille, T., and F. Schweppe [1], "On Line Identification of Interconnected Network Equivalents from Operating Data," *IEEE Trans. PAS*, Summer Power Meeting, 1972.

Deyst, J., and R. Price [1], "Conditions for Asymptotic Stability of the Discrete Minimum Variance Linear Estimator," *IEEE Trans. Autom. Control, AC-13*, Dec. 1968, 702–705.

Doob, J. [1], *Stochastic Processes.* New York: Wiley, 1953.

Dubes, R. [1], *Theory of Applied Probability.* Englewood Cliffs, N. J.: Prentice-Hall, 1968.

Duncan, T. [1], "Probability Densities for Diffusion Processes with Applications to Non-Linear Filtering Theory and Detection Theory," *Inform. Control*, 13, July 1968, 62–74.

Dynkin, E., *Markov Processes.* New York: Academic Press, 1965.

Feldbaum, A. [1], "Dual Control Theory—I," *Automation Remote Control, 21*, 1960, 1240–1249.

Feller, W. [1], *Probability Theory and Its Applications.* New York: Wiley, 1950.

Ferguson, T. [1], "A Method of Generating Best Asymptotically Normal Estimates," *Ann. Math. Stat., 29*, 1959, 1046–1062.

Fisher, J., and E. Stear [1], "Optimal Nonlinear Filtering for Independent Increment Processes," *Parts I and II, IEEE Trans. Inform. Theory, IT-13*, Oct. 1967, 558—578.

Fisher, R. A., "On the Mathematical Foundations of Theoretical Statistics," *Phil. Trans. Roy. Soc. London, 222*, 1922, 309.

Fitzgerald, R. [1], "Divergence of the Kalman Filter," *IEEE Trans. Autom. Control, AC-16*, Dec. 1971, 736–748.

Forsythe, G, and C. Moller [1], *Computer Solution of Linear Algebraic Systems.* Englewood Cliffs, N. J.: Prentice-Hall, 1967.

Franklin, J. [1], *Matrix Theory.* Englewood Cliffs, N. J.: Prentice-Hall, 1968.

Fraser, D., and J. Potter [1], "The Optimum Linear Smoother as a Combination of Two Optimum Linear Filters," *IEEE Trans. Autom. Control, AC-14*, Aug. 1969, 387–390.

Freund, J. [1], *Mathematical Statistics*, 2nd ed. Englewood Cliffs, N. J.: Prentice-Hall, 1971.

Friedland, B. [1], "Treatment of Bias in Recursive Filtering," *IEEE Trans. Autom. Control, AC-14*, Aug. 1969, 359–367.

Galiana, F., and F. Schweppe [1], "A Weather Dependent Probabilistic Model for Short Term Load Forecasting," *IEEE Trans. PAS*, Winter Power Meeting, 1972 (C72-075-5).

Gardner, L. [1], "Adaptive Predictors," in *Transactions of the Third Prague Conference on Information Theory, Statistics, Decision Functions, Random Processes.* Prague: Publishing House of Czechoslovak Academy of Sciences, 1964.

Glover, J. [1], "Linear Estimation of Completely Unknown Signals," *IEEE Trans. Autom. Control, AC-14*, Dec. 1969, 766–767.

Glover, J., and F. Schweppe [1], "Control of Linear Dynamic Systems with Unknown but Bounded Disturbances," *IEEE Group Autom. Control, AC-16*, Oct. 1971, 411–423.

Glover, J., and F. Schweppe [2], "Advanced Load Frequency Control," *IEEE Trans. PAS*, Oct. 1972.

Grenander, U., and M. Rosenblatt [1], *Statistical Analysis of Time Series.* New York: Wiley, 1957.

Guntmacher, F. [1], *The Theory of Matrices.* New York: Chelsea Publishing Co., 1959.

Heffes, H., [1], "Effect of Erroneous Models on the Kalman Filter Response," *IEEE Trans. Autom. Control, AC-11*, No. 3, July 1966, 541–543.

Helstrom, C. [1], *Statistical Theory of Signal Detection.* Elmsford, N. Y.: Pergamon, 1960.

Hnyilicza, E. [1], "A Set-Theoretic Approach to State Estimation," M.S. thesis, Cambridge, Mass.: Department of Electrical Engineering, M.I.T., June 1969.

Hofstetter, E., and R. McAulay [1], "Barankin Bounds on Parameter Estimation," *IEEE Trans. Inform. Theory*, 17, Nov. 1971, 669–676.

Householder, A. [1], *Principles of Numerical Analysis.* New York: McGraw-Hill, 1953.

Ito, K., and H. McKean [1], *Properties of Diffusion Processes and Their Sample Paths.* Berlin: Springer, 1965.

Jazwinski, A. [1], "Filtering for Nonlinear Dynamical Systems," *IEEE Trans. Autom. Control, AC-11*, Oct. 1966, 765–766.

Jazwinski, A. [2], "Limited Memory Optimal Filtering," *IEEE Trans. Autom. Control, AC-13*, Oct. 1968, 558–563.

Jazwinski, A. [3], *Stochastic Processes and Filtering Theory.* New York: Academic Press, 1970.

Jones, L., and F. Schweppe [1], "A Random Acceleration Model for Filtering Polynomial-Like Signals," 28G-3, Lincoln Lab., M.I.T., Lexington, Mass., Oct. 1961.

Joseph, P. D., and J. Tou [1], "On Linear Control Theory," *Trans. AIEE, 80*, Part II, 1961, 193–196.

Kahan, W. [1], "Circumscribing an Ellipsoid About an Inter-Section of Two Ellipsoids," *Can. Math. Bull., 2*, No. 3, 1968, 437–441.

Kailath, T. [1], "The Divergence and Bhattacharyya Distance Measures in Signal Selection," *IEEE Trans. Commun. Technol.*, Feb. 1967.

Kailath, T. [2], "An Innovations Approach to Least Squares Estimation, Part I: Linear Filtering in Additive White Noise," *IEEE Trans. Autom. Control, AC-13*, Dec. 1968, 646–655.

Kailath, T. [3], "A General Likelihood-Ratio Formula for Random Signals in Gaussian Noise," *IEEE Trans. Inform. Theory*, *IT-15*, May 1969, 350–361.

Kailath, T. [4], "The Innovations Approach to Detection and Estimation Theory," *Proc. IEEE*, *58*, No. 5, May 1970, 680–695.

Kailath, T., and P. Frost [1], "An Innovations Approach to Least-Squares Estimation, Part II: Linear Smoothing in Additive White Noise," *IEEE Trans. Autom. Control*, *AC-13*, Dec. 1968, 655–660.

Kalman, R. [1], "A New Approach to Linear Filtering and Prediction Problems," *J. Basic Eng.*, *Ser. D*, *82*, March 1960, 35–45.

Kalman, R. [2], "New Methods and Results in Linear Prediction and Filtering Theory," in *Proceedings of First Symposium on Engineering Applications of Random Function Theory and Probability*, J. Bogdanoff and F. Kozin, Eds. New York: Wiley, 1963, 270–388.

Kalman, R. and R. Bucy [1], "New Results in Filtering and Prediction Theory," *J. Basic Eng.*, *Ser. D*, *83*, March 1961, 95–108.

Kaminski, P., A. Bryson, and S. Schmidt [1], "Discrete Square Root Filtering: A Survey of Current Techniques," *IEEE Trans. Autom. Control*, *AC-16*, No. 6, Dec. 1971, 727–736.

Kullbeck, S. [1], *Information Theory and Statistics*. New York: Wiley, 1959.

Kushner, H. [1], "On the Differential Equations Satisfied by Conditional Probability Densities of Markov Processes, with Applications," *J. SIAM Control*, Ser. A, *2*, No. 1, 1964, 106–119.

Kushner, H. [2], "On the Dynamical Equations of Conditional Probability Density Functions of Markov Processes, with Application to Optimal Stochastic Control Theory," *J. Math. Anal. Appl.*, *8*, No. 2, April 1964, 332–344.

Kushner, H. [3], "On the Optimum Timing of Observations for Linear Control Systems with Unknown Initial State," *IEEE Trans. Autom. Control*, *AC-9*, No. 2, April 1964, 144–150.

Kushner, H. [4], "Approximations to Optimal Nonlinear Filters," *IEEE Trans. Autom. Control*, *AC-12*, Oct. 1967, 546–556.

Kushner, H. [5], *Stochastic Stability and Control*. New York: Academic Press, 1967.

Kushner, H. [6], *Introduction to Stochastic Control*. New York: Holt, Reinhart, & Winston, 1971.

Kyskernask, H. [1], "Optimal Filtering in Linear Systems with Time Delays," *IEEE Trans. Autom. Control*, *AC-12*, No. 2, April 1967, 169–173.

Lanczos, C. [1], *Applied Analysis*. Englewood Cliffs, N. J.: Prentice-Hall, 1956.

Laning, J., and R. Battin [1], *Random Processes in Automatic Control*. New York: McGraw-Hill, 1956.

Larson, R., and J. Peschon [1], "A Dynamic Programming Approach to Trajectory Estimation," *IEEE Trans. Autom. Control*, AC-11, No. 3, July 1966, 537–541.

Lavi, N. E. [1], *Estimation Theory and Applications.* New York: Wiley, 1969.

Lelam, L. [1], "On Some Asymptotic Properties of Maximum Likelihood Estimates and Related Bayes Estimates," *Univ. Calif. Pub. Stat., 1*, No. 11, 277–330.

Levin, J. [1], "On the Matrix Riccatti Equation," *Proc. Am. Math. Soc., 10*, 1959, 519–524.

Lewis, T. O., and P. L. Odell [1], *Estimation in Linear Models.* Englewood Cliffs, N. J.: Prentice-Hall, 1971.

Liebelt, P. [1], *An Introduction to Optimal Estimation.* Reading, Mass.: Addison-Wesley, 1967.

Luenberger, D. G. [1], "Observing the State of a Linear System." *IEEE Trans. Mil. Elec. MIL-8*, No. 2, April 1966, 190–197.

Magil, D. T. [1], "Optimal Adaptive Estimation of Sampled Stochastic Processes," *IEEE Trans. Autom. Control, AC-10*, No. 4, Oct. 1965, 434–439.

Magness, T., and J. McGuire [1], "Comparison of Least Squares and Minimum Variance Estimates of Regression Parameters," *Ann. Math. Stat., 33*, June 1962, 462–470.

Masiello, R., and F. Schweppe [1], "A Tracking Static State Estimator," *IEEE Trans. on PAS, 90*, June 1971, 1025–1033.

Meditch, J. [1], "On Optimal Linear Smoothing Theory," *Inform. Control, 10*, 1967, 598–615.

Meditch, J. [2], "On Optimal Fixed-Point Linear Smoothing," *Int. J. Control, 6*, 1967, 189–199.

Meditch, J. [3], *Stochastic Optimal Linear Estimation and Control.* New York: McGraw-Hill, 1969.

Mehra, R. [1], "On the Limiting Procedure in Linear Recursive Estimation Theory," *J. Math. Ann. Appl., 26*, April 1969, 1–8.

Meier, L. III, J. Peschon, and R. M. Dressler [1], "Optimal Control of Measurement Subsystems," *IEEE Trans. Autom. Control, AC-12*, Oct. 1967, 528–535.

Melsa, J., and A. Sage [1], *Introduction to Probability and Stochastic Processes.* Englewood Cliffs, N. J.: Prentice-Hall, 1972.

Mendel, J. and D. Gieseking [1], "Bibliography on the Linear Quadratic—Gaussian Problem," *IEEE Trans. Autom. Control, AC-16*, Dec. 1971, 847–869.

Middleton, D. [1], *An Introduction to Statistical Communication Theory.* New York: McGraw-Hill, 1960.

Moore, R., and F. Schweppe [1], "Adaptive Control for Nuclear Power Plant Load Changes," Fifth World Congress of the International Federation of Automatic Control, Paris, 1972.

Mowerey, V. O. [1], "Least Squares Recursive Differential-Correction Estimation in Nonlinear Problems," *IEEE Trans. Autom. Control, AC-10*, No. 4, Oct. 1965, 399–407.

Nahi, N. [1], *Estimation Theory and Applications.* New York: Wiley, 1969.

Nash, R. A. Jr., and F. B. Tuteur [1], "The Effect of Uncertainties in the Noise Covariance Matrices on the Maximum Likelihood Estimate of a Vector," *IEEE Trans. Autom. Control, AC-13,* Feb. 1968, 86–88.

Neal, S. [1], "Linear Estimation in the Presence of Errors in Assumed Plant Dynamics," *IEEE Trans. Autom. Control, AC-12,* Oct. 1967, 592–594.

Nishimina, T. [1], "On the A priori Information in Sequential Estimation Problems," IEEE Trans. Autom. Control, AC-11, No. 2, April 1966, 199–205.

Ogata, K. [1], *State Space Analysis of Control Systems.* Englewood Cliffs, N. J.: Prentice-Hall, 1967.

Ohap, R., and A. Stubberand [1], "A Technique for Estimating the State of a Nonlinear System," *IEEE Trans. Autom. Control, AC-10,* No. 2, April 1965, 150–155.

Parzen, E. [1], *Stochastic Processes.* San Francisco: Holden-Day, Inc., 1962.

Penrose, R. [1], "On the Generalized Inverse of A Matrix," *Proc. Cambridge Phil. Soc., 51,* 1956, 17–19.

Price, C. [1]. "An Analysis of Divergence Problem in the Kalman Filter," *IEEE Trans. Autom. Control, AC-13,* Dec. 1968, 699–702.

Rauch, F., F. Tung, and C. Striebel [1], "Maximum Likelihood Estimates for Linear Dynamic Systems," *AIAA J., 3,* No. 8, Aug. 1965, 1445–1450.

Sage, A. [1], *Optimum System Controls.* Englewood Cliffs, N. J.: Prentice-Hall, 1968.

Sage, A., and J. Melsa [1], *Estimation Theory with Applications to Communications and Control.* New York: McGraw-Hill, 1970.

Sakrison, D. [1], "Iterative Design of Optimum Filters for Non-Mean-Square Error Performance Criteria," *IEEE Trans. Inform. Theory, IT-9,* July 1963, 161–167.

Sakrison, D. [2], "Estimation of Signals Containing Unknown Parameters: Comparison of Linear and Arbitrary Unbiased Estimates," *J. Soc. Ind. Appl. Math., 13,* No. 3, 1965, 706–719.

Schlaepfer, F. [1], "Set-Theoretic Estimation of Distributed Parameter Systems,": Report ESL-R-413, M.I.T., Cambridge, Mass., Jan. 1970.

Schlaepfer, F., and F. Schweppe [1], "Continuous Time State Estimation Under Disturbance Bounded by Convex Sets," *IEEE Trans. Autom. Control, AC-17,* April 1972, 197–206.

Schmidt, S. [1], "State Space Techniques Applied to the Design of a Space Navigation System," *JACC Preprints,* 1962, paper 11–3.

Schwartz, L., and E. B. Stear [1], "A Computational Comparison of Several Nonlinear Filters," *IEEE Trans. Autom. Control, AC-13,* Feb. 1968, 83–86.

Schweppe, F. [1], "On the Linear Smoothing of Redundant Radar Data from Satellites," 22G-0021, Lincoln Lab., Lexington, Mass., M.I.T., Aug. 1960.

Schweppe, F. [2], "Optimization of Signals," GR 1964–4, Lincoln Lab., Lexington, Mass, M.I.T., Jan. 1964.

Schweppe, F. [3], "Algorithms for Estimating a Re-Entry Body's Position, Velocity, and Ballistic Coefficient in Real Time or for Post Flight Analysis," 1964–4, Lincoln Lab., Lexington, Mass., M.I.T., Jan. 1965.

Schweppe, F. [4], "Evaluation of Likelihood Functions for Gaussian Signals," IEEE Trans Inform. Theory, IT-11, No. 1, Jan. 1965, 61–70.

Schweppe, F. [5], "Radar Frequency Modulations for Accelerating Targets Under a Bandwidth Constraint," *IEEE Trans. Mil. Elec.*, *MIT-9*, No. 1, Jan. 1965, 25–32.

Schweppe, F. [6], "On the Accuracy and Resolution of Radar Signals," *IEEE Trans. Aerospace Electron. Systems*, *AES-1*, No. 3, Dec. 1965, 235–245.

Schweppe, F. [7], "State Space Evaluation of the Bhattacharyya Distance Between Two Gaussian Processes," *Inform. Control*, *11*, *No. 3*, Sept. 1967, 352–372.

Schweppe, F. [8], "On the Bhattacharyya Distance and the Divergence Between Gaussian Processes," *Inform. Control*, *11*, No. 4, Oct. 1967, 373–395.

Schweppe, F. [9], "Recursive State Estimation: Unknown but Bounded Errors and System Inputs," *IEEE Trans.*, *AC-13*, Feb. 1968, 22–29.

Schweppe, F. [10], "Sensor Array Data Processing for Multiple Signal Sources," *IEEE Trans. Inform. Theory*, *IT-14*, No. 2, March 1968, 297–305.

Schweppe, F. [11], "Power System Static State Estimation, Part III—Implementation," *IEEE Trans. PAS*, *PAS-89*, No. 1, Jan, 1970, 130.

Schweppe, F., and D. Gray [1], "Radar Signal Design Subject to Simultaneous Peak and Average Power Constraint," *IEEE Trans. Inform. Theory*, *IT-12*, No. 1, Jan. 1966, 13–25.

Schweppe, F., and L. Jones [1], "A Random Acceleration Model for Filtering Polynomial Like Signals," 28G–3, Lincoln Lab., Lexington, Mass., M.I.T., Oct. 1963.

Schweppe, F., and H. Knudsen [1], "The Theory of Amorphous Cloud Trajectory Prediction," *IEEE Trans. Inform. Theory*, *IT-14*, No. 3, May 1968, 415–427.

Schweppe, F., and D. Rom [1], "Power System Static State Estimation, Part II—Approximate Model," *IEEE Trans. PAS*, *PAS-89*, No. 1, Jan. 1970, 125.

Schweppe, F., and J. Wildes [1], "Power System Static State Estimation, Part I—Exact Model," *IEEE Trans. PAS*, *PAS-89*, No. 1, Jan. 1970, 120.

Sherman, S. [1], "Non Mean Square Error Criteria," *IRE Trans. Inform. Theory*, *IT-4*, 1958, 125–126.

Shinbrot, M. [1], "Optimization of Time-Varying Linear Systems with Nonstationary Inputs," *Trans. ASME*, *80*, 1958, 457–462.

Singer, R. A., and P. A. Frost [1], "On the Relative Performance of the Kalman and Wiener Filters," *IEE Trans. Autom. Control*, *AC-14*, Aug. 1969, 390–394.

Sklar, J., and F. Schweppe [1], "On the Angular Resolution of Multiple Targets," *Proc. IEEE* (correspondence), *52*, No. 9, Sept. 1964.

Smith, G. [1], "Multivariable Linear Filter Theory Applied to Space Vehicle Guidance," *SIAM J. Control*, *2*, 1964, 19–32.

Snyder, D. L. [1], *The State-Variable Approach to Continuous Estimation with Applications to Analog Communication Theory*. Cambridge, Mass.: M.I.T. Press, 1968.

Sorenson, H. W. [1], "On the Error Behavior in Linear Minimum Variance Estimation Problems," *IEEE Trans. Autom. Control*, *AC-12*, Oct. 1967, 557–562.

Sridhar, R., and D. Detchmendy [1], "Sequential Estimation of States and Parameters in Noisy Nonlinear Dynamical Systems," *JACC Preprints*, 1965, 56–63.

Stratonovich, R. [1], "On the Theory of Optimal Nonlinear Filtering of Random Functions," *Theory Probability Appl.*, *4*, 1959, 223–225.

Stratonovich, R. [2], "Conditional Markov Processes," *Theory Probability Appl.*, *5*, No. 2, 1960, 156–178.

Stratonovich, R. [3], *Conditional Markov Processes and Their Application to Problems of Optimal Control*. New York: American Elsevier, 1968.

Stratonovich, R. [4], "A New Representation for Stochastic Integrals and Equations," *J. SIAM Control*, *4*, No. 2, 1966. See also Stanford Electronics System Theory Lab Report 7050–9, Stanford, Calif., by P. Frost, Aug. 1966.

Swerling, P. [1], "Optimum Linear Estimation for Random Processes as Limit of Estimates Based on Sampled Data," *IRE Wescon. Conv. Rec.*, Part 4, 1958, 158–163.

Swerling, P. [2], "A Proposed Stagewise Differential Correction Procedure for Satellite Tracking and Prediction," The Rand Corp., Paper P-1292, Santa Monica, Calif., Jan. 1958.

Swerling, P. [3], "First Order Error Propagation in a Stagewise Smoothing Procedure for Satellite Observation," *J. Astron. Sci.*, *6*, 1959, 46–52.

Timothy, L., and B. E. Bona [1], *State Space Analysis*. New York: McGraw-Hill, 1968.

Valentine, F. [1], *Convex Sets*. New York: McGraw-Hill, 1964.

Van Trees, H. [1], *Detection, Estimation and Modulation Theory*. New York: Wiley, 1961.

Van Trees, H. [2], "Applications of State-Variable Techniques in Detection Theory," *Proc. IEEE*, *58*, May 1970, 653–669.

Wald, A. [1], *Sequential Analysis*. New York: Wiley, 1947.

Wiener, N. [1], *The Extrapolation, Interpolation, and Smoothing of Stationary Time Series*. New York: Wiley, 1949.

Wilde, D., and C. Beightler [1], *Foundations of Optimization*. Englewood Cliffs, N. J.: Prentice-Hall, 1967.

Wilks, S. [1], *Mathematical Statistics.* New York: Wiley, 1962.

Witsenhausen, H. S. [1], "Minimax Control of Uncertain Systems," Report ESL-R-269. Cambridge, Mass., Electronic Systems Lab., M.I.T., May 1966.

Witsenhausen, H. S. [2], "A Minimax Control Problem for Sampled Linear Systems," *IEEE Trans. Autom. Control, AC-13,* Feb. 1968, 5–21.

Witsenhausen, H. S. [3], "Set of Possible States of Linear Systems Given Perturbed Observations," *IEEE Trans., AC-13,* Oct. 1968, 556–558.

Witsenhausen, H. S. [4], "Separation of Estimation and Control for Discrete Time Systems," *Proc. of IEEE, 59,* Nov. 1971, 1557–1566.

Wong, E. [1], *Stochastic Processes in Information and Dynamical Systems.* New York: McGraw-Hill, 1971.

Wong, E., and M. Zakai [1], "On the Convergence of Ordinary Integrals to Stochastic Integrals," *Ann. Math. Stat., 36,* 1965, 1560–1564.

Wong, E., and M. Zakai [2], "On the Relation Between Ordinary and Stochastic Differential Equations," *Inst. J. Eng. Sci.,* 3, 1965, 213–229.

Wonham, W. [1], "Some Applications of Stochastic Differential Equations to Optimal Nonlinear Filtering," *J. SIAM Control,* Ser. A, 2, No. 3, 1965, 347–369.

Wonham, W. [2], "Random Differential Equations in Control Theory," Chap. 2, *Probabilistic Methods in Applied Mathematics,* A. Bharucha-Reid, Ed., New York: Academic Press, 1969.

Wonnacott, R., and T. Wonnacott [1], *Econometrics.* New York: Wiley, 1970.

Woodward, P. [1], *Probability and Information Theory, with Applications to Radar.* New York: McGraw-Hill, 1953.

Zadeh, L., and C. Desoer [1], *Linear System Theory, the State Space Approach.* New York: McGraw-Hill, 1963.

Zakai, M. [1], "General Error Criteria," *IEEE Trans. PGIT, IT-10,* Jan. 1964, 94–95.

Index

C

D

E

N

O

P

Q

R

S

T

U